THE PHYSIOLOGY OF FISHES

Edited by

David H. Evans, Ph.D.
Professor of Zoology
Coordinator of Biological Sciences Program
University of Florida
Gainesville, Florida
and
Mt. Desert Island Biological Laboratory
Salsbury Cove, Maine

CRC Press
Boca Raton Ann Arbor London Tokyo

Library of Congress Cataloging-in-Publication Data

The Physiology of fishes / by David H. Evans.
p. cm. -- (CRC series in marine science)
Includes bibliographical references and index.
ISBN 0-8493-8042-1
1. Fishes--Physiology. I. Evans, David H. (David Hudson), 1940– . II. Series.
QL639.1.P49 1993
597′.01—dc20 93-18071
CIP

This book represents information obtained from authentic and highly regarded sources. Reprinted material is quoted with permission, and sources are indicated. A wide variety of references are listed. Every reasonable effort has been made to give reliable data and information, but the authors, editor, and the publisher cannot assume responsibility for the validity of all materials or for the consequences of their use.

Direct all inquiries to CRC Press, Inc., 2000 Corporate Blvd., N.W., Boca Raton, Florida 33431.

International Standard Book Number 0-8493-8042-1

Library of Congress Card Number 93-18071
Printed in the United States of America 3 4 5 6 7 8 9 0
Printed on acid-free paper

The CRC Marine Science Series provides publications that synthesize recent advances in Marine Science. Marine Science is at an exciting new threshold where new developments are providing fresh perspectives on how the biology of the ocean is integrated with its chemistry and physics.

CRC MARINE SCIENCE SERIES

Series Editors

Michael J. Kennish, Ph.D.
Peter L. Lutz, Ph.D.

Published Titles

Ecology of Estuaries: Anthropogenic Effects, Michael J. Kennish
The Physiology of Fishes, David H. Evans

Forthcoming Titles

Morphodynamics of the Inner Continental Shelf, L. Donelson Wright
Major Marine Ecological Disturbances, Ernest H. Williams, Jr. and Lucy Bunkley-Williams
Marine Bivalves and Ecosystem Processes, Richard F. Dame
Practical Handbook of Marine Science, 2nd Edition, Michael J. Kennish
Ecology of Marine Invertebrate Larvae, Larry McEdward
Benthic Microbial Ecology, Paul F. Kemp
Physical Oceanography of the Great Barrier Reef Processes, Eric Wolanski
Sediment Studies of River Mouths, Tidal Flats, and Coastal Lagoons, Doeke Eisma
Seabed Instability, M. Shamim Rahman

The Physiology of Fishes describes how the dominant physical and chemical features of freshwater environments and the ocean influence and even determine the nature and operation of various systems within fishes, including the sensory systems, mechanisms for salt and gas exchange, locomotion, reproduction, and others.

PREFACE

The idea for this volume dates from 1969, when I taught my graduate course in Fish Physiology for the first time at the University of Miami and found that no suitable text existed. In the ensuing years, references to chapters in Hoar and Randall's *Fish Physiology* (Academic Press), as well as the primary literature, have been useful for this course, but a single-volume, in-depth text was still lacking. CRC Press has provided a solution to this problem, and hopefully, one for other fish biologists who may be seeking a text for a course or a readily accessible source of information on various aspects of fish physiology. The authors for this volume have been carefully selected from the tanks of active researchers who are best able to summarize the state-of-the-art as of the end of 1992. The chapters, by definition, will be somewhat dated once they are published, but my hope is that they will form bases for historical reference and future research in the areas covered. Some areas of interest are missing, most notably aspects of brain structure and function, but I have tried to select the areas of greatest general interest, based upon 25 years of reading, writing, and lecturing in fish physiology and comparative physiology. Any omissions are my own. I hope that these topics will be of some interest to marine biologists and ichthyologists ranging from fisheries biologists to physiologists, and of course, to comparative physiologists who may want to learn more about the physiological strategies that are unique to fishes, and those shared with other organisms.

THE EDITOR

David H. Evans, Ph.D., is Professor of Zoology and Coordinator of the Biological Sciences Program at the University of Florida.

Dr. Evans received his A.B. (cum laude) in Zoology from DePauw University, Indiana in 1962, and his Ph.D. in Biological Sciences from Stanford University, California in 1967. He held postdoctoral positions in the Biological Sciences Department of the University of Lancaster, U.K., and the Groupe de Biologie Marine du C.E.A., Villefranche-sur-mer, France during 1967 and 1968. In 1969 he joined the Department of Biology at the University of Miami, Florida, as Assistant Professor and served as Professor and Chair from 1978 to 1981, when he became Professor of Zoology at the University of Florida. He also served as Chair of that department from 1982 to 1985.

Dr. Evans also served as Director of the Mt. Desert Island Biological Laboratory, Salsbury Cove, Maine from 1983 to 1992, as well as Director of the MDIBL's Center for Membrane Toxicity Studies from 1985 to 1992. He has served on a White House, Office of Science and Technology Policy, Acid Rain Peer Review Panel in 1982 to 1984 and is currently serving on the Physiology and Behavior Panel of the National Science Foundation. He is a member of the American Society of Zoologists, American Physiological Society, the Society for Experimental Biology, and Sigma Xi. Currently he is serving on the editorial boards of the *Journal of Experimental Biology*, *Journal of Comparative Physiology*, and the *American Journal of Physiology*. He has also served on the editorial boards of the *The Biological Bulletin* and the *Journal of Experimental Zoology*.

Dr. Evans received the University of Miami, Alpha Epsilon Delta, Premedical Teacher of the Year Award in 1974. He was also awarded the University of Florida and the College of Liberal Arts and Sciences Outstanding Teaching Awards in 1991; in 1993, he was named Teacher-Scholar of the Year.

Dr. Evans has presented 20 invitational lectures at international meetings and has published over 100 papers and book chapters. He has been the recipient of research grants from the National Science Foundation and the National Institute of Environmental Health Sciences. His current research interest is the role of natriuretic hormones in fish physiology and the effect of heavy metals on smooth muscle contraction.

CONTRIBUTORS

Robert McNeill Alexander, Ph.D., D.Sc., F.R.S.
Department of Pure & Applied Biology
University of Leeds
Leeds, England

David H. Evans, Ph.D.
Department of Zoology
University of Florida
Gainesville, Florida
and
Mt. Desert Island Biological Laboratory
Salsbury Cove, Maine

Anthony P. Farrell, Ph.D.
Department of Biological Sciences
Simon Fraser University
Burnaby, British Columbia, Canada

Russell D. Fernald, Ph.D.
Department of Psychology
Stanford University
Stanford, California

Ryozo Fujii, Ph.D.
Department of Biomolecular Science
Faculty of Science
Toho University
Funabashi, Chiba, Japan

Carter R. Gilbert, Ph.D.
Florida Museum of Natural History
University of Florida
Gainesville, Florida

Toshiaki J. Hara, Ph.D.
Department of Fisheries & Oceans
Freshwater Institute
and
Department of Zoology
University of Manitoba
Winnipeg, Manitoba, Canada

Jeffrey R. Hazel, Ph.D.
Department of Zoology
Arizona State University
Tempe, Arizona

Walter Heiligenberg, Ph.D.
Marine Biology Research Division
Scripps Institution of Oceanography, UCSD
La Jolla, California

Norbert Heisler, M.D.
Department of Animal Physiology
Humboldt University
Berlin, Germany
and
Max Planck Institute for Experimental Medicine
Göttingen, Germany

Susanne Holmgren, Ph.D.
Department of Zoophysiology
University of Göteborg
Göteborg, Sweden

Gordon McDonald, Ph.D.
Department of Biology
McMaster University
Hamilton, Ontario, Canada

Stefan Nilsson, Ph.D.
Department of Zoophysiology
University of Göteborg
Göteborg, Sweden

Reynaldo Patiño, Ph.D.
Texas Cooperative Fish & Wildlife Research Unit
Texas Technological University
Lubbock, Texas

Steve F. Perry, Ph.D.
Department of Biology
University of Ottawa
Ottawa, Ontario, Canada

Christopher Platt, Ph.D.
Department of Zoology
University of Maryland
College Park, Maryland

Arthur N. Popper, Ph.D.
Department of Zoology
University of Maryland
College Park, Maryland

J. Michael Redding, Ph.D.
Department of Biology
Tennessee Technological University
Cookeville, Tennessee

Paul W. Webb, Ph.D.
School of Natural Resources and Environment
Department of Biology
University of Michigan
Ann Arbor, Michigan

Sjoerd E. Wendelaar Bonga, Ph.D.
Department of Animal Physiology
University of Nijmegen
Nijmegen, The Netherlands

Chris M. Wood, Ph.D.
Department of Biology
McMaster University
Hamilton, Ontario, Canada

TABLE OF CONTENTS

Dedicated

to my family: Jean, Drew and Matt, and my parents in their 80th year

to my mentors: Russ Poppenhager, Forst Fuller, Loren Woods, Marion Grey, Howard Bern, Bill Potts, and Jean Maetz

1 Evolution and Phylogeny

Carter R. Gilbert

INTRODUCTION

The following represents a summary of the evolution and phylogeny of fishes, information for which has come from various sources. It has been necessary to condense a considerable amount of factual information because of space constraints, and a number of details that some might consider important have necessarily been omitted. For those readers who wish to obtain additional information on the subject, J. S. Nelson's (1984) *Fishes of the World* (the first edition of which appeared in 1976) is especially recommended. Included in the book are one or more excellent line drawings of each living fish family (and some extinct families), as well as a complete list, in phylogenetic sequence (Appendix 1), of the 50 living orders and 445 families recognized at that time. Nelson's book represents the most modern classification of fishes, and thus is the one followed throughout this chapter (Figure 1). In this state-of-the-art work (which involves both recent and fossil groups) are discussions of changing concepts of evolutionary relationships at all taxonomic levels and an extensive list of pertinent references. Although not all these references can be listed here, the reader is especially referred to the important treatises by Greenwood et al. (1966, 1973). Nelson's book appeared too early, however, to incorporate certain important changes emanating from chapters appearing in the book edited by Moser et al. (1984). References of a more general nature include Bailey and Cavender's (1971) review of fish classification and Carroll's (1988) *Vertebrate Paleontology and Evolution*. The latter is essentially an updated version of Romer's (1966) classic *Vertebrate Paleontology* (the last [third] edition of which was published in 1966), and contains a wealth of anatomical information relating to evolution of the vertebrates as a whole. Finally, Pough et al. (1989), in the third edition of their book *Vertebrate Life*, have done an excellent job of integrating the entire vertebrate story, with evolution being the connecting thread throughout.

JAWLESS FISHES (AGNATHANS)

FOSSIL HISTORY

Although the first fragmentary fish remains appear in marine fossil deposits of the Upper Cambrian period, over 550 million years ago (Figure 2), the earliest complete remains date from the Silurian period, at least 425 million years ago (Repetski, 1978). These were not conventional fishes of the type seen

0-8493-8042-1/93/$0.00+$.50

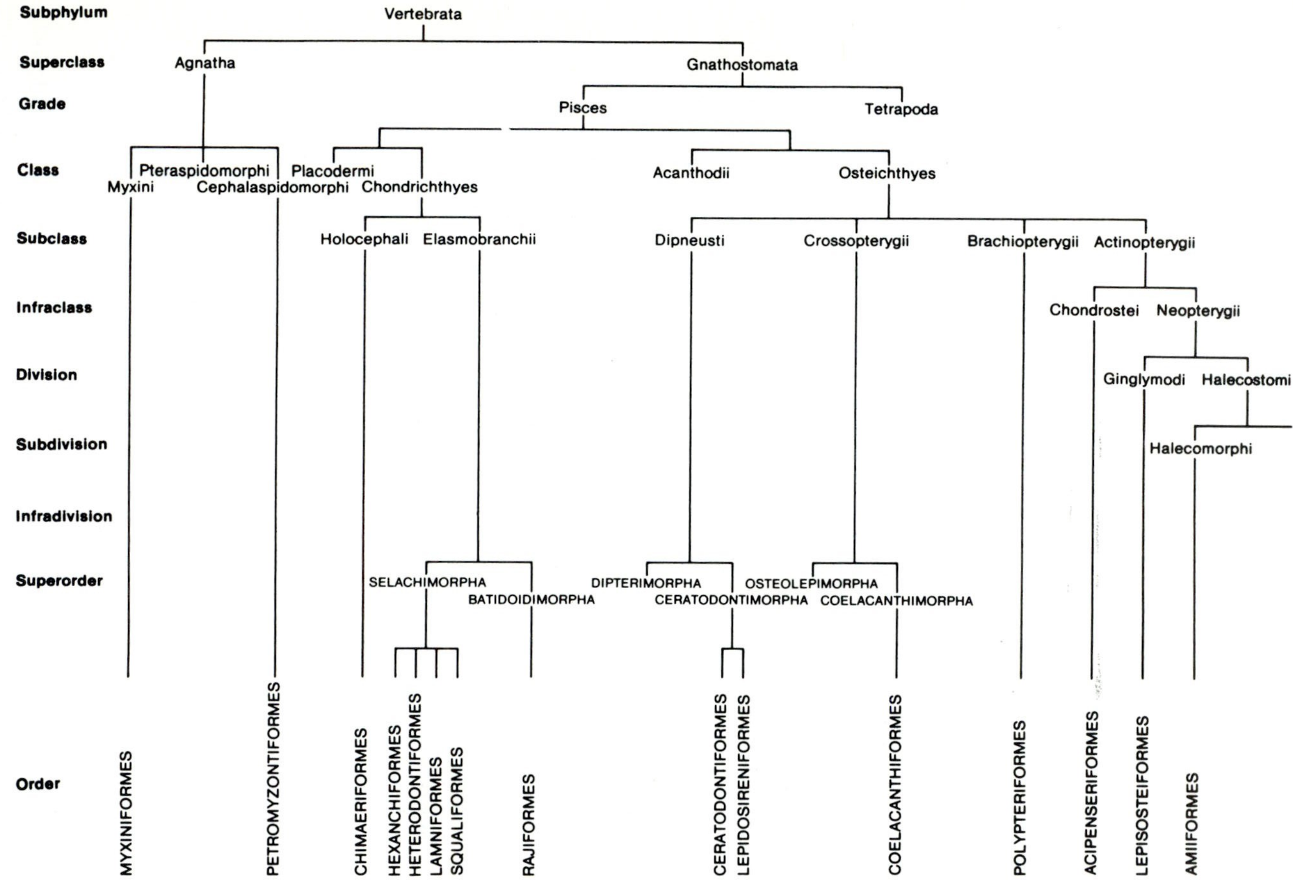
Subphylum
Vertebrata
Superclass
Agnatha
Gnathostomata
Grade
Pisces
Tetrapoda
Class
Pteraspidomorphi
Placodermi
Myxini
Cephalaspidomorphi
Chondrichthyes
Acanthodii
Osteichthyes
Subclass
Holocephali
Elasmobranchii
Dipneusti
Crossopterygii
Brachiopterygii
Actinopterygii
Infraclass
Chondrostei
Neopterygii
Division
Ginglymodi
Halecostomi
Subdivision
Halecomorphi
Infradivision
Superorder
SELACHIMORPHA
BATIDOIDIMORPHA
DIPTERIMORPHA
CERATODONTIMORPHA
OSTEOLEPIMORPHA
COELACANTHIMORPHA
Order
MYXINIFORMES
PETROMYZONTIFORMES
CHIMAERIFORMES
HEXANCHIFORMES
HETERODONTIFORMES
LAMNIFORMES
SQUALIFORMES
RAJIFORMES
CERATODONTIFORMES
LEPIDOSIRENIFORMES
COELACANTHIFORMES
POLYPTERIFORMES
ACIPENSERIFORMES
LEPISOSTEIFORMES
AMIIFORMES

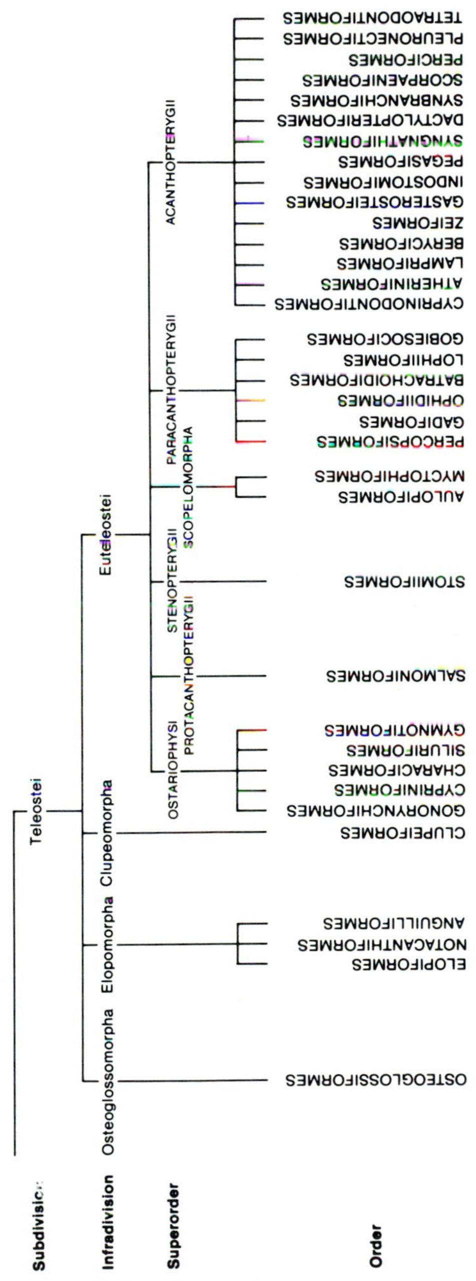

FIGURE 1. Hierarchy of higher categories of fishes. (From Nelson, J. S., *Fishes of the World*, 2nd ed., John Wiley & Sons, New York, 1984. With permission from copyright owner©, John Wiley & Sons.)

ERA	PERIOD	EPOCH	DURATION IN MILLIONS OF YEARS	TIME FROM BEGINNING OF PERIOD TO PRESENT (MILLIONS OF YEARS)	GEOLOGIC CONDITIONS	PLANT LIFE	ANIMAL LIFE
Cenozoic (Age of Mammals)	Quaternary	Recent	0.025	0.025	End of last ice age; climate warmer	Decline of woody plants; rise of herbaceous ones	Age of man
		Pleistocene	1	1	Repeated glaciation; 4 ice ages	Great extinction of species	Extinction of great mammals; first human social life
	Tertiary	Pliocene	11	12	Continued rise of mountains of western North America; volcanic activity	Decline of forests; spread of grasslands; flowering plants, monocotyledons developed	Man evolved from manlike apes; elephants, horses, camels almost like modern species
		Miocene	16	28	Sierra and Cascade mountains formed; volcanic activity in northwest U. S.; climate cooler		Mammals at height of evolution; first manlike apes
		Oligocene	11	39	Lands lower; climate warmer	Maximum spread of forests; rise of monocotyledons, flowering plants	Archaic mammals extinct; rise of anthropoids; forerunners of most living genera of mammals
		Eocene	19	58	Mountains eroded; no continental seas; climate warmer		Placental mammals diversified and specialized; hoofed mammals and carnivores established
		Paleocene	17	75			Spread of archaic mammal
	Rocky Mountain Revolution (Little Destruction of Fossils)						
Mesozoic (Age of Reptiles)	Cretaceous		60	135	Andes, Alps, Himalayas, Rockies formed late; earlier, inland seas and swamps; chalk, shale deposited	First monocotyledons; first oak and maple forests; gymnosperms declined	Dinosaurs reached peak, became extinct; toothed birds became extinct; first modern birds; archaic mammals common
	Jurassic		30	165	Continents fairly high; shallow seas over some of Europe and western U.S.	Increase of dicotyledons; cycads and conifers common	First toothed birds; dinosaurs larger and specialized; insectivorous marsupials
	Triassic		40	205	Continents exposed; widespread desert conditions; many land deposits	Gymnosperms dominant, declining toward end; extinction of seed ferns	First dinosaurs, pterosaurs and egg-laying mammals; extinction of primitive amphibians

	Appalachian Revolution (Some Loss of Fossils)						
Paleozoic (Age of Ancient Life)	Permian		25	230	Continents rose; Appalachians formed; increasing glaciation and aridity	Decline of lycopods and horsetails	Many ancient animals died out; mammal-like reptiles, modern insects arose
	Pennsylvanian		25	255	Lands at first low; great coal swamps	Great forests of seed ferns and gymnosperms	First reptiles; insects common; spread of ancient amphibians
	Mississippian		25	280	Climate warm and humid at first, cooler later as land rose	Lycopods and horsetails dominant; gymnosperms increasingly widespread	Sea lilies at height; spread of ancient sharks
	Devonian		45	325	Smaller inland seas; land higher, more arid; glaciation	First forests; land plants well established; first gymnosperms	First amphibians; lungfishes, sharks abundant
	Silurian		35	360	Extensive continental seas; lowlands increasingly arid as land rose	First definite evidence of land plants; algae dominant	Marine arachnids dominant; first (wingless) insects; rise of fishes
	Ordovician		65	425	Great submergence of land; warm climates even in Arctic	Land plants probably first appeared; marine algae abundant	First fishes, probably freshwater; corals, trilobites abundant; diversified molluscs
	Cambrian		80	505	Lands low, climate mild; earliest rocks with abundant fossils	Marine algae	Trilobites, brachiopods dominant; most modern phyla established
	Second Great Revolution (Considerable Loss of Fossils)						
Proterozoic			1500	2000	Great sedimentation; volcanic activity later; extensive erosion, repeated glaciations	Primitive aquatic plants—algae, fungi	Various marine protozoa; towards end, molluscs, worms, other marine invertebrates
	First Great Revolution (Considerable Loss of Fossils)						
Archeozoic			???	???	Great volcanic activity; some sedimentary deposition; extensive erosion	No recognizable fossils; indirect evidence of living things from deposits of organic material in rock	

FIGURE 2. Geologic timetable. Beginning times indicated for some geologic periods may differ somewhat from those given in text. (From Dodson, E. O., *A Textbook of Evolution*, W. B. Saunders, Philadelphia, 1952. With permission.)

today, but rather were members of a long-extinct group known as the ostracoderms (Figure 3). For many years the exact relationships of the ostracoderms to living fishes were not understood. This situation changed in 1927, when Stensio, in a classic study of the abundant ostracoderm fossil material from Spitzbergen Island (off the coast of Norway), concluded that these animals share their closest phylogenetic relationships with the living lampreys and hagfishes (the so-called cyclostomes). In outward appearance these groups seem to have little in common, the more or less fish-shaped body and bony external covering of the ostracoderms being radically different from the eel-like body and absence of a hard external body covering characteristic of the cyclostomes. Stensio (1927) nevertheless showed that these two groups share a number of basic anatomical similarities found in no other vertebrates, including absence of jaws and true teeth, complete absence of paired fins and limb girdles, a single nare situated medially on top of the head, either one or two chambers in the semicircular canal of the inner ear, a persistent and complete notochord in adults, and multiple (i.e., more than seven) independent external gill openings. He concluded that all belong to the same basic group, which has been termed the class Agnatha (Nelson [1984] elevated the Agnatha to a superclass), in reference to the absence of jaws.

Lampreys and hagfishes have been assumed to have had a long evolutionary history, dating back well into the Paleozoic era. The primary basis for this belief was the disappearance of other agnathans from the fossil record after the Devonian period, around 360 million years ago. Unfortunately, the cartilaginous endoskeleton and absence of other hard body parts makes the lampreys and hagfishes very poor candidates for fossilization, so that direct evidence relating to the evolution of these animals was consequently missing until recently. In 1968, however, Bardack and Zangerl described a fossil lamprey, *Mayomyzon pieckoensis*, from coal deposits from the Pennsylvanian (Middle) period (dating back about 300 million years) near Chicago. The 14 known specimens (8 of which were discovered after the original description [Bardack and Zangerl, 1971]) are in good condition (some are whole-animal casts), with 11 having been preserved in lateral aspect and 3 in a dorso-ventral plane. A fossil hagfish (still undescribed) was subsequently found in the same deposits as *Mayomyzon* (Carroll, 1988; Pough et al., 1989). These fossils show conclusively that lampreys and hagfishes were well differentiated by the mid-Pennsylvanian, and had split off from other agnathans by the Devonian period, if not earlier.

Classes of Agnathans

Lampreys and Hagfishes

There are over 70 living species of lampreys and hagfishes, of which about 60% are lampreys (Nelson, 1984). As a group the lampreys are euryhaline, although the majority of species live entirely in fresh water. They are about

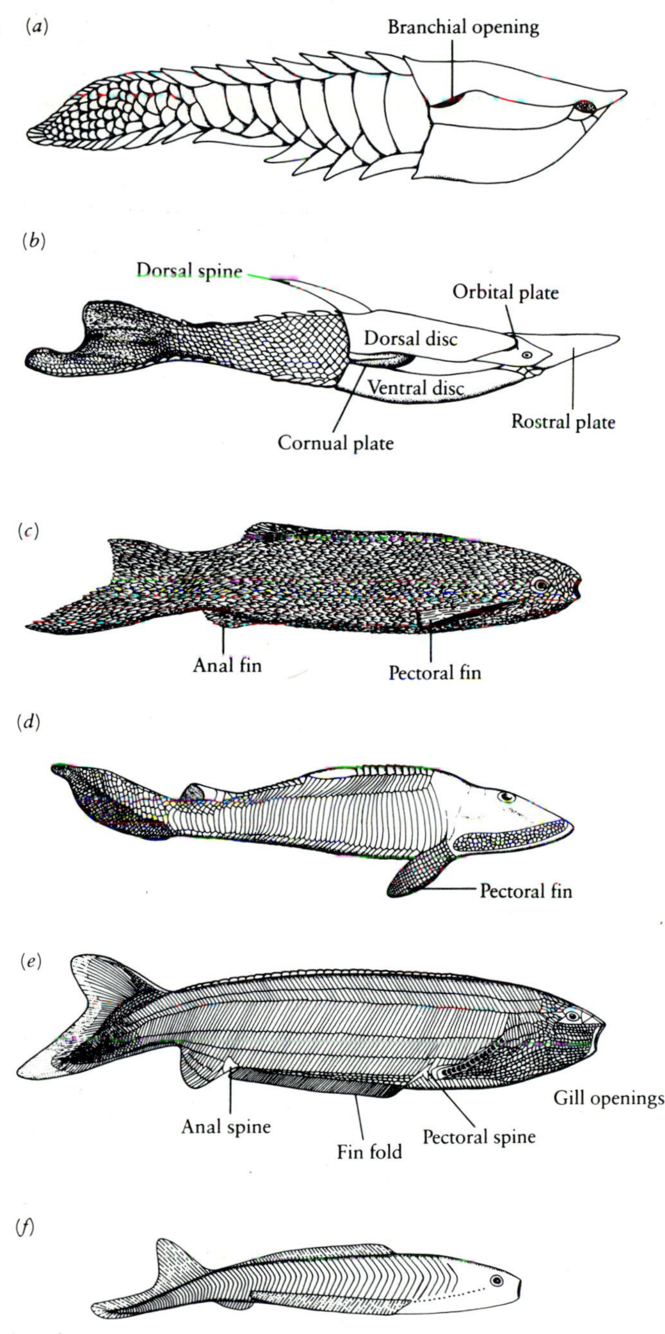

FIGURE 3. Representative ostracoderms. (a) *Anglaspis*; (b) *Pteraspis*; (c) *Phlebolepis*; (d) *Hemicyclaspis*; (e) *Pharyngolepis*; (f) *Jamoytius*. (From Moy-Thomas, J. A. and Miles, R. S., *Palaeozoic Fishes*, 2nd ed., Chapman and Hall, London, 1971. With permission.)

equally divided between parasitic and nonparasitic forms, of which the latter feed only during the larval (or ammocoete) stage and the adults live only long enough to reproduce. Hagfishes are completely stenohaline, and are the only vertebrates in which the body fluids are isosmotic with seawater. Unlike lampreys, the hagfishes lack a larval stage. Lampreys have an antitropical, or bipolar, distribution, being absent from the tropics but displaying a broad distribution in temperate areas of the northern hemisphere and a widely disjunct distribution in the southern hemisphere (South America, Australia, and New Zealand) (Nelson, 1984). Hagfishes are distributed in temperate marine waters throughout the world.

Ostracoderms

Ostracoderms comprise at least two distinct phyletic groups, the pteraspids (or heterostracans) (Figure 3a–c) and cephalaspids (including the anaspids) (Figure 3d–f), each of which can be further subdivided into recognizable subdivisions (Pough et al, 1989). One of the main distinctions between pteraspids and cephalaspids relates to the body covering, which consists of large bony plates in the former group and smaller units (sometimes only denticles) in the latter. Stensio (1968) believed that lampreys and hagfishes evolved independently from the cephalaspid and pteraspid lines, respectively, a conclusion based primarily on differences in mode of formation of the snout. Other paleontologists have come to the opposite conclusion, contending that cyclostomes are of monophyletic origin and probably were derived from the cephalaspid line, specifically from an anaspid or anaspid-like ostracoderm (see numerous references in Bardack and Zangerl [1971]). Among other things, they believed that the complex reconstructions of the snout on which Stensio (1968) based his conclusions are too hypothetical to justify the conclusions reached. Bardack and Zangerl (1971), however, concluded that since *Mayomyzon pieckoensis* shows no hagfish characters, and since hagfish and lamprey remains have been found in the same mid-Pennsylvanian fossil deposits, it seems reasonable to assume that the two groups of living cyclostomes are not intimately related and have had a long, separate phyletic history. Other recent research seems to support this idea. Nelson (1984) pointed out that lampreys differ from hagfishes in a number of characters that are, in turn, shared with jawed vertebrates, including highly differentiated kidney tubules, absence of a persistent pronephros, more than one semicircular canal (lampreys have two, higher vertebrates have three), a large exocrine pancreas, a photosensory pineal organ, vertebral elements, histology of the adenohypophysis, and composition of the body fluids. He considered these differences to be of sufficient magnitude to justify placement of the lampreys and hagfishes in two separate classes, the Cephalaspidomorphi and Myxini, respectively.

Ritchie (1968) showed that *Jamoytius kerwoodi* (Figure 4), an anaspid-like ostracoderm from Silurian (Middle) deposits of Scotland, resembles lampreys in several important particulars, including a subterminal, circular mouth (per-

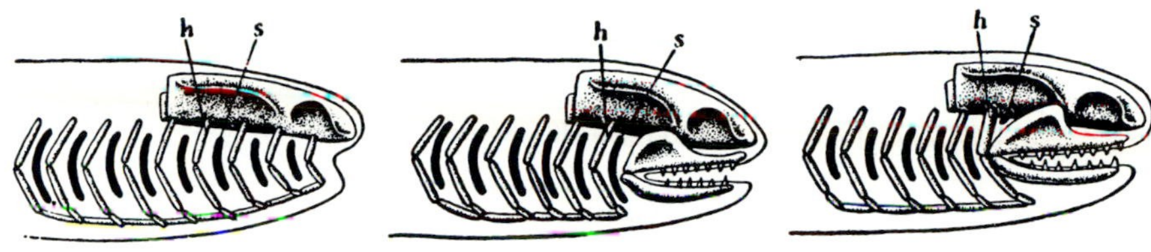

FIGURE 4. Diagrams to show stages in evolution of jaw structure in vertebrates. **Left**, jawless stage found in the Agnatha; **center**, the hypothetical autostylic condition, in which jaw support is limited to the overlying skull; **right**, the amphistylic condition found in all living jawed vertebrates, in which the hyoid arch is specialized as the hyomandibular bone (h) and the spiracle (s) is reduced. (From Romer, A. S., *Vertebrate Paleontology,* 2nd ed., University of Chicago Press, Chicago, 1945. With permission.)

haps with an annular cartilage) and a branchial basket beginning immediately behind the orbit. Although *Jamoytius* retains various ostracoderm features, including lateral fin folds and body scales, both Ritchie (1968) and Bardack and Zangerl (1971) nevertheless concluded that it belongs to the group of ostracoderms from which lampreys most likely evolved.

Although the ostracoderms continued to thrive until the end of the Devonian period, several major developments had begun to occur earlier that were to be of immense significance in the evolution of the vertebrates. Most important of these was the appearance of jaws, which have been identified in Silurian (Middle) fossils. Also appearing shortly thereafter in some of these jawed vertebrates were paired fins attached to internal fin girdles. The hinged jaws (accompanied by teeth) and paired fins together provided several important selective advantages: (1) an increased range of available food, (2) increased ability to utilize food (i.e., large food items could now be broken down into smaller particles), and (3) greater ease of acquisition. The fossil agnathans, by contrast, are presumed to have been mostly, if not entirely, filter feeders, and thus were limited to types of food small enough to be taken in through the oral opening, with no further reduction in size. Although some ostracoderms may have been parasitic, as is the case with many living agnathans, physical limitations (rounded or flattened body, heavy body armor, and limited mobility), coupled with the probable scarcity of suitable hosts, seem to argue against this. Despite the above disadvantages, the ostracoderms continued to survive for many millions of years while living side by side with primitive jawed vertebrates. As Pough et al. (1989) have noted, "It may seem strange that a major new morphological feature like jaws should have arisen before the extensive radiation of agnathans had occurred, instead of arising from some later product of that radiation. This pattern of evolution, however, is seen over and over again in an examination of vertebrate life: major new innovations arise from less specialized members of a lineage."

One may ask why lampreys and hagfishes alone were able to perpetuate the agnathan line, 360 million years after the disappearance of all other members of this group. The standard (if overly simplistic) explanation for their survival is that "the ecological niches occupied by these animals were so suited to their body plan that no evolutionary advance yet realized has been able to replace it" (Pough et al., 1989).

EVOLUTIONARY ADVANCEMENTS

Jaws are derived from the anterior branchial arches (or bars), which are located between the gill openings. Initially (i.e., in agnathans), these arches formed a united set of structures fused dorsally to the head region that served to stiffen the gill (or branchial) region and afford support for the muscles involved in opening and closing the gill slits (Figure 4 [left]). Subsequently, the branchial arches became separated from the skull and divided into upper and lower halves, which were bent somewhat on each other. This was accompanied by modification of one of the three anteriormost bars (most likely the third) into jaws, with the upper part becoming the palatoquadrate and the lower the mandible (Figure 4 [right]). Initially, the only support for the jaws came from the overlying skull (the so-called autostylic condition), in which case the articulation would have been a movable one and jaw function therefore not particularly effective (see Figure 4, center figure). This condition was essentially transitory and largely presumptive, since, contrary to previous belief, it is absent from most of the earliest jawed vertebrates. (In some groups of living fishes, notably the chimaerids and lungfishes, there does exist an autostylic condition, but this is a highly specialized situation [sometimes termed "holostyly"] involving actual fusion of the upper jaws with the braincase.) Subsequently, the upper part of the fourth branchial arch became enlarged and modified to form the hyoid arch or hyomandibular bone (hyomandibular cartilage in chondrichthyan fishes) (Figure 4 [right]), which provided additional support for propping the jaws against the braincase. This condition is termed amphistyly, and was already present in most of the earliest known jawed fossil vertebrates, including all of the acanthodians and at least some of the placoderms. Later, in higher actinopterygian fishes and modern sharks, the hyomandibular bone (or cartilage) came to bear the main burden of jaw support, a condition termed hyostyly. In conjunction with these changes was a reduction in size of the fourth gill slit, which in some fish groups (most notably the bottom-dwelling chondrichthyans) is retained as a spiracle and serves as a duct for transmission of water down to and across the gills.

JAWED FISHES (GNATHOSTOMES)

Although evolution of jawed fishes commenced in the Silurian period, it was during the Devonian that the proliferation of these animals becomes particularly evident in the fossil record, a circumstance that has resulted in this period being called the "Age of Fishes". In addition to hinged jaws and paired fins, these more advanced fishes differed from agnathans in a number of other major features, including presence of paired lateral nares and a three-chambered semicircular canal. Once acquired, these characters were not only retained in the fishes but in all other subsequent vertebrate groups as well. Nelson (1984) placed all fishes possessing the above features in the superclass Gnathostomata, within which he recognized four separate classes: Placodermi, Chondrichthyes, Acanthodii, and Osteichthyes. Although surely derived from agnathans, these four groups appear in the fossil record fully developed, without intermediate connecting links. The osteichthyans (bony fishes) and chondrichthyans (sharks, skates, rays, and chimaerids) survive today, whereas the placoderms and acanthodians became extinct in the Mississippian and Permian (Lower) periods, respectively.

ACANTHODII

The first of the four gnathostome classes to appear in the fossil record were the acanthodians (Figure 5), which are known from isolated spines from Silurian (Early) marine deposits. Were these spines the only evidence available, their determination would be extremely difficult, but subsequent finds of associated teeth and scales from later in the Silurian, and of complete specimens from Devonian to early Permian freshwater deposits, provide positive identification. As might be expected for such an early group, acanthodians possess a suite of characters that are primitive for gnathostomes, and for this reason they have at various times been considered to bear their closest relationships to each of the other three classes of fish-like vertebrates. During the 1930s to 1950s, for example, the acanthodians were included in the class Placodermi in a number of authoritative references (Romer, 1945). Superficially, however, the acanthodians most closely resembled the bony fishes (osteichthyans), especially in the presence of large anterior eyes and a terminal or near-terminal mouth. Conventional fins were absent, but instead were represented by a series of hard, immovable spines, often with attached membranes, which were located in the same approximate positions on the body as fins and presumably functioned in a similar, though much less efficient, way. Although there is no evidence that these spines evolved directly into the fins of modern fishes, the "fin-fold theory" of the origin of paired appendages has long been held in high regard (recently this theory has come under attack). Some have advocated a close relationship to the chondrichthyans (Nelson, 1968), but the two groups exhibit such fundamental differences in dental histology and development (the teeth in acanthodians lack an enamel-like surface and show no evidence of regular replacement) that this idea has gained little support (Carroll, 1988). Recent studies indicate a number of functional and morphological similarities between acanthodians and osteichthyans, especially involving the cranium and jaws, and these provide strong evidence that the acanthodians are the sister group to the osteichthyans. In particular, these

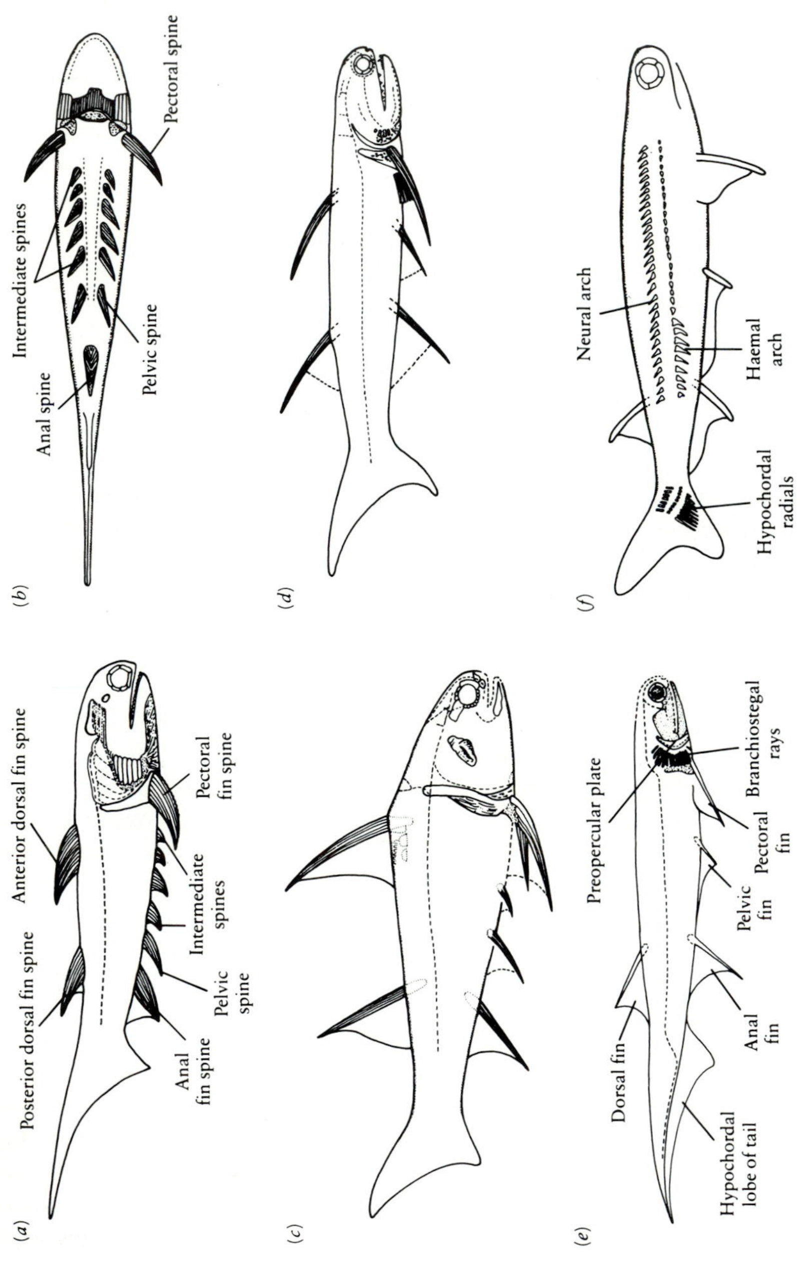

FIGURE 5. Representative acanthodians. (a) *Climatius*; (b) *Euthacanthus*; (c) *Diplacanthus*; (d) *Ischnacanthus*; (e) *Homalacanthus*; (f) *Acanthodes*. (From Moy-Thomas, J. A. and Miles, R. S., *Palaeozoic Fishes*, 2nd ed., Chapman and Hall, London, 1971. With permission.)

two groups are defined by a unique mechanism for opening the mouth, as discussed by Lauder and Liem (1983) and summarized by Pough et al. (1989).

PLACODERMI

The placoderms (or armored fishes) are first known from Devonian (Early) deposits, but had disappeared by the early Mississippian period. They probably can best be described as looking like ostracoderms with jaws (Figure 6). Although placoderms share an impressive list of derived characters with other gnathostomes, several fundamental aspects of their morphology appear to isolate them from all other jawed vertebrates. The most significant of these are the positions of the jaw muscles, which in all other gnathostomes lie external to the jaw's skeletal elements but in the placoderms are situated medially. This difference is sufficiently fundamental as to suggest that jaws may have evolved more than once among primitive fishes. Placoderms also lacked teeth that correspond to those of any other gnathostomes, and they had a hyoid arch that was distinctively different in arrangement and number of elements from that of other vertebrates. They have no modern analogs, and an increasing number of paleontologists have come to believe that the placoderms likely represent an evolutionary dead-end, with no known descendants (Pough et al., 1989). This represents a major conceptual change from fish classifications appearing from the 1930s through the 1950s (Romer, 1945). According to those arrangements,

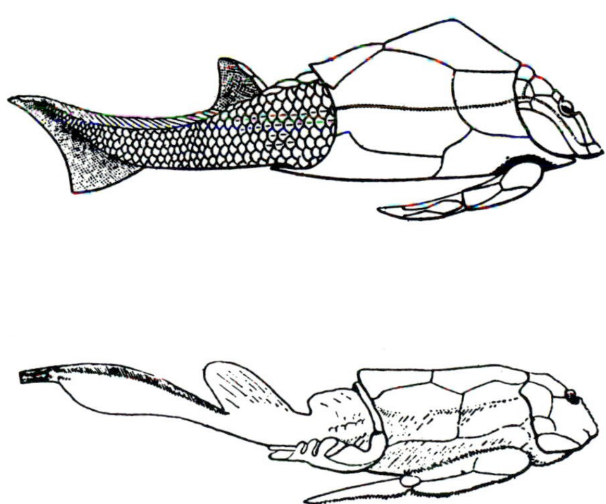

FIGURE 6. Representative placoderms. **Top**, *Pterichthyodes*; **bottom**, *Bothriolepis*. (From Romer, A. S., *Vertebrate Paleontology*, 2nd ed., University of Chicago Press, Chicago, 1945. With permission.)

the acanthodians were included in an expanded class Placodermi, which in turn was considered a transitional stage that was directly ancestral to all other classes of vertebrates. Considering this, it is interesting to note that Nelson (1984) has suggested (see following discussion) that the chondrichthyans may be more closely related to the placoderms than to any of the other groups under consideration.

Chondrichthyes

Although the earliest fossil remains attributed to the class Chondrichthyes (the literal translation of which is "cartilaginous fishes") are denticles from the Silurian (Late) period (Zangerl, 1981), the first complete remains are from the Devonian (Early). Chondrichthyans are characterized by a cartilaginous endoskeleton, a unique dental histology and development, the usual presence of urea in the blood and tissues, the inner margins of the pelvic fins in males modified into reproductive structures called "claspers", no esophageal diverticulum that functions either as a respiratory organ or buoyancy device, and an unusually well-developed electrosensory system. Living forms can be divided into two distinct groups. One is characterized by four pairs of gill arches and a single external gill opening on each side of the head, and is placed in the subclass Holocephali (rattails or chimaerids). The other has five to seven separate gill openings on each side and comprises the subclass Elasmobranchii (sharks, skates, and rays). In addition to the above, holocephalans differ from elasmobranchs in having a smooth epidermis, a long flexible tail, large eyes, absence of a cloaca, absence of ribs, and the upper jaw fused with the braincase (hyostyly). All chimaerids have the teeth consolidated into grinding plates, an adaptation for feeding on the molluscs that comprise their chief food. Although divergence of these two subclasses is thought to have begun in the Pennsylvanian period, the first undoubted holocephalans are actually of Jurassic age (Pough et al., 1989).

Most classifications of the subclass Elasmobranchii involve a dichotomous arrangement, in which the sharks are included in one group (the superorder Selachimorpha) and the skates and rays (or batoids) are in a second (the superorder Batidoidimorpha) (see Appendix). Compagno (1973, 1977), however, proposed several changes in this traditional scheme, which is centered around a more intimate relationship of the skates and rays to sharks in general. In his new classification, Compagno recognized four superorders of elasmobranchs, three of which are comprised of sharks and the fourth the skates and rays. This arrangement has not been universally accepted, however, and Nelson (1984) has indicated that further changes may be in order when research now underway is published.

The ancestral group of the chondrichthyans is unknown. According to Nelson (1984), there is no good evidence to indicate derivation from either placoderms or acanthodians, although he postulated this group may be more closely related to the placoderms (Figure 1). A factor presumably involved in this assumption is the presence of pelvic claspers in both groups (Moy-Thomas and Miles, 1971).

There are almost 800 living species of chondrichthyans, including about 425 skates and rays, 340 sharks, and 30 chimaerids (Nelson, 1984). The basic body form appears to have changed relatively little during their evolutionary history, although several bizarre forms (none of which survive today) are known from the fossil record. All available evidence suggests that the group had a marine origin, and with few exceptions all living species live in a marine environment. The most notable exceptions are the freshwater stingrays of South America (comprising an endemic family and about 14 species), in which the urea concentration of the blood and tissues has been reduced to the point that the group is now incapable of living in salt water.

Osteichthyes

The fourth major line of gnathostomes, the class Osteichthyes (which, translated, means "bony fishes"), comprises by far the greatest number of living species of any class of vertebrates, with best estimates placing the number at around 21,000. Fragmentary remains are known from the Silurian (Late), and more complete remains from the Devonian (Early), periods. Since these animals resemble acanthodians in details of head structure, it is thought these similarities may indicate a common ancestor of the two groups in the Silurian (Early). As might be expected for a group so large and varied, and whose phylogenetic interrelationships are still being studied by a large number of scientists, concepts are constantly changing.

Within the class Osteichthyes, Nelson (1984) recognized four subclasses (Figure 1), each of which contains one or more living species; of these, the Actinopterygii (ray-finned fishes) (Figure 7) comprises all but 18 of the total number of living osteichthyans. The other subclasses include the Dipneusti (lungfishes, six living species [Figure 7]); Crossopterygii (lobe-finned fishes, one living species [Figure 7]); and Brachiopterygii (bichirs, 11 living species [Figure 7]). Some classifications recognize only two subclasses, the Sarcopteryii, to include the lungfishes and lobe-finned fishes; and the Actinopteryii, which combines the bichirs with the remaining ray-finned fishes (Figure 7). For the sake of convenience, we will continue to use the term "sarcopterygian" in the following discussion.

Remains of the bony fishes indicate that radiation already was well underway by Early to Middle Devonian, with three of the four subclasses recognized by Nelson (1984) being known from fossils of that age (Dipnesti, Crossopterygii, and Actinopterygii). Interestingly, the earliest fossils of the fourth subclass (Brachiopterygii) date back only to the Eocene epoch of Africa, about 60 million years ago (Romer, 1966; Carroll, 1988).

Despite the class name, bone is not uniformly present in all species of osteichthyans. The skeleton of the great majority of species is comprised mostly of this material, but cartilage is always present to some degree and in

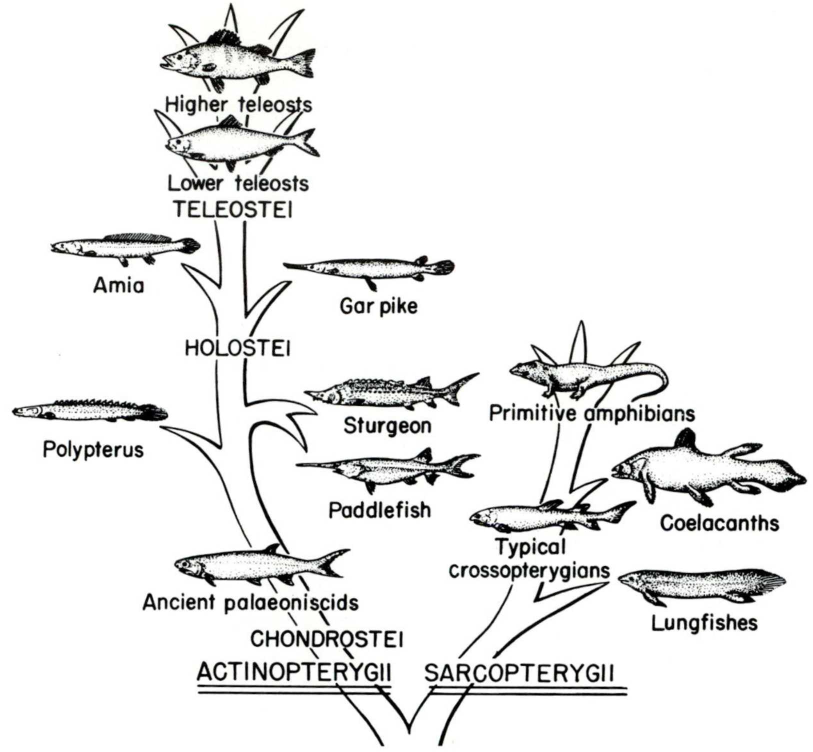

FIGURE 7. An earlier simplified concept of relationships of the bony fishes, including illustrations of various major fish groups discussed in text. (From Romer, A. S., *Man and the Vertebrates,* University of Chicago Press, Chicago, 1959. With permission.)

some cases bone has been lost entirely. Also, the presence of bone is not unique to this group, since agnathans, placoderms, and acanthodians also possessed true bone in diverse regions of their bodies. All four osteichthyan subclasses are characterized by similarly derived patterns of the lateral-line canals, similar bony elements in the operculum and pectoral girdle, fin webs supported by bony dermal rays, and a gas-filled diverticulum of the esophagus that functions as an accessory respiratory organ and buoyancy device. Many forms have two dermal bones tightly associated with the palatoquadrate bone, where they form the upper biting edge of the mouth (the premaxilla and maxilla), and to which are typically fused enamel-coated teeth. A neurocranium with anterior and posterior ossified sections separated by a fissure allows movement between the two halves of the skull in many forms.

CLASSIFICATION OF OSTEICHTHYANS

Carroll (1988) and Pough et al. (1989), who have adopted the more traditional scheme of osteichthyan classification (i.e., recognition of two subclasses, Sarcopterygii and Actinopterygii), listed as major unifying characters of primitive sarcopterygians (1) cosmoid scales, in which the scales were comprised of a substance called cosmine and were invaded and nurtured by the vascular system; (2) two dorsal fins with separate girdles; (3) an epichordal lobe on a heterocercal caudal fin; and (4) paired fins with a fleshy, scaled, and bony central axis (Figure 7). The paired-fin rays extend in a feather or compound leaf-like manner, in contrast to the fan-like form of paired fins in actinopterygians. Open passages from the nares to the throat (choanae) were present in some groups, including the lungfishes and one extinct line of crossopterygian fishes. As discussed subsequently, inclusion of the dipnoans with the sarcopterygians is somewhat controversial, since lungfishes differ in several important respects from the situation described above and their exact relationships to other fishes remain the subject of debate.

Actinopterygian fishes (1) lack cosmoid scales, (2) have a single dorsal fin with a single girdle (the fin may be subdivided into two or more sections in some species); (3) have a heterocercal or homocercal caudal fin with no epichordal lobe, and (4) have fins that are comprised only of rays, without a bony central axis (Figure 7). Choanae are invariably absent. Scales are usually present but may be secondarily lost; when present, they may contain a hard substance known as ganoin, but in most species this substance is lacking and the scales are consequently lighter and more flexible, with the edges either smooth or containing numerous comb-like teeth or ctenii. At no time does the vascular system impinge on the scales.

DIPNEUSTII

The earliest dipnoans, which are found in marine fossil deposits from the Devonian (Lower), already possessed characters distinct from other osteichthyans. In additional to functional lungs, lungfishes have the premaxilla and maxilla absent from the upper jaw, the palatoquadrate fused to an undivided cranium, and the teeth scattered over the palate and fused into tooth ridges along the lateral palatal margins. Although these basic features are retained today, the living species have undergone several changes, including loss of the anterior dorsal fin; loss of the branchiostegal rays and gular plate; modification of the caudal fin from a heterocercal to a symmetrical (diphycercal) condition; fusion of the dorsal and anal fins with the caudal into a single undivided unit around the posterior third of the body; and reduction in number of dermal bones in the skull, together with loss of the cosmine cover. Finally, in five of the six living species the pectoral and pelvic fins are greatly reduced in size and bony elements are lacking in the fins.

Living lungfishes occur in three geographically isolated regions (Australia, South America, and Africa) and comprise two distinct phyletic groups. The Australian species (*Neoceratodus forsteri*) is characterized by a relatively heavy body, large paddlelike pectoral and pelvic fins, large scales, an unpaired lung, larvae without external gills, and lack of estivation in the adult. The South American and African lungfishes (the latter group with four species) are obviously closely related, and are distinguished by a more slender body, slender (often threadlike) fins, small scales, paired lungs, the larvae with external gills, and estivation in the adults. Recently, the idea embraced by many early workers that dipnoans were directly ancestral to amphibians was revived by Gardiner (1980) and by Rosen et al. (1981), who at the same time rejected the generally accepted view that the crossopterygian genus *Eusthenopteron* (Figure 8) is close to the ancestral tetrapods. They based this partly on the conclusion that the internal (excurrent) nostril of Recent lungfishes is a true choana. Nelson (1984) noted that resolution of this problem is not easily reached, and probably can be resolved only through the discovery and study of additional fossil material. He concluded, however, with the view that the present weight of evidence still supports the concept that crossopterygians are the ancestral group to the amphibians.

Despite their frequent inclusion with the sarcopterygian fishes, there is disagreement regarding the exact relationships of the lungfishes to other osteichthyans, as well as to the amphibians. This is reflected in analyses of osteichthyan relationships, which invariably show dipnoans branching off near the base of the phylogenetic tree (Figure 7). It is generally thought that the many similarities between dipnoans and amphibians are the result of convergent evolution, and that amphibians evolved from rhipidistian crossopterygian fishes, specifically certain of those in the extinct order Osteolepiformes.

Crossopterygii

The crossopterygian fishes (also called "fringe-finned" or "tassel-finned" fishes) comprise two distinct phyletic lines. Most workers now consider them to be polyphyletic, and comprised of two equally ancient sister groups. One of these (the superorder Osteolepimorpha, or rhipidistians) (Figure 8) has been extinct since the Permian (Lower) period, but is important because most believe it gave rise directly to the amphibians and ultimately to all other

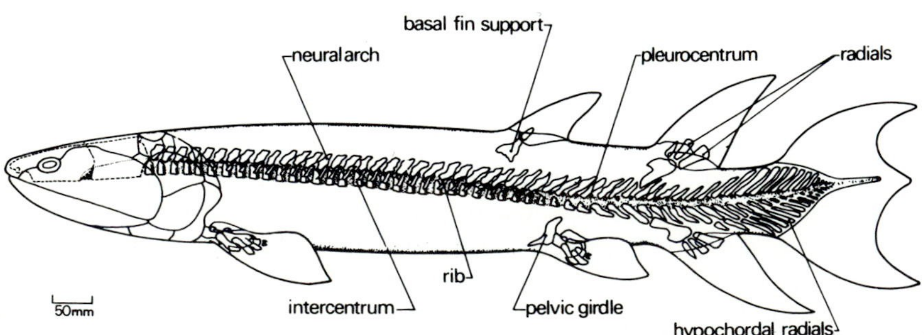

FIGURE 8. The rhipidistian crossopterygian fish, *Eusthenopteron*, with scales omitted to show post-cranial skeleton. (From Moy-Thomas, J. A. and Miles, R. S., *Palaeozoic Fishes*, 2nd ed., Chapman and Hall, London, 1971. With permission.)

vertebrate animals. They are characterized by internal nostrils (choanae), the presence of branchiostegal rays (4 to 13 in number), the fin rays (lepidotrichia) branched, many more lepidotrichia than radials in the caudal fin, and the first dorsal fin situated posterior to the middle of the body (see Carroll [1988] for additional details). Externally, the skull of rhipidistians resembles that of early actinopterygian fishes in its general features, a major difference being that in the latter group the bones of the skull roof hinge on one another above the division between the anterior and posterior units of the braincase. The pattern of dermal bones in osteolepiform rhipidistians such as *Eusthenopteron* is very similar to that of early amphibians, and is one of the reasons why that genus is believed to belong to the line of fishes that gave rise to the amphibians during the Devonian period.

The rhipidistians were the dominant freshwater predators among the bony fish during the late Paleozoic, and many specializations of the group may be attributed to their living in shallow waters. Although drying of the continents has been the reason traditionally given as the impetus that led to evolution of amphibians from rhipidistian fishes, many contend that the reverse (i.e., gradual flooding) offers an even more plausible explanation.

The other major line of crossopterygian fishes (the superorder Coelacanthimorpha, or actinistians) thrived in fresh waters from the Devonian to the Cretaceous periods. No Cenozoic fossils are known, and it was long believed that this group had disappeared about 75 million years ago. Thus, the discovery in 1938 of a living coelacanth from marine waters off southeast Africa remains probably the most startling scientific find of this or any other century. The living form is very similar morphologically to reconstructions of extinct coelacanths based on fossil remains. The circumstances leading to the initial discovery and subsequent rediscovery of this fish, which was named *Latimeria chalumnae*, have been recounted in a book written by Smith (1956), the fish's describer. Actinistians differ from rhipidistians in having a diphycercal tail consisting of three lobes, absence of choanae, no branchiostegal rays, lepidotrichia never branched, the lepidotrichia in the tail either equal to the number of radials or somewhat more numerous, and the anterior dorsal fin in advance of the center of the body.

As mentioned earlier, controversy surrounds the interrelationships of the crossopterygians. Most agree that these fishes are polyphyletic, the basis of which mostly centers around details of anatomy of the head and paired fins. Wiley (1979) considered *Latimeria* to be the primitive sister group of all other osteichthyans, and some have even gone so far as to suggest that genus might be more closely related to chondrichthyans.

Brachiopterygii

Nelson's (1984) subclass Brachiopterygii (or Cladistia) is a small group comprising a single order and family of freshwater fishes (Polypteriformes: Polypteridae), the living bichirs of central and western Africa. There is no fossil evidence to suggest that this family has ever occurred outside of Africa. The group is characterized by a large suite of primitive features, including a ventrally situated, two-lobed vascularized air bladder, intestine with a spiral valve, presence of a spiracle, a pair of gular plates situated ventrally between the jaws, absence of an interopercle, no branchiostegal rays, the maxillary firmly united to the skull, and rhombic ganoid scales. Considering the extremely primitive morphology of these fishes, the absence of pre-Cenozoic fossil remains is surprising, particularly when compared to the antiquity of the other three subclasses, whose fossil histories all date back to the Devonian. The bichirs have been allied to the dipnoans by some (Nelson, 1969; Mok, 1981), but in most classifications (including Romer [1966] and Carroll [1988]) they have been regarded as among the most primitive of living actinopterygian fishes. Schaeffer (1973) believed that the brachiopterygians are derived from an early line of palaeoniscoid fishes (see subsequent discussion), and Rosen et al. (1981) and Patterson (1982) regarded them as the primitive sister group to other actinopterygian fishes.

Actinopterygii

The subclass Actinopterygii comprises by far the largest group of bony fishes, including all but a tiny fraction of the estimated 21,000 living species of osteichthyans. Since their initial appearance in the Devonian period, this group has undergone a progression of morphological changes. Comparisons of fossil and recent actinopterygians permit identification of primitive vs. advanced morphological characters, the combinations of which resulted in recognition of three superorders (Chondrostei, Holostei, and Teleostei) in early classifications (Romer, 1945, 1959) (Figure 7). Of these, the chondrosteans include the most primitive combination of characters and the teleosteans the most advanced, with the holosteans occupying an intermediate position. Among the more obvious of these are a heterocercal tail (with an unrestricted notochord) vs. a homocercal tail (with a restricted notochord); the air bladder a two-lobed, ventrally situated vascularized organ vs. a dorsally situated, nonvascularized structure with a single lobe; presence (vs. absence) of a spiracle; hard scales containing ganoin vs. softer, more flexible scales in which the ganoin has been lost; and changes in size and relative position of skull bones, together with an overall reduction in number of bones. As conceived in earlier classifications, the chondrosteans were considered to include the Polypteriformes (bichirs) (Figure 7) and Acipenseriformes (sturgeons and paddlefishes) (Figure 7), the holosteans the Amiiformes and Lepisosteiformes (bowfins and gars, respectively) (Figure 7), and the teleosteans all remaining orders. These superorders are now known to be artificial phyletic units, and merely reflect a progression of primitive to more advanced morphological character states through time. Greenwood et al. (1966), for example, postulated that holostean-level fishes independently gave rise to higher-level teleostean-level groups on at least three different occasions. Patterson (1973) was the one primarily responsible for the breakup of the holosteans as a natural phylogenetic unit, but continued to use the term "Holostei" informally.

Although the above superorders are no longer recognized, scientists still find it convenient to use the names Chondrostei, Holostei, and Teleostei in describing the overall morphological organization of particular groups of fishes. Some of the names are also retained in formal use in classification, albeit at different phyletic levels than previously. In Nelson's (1976) earlier classification, they were employed for three of his four infraclasses of the subclass Actinopterygii (the fourth, the Halecostomi, included only certain extinct orders). By 1984, however, belated adoption by Nelson of the phylogenetic scheme proposed earlier by Patterson (1973) resulted in reduction of the number of infraclasses from four to two. Of these, one (the Chondrostei) remained unchanged from the 1976 classification, whereas the other (the Neopterygii) comprised the remaining three infraclasses included earlier. In this revised scheme, the name Teleostei was downgraded to a lower phyletic level and the name Holostei was dropped entirely. Included within the infraclass Neopterygii are two divisions, one of which (Ginglymodi) includes only the living gars (order Lepisosteiformes) (Figure 7) and the other (Halecostomi) all remaining groups.

Infraclass Chondrostei

The infraclass Chondrostei is characterized by having the interoperculum absent, premaxilla and maxilla rigidly attached to the ectopterygoid and palatine, a spiracle usually present, and the myodome usually absent. It includes a single living order, Acipenseriformes (the sturgeons and paddlefishes) (Figure 7), and 13 extinct orders. Living sturgeons comprise four genera and 24 species living in fresh and marine waters in temperate regions of the northern hemisphere, whereas the paddlefishes include two genera and two species occurring in fresh waters of eastern North America and eastern China. Among other things, these fishes are characterized by a pronounced heterocercal tail, a largely cartilagenous endoskeleton, and the intestine with a spiral valve.

Most notable of the extinct chondrosteans is the order Palaeonisciformes. It includes the earliest and most primitive group of actinopterygians, and numerically probably was the dominant group of fishes from the Devonian (Lower) to Cretaceous (Lower) periods (some classifications recognize as many as 42 families in the order). It should be noted, however, that experts disagree regarding monophyly of the paleoniscoids, some believing these fishes to be a polyphyletic group united by a host of primitive characters. Probably the best-studied palaeoniscoid is the Devonian genus *Cheirolepis* (family Cheirolepididae) (Figure 9) (Pearson and Westoll, 1979), which Patterson (1982) considered to be one of the stem group forms of the subclasses Brachiopterygii and Actinopterygii. Several primitive morphological features characteristic of this group were mentioned earlier (structure and morphology

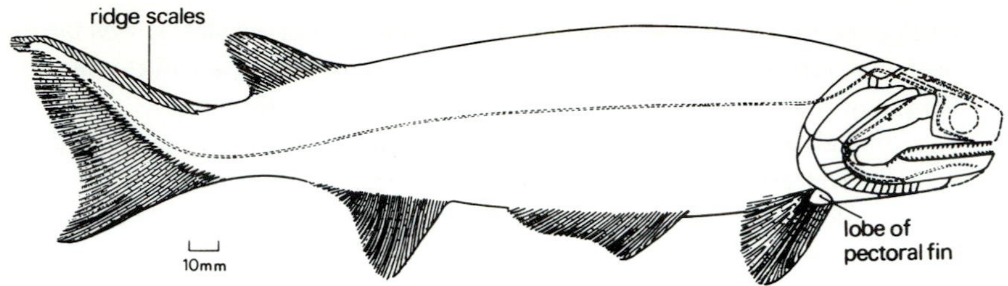

FIGURE 9. The palaeoniscoid fish, *Cheirolepis*. (From Moy-Thomas, J. A. and Miles, R. S., *Palaeozoic Fishes,* 2nd ed., Chapman and Hall, London, 1971. With permission.)

of tail, structure and position of the air bladder, presence of a spiracle, hard scales with ganoin). Other important features involve the position and structure of the mouth, as well as structural support of the jaws. The gape of the mouth was quite broad, and extended posteriorly at least three quarters of the length of the head, with the posterior ends of the dentary and maxillary bones in close proximity to the suboperculum. The hyomandibular, quadrate, and epipterygoid bones, which served as jaw supports, were relatively small. Finally, the cheek bones were firmly connected, which in turn permitted only limited expansion of the orobranchial chamber.

The skull of *Cheirolepis* (Figure 10) serves to demonstrate other primitive features of cranial osteology and organization. Included among the dermal bones of the skull are (1) schlerotic bones (**sc**) ringing the eye; (2) a large preopercular bone (**pop**) (larger than the opercular bone), which extends anteriorly well forward (almost reaching the eye) and forms the dorsal border of the maxilla (**m**) over much of its length; (3) a large maxilla that in turn forms almost the entire upper margin of the mouth and whose posterior end is attached to the bones of the cheek; (4) presence of a clavicle (**cl**); and (5) presence of paired gular plates (**g**) on the undersurface of the head.

Infraclass Neopterygii

Neopterygian fishes were well differentiated by the Upper Permian. Presumably derived from primitive chondrosteans, they lived side by side with the palaeoniscoids until the Lower Cretaceous, when the latter group disappeared. Many changes are evident in skull organization and morphology from that seen in the chondrosteans. Some of these are shown in Figure 11, which is a neopterygian fish from the Upper Permian. The gape of the mouth is smaller, and the general position of the mouth much farther forward. The maxilla is reduced in size, and is no longer attached posteriorly to the cheek bones. Increase in size of the hyomandibular, quadrate, and epipterygoid bones permits increased jaw support, and in addition a new bone, the symplectic, has arisen which serves to unite the hyomandibular and quadrate. The overall size and number of bones in the skull are reduced, resulting in a looser and less completely fused skull. Because the cheeks are no longer solid, lateral movement of the hyomandibular bone is possible, which permits a rapid increase in volume of the orobranchial chamber and permits a powerful suction, useful in capturing prey. Other changes evident in Figure 11b include (1) change in position and reduction in size of the preopercle (**pop**), now reduced to a crescent-shaped bone forming the anterior margin of the gill cover (comprised in turn of the operculum [**o**] and suboperculum [**sop**]); (2) loss of the clavicle (**cl** in Figure 10c); (3) appearance of the interopercular bone; and (4) increase in size of the premaxillary bone (**pm**). Retained in some neopterygians are such primitive features as ganoid scales (gars), a well-developed gular plate (bowfins and elopiform fishes), a functional lung (gars and bowfins), and an abbreviated heterocercal tail (gars and bowfins).

Within the infraclass Neopterygii are two divisions. One (the Ginglymodi) comprises the living gars (order Lepisosteiformes; family Lepisosteidae) (Figure 7), which are regarded as the most primitive group of neopterygians and which have a fossil history dating back at least to the Cretaceous period (Wiley, 1976). Formerly present in fresh and brackish waters of both North and South America (including Central America), Africa, Europe, and Asia, two genera and seven species of gars survive today only in eastern North America, Central America, and Cuba. The other division (the Halecostomi) includes all remaining fishes. These two groups differ in several features (characters of the Ginglymodi are indicated first): interoperculum absent (vs. present); two or more supratemporal bones present on each side (vs. one supratemporal bone on each side); maxilla immobile (vs. mobile); supramaxilla absent (vs. present); and myodome absent (vs. present).

Nelson (1984) recognized two subdivisions of the Halecostomi, in addition to a residue of extinct forms to which he attached no formal name. One subdivision (the Halecomorphi) includes the order Amiiformes, with several extinct families and one extant family (Amiidae). Although amiids once occurred throughout much of the world (North America, South America, Europe, Africa, and Asia), and marine and freshwater fossils are known from as far back as the Jurassic period (Patterson, 1973; Wilson, 1982, 1983), they are today represented by a single living species (the bowfin, *Amia calva*) (Figure 7) confined to eastern North America. The second subdivision (the Teleostei) (Figure 7) comprises over 99% of all living osteichthyans and about 96% of all living fishes (i.e., including chondrichthyans and agnathans), Nelson (1984) having listed 20,812 extant species in 35 orders, 409 families, and 3876 genera. Although modern teleosts can usually be distinguished superficially from the small number of living nonteleosts by a combination of characters, a truly definitive definition apparently is not easy. Patterson and Rosen (1977) defined the teleosts as a group of halecostomes with the ural neural arches elongated as uroneurals, basibranchial toothplates unpaired, and the premaxilla mobile.

Present classification of the bowfins (Amiidae) and gars (Lepisosteidae) demonstrates how certain concepts of evolutionary relationships have changed

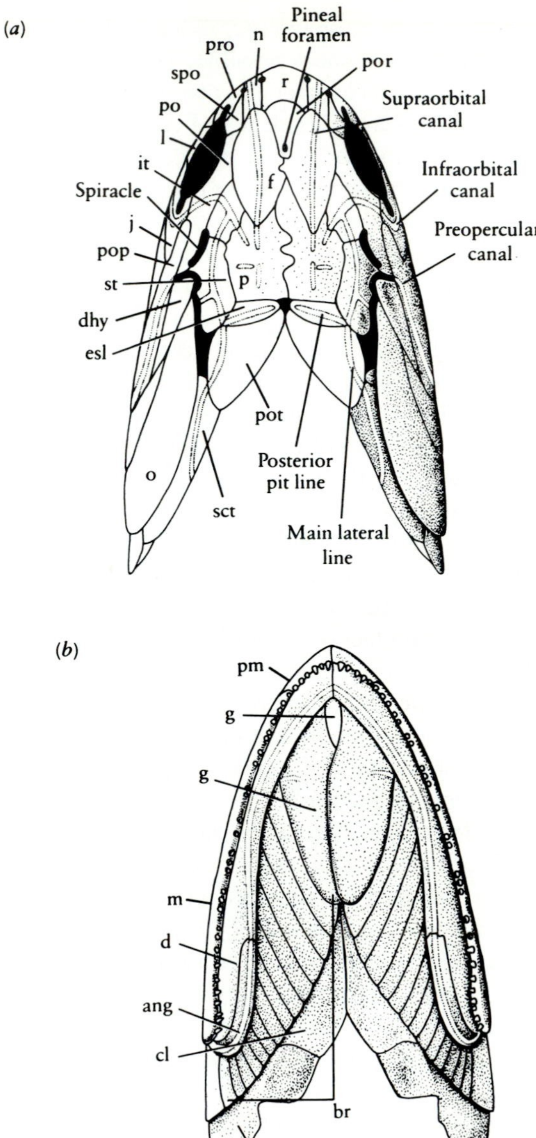

FIGURE 10. Dermal bones of the head region of a primitive palaeoniscoid fish, *Cheirolepis*. Following abbreviations limited to more prominent bones and to skull elements discussed in text: br, branchiostegal rays; cl, clavicle; ct, cleithrum; d, dentary; ec, ectopterygoid; en, external naris; ep, epipterygoid; g, gulars; hy, hyomandibular; iop, interopercular; j, jugal; l, lacrimal (infraorbital); m, maxilla; n, nasal; o, opercular; p, parasphenoid; pa, parietal (frontal); pf, postfrontal (supraorbital); pl, palatine; pm, premaxilla; pop, preopercular; ps, parasphenoid; pt, pterygoid; q, quadrate; s, suprapterygoid(s); sc, scletoric ring; sm, supramaxilla; smp, symplectic; sop, subopercular; sp, spiracular cleft; st, supratemporal (dermal sphenotic); v, vomer. (From Pearson, D. M. and Westoll, T. S., *Trans. R. Soc. Edinburgh,* 70, 337, 1979. With permission of Royal Society of Edinburgh.)

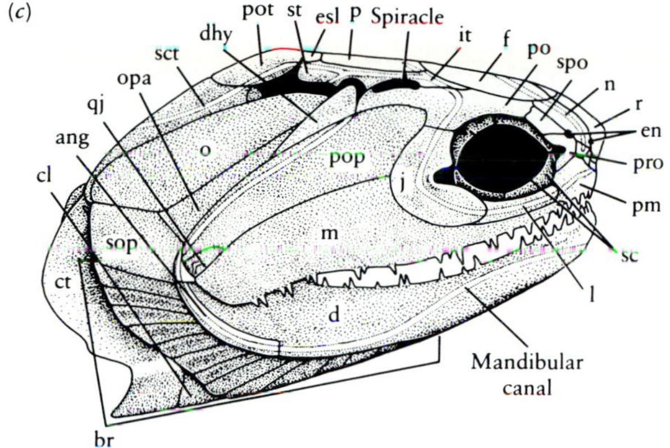

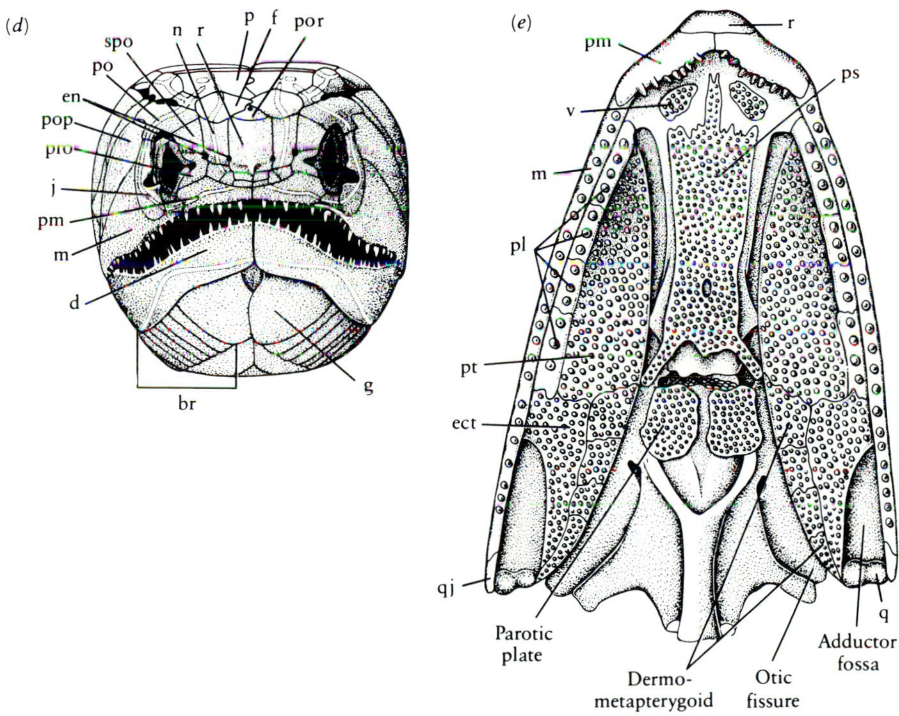

FIGURE 10. (continued)

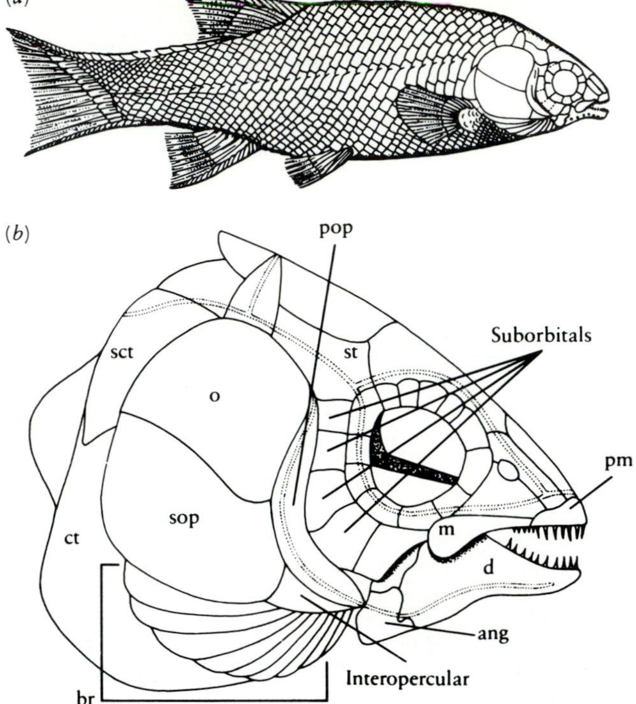

FIGURE 11. The semionotid fish, *Acentrophorus*; a, whole individual; b, skull. Abbreviations as in Figure 10. (From Moy-Thomas, J. A. and Miles, R. S., *Palaeozoic Fishes,* 2nd ed., Chapman and Hall, London, 1971. With permission.)

in recent years. Gars and bowfins share a suite of primitive features, including an abbreviated heterocercal tail and a compartmentalized air bladder that retains its original functional in aerial respiration. They were once classified as "holosteans", and although not necessarily regarded as intimately related, they were placed at the same approximate level in the evolutionary hierarchy (Figure 7). The amiids are now considered to be more closely related to the teleosts than formerly, whereas the lepisosteids are regarded as the most primitive of all neopterygians. The order Semionotiformes (Figure 11), to which the gars were once allocated, now comprises only extinct Mesozoic forms, and is even placed in a different division (Halecostomi).

TELEOSTEAN FISHES

Teleostean evolution was largely a Mesozoic event, with all major phyletic lines having become established by the Cretaceous period. Prominent but now-extinct Mesozoic orders include the Aspidorhynchiformes, Pholidophoriformes, Leptolepidiformes, and Ichthyodectiformes, and it is possible that one or more of these groups is (are) directly ancestral to the four main lines of modern teleosts. Of these, the pholidophoroids deserve special mention, since they are thought to have given rise independently to the living osteoglossomorphs and elopomorphs during the Triassic or Jurassic periods (Nelson, 1984).

Morphological Changes During Teleostean Evolution

It is appropriate here to review some of the more obvious morphological changes seen during teleostean evolution (i.e., from primitive to more advanced conditions) (some can be seen in Figures 7, 10, and 11), since reference will be made to these in later discussions. Although characteristic of teleosts in general, these changes are best observed among the various taxa included in the large infradivision Euteleostei (see below). Among these are

1. Loss of the duct connecting the air bladder to the alimentary canal (i.e., change from a physostomous to a physoclistous condition)
2. Forward migration of the pelvic girdle and fins from near the middle of the belly to a position beneath (or even anterior to) the pectoral girdle and fins
3. Progressive reduction in size and significance of the maxilla, which is replaced by the premaxilla as the principle bone in the upper jaw
4. Increased protrusibility of the upper jaw
5. Change in scale morphology from the cycloid (with smooth margins) to ctenoid condition (with numerous teeth on the margins)
6. Change from a thin, shallowly embedded, deciduous scale to a thicker, more deeply embedded, less deciduous scale
7. Fusion of the two halves of the anteriormost rays of the dorsal, pelvic, and anal fins into a single unit, which frequently is hard and stiff.

Combinations of these characters may vary considerably; it is not uncommon, for example, for individuals of certain species to have both cycloid and ctenoid scales on different parts of the body. In addition, loss of the adipose fin in certain teleosts should also be mentioned. (This fin is a small fleshy structure located on the dorsal midline of the back between the dorsal and caudal fins.) It is present in certain of the more primitive species of euteleosts, but in none of the other three teleostean lineages (see discussion below); however, it is invariably absent in more advanced euteleosts. It is highly variable in terms of its presence or absence, sometimes even among different individuals of the same species. Species of the exclusively North American order Percopsiformes (most notably the two living species of the family Percopsidae) are of particular interest since they display, to an unusual degree, various combinations of the primitive and advanced characters described above.

Infradivisions of Teleostean Fishes

Recent research has resulted in recognition of four main lineages (or infradivisions) of recent teleosts (Nelson, 1973a, 1973b; Patterson and Rosen,

1977): Osteoglossomorpha, Elopomorpha, Clupeomorpha, and Euteleostei. Of these, the last group comprises an estimated 19,642 species, or about 94% of all living teleosts. Recognition of these lineages is often based on obscure but significant internal anatomical characters (usually osteological or myological) that are difficult to describe or figure. For the sake of completeness, the most important of these are mentioned in the following text, even though they may have little meaning except to a specialist. Overall, however, the listing and discussion of such complex characters has usually been abbreviated.

Osteoglossomorpha

The infradivision Osteoglossomorpha comprises a single order, the Osteoglossiformes, in which are included six living families, 26 genera, and about 206 species. Of these, all but 16 species belong to the African elephantfish family (Mormyridae). All families are strictly freshwater in their orientation, with no living or fossil species being known from marine habitats. Osteoglossiforms are distinguished by a host of characters, including well-developed teeth usually present on both the parasphenoid and tongue bones, and the intestine usually passing posteriorly to the left of the esophagus and stomach, rather than to the right. The two species of mooneyes (family Hiodontidae) are of conventional appearance (unlike other osteoglossiform fishes), superficially resembling herrings, and for that reason were included by Berg (1940) in the order Clupeiformes. Placement of the mooneyes in the Osteoglossiformes follows the work of Greenwood (1970, 1973).

Elopomorpha

The infradivision Elopomorpha comprises three orders, 25 families, 157 genera, and 637 species. The three orders included are the Elopiformes (tarpons, ladyfishes, and bonefishes: three families, four genera, and 15 species); Notacanthiformes (spiny eels and halosaurs: three families, six genera, and 25 species); and Anguilliformes (true eels: 19 families, 147 genera, and 597 species). In contrast to the osteoglossomorph fishes, elopomorphs are basically marine, with only a few individual species spending portions of their lives in fresh water. Adults differ considerably in physical appearance, ranging from the conventional body form of elopiforms to the thin and elongated bodies characteristic of eels. Despite this, all possess a translucent, laterally compressed, ribbon-like larva known as a leptocephalous. Leptocephala are so totally different in appearance from adults that even today the leptocephalous and adult stages of many elopomorph species have yet to be properly associated. Leptocephala characteristically are passively transported long distances by ocean currents, which is a major factor in the unusually wide distributions of many species. Despite this unique unifying character, only recently have the elopomorphs come to be recognized as a natural phyletic unit (Greenwood et al., 1966).

The eels, in association with their distinctive body form, are characterized by other unusual morphological features (usually reductions) that have occurred in response to life in the restricted physical habitats in which most eels live. Among these are (1) complete loss of pelvic fins and loss or reduction of other fins; (2) loss or reduction of body scales (present [but deeply embedded] only in family Anguillidae); (3) gill openings extremely restricted; and (4) loss or fusion of various bones in and immediately posterior to the skull. Although all eels are basically marine, species of the family Anguillidae are unusual in that adults spend their lives in fresh or brackish water, returning to the sea only to spawn.

Clupeomorpha

The infradivision Clupeomorpha includes one living order, Clupeiformes. This order comprises four living families, 68 genera, and about 331 species. Two of these families are obscure (each containing a single species), whereas the other two (the Clupeidae [herrings, shads, sardines, and menhadens] and Engraulididae [anchovies]) include a number of commercially important species. Clupeomorphs are generalized silvery fishes, usually with laterally compressed bodies and scutes on the ridge of the belly, which display most of the primitive teleostean features listed earlier. Other distinguishing features include extension of the cephalic canal system onto the operculum, an undeveloped lateral line (except in one monotypic family), and nonprotrusible jaws. The feature unique to clupeomorphs, however, is the otophysic connection involving a diverticulum of the swim bladder that penetrates the exoccipital bone and extends into the prootic within the lateral wall of the brain case.

Euteleostei

The infradivision Euteleostei comprises by far the largest percentage of teleostean species. Among the families included are found the entire range of primitive and advanced morphological characters found in teleosts (Figure 7). Because of its size and morphological diversity it is very poorly characterized, and there is no unique feature that distinguishes euteleosts from the other three teleostean lines. For these reasons, there is no convincing evidence that the group is monophyletic, and considerable work remains to be done before a sound classification of the euteleosts can be erected. Many changes in alignment and composition of the major groups included have already occurred since it was first established by Greenwood et al. (1966). Nelson (1984) divided the euteleosts into six superorders, 30 orders, 374 families, 3625 genera, and 19,642 species. Only one of the six superorders (the Ostariophysi), however, is united by a single, well-defined and totally unique character. Of the nearly 20,000 species in the infradivision Euteleostei, about 4100 species (almost 20%) belong to just two families (Cyprinidae and Gobiidae), and three other families (Characidae, Cichlidae, and Labridae) contain another 2000 or more species. These five families combined thus have around 6100 species, or about 31% of the total number of euteleosts.

Because of the large number of euteleostean taxonomic groups, a detailed discussion of the various orders and families is not possible here. Consequently, the following discussion is confined mostly to a general analysis of the six superorders included in Nelson's (1984) classification: Ostariophysi, Protacanthopterygii, Stenopterygii, Scopelomorpha, Paracanthopterygii, and Acanthopterygii.

Ostariophysi

The ostariophysans comprise five orders, 57 families, 938 genera, and 6050 species, and include such well-known fishes as minnows and suckers (order Cypriniformes), catfishes (order Siluriformes), and characins or tetras (order Characiformes) (Fink and Fink, 1981). The vast majority of ostariophysan species live in fresh water, although two catfish families are largely marine. Ostariophysans possess the primitive teleostean features discussed earlier, together with a number of other characters listed by Nelson (1984). The character serving to unite all ostariophysans, however, is the unique modification of the anterior vertebrae, which in all orders but one involves fusion of the first four to seven vertebrae (rarely more than five) into bony ossicles. This structure, known as the Weberian apparatus, connects the air bladder to the inner ear and is believed to enhance sound transmission. In the order Gonorynchiformes only the first three vertebrae are specialized, and these are associated with one or more cephalic ribs; Rosen and Greenwood (1970) considered this to represent a primitive Weberian apparatus. Other characters unique to ostariophysans are (1) the minute, unicellular, horny projections (termed unculi) that are commonly present on various parts of the body (e.g., mouth region or ventral surface of the paired fins), and (2) the multicellular horny tubercles (nuptial tubercles) with a well-developed keratinous cap, in contrast to the thin cuticle found in other euteleosts (Roberts, 1982). Most ostariophysans have well-defined scales on the body, but in the catfishes these have either been lost or modified into bony plates.

The minnow family (Cyprinidae) is probably the largest family of fishes (over 2000 species), occurs natively in fresh waters of all major continents except South America and Australia, and has the greatest latitudinal range of any family of freshwater fishes. It is characterized by the absence of teeth in the mouth, together with modification of the fifth branchial arch into one to three rows of pharyngeal teeth, which in turn assume the function normally associated with the mouth teeth. Most species are small, but some can reach a large size, as for example certain Asiatic species that may reach lengths of up to 10 feet. Although the term "minnow" is commonly applied to small fish belonging to a large number of marine and freshwater families, from a scientific standpoint it should be limited only to species of the family Cyprinidae.

Protacanthopterygii

The superorder Protacanthopterygii includes a single order (Salmoniformes), four suborders, 15 families, 90 genera, and 320 species of both freshwater and marine fishes. The studies on which present ideas on salmoniform classification are mainly based are by Rosen and Patterson (1969), Rosen (1973, 1974), and Fink and Weitzman (1982). Included in this group are such well-known fish as salmon, trout, and pike, together with certain obscure deep-sea groups. Members of the salmon and trout family (Salmonidae) are of special scientific interest because of the remarkable speed and facility with which speciational processes can occur. Salmoniform fishes display the same primitive characters as ostariophysans, and in many cases the species bear a strong superficial resemblance; however, they differ in lacking the Weberian apparatus that distinguishes all ostariophysans. Nelson (1984) otherwise presented no specific features to characterize the group. The size of the order Salmoniformes has been greatly reduced from that formerly recognized, with several groups now placed in other superorders, and Fink and Weitzmann (1982) presented evidence to indicate that it should be reduced still further. Specifically, they found no indication that the esocoids (families Esocidae and Umbridae) were closely related to other members of the order; they suggested instead that they are the most primitive euteleosts, with both ostariophysans and other salmoniforms being derived groups.

Stenopterygii

The superorder Stenopterygii also includes a single order (Stomiiformes), which also retains primitive salmoniform characters and was formerly included in that order. Classification of the group is based largely on work by Weitzman (1974), who proposed several changes in relationships from those recognized previously, including transfer of certain species of the family Gonostomatidae into two other families. The stomiiforms comprise nine families, 53 genera, and about 248 species of exclusively deep-sea fishes, all of which have luminescent organs (photophores) and other features typical of fish living in the deep ocean (large mouths, usually extending past the eye, and often with prominent teeth on both the maxilla and premaxilla; soft bodies; weakly developed skeletons; brown or black bodies). Some species (specifically certain of those in the family Gonostomatidae) may occur in tremendous numbers and are important as a source of food for predaceous deep-sea species.

Scopelomorpha

The superorder Scopelomorpha is another group of exclusively marine fishes that retains most of the primitive teleostean characters shared by the preceding groups, including abdominal pelvic fins, a nonprotrusible upper jaw, and usual presence of an adipose fin. More advanced features include exclusion of the maxilla from the gape of the mouth (now occupied exclusively by the premaxilla) and loss of the duct to the air bladder (physoclistic condition). The scopelomorphs comprise two orders, 14 families, 75 genera, and about 429 species of mostly deep-sea fishes. All were formerly included in a single order (Myctophiformes), but a second order (Aulopiformes) was recognized by Rosen (1973) based on unique specialization of the gill arches (Nelson, 1984). A few aulopiforms inhabit shallow benthic waters (e.g., lizardfishes of the family Synodontidae), but most are mid-water predators of the deep ocean, and usually possess morphological features (e.g., large mouths and prominent stiletto-like teeth) characteristic of such fishes. All aulopiform species apparently lack melanophores, in contrast to the great majority of myctophiform species (e.g., all species of the family Myctophidae), which are noteworthy because of their extensive night-time migrations to the surface.

Paracanthopterygii

The superorder Paracanthopterygii comprises a large assemblage of six living orders, 33 families, 287 living genera, and about 1160 living species. Also included in the paracanthopterygians is the extinct order Ctenothrissiformes, a group of marine fishes of the Upper Cretaceous period. The significance of the ctenothrissiforms relates to the fact that Rosen (1973) considered them possibly to be the sister group of the vast assemblage of paracanthopterygian-acanthopterygian fishes. Rosen and Patterson (1969), who redefined the paracanthopterygians, used, as their basis for recognition, features of both the caudal skeleton and jaw muscles (particularly the levator maxillae superioris). The reader is referred to papers by Fraser (1972) and Rosen (1973), who discussed the problem and critically evaluated earlier works. Included among the paracanthopterygians are the orders Percopsiformes (3 families), Gadiformes (7 families), Ophiidiiformes (4 families), Batrachoidiformes (1 family), Lophiiformes (16 families), and Gobiesciformes (2 families). As mentioned earlier, the percopsiforms are remarkable for the combination of primitive and advanced features, particularly in the family Percopsidae. The other five orders, however, display advanced morphological characters almost exclusively, some of which are among the most bizarre seen in any fishes. Probably the most notable of these is the forward migration, in lophiiforms, of the anteriormost dorsal-fin ray to the tip of the snout, where it becomes modified into a flexible fishing lure.

Acanthopterygii

The final superorder, the Acanthopterygii, represents the largest assemblage of euteleosts, Nelson (1984) having listed a total of 15 orders, 246 families, and 11,435 species. These orders include the Cyprinodontiformes (13 families), Atheriniformes (5 families), Lampriformes (11 families), Beryciformes (14 families), Zeiformes (6 families), Gasterosteiformes (3 families), Indostomiformes (1 family), Pegasiformes (1 family), Syngnathiformes (6 families), Dactylopteriformes (1 family), Synbranchiformes (1 family),

Scorpaeniformes (20 families), Perciformes (150 families), Pleuronectiformes (6 families), and Tetraodontiformes (8 families). According to Rosen (1973), the group can be defined as having the retractor arcuum branchialum muscle inserted prinicipally or entirely on the third pharyngobranchial, the articular surface of the fourth epibranchial reduced, and the second and third epibranchials enlarged as the principal support for the upper pharyngeal dentition. The acanthopterygians include the typical deep-bodied, spiny-rayed fishes (sunfishes and "perches") so familiar to laymen. Also included are a wide range of primitive and advanced morphological features, including such bizarre adaptations as modification of the anterior (spinous) portion of the dorsal fin into a suction disc located on top of the head (in the family Echeneididae); the extraordinarily thin and ribbonlike body of the several families of oceanic lampriform fishes; the flattened body and twisted head characteristic of the various flatfish families; the ability of certain families to swell to several times their normal size by ingestion of water (Tetraodontidae and Diodontidae); and modifications of the mouth and snout region into a suction tube and of the tail and posterior part of the body into a prehensile grasping structure in the seahorses (family Syngnathidae). Equally unusual are the life-history strategies of certain groups, as for example the ability of the eggs in some species of the families Atherinidae and Cyprinodontidae to survive in a completely dry environment, with hatching triggered immediately by exposure to water.

The family Gobiidae should also be mentioned here, since it is one of the two most speciose families of fishes (along with the Cyprinidae). Best estimates place the total at around 2000 (which is very close to the 2070 estimated species of Cyprinidae), but this may be a conservative figure considering the large number of undescribed forms. The family also is noteworthy because it includes the smallest known fishes, some species maturing at only about 6 mm standard length.

REFERENCES

Bailey, R. M. and Cavender, T. M., Fishes, in *Encyclopedia of Science and Technology*, McGraw-Hill, New York, 1971.

Bardack, D. and Zangerl, R., First fossil lamprey: a record from the Pennsylvanian of Illinois, *Science,* 162, 1265, 1968.

Bardack, D. and Zangerl, R., Lampreys in the fossil record, in *The Biology of Lampreys*, Vol. 1, Hardisty, M. W. and Potter, I. C., Eds., Academic Press, New York, 1971, 13.

Berg, L. S., Classification of fishes, both Recent and fossil, *Trav. Inst. Zool. Acad. Sci. URSS*, 5, 87, 1940. (Reprinted in English by J. W. Edwards, Ann Arbor, Michigan, 1947.)

Carroll, R. L., *Vertebrate Paleontology and Evolution,* W. H. Freeman, New York, 1988.

Compagno, L. J. V., Interrelationships of living elasmobranches, in *Interrelationships of Fishes,* Greenwood, P. H. et al., Eds.; *J. Linnaean Soc. (Zool.),* 53 (Suppl. 1), 15, 1973. (Published for Linnaean Society by Academic Press, New York).

Compagno, L. J. V., Phyletic relationships of living sharks and rays, *Am. Zool.,* 17, 303, 1977.

Dodson, E. O., *A Textbook of Evolution,* W. B. Saunders, Philadelphia, 1952.

Fink, S. V. and Fink, W. R., Interrelationships of the ostariophysan fishes (Teleostei), *J. Linnaean Soc. (Zool.),* 72, 297, 1981.

Fink, W. E. and Weitzman, S. H., Relationships of the stomiiform fishes (Teleostei), with a description of *Diplophos, Bull. Mus. Comp. Zool.,* 150, 31, 1982.

Fraser, T. H., Some thought about the teleostean fish concept — the Paracanthopterygii, *Jpn. J. Ichthy.,* 19, 232, 1972.

Gardiner, B. G., Tetrapod ancestry: a reappraisal, in *The Terrestrial Environment and the origin of Land Vertebrates,* Panchen, A. L., Ed., Academic Press, London, 1980.

Greenwood, P. H., On the genus *Lycoptera* and its relationships with the family Hiodontidae (Pisces, Osteoglossomorpha), *Bull. Br. Mus. Nat. (Zool.),* 19, 257, 1970.

Greenwood, P. H., Interrelationships of osteoglossomorphs, in *Interrelationships of Fishes,* Greenwood, P. H. et al., Eds.; *J. Linnaean Soc. (Zool.),* 53 (Suppl. 1), 307, 1973. (Published for Linnaean Society by Academic Press, New York).

Greenwood, P. H., Rosen, D. E., Weitzman, S. H., and Myers, G. S., Phyletic studies of teleostean fishes, with a provisional classification of living forms, *Bull. Am. Mus. Nat. Hist.,* 131, 339, 1966.

Greenwood, P. H., Miles, R. S., and Patterson, C., Eds., *Interrelationships of Fishes, J. Linnaean Soc. (Zool.),* 53 (Suppl. 1), 1973. (Published for Linnaean Society by Academic Press, New York).

Lauder, G. V. and Liem, K. F., The evolution and interrelationships of the actinopterygian fishes, *Bull. Mus. Comp. Zool.,* 150, 95, 1983.

Mok, H. K., The posterior cardinal veins and kidneys of fishes, with notes on their phylogenetic significance, *Jpn. J. Ichthy.,* 27, 281, 1981.

Moser, H. G., Richards, W. J., Cohen, D. M., Fahay, M. P., Kendall, A. W., Jr., and Richardson, S. L., Eds., *Ontogeny and Systematics of Fishes,* American Society of Ichthyologists and Herpetologists, Allen Press, Lawrence, Kansas, 1984.

Moy-Thomas, J. A. and Miles, R. S., *Palaeozoic Fishes,* 2nd ed., Chapman and Hall, London, 1971.

Nelson, G. J., Gill-arch structure in *Acanthodes,* in *Current Problems of Lower Vertebrate Phylogeny,* Proc. 4th Nobel Symposium (1967), Orvig, T., Ed., Wiley-Interscience, New York, 1968, 129.

Nelson, G. J., Gill arches and the phylogeny of fishes, with notes on the classification of vertebrates, *Bull. Am. Mus. Nat. Hist.,* 141, 75, 1969.

Nelson, G. J., Notes on the structure and relationships of certain Cretaceous and Eocene teleostean fishes, *Am. Mus. Novit.,* 2524, 1, 1973a

Nelson, G. J., Relationships of clupeomorphs, with remarks on the structure of the lower jaw in fishes, in *Interrelationships of Fishes,* Greenwood, P. H., et al., Eds.; *J. Linnaean Soc. (Zool.),* 53 (Suppl. 1), 333, 1973b. (Published for Linnaean Society by Academic Press, New York).

Nelson, J. S., *Fishes of the World,* 1st ed., John Wiley & Sons, New York, 1976.

Nelson, J. S., *Fishes of the World,* 2nd ed., John Wiley & Sons, New York, 1984.

Patterson, C., Interrelationships of holosteans, in *Interrelationships of Fishes,* Greenwood, P. H., et al., Eds.; *J. Linnaean Soc. (Zool.),* 53 (Suppl. 1), 233, 1973. (Published for Linnaean Society by Academic Press, New York).

Patterson, C., Morphology and interrelationships of primitive actinopterygian fishes, *Am. Zool.,* 22, 241, 1982.

Patterson, C. and Rosen, D. E., Review of the ichthyodectiform and other Mesozoic teleost fishes and the theory and practice of classifying fossils, *Bull. Am. Mus. Nat. Hist.,* 158, 81, 1977.

Pearson, D. M. and Westoll, T. S., The Devonian actinopterygian *Cheirolepis* Agassiz, *Trans. R. Soc. Endinburgh,* 70, 337, 1979.

Pough, F. H., Heiser, J. B., and McFarland, W. N., *Vertebrate Life*, Macmillan, New York, 1989.

Repetski, J. E., A fish from the Upper Cambrian of North America, *Science*, 200, 529, 1978.

Ritchie, A., New evidence on *Jamoytius kerwoodi* White, an important ostracoderm from the Silurian of Lanarkshire, Scotland, *Paleontology,* 11, 21, 1968.

Roberts, T. R., Unculi (horny projections arising from single cells), an adaptive feature of the epidermis of ostariophysan fishes, *Zool. Scr.,* 11, 55, 1982.

Romer, A. S., *Vertebrate Paleontology,* 2nd ed., University of Chicago Press, Chicago, 1945.

Romer, A. S., *Man and the Vertebrates,* University of Chicago Press, Chicago, 1959.

Romer, A. S., *Vertebrate Paleontology,* 3rd Ed., University of Chicago Press, Chicago, 1966.

Rosen, D. E., Interrelationships of higher euteleostean fishes, in *Interrelationships of Fishes,* Greenwood, P. H., et al., Eds.; *J. Linnaean Soc. (Zool.),* 53 (Suppl. 1), 97, 1973. (Published for Linnaean Society by Academic Press, New York).

Rosen, D. W., Phylogeny and zoogeography of salmoniform fishes and relationships of *Lepidogalaxias salmandroides, Bull. Am. Mus. Nat. Hist.,* 153, 265, 1974.

Rosen, D. E., Forey, P. L., Gardiner, B. G., and Patterson, C., Lungfishes, tetrapods, paleontology, and plesiomorphy, *Bull. Am. Mus. Nat. Hist.,* 167, 159, 1981.

Rosen, D. E. and Greenwood, P. H., Origin of the Weberian apparatus and the relationships of the ostariophysan and gonorynchiform fishes, *Am. Mus. Novit.,* 2428, 1, 1970.

Rosen, D. E. and Patterson, C., The structure and relationships of the paracanthopterygian fishes, *Bull. Am. Mus. Nat. Hist.,* 141, 357, 1969.

Schaeffer, B., Interrelationships of chondrosteans, in *Interrelationships of Fishes,* Greenwood, P. H. et al., Eds., J. Linnaean Soc. (Zool.), 53 (Suppl. 1), 207, 1973. (Published for Linnaean Society by Academic Press, New York).

Smith, J. L. B., *The Search Beneath the Sea.* Henry Holt, New York, 1956.

Stensio, E. A., The Downtonian and Devonian vertebrates of Spitzbergen. I. Family Cephalaspidae, *Skr. Svalbard Ishavet.,* 12, 1, 1927.

Stensio, E. A., The cyclostomes with special reference to the diphyletic origin of the Petromyzontida and Myxinoidea, in *Current Problems of Lower Vertebrate Phylogeny,* Proc. 4th Nobel Symposium (1967), Orvig, T., Ed., Wiley-Interscience, New York, 1968, 13.

Weitzman, S. H., Osteology and evolutionary relationships of the Sternoptychidae, with a new classification of stomiatoid families, *Bull. Am. Mus. Nat. Hist.,* 153, 327, 1974.

Wiley, E. O., The phylogeny and biogeography of fossil and Recent gars (Actinopterygii: Lepisosteidae), *Misc. Publ. Mus. Nat. Hist. Univ. Kansas*, 64, 1, 1976.

Wiley, E. O., Ventral gill arch muscles and the interrelationships of gnathostomes, with a new classification of the Vertebrata, *J. Linnaean Soc. (Zool.),* 67, 149, 1979.

Wilson, M. V. H., A new species of the fish *Amia* from the Middle Eocene of British Columbia, *Palaeontology*, 25, 413, 1982.

Wilson, M. V. H., Paleocene amiid fish from Jabal Umm Himar in the Harrat Hadan Area, at Taif Region, Kingdom of Saudi Arabia. U.S. Geol. Surv., Saudi Arabian Project Rep., Jiddah, Saudi Arabia, 1983.

Zangerl, R., Chondrichthyes, 1. Paleozoic Elasmobranchii, *Handbook of Paleoichthyology,* Vol. 3A. Fischer, G., Ed., Verlag, Stuttgart, 1981.

APPENDIX

Selected List of Common and Representative Families of Living Fishes

Superclass Agnatha (Cyclostomata, Marsipobranchii)
Class Myxini
Order Myxiniformes (Hyperotreti)
Family Myxinidae—hagfishes
Class Cephalaspidomorphi (Monorhina)
Order Petromyzontiformes
Family Petromyzontidae—lampreys
Superclass Gnathostomata
Class Chondrichthyes—cartilaginous fishes
Subclass Holocephali
Order Chimaeriformes
Family Chimaeridae—shortnose chimaeras or ratfishes
Family Callorhynchidae—plownose chimaeras
Family Rhinochimaeridae—longnose chimaeras
Subclass Elasmobranchii
Superorder Selachimorpha (Pleurotremata)—sharks
Order Hexanchiformes
Family Chlamydoselachidae—frill shark
Family Hexanchidae—cow sharks
Order Heterodontiformes
Family Heterodontidae—bullhead or horn sharks
Order Lamniformes (Galeoidea)
Suborder Lamnoidei
Family Lamnidae—mackerel, thresher and basking sharks
Family Rhincodontidae—whale shark
Family Orectolobidae—carpet or nurse sharks
Family Odontaspididae—sand tigers
Suborder Scyliorhinoidei
Family Scyliorhinidae—cat sharks
Family Carcharhinidae—requiem sharks
Family Sphyrnidae—hammerhead sharks
Order Squaliformes (Tectospondyli)
Suborder Squaloidei
Family Squalidae—dogfish sharks
Suborder Pristiophoroidei
Family Pristiophoridae—saw sharks

Suborder Squatinoidei
Family Squatinidae—angel sharks
Superorder Batidoidimorpha (Hypotremata)—rays
Order Rajiformes
Suborder Rajoidei
Family Rajidae—skates
Suborder Pristoidei
Family Pristidae—sawfishes
Suborder Torpedinoidei
Family Torpedinidae—electric rays
Family Rhinobatidae—guitarfishes
Suborder Myliobatidoidei
Family Myliobatididae—eagle rays
Family Dasyatidae—stingrays
Family Potamotrygonidae—river stingrays
Family Mobulidae—manta rays and devil rays
Class Osteichthyes—bony fishes
Subclass Dipneusti—lungfishes
Order Ceratodontiformes
Family Ceratodontidae—Australian lungfish
Order Lepidosireniformes
Family Lepidosirenidae—South American lungfish
Family Protopteridae—African lungfishes
Subclass Crossopterygii—fringe-finned or tassel-finned fishes
Order Coelacanthiformes—coelacanths
Family Latimeriidae—gombessa
Subclass Brachiopterygii (Cladistia)
Order Polypteriformes
Family Polypteridae—bichirs
Subclass Actinopterygii—ray-finned fishes
Infraclass Chondrostei
Order Acipenseriformes
Suborder Acipenseroidei
Family Acipenseridae—sturgeons
Suborder Polyodontoidei
Family Polyodontidae—paddlefishes
Infraclass Neopterygii
Division Ginglymodi
Order Lepisosteiformes
Family Lepisosteidae—gars
Division Halecostomi
Subdivision Halecomorphi

Order Amiiformes
Family Amiidae-bowfin
Subdivision Teleostei
Infradivision Osteoglossomorpha
Order Osteoglossiformes
Suborder Osteoglossoidei
Family Osteoglossidae—osteoglossids or bonytongues
Family Pantodontidae—African butterflyfish
Suborder Notoperoidei
Superfamily Notopteroidea
Family Notopteridae—featherbacks
Superfamily Hiodontoidea
Family Hiodontidae—mooneyes
Suborder Mormyroidei
Family Mormyridae—elephantfishes
Family Gymnarchidae—gymnarchids
Infradivision Elopomorpha
Order Elopiformes
Suborder Elopoidei
Family Elopidae—ladyfishes or tenpounders
Family Megalopidae—tarpons
Suborder Albuloidei
Family Albulidae—bonefishes
Order Notacanthiformes (Lyopomi and Heteromi)
Suborder Notacanthoidei
Family Notacanthidae—spiny eels
Suborder Halosauroidei
Family Halosauridae—halosaurs
Order Anguilliformes
Suborder Anguilloidei (Apodes)
Family Anguillidae—freshwater eels
Family Moringuidae—spaghetti eels
Family Muraenidae—moray eels
Family Congridae—conger eels
Family Ophichthidae—snake eels
Infradivision Clupeomorpha
Order Clupeiformes
Suborder Denticipitoidei
Family Denticipitidae—denticle herring
Suborder Clupeoidei
Family Clupeidae—herrings
Family Engraulididae—anchovies

Infradivision Euteleostei
Superorder Ostariophysi
Order Gonorynchiformes
Family Gonorynchidae—gonorynchids
Family Chanidae—milkfish
Order Cypriniformes
Family Cyprinidae—minnows and carps
Family Catostomidae—suckers
Family Cobitididae—loaches
Order Characiformes
Family Characidae—characins and tetras
Family Curimatidae—curimatids
Family Anostomidae—anostomids
Family Erythrinidae—trahiras
Family Gasteropelecidae—freshwater hatchetfishes
Family Ctenoluciidae—pike-characins
Family Citharinidae—citharinids
Order Siluriformes
Family Siluridae—sheatfish catfishes
Family Ictaluridae—North American freshwater catfishes
Family Bagridae—bagrid catfishes
Family Pangasiidae—pangasid catfishes
Family Sisoridae—sisorid catfishes
Family Clariidae—airbreathing catfishes
Family Malapteruridae—electric catfishes
Family Ariidae—sea catfishes
Family Plotosidae—eeltail catfishes
Family Mochokidae—upside-down catfishes
Family Auchenipteridae—auchenipterid catfishes
Family Ageneiosidae—barbelless catfishes
Family Pimelodidae—long-whiskered catfishes
Family Aspredinidae—banjo catfishes
Family Trichomycteridae—pencil catfishes
Family Callichthyidae—callichthyid armored catfishes
Family Loricariidae—suckermouth armored catfishes
Order Gymnotiformes
Family Gymnotidae—naked-back knifefishes
Family Sternopygidae—sternopygid knifefishes
Family Apteronotidae—freefin knifefishes

Family Electrophoridae—electric knifefish
Superorder Protacanthopterygii
Order Salmoniformes
Suborder Esocoidei (Haplomi)
Family Esocidae—pikes
Family Umbridae—mudminnows
Suborder Salmonoidei
Family Salmonidae—salmons, trouts and whitefishes
Family Osmeridae—smelts
Family Galaxiidae—galaxiids and whitebaits
Superorder Stenopterygii
Order Stomiiformes
Family Stomiidae—scaly dragonfishes
Family Gonostomatidae—lightfishes and bristlemouths
Family Sternoptychidae—deepwater hatchetfishes
Family Chauliodontidae—viperfishes
Family Idiacanthidae—black dragonfishes
Superorder Scolepomorpha
Order Aulopiformes
Suborder Aulopoidei
Family Aulopodidae—aulopids
Family Chlorophthalmidae—greeneyes
Suborder Alepisauroidei
Family Alepisauridae—lancetfishes
Family Synodontidae—lizardfishes
Order Myctophiformes
Family Myctophidae—lanternfishes
Superorder Paracanthopterygii
Order Percopsiformes
Suborder Percopsoidei
Family Percopsidae—trout-perches
Suborder Aphredoderoidei
Family Aphredoderidae—pirate perch
Family Amblyopsidae—cavefishes
Order Gadiformes
Suborder Gadoidei
Family Gadidae—cods
Family Moridae—morid cods
Family Merlucciidae—hakes
Suborder Macrouroidei

Family Macrouridae—grenadiers or rattails
Order Ophidiiformes
Suborder Ophidioidei
Family Ophidiidae—cusk-eels
Family Carapidae—pearlfishes
Suborder Bythitoidei
Family Bythitidae—viviparous brotulas
Order Batrachoidiformes
Family Batrachoididae—toadfishes and midshipmen
Order Lophiiformes
Suborder Lophioidei
Family Lophiidae—goosefishes
Suborder Antennarioidei
Family Antennariidae—frogfishes
Family Ogcocephalidae—batfishes
Suborder Ceratioidei
Family Ceratiidae—seadevils
Family Himantolophidae—footballfishes
Order Gobiesociformes
Family Gobiesocidae—clingfishes
Order Cyprinodontiformes
Suborder Exocoetoidei
Family Exocoetidae—flyingfishes and halfbeaks
Family Belonidae—needlefishes
Suborder Cyprinodontoidei
Family Cyprinodontidae—killifishes and topminnows
Family Aplocheilidae—rivulines
Family Anablepidae—four-eyed fishes
Family Poeciliidae—livebearers
Order Atheriniformes
Family Atherinidae—silversides
Family Melanotaeniidae—rainbowfishes
Order Lampriformes
Suborder Lamproidei
Family Lampridae—opah
Suborder Trachipteroidei
Family Trachipteridae—ribbonfishes
Family Regalecidae—oarfishes
Order Beryciformes
Suborder Berycoidei
Family Berycidae—alfonsinos

Family Trachichthyidae—slimeheads and roughies
Family Holocentridae—squirrelfishes
Suborder Polymixioidei
Family Polymixiidae—beardfishes
Order Zeiformes
Family Zeidae—dories
Family Caproidae—boarfishes
Order Gasterosteiformes (Thoracostei)
Family Gasterosteidae—sticklebacks
Order Indostomiformes
Family Indostomidae—indostomids
Order Pegasiformes
Family Pegasidae—seamoths
Order Syngnathiformes (Solenichthys)
Suborder Syngnathoidei
Family Syngnathidae—pipefishes and seahorses
Suborder Aulostomoidei
Family Aulostomidae—trumpetfishes
Family Fistulariidae—cornetfishes
Family Macrorhamphosidae—snipefishes
Order Dactylopteriformes
Family Dactylopteridae—flying gurnards
Order Synbranchiformes
Family Synbranchidae—swamp-eels
Order Scorpaeniformes
Suborder Scorpaenoidei
Family Scorpaenidae—scorpionfishes and rockfishes
Family Synanceiidae—stonefishes
Family Triglidae—searobins
Suborder Anoplopomatoidei
Family Anoplopomatidae—sablefishes
Suborder Hexagrammoidei
Family Hexagrammidae—greenlings
Suborder Cottoidei
Family Cottidae—sculpins
Family Agonidae—alligatorfishes and poachers
Family Cyclopteridae—lumpfishes and snailfishes
Order Perciformes
Suborder Percoidei
Family Percidae—perches
Family Centropomidae—snooks

Family Percichthyidae—temperate basses
Family Serranidae—sea basses
Family Grammidae—basslets
Family Teraponidae—grunters and tigerperches
Family Centrarchidae—sunfishes and black basses
Family Priacanthidae—bigeyes or catalufas
Family Apogonidae—cardinalfishes
Family Malacanthidae—tilefishes
Family Pomatomidae—bluefishes
Family Rachycentridae—cobias
Family Echeneididae—remoras
Family Carangidae—jacks and pompanos
Family Coryphaenidae—dolphins
Family Bramidae—pomfrets
Family Lutjanidae—snappers
Family Lobotidae—tripletails
Family Gerreidae—mojarras
Family Haemulidae—grunts
Family Sparidae—porgies
Family Sciaenidae—croakers and drums
Family Mullidae—goatfishes
Family Pempherididae—sweepers
Family Toxotidae—archerfishes
Family Kyphosidae—sea chubs
Family Ephippididae—spadefishes
Family Chaetodontidae—butterflyfishes
Family Pomacanthidae—angelfishes
Family Nandidae—leaffishes
Family Cichlidae—cichlids
Family Embiotocidae—surfperches
Family Pomacentridae—damselfishes
Family Cirrhitidae—hawkfishes

Suborder Mugiloidei
Family Mugilidae—mullets

Suborder Sphyraenoidei
Family Sphyraenidae—barracudas

Suborder Polynemoidei
Family Polynemidae—threadfins

Suborder Labroidei
Family Labridae—wrasses
Family Scaridae—parrotfishes

Suborder Zoarcoidei
Family Zoarcidae—eelpouts

Family Stichaeidae—pricklebacks
Family Pholididae—gunnels
Family Anarhichadidae—wolffishes
Suborder Notothenioidei
Family Nototheniidae—cod icefishes
Family Channichthyidae—crocodile icefishes
Suborder Trachinoidei
Family Trachinidae—weeverfishes
Family Opistognathidae—jawfishes
Family Uranoscopidae—stargazers
Family Percophidae—duckbills
Family Mugiloididae—sandperches
Suborder Blennioidei
Family Blenniidae—combtooth blennies
Family Tripterygiidae—triplefins
Family Dactyloscopidae—sand stargazers
Family Labrisomidae—labrisomids
Family Clinidae—clinids
Family Chaenopsidae—pikeblennies and flagblennies
Suborder Icosteoidei
Family Icosteidae—ragfish
Suborder Schindlerioidei
Family Schindleriidae—neotenicfishes
Suborder Ammodytoidei
Family Ammodytidae—sand lances
Suborder Callionymoidei
Family Callionymidae—dragonets
Suborder Gobioidei
Family Gobiidae—gobies
Family Eleotrididae—sleepers
Family Microdesmidae—wormfishes
Suborder Kurtoidei
Family Kurtidae—nurseryfishes
Suborder Acanthuroidei
Family Acanthuridae—surgeonfishes
Family Trichiuridae—cutlassfishes
Family Scombridae—mackerels and tunas
Family Xiphiidae—swordfish
Family Istiophoridae—billfishes
Suborder Stromateoidei
Family Nomeidae—driftfishes
Family Stromateidae—butterfishes

Suborder Anabantoidei
Family Anabantidae—climbing gouramies
Family Helostomatidae—kissing gourami
Suborder Luciocephaloidei
Family Luciocephalidae—pikehead
Suborder Channoidei
Family Channidae—snakeheads
Suborder Mastacembeloidei
Family Mastacembelidae—spiny eels
Order Pleuronectiformes (Heterosomata)
Suborder Pleuronectoidei
Family Pleuronectidae—righteye flounders
Family Bothidae—lefteye flounders
Suborder Soleoidei
Family Soleidae—soles
Family Cynoglossidae—tonguefishes
Order Tetraodontiformes (Plectognathi)
Suborder Balistoidei
Family Balistidae—triggerfishes and filefishes
Family Triacanthodidae—spikefishes
Family Ostraciidae—boxfishes
Suborder Tetraodontoidei
Family Tetraodontidae—pufferfishes
Family Diodontidae—porcupinefishes
Family Molidae—molas or ocean sunfishes

(Modified and abbreviated from Nelson, J. S., *Fishes of the World,* 2nd ed., John Wiley & Sons, New York, 1984.)

2 Swimming

Paul W. Webb

INTRODUCTION

By definition, fish swim, but "swimming" is a loose term for a wide and complex set of adaptive movements whereby fish perform the numerous activities necessary to survive in diverse habitats. As a result, physiological studies pertinent to swimming are legion, necessitating discussion of only a few selected topics in this chapter, which focuses on the nature and properties of the propulsion system. This is comprised of (1) a propulsor that transfers momentum from the fish to the water, thereby generating thrust, and (2) the muscles that drive those propulsors. Neural control systems are not discussed. Swimming energetics are discussed because driving the propulsors is a major expense affecting design criteria for many other physiological systems as well as the impact of fish on their ecological resource base. In addition, the amount of energy available for swimming is often constrained by environmental factors. Scale effects are omitted. These have been reviewed recently by Goolish (1991a).

PROPULSION SYSTEMS

PROPULSORS

Diversity and Classification

High water density allows many fish to be at or close to neutral buoyancy (Bone and Marshall, 1982; Gee, 1983; Alexander, 1990), so that axial and appendicular systems can be specialized for swimming (Figure 1). The resultant diversity of propulsors is usually classified on the basis of functional similarities defined by kinematics and associated forces for separate median and paired fin propulsors (MPF propulsion) and body and caudal fin propulsors (BCF propulsion) (Lindsey, 1978; Braun and Reif, 1985; Webb and Blake, 1985; Daniel and Webb, 1987; Webb, 1988a). Of particular importance is the degree of undulation, measured by the number of half waves within the length of a propulsor, and the rate at which the amplitude of lateral movements increases along the body. These two factors largely determine the forces that dominate thrust production and efficiency (Lighthill, 1975; Daniel and Webb, 1987).

Anguilliform BCF undulation (Figure 1) involves more than two half waves within the body length and large amplitudes along most of that length. Early fishes probably had a well-developed head skeleton and probably swam with

0-8493-8042-1/93/$0.00+$.50

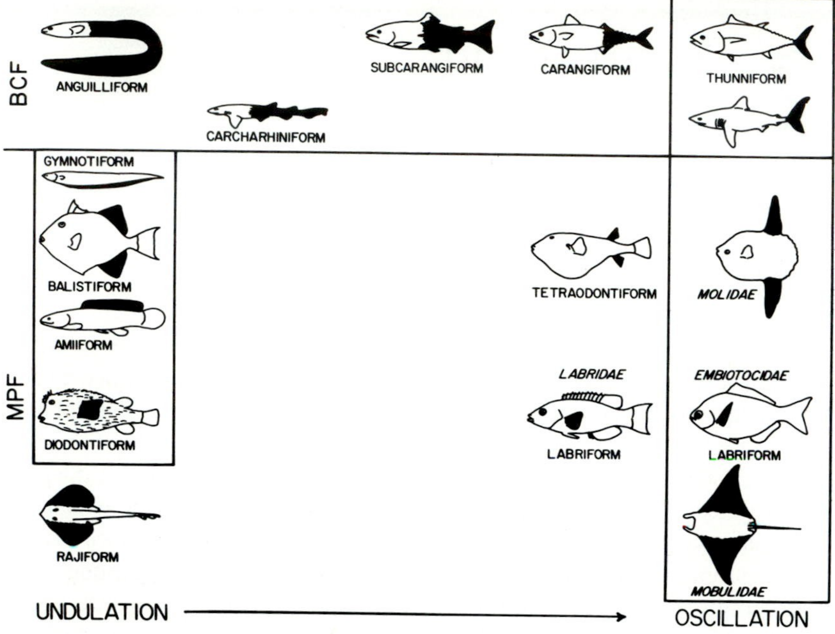

FIGURE 1. The diversity of propulsors (shaded regions) among fishes is large, as all possible axial and appendicular systems are used. In recognizing this diversity, Breder (1926) proposed names for various morphologies based on the familial name of an exemplary family and the addition of the suffix "form"; e.g., eel-like = anguilliform. This approach does not recognize similarities in kinematics and physical mechanisms whereby the propulsors generate thrust (Webb and Blake, 1985). As a result, several probably distinct propulsion variants are unnamed (e.g., *Mola*) or the same names are used for mechanically different systems (e.g., labriform mode). Homologous propulsors often function in different ways (rows). BCF propulsors have been studied most extensively , and a succession of kinematic patterns is recognized from undulation to oscillation. This is associated with decreasing use of drag as a thrust force and increasing use of acceleration reaction and, for oscillating propulsors, lift. Kinematic trends from undulation to oscillation are often associated with higher thrust and efficiency (Lighthill, 1975; Daniel and Webb, 1987; Webb, 1988a). Morphologically different propulsors often function in the same way. The oscillatory mechanisms boxed on the right are probably all lift-based (Daniel and Webb, 1987; Webb, 1984b). The undulatory fins with long fin-bases (boxed left) generate thrust using drag and acceleration reaction, enhanced by body-fin interactions (Lighthill and Blake, 1989; Lighthill, 1990a, 1990b, 1990c). Undulatory rajiform propulsion is probably distinct as significant bending along the span of the fin results in different relationships between kinematics and thrust and efficiency (Daniel, 1988). Similar trends undoubtedly occur for MPF propulsors, but there are insufficient documentary data. Oscillation by fins in the tetraodontiform mode and labrids using the labriform mode may be more drag-based than lift-based. Oscillatory ostraciiform propulsion, believed to be of low efficiency, is not shown.

anguilliform undulation of the posterior of the body, probably like modern anuran tadpoles (Wassersug, 1989). When the posterior body was armored, it was articulated to permit undulation (Moy-Thomas and Miles, 1971). These anguilliform motions probably generated thrust through a combination of drag and acceleration reaction forces, similar to modern eels (Lighthill, 1975).

The early fish were probably quite sluggish (Webb and Smith, 1980). Their small bodies and fin depths and anguilliform motions would generate little thrust (Lighthill, 1975; Webb, 1984a) while armor increases acceleration resistance (Webb et al., 1992). Not surprisingly, a key feature of ostracoderm evolution was improved performance through the concomitant enhancement of body depth and loss of armor (Webb and Smith, 1980).

Larger body and tail depth increases the mass of water entrained by BCF motions so that acceleration reaction forces dominate, increasing thrust and improving speed and acceleration. Modern actinopterygian fishes concentrate lateral movements towards the posterior half to third of the body, which further improves thrust and efficiency (Lighthill, 1975). Among undulatory swimmers, the carangiform pattern of swimming is most advanced, with little more than half a wave within the body length (Breder, 1926). Most fish, however, are sub-carangiform (Breder, 1926), retaining more of a wave within the body length but concentrating increases in amplitude towards the tail (Figure 1). Sub-carangiform swimmers are presumed to be less efficient and powerful than carangiform swimmers.

The morphology and swimming patterns of many selachians are carchariniform, differing substantially from bony fishes, but providing thrust enhancement for undulatory swimmers (Webb, 1988b). In carchariniform swimming, the body length typically includes more than two half waves within its length (Gray, 1933a, 1933b; Webb and Keyes, 1982). Amplitude is often large over the posterior half of the body, resulting in the first dorsal fin making large lateral movements. The timing between the motion of the first dorsal fin and of the tail is such that the wake shed by the former fin enhances thrust when that wake reaches the tail (Lighthill, 1975; Webb and Keyes, 1982; Weihs, 1989).

Fish larvae, hatching at lengths of ≥2 to 3 mm, also swim with undulatory anguilliform motions in which drag forces are used to produce thrust (Vlymen, 1974; Jordan, 1992). Development of adult propulsor morphology and function changes occur through a number of metamorphoses or saltatory stages (Balon, 1984). The functional importance of these changes remains to be fully explored, although some general features are recognized. These are largely attributable to changes in Re* during development. For example, energy-saving burst-and-coast behavior of alternating swimming bouts with unpowered glides is delayed until Re >20 (Weihs, 1973a, 1974), while smaller larvae are constrained by their hydrodynamic environment to making small-radius turns (Fuiman and Webb, 1988). Fin development is delayed until the hydrodynamic environment is appropriate for effective development of large thrust forces and

---

*Reynolds number, Re, is an indicator of the relative importance of viscous and inertial forces. Re = Lu/ν, where L is total length, u is swimming speed, and ν is the kinematic viscosity of water. Viscous forces dominate when Re is small (L and u are small). Inertial forces become important for swimming at Re of about 20.

muscle stress is large enough to power deeper fins (Webb and Weihs, 1986; Jordan, 1992).

Some highly derived thunniform selachians and actinopterygians swim with large-amplitude oscillations of a deep half-moon shaped caudal fin (Figure 1). The caudal fin acts as a hydrofoil, using lift to generate high thrust, and the body is streamlined to minimize resistance (Lighthill, 1975). These animals are the fastest representatives of the phylogenetic groups to which they belong.

In contrast to evolutionary trends towards oscillatory BCF motions, those in MPF propulsion are from oscillation to undulation with recurrence of oscillation in some groups. Appendages probably originally worked as planes controlling directional stability similar to modern selachians (Harris, 1936, 1938; Aleyev, 1977; Weihs, 1989). Articulation of the fin base would have improved control, leading to the use of these fins as propulsors. Subsequently, flexible fins arose and are common in modern actinopterygian and batoidimorph fishes (Gosline, 1971). In bony fishes, these fins produce a wide range of undulatory and oscillatory motions used especially for slow swimming. Some fish derived from those with flexible fins have stiff median and paired fins as control surfaces, for example tuna (Magnuson, 1978). Others also use relatively stiff but mobile appendages as oscillating hydrofoils (e.g., *Mola*, the ocean sunfish; Figure 1) for fast, efficient swimming.

Median and paired fins are used for lower levels of swimming performance where their efficiency is higher than BCF propulsors. Performance can also be increased because body drag is reduced when the body is held stretched-straight, while thrust may be enhanced from interactions between the fins and the body (Geerlink, 1986; Lighthill, 1975, 1990a, 1990b, 1990c; Lighthill and Blake, 1989; Weihs, 1989). Nevertheless, the amount of muscle associated with MPF systems is relatively small so that BCF propulsion is used at high speeds and acceleration rates. High speed and acceleration are important for predator avoidance, so fish that rely on MPF swimming typically minimize such risks by being large (e.g., *Mola*, skates, and rays) or by using additional defenses such as spines or toxins (e.g., lionfish).

Propulsor Recruitment

Most studies of propulsors have focused on differences among species. In practice, no single propulsor can function efficiently over the entire swimming performance range, so that a variety of propulsors are required by most individuals (Webb, in press). These are recruited in an orderly sequence to match power generation with energy needs for swimming at different speeds and acceleration rates (Table 1). Undulation of the paired and/or median fins is commonly used at the lowest speeds and acceleration rates. At slightly higher speeds these MPF propulsors tend to shift to drag and/or lift-based oscillations. As speed increases, BCF propulsors first supplement and then succeed MPF propulsors (Alexander, 1989; Webb, in press). Intermittent burst-and-coast swimming may be used to increase endurance as part of the BCF swimming

TABLE 1
A Generalized Summary of Recruitment Patterns for Muscles, Propulsors and Swimming Behavior Which Define Various Gaits as Performance Level Increases

	Propulsor Recruitment Sequence	Muscle Recruitment Sequence	Behavior Recruitment Sequence
Low performance			
↓	MPF undulatory propulsion	MPF red fibers	
↓	MPF undulatory/ oscillatory propulsion	MPF red fibers	
↓	MPF plus BCF propulsors	MPF and BCF red fibers	
↓	BCF propulsor	BCF red fibers	Burst-and-coast swimming, drag enhancement
↓	BCF propulsor	BCF red fibers	Steady swimming
↓	BCF propulsor	BCF red and white fibers	Steady swimming
↓	BCF propulsor	BCF red and white fibers	Burst-and-coast swimming
↓	BCF propulsion	BCF white fibers	Steady swimming (sprints)
↓	BCF propulsion	BCF white fibers	Fast-starts
High performance			

Adapted from Webb, P. W., in *Comparative Vertebrate Exercise Physiology,* Jones, J. H., Ed., Academic Press, Orlando, FL, in press.

repertoire. Finally, large-amplitude transient movements are used in fast-starts to accelerate at high rates (Weihs, 1972, 1973b; Rome et al., 1988; Rome and Sosnicki, 1991; Webb, in press).

Muscle

Fiber Types

Fish, like other vertebrates, have several types of muscle fiber with different properties. Two fiber types are most important for locomotion (Bone, 1975, 1978a, 1978b); red or slow oxidative fibers and white or fast glycolytic fibers (Table 2). A third type of muscle, comprised of pink or fast oxidative fibers, is also used for locomotion in at least some fish (Johnston, 1981; 1991). Most studies have been performed on myotomal muscle fibers, but more recent observations on fin muscles show they have similar mechanical and physiological properties (Johnson and Johnston, 1991a).

Speed of shortening is one of the principal diagnostic features of fiber type and a critical determinant of power output. The maximum shortening speed, V_{max}, appears to be determined by the rate of cross-bridge detachment which in turn is set by the properties of the myosin molecule and the rate of removal of Ca^{2+}. Both red and white fibers generate maximum power and maximum efficiency at intermediate shortening speeds, V, at strain rates V/V_{max} of 0.15 to 0.4

TABLE 2
A Comparison of Some Principal Characteristics of Red and White Muscle of Fishes

Red Muscle	White Muscle
SO (slow oxidative fibers) type IIb	FG (fast glycolytic fibers) type Ia
0.5–30% myotome mass	
60–150 mm in diameter	Up to about 300 mm in diameter
Metabolic Properties	
Aerobic	Anaerobic
1.9–2.5 capillaries/fiber	0.2–0.9 capillaries/fiber
15–35% of fiber volume	0.5–4% fiber volume
Lipid and glycogen stores	Usually glycogen stores
Usually high (myoglobin)	Low (myoglobin)
Enzyme activities $\alpha\ M^{0.9-1.0}$	Enzyme activities $\alpha\ M^{1.2-1.5}$
Electromechanical Coupling	
Tonic	Phasic
Multiterminal distributed innervation	Focal innervation in less derived species. Polyneuronal distributed innervation in more derived actinopterygians
Sarcoplasmic reticulum	
0.1–0.6% fiber volume	0.3–0.9% fiber volume
T-tubule system	
3–5% fiber volume	5–14% fiber volume
	Parvalbumims approx. 15% of soluble proteins
Mechanics	
V_{max} 4.5 muscle lengths (ML)/s	V_{max} 13 ML/s
In situ V/V_{max} 0.17–0.36	$V/V_{max} < 0.38$
	Gear ratio 4 times greater
Maximum power about 150 W/kg at 20°C	Maximum power 25–35 W/kg
Maximum power about 5–8 W/kg at 5°C during oscillatory work	

Note: Pink muscle (FOG, type IIa) is intermediate, although detailed information is lacking.

Data from Bone, 1975, 1978a, 1978b; Bone and Marshall, 1982; Somero and Childress, 1984; Johnston, 1981, 1991; Rome, in press.

(Altringham and Johnston, 1988; Rome, in press). Different isoforms of myosin and troponin, together with high concentrations of parvalbumin, result in higher V_{max} for white than red fibers.

A second major discriminating feature of red and white muscle is the metabolic basis for energy production (Table 2). Red muscle is a highly vascularized, aerobic, non-fatiguing low-power system. White muscle is a primarily anaerobic, capacity-limited, high-power system. Pink muscle is intermediate. Therefore, red muscle is adapted to function in low-level sustainable activities of routine swimming while white fibers are used for bursts of activity, commonly in fitness critical situations where survival is important.

Anatomy

Separate muscle systems move the fin blade of the median caudal, anal, and dorsal fins and the paired pectoral and pelvic fins (Figure 2), while additional muscle systems and the spring-like fin rays control the shape of the fin blade (McCutcheon, 1970; Geerlink and Videler, 1974; Videler, 1977).

The principal muscles in the pectoral fins (Figure 2C) are deep and superficial adductor and abductor muscle blocks inserting on expanded T-shaped bases of the paired fin rays (Harden, 1964; Alexander, 1967; Gosline, 1971; Geerlink, 1986; Langfeld et al., 1991). The first ray, which would be subject to the greatest fluid forces, also has arrector and flexor muscles (Geerlink, 1979). Similarly, erector and depressor muscles of median fins (Figure 2B) link lepidotrichia or fin rays to the pterygiophores. Various oblique muscles control fin area (Figure 2B, C, D), for example the inclinator muscles of the dorsal and anal fins and arrector muscles of the pectoral fins (Harden, 1964; Geerlink, 1979).

A major trend in actinopterygian caudal fin evolution is towards increasing complexity of muscles to control the orientation and shape of the fin blade (Lauder, 1989). In selachians and primitive actinopterygians, caudal fin bending is achieved via the contraction of myotomal muscle, with the lateral superficial muscles inserting directly onto the caudal fin rays. Caudal fin ray movement independent of the myotomal muscles occurred first in the Ginglymodi, primarily for hypaxial rays as a result of development of an initially weak hypochordal longitudinalis muscle and an extensive ventral flexor muscle. Antagonistic interradialis and supracarinalis muscles originated in the Halecostomes and provide increased control over fin area, in this case for epaxial fin rays. Teleost clades have both epaxial and hypaxial interradial and carinal muscles (Figure 2D). The hypochordal longitudinalis muscle is also well developed and is probably critical in stiffening and stabilizing the epaxial portion of the caudal fin during fast-starts (Lauder, 1989).

Actinopterygian fin muscle bundles tend to be small and fusiform or moderately pinnate (Harden, 1964; Geerlink, 1983) while those of batoidimorph fish tend to be more extensive and organized in sheets (Bone, 1975, 1978a). Superficial red fiber bundles in the pectoral fins of rays are oriented parallel to and insert about halfway along the ceratotrichia. White muscle fibers are arranged obliquely to the ceratotrichia with a pinnate-like arrangement, which probably increases power output (Bone, 1975, 1978a).

The anatomy and fiber arrangements of myotomal muscle are different from those of the median fins, and the way the muscle, myosepta, and skin drive the

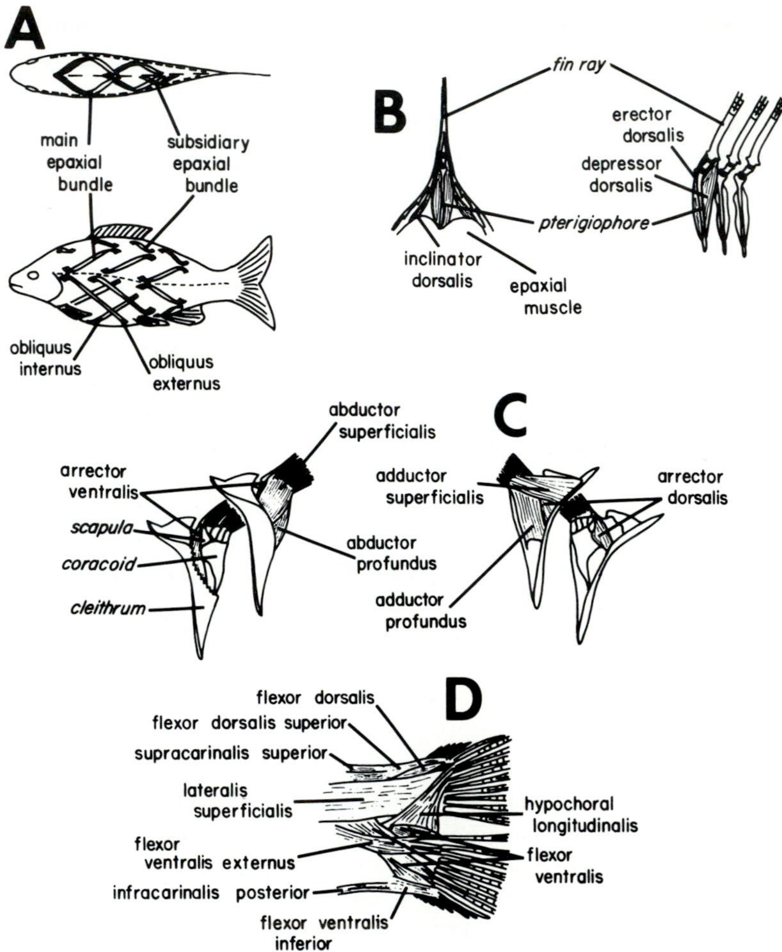

FIGURE 2. Muscle systems controlling actinopterygian propulsors. (A) Helical trajectories in the myotomal muscle of teleostean fishes seen from above and in lateral view. The location of red muscle is illustrated by the dotted lines. The lateral red muscle lies in a strip above and below the dotted line seen in the lateral view. (B) Transverse and lateral views of the dorsal fin skeleton and principal musculature moving the fin rays. Carinalis muscles are omitted for simplicity. The anal fin has similar anatomy. (C) Major muscles moving the fin rays of the left-hand pectoral girdle seen from the left (on the left) and from the right. (D) Major muscle moving the caudal fin rays and controlling the spread of the caudal fin of an acanthopterygian fish. The ventral lateralis superficialis is omitted to show the deeper muscles. (Based on Harden, 1964; Alexander, 1967, 1969; Gosline, 1971; Cailliet et al., 1986; Lauder, 1989.)

tail is still not understood (Wainwright, 1983, 1988; Long, 1992; Hebrank et al., 1990). Segmental myomeres in the least derived chordates and larval fishes are folded into V shapes, with the apex pointing rostrally. The flexures become increasingly W-shaped during development as well as in more derived species

(Bone, 1989). Myosepta between the myomeres are flexible in bending but stiff under tension and undoubtedly are key to force transmission from the muscle to the body (Wainwright, 1983). Myosepta are often stiffened in teleosts, probably to control torsional moments due to muscle contraction (Bone and Marshall, 1982).

Fiber organization within the myotome is also complex (Figure 2a) compared with that of median and paired fins. By following adjacent white fibers from myomere to myomere, Alexander (1969) showed they traced complex spiral patterns, or trajectories, through the myotome. Alexander (1969) identified two trajectory patterns. One is associated with tendons or equivalent functional components and is found in the myosepta of selachians, less derived actinopterygians (*Acipenser, Protopterus, Amia, Anguilla,* and *Salmo*), and the caudal regions of more derived teleosts. The second pattern, with muscle trajectories following a helical pattern, occurs in the anterior myomeres of more derived teleosts where tendon-like structures are absent.

All myotomal red fibers are approximately equidistant from, and parallel to, the medial axis of body bending. All sarcomeres shorten by the same amount in the length range where force generation is >96% of the maximum (Rome et al., 1990; Rome and Sosnicki, 1990, 1991; Rome, in press). In contrast, myotomal white muscle fibers vary in distance from the median plane, and subtend angles from 10 to nearly 40° to the body axis. This organization ensures that all white fibers contract by similar proportions of their length (Alexander, 1969), developing forces >85% of the maximum, while working in the portion of the stress-strain curve where power and efficiency are high (Rome, in press). The selachian pattern provides greater mechanical advantage, which probably explains its retention in the narrow caudal peduncle region in more derived teleosts (Alexander, 1969).

The helical trajectories allow high bending rates compared with the red fiber organization. This has important consequences for the gear ratio of the fibers, where gear ratio is the change in body position for a given change in sarcomere length. The gear ratio of white fibers is four times higher than for red fibers (Rome et al., 1992a, 1992b), which, combined with higher V_{max}, results in white fibers producing movements that are an order of magnitude faster than those due to red fibers.

Muscle Recruitment

The characteristics of the different muscle fiber types suit them for different levels of work (Table 2). Thus, as locomotor power requirements increase, red, then pink, and finally white fibers in increasing numbers of motor units are recruited (Bone, 1978a, 1978b; Johnston, 1981; Eaton and Emberley, 1991). Fast-starts traditionally have been thought to be maximum all-or-none events using the entire muscle mass first on one side of the body and then on the other. Recently it has been shown that a variable number of motor units is used, depending on turning angle (Eaton et al., 1988; Foreman and Eaton,

1990; Foreman, 1991), and hence on the amount of work performed (Webb, in press).

Recruitment patterns correlate with innervation patterns. Red muscle is comprised of tonic fibers with distributed multiterminal innervation. Power requirements are probably matched to swimming resistance by graded contractions and by varying the number of active motor units. White fibers of elasmobranchs, sarcopterygians, and primitive actinopterygians (Halecostomes, Ginglymodi, and basal elopomorph and clupeomorph teleosts) are focally innervated (Fetcho, 1987). These fish use red muscle at low aerobic swimming speeds, and recruit white muscle only for sprints and rapid acceleration. White muscle fibers of more derived teleosts are unique in having multiterminal innervation like red muscle fibers (Bone, 1978a; Johnston, 1981; Fetcho, 1987). This white muscle is also believed to show graded contractions permitting white muscle recruitment at relatively lower speeds when it can support red muscle in sustainable swimming.

Red and White Muscle Proportions

The amount of red and white muscle varies markedly among fishes. White muscle comprises about 50 to 60% of the body mass of fish that swim fast and/or continuously, such as the thunniform fishes, salmonids, clupeids, etc. (Webb, 1975a, 1978a; Graham, 1983; Block, 1991) and also in sit-and-wait ambushers that use whole-body accelerations and sprints to capture elusive prey. Whole-body acceleration, however, is not very efficient, and has been replaced by inertial suction feeding in many clades (Lauder and Liem, 1983). Many such fish have reduced proportions of white muscle, for example cottids (Webb, 1990). In food-poor habitats, contractile muscle proteins may be reduced in suction feeding fishes (Bone and Marshall, 1982; Hochachka and Somero, 1984).

More mobile fish have higher proportions of red muscle in the myotome, as assayed at the caudal peduncle. It is believed that the proportions of red and white muscle relate to foraging patterns, with high red muscle proportions being associated with wide-ranging foragers and low proportions occurring in sit-and-wait ambushers. McLaughlin and Kramer (1991) analyzed data from the literature on muscle proportions and natural history reports on foraging behavior and did not support this idea. Instead they found overlap, with many fish having intermediate amounts of muscle, implying mixed foraging strategies. This situation appears to be different from that for terrestrial animals, implying fish are more opportunistic or more plastic in their foraging behavior.

A number of factors contribute to the plasticity and attendant opportunism in locomotor muscle patterns and reported feeding behaviors. First, most of the body weight is supported by the water so that a diversity of propulsors has evolved which can be used for different swimming activities (McLaughlin and Kramer, 1991). Second, inertial suction feeding can reduce dependence on white muscle for attacks. Then there is no need for red muscle to be displaced to increase white. Third, the distributed multiterminal innervation of white muscle of many fish (Fetcho, 1987) probably provides a graded thrust capability so that fish with little red muscle may still have considerable cruising capability (Bone, 1975).

In addition, lumping observations from diverse groups of fishes may have obscured actual relationships between muscle proportions and foraging behavior. More complete data sets and comparisons are desirable for different groups, such as less derived actinopterygians with focally innervated muscle, elasmobranchs, and clades with well-developed suction feeding (Brooks and MacLennan, 1991).

Interactions Between the Muscles and Propulsors

The mechanisms translating muscle contraction into propulsive movements are poorly understood. The presence of discrete muscle bundles rotating fin rays like levers, clearly definable force trajectories (Geerlink, 1986), and the ability for manipulation (Lauder, 1989) should make studies on MPF systems comparatively simple. Instead, research has focused on the more complex BCF system.

The simplest situation for BCF swimming is alternate and complete muscle contractions on opposite sides of the body. This is probably the situation in prehatching and early larvae, involving only the anterior myomeres (Blight, 1976, 1977; Eaton et al., 1977) and in high-speed sprints (Hess and Videler, 1984). Using amphibian larvae, Blight (1977) has shown that such alternating contractions of anterior muscles attached to a passive tail region result in a travelling wave analogous to that seen during swimming.

During continuous swimming by larvae and elongate fishes, the head and anterior of the body make large lateral movements. These movements not only waste energy but also reduce the effectiveness of sensory organs concentrated rostrally (Blight, 1977). During ontogeny, and with the onset of steady swimming after a fast-start, the head of fusiform fishes is stabilized relative to the tail. A feature of this transition appears to be the faster caudal propagation of waves of muscular activity compared with the speed of the propulsive wave (Blight, 1977; Sigvardt, 1989; Williams et al., 1989; Frolich and Biewener, 1992). This has three effects: (1) the body is stiffened rostrally, (2) bending moments are generated at the tail, and (3) the posterior body and tail are stiffened such that the curvature induced by resistive forces at various speeds results in appropriate angles to orient momentum changes close to the axis of progression of the body (Alexander, 1969; Blight, 1977; Williams et al., 1989).

Performance Range Fractionation by Swimming Gaits

Animals use various gaits to move at different speeds. A gait is a locomotor pattern typical of a limited performance range, and characterized by discontinuities in one or more quantities describing the locomotor pattern

(Alexander, 1989). Although fish swimming patterns are very variable, swimming gaits are readily defined in terms of muscle fiber type, propulsor type, and propulsor kinematics (Table 1). These gaits divide a large performance range into small parts in which thrust power and muscle power are generated efficiently (Videler, 1981; Rome et al., 1990; Webb, in press).

Maximum swimming speeds achieved in different gaits vary considerably among species, reflecting interspecific differences in muscle and propulsor characteristics. For example, higher temperature increases muscle V_{max} and hence power output (Rome, in press; Rome et al., 1992a, 1992b). This is probably critical for thunniform fish in achieving their high swimming speeds. Similarly, studies of carp, *Cyprinus carpio*, and scup, *Stenitomus chrysops,* have shown that the maximum isometric force generated by carp myotomal muscle is about half that of scup (Rome et al., 1992a, 1992b). Scup, therefore, swim over a range of higher speeds than carp. However, carp are able to swim efficiently at lower speeds.

In spite of the differences in swimming speed between carp and scup, V/V_{max} of the myotomal red muscle is similar over the cruising speed ranges of each species. This is achieved by scup having a longer body wave, which increases gear ratio (Rome et al., 1992a, 1992b). However, Lighthill (1975) has pointed out that effective propulsion is a balance between opposing effects of wavelength on thrust and efficiency. Larger wavelengths increase thrust. Thus (Wu, 1977):

$$T \propto F_{tb}^{2}H^{2}B^{2}\,(1 - u/F_{tb}\lambda) \qquad (1)$$

where T = thrust, F_{tb} = tail beat frequency, H = tail beat amplitude, B = tail trailing edge span, and λ = wavelength.

Therefore, if all else is equal, larger wavelengths increase $1 - u/F_{tb}\lambda$ and hence T. Within the range of wavelengths seen among fishes, Froude efficiency, the mechanical efficiency of the propulsor, is related to wavelength as (Lighthill, 1975):

$$\text{Froude efficiency} = 0.5 + 0.5\ u/F_{tb}\lambda \qquad (2)$$

and hence tends to decrease with increasing wavelength.

Therefore, high swimming speed requires as long a wavelength as possible within constraints set by efficiency considerations, not only to maximize thrust but also to maximize gear ratio for the muscles (Lighthill, 1975; Rome et al., 1992a, 1992b; Rome, in press). In contrast, slow swimming is facilitated by smaller wavelengths, or a shift to an alternative MPF propulsion system.

Behavioral factors also affect speed ranges over which various gaits are used. The burst-and-coast gait allows cod and carp to extend their swimming ranges to lower speeds (Videler, 1981; Rome et al., 1990).

ENERGETICS

Oxygen Consumption

Relationships Between Metabolic Rate and Swimming Speed

Aerobic metabolism is used to provide energy for sustained activities. One energy cost, the standard metabolic rate, is a basic maintenance requirement measured as the minimum rate of oxygen consumption of a postprandial, unstressed fish at rest (Brett and Groves, 1979). Long-term energy demands for swimming, food acquisition and treatment, regulation due to environmental perturbation, and reproduction are additional to standard metabolism. These demands are met within a constraint of the maximum rate of aerobic metabolism, defined as the active metabolic rate.

Active metabolic rate traditionally has been measured using swimming as a means to elicit high rates of metabolism. This approach assumes performance is limited by oxygen delivery to the locomotor muscles (Beamish, 1978). However Goolish (1991a) has pointed out that higher rates of metabolism than those induced by swimming are achieved in young growing fish, while ionosmoregulation can add to swimming costs even at high levels of aerobic swimming performance (Rao, 1968). Therefore, it appears that muscle oxygen demand, rather than oxygen supply, limits aerobically supported performance for many fish, certainly for small fish. Thus the phrase "active metabolism" is used here for the maximum rate of aerobic metabolism inducible by swimming.

The relationship between metabolic rate and swimming speed is usually described after Brett (1964) by an exponential function:

$$\log Q_{O2} = \log Q_S + \alpha u \tag{3}$$

$$Q_{O2} = Q_S e^{\alpha u} \tag{4}$$

where Q_{O2} = total metabolic rate, Q_S = standard metabolic rate, α = regression coefficient.

There is no reason to expect an exponential relationship (Smit et al., 1971; Gordon et al., 1989). Indeed, Fry (1957) suggested metabolic rate was related to swimming speed by:

$$Q_{O2} = Q_S + \beta u^2 = Q_S + \beta u^\alpha \tag{5}$$

while Smit et al. (1971) found a linear relationship:

$$Q_{O2} = Q_S + \beta u \tag{6}$$

These three model forms were compared (R. Brazee and P.W. Webb, unpublished observations) for a selection of data on fish that swim in different ways (Brett, 1964; Smit et al., 1971; Webb, 1975b; Gordon et al., 1989) and for squid (D'Orr, 1982). It was found that the linear model did not usually give as good a fit to the data as the exponential and power functions. Examination of the residuals for the two nonlinear models showed them to be almost identical, as were mean squared errors, while predicted values of rates of oxygen consumption at a given speed were similar. Thus, the exponential function should not be assumed, but rather a model should be used that best describes a given data set. The power function, however, has some attraction since composite efficiency changes in the metabolic, muscle, and propulsor systems emerge as a simple power function of swimming speed.

Relationships between metabolic rate and speed (using either nonlinear model) are very variable. For example, Gordon et al. (1989) and Facey and Grossman (1990) found metabolic rate increased little with increasing swimming speed for two species of labriform swimmers, and some benthic and bentho-pelagic fishes. Variations presumably reflect interspecific differences in thrust, resistance, and efficiency. For example, negatively buoyant fish have U-shaped energy-speed relationships (Blake, 1979; Bunker and Machin, 1992). Such fish use hydrodynamic lift to control vertical position, resulting in an additional energy cost measured as induced drag. The U-shaped curve then occurs because induced drag decreases as speed increases, while parasite drag increases (Hoerner, 1965, 1975; Bone and Marshall, 1982).

Cost of Transport

Energy used for swimming has major impacts on ecological resource bases, net energy gain from different foraging strategies, and reserves required for migrations. The overall impact of swimming speed and energy costs is often measured and compared among species using the gross cost of transport, GCOT, the energy required to move unit mass through unit distance (Tucker, 1975; Videler and Nolet, 1990). There is a U-shaped relationship between GCOT and swimming speed because standard metabolism makes large contributions to total energy expenditure in traversing any distance at low speeds, while activity metabolism dominates at high speeds.

An optimum speed can be defined at the minimum GCOT (Weihs, 1973c). However, the slope, α, of the metabolic rate-speed relationship (exponential model, Equation 3) is a major determinant of this optimum speed which increases as α decreases (Figure 3a). Strongly U-shaped GCOT-speed relationships are found when α is large. If α is small, the optimum speed is outside the aerobic speed range. Thus, any factor that affects the slope (e.g., endothermy, negative buoyancy, environmental factors such as temperature or toxicants), will predictably increase optimum speed. Similarly, Q_S influences the shape of the GCOT-speed relationship (Figure 3b). Strongly U-shaped relationships are found only when Q_S is large, but the optimum speed is not affected. Physiological differences among species, for example size effects and responses to environmental factors, strongly influence Q_S and

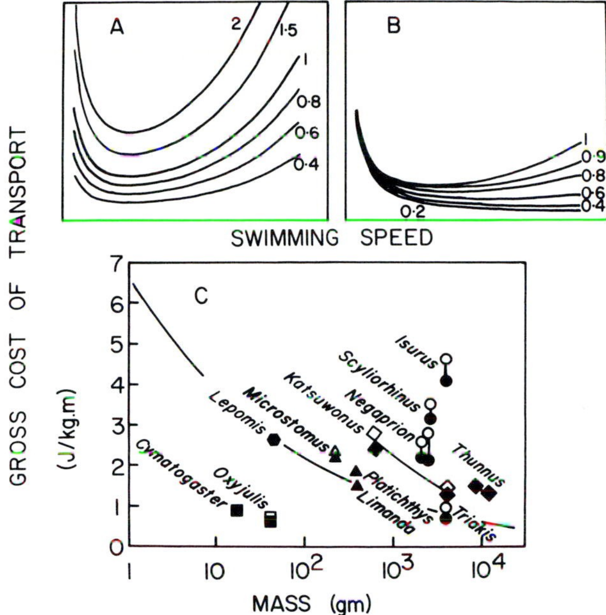

FIGURE 3. Illustration of some factors that affect gross cost of transport, GCOT. Panels A and B show how the magnitude of regression coefficients for the exponential model affects the general shape of the GCOT-swimming speed curve, based on data for sockeye salmon from Brett (1964), where $Q_s = 0.15M^{0.846}$, active metabolic rate = $0.95M^{0.963}$, and log u_{crit} = 1.29 + log · L. In panel A, the slope of the relationship is multiplied by factors of 0.4, 0.6, 0.8, 1, 1.5, and 2. Strongly U-shaped curves occur only with large values for the slope. In panel B, the intercept for standard metabolism is multiplied by factors of 0.2, 0.4, 0.6, 0.8, 0.9, and 1. Low rates of standard metabolism flatten the GCOT-swimming speed curve. In panel C, the effects of size are illustrated with respect to a convenience reference (solid line) for sockeye salmon (Brett and Glass, 1973) for a selection of species. Solid symbols show GCOT at the optimum speed. This speed is sometimes outside the aerobic swimming performance range, when GCOT at maximum reported aerobic speeds is shown by an open symbol. Squares, labriform swimmers (Gordon et al., 1989); hexagon, subcarangiform swimming by a gibbose fish, *Leponis gibbosus* (Brett and Sutherland, 1965); triangles, pleuronectiforms (Duthie, 1982); circles, thunniform fishes; and diamonds, sharks (summary data in Graham et al., 1990).

hence the speed range over which transport costs can be minimized. Costs of transport are believed to be ecologically important. For example, fish are believed to swim close to optimum speeds during routine movements such as foraging (Ware, 1978; Weihs and Webb, 1983; Webb, 1991) when changing α or Q_s should affect net food gains and habitat choice. Therefore, variation in α and Q_S should affect behavior and habitat choices, but such interactions remain to be explored.

Some consequences of different physiological and morphological factors on GCOT are illustrated in Figure 3c. The specific optimum speed, measured as body lengths/second (L/s), decreases with increasing size, reflecting α increasing with L (Brett and Glass, 1973). Using sockeye salmon as a convenience reference, skipjack tuna and albacore have higher GCOT for their size, presumably due to endothermy. Weight support for the negatively buoyant pleuronectiformes gives most species high minimum GCOT. The elasmobranchs studied to date also have high GCOT, perhaps because they too tend to be negatively buoyant. In contrast, labriform swimmers that hold the body straight while swimming achieve low GCOT.

Anaerobic Metabolism

Power for bursts of activity (sprints and fast-starts) is provided by white muscle via anaerobic pathways. Anaerobic pathways also fuel intermediate or prolonged speeds between cruising and sprints that can be sustained for tens of minutes, but nevertheless finally result in fatigue. Some cyprinids use anaerobic metabolism at low sustained speeds, but this appears to be the exception among fishes (Bone, 1975).

The capacity for anaerobic energy production has been estimated from decreases in substrate reserves, endproduct accumulation, and "oxygen debt" (Weiser et al., 1985; Goolish, 1991a; Puckett and Dill, 1984, 1985; Kauffman, 1990). For example, phosphocreatine decreases by about 75% from 5.8 μmol/g muscle to 1.5 μmol/g muscle within 30 s when young rainbow trout are chased; white muscle glycogen reserves of ≤1% can be depleted by 50% in only 2 min of forced activity (Wardle, 1975; Goolish, 1991a). Energy expenditure during foraging lunges averaging a speed of 9 L/s has been measured from oxygen repayment by juvenile coho salmon, *Oncorhynchus kisutch* (mean mass, 1.2 g). Metabolic rates were estimated to be equivalent to 38,000 mg $O_2 \cdot kg^{-1} \cdot h^{-1}$ (Puckett and Dill, 1984, 1985). Wu (1977) suggested that measurements of oxygen consumption at cruising and prolonged swimming speeds could be extrapolated to bursts to estimate energy needs. Since both muscle and Froude efficiencies increase to a plateau at prolonged and sprint speeds (Lighthill, 1975; Rome, in press), this procedure is not theoretically defensible. Puckett and Dill (1984, 1985) measured metabolic rates at sustained speeds of their coho, and found the rate extrapolated from these data to the mean burst speed in lunges was 16 times lower than observed.

Anaerobic energy production, like that of ectothermic tetrapods (Bennett and Licht, 1972, 1974), varies among species and with size, temperature, and season, but studies to fully characterize patterns are lacking (Goolish, 1991a). In addition, relationships between anaerobic metabolic rate and performance level are unknown and maximum performance levels supportable by observed anaerobic capacity are largely crude estimates based on resistance relationships (e.g., Bainbridge, 1961; Webb, 1975a; Blake, 1983). Thus, considerable research is desirable in this area.

Variations in Aerobic and Anaerobic Capacity among Species

Fish differ in the relative development of aerobic and anaerobic capabilities. Active species appear to have high capacities for both aerobic and anaerobic metabolism, while sedentary (sluggish) species have low capacities for both. For example, active pelagic and endothermic species typically have higher lactate dehydrogenase activities, buffering capabilities, and rates of oxygen consumption than more sedentary species (see Goolish, 1991a).

Within a spectrum from more to less active fish, there may be inverse relationships in the relative development of aerobic and anaerobic capacity (Hochachka and Somero, 1984). Goolish (1991b) found that mudminnow, *Umbra limi,* had an 87% higher rate of lactate production and roughly a 50% higher anaerobic capacity than creek chub, *Semotilus atromaculatus*. Mudminnow also more rapidly metabolized lactate, resynthesizing glycogen, while creek chub appeared to use the lactate as a substrate for aerobic metabolism. However, mudminnow used anaerobic metabolism at much lower swimming speeds than creek chub. Mudminnows are sit-and-wait predators, with better-developed anaerobic capabilities, while high aerobic capacity is not necessary. In contrast, creek chub are active foragers reliant on aerobic metabolism.

ENVIRONMENTAL EFFECTS ON METABOLISM AND PERFORMANCE

Sustained locomotion is powered ultimately by aerobic metabolism. The energy available for aerobic swimming activities is the difference between the standard and active metabolic rate (Brett and Groves, 1979), traditionally called the metabolic scope. Numerous environmental elements affect locomotor metabolic scope and hence sustainable speeds. Fry (1971) recognized that many of these environmental elements, or physical components, had similar metabolic consequences, and therefore developed a classification based on common effects of elements. This approach is known as the Fry Paradigm (Fry, 1971; Webb, 1978b; Brett, 1979; Kerr, 1990). Within the Fry Paradigm, there are four categories of factors causing nonlethal effects due to controlling, limiting, masking (regulatory), and directive factors.

Controlling Factors

The criterion defining controlling factors (temperature and pressure) is a direct effect on the kinetics of enzyme-catalyzed reactions (Fry, 1971). For example, temperature affects the kinetic energy of substrates and hence rates of reactions, while pressure resists volume changes of reactants, products, or enzymes during reactions (Hochachka and Somero, 1984). Many molecules in the cell also affect enzyme kinetics, and many serve regulatory functions essential for homeokinesis. All such regulatory functions are excluded from the controlling factor category as they do not meet the basic criteria of Fry (1971). However, Brett (1979) considers pH, which is a potent regulatory factor and/or waste product, to be a controlling factor.

Temperature

Fish experience large temperature ranges, both among species from different latitudes and within species from season to season. Muscle performance varies with temperature, but because of the importance of swimming in all aspects of behavior, a variety of species-specific compensatory responses increase swimming capacity as temperature decreases (Guderley and Blier, 1988); responses affect twitch time, V_{max}, maximum force, enzyme concentrations, ultrastructure, relative proportions of fiber types, and recruitment patterns (Moerland and Sidell, 1986a, 1986b; Guderley and Blier, 1988; Wardle et al., 1989; Johnston et al., 1990; Langfeld et al., 1989, 1991)

Fish from different latitudes have various isoforms of contractile proteins which facilitate development of similar maximum muscle stress of 200 to 300 kN/m^2 at normal environmental temperatures (Johnson and Johnston, 1991a). However, times required to activate and relax muscle increase among species as: antarctic > temperate > tropical. Consequently, V_{max} is lower in antarctic species than in tropical species. In addition, the stress-strain relationship is more deeply curved for species from higher latitudes, resulting in lower power output for a given stress. Therefore, maximum beat frequencies for propulsors and muscle power output are lower for fish resident in colder habitats. Acute temperature changes also reduce muscle power output for a given species, but acclimatization ameliorates the reduction, mirroring the changes seen in species from different latitudes (Johnston et al., 1985; Crockford and Johnston, 1990; Johnson and Johnston, 1991b).

While temperature affects muscle power and V_{max}, kinematics vary little with temperature (Stevens, 1979; Rome et al., 1990, 1992a). As a result, the ability of muscle to meet power requirements to swim at a given speed decreases at lower temperatures (Rome, 1992). For example, Rome and Sosnicki (1990) found that V_{max} for carp red muscle decreased by a factor of 1.6 from 20 to 10°C resulting in a decrease in the maximum swimming speed at which red muscle was used exclusively from 45 to 30 cm/s. The lowest speed at which these fish could swim steadily was similarly reduced. Red muscle still contracted over the same V/V_{max} range at both temperatures so that efficiency and power output were maximized within the constraints imposed by temperature. Similar results have been obtained for scup (Rome et al., 1992a). White muscle was also recruited to support red muscle at lower swimming speeds (Rome et al., 1992a), thus compressing the motor unit recruitment pattern into a smaller range of speeds at lower temperatures (see Rome, in press; Rome et al., 1992a).

Locomotor metabolic scope and maximum swimming speed increase with temperature, although there is often an optimum temperature above which they decline (Beamish, 1978; Brett, 1979). This optimum temperature is typically close to the preferred temperature chosen by fish in a temperature gradient (Bryan et al., 1990). It has been postulated that the optimum temperature occurs because temperature-reduced water oxygen levels limit metabolic scope (Brett, 1964). However, the occurrence of similar metabolic rate-temperature relationships in terrestrial ectotherms (see Gordon, 1977) suggests this hypothesis is incomplete at best, and invites other (currently unknown) explanations.

Pressure

Pressure effects all arise from volume changes during reactions. For example, if the volume of reactants increases, even if the change is transitory in the formation of enzyme-substrate complex, pressure will inhibit the reaction. Substantial volume-changing hydrations and dehydrations are common in enzyme-catalyzed reactions, including enzyme activation and inhibition. These make water the most important contributor to volume changes. Volume changes occur because water molecules structured through interactions with organic molecules tend to occupy less volume than free molecules. Enzymatic adaptation occurs, measurable as changes in the apparent Michaelis constant, a measure of the effectiveness with which enzyme-substrate complexes are formed. However, the molecular stability conferring low pressure sensitivity reduces efficiency and the pressure compensation changes in essential proteins are sufficiently large that they tend to restrict ranges of both shallow-water and deep-water fishes (see Hochachka and Somero, 1984).

The metabolic rate of deep-water pelagic fishes is lower than expected from observations on shallow-water fish. However, this is not due to incomplete adaptation to pressure. Over most of the oceanic depths, pressure covaries with temperature, light, and, in particular, with productivity. Deep-sea fishes tend to have reduced capacities for locomotion, including reduction in muscle protein. This reduces maintenance costs, permitting conversion of higher proportions of rare meals to growth and reproduction. Although predators, these fish mostly float in the water column waiting for food, or at most swim very slowly when food is detected (Sullivan and Somero, 1980; Hochachka and Somero, 1984). Fishes near hydrothermal vents, where productivity is not limited by the absence of light, tend to have energetic profiles reminiscent of shallow-water fishes.

Limiting Factors

Metabolic rate and performance levels possible for given controlling factor levels are only achieved in the absence of limiting factors. Limiting factors reduce active metabolism and hence performance (Figure 4a) by interfering with the delivery of oxygen and substrate and the removal of wastes. Common

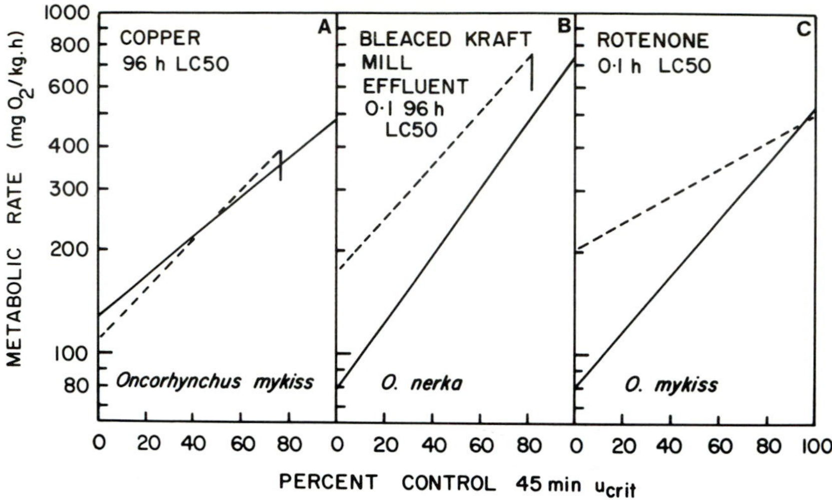

FIGURE 4. Effects of limiting and regulatory environmental factors (dotted lines) on swimming metabolism and performance. Solid lines show the relationship for control fish. Toxicants used as limiting and regulatory factors are shown as proportions of concentrations at which 50% of fish died in 96 h, as determined in a bioassay. Swimming speeds are shown as percentages of the 45-min u_{crit}, the speed at which 50% of the control group became exhausted in an increasing velocity test in which the speed was increased in increments of approximately 10% u_{crit} every 45 min. (A) The response of rainbow trout, *Oncorhynchus mykiss*, to copper illustrates a limiting factor effect, reducing active metabolic rate and hence locomotor metabolic scope and swimming speed. (B) The response of sockeye salmon, *O. nerka*, to bleached Kraft mill effluent shows a regulatory factor where standard metabolism is elevated, reducing locomotor scope and swimming speed. (C) A modulated regulatory factor response is illustrated for *O. mykiss* exposed to rotenone. Standard metabolism is increased, reducing locomotor scope, but active metabolic rate and swimming speed are not affected.

limiting factors are increased carbon dioxide and nitrogenous products of metabolism, and reduced dissolved oxygen. Many environmental elements act as limiting factors by indirectly reducing water oxygen concentration, including altitude, dissolved substances (e.g., salt in seawater), suspended material (e.g., silt, fiber), and perhaps temperature.

Fish have substantial regulatory abilities to compensate for variation in the availability of essential metabolites. As a result, there are usually thresholds below which physiological compensation is no longer sufficient to supply energy for high swimming speeds, so that swimming performance begins to decline (Skadsen et al., 1980; Randall and Brauner, 1991). This threshold varies among species according to abilities to regulate metabolite delivery and removal (Randall and Brauner, 1991).

Regulatory Factors

Regulation for any environmental element that perturbs a controlled system will require the expenditure of metabolic energy which may reduce metabolic scope and swimming speed by competing for energy (Figure 4b). Fry (1971) called such elements masking factors, but the term "regulatory factors" is more descriptive (Webb, 1978b). Regulatory factors include pH, salinity, toxicants, and, for endothermic fishes, temperature. Regulation may be deferrable with short-term allocation of energy to swimming. Then sustained swimming performance may not be affected, at least in the time frame of typical experimental studies (Figure 4c). Fry (1971) called this phenomenon a modulated response to regulatory factors. Because of the possibility of modulation, swimming performance is an unreliable method for quantifying or screening effects of pollutants on fishes.

DIRECTIVE FACTORS

Directive factors, among other effects, alter physiological profiles in time, for example in seasonal acclimation, ontogenetic changes with niche shifts (e.g., alternation of freshwater and marine stages in salmonids and eels), and sexual maturation. Locomotor metabolic scope may be expanded or contracted by these events as a result of changes at the enzyme-substrate level (e.g., Johnston and Harrison, 1985) and metabolite delivery systems (Cech et al., 1976). Thus, the thermal acclimation in muscle that minimizes seasonal controlling factor effects on performance is a result of directive factors.

ACKNOWLEDGMENTS

This analysis draws on research supported by the National Science Foundation and was completed with support from NSF Grant DCB9017817.

REFERENCES

Alexander, D. E., Drag coefficients of swimming animals: effects of using different reference areas, *Biol. Bull.,* 179, 186, 1990.

Alexander, R. McN., *Functional Design in Fishes,* Hutchinson, London, 1967.

Alexander, R. McN., Orientation of muscle fibres in the myomeres of fishes, *J. Mar. Biol. Assoc., U.K.,* 49, 263, 1969.

Alexander, R. McN., Optimization and gaits in the locomotion of vertebrates, *Physiol. Rev.,* 69, 1199, 1989.

Aleyev, Y. G., *Nekton,* Junk, The Haque, 1977.

Altringham, J. D. and Johnston, I. A., The mechanical properties of polyneuronally innervated, myotomal muscle fibres isolated from a teleost fish (*Myoxocephalus scorpius*), Pfluegers Arch., 412, 524, 1988.

Bainbridge, R., Problems of fish locomotion, *Symp. R. Soc. London,* 40, 23, 1961.

Balon, K. F., Saltatory processes and altricial to precocial forms in the ontogeny of fishes, *Am. Zool.,* 21, 573, 1984.

Beamish, F. W. H., Swimming capacity, in *Fish Physiology,* Vol. 7, Locomotion, Hoar, W. S. and Randall, D. J., Eds., Academic Press, New York, 1978, chap. 2.

Bennett, A. F. and Licht, P., Anaerobic metabolism during activity in lizards, *J. Comp. Physiol.,* 81, 277, 1972.

Bennett, A. F and Licht, P., Anaerobic metabolism during activity in amphibians, *Comp. Biochem. Physiol.,* 48A, 319, 1974.

Blake, R. W.,The energetics of hovering in the mandarin fish (*Synchropus picturatus*) *J. Exp. Biol.,* 82, 25, 1979.

Blake, R. W., Fish Locomotion, Cambridge University Press, Cambridge, 1983.

Blight, A. R., Undulatory swimming with and without waves of contraction, *Nature (London),* 264, 352, 1976.

Blight, A. R., The muscular control of vertebrate swimming movements, *Biol. Rev.,* 52, 181, 1977.

Block, B. A., Evolution novelties: how fish have built a heater out of muscle, *Am. Zool.,* 31, 726, 1991.

Bone, Q., Muscular and energetic aspects of fish swimming, in *Swimming and Flying in Nature,* Wu, T. Y., Brokaw, C. J., and Brennen, C., Eds., Plenum Press, New York, 1975, 493.

Bone, Q., Locomotor muscle, in *Fish Physiology,* Vol. 7, *Locomotion,* Hoar, W. S. and Randall, D. J., Eds., Academic Press, New York, 1978a, chap. 6.

Bone, Q., Myotomal muscle fibre type sin *Scomber* and *Katsuwonus* in *Physiological Ecology of Tunas,* Sharp, G. D. and Dizon, A. E., Eds., Academic Press, New York, 1978b, 183.

Bone, Q., Evolutionary patterns of axial muscle systems in some invertebrates and fish, *Am. Zool.,* 29, 5, 1989.

Bone, Q. and Marshall, N. B., *Biology of Fishes,* Blackie, London, 1982, chap. 3.

Braun, J. and Reif, W., A survey of aquatic locomotion in fishes and tetrapods, *Neues Jahrb. Geol. Palaeontol. Abh.,* 169, 307, 1985.

Breder, C. M., The locomotion of fishes, *Zoologica (NY)* 4, 159, 1926.

Brett, J. R., The respiratory metabolism and swimming performance of young sockeye salmon, *J. Fish. Res. Bd. Can.,* 21, 1183, 1964.

Brett, J. R., Environmental factors and growth, in *Fish Physiology, Bioenergetics and Growth,* Hoar, W. S., Randall, D. J., and Brett, J. R., Eds., Academic Press, New York, 1979, 599.

Brett, J. R. and Glass, N. R., Metabolic rate and critical swimming speeds of sockeye salmon (*Oncorhynchus nerka*) in relation to size and temperature, *J. Fish. Res. Bd. Can.,* 15, 587, 1973.

Brett, J. R. and Groves, T. D. D., Physiological energetics, in *Fish Physiology, Bioenergetics and Growth,* Hoar, W. S., Randall, D. J., and Brett, J. R., Eds., Academic Press, New York, 1979, 279.

Brett, J. R. and Sutherland, D. B., Respiratory metabolism of pumkinseed (Lepomis gibbosus) in relation to swimming speed, *J. Fish. Res. Bd. Can.,* 22, 405, 1965.

Brooks, D. R. and McLennan, D. A., *Phylogeny, Ecology, and Behavior,* University of Chicago Press, Chicago, 1991.

Bryan, J. D., Kelsch, S. W., and Neill, W. H., The maximum power principle in behavioral thermoregulation by fishes, *Trans. Am. Fish. Soc.,* 119, 611, 1990.

Bunker, S. J. and Machin, K. E., The hydrodynamics of cephalaspids, *in Biomechanics in Evolution,* Rayner, J. M. V. and Wooton, R. J., Eds., Cambridge University Press, Cambridge, 1992, 113.

Cailliet, G. M., Love, M. S., and Ebeling, A. W., *Fishes: A Field and Laboratory Manual on their Structure, Identification, and Natural History,* Wadsworth Publishing, Belmont, CA, 1986.

Cech, J. J., Bridges, D. M., Rowell, D. M., and Balzer, P. J., Cardiovascular responses of winter flounder *Pseudopleuronectes americanus* (Walbaum) to acute temperature increase, *Can. J. Zool.,* 54, 1383, 1976.

Crockford, T. and Johnston, I. A., Temperature acclimation and the expression of contractile protein isoforms in the skeletal muscles of the common carp (*Cyprinus carpio* L.), *J. Comp. Physiol.,* 160B, 23, 1990.

D'Orr, R. K., Respiratory metabolism and swimming performance of the squid, *Liligo opalescens, Can. J. Fish. Aquat. Sci.,* 39, 580, 1982.

Daniel, T. L., Forward flapping flight from flexible fins, *Can. J. Zool.,* 66, 630, 1988.

Daniel, T. L. and Webb, P. W., Physics, design and locomotor performance, in *Comparative Physiology: Life in Water and on Land,* Dejours, P., Bolis, L., Taylor, C. R., and Weibel, E. R., Eds., Liviana Press, New York, 1987, 343.

Duthie, G. G., The respiratory metabolism of temperature-adapted flatfish at rest and during swimming activity and the use of anaerobic metabolism at moderate swimming speeds, *J. Exp. Biol.,* 97, 359, 1982.

Eaton, R. C. and Emberley, D. S., How stimulus direction determines the trajectory of the Mauthner-initiated escape response in a teleost fish, *J. Exp. Biol.,* 161, 469, 1991.

Eaton, R. C., DiDomenico, R., and Nissanov, J., Flexible body dynamics of the goldfish C-start: implications for reticulospinal command mechanisms, *J. Neurosci.,* 8, 2758, 1988.

Eaton, R. C., Farley, R. D., Kimmel, C. B., and Schabtach, E., Functional development in the Mauthner cell system of embryos and larvae of the zebra fish, *J. Neurobiol.,* 8, 151, 1977.

Facey, D. E. and Grossman, G. D., The metabolic cost of maintaining position for four North American stream fishes: effects of season and velocity, *Physiol. Zool.,* 63, 757, 1990.

Fetcho, J. R., A review of the organization and evolution of motoneurons innervating the axial musculature of vertebrates, *Brain Res. Rev.,* 12, 243, 1987.

Foreman, M. B., The Kinematics and Neuroethology of the Mauthner-Initiated Escape Response, Ph.D. dissertation, University of Colorado, Boulder, 1991.

Foreman, M. B. and Eaton, R. C., EMG and kinematic analysis of the stages of the Mauthner-initiated escape response, *Soc. Neurosci. Abstr.,* 16, 1328, 1990.

Frolich, L. M. and Biewener, A. A., Kinematic and electromyographic analysis of the functional role of the body axis during terrestrial and aquatic locomotion in the salamander *Ambystoma tigrinum, J. Exp. Biol.,* 162, 107, 1992.

Fry, F. E. J., The aquatic respiration of fish, in *Fish Physiology,* Brown, M. E., Ed., Academic Press, New York, 1957, 1.

Fry, F. E. J., The effect of environmental factors on the physiology of fish, in *Fish Physiology,* Vol. 6, Hoar, W. S. and Randall, D. J., Eds., Academic Press, Orlando, 1971, 1.

Fuiman, L. A. and Webb, P. W., Ontogeny of routine swimming activity and performance in *Zebra danois* (Teleostei: Cyprinidae), *Anim. Behav.,* 36, 250, 1988.

Gee, J. H., Ecological implications of buoyancy control in fish, in *Fish Biomechanics,* Webb, P. W. and Weihs, D., Eds., Praeger, New York, 1983, 140.

Geerlink, P. J., The anatomy of the pectoral fins in *Sarotherodon niloticus* Trewavas (Cichlidae), *Neth. J. Zool.,* 29, 9, 1979.

Geerlink, P. J., Pectoral fin kinematics of *Coris formosa* (Teleostei, Labridae), *Neth. J. Zool.,* 33, 515, 1983.

Geerlink, P. J., Pectoral Fins. Aspects of Propulsion and Braking in Teleost Fishes, Ph.D. thesis, University of Groningen, The Netherlands, 1986.

Geerlink, P. J. and Videler, J. J., Joints and muscles of the dorsal fin of *Tilapia nilotica* L. (Fam. Cichlidae), *Neth. J. Zool.,* 24, 279, 1974.

Goolish, E. M., Aerobic and anaerobic scaling in fish, *Biol. Rev.,* 66, 33, 1991a.

Goolish, E. M., Anaerobic swimming metabolism of fish: sit-and-wait versus active forager, *Physiol. Zool.,* 64, 485, 1991b.

Gordon, M. S., *Animal Physiology,* Macmillan, New York, 1977.

Gordon, M. S., Chin, H. G., and Vojkovich, M., Energetics of swimming in fishes using different methods of locomotion. I. Labriform swimmers, *Fish Physiol. Biochem.,* 6, 341, 1989.

Gosline, W. A., *Functional Morphology and Classification of Teleostean Fishes,* University Press of Hawaii, Honolulu, 1971.

Graham, J. B., Heat transfer, in *Fish Biomechanics,* Webb, P. W. and Weihs, D., Eds., Praeger, New York, 1983, 248.

Graham, J. B., Dewar, H., Lai, N. C., Lowell, W. R., and Arce, S. M., Aspects of shark swimming performance using a large water tunnel, *J. Exp. Biol.,* 151, 175, 1990.

Gray, J., Studies in animal locomotion. I. The movement of fish with special reference to the eel, *J. Exp. Biol.,* 10, 88, 1933a.

Gray, J., Studies in animal locomotion. II. The relationship between the eaves of muscular contraction and the propulsive mechanism of the eel, *J. Exp. Biol.,* 10, 386, 1933b.

Guderley, H. and Blier, P., Thermal acclimation in fish: conservative and labile properties of swimming muscle, *Can. J. Zool.,* 66, 1105, 1988.

Harden, W., Anatomie der Fische, *Handbuch der Binnenfischerei Mitteleuropas,* 2A, 1, 1964.

Harris, J. E., The role of fins in the equilibrium of swimming fish. I. Wind-tunnel tests on a model of *Mustelus canis* (Mitchell), *J. Exp. Biol.,* 13, 476, 1936.

Harris, J. E., The role of fins in the equilibrium of swimming fish. II. The role of the pelvic fins, *J. Exp. Biol.,* 15, 32, 1938.

Hebrank, J. H., Hebrank, M. R., Long, J. H., Block, B. A., and Wainwright, S. A., Backbone mechanics of the blue marlin Makaira nigricans *(Pisces, Istiophoridae), J. Exp. Biol.,* 148, 449, 1990.

Hess, F. and Videler, J. J., Fast continuous swimming of saithe (*Pollachius virens*): a dynamic analysis of bending moments and muscle power, *J. Exp. Biol.,* 109, 229, 1984.

Hochachka, P. W. and Somero, G. N., *Biochemical Adaptation,* Princeton University Press, Princeton, 1984.

Hoerner, S. F., *Fluid-Dynamic Drag,* Hoerner Fluid Dynamics, Brick Town, NJ, 1965.

Hoerner, S. F., *Fluid-Dynamic Lift,* Hoerner Fluid Dynamics, Brick Town, NJ, 1975.

Johnson, T. P. and Johnston, I. A., Temperature adaptation and the contractile properties of live muscle fibres from teleost fish, *J. Comp. Physiol.,* 161B, 27, 1991a.

Johnson, T. P. and Johnston, I. A., Power output of fish muscle fibres performing oscillatory work: effects of acute and seasonal temperature change, *J. Exp. Biol.,* 157, 409, 1991b.

Johnston, I. A., Structure and function of fish muscle, *Symp. R. Soc. London,* 48, 71, 1981.

Johnston, I. A., Muscle action during locomotion: a comparative perspective, *J. Exp. Biol.,* 160, 167, 1991.

Johnston, I. A. and Harrison, P., Contractile and metabolic characteristics of muscle fibres from antarctic fish, *J. Exp. Biol.,* 116, 223, 1985.

Johnston, I. A., Fleming, J. D., and Crockford, T., Thermal acclimation and muscle contractile properties in cyprinid fish, *A. J. Physiol.,* 259, R231, 1990.

Johnston, I. A., Sidell, B. D., and Driedzic, W. R., Force-velocity characteristics and metabolism of carp muscle fibres following temperature acclimation, *J. Exp. Biol.,* 119, 239, 1985.

Jordan, C. E., A model of rapid-start swimming at intermediate Reynolds number: undulatory locomotion in the chaetognath *Sagitta elegans, J. Exp. Biol.,* 163, 119, 1992.

Kaufmann, R., Respiratory cost of swimming in larval and juvenile cyprinids, *J. Exp. Biol.,* 150, 343, 1990.

Kerr, S. R., Niche theory in fisheries ecology, *Trans. Am. Fish. Soc.,* 109, 254, 1980.

Langfeld, K. S., Altringham, J. D., and Johnston, I. A., Temperature and the force-velocity relationship of live muscle fibres from the teleost Myoxocephalus scorpius, *J. Exp. Biol.,* 144, 437, 1989.

Langfeld, K. S., Crockford, T., and Johnston, I. A., Temperature acclimation in the common carp: force-velocity characteristics and myosin subunit composition of slow muscle fibers, *J. Exp. Biol.,* 155, 291, 1991.

Lauder, G. V., Caudal fin locomotion in ray-finned fishes: historical and functional analysis, *Am. Zool.,* 29, 85, 1989.

Lauder, G. V. and Liem, K. F., The evolution and interrelationships of the actinopterygian fishes, *Bull. Mus. Comp. Zool. Harv. Univ.,* 150, 95, 1983.

Lighthill, J., *Mathematical Biofluiddynamics,* Society for Industrial and Applied Mathematics, Philadelphia, 1975.

Lighthill, J., Biofluiddynamics of balistiform and gymnotiform locomotion. II. The pressure distribution arising in two-dimensional irrotational flow from a general symmetrical motion of a flexible flat plate normal to itself, *J. Fluid Mech.,* 213, 1, 1990a.

Lighthill, J., Biofluiddynamics of balistiform and gymnotiform locomotion. III. Momentum enhancement in the presence of a body of elliptical cross-section, *J. Fluid Mech.,* 213, 11, 1990b.

Lighthill, J., Biofluiddynamics of balistiform and gymnotiform locomotion. IV. Short-wavelength limitations on momentum enhancement, *J. Fluid Mech.,* 213, 21, 1990c.

Lighthill, J. and Blake, R. W., Biofluidynamics of balistiform and gynotiform locomotion. I. Biological background, and analysis by elongated-body theory, *J. Fluid Mech.,* 212, 183, 1989.

Lindsey, C. C., Form, function, and locomotory habits in fish, in *Fish Physiology,* Vol. 7, *Locomotion,* Hoar, W. S. and Randall, D. J., Eds., Academic Press, New York, 1978, chap. 1.

Long, J. H., Stiffness and damping forces in the intervertebral joints of blue marlin (*Makaira nigricans*), *J. Exp. Biol.,* 162, 131, 1992.

Magnuson, J. J., Locomotion by scombroid fishes: hydrodynamics, morphology and behavior, in *Fish Physiology,* Vol. 7, Locomotion, Hoar, W. S. and Randall, D. J., Eds., Academic Press, New York, 1978, chap. 4.

McCutcheon, C. W., The trout tail fin: a self-cambering hydrofoil, *J. Biomech.,* 3, 271, 1970.

McLaughlin, R. L. and Kramer, D. L., The association between amount of red muscle and mobility in fishes: a statistical analysis *Environ. Biol. Fishes,* 30, 369, 1991.

Moerland, T. S. and Sidell, B. D., Biochemical responses to temperature in the contractile protein complex of striped bass, *Morone saxalis, J. Exp. Zool.,* 238, 287, 1986a.

Moerland, T. S. and Sidell, B. D., Contractile responses to temperature in the locomotory musculature of striped bass *Morone saxalis, J. Exp. Zool.,* 240, 25, 1986b.

Moy-Thomas, J. A. and Miles, R. S., *Palaeozoic Fishes,* W. B. Saunders, Philadelphia, 1971.

Puckett, K. J. and Dill, L. M., Cost of sustained and burst swimming to juvenile coho salmon (*Oncorhynchus kisutch*), *Can. J. Fish. Aquat. Sci.,* 41, 1546, 1984.

Puckett, K. J and Dill, L. M., The energetics of feeding territorially in juvenile coho salmon (*Oncorhynchus kisutch*), *Behaviour,* 92, 97, 1985.

Randall, D. J. and Brauner, C., Effects of environmental factors on exercise in fish, *J. Exp. Biol.,* 160, 113, 1991.

Rao, G. M. M., Oxygen consumption of rainbow trout (*Salmo gairdneri*) in relation to activity and salinity, *Can. J. Zool.,* 46, 781, 1968.

Rome, L. C., The mechanical design of the muscular system, in *Comparative Vertebrate Exercise Physiology,* Jones, J. H., Academic Press, Orlando, FL, in press.

Rome, L. C. and Sosnicki, A. A., The influence of temperature on mechanics of red muscle in carp, *J. Physiol.,* 427, 151, 1990.

Rome, L. C. and Sosnicki, A. A., Myofilament overlap in swimming carp. II. Sarcomere length changes during swimming, *Am. J. Physiol.,* 260, C289, 1991.

Rome, L. C., Choi, I., Lutz, G., and Sosnicki, A., The influence of temperature on muscle function in the fast-swimming scup. I. Shortening velocity and muscle recruitment during swimming, *J. Exp. Biol.,* 163, 259, 1992a.

Rome, L. C., Choi, I., Lutz, G., and Sosnicki, A., The influence of temperature on muscle function in the fast-swimming scup. II. The mechanics of red muscle, *J. Exp. Biol.,* 163, 279, 1992b.

Rome, L. C., Funke, R. P., and Alexander, R. McN., The influence of temperature on muscle velocity and sustained performance in swimming carp, *J. Exp. Biol.,* 154, 163, 1990.

Rome, L. C., Funke, R. P., Alexander, R. McN., Lutz, G., Aldridge, H., Scott, F., and Freadman, M., Why animals have different muscle fibre types, *Nature,* 335, 824, 1988.

Sigvardt, K. A., Spinal mechanisms in the control of lamprey swimming, *Am. Zool.,* 29, 19, 1989.

Skadsen, J. M., Webb, P. W., and Kostecki, P. T., Measurement of sublethal metabolic stress in rainbow trout (*Salmo gairdneri*) using automated respirometry, *J. Environ. Sci. Health B,* 15, 193, 1980.

Smit, H., Amelink-Koutstaal, J. M., Vijverberg, J., and von Vaupel-Kelin, J. C., Oxygen consumption and efficiency of swimming goldfish, *Comp. Biochem. Physiol. A,* 39, 1, 1971.

Somero, G. N. and Childress, J. J., A violation of the metabolism-scaling paradigm: activities of glycolytic enzymes increase in larger-sized fish, *Physiol. Zool.,* 53, 322, 1980.

Stevens, E. D., The effect of temperature on tail beat frequency of fish swimming at constant velocity, *Can. J. Zool.,* 57, 1628, 1979.

Sullivan, K. M. and Somero, G. N., Enzyme activities of fish skeletal muscle and brain as influenced by depth of occurrence and habits of feeding and locomotion, *Mar. Biol.,* 60, 91, 1980.

Tucker, V. A., The energetic cost of moving about, *Am. Sci.,* 63, 413, 1975.

Videler, J. J., Mechanical properties of fish tail joints, *Fortschr. Zool.* 24, 183, 1977.

Videler, J. J., Swimming movements, body structure and propulsionin cod Gadus morhua, *Symp. Zool. Soc. London,* 48, 1, 1981.

Videler, J. J. and Nolet, B. A., Costs of swimming measured at optimum speed: scale effects, differences between swimming styles, taxonomic groups and submerged and surface swimming, *Comp. Biochem. Physiol. A,* 97, 91, 1990.

Vlymen, W. J., Swimming energetics of larval anchovy, *Engraulis mordax, Fish. Bull. U.S.,* 72, 885, 1974.

Wainwright, S. A., To bend a fish, in *Fish Biomechanics,* Webb, P. W. and Weihs, D., Eds., Praeger, New York, 1983, 68.

Wainwright, S. A., *Axis and Circumference,* Harvard University Press, Cambridge, 1988.

Wardle, C. S., Limit of fish swimming speed, *Nature (London),* 255, 725, 1975.

Wardle, C. S., Videler, J. J., Arimoto, T., Franco, J. M., and He, P., The muscle twitch an the maximum swimming speed of giant bluefin tuna, *Thunnus thynnus* L, *J. Fish Biol.,* 35, 129, 1989.

Ware, D. M., Bioenergetics of pelagic fish: theoretical change in swimming speed and ratio with body size, *J. Fish. Res. Board Can.,* 35, 220, 1978.

Wassersug, R. J., Locomotion in amphibian larvae (or "Why aren't tadpoles built like fishes?"), *Am. Zool.,* 29, 65, 1989.

Webb, P. W., Hydrodynamics and energetics of fish propulsion, *Bull. Fish. Res. Board Can.,* 190, 1, 1975a.

Webb, P. W., Efficiency of pectoral-fin propulsion in Cymatogaster aggregata, in *Swimming and Flying in Nature,* Wu, T. Y., Brokaw, C. J., and Brennen, C., Eds., Plenum Press, New York, 1975b, 573.

Webb, P. W., Fast-start performance and body form in seven species of teleost fish, *J. Exp. Biol.,* 74, 211, 1978a.

Webb, P. W., Partitioning of energy into metabolism and growth, in *Ecology of Freshwater Fish Production,* Gerking, S. D., Ed., Blackwell Scientific, Cambridge, England, 1978b, 184.

Webb, P. W., Form and function in fish swimming, *Sci. Am.,* 251, 72, 1984a.

Webb, P. W., Body form, locomotion and foraging in aquatic vertebrates, *Am. Zool.,* 24, 107, 1984b.

Webb, P. W., Simple physical principles and vertebrate aquatic locomotion, *Am. Zool.,* 28, 709, 1988a.

Webb, P. W., "Steady" swimming kinematics of tiger musky, an esociform accelerator, and rainbow trout, a generalist cruiser, *J. Exp. Biol.,* 138, 51, 1988b.

Webb, P. W., How does benthic living affect body volume, tissue composition and density of fishes?, *Can. J. Zool.,* 68, 1250, 1990.

Webb, P. W., Composition and mechanics of routine swimming of rainbow trout, *Oncorhynchus nerka, Can. J. Fish. Aquat. Sci.,* 48, 583, 1991.

Webb, P. W., Exercise performance of fish, in *Comparative Vertebrate Exercise Physiology,* Jones, J. H., Ed., Academic Press, Orlando, FL, in press.

Webb, P. W. and Blake, R. W., *Swimming, in Functional Vertebrate Morphology,* Hildebrand, M., Bramble, D. M., Liem, K. F., and Wake, D. B., Eds., Harvard University Press, Cambridge, 1985, 110.

Webb, P. W. and Keyes, R. S., Swimming kinematics of sharks, *Fish.Bull. U.S.,* 80, 803, 1982.

Webb, P. W. and Smith, G. R., Function of the caudal fin in early fishes, *Copeia,* 1980, 559, 1980.

Webb, P. W. and Weihs, D., Functional locomotor morphology of early life history stages of fishes, *Trans. Am. Fish. Soc.,* 115, 115, 1986.

Webb, P. W., Hardy, D. G., and Mehl, V. L., The effect of armored skin on the swimming of longnose gar, *Lepisosteus osseus, Can. J. Zool.,* 70, 1173, 1992.

Weihs, D., A hydrodynamic analysis of fish turning maneuvers, *Proc. R. Soc. London Ser. B,* 182, 59, 1972.

Weihs, D., Mechanically efficient swimming techniques for fish with negative buoyancy, *J. Mar. Res.,* 31, 194, 1973a.

Weihs, D., The mechanism of rapid starting of slender fish, *Biorheology,* 10, 343, 1973b.

Weihs, D., Optimal fish cruising speed, *Nature (London),* 245, 48, 1973c.

Weihs, D., The energetic advantages of burst swimming, *J. Theor. Biol.,* 49, 215, 1974.

Weihs, D., Design features and mechanics of axial locomotion in fish, *Am. Zool.,* 29, 151, 1989.

Weihs, D. and Webb, P. W., *Optimization of locomotion, in Fish Biomechanics,* Webb, P. W. and Weihs, D., Eds., Praeger, New York, 1983, 339.

Weiser, W., Platzer, U., and Hinterleitner, S., Anaerobic and aerobic energy production of young rainbow trout (Salmo gairdneri) during and after bursts of activity, *J. Comp. Physiol.,* 155B, 485, 1985.

Williams, T. L., Grillner, S., Smoljaninov, V. V., Wallen, P., Kashin, S., and Rossignol, S., Locomotion in lamprey and trout: the relative timing of activation and movement, *J. Exp. Biol.,* 143, 559, 1989.

Wu, T. Y., Introduction to scaling of aquatic animal locomotion, in *Scale Effects of Animal Locomotion,* Pedley, T. J., Ed., Academic Press, New York, 1977, 203.

3 Buoyancy

R. McNeill Alexander

INTRODUCTION

This chapter shows that fishes are denser than the water they live in, unless they have adaptations to give them buoyancy. Some fishes tolerate their excess density, relying on hydrodynamic forces to keep them off the bottom when they swim and sinking if they stop swimming. Others depend on lipids or low-density aqueous fluids to reduce their density to that of the water. Yet others are given buoyancy by gas-filled swim bladders, into which they may have to secrete gases at very high pressures if they live at great depths. These buoyancy strategies are reviewed, and their merits are compared, for fishes with different ways of life.

DENSITIES OF FISHES AND TISSUES

Most fishes live either in fresh water with a density of 1000 kg/m^3, or in seawater of about 1026 kg/m^3. However, their bodies consist largely of tissues that are denser than either, as Table 1 shows. Muscle generally has densities close to 1060 kg/m^3, both in selachians and in teleosts. Other soft tissues such as gut, uncalcified cartilage, and scale-less skin have similar densities but calcified cartilage and scaly skin are denser. Bone is denser still, but there are striking differences between fish bone densities determined by different investigators. These may be due to differences in the extent to which other tissues were cleaned off the bones, before the density was determined. The cyprinid bones in the table had been thoroughly cleaned with hydrogen peroxide, but the bones of the marine fishes had simply been brushed clean after boiling. Though all these tissues are denser than water, liver may be less dense, if it contains a large proportion of oil. The triglyceride oils of fishes generally have densities of about 920 kg/m^3 (Lewis, 1970).

The tissues of most fishes probably have densities close to those of Table 1, but we will see in later sections that in some fishes tissues have unusual compositions and are much less dense, apparently as adaptations for buoyancy.

Since they consist mainly of tissues that are denser than seawater, fishes without buoyancy adaptations are themselves denser than seawater, with densities ranging from about 1050 to 1090 kg/m^3 (Jones and Marshall 1953; Magnuson, 1978; Davenport and Kjorsvik, 1986; Webb, 1990). Other fishes from which the buoyancy organs have been removed also have densities in this range. For

0-8493-8042-1/93/$0.00+$.50

TABLE 1
Densities of Various Fishes (kg/m^3)

	Selachians	Marine Teleosts	Freshwater Teleosts
Muscle	1038–1081	1055	1046–1063
Cartilage	1061–1183	—	—
Bone	—	1300–1500	1570–2040
Skin	1079–1188	1054–1066	—
Gut	—	1038	—
Liver	893–1069	986–1050	—

Note: Densities (kg/m^3) of tissues of five species of selachian (Bone and Roberts, 1969), three species of marine teleosts (Webb, 1990), and three species of freshwater teleosts (Cyprinidae; Alexander, 1959a). All data refer to typical tissues without special buoyancy adaptations.

example, intact haddock (*Merlangus*) have densities close to seawater, but when the swim bladder is deflated the density rises to 1060 kg/m^3 (Webb, 1990). Some deep-sea squalid sharks also have densities close to seawater, but when their huge oily livers are removed the density of the rest of the fish is about 1075 kg/m^3 (Corner et al., 1969).

Even with the buoyancy organs removed, there are differences in density between fishes due to differences in the proportions and densities of their tissues. For example, the thick bony scales of *Polypterus* make up 10% of its body mass and increase the density of the fish (with lung deflated) to about 1130 kg/m^3 (Alexander, 1966a). In contrast, the scales of cyprinid fishes are only 2 to 4% of body mass (Alexander, 1959a).

The densities of fishes may be altered temporarily by their food, especially by the shells of molluscs, if these are ingested (Hoogerhoud, 1987).

The densities of many pelagic fishes are reduced by buoyancy organs to within 1% of the density of the water they live in, whether it be fresh or salt (Jones and Marshall, 1953; Alexander, 1959a, 1959b; Corner et al., 1969; and many other papers). The buoyancy organs involved are reviewed in later sections. Many fishes, however, have no buoyancy organ, either because their ancestors apparently did not have one (as in the case of typical selachians) or because a swim bladder possessed by ancestors has been lost in the course of evolution. Bottom-living teleosts such as plaice (*Pleuronectes*) have lost their swim bladders (Webb, 1990), but so also have some pelagic fishes, including many scombrids (mackerel and tunnies: Magnuson, 1978). The relative merits of having a buoyancy organ or doing without one are discussed in a later section.

Archimedes' Principle tells us that an upthrust acts on a submerged body, equal to the weight of an equal volume of the surrounding fluid. If the body has the same density as the fluid, its weight and the upthrust are, of course, equal. Just as the weight can be thought of as a single force acting at the center of gravity, the upthrust can be thought of as acting at the center of buoyancy. If fishes were made of material of uniform density, the centers of gravity and buoyancy would coincide, but because tissues of different densities are distributed nonuniformly in the body, the two centers are generally slightly separated. The centers of buoyancy of fishes with swim bladders are generally slightly below the center of gravity, making the equilibrium of such fish unstable (Ohlmer, 1964); that is why dead fish float upside-down.

SWIMMING OF DENSE FISHES

Fishes that are denser than the water in which they live can, nevertheless, swim. To do this they must generate upward hydrodynamic forces equal to the difference between the weight and the upthrust.

Selachians and sturgeons have fins that cannot be folded, but project permanently from the body. The pectoral fins project like airplane wings, and like these, produce upward lift forces if they are tilted to an appropriate angle as the fish swims. These fishes also have asymmetrical (heterocercal) tails that produce an upward component of force, as well as the forward thrust needed to propel the animal (Alexander, 1965, 1966b; Simons, 1970). Lift on the pectoral fins tends to tilt the fish head-up and lift on the tail to tilt it head-down, so for level swimming they must be balanced. By taking account of the relative positions of the fins and the centers of gravity and buoyancy, Alexander (1965) estimated that 71% of the upward lift required by a dogfish (*Scyliorhinus*) must be supplied by the pectoral fins, and only 29% by the tail.

Tunnies that are denser than seawater also use their pectoral fins as hydrofoils, but their tails are symmetrical and the posterior upward lift that is needed for equilibrium seems to be supplied not by them but by the caudal peduncle (Magnuson, 1978). As in dogfish, the pectoral fins must supply most of the upward lift — 80% according to Magnuson's (1970) calculations, in the case of *Euthynnus*.

For some fishes, the body rather than the fins may supply the lift that prevents sinking. Bunker and Machin (1991) made wind tunnel tests on a model of a long-extinct cephalaspid fish whose broad head shield was flat below and convex above. They concluded that if it swam with its body slightly tilted, lift on the head supplemented by a little from the heterocercal tail would be enough to prevent sinking.

In a later section we will want to compare the energy costs in different means of preventing sinking. The fishes that we have been discussing use fins as fixed hydrofoils to drive water downward, to obtain upward lift. The power needed for this (known as the induced power) can be calculated by equating it to the kinetic energy given to the water that is driven downward. The induced power (P_{ind}) needed to generate lift (L) by means of a hydrofoil of span (s) traveling at speed (u) in water of density (ρ_w) is

$$P_{ind} = 4L^2/\pi\rho_w us^2 \quad (1)$$

(see Alexander, 1990 or a textbook of aerodynamics).

If the fish has other reasons for swimming, Equation 1 is a reasonable estimate of the additional power needed to prevent sinking. If the fish has to keep swimming simply to produce lift, the whole energy cost of swimming should be included in the energy cost of keeping afloat. Like airplanes, fishes that use fins as fixed hydrofoils have a minimum power speed, at which the energy cost of swimming is least. Bunker and Machin (1991) estimated that the minimum power speed for a 0.4-m cephalaspid was 0.3 m/s.

Again like an aircraft, a dense fish that uses its fins as fixed hydrofoils must have a minimum speed (the stalling speed) below which the fins cannot generate the required lift. This speed has been estimated as 0.24 m/s for dogfish (*Scyliorhinus*; Alexander, 1965) and 0.6 m/s for Skipjack tuna (*Euthynnus*; Magnuson, 1970).

An alternative strategy for a dense fish is to hover like a helicopter, keeping the body stationary but beating the pectoral fins backward and forward to drive water downward. The mandarin fish *Synchiropus* is a small tropical marine teleost that is able to do this, although its density is unusually high, 1120 kg/m^3 (Blake, 1979). However, it hovers very close to the bottom where the mechanical power required is greatly reduced by ground effect, the effect that makes hovercraft so much more economical of fuel than helicopters.

Teleosts with densities only a little different from the water may be able to hover well clear of the bottom. For example, if a fish with a swim bladder swims up from the depth to which it is adapted, its swim bladder will expand as the pressure falls, and if it swims down the swim bladder will be compressed. Either change makes the density of the fish a little different from that of the water, but the fish may be able to compensate by beating its pectoral fins. Jones (1952) subjected perch (*Perca*) to reduced pressures and found that they could hover comfortably only when the pressure reduction was 16% or less, reducing the density of these fish by no more than about 10 kg/m^3 below the density of the fresh water that they live in. This is very much less than the density difference of 90 kg/m^3 for which *Synchiropus* (aided by ground effect) compensates by hovering.

SIZES OF BUOYANCY ORGANS

Consider a fish which, without any buoyancy organ, would have volume (V) and density (ρ). How large a buoyancy organ does it need to reduce its density to match the density of the water (ρ_w) that it lives in? The answer depends on the density of the buoyancy organ (ρ_b).

If the volume of the buoyancy organ is V_b, the total volume of the fish including it is $(V + V_b)$ and the total mass is $(\rho V + \rho_b V_b)$. Thus, if its overall density is matched to the density of the water:

$$\rho_w = (\rho V + \rho_b V_b)/(V + V_b)$$
$$\text{whence } V_b/V = (\rho - \rho_w)/(\rho_w - \rho_b) \tag{2}$$

This gives the volume of the buoyancy organ as a fraction of the initial volume, but we may want it as a fraction of the total volume of the intact fish: we may want $V_b/(V + V_b)$. It can be shown by manipulating Equation 2 that this is

$$V_b/(V + V_b) = (\rho - \rho_w)/(\rho - \rho_b) \tag{3}$$

Alternately, we may want to know the mass of the required buoyancy organ (m_b) as a fraction of the total mass of fish and buoyancy organ ($m + m_b$). Because $m_b = \rho_b V_b$ and $(m + m_b) = \rho_w (V + V_b)$, Equation 3 gives us

$$m_b/(m + m_b) = (\rho_b/\rho_w)(\rho - \rho_w)/(\rho - \rho_b) \tag{4}$$

Table 2 shows how large different buoyancy aids have to be, according to Equations 3 and 4. Gas-filled swim bladders can be much smaller than alternative buoyancy aids because the densities of gases are so low. The other buoyant materials in the table will be described in the next section. Notice that a modest difference in density between 930 and 860 kg/m³ makes a very large difference to the quantity required.

OILY FISHES

This section is about oceanic fishes which are either mesopelagic (swimming in midwater) or benthopelagic (swimming just above the bottom). These habits can often be inferred from the depths at which the fishes are caught but

TABLE 2
Fractional Volumes and Masses of Buoyancy Aids Required to Match the Density of Typical Fish to Fresh Water or Seawater

		Fresh Water		Seawater	
	Density of Buoyancy Aid	Volume	Mass	Volume	Mass
Gas	Negligible	0.07	—	0.05	—
Squalene	860 kg/m³	0.35	0.30	0.23	0.19
Wax esters	860 kg/m³	0.35	0.30	0.23	0.19
Triglycerides	930 kg/m³	0.52	0.48	0.34	0.31

Note: Fresh water of density 1000 kg/m³, seawater of density 1026 kg/m³. Density of the fish without buoyancy aids is assumed to be 1075 kg/m³. Fractional volumes are $V_b/(V + V_b)$ and fractional masses are $m_b/(m + m_b)$, as given by Equations 3 and 4.

have in some cases been confirmed by direct observation from bathyscaphes or other research submarines.

Most selachians are considerably denser than seawater, but Corner et al. (1969) found that four benthopelagic species of the family Squalidae were 0 to 0.3% less dense than seawater. These densities were measured after the fishes had been brought to the surface but they were caught at depths of several hundred meters where both their densities and the density of the water would be altered by the low temperature and high pressure. Corner et al. calculated that at their natural depth the densities of these fishes would have matched the water even more closely. The huge basking shark (*Cetorhinus*), which swims near the surface, also has about the same density as seawater, as does the benthopelagic chimeroid *Hydrolagus* (Corner et al., 1969).

All these fishes have huge livers. The livers of the squalids made up 20 to 30% of body mass (compared to about 5% in typical sharks) and over 80% of its mass was oil with a density of 870 to 880 kg/m^3. Thus, low-density oil made up around 20% of the mass of these fishes, enough to explain their being no denser than seawater (Table 2). With the liver removed, the density of the remainder of the fish was 1070 to 1080 kg/m^3 or about the same as for most fishes without buoyancy organs. A large proportion of the oil (40 to 80%) was the hydrocarbon squalene, which has a density of 860 kg/m^3, considerably lower than the density of the triglycerides that predominate in most fish oils: about 920 kg/m^3 (Lewis, 1970). Squalene is also plentiful in the liver oil of the basking shark (Bone and Roberts, 1969). In contrast to these findings, Van Vleet et al. (1984) analyzed two other benthopelagic shark species and found that in their liver oil wax esters were plentiful and squalene sparse. Wax esters have about the same density as squalene (860 kg/m^3; Lewis, 1970) so are equally effective as a buoyancy aid.

Other oily fishes do not depend on a huge liver for their buoyancy but on oil in other tissues. One of the most remarkable and best studied is the castor oil fish *Ruvettus*, a large benthopelagic teleost that is fished commercially in the Pacific. Its density is almost exactly that of seawater, due to huge quantities of oil in the muscles, the skin, and even in cavities in the bones (Bone, 1972). Ordinary bones are considerably denser than water but the vertebrae and skull bones of this fish float. Nevenzel et al. (1965) found that the muscles contained 15% of lipid, of which more than 90% was wax esters. A similar proportion of wax esters is present in the oil of the adipose tissues and extraordinarily oily muscles of the coelacanth (*Latimeria*: Nevenzel et al., 1966). This primitive bony fish has about the same density as seawater although (like *Ruvettus*) it has no functional swim bladder (Fricke et al., 1987).

The notothenioids are a group of marine teleosts that are remarkably successful in the Antarctic. They have no swim bladder, and most of them are denser than the water and live on the bottom. However, Eastman (1988) described two midwater species that have about the same density as seawater. *Pleuragramma* has small sacs of oil in every segment, and *Dissostichus* has adipose cells under its skin and permeating its muscle, but in both the oil is predominantly triglyceride with a density of about 930 kg/m^3. Also, the oil is less plentiful than in the other oily fish that we have been discussing (about 9% of body mass in *Dissostichus*) and would not by itself be enough to explain the low densities of these fishes. Their skeletons are remarkably lightly mineralized.

The lantern fishes (Myctophidae) are small mesopelagic fishes, extremely plentiful in the oceans. Some have well-developed swim bladders, for example, *Myctophus* spp. caught near the surface at night have swim bladders occupying 3.2 to 5.5% of body volume (Kanwisher and Ebeling, 1957), and one had a density of 1040 kg/m^3 (Neighbors and Nafpaktatis, 1982). Others, such as *Gonichthys* (1085 kg/m^3), lose their swim bladders when adult and have no alternative buoyancy aid. Yet others, such as *Lampanyctus mexicanus,* also lose their swim bladders but accumulate wax esters around the remnant of the swim bladder and in their muscles. With lipid amounting to about 15% of body mass, the adult fish are only a little denser than the water (1025 to 1037 kg/m^3: Capen, 1967). Three other species of *Lampanyctus* with a small or degenerate swim bladder contain only 1 to 6% lipid but nevertheless have densities of 1021 to 1029 kg/m^3, apparently due to adaptations of the kind described in the next section: their bodies contain unusually high proportions of water (Neighbors and Nafpaktatis, 1982).

WATERY FISHES

The fishes described in this section have their densities reduced by watery tissues and poorly ossified bone. In the previous section we described fishes as mesopelagic (swimming in midwater) or benthopelagic (swimming close above the bottom). Our first examples in this section are generally described as bathypelagic, meaning that they swim many hundreds of meters below the surface, but well clear of the bottom.

Denton and Marshall (1958) found that two bathypelagic species, *Gonostoma elongatum* and *Xenodermichthys copei,* had densities (at the surface) of 1032 and 1039 kg/m^3. They had no swim bladders and contained only modest quantities of oil, and owed their low densities to watery tissues and poorly ossified bone. They contained 87 to 90% water, compared to 72% for *Ctenolabrus*, a typical coastal teleost. Protein made up 7% or less of their body masses, compared to over 16% for two coastal species. When *Xenodermichthys* was reduced to ash by heating, it yielded only about one third as much ash, per unit body mass, as *Ctenolabrus*. The poorly ossified bones of *Gonostoma* and *Xenodermichthys* showed far more faintly in X-ray pictures, than did the bones of other teleosts of similar size. Denton and Marshall (1958) report that a thick gelatinous layer under the skin of the bathypelagic teleost *Chauliodus* is denser than seawater but Yancy et al. (1989) found that this layer was less dense than seawater, in *Chauliodus* and two other teleosts from similar depths. It may contribute to the buoyancy of these fishes.

Similar adaptations are found, rather surprisingly, in the lumpsucker *Cyclopterus*, a coastal teleost without a swim bladder (Davenport and Kjorsvik, 1986). Females have densities of only about 1030 kg/m^3, very close to the density of seawater. Males are slightly denser, and the data that follow refer to females. In them, 18% of body mass is a gelatinous layer close under the skin, with almost the same density as seawater. Another 7% in ripe females is ovarian fluid, with an osmotic concentration only 36% that of seawater, and a density of 1010 kg/m^3. There are dorsal muscles making up 7% of body mass which have a density of only 1011 kg/m^3 (compared to 1060 kg/m^3 for typical teleost muscle). This is because (like the ovarian fluid) they have a low osmotic concentration, they contain only half as much protein per unit mass as plaice (*Pleuronectes*) muscle, and they contain 8% oil. The skeleton is almost uncalcified, with a density of only 1040 kg/m^3.

The electric ray (*Torpedo nobiliana*) has a mean density of 1030 kg/m, almost the same as the seawater that it lives in (Roberts, 1969). Like the buoyant selachians described in the previous section, it has a large oily liver, but this is not enough to explain its low density: with the liver removed, the density of the remainder of the fish is still only 1043 kg/m^3. Another part of the explanation is that the electric organ which makes up 19% of the body mass has a much lower protein content than the ordinary muscle. Accordingly, its density is much lower (1030 kg/m^3, compared to 1055 kg/m^3).

SWIM BLADDERS

Since densities of gases are so low, relatively small gas-filled floats can match the densities of fishes to that of the water in which they swim. Table 2 suggests that a swim bladder need occupy only about 7% of the volume of a freshwater fish or 5% of a marine one. Swim bladders of about these sizes are indeed commonly found, and fishes with swim bladders commonly have densities within 0.5% of the density of the water (Jones and Marshall, 1953; Alexander, 1959b).

If a fish has an air-breathing organ in addition to a swim bladder, the gas in that will give buoyancy and the swim bladder should be correspondingly smaller (see, for example, Gee, 1976). Neither such air-breathing organs, nor the lungs possessed by some fishes, are wholly satisfactory as buoyancy aids, because the oxygen extracted from them is not replaced by carbon dioxide (instead, carbon dioxide diffuses from the gills into the water) so the volume of gas and the buoyancy of the fish diminish between breaths (Gee, 1981).

I will argue in a later section that small size is advantageous for buoyancy organs, because if a fish has to be bigger to contain a large buoyancy organ it will need more energy for swimming. Swim bladders have the great advantage of being much smaller than equivalent nongas buoyancy aids, but they have some severe disadvantages, as the rest of this section will show.

Swim bladders expand when the fish swims nearer the surface of the water, where the pressure is less, and are compressed when it swims deeper. The pressure is 1 atm at the surface and increases by 1 atm for every 10 m of depth. Thus, a swim bladder that had the required volume at the surface would be compressed to half that volume by the 2 atm total pressure at 10 m depth, and to one tenth by the 10 atm at 90 m. Most swim bladders seem to be compressed like this, in almost exact accord with Boyle's Law, at increased pressures. At decreased pressures, however, expansion is rather less than Boyle's Law predicts, because it is restricted by the taut swimbladder wall, and body wall. Cypriniform fishes are unusual in having swim bladders with rather stout walls, inflated with gas at pressures up to 0.14 atm above ambient. These change volume less, for small changes of depth, than do other swim bladders (Alexander, 1959b).

Thus the density of a fish with a swim bladder matches that of the water only at one depth, unless the quantity of gas in it is adjusted. The fish may be able to hover in midwater only in a restricted range of depths. Perch (*Perca*) cannot hover by fin movements alone at pressures more than 16% below the pressure to which they are adapted (Jones, 1952). Thus a perch adapted to 20 m depth (pressure 3 atm) could not hover at depths less than 15 m (2.5 atm).

Further, even at the depth to which it is adapted a fish with a swim bladder is in unstable equilibrium. If it rises a little its density will fall, making it tend to rise further. If it sinks a little its density will increase and it will sink more. To hover at constant depth, a fish must continually use its fins to make small adjustments, just as circus performers must make small movements to keep their balance on a tightrope.

Another disadvantage of swim bladders is that they tend to lose gas by diffusion to the blood and, eventually, through the gills to the water. Consider a fish living at a depth at which the pressure is p atm. The partial pressures of the swimbladder gases must total p atm, and are commonly found to be about 0.8 atm nitrogen and $(p - 0.8)$ atm oxygen (Scholander et al., 1953). The atmosphere contains 0.8 atm nitrogen and 0.2 atm oxygen, and the dissolved gases in the sea tend to equilibrate with it. Thus, however deep the fish swims, the partial pressure of dissolved nitrogen in the water around it will be 0.8 atm and that of oxygen 0.2 atm (or less, because dissolved oxygen is removed by the respiration of marine organisms). Thus, the partial pressure of oxygen can be expected to be $(p - 0.8)$ atm in the swim bladder and 0.2 atm (or less) in the water. The difference of $(p - 1)$ atm will tend to drive oxygen out of the swim bladder. This is because the swimbladder gases tend to equilibrate with the blood flowing through nearby blood vessels and the blood equilibrates with the water at the gills. This partial pressure difference would be 1 atm at a depth of 10 m, and 100 atm at 1000 m.

That argument assumed a particular rule for the composition of swim-bladder gases, but salmonids and fish from great depths tend to have more nitrogen in their swim bladders than the rule suggests (see Alexander, 1966a). It is easy to show by similar arguments that whatever the composition, there must be a partial pressure difference tending to drive at least one gas out.

Losses by diffusion from the swim bladder are inevitable, but the rate of loss can be minimized by making the swimbladder wall as impermeable as possible. Swimbladder walls generally have three layers, of which the middle one, the submucosa, has a silvery or pearly appearance due to being packed with crystals of guanine. These crystals are flexible sheets, exceedingly thin (about 20 nm) but at least 90 μm long (Lapennas and Schmidt-Nielsen, 1977). Their presence makes the submucosa up to 100 times less permeable to diffusing gases than an equal thickness of frog muscle or connective tissue (Denton et al., 1970; Lapennas and Schmidt-Nielsen, 1977). The submucosa was thicker and had a lower diffusion constant in conger eels (*Conger*), which inhabit depths down to 300 m, than in a selection of teleosts from shallower water.

Diffusion losses from the swim bladder must be replaced by secretion (the mechanism is discussed in the next section). The greater the depth at which a fish lives, the faster will be the diffusion losses, and if the depth is too great the fish will be unable to keep its swim bladder inflated. Lapennas and Schmidt-Nielsen (1977) estimated that the swimbladder wall of *Conger* was impermeable enough to enable the eel to live at depths down to 1000 m, considerably beyond its natural depth limit of about 300 m. They had no measurements of gas secretion rates for *Conger* but assumed that it would secrete as fast as *Anguilla*, which had been shown to be able to refill its emptied swim bladder in 10 h. Swim bladders from fishes caught at depths greater than 1000 m contained very much more guanine, per unit area of wall, than *Conger* and other fish from lesser depths (Denton et al., 1970).

GAS SECRETION AND RESORPTION

The rates at which fishes can fill their swim bladders or remove gas from them have been measured many times since the pioneering experiments of Moreau (1874). One of the most informative experiments was that of Jones and Scholes (1985), who kept cod (*Gadus morhua*) in a pressure tank with a viewing window. They could adjust the pressure to any value between 1 and 7.5 atm, simulating depths from the surface to 65 m. They allowed the fish to adapt to one pressure, and then changed to another. After an increase of pressure the fish were denser than water and corrected this by secreting into the swim bladder, but after a decrease they were too buoyant and had to correct by removing gas. Jones and Scholes (1985) checked the progress of secretion or resorption by occasional brief tests in which they adjusted the pressure, to find the one at which the fish just floated. They found that when the pressure was increased to simulate increasing depth, the cod compensated at a rate of only 1 m/h at 12°C, or less at lower temperatures. When it was decreased to simulate depth reduction, the rate of compensation depended on the depth but not on the temperature. At 5 m the fish compensated at only 1 m/h, but at 65 m they compensated at 20 m/h.

Some other fishes have been shown to be able to fill their swim bladders faster. For example, bluefish (*Pomatomus*) secrete fast enough to compensate for depth changes of 2.5 m/h (Wittenberg et al., 1964).

The more primitive teleosts have connections from the swim bladder to the gut and can spit excess gas out or (if they are at the surface) gulp air in. Higher teleosts (Paracanthopterygii and Acanthopterygii) have lost the pneumatic duct and can only add or remove gas by secretion from or resorption into the blood. However, even fishes that possess the duct are generally capable of secreting and resorbing gases. Indeed, later paragraphs show that much of our knowledge of the physiology of secretion and resorption comes from experiments on the eel *Anguilla*, a teleost with a pneumatic duct.

Secretion and resorption generally occur in specialized regions of the swim bladder. Secretion occurs at the gas gland, described below. Resorption occurs in a swollen pneumatic duct in eels, in a dorsal pocket called the oval in Gadiformes (cod, etc.), and in other specialized regions in other groups. To resorb gases all that is needed is to expose them to the blood, because their pressure makes them tend to diffuse into the blood, as we have already seen. The greater the depth, the larger the partial pressure difference to drive diffusion; that is presumably why the cod could absorb faster at greater depths in the experiments of Jones and Scholes (1985). When gas is to be resorbed, the resorbent region of the swim bladder is expanded and its blood vessels are dilated, but at other times the resorbent region is contracted and very little blood is allowed to flow through it, to minimize unwanted losses by diffusion (Fänge, 1983). Steen (1963a) performed experiments on eels in which he showed that the rate of resorption depends on the differences of gas partial pressure between the swim bladder and the blood, just as would be expected for passive diffusion. It is even possible to make oxygen diffuse *into* the swim bladder in the resorbent region, by filling the swim bladder with nitrogen and having the blood well oxygenated. The passive nature of resorption is also confirmed by Jones and Scholes' (1985) finding that the maximum rate of resorption is independent of temperature.

Resorption is passive, but gas secretion has to be active if gases are being moved from the water, where their partial pressures are relatively low, to the swim bladder, where their partial pressures are higher. Jones and Scholes (1985) found that the maximum rate of secretion increases with increasing temperature, as would be expected for a process driven by metabolism.

Gas is secreted into the swim bladder at the gas gland, a region of the swimbladder wall that is generally reddish in color. In favorable conditions, gas bubbles can be seen forming at the surface of the gas gland of an anesthetized fish whose swim bladder has been opened (Scholander, 1956). The cells of the gas gland are capable of producing lactic acid rapidly by anaerobic metabolism, even in the presence of high partial pressures of oxygen which would inhibit anaerobic metabolism in other cells (Boström et al., 1972). The secretion of gases into the swim bladder is driven by release of lactic acid into the blood passing through the gas gland.

The lactic acid entering the blood drives gases out by several mechanisms. First, it reduces the pH, and thereby releases oxygen from hemoglobin by the Bohr and Root effects. The Bohr effect is a reduction, at low pH, of the degree to which the hemoglobin becomes saturated with oxygen, at any given partial pressure of oxygen. The Root effect is a reduction, at low pH, of the oxygen capacity of the blood. Pelster and Weber (1990) used electrophoresis to separate teleost hemoglobin into two components, and found that only one of these two is subject to the Root effect. Second, again by reducing the pH, lactic acid releases carbon dioxide from bicarbonate in the blood. This seems relatively unimportant as a mechanism of gas secretion, for carbon dioxide secreted into the swim bladder tends to be lost again rapidly because of its high diffusion constant. Finally, like any other electrolyte, lactic acid reduces the solubilities of all gases in the blood by the salting out effect (the effect that makes gases less soluble in seawater than in fresh water). This is the only known mechanism capable of explaining secretion of nitrogen and inert gases, though Wittenberg (1958) has shown how these may be carried passively into the swim bladder with secreted oxygen. It had been claimed that the solubility of nitrogen in eel blood falls sharply as pH falls, an effect that seemed potentially important for nitrogen secretion. However, Pelster et al. (1988) could not demonstrate any effect of pH on nitrogen solubility in eel blood.

Carbon dioxide is also produced in the gas gland, mainly, it seems, by anaerobic metabolism of glucose by way of the pentose phosphate shunt (Pelster et al., 1989). This would produce NADPH which would have to be reoxidized. Pelster and his colleagues suggest that the oxygen may be provided by lengthening fatty acid chains. Carbon dioxide produced in the gas gland may contribute directly to the swimbladder gases, or release oxygen from oxyhemoglobin by the Bohr and Root effects.

The effects described in the above two paragraphs cannot by themselves explain gas secretion at any substantial depth. Even if all the oxyhemoglobin could be dissociated, the oxygen would go into physical solution in the blood instead of being released as bubbles, at depths at which the pressure was more than about 4 atm (Alexander, 1966a).

Gas secretion is possible at substantial depths, only because the effects described above are amplified by a process of countercurrent multiplication in the rete mirabile, an assembly of parallel blood capillaries (Figure 1). The artery approaching the gas gland and the vein leaving it are broken up into capillaries which run parallel to each other, the venous capillaries interspersed among the arterial ones. The capillaries are packed tightly together, allowing rapid diffusion of substances between them. Typical capillaries in teleost muscle are only about

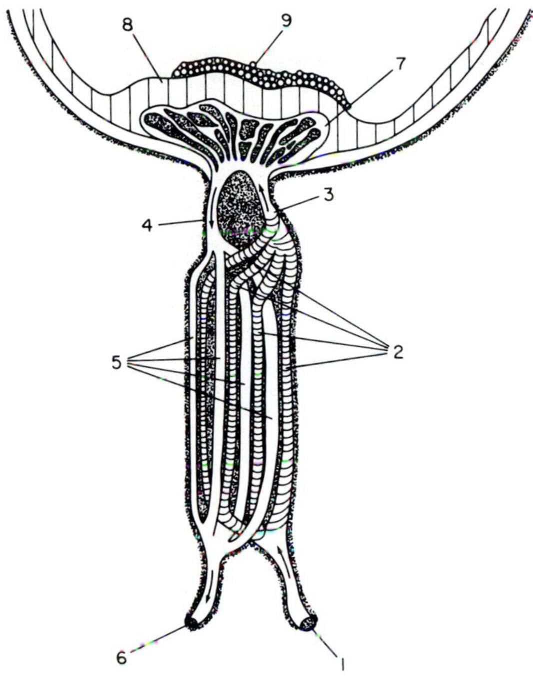

FIGURE 1. A diagram of the rete mirabile and gas gland of an eel: 1, artery; 2, arterial capillaries; 3, 4, artery and vein connecting rete to gas gland; 5, venous capillaries; 6, vein; 7, capillaries of gas gland; 8, epithelium of gas gland; 9, gas foam. The few capillaries shown here represent thousands in the actual structure. (From Blaxter, J. H. S. and Tytler, P., *Adv. Comp. Physiol. Biochem.*, 7, 311, 1978. With permission.)

0.5 mm long but the capillaries of the rete are very much longer: 10 mm in large eels (*Anguilla*) and up to 25 mm in some deep-sea teleosts (Marshall, 1972).

Kuhn et al. (1963) showed theoretically how countercurrent multiplication could work in the rete, and Steen (1963b) demonstrated it in experiments on eels (*Anguilla*). Eels have played an intensely important part in research on swimbladder physiology, because of a small but crucial anatomical peculiarity: the capillaries of the rete re-form into large blood vessels (at positions 3 and 4 in Figure 1) between the rete and the gas gland. In other teleosts the arterial capillaries continue into the gas gland and loop back to form the venous capillaries. The advantage to physiologists of the arrangement in eels is that it enables them to take blood samples from anesthetized, opened eels at positions 3 and 4, as well as 1 and 6 (Figure 1). With samples from those four positions, the countercurrent multiplication principle can be demonstrated clearly.

Instead of the classic experiments of Steen (1936b), Figure 2 shows results from very similar experiments by Kobayashi et al. (1989a, 1990), in which more quantities were measured. These recent experiments have substantially modified our understanding of gas secretion.

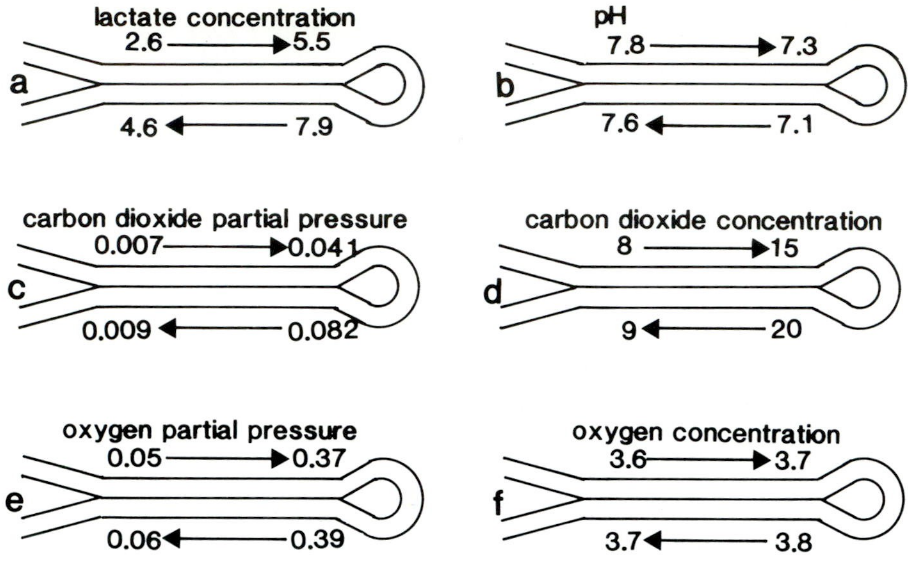

FIGURE 2. Changes in the composition of the blood as it passes through the rete mirabile and gas gland of eels (*Anguilla*) during gas secretion. (a) lactate concentration, mmol l^{-1}; (b) pH; (c) partial pressure, atm, and (d) concentration, mmol l^{-1}, of carbon dioxide; (e), partial pressure, atm, and (f) concentration, mmol l^{-1}, of oxygen. (a) Shows mean values from six experiments by Kobayashi et al. (1989a) and (b) to (f) show mean values from nine experiments by Kobayashi et al. (1990).

Each of the diagrams in Figure 2 represents the path of blood along the arterial capillaries of the rete, through the gas gland and back along the venous capillaries. The numbers show the results of analysis of blood samples taken from positions 1, 3, 4, and 6 (Figure 1). Samples were taken from these four positions in as rapid sequence as possible, at times when the preparation was believed to be in a steady state.

Figure 2a shows that the lactate concentration in the blood increased (from a mean of 5.5 to 7.9 mmol l^{-1}) as the blood passed through the gas gland. An increase of about the same amount (from 2.6 to 4.6 mmol l^{-1}) occurred between the blood's first entering the rete and its finally leaving — as expected if lactate is neither produced nor broken down in the capillaries of the rete. However, the lactate concentration increases markedly as the blood travels along the arterial capillaries and decreases as it travels along the venous ones. This must be due to diffusion from the venous capillaries (in which the blood has a higher lactate concentration, due to addition of lactate in the gas gland) to adjacent parts of arterial capillaries (in which the lactate concentration is lower). The concentration of lactate in the blood leaving the gas gland is very much higher than it would be if the blood entered the gland directly, instead of by way of the rete.

Figures 2c and d show, similarly, that carbon dioxide is added to the blood as it passes through the gas gland, and that both the partial pressure and the concentration of carbon dioxide increase towards the gas gland end of the rete. Carbon dioxide as well as lactate must diffuse from the venous to the arterial capillaries. Figure 2b shows that the higher concentrations of lactate and carbon dioxide, towards the gas gland end, are accompanied as expected by lower pH.

Figures 2e and f show that partial pressures of oxygen increased towards the gas gland end of the rete, but that the concentration of oxygen remained almost constant. There is no paradox here: the lower pH at the gas gland end would be expected to cause higher partial pressures by the Root effect, even if concentrations remain unchanged. Oxygen was being secreted into the swim bladder so the oxygen concentration in the blood was presumably falling, as the blood passed through the gas gland. However, the expected fall (0.1 mmol l^{-1}) was too small to be detectable in this experiment.

The classic theory of the rete mirabile (Kuhn et al., 1963) allows diffusion of gases, but not of lactate, between the venous and arterial capillaries. The experiments illustrated in Figure 2 show little or no oxygen diffusion, but substantial diffusion of lactate. Kobayashi et al. (1989b) have formulated a new version of the theory, that allows lactate diffusion. It predicts (as the original theory also did) that the maximum partial pressure that can be developed in the gas gland increases very sharply as the conductance ratio increases

$$\text{conductance ratio} = \frac{\text{gas diffusion conductance}}{\text{blood flow rate} \times \text{gas solubility}} \qquad (5)$$

Larger retia mirabilia (ones with longer capillaries, or with larger numbers of capillaries in parallel) have higher gas diffusion conductances, so should be able to secrete into the swim bladder at higher pressures. Deep-sea fishes that have swim bladders generally have large retia (Marshall, 1972). The theory also predicts that higher gas secretion rates, at given pressure, require higher conductance ratios, so we may expect to find large retia in fishes that secrete fast, as well as in ones that live at high pressures.

It had been thought that any diffusion of lactate from the venous to the arterial capillaries would make the rete less effective, but the analysis of Kobayashi et al. (1989b) seems to show on the contrary that it makes the rete more effective, by enabling higher lactate concentrations to build up in the gas gland, for any given rate of lactate secretion.

In the experiments shown in Figure 2, there was no appreciable increase of oxygen concentration towards the gas gland end of the rete. It seems clear, however, that when oxygen is being secreted at substantial depths, oxygen must diffuse from the venous to the arterial capillaries, as envisaged in the classic theory (Kuhn et al., 1963), and the concentration as well as the partial pressure of oxygen must increase towards the gas gland. The reason is that even if all the oxyhemoglobin were dissociated, a given concentration of oxygen could exert only a limited partial pressure.

BUOYANCY STRATEGIES

We have seen in previous sections that some fishes are denser than the water they live in while others have buoyancy aids that match their densities to the water. Among those with buoyancy aids, some depend on oils, some on watery tissues and poorly ossified skeletons, and some on swim bladders. Can we explain this diversity in terms of the fishes' ways of life?

It has long been recognized that bottom-living fishes may benefit from being denser than the water, so that friction tends to keep them in place when they rest on the bottom (see, for example, Alexander, 1966a). Bottom-living selachians, such as *Scyliorhinus* and *Raia*, and bottom-living teleosts, such as *Pleuronectes* and *Myoxocephalus*, are considerably denser than seawater. Webb (1990) wondered whether bottom-living teleosts might tend to have more or denser bone than pelagic ones, but found no evidence for this in the species he chose for study. In other experiments, Webb (1989) showed that the lift that acts on *Raia* and *Pleuronectes* when they rest on the bottom in even quite a slow current may be enough to make them lose their frictional grip on the bottom.

In a series of studies, Gee and his colleagues have shown that freshwater fishes living in fast currents tend to have reduced swimbladder volumes, compared to fishes of the same species living in slower water (reviewed by Gee, 1983). Some of the species they investigated habitually rest on the bottom. For them, the advantage of a reduced swim bladder (and increased density) is presumably improved frictional grip on the bottom. Others are more pelagic in their habits, making the advantage of swimbladder reduction less obvious.

Many other pelagic fishes lack buoyancy aids. They include sharks such as *Lamna* (Bone and Roberts, 1969) and many scombrids which seldom or never rest on the bottom, at least in aquaria (Magnuson, 1978). The following argument (Alexander 1966a, 1990) suggests that buoyancy aids may be disadvantageous for fast-swimming fishes. Imagine a fish that is initially denser than water, which swims at speed u. The power required to overcome the drag on its body is approximately proportional to $u^{2.5}$. If it uses fins as hydrofoils to supply the lift needed to prevent sinking, there is an additional (induced) power requirement: by Equation 1, this is proportional to u^{-1} Alternatively, if it acquires a buoyancy aid, this will increase the volume of the body, as shown by Equation 2. If the volume is increased, the power needed to propel the body will be increased by an appropriate factor — and we have already seen that this power is proportional to $u^{2.5}$. Thus, the extra power needed to prevent sinking is proportional to u^{-1} if hydrofoils are used, and to $u^{2.5}$ if there is a buoyancy aid. Buoyancy aids must be the more economical for slow-swimming fish and hydrofoils for fast ones. The critical speed, above which hydrofoils are predicted to be more economical, is faster for large fish than for small ones. It is also higher if the available buoyancy aid is a swim bladder than if it is a bulkier alternative such as squalene (see Table 2). Figure 3 shows estimates of critical speeds from Alexander (1990), together with masses and typical swimming speeds of a few species. The buoyant species (open symbols) fall well below the lines, so buoyancy aids would be expected to be the better option for them. The dense pelagic teleosts (filled symbols) lie on or just below the "swimbladder" line, indicating that even for them a swim bladder might be more economical. However, there is a good deal of uncertainty about the estimates of energy costs, and no account has been taken of the cost of secreting gas to replace diffusion losses from the swim bladder.

Swim bladders are the smallest effective buoyancy aids (Table 2) so add least to the volume of the body and the power needed for swimming. They are found only in teleosts (they seem to have evolved from the lungs of primitive bony fishes), but we might suppose that they would be the best option for every

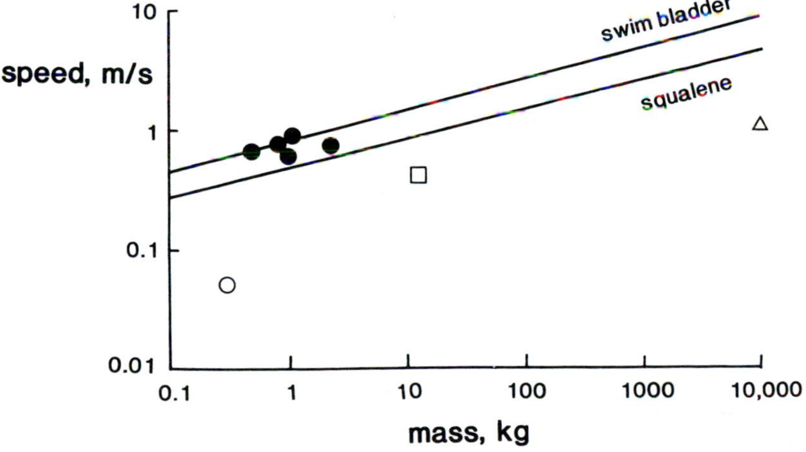

FIGURE 3 . The speeds at which fishes using fins as hydrofoils to prevent themselves from sinking are estimated to become more economical of energy than if they used a swim bladder or squalene as a buoyancy aid, plotted against body mass. The points show typical swimming speeds for (●) various scombroids which depend on the hydrofoil action of fins, and fishes (open symbols) with buoyancy aids: (○) trout (*Salmo*) and (□) wahoo (*Acanthocybium*), which have swim bladders, and (Δ) basking shark (*Cetorhinus*), which is made buoyant by squalene. (From Alexander, R. McN., *Am. Zool.*, 30, 189, 1990. With permission.)

teleost that swims slowly enough for buoyancy aids to be advantageous. However, there are many pelagic fishes that have lost the swim bladder and rely on other buoyancy aids, either low-density oils, as in some lantern fishes, or watery tissues, as in *Gonostoma* and *Xenodermichthys*.

One of the disadvantages of swim bladders is that they change volume as the fish changes depth — and many fishes make substantial daily changes of depth (Alexander, 1972). A community of small fishes, crustaceans, and siphonophores spends the night near the surface of the oceans and the day at depths of a few hundred meters. They are so plentiful that they form a "deep-scattering layer" that is quite conspicuous in echo-sounder traces, and their movements have also been observed from submersible vehicles (Barham, 1966). Lantern fishes are the most plentiful of the fish in the deep-scattering layer. As we have seen, some of them have swim bladders which are large enough to give useful buoyancy when the fish is at the surface. To maintain their buoyancy as they swim down to their daytime depths, they would have to secrete gas into the swim bladder at enormous rates. The lantern fishes observed by Barham commuted between the top 50 m of the sea and depths of around 300 m, at rates of the order of 100 m/h, but no fish has been shown to secrete gas fast enough to compensate for depth changes at more than 2.5 m/h (Wittenberg et al., 1964). It seems inevitable that at their daytime depths, the swim bladders of lantern fishes are compressed to a tiny fraction of their night-time size. A swim bladder is an effective buoyancy organ only at night, for a vertically migrating lantern fish. Wax esters are bulkier, but are effective day and night. This is probably why some lantern fishes have lost their swim bladders and depend on wax esters for buoyancy.

Some other teleosts spend the day resting on the bottom and the night feeding at lesser depths. For them, the volume changes of the swim bladder may be no disadvantage if they keep enough gas in the swim bladder to match their density to the water at night. When they descend to the bottom for the day their swim bladders will be compressed, but if they spend the day actually resting on the bottom it may be an advantage to be denser than the water there, so that friction will tend to hold them in place. Swordfish (*Xiphius*) have swim bladders which give them about the same density as the water when they are swimming at the surface. Observers in submersibles have seen them resting on the bottom, and it has been shown by acoustic telemetry that they swim up from 100 m to the surface in less than 5 min (Carey and Robinson, 1981). A swim bladder seems a very satisfactory buoyancy aid for that style of life.

We have seen that, at substantial depths, gases inevitably diffuse out of swim bladders and must be replaced by diffusion. Just as work is needed to compress gases in a pump, energy is needed to secrete gases into a swim bladder against partial pressure gradients (Alexander, 1972). It seems possible that for fishes that live at great depths, keeping a swim bladder filled may entail a substantial energy cost. This may tip the balance of advantage to other buoyancy aids, especially if the fish is a sluggish one for which the increased energy cost of swimming, due to the extra bulk of alternative buoyancy aids, may not be very important.

Many bathypelagic teleosts such as *Gonostoma* and *Xenodermichthys* have lost their swim bladders, although many teleosts that live close to the bottom at even greater depths have retained theirs (Marshall, 1972). The reason for the difference may be that the fishes close to the bottom may be more active, because food is more plentiful on the bottom (where detritus sinking from above accumulates) than at the depths inhabited by the bathypelagic fishes.

Bathypelagic fishes tend to get their buoyancy from water tissues and poorly ossified bone, rather than from oils. A possible reason is the high heats of combustion of all the oily materials used by fishes as buoyancy aids (Alexander, 1990). The energy cost of depositing the required quantity of oil, as the fish grows, may make watery tissues the better alternative for these fishes, although fishes with watery muscles presumably cannot swim fast.

CONCLUSION

We have seen that fishes without buoyancy aids are denser than either fresh or salt water. Excess density may be an advantage for bottom-living fishes, especially for those that live in fast-flowing water, but for other fishes it presents a problem that has been met in four principal ways.

- Pelagic sharks and many scombrids tolerate excess density, using fins as hydrofoils to prevent themselves from sinking. This seems to be the most economical option for fast-swimming fishes.
- Most pelagic teleosts have swim bladders, which match their density to that of the water. The volume of the swim bladder changes as the fish changes depth, but gas can be removed from the swim bladder by allowing it to diffuse into the blood, or added by secretion from the gas gland. Secretion at depths where the presssure is high is made possible by counter-current multiplication in the rete mirabile. Diffusion losses from swim bladders are inevitable, but are kept low by a layer of guanine crystals in the swim-bladder wall. The swim bladder is the least bulky buoyancy aid, and seems generally to be the best option for slow-swimming fishes, but is not available to selachians because it has evolved only in bony fishes.
- A variety of fishes contain low-density lipids (often squalene or wax esters) in sufficient quantity to match their density to that of the water. This seems to be the best option for slow-swimming selachians, and has been adopted by *Cetorhinus* and deep-sea squalids. It has also been adopted by some

lantern fishes that make very large daily depth changes, so could benefit from the buoyancy given by a swim bladder for only part of each day. The use of oil rather than a swim bladder by the pelagic notothenioids may be a consequence of their evolutionary history: other notothenioids live on the bottom and the group as a whole lacks the swim bladder.

- Some bathypelagic teleosts have watery muscle and poorly ossified bone to keep their density low. This may be the best option in an environment where hydrostatic pressure is high and food is sparse. The energy cost of secreting gas into the swim bladder to replace diffusion losses might be high (Alexander, 1972), as would the energy cost of accumulating large quantities of low-density oils (Alexander, 1990). It is not clear why the lumpsucker *Cyclopterus* has adopted the same buoyancy strategy.

Though this chapter has ranged widely, it has not exhausted the subject of fish buoyancy. The morphology and physiology of the swim bladder are discussed more fully in other reviews, such as Fänge (1983). I have entirely omitted the important topic of the buoyancy of fish eggs (on which, see Craik and Harvey, 1987).

REFERENCES

Alexander, R. McN., The densities of Cyprinidae, *J. Exp. Biol.,* 36, 333, 1959a.

Alexander, R. McN., The physical properties of the swimbladders of fish other than Cypriniformes, *J. Exp. Biol.,* 36, 347, 1959b.

Alexander, R. McN., The lift produced by the heterocercal tails of Selachii, *J. Exp. Biol.,* 43, 131, 1965.

Alexander, R. McN., Physical aspects of swimbladder function, *Biol. Rev.,* 41, 141, 1966a.

Alexander, R. McN., Lift produced by the heterocercal tail of *Acipenser, Nature,* 210, 1049, 1966b.

Alexander, R. McN., The energetics of vertical migration by fishes, *Symp. Soc. Exp. Biol.,* 26, 273, 1972.

Alexander, R. McN., Size, speed and buoyancy adaptations in aquatic animals, *Am. Zool.,* 30, 189, 1990.

Barham, E. G., Deep scattering layer migration and composition: observations from a diving saucer, *Science,* 151. 1399, 1966.

Blake, R. W., The energetics of hovering in the mandarin fish *(Synchropus picturatus), J. Exp. Biol.,* 82, 25, 1979.

Blaxter, J. H. S. and Tytler, P., Physiology and function of the swimbladder, *Adv. Comp. Physiol. Biochem.,* 7, 311, 1978.

Bone, Q., Buoyancy and hydrodynamic functions of integument in the castor oil fish *Ruvettus pretiosus* (Pisces: Gempylidae), *Copeia,* 1972, 78.

Bone, Q. and Roberts, B. L., The density of elasmobranchs, *J. Mar. Biol. Assoc. U.K.,* 49, 913, 1969.

Boström, S.-L., Fänge, R., and Johansson, R. G., Enzyme activity patterns in gas gland tissue of the swimbladder of the cod (*Gadus morrhua), Comp. Biochem. Physiol.* 43B, 473, 1972.

Bunker, S. J. and Mahin, K. E., The hydrodynamics of cephalaspids, in *Biomechanics in Evolution,* Rayner, J. M. V. and Wootton, R. J., Eds., Cambridge University Press, New York, 1991, 113

Capen, R. L., Swimbladder morphology of some mesopelagic fishes in relation to sound scattering, Report 1447, U. S. Navy Electronics Laboratory, San Diego CA, 1967, 1.

Carey, F. G. and Robinson, B. H., Daily patterns in the activities of swordfish, *Xiphius gladius,* observed by acoustic telemetry, *Fish. Bull. U.S.,* 79, 277, 1981.

Corner, E. D. S., Denton, E. J., and Forster, G. R., On the buoyancy of some deep-sea sharks, *Proc. R. Soc. London, Ser. B,* 171, 415, 1969.

Craik, J. C. A. and Harvey, S. M., The causes of buoyancy in eggs of marine teleosts, *J. Mar. Biol. Assoc. U.K.,* 67, 169, 1987.

Davenport, J. and Kjorsvik, E., Buoyancy in the lumpsucker *Cyclopterus lumpus, J. Mar. Biol. Assoc. U.K.,* 66, 159, 1986.

Denton, E. J. and Marshall, N. B., The buoyancy of bathypelagic fishes without a gas-filled swim bladder, *J. Mar. Biol. Assoc. U.K.,* 37, 753, 1958.

Denton, E. J., Liddicoat, J. D., and Taylor, D. W., Impermeable "silvery" layers in fishes, *J. Physiol.,* 207, 64P, 1970.

Eastman, J. T., Lipid storage systems and the biology of two neutrally buoyant Antarctic notothenioid fishes, *Comp. Biochem. Physiol.,* 90B, 529, 1988.

Fänge, R., Gas exchange in fish swim bladder, *Rev. Physiol. Biochem. Pharmacol.,* 97, 111, 1983.

Fricke, H., Reinicke, O., Hofer, H., and Nachtigall, W., Locomotion of the coelacanth *Latimeria chalumnae* in the natural environment, *Nature,* 329, 331, 1987.

Gee, J. H., Buoyancy and aerial respiration: factors influencing the evolution of reduced swimbladder volume of some Central American catfishes (Trichomycteridae, Callichthyidae, Loricariidae, Astroblepidae), *Can. J. Zool.,* 54, 1030, 1976.

Gee, J. H., Coordination of respiratory and hydrostatic functions of the swimbladder in the Central American mudminnow, *Umbra limi,* J. Exp. Biol., 92, 37, 1981.

Gee, J. H., Ecologic implications of buoyancy control in fish, in *Fish Biomechanics,* Webb, P. W. and Weihs, D., Eds., Praeger, New York, 1983, 140.

Hoogerhoud, R. J. C., The adverse effects of shell ingestion for molluscivorous cichlids, a constructional morphological approach, *Neth. J. Zool.,* 37, 277, 1987.

Jones, F. R. H., The swimbladder and the vertical movements of teleostean fishes. II. The restriction to rapid and slow movements, *J. Exp. Biol.,* 29, 94, 1952.

Jones, F. R. H. and Marshall, N. B., The structure and functions of the teleostean swimbladder, *Biol. Rev.,* 28, 16, 1953.

Jones, F. R. H. and Scholes, P., Gas secretion and resorption in the swimbladder of the cod *Gadus morhua, J. Comp. Physiol. B,* 155, 319, 1985

Kanwisher, J. and Ebeling, A., Composition of the swimbladder gas in bathypelagic fishes, *Deep Sea Res.,* 4, 211, 1957.

Kobayashi, H., Pelster, B., and Scheid, P., Water and lactate movement in the swimbladder of the eel, *Anguilla anguilla, Resp. Physiol.,* 78, 45, 1989a.

Kobayashi, H., Pelster, B., and Scheid, P., Solute back-diffusion raises the gas-concentrating efficiency in counter-current flow, *Resp. Physiol.,* 78, 59, 1989b.

Kobayashi, H., Pelster, B., and Scheid, P., CO_2 back-diffusion in the rete aids O_2 secretion in the swimbladder of the eel, *Resp. Physiol.,* 79, 231, 1990.

Kuhn, W., Ramel, A., Kuhn, H. J., and Marti, E., The filling mechanism of the swimbladder, *Experientia,* 19, 497, 1963.

Lapennas, G. N. and Schmidt-Nielsen, K., Swimbladder permeability to oxygen, *J. Exp. Biol.,* 67, 175, 1977.

Lewis, R. W., The densities of three classes of marine lipids in relation to their possible role as hydrostatic agents, *Lipids,* 5, 151, 1970.

Magnuson, J. J., Hydrostatic equilibrium of *Euthynnuss affinis,* a pelagic teleost without a gas bladder, *Copeia,* 1970, 56.

Magnuson, J. J., Locomotion by scombrid fishes: hydromechanics, morphology and behaviour, in *Fish Physiology,* Vol. 7, Hoar, W. S. and Randall, D. J., Eds., Academic Press, New York, 1978, 239.

Marshall, N. B., Swimbladder organization and depth ranges of deep-sea teleosts, *Symp. Soc. Exp. Biol.,* 26, 261, 1972.

Moreau, A., Mémoire sur la vessie natatoire, au point de vue de la station et de la locomotion des poissons, *C. R. Acad. Sci. Paris,* 78, 541, 1874.

Neighbors, M. A. and Nafpaktatis, B. G., Lipid compositions, water contents, swimbladder morphologies and buoyancies of nineteen species of midwater fishes, *Mar. Biol.,* 66, 207, 1982.

Nevenzel, J. C., Rodegker, W., and Mead, J. F., The lipids of *Ruvettus pretiosus* muscle and liver, *Biochemistry,* 4, 1589, 1965.

Nevenzel, J. C., Rodegker, W., Mead, J. F., and Gordon, M. S., Lipids of the living coelacanth, *Latimeria chalumnae, Science,* 152, 1753, 1966.

Nevenzel, J. C., Rodegker, W., Robinson, J. S., and Kayama, M., The lipids of some lantern fishes (family Myctophidae), *Comp. Biochem. Physiol.,* 31, 25, 1969.

Ohlmer, W., Untersuchungen uber de Beziehungen zwischen Korperform and Bewegungsmedium bei Fishchen aus stehenden Binnengewassern, *Zool. Jahrb. (Anat.),* 81, 151, 1964.

Pelster, B., Kobayashi, H., and Scheid, P., Solubility of nitrogen and argon in eel whole blood and its relationships to pH, *J. Exp. Biol.,* 135, 243, 1988.

Pelster, B., Kobayashi, H., and Scheid, P., Metabolism of the perfused swimbladder of European eel: oxygen, carbon dioxide, glucose and lactate balance, *J. Exp. Biol.,* 144, 495, 1989.

Pelster, B. and Weber, R. E., Influence of organic phosphates on the Root effect of multiple fish haemoglobins, *J. Exp. Biol.,* 149, 425, 1990.

Roberts, B. L., The buoyancy and locomotory movements of electric rays, *J. Mar. Biol. Assoc. U.K.,* 49, 621, 1969.

Scholander, P. F., Observations on the gas gland in living fish, *J. Cell. Comp. Physiol.,* 49, 523, 1956.

Scholander, P. F. and van Dam, L., Composition of the swimbladder gas in deep sea fishes, *Biol. Bull. Mar. Biol. Lab. Wood's Hole,* 107, 247, 1953.

Simons, J. R., The direction of the thrust produced by the heterocercal tails of two dissimilar elasmobranchs: the Port Jackson shark, *Heterodontus portusjacksoni* (Meyer) and the Piked dogfish, *Squalus megalops* (Macleay), *J. Exp. Biol.,* 52, 95, 1970.

Steen, J. B., The physiology of the swimbladder of the eel, *Anguilla vulgaris.* II. The reabsorption of gases, *Acta Physiol. Scand.,* 58, 138, 1963a.

Steen, J. B., The physiology of the swimbladder of the eel, *Anguilla vulgaris.* III. The mechanism of gas secretion, *Acta Physiol. Scand.,* 59, 221, 1963b.

Stickney, D. G. and Torres, J. J., Proximate composition and energy content of mesopelagic fishes from the eastern Gulf of Mexico, *Mar. Biol.,* 103, 13, 1989.

Van Vleet, E. S., Candileri, S., McNellie, J., Reinhardt, S. B., Conkright, M. E., and Zwissler, A., Neutral lipid components of eleven species of Caribbean sharks, *Comp. Biochem. Physiol. B,* 79, 549, 1984.

Webb, P. W., Station holding by three species of benthic fish, *J. Exp. Biol.,* 145, 303, 1989.

Webb, P. W., How does benthic living affect body volume, tissue composition and density of fishes?, *Can. J. Zool.,* 68, 1250, 1990.

Wittenberg, J. B., The secretion of inert gas into the swim bladder of fish, *J. Gen. Physiol.,* 41, 783, 1958.

Wittenberg, J. B., Schwend, M. J., and Wittenberg, B. A., The secretion of oxygen into the swim-bladder of fish. III. The role of carbon dioxide, *J. Gen. Physiol.,* 48, 337, 1964.

Yancey, P. H., Lawrence-Berrey, R., and Douglas, M. D., Adaptations in mesopelagic fishes. I. Buoyant glycosaminoglycan layers in species without diel vertical migration, *Mar. Biol.,* 103, 453, 1989.

4 Inner Ear and Lateral Line

Arthur N. Popper and Christopher Platt

INTRODUCTION

Though the components of the octavolateralis system have not been reviewed as a single entity in over 25 years (Lowenstein, 1967), considerable information is available on various aspects of the system in prior reviews on the combined ear and lateral line (Lowenstein, 1967; Kalmijn, 1988), or on components of the system such as inner ear (Lowenstein, 1971), labyrinthine/vestibular system (Pfeiffer, 1964; Platt, 1983, 1988), auditory system (Platt and Popper, 1981; Schellart and Popper, 1992; Popper and Fay, 1993), and lateral line (Flock, 1971; Coombs et al., 1992). There also have been several edited volumes on fish hearing (Schuijf and Hawkins, 1976; Tavolga et al., 1981) and on the lateral line (Coombs et al., 1989) that contain chapters on these topics.

In order to limit the length of the chapter, we have chosen not to cite all relevant papers on a particular topic. Instead, we often will cite one of these recent reviews or very recent papers that are not yet cited much, or one or two pertinent primary references when we feel that they are needed to understand a topic or they are of particular historic significance.

This chapter is directed towards the octavolateralis mechanosensory system of bony fishes (Osteichthyes). While there is an important literature on Agnatha and Chondrichthyes, we do not cover those groups except in the most cursory fashion, nor do we deal with the electrosensory system despite its relationship to the lateral line. Discussions of the octavolateralis system of Chondrichthyes can be found in the references cited in the preceding paragraphs (also see Corwin, 1989).

OCTAVOLATERALIS TERMINOLOGY

The term "octavolateralis" refers to the inner ear and the lateral line sensory systems, which for a long period of time were collectively called the acousticolateralis system (see Lowenstein, 1967). The older name came from early conceptions that both systems were primarily for acoustic reception, that they shared the same embryonic anlage, and that the inner ear was derived evolutionarily from a lateral line precursor containing hair cells (reviewed in Popper et al., 1992). However, as was pointed out first by Wever (1974), the ear and lateral line are each distinct with regard to their embryonic placodes, innervation, and central projections, and there are reasons to believe that neither one is evolutionarily derived from the other (see Popper et al., 1992). We have chosen to use the term octavolateralis (see

0-8493-8042-1/93/$0.00+$.50

Nieuwenhuys, 1967; Northcutt, 1980; McCormick, 1981, 1989), because it simply notes these organs are innervated by the eighth and lateral line cranial nerves.

The Role of the Inner Ear

The vertebrate inner ear is involved both with auditory and postural senses. The ear of fishes was believed to have one or two of its endorgans (saccule and lagena, see below) as the acoustic receptors, but the current view is that at least three, and possibly four, of the endorgans may have at least some involvement in hearing (reviewed in Schellart and Popper, 1992). Likewise, the semicircular canals and the utricle were believed to be the acceleration receptors for the vestibular senses controlling posture, locomotion, and visual stability (see Lowenstein, 1971; Platt, 1988), but we now suspect that additional inner ear endorgans may contribute to the vestibular sense. In short, our current view is that, while certain endorgans provide major inputs to different systems, many of the endorgans in fishes are multifunctional.

The essence of the argument is that the inner ear of fishes cannot be divided as neatly as that of mammals into a "vestibular" part for postural control and an "auditory" part for hearing. The semicircular canal organs respond to angular accelerations of the head, and the otolithic organs and the nonotolithic macula neglecta respond to linear accelerations. Accelerations can be produced by gravity, locomotion, underwater displacement waves, and sound waves. The geometry of the fluids, partitions, inertial masses, and other accessory structures affect the frequency, sensitivity, phase angle, and other parameters of response of the mechanosensory hair cells in a given endorgan.

The Role of the Lateral Line

There has been increasing evidence supporting the function of the lateral line as a hydrodynamic receiver (see papers in Coombs et al., 1989) or a detector of low-frequency motions that originate near (within a body length or two) of the receiver. Several investigators (e.g., Dijkgraaf, 1960) consider these motions to be distinct from those usually associated with "hearing" under water (although that remains an elusive definition; see Webster et al., 1992). These functions make the lateral line an important sensory system initiating appropriate behavioral responses for avoiding obstacles while swimming, for schooling, for capturing prey at the surface and below the surface, and for predator avoidance.

STRUCTURES OF THE OCTAVOLATERALIS SYSTEM

Sensory Hair Cells

The receptor component for each part of the inner ear and for the mechanosensory lateral line is the sensory hair cell, a mechanoreceptive cell with characteristic ultrastructural features (Wersäll, 1961). The receptor mechanism of each hair cell involves an apical array of microvillar processes called stereocilia and a single true cilium called the kinocilium (Figure 1A); this cluster is known as the ciliary bundle. There may be only a few tens of stereocilia in the bundle, or more than 100.

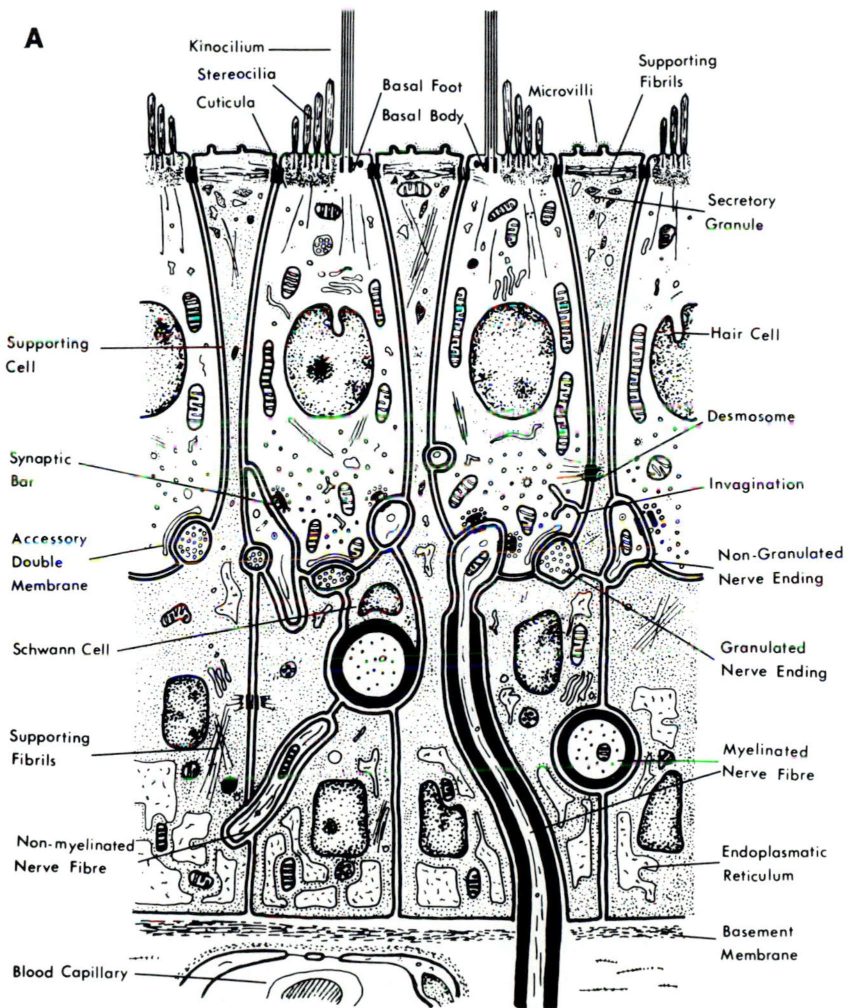

FIGURE 1. Sensory hair cells similar to those found in each of the endorgans of the octavolateralis system. (A) Schematic illustration of the ultrastructure of hair cells of the sensory epithelium of lateral line canal organs of the burbot, *Lota vulgaris*. Note that the sensory hair cell on the left is oriented to the right and the hair cell on the right is oriented towards the left (see text and Figure 5). (From Flock, Å., *Cold Spring Harbor Symp.*, 30, 135, 1965. With permission.)

FIGURE 1 (continued). (B) Scanning electron micrograph of a field of hair cells from the lagenar epithelium of a zebrafish (*Brachydanio rerio*). The dashed dividing line separates two "groups" of hair cells that are oriented in opposite directions (as defined in the text). Arrows show the orientation direction of each region. (Width of field, approximately 20 μm.)

The ciliary bundle has a morphological orientation that correlates with a physiological directional sensitivity to bending this bundle (reviewed by Hudspeth, 1985). Orientation is determined by the eccentric location of the kinocilium at one side of the bundle, and the gradation of heights of the stereocilia, with the tallest row of stereocilia closest to the kinocilium and the shortest stereocilia at the opposite end of the ciliary array from the kinocilium (Figure 1). Bending the bundle towards the tall side causes excitatory depolarization of the intracellular resting potential voltage of the hair cell, while bending in the opposite direction causes inhibitory hyperpolarization. Response magnitude depends on the degree of bending, and measurable responses occur to displacements of nanometer proportions. Bending the bundle in directions other than along the major axis gives a smaller maximal response magnitude that shows a cosine relation to the direction of bending. This vectorial response property gives the hair cell a potential mechanism for initiating directional motor responses and for directional localization of underwater sound sources.

In each endorgan, the sensory epithelium (Figure 1B) contains an array of hair cells oriented in distinct patterns of directional polarity based on the morphological polarization of the ciliary bundles (reviewed by Platt and Popper, 1981; Platt, 1983; Coombs et al., 1988).

Anatomy of the Inner Ear and Accessory Structures

The classic description of the vertebrate labyrinth divides it into the *pars superior* and the *pars inferior*. In cartilaginous and bony fishes, the *pars superior* includes the three semicircular canals and one of the three otolith organs, the utriculus ("little pouch") or utricle with a roughly horizontal floor; the *pars inferior* includes two additional otolith organs, the sacculus ("little sac") or saccule, and the lagena ("flask") which lie nearly vertically (Figure 2). Many, but not all, fishes have an additional endorgan, the macula neglecta, often located near the utricle or posterior semicircular canal ampulla. Each of the endorgans is innervated by branches of cranial nerve VIII.

Semicircular Canals

The three semicircular canal ducts (anterior, posterior, and horizontal [or lateral]) extend outward from the utricle, and are filled with endolymph. The anterior and posterior canals share a common vertical section, the *crus commune* (Figure 2). A dome-like enlargement, the ampulla, lies at the base of each canal. Within each ampulla, the crista ampullaris is a high, narrow ridge lying transversely across the duct and covered with sensory hair cells. A gelatinous structure, the cupula, extends like a thick curtain from the surface of the crista to the roof of the ampulla, so the crista/cupula partition acts as a deformable diaphragm for detecting fluid movement in the canal.

Otolith Organs

The three otolith organs are fluid-filled pouches that each contain a dense mass of some crystalline forms of calcium minerals (Schultze, 1990). Nonteleost vertebrates usually have otolith organs containing a somewhat pasty collection of tiny crystals called otoconia or statoconia, bound together in a collagenous or other matrix, in which case it is not strictly an otolith, but is called an otoconial or statoconial mass (Carlström, 1963). In contrast, most ray-finned fishes, and particularly teleosts, have the crystalline matrix developed into a much more rigid, stone-like structure, often with well-defined elaborate grooves and protrusions that are species specific, and these rigid structures are correctly called otoliths.

The otolith organs have a sensory epithelium that forms a plate-like patch, the macula, on one wall of the pouch. The otolithic mass lies close to the sensory epithelium, and they are mechanically coupled together by a gelatinous sheet or plate called the otolithic membrane (see Platt and Popper, 1981). Viscoelastic properties of the otolith membrane, the endolymph, and the epithelial surface, including ciliary bundles, are important parameters affecting the response properties (Mayne, 1974).

The utricle generally has a region where the epithelium is thicker and cells larger. This region forms a prominent stripe, or "striola", around the dorsolateral

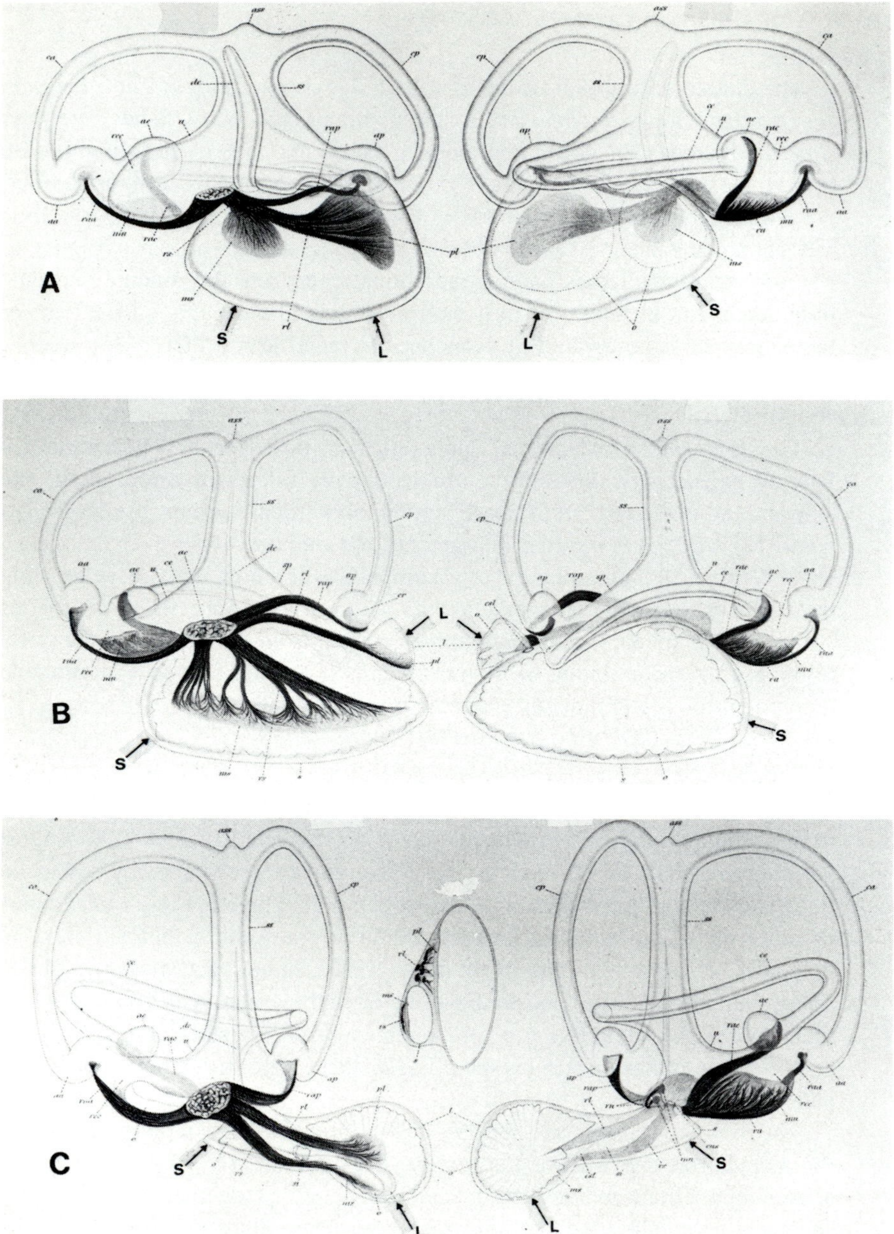

FIGURE 2. Illustrations from Retzius (1881) showing medial (on the left) and lateral (on the right) views of the ears from a few bony fish including (A) the bowfin, *Amia calva*, representing the halecomorphs, the sister group to teleosts; (B) a cod, *Gadus morhua*, a teleost that is not a hearing specialist; and (C) a carp, *Cyprinus idus*, a teleost that is a hearing specialist. Note the differences in relative sizes and in separation between the saccule (S) and lagena (L).

curve of the utricular macula. Other epithelia may sometimes show a striolar-like region as well. The cells in the striolar region often have ciliary bundles that are distinct from those in surrounding regions (see Platt and Popper, 1981; Platt, 1983).

A puzzling endorgan of the ear is the macula (or papilla) neglecta, which was first described by Retzius (1881). The size, structure, and position of the neglecta varies in different species (Figure 2). When present, the neglecta has one or two sensory patches that are overlain by a gelatinous cupula-like structure with otoconia or otolith. Although the neglecta can be huge in some elasmobranch species, where it is considered an auditory endorgan (Corwin, 1989), it is generally diminutive in bony fishes.

Ciliary Bundles

Scanning electron micrographic (SEM) examination of the sensory epithelia of inner ear endorgans reveals substantial variability in the forms of the ciliary bundles, both between different macular epithelia and even within a single endorgan epithelium. Different "classification" schemes have been proposed to describe the different-length ciliary bundles in fishes (Popper, 1977; Platt and Popper, 1981; Platt, 1983).

Different lengths of ciliary bundles are located on different regions of various epithelia. The longest cilia (up to 100 μm — Figure 3G) are found on sensory cells in the cristae of the semicircular canals, and these bundles extend well up into the cupula. Three of the most common forms in fish otolith organs have been termed F1, F2, and F3 (Popper, 1977). Type F1 ciliary bundles have kinocilia that are just a bit longer than the longest stereocilia (Figure 3E), with a maximum height usually about 5 to 6 μm. The F1 bundles often are found in the central regions of saccular epithelia in many fishes, as well as in the striolar region of the utricle and the lagena of many species. Type F2 bundles have long kinocilia (to over 10 μm) but short stereocilia (Figure 3B) and are most often seen at the margins of the sensory epithelia. Type F3 bundles (up to about 8 μm long) are similar to, but longer than, F1 bundles. The F3 bundles are often found just inside the margins of the epithelia, but they often make up the most caudal group of bundles in the saccular macula of otophysan fishes.

It is not yet clear what functions such differences in bundle length may have in fishes. In lizards and birds, regions having longer ciliary bundles detect lower-frequency signals while shorter bundles detect higher frequencies (e.g., Saunders and Dear, 1983). Indirect evidence in fish raises the possibility of a similar correlation. Recordings from afferent neurons and from hair cells from different saccular regions of the goldfish, *Carassius auratus* (Furukawa and Ishii, 1967; Sugihara and Furukawa, 1989), suggest that hair cells from the different regions of the saccular epithelium in otophysans respond to different frequencies. Since rostral ciliary bundles are significantly shorter than the caudal bundles in these species (Platt and Popper, 1984), shorter bundles might be correlated with detection of higher-frequency signals. However, considerably more data are needed to establish firm correlations between frequency and ciliary bundle length.

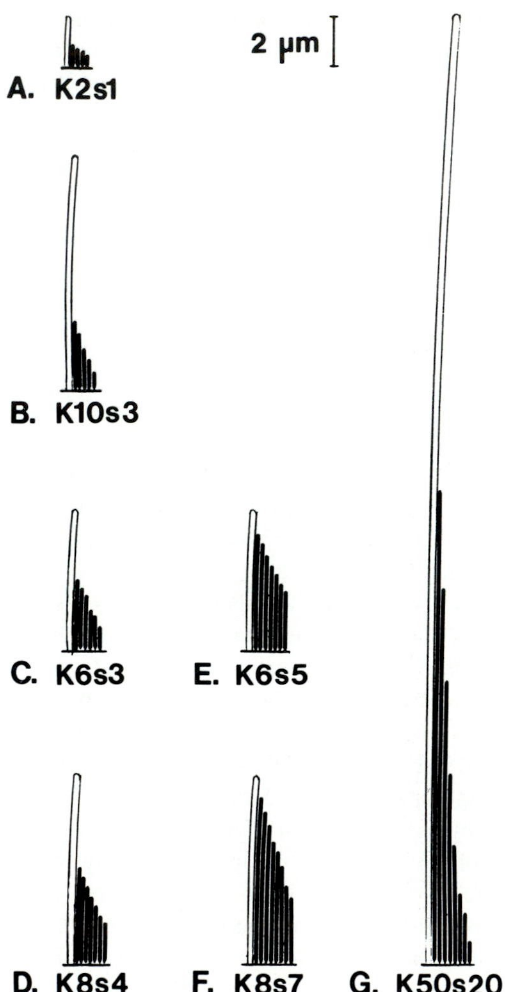

FIGURE 3. Schematic representation of the ciliary bundles from different inner ear endorgans of fish showing the wide range of lengths found in different regions. A to F are generally found on the otolith maculae, while G is only found on the cristae of the semicircular canals. (From Platt, C. and Popper, A. N., in *Hearing and Sound Communication in Fishes,* Tavolga, W. N. et al., Eds., Springer-Verlag, New York, 1981, 21. With permission.)

Hair Cell Orientation Patterns

Neighboring hair cells in local epithelial areas often share a common morphological orientation (Figure 1b), which has important implications for directional sensitivity (e.g., Flock, 1971; Lowenstein, 1971). Orientation maps are now known for tissues from a wide range of species, including agnathans, elasmobranchs, and bony fishes (see reviews in Lowenstein, 1971; Platt and Popper, 1981; Platt, 1983, 1988; Corwin, 1989; Schellart and Popper, 1992).

The semicircular canal organs have the most conservative pattern, which is not known to vary among all jawed vertebrates (gnathostomes). All of the ciliary bundles within a crista are oriented in the same direction, along the axis of the canal. The crista of the horizontal canal ampulla has all its hair cells oriented toward the utricle, while the cristae of both the anterior and posterior ampullae always have all their cells oriented away from the utricle (see Lowenstein, 1971; Platt, 1983).

In the otolith organs the hair cells usually can be divided into two groups (Figure 4). Often there is a gradual shift of orientation over a broad area, without an abrupt change in orientation; for example, cells oriented consistently "inward" from the periphery of a circular macula would have a gradual shift in orientation that changes from facing caudally at the front edge to facing medially at the lateral edge. All of these cells would still be considered one orientation group (Platt, 1977, 1983).

The utricle appears to have a fairly stable orientation pattern throughout the vertebrates, although it may be less tightly organized in elasmobranchs than in bony fishes. In bony fish (Figure 4A) and all tetrapods, the pattern has cells of opposite orientation divided into just two groups that are directed towards one another. Since the utricular macula usually lies near the horizontal plane, these orientations are in the rostral/caudal or medial/lateral directions.

The lagenar pattern in bony fishes has some similarities to that of the utricle. Usually two hair cell orientation groups are separated by a dividing line with the predominant cellular orientations in the dorsoventral direction (Figure 4B). Often over much of the macula, the cells are oriented inward from the margin, so at the dividing line the orientations are facing towards each other. However, it is not uncommon in many species to have a region where the opposing orientations are more "antiparallel" than "facing" opposition.

The saccular macula has more variety between species in orientation patterns than either the utricle or lagena (Popper and Coombs, 1982). Basically there appear to be either two or four hair cell orientation directions, rostral/caudal and dorsal/ventral, but the cells with those orientations sometimes form complex grouping patterns (Figure 4C).

The most common pattern in teleost saccules, the "standard" pattern (Figure 4C) (Popper and Coombs, 1982), has four orientations in two pairs of opposing groups of hair cells, dividing the macula into four unequal quadrants. In the rostral part of the macula, the dorsal half has orientations facing caudally, while the ventral half has cells facing rostrally. In the caudal part of the macula, the dorsal half has orientations facing dorsally, while the ventral half has cells facing ventrally. The standard pattern is found in virtually all teleost taxonomic orders so far examined, other than the otophysans (ostariophysans).

Other patterns are found among different taxa (Figure 4C). The "vertical" pattern has only two orientation groups, one directed dorsally and the other ventrally. Among teleosts, this pattern occurs in all representatives examined from the four orders in the Otophysi (superorder Ostariophysi) (Popper and

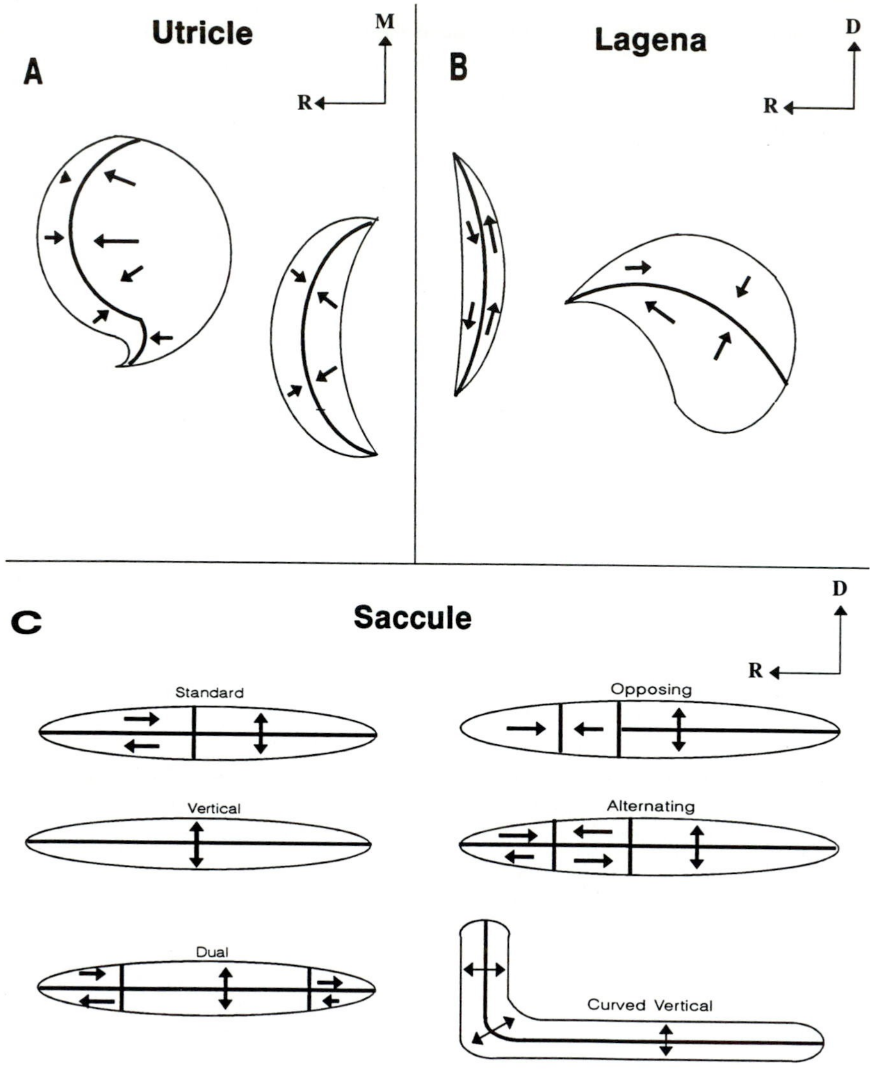

FIGURE 4. Schematic illustrations of the sensory hair cell orientation patterns of each of the otolithic endorgans in different species of fish. In each case, the arrows indicate the orientation of the bulk of the hair cells in each "orientation group", based on ciliary bundle structure (see text and Figure 1). (A) Two different utricular patterns. The pattern on the left is the most common pattern found among bony fish, while that on the right is found in a few species that may use their utricles for sound detection. (B) The lagenar epithelial pattern on the left is found in most teleosts, while that on the right is found in otophysans. (C) Six different saccular patterns have been identified (Popper and Coombs, 1982). See text for discussion of the patterns. D, Dorsal; M, medialateral; R, rostral.

Platt, 1983), as well as among the taxonomically distant mormyrids (order Osteoglossiformes) (Popper, 1981). A variant on the vertical pattern, called the "curved vertical", is found among all of the nonteleost bony fishes and in the cartilaginous fishes (e.g., Popper, 1978; Corwin, 1989). This pattern has only two hair cell orientation groups facing away from the midline, so the caudal part has dorsoventral orientations. As a result of a curvature in the epithelium, the cells at the rostral end of the epithelium are shifted to the rostral-caudal orientations (Popper, 1978; Popper and Northcutt, 1983; Popper, unpublished).

Hair Cell Differences

Ultrastructural investigations of the inner ear using transmission electron microscopy (TEM) indicated type I and type II hair cells in amniote vertebrates (Wersäll, 1961). Anamniotes were supposed to have only the type II hair cell, but recent evidence indicates that a cichlid fish, the oscar (*Astronotus ocellatus*), has, in addition to type II cells, another type of hair cell that is very similar to the amniote type I cell (and called type I-like by Chang et al., 1992). The presence of two physiologically different types of sensory hair cells in fish (Saidel et al., 1990a, 1990b; Yan et al., 1991; Chang et al., 1992) may parallel findings for the toadfish (*Opsanus tau*), where there are differences in the ion channels of different saccular hair cells (Steinacker and Romero, 1992). These data, along with the observations of ciliary bundle morphology, show differences among different epithelial regions of the saccule, lagena, and utricle, suggesting that different regions of the otolithic endorgans may have different functions. The possibility that the different cells respond to different modalities (vestibular or auditory information), or to different aspects of the same modality (such as different frequencies of an acoustic stimulus), means that, functionally, the otolithic endorgans of fish may be more complex than previously appreciated.

Auxiliary Structures and Auditory Specializations

A mechanism to increase sensitivity to underwater sound is to exploit a compressible chamber or bubble of gas, because a sound pressure wave in water produces volume changes of the gas that can be linked to displacement stimuli for the inner ear.

The teleost gas bladder (or swim bladder) is located in the abdominal cavity and serves a variety of functions in hydrodynamics, sound production, and hearing. Actual data on the contributions of the gas bladder to hearing are surprisingly scanty, with only a few studies demonstrating that removal of the gas bladder adversely affects hearing sensitivity (Poggendorf, 1952; Kleerekoper and Roggenkamp, 1959) and that "addition" of a bubble of air near the head enhances sensitivity of species that do not have a gas bladder (Chapman and Sand, 1974).

In the otophysans (formerly called ostariophysans) a series of bones, the Weberian ossicles, physically connect the rostral end of the gas bladder to the fluid system of the inner ear at the midline, between the saccules. Although never tested experimentally, it has been proposed that movement of the gas bladder walls is transmitted to the fluid system of the inner ear via the hinged Weberian ossicles (see Alexander, 1962; van Bergeijk, 1967). Fishes with Weberian ossicles clearly detect a wider range of frequencies (bandwidths) and have better sensitivity (lower thresholds) than fishes without the bones (reviewed in Fay, 1988; Schellart and Popper, 1992). Removal of the ossicles causes loss of sensitivity and bandwidth (Poggendorf, 1952), although this does not prove that the ossicles themselves move.

A second form of adaptation for improvement of coupling involves rostral projections of the gas bladder to carry motions directly to the ear. Specializations of this type occur in a wide range of species. In some squirrelfish (Holocentridae), the rostral end of the gas bladder terminates on the wall of the saccule, and these species hear considerably better than species without it (Coombs and Popper, 1979). Similarly, rostral projections of the gas bladder in clupeids end in a bubble of gas as part of the utricle, supporting other evidence that the utricle is a hearing endorgan in these species (reviewed in Blaxter et al., 1981).

The third major specialization to enhance hearing is the presence of a separate gas bubble in the head and intimate with the ear. Such bubbles are found in the mormyrids, a group with excellent hearing (McCormick and Popper, 1984), and in the anabantids (bubble-nest builders), another group that seems to hear well (Saidel and Popper, 1987).

Anatomy of the Lateral Line Organs

Two structurally distinct kinds of endorgans of the mechanosensory lateral line system are the free neuromasts and the canal neuromasts (see chapters in Coombs et al., 1989). The free neuromasts are small patches on the skin, often forming groups or lines called “stitches”. The canal neuromasts are similar patches, but located within the fluid-filled lateral line canals lying under the skin. Each neuromast contains from a few to a few hundred hair cells, usually with a gelatinous cupula enclosing the ciliary bundles and extending upward from the epithelium. Neuromasts consistently have hair cells oriented in two opposing polarities, with the opposing hair cells fully intermingled. Hair cells of the lateral line also show a variety of ciliary bundle sizes, but it is not yet clear whether there are categories or types like those of the inner ear (Coombs et al., 1988, 1992).

The structure of accessory tissues is important for the mechanosensory function of lateral line endorgans. For example, the shape and size of the cupula varies considerably, depending upon the location of the neuromast on the body

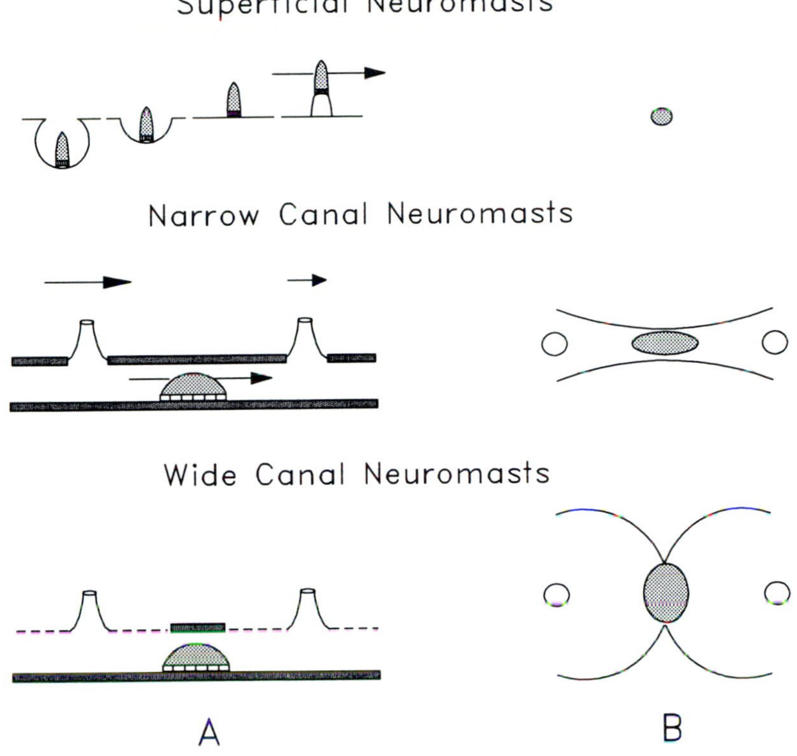

FIGURE 5. Schematic illustration of the variation in the lateral line system of various fishes. (A) Side views of the cupula over neuromast of various forms, with water flow shown by arrows; in the narrow canal the pressure flow inside the canal is induced from the water pressure difference above the two pores. The heavy lines indicate firm structures, the dashed lines indicate flexible soft tissue. (B) Top views looking directly down on the stippled cupulae for the various forms shown in (A), and the canal shapes with the pores indicated as open circles. (From Coombs, S. et al., in *The Evolutionary Biology of Hearing,* Webster, D. B. et al., Eds., Springer-Verlag, New York, 1992, 268. With permission.)

and depending upon species (Figure 5). The cupula of a canal neuromast may be dome-like, or in narrow canals it may be somewhat like a tall "keel" with its long axis parallel to the axis of the canal, or in wide canals it may lie transversely, similar to that in the semicircular canal of the inner ear (Coombs et al., 1988). The canals also have different degrees of opening to the outside water, but the most common situation is to have pores at regular intervals, with neuromasts lying between the pores.

Lateral line systems range from being fairly simple to those systems having extensive branching in different species (see Coombs et al., 1988). There often are three major canals on the head and a single major trunk canal running along the lateral side of the body. Sometimes the trunk canal is very short, sometimes it extends out into the tail fin itself. Moreover, it may be relatively straight, or with a pronounced curve, which Dijkgraaf (1963) noted as common where the pectoral fin was high on the side of the body, presumably affecting local water flow patterns. Diameters of the canals vary widely, and may be as large as 7 mm, and the covering over the canal may be fairly rigid, or fairly compliant. Sizes of the neuromast cover a wide range, even within a single species (Figure 6).

Central Projections of Octavolateralis Nerves

The sensory ganglion cells of the eighth nerve and the lateral line nerves project centrally to synapse on cells in octavolateralis nuclei in the medulla oblongata in cartilaginous and bony fishes (Figure 7). These octavolateralis nuclei form longitudinal columns extending roughly from the level of the cerebellar crest to near the level of the vagus nerve (Northcutt, 1981; McCormick, 1981, 1983, 1992). Early work considered the lateral line nerves to be branches of cranial nerves V, VII, and IX, but now we know that is not true. Instead, the lateral line nerves are cranial sensory nerves unique to aquatic vertebrates (see McCormick, 1983).

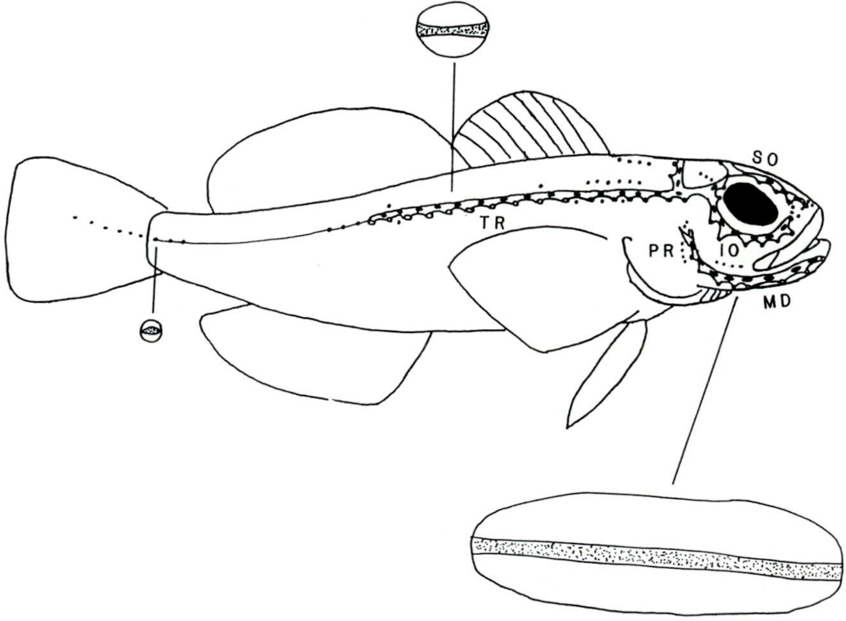

FIGURE 6. Schematic illustration of the distribution of canal and surface neuromasts on the body of the mottled sculpin. The enlarged drawings show the dorsal surface of neuromasts found on the mandible (bottom), and trunk (top) canal organs and a superficial neuromast (left). These illustrate the extent of size variation of neuromasts found at different locations on the body. The stippled areas represent the region of the epithelium containing sensory hair cells. MD, Mandibular canal; SO, supraorbital canal; IO, infraorbital canal; PR, preopercular canal; TR, trunk canal. (From Coombs, S. et al., *The Mechanosensory Lateral Line: Neurobiology and Evolution,* Springer-Verlag, New York, 1989, 301. With permission.)

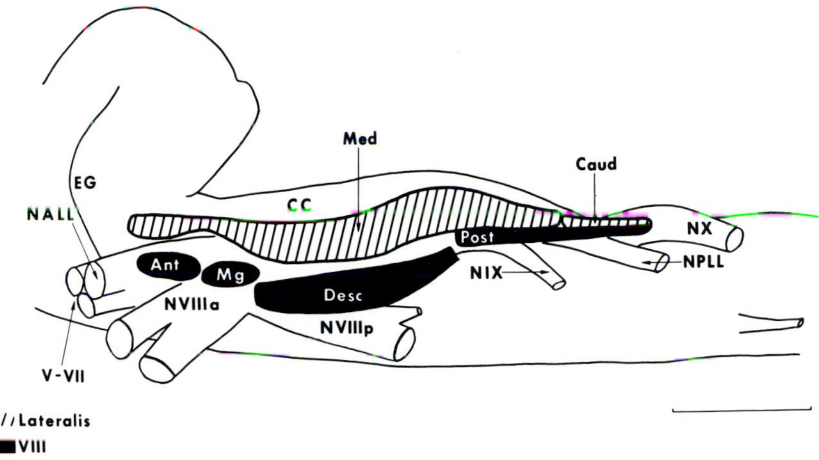

FIGURE 7. Schematic illustration of a lateral view of the brainstem of the bowfin, *Amia calva*, illustrating the positions of the six octavolateralis nuclei found in this species (rostral to the left, dorsal to the top). Each nucleus is projected onto the lateral surface of the brain. Ant, anterior octavus nucleus; CC, cerebellar crest; Caud, nucleus caudalis; Desc, descending octavus nucleus; EG, eminentia granularis; Med, nucleus medialis; Mg, nucleus magnocellularis; N ALL, anterior lateral line nerve; N IX, glossopharyngeal nerve; N PLL, posterior lateral line nerve; N VIIIa, anterior ramus of the eighth nerve; N VIIIp, posterior ramus of the eighth nerve; N X, vagus nerve; Post, posterior octavus nucleus; V-VII, trigeminal and facial nerves. Scale bar = 1 mm. (From McCormick, C. A., *J. Comp. Neurol.*, 197, 3, 1981. With permission of Wiley-Liss, copyright owner, a division of John Wiley & Sons.)

Vestibular and Auditory Pathways

The more ventral of the columns contains four or five main octaval nuclei which receive afferents from the semicircular canal cristae and from the otolith organs, representing vestibular and auditory input (Figure 7) (see McCormick, 1992).

It is supposed, but not yet clear, that there may be a functional division in the brainstem between vestibular and auditory input. Such a division would clarify a central question in fish inner ear physiology, that of how fishes separate auditory from gravitational/locomotor signals, given that they use the otolith organs for both functions. The inputs from the semicircular canals and the utricle terminate on more ventral parts of the descending nucleus, while the saccular and lagenar inputs terminate more dorsally in this nucleus (McCormick, 1992). The overlaps of anatomical input, and possible multifunctional properties of the endorgans themselves, make the question difficult to resolve.

From the medulla, octaval projections ascend to the cerebellum and to the midbrain. A pathway believed to contain auditory information carries input from the octaval column up to the torus semicircularis of the midbrain. There are further thalamic and forebrain projections that are still being worked out, and show some similarity to tetrapod auditory paths (see Striedter 1991; McCormick, 1992).

Lateral Line Mechanosensory Pathways

The mechanosensory lateral line nuclei are located above the octaval column and occupy roughly the same rostrocaudal extent (Figure 7) receiving mechanosensory input from branches of lateral line nerves (see reviews by Boord and Montgomery, 1989; McCormick, 1989). In bony fishes, the mechanosensory column is subdivided rostrocaudally into two nuclei, both of which receive lateral line inputs.

Precise topographic projections of the mechanosensory inputs have not been discovered. Both the anterior and posterior lateral line nerve trunks project to the nucleus medialis and n. caudalis, without obvious somatotopy, except that the input from the head projects to more ventromedial parts of the nuclei than the inputs from the body (McCormick, 1989). There also is no evidence as yet that free neuromasts and canal neuromasts have differential projections. It needs to be pointed out, however, that data are very limited and somatotopy cannot be excluded until studies are done with finer regional localization.

Efferent System

Both the ear and lateral line of fishes receive efferent projections from the octavolateralis efferent nucleus located in the medulla (reviewed in Highstein, 1991; Roberts and Meredith, 1992). The neurons projecting to the ear and lateral line are intermingled in this nucleus, and there is evidence that individual efferent neurons might project to both the lateral line and inner ear (Claas et al., 1981; Meredith and Roberts, 1987). It is not clear whether hair cells in all endorgans receive efferent input, and the role of the efferent system is still being clarified. The efferent system functions to modulate the response of afferent hair cells or afferent neurons in various endorgans (see reviews by Highstein, 1991; Roberts and Meredith, 1992). For example, stimulation of the efferent system will modulate the responses recorded from afferent neurons in goldfish (Furukawa, 1981) and *Opsanus* (Highstein and Baker, 1985; Boyle and Highstein, 1990) and there is a decrease in the microphonic responses from the saccule of *Carassius* when the efferent system is stimulated centrally (Piddington, 1971). There are also data from elasmobranchs and *Opsanus* showing that stimulation of the efferent system modifies the response of the trunk lateral line (Russell and Roberts, 1972; Tricas and Highstein, 1990, 1991).

FISH BEHAVIOR AND THE OCTAVOLATERALIS SYSTEM

Responses to Gravity and Motion

Fish move about in a fully three-dimensional world, and the vestibular system provides information for controlling postural attitude, for locomotion, and for stabilization of gaze.

Postural Behavior

For most fish in their normal postural attitude, the head and body are held with the rostrocaudal axis horizontal, and the dorsoventral axis vertical relative to gravity, with a few remarkable exceptions which will be discussed below. Early work using surgical manipulations on elasmobranchs and teleosts (reviewed in Lowenstein, 1936) developed the concept that certain parts of the ear were more necessary than others for normal postural and locomotor control.

Evaluating vestibular inputs to behavior was strikingly facilitated by the discovery of the "dorsal light reflex" by von Holst (1935). Many fish, when subject to strong light from one side, adopt a tilted posture with the back turned toward the light. This tilted "equilibrium" posture is produced when the effect of the light to tilt the fish is exactly balanced by the effect of gravity to restore the fish to an upright position. Measurements can be made of tilting behavior following experimental manipulations on the ear, thus giving quantitative data on the relative importance of individual otolith organs.

Using surgical extirpations of individual otolithic organs, and quantitative tests in a large centrifuge to increase *g* forces, the "adequate stimulus" for gravitational response was found to be the shearing force produced across the sensory epithelium when it is tilted (von Holst, 1950). The utricle was identified as the major otolith organ for postural control in several freshwater species, including minnows (Cyprinidae), tetras (Characidae), and angelfish (Cichlidae) (von Holst, 1935, 1950). The lagena also contributes to the gravitational response in some species (Schoen and von Holst, 1950). As earlier workers had found, the saccule apparently contributes little, if any, input to postural behavior in the species tested. (A comprehensive account in English of these German papers is given in Pfeiffer, 1964.)

Locomotion

The utricle and the semicircular canals all contribute to locomotor control. Swimming shows serious disorientation after bilateral removal of the canals and utricle, and asymmetrical effects are produced by unilateral labyrinthine lesions. Central compensation for such loss can be quite rapid, as in goldfish, where substantial recovery after unilateral partial labyrinthectomy sometimes occurs within hours (Ott and Platt, 1988a, 1988b). The potential problems of interpreting canal and otolith function with regard to static and dynamic functions are discussed critically by Lowenstein (1936). His major caution remains true, that the most reliable manipulation is probably severance of the appropriate nerve branch, since a variety of residual functions may remain after attempts at otolith removal or canal blockage.

Visual Stabilization

A major mechanism for visual stabilization is the vestibulo-ocular reflex (VOR), which actually involves a whole suite of oculomotor responses. When the head rotates from side to side, for example, the horizontal (lateral) semicircular canal organs are stimulated. This sensory stimulation initiates compensatory eye movements so that both eyes show conjugate motions opposite to the head motion, keeping the direction of gaze constant. Similar reflexes drive head movements around the roll and tilt axes, involving inputs from the vertical canals. It is important to note that all three canals are stimulated by most head movements, and eye movements rarely involve only one pair of muscles. There are similar static responses of maintained eye deflection during maintained tilt, attributed to otolithic input rather than canals (see Lowenstein, 1936; von Holst, 1950). VOR studies have been done in lampreys (Rovainen, 1976), sharks, and several teleosts (see reviews by Lowenstein, 1936, 1971; Platt, 1983). In elasmobranchs the lateralization and crossing of vestibulo-ocular pathways appear different from the pathways in other vertebrates (Graf and Brunken, 1984).

Acoustic Communication: Sound Production and Behavior

Acoustic communication has been well documented for a number of fish species (reviewed in Tavolga, 1971; Demski et al., 1973; Fine et al., 1977; Myrberg, 1981). As pointed out by Tavolga (1971), sound is a particularly useful channel for underwater communication since acoustic signals are not affected by murkiness or darkness of the environment, sound travels rapidly over long distances, is highly directional, and is not particularly affected by rocks or coral reefs. While sound is important for communication behavior in a variety of species, sounds also are produced as a side effect of other behavior such as feeding or locomotion (see Tavolga, 1971).

Sound Production Mechanisms

Sounds are produced intentionally in a large number of species, although the emission mechanisms and the behavioral context of the sounds varies interspecifically. "Stridulatory" sounds are produced by moving or grinding of body parts against one another such as using pharyngeal teeth or other hard body parts. Other species produce sounds by directly or indirectly involving the swim bladder. Removal of the swim bladder results in a striking loss of acoustic energy from the sound (Schneider, 1967), although the sound spectrum does not change noticeably. In what is known as the "extrinsic" swimbladder system, sounds are produced by moving body parts or muscles that contact the swim bladder, which then amplifies the sound. For example, in some marine catfish, a muscle from the skull activates a spring-like mechanism that hits the swim bladder (Tavolga, 1962). Muscles involved with sound production may also be "intrinsic" to the swim bladder. In this case, as found in the toadfish *Opsanus*, the muscle has its origin and insertion on the swim bladder, and the contraction rate of the muscle is the fundamental frequency of the sound.

Examples of Acoustic Communication

Fish sounds are quite variant, depending upon the species. Individual species may have more than one type of sound (Tavolga, 1971; Fine et al., 1977). Some sounds consist of short bursts of broad-band noise (especially those produced using stridulatory mechanisms), while others are tonal and contain a fundamental frequency and multiple harmonics. One of the best-known examples of fish sounds is found in toadfish, *Opsanus tau*, where males and females produce a chirping pulsed sound using an intrinsic swimbladder system (e.g., Winn, 1967). During reproductive season, the male also produces a very intense and long-duration "boatwhistle" sound that can be heard for great distances, and females approach the sound produced by courting males (Winn, 1967).

Perhaps the most elegant study of the use of sound in behavior was done by Myrberg and colleagues on damselfish (family Pomacentridae) (Spanier, 1979; Myrberg and Spires, 1980). Several species live conspecifically on reefs and use sounds to defend territories, including nests. Each species has a sound that differs in frequency and in pulse repetition rate. Behavioral studies showed that species can discriminate between sounds of conspecifics and heterospecifics based on differences in pulse rate.

Lateral Line in Behavior

It now seems clear that the lateral line is primarily used in hydrodynamic interactions at very short distances, on the order of the body length of the receiver (Kalmijn, 1989). For a predator, the lateral line organs may be important for accurate localization of moving prey and to supplement more distant chemoreception and as an aid in darkness or murky water where vision may not be useful (Coombs et al., 1992). For a prey animal, the system may provide warning of imminent capture, allowing a quick movement, just at the moment of the strike, that may be sufficient to move the target safely away from the predator (Blaxter et al., 1981).

All the lateral line organs studied so far in fishes appear to function over the range of about 10 to 200 Hz (Coombs et al., 1992). This range is below the best sensitivity for the acoustic organs of the inner ear for most fish yet tested, though it overlaps with the low end of the hearing range for many fishes (see Figure 8 below). At this low end, the hydrodynamic forces detected by the lateral line overlap with the forces that can accelerate the whole body and be detected by the inner ear. The systems are in a sense complementary, because the lateral line detects hydrodynamic sources that cause local disturbances over a part of the body surface, while the inner ear, which does not compare effects of the stimulus on different parts of the body, may detect sources at much greater distances from the animal (Sand, 1984; Kalmijn, 1989; Coombs et al., 1992). By analogy to "hearing", the term "svenning" has been proposed as the behavioral use of the lateral line for this sense of "distant touch" (Dijkgraaf, 1963), to honor Professor Sven Dijkgraaf (Platt et al., 1989).

Obstacle Detection

If the water is moving, a flow field will occur around a stationary object. Conversely, when a fish is swimming, it generates a flow field that can be reflected back by nearby objects, stationary or moving. Some data on discrimination of the grid size of an array of bars by blind cave fish (*Anoptichthys jordani*) indicate that the swimming fish may be coding some phase information by the lateral line sensors (see Hassan, 1989).

Feeding

Studies on feeding have shown that both predators and prey may use lateral line information. Sculpin (*Cottus bairdi*) respond to a vibrating ball stimulus as though it were food, and will turn to snap at the ball when it stimulates the lateral line (Coombs and Janssen, 1990). Bluegill sunfish (*Lepomis macrochirus*) that normally attack food fish or an artificial prey in the dark are unable to when the lateral line function is blocked (Enger et al., 1989). Lateral line capabilities have been implicated in other fish that feed on zooplankton (Montgomery, 1989). Some surface-feeding fishes use the lateral line to detect surface waves, and can apparently make some judgement of both direction and distance of a local stimulus (Bleckmann et al., 1989).

The lateral line may also give information about predatory strikes to potential prey. Larval fish have been filmed making avoidance movements to a mechanical probe in the dark several milliseconds before they are actually touched (Blaxter and Fuiman, 1989).

Schooling

Lateral line information has been implicated in maintaining the structure of fish schools (Partridge, 1981). Maintaining position and velocity relative to nearest neighbors clearly has a visual component, but a series of experiments on blinded animals that also had selective cutting of lateral line nerves, established the use of the posterior (trunk) lateral line in maintaining school structure. It remains unclear how these results can be generalized to schooling fish lacking a trunk lateral line, such as herring and other clupeids (Sand, 1984).

FUNCTION OF THE OCTAVOLATERALIS SYSTEM

Locomotion and Posture

As discussed above, a main function of the vestibular system is regulation of body orientation, including both postural control when relatively still, and dynamic positioning and movement of the body during locomotion. Aside from controlling the upright position seen in most fishes, there are a few cases of unusual posture that have been studied.

Flatfish (order Pleuronectiformes — the flounders and allies) all metamorphose into adults that lie and swim rolled on one side. One eye migrates across the top of the head, leaving a "blind side" to lie on the bottom, while the "eyed" side is uppermost. The vestibular system does not change in orientation relative to the fish, so it functions rotated 90° relative to that in other fish. Behavioral studies suggest that the saccule has replaced the utricle as the major otolith controlling posture (Schöne, 1964; Platt, 1973). Anatomical tracing of fibers shows the central circuits for the VOR are modified in flatfish compared to other vertebrates (Graf and Baker, 1985). So adult flatfish apparently depend chiefly on central reorganization of some neural pathways rather than peripheral reorganization for postural control.

A second case is that of the "upside-down" catfish, a group that includes the small *Synodontis nigriventris* of Africa. This fish often swims with the dorsal side down in open water, and feeds on the underside of floating vegetation. When it's near the bottom, or near a vertical wall, this fish shows a "ventral substrate response" turning its belly toward the substrate while it is close to it (Meyer et al., 1976). This species shows no unusual peripheral structure in the inner ear, and the unusual behavior also is inferred to be a central phenomenon (Meyer et al., 1976).

Some fishes show an unusual posture in the pitch axis instead of the roll axis. Some species of tetra (Family Characidae) are known as "head-standers" or "tail-standers" because they have a usual posture with the body and head tilted upward or downward as much as 30°. While many of these show no vestibular specializations, one species of tail-stander, the penguin fish (*Thayeria boehlkei*), has the utricular otolithic organ modified so that the sensory macula is most level not when the fish is level, but when the head is tilted upward roughly 25° (Braemer and Braemer, 1958).

Important physiological studies on vestibular function exploited relatively robust preparations from elasmobranchs to reveal the presence of a resting discharge in the afferent nerves, with increases or decreases to acceleration forces in specific directions, for semicircular canals and otolith organs (see reviews by Lowenstein, 1971; Platt, 1983).

It is now known that the cupula and crista of each semicircular canal organ do not act like the "swinging door" model that was influential for decades, but more like a diaphragm that deforms like a drumhead (McLaren and Hillman, 1979). The differential equations that utilize terms for fluid inertia, viscosity, and compliance to describe quantitatively the acceleration, velocity, and displacement terms of semicircular canal fluid dynamics are not fundamentally changed by this model, and the mathematical approach is very useful for examining the relation of structure to rotational sensitivity (see Mayne, 1974).

Teleost semicircular canal work has been extensive on toadfish, *Opsanus tau*, by Highstein and co-workers. They found three types of afferent activity based on phase and gain relations to sinusoidal stimuli. Two types were primarily responsive to angular velocity of the head, and one type primarily to acceleration. Of the velocity-sensitive units, one group showed a resting discharge that was regular and had a low gain, in terms of added spikes per degree per second of stimulus; the other group was irregularly firing, and showed a high gain (Boyle and Highstein, 1990). Stimulation of the efferent system caused an increased discharge rate and decrease in sensitivity of the afferent units (see review by Highstein, 1991).

Very few studies have been done on primary afferent fibers to examine responses to tilt from otolith organs, in contrast to the extensive studies on auditory fibers from these organs, particularly in goldfish (see review by Popper and Fay, 1993). Again, the definitive work remains that by Lowenstein and co-workers on the isolated head preparation of the ray (*Raja clavata*), showing specialized units for position, for "change in position", and for vibration sensitivity, even within a single endorgan (see Lowenstein, 1971). Some more recent work on elasmobranchs also has noted the multimodal responses to tilt and vibration (Budelli and Macadar, 1979; Plassmann, 1983). Flatfish are apparently the only teleost in which primary afferent responses to tilts have been reported (Platt, 1973), and although their posture is a special case, the peripheral responses showed properties basically similar to those in elasmobranchs.

Hearing

It is apparent from ethological studies that many fish species are capable of detecting sounds, although most such studies tell little about the limits of hearing and discrimination capabilities. Instead, a variety of data have been obtained using various behavioral paradigms (see Fay, 1988 for a review) to essentially "ask" fish what they can hear.

How Do Fish Hear?

Sound in water has some particularly important characteristics of the pressure and particle displacement components that are not prominent in airborne sound (see Kalmijn 1988, 1989; Rogers and Cox, 1988 for excellent reviews of underwater acoustics). In air, which is quite compressible, the particle displacement components of sound are relatively inconsequential within a very short distance from the source since they attenuate rapidly over distance. Water is far less compressible than air. As a result, sound attenuation is not as great as in air, and the displacement component of the sound field remains considerable for a distance extending to several wavelengths from the source. The decline in magnitude depends on the sound frequency, but tails off asymptotically to eventually approach zero. Since the speed of sound in water is about 1500 m/s, the wavelength of a 100-Hz signal is on the order of 15 m. Thus, at low frequencies, particle displacement may still be detectable by a fish many meters from a sound source.

Much of the literature often talks of the acoustic nearfield and acoustic farfield (see van Bergeijk, 1967 for one of the classic discussions). In the nearfield, particle motion (which is vectorial and thus has a directional component) includes a hydrodynamic flow of the water as a consequence of the motion of the source. The farfield particle motion is only that molecular motion accompanying the pressure component of the signal. The measure often used for the transition between near- and farfield is wavelength/2π, which is approximately 1/6 of a wavelength of the sound source, but its precise distance depends upon whether the source is a monopole, dipole, or more complex source (van Bergeijk, 1967; Kalmijn, 1989). It must be emphasized that near- and farfield components do not suddenly change at this point, and that some fishes may be sensitive to hydrodynamic motions well past this point, while others may be insensitive well inside that point. In essence, the old concept of differences in detection capabilities for near- and farfields, as described by van Bergeijk (1967), is irrelevant when we consider what fish can detect at different distances from a sound source.

The significance of the presence of pressure and particle motion for a fish is that many species can detect both forms of energy, and both may be important for fish behavior. In fact, it has been argued that fishes may have two different sound detection "pathways", the "direct" and "indirect" (Fay and Popper, 1975; Popper and Fay, 1993). The "direct" pathway is for detection of particle motion through direct stimulation of the inner ear. Keeping in mind that the body of a fish is essentially acoustically transparent (since the density is about the same as that of water), the body moves in a sound field along with the rest of the water mass. However, the otoliths are far denser than the rest of the body and, as a consequence, move out of phase with the body and the sensory epithelium of the ear. This differential motion results in a shearing action on the cilia of the sensory hair cells, and stimulation of the ear.

The direct pathway is thought to be operative in most species, and can affect any of the otolithic endorgans. However, fishes that only use the direct pathway are thought to detect sounds only to a few hundred Hz, and with fairly poor sensitivity (Fay and Popper, 1975; Schellart and Popper, 1992). Examples of species that must use only the indirect pathway are fishes without a swim bladder (e.g., *Euthynnus*, Figure 8).

The "indirect" pathway involves detection of the pressure component of the acoustic signal. Pressure does not stimulate the inner ear directly, but it does set gas bubbles, such as the swim bladder, into motion. The gas bubble then becomes a signal source and re-radiates the energy induced in it, generating its own near- and farfields. If the bubble is sufficiently close, or otherwise acoustically coupled to the ear, there is little attenuation of the particle motion, which then can cause otolith motion relative to the sensory epithelium. Without such coupling, the energy from the gas bubble attenuates sufficiently rapidly so that it does not affect the inner ear.

Fishes which have specializations for acoustically coupling a gas bubble to the inner ear are called "hearing specialists" and all appear to have better bandwidth and sensitivity than fishes without such coupling, or "nonspecialists". Specialists include the otophysans (e.g., *Carassius* in Figure 8), in which the swim bladder

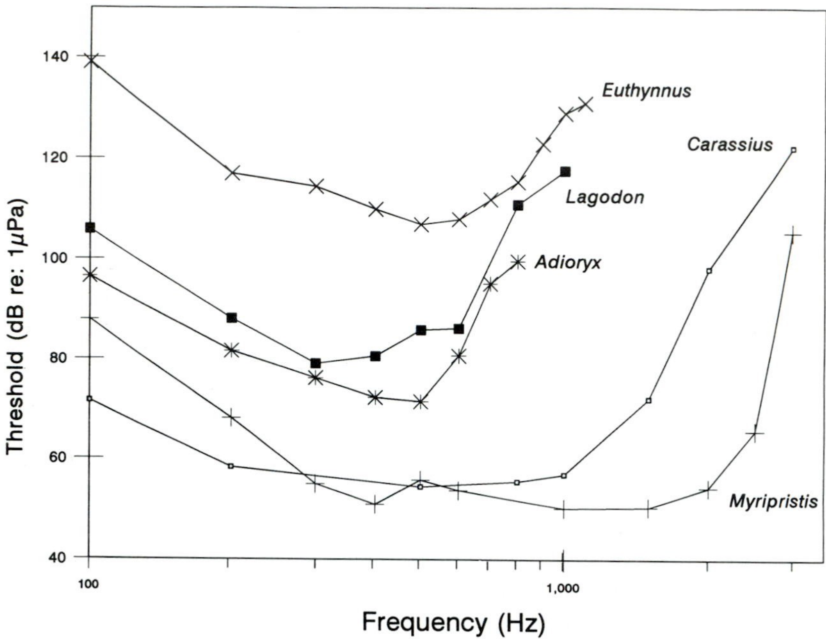

FIGURE 8. Hearing thresholds for representative teleost species. Best hearing is for *Carassius auratus* (goldfish — Jacobs and Tavolga, 1967) and *Myripristis kuntee* (squirrelfish — Coombs and Popper, 1979), two hearing specialists. *Adioryx xantherythrus* (another squirrelfish — Coombs and Popper, 1979) and *Lagodon rhomboides* (pinfish — Tavolga, 1974), are hearing nonspecialists having a swim bladder. *Euthynnus affinis* (tuna — Iversen, 1969) has no swim bladder.

is coupled acoustically to the inner ear via the Weberian ossicles (Alexander, 1962; van Bergeijk, 1967). This motion is carried at least to the saccular otolith which moves relative to the sensory epithelium. (There is some question as to whether the lagenar otolith in otophysans also is stimulated "indirectly" or whether it serves as a direct detector of particle motion.) In addition to the otophysans, species using special acoustic coupling include the squirrelfish, *Myripristis kuntee*, a species that has an anterior projection of the swim bladder intimately tied to the saccular wall and that hears better than *Carassius* (Figure 8). Other "specialist" species, such as the anabantids and mormyrids, have gas bubbles other than the swim bladder near the ear (reviewed in Popper and Coombs, 1982; Schellart and Popper, 1992).

Hearing nonspecialists include such fish as the squirrelfish, *Adioryx xanthyrythrus,* and the pinfish, *Lagodon rhomboides*, in addition to fishes that do not have a swim bladder (Figure 8). The rostral end of the swim bladder in both species is some distance from the inner ear, and neither species hears as well as the specialists. Interestingly, *Adioryx* is closely related to *Myripristis* and this is one of the classic examples where the same taxonomic group has both hearing specialists and nonspecialists (Coombs and Popper, 1979).

Pure Tone Thresholds

Threshold data have been summarized by Fay (1988). Figure 8 presents a sample of these data for a number of different fishes selected to illustrate the variability found in overall hearing capabilities. There is little data below 50 Hz because that is generally the lowest frequency that can be produced easily in behavioral experiments, though not necessarily the lower limit of the hearing capability of fishes.

The species most widely studied with regard to hearing is the goldfish, *Carassius auratus*, which can detect tones from below 50 to about 3000 Hz (Figure 8); best hearing is from about 200 to 1000 Hz. Similar data are found for other otophysans, although data are limited to just a few species (see Fay, 1988). A similar bandwidth and sensitivity is found for a number of other hearing specialists, including *Myripristis*. In contrast, nonspecialists generally have poorer hearing capabilities than specialists. For example, *Adioryx* and *Lagodon* can only detect sounds to about 800 Hz, with best sensitivity from 200 to 500 Hz.

A number of species without swim bladders have been studied behaviorally (reviewed in Fay, 1988). These species, such as the tuna, *Euthynnus affinus* (Figure 8), have poor sensitivity. However, it should be noted that the data for *Euthynnus* is presented in the figure in terms of pressure, although the data should be in terms of particle displacement since such species are not likely to be able to detect pressure.

The ability to detect pure tones tells something about the absolute sensitivity and bandwidth of an auditory system, but such signals are not much like natural signals. Moreover, under natural conditions the role of the auditory system is to detect signals in the presence of other signals (e.g., background noise) and to discriminate between signals, as is done by the pomacentrids (Myrberg and Spires, 1980). There have been extensive studies on behavioral capabilities of *Carassius* to detect more complex signals; these have been reviewed by Fay (1988).

Masking and Discrimination

Behavioral studies are used to measure detection of signal in noise (often called masking experiments) and discrimination between signals that differ in frequency and/or intensity. However, while data on pure tone sensitivity are available for over 50 species, data on detection of signals in noise and discrimination are available primarily for *Carassius*, while far fewer data are available for a limited number of other species. Those data that are available (reviewed in Fay, 1988) suggest that nonspecialists may not be able to discriminate as well as specialists.

Sound Source Localization

One of the most important roles for the auditory system is to detect the presence and location of a predator or prey. Determination of sound source location, or sound localization, is highly refined in virtually all terrestrial vertebrates. Early studies of localization by fishes were ambiguous as to whether fish could localize sounds (reviewed in van Bergeijk, 1967; Schuijf, 1975). Van Bergeijk (1967) suggested, in fact, that fishes should not be able to localize sounds in the same way as terrestrial vertebrates, where two ears are used to provide interaural differences in information about the sound location. He argued that since fish have a single sound receptor, the swim bladder, the information received by the two ears must be identical, thereby not allowing for the interaural cues needed to perform localization.

Despite the caveat presented by van Bergeijk, recent studies have demonstrated that sound source localization probably occurs using a mechanism that is completely different from that in terrestrial vertebrates. Behavioral studies, albeit on a very limited number of species, show that the cod, *Gadus morhua*, can discriminate between two sounds that are separated by 10 to 20 degrees in azimuth (Schuijf, 1975; Hawkins and Sand, 1977) or in the median vertical plane (Hawkins and Sand, 1977). It also appears that *Gadus* may also be able to discriminate between two sounds that are at different distances (Schuijf and Hawkins, 1983). Very limited data for an otophysan (goldfish) suggest that these hearing specialists can determine sound source position (Schuijf et al., 1977), but these experiments need replication and extension before a definitive statement about localization in these species can be made.

It is not yet clear how localization could be performed without the availability of binaural information, as argued by van Bergeijk (1967). Some answers may well involve the functional morphology of the auditory periphery. The peripheral organization of the hair cells into "orientation groups" reflects functional polarity to vectorial stimuli. The stimulus of particle motion should produce different movements of the otolith relative to the epithelia for different sound source directions. It has been proposed (e.g., Schuijf, 1975, 1981; Buwalda, 1981; Fay, 1984; Rogers et al., 1988; Popper et al., 1988; Schellart and Popper, 1992) that this differential motion in the various endorgans provides unique information to the central nervous system (CNS) for stimuli from each different direction, with the exception of exactly opposite stimuli. Schuijf (1975; also Buwalda, 1981) dealt with this 180° ambiguity issue by demonstrating that specialist fish not only use the particle motion to determine direction, but that they also compare the phase information of the particle motion with the phase of the motion obtained from pressure. Of course, nonspecialists do not detect pressure and it is not yet clear whether they can resolve the ambiguity, or whether they need other cues to fully resolve direction. One additional problem with the Schuijf model is that it primarily deals with detection of pure tone sounds that contain clear phase information, but most signals of consequence to fish are broadband noises without such phase information. This problem has, however, been dealt with in another model that still requires direct and indirect information, but which eliminates the need for phase comparison *per se* (Popper et al., 1988; Rogers et al., 1988).

Why Do Fish Hear?

An enigma of fish audition is that while many species hear very well, the very same species often are not known to produce sounds or to use sound to communicate. Clearly, the best-known example of this is the goldfish, which detects sounds to over 2000 Hz with excellent sensitivity (Figure 8). While we know little about the natural history of goldfish, there is no evidence that this species produces sounds. When the initial data on hearing in a mormyrid (*Gnathonemus*) showed that this species also hears well (McCormick and Popper, 1984), the reason for the hearing capability was not understood, but we now know that mormyrids use sound for communication (Crawford, 1991). Still, despite years of trying we have never detected sounds from goldfish and a number of other species that can hear, so the question that must still be asked is: what does the goldfish listen to?

While answers to this question have not been obtained directly, Rogers and colleagues (Rogers, 1986; Lewis and Rogers, 1992) have argued that goldfish, and other species, are listening to their "environment". Along with hearing turbulence, rain, and possible predators, Rogers argues that fish learn a good deal about their world through detection of ambient noises and that this can tell a fish (as it does a blind human) about the presence of neighbors, the extent of open space, etc. (also see Popper and Fay, 1993). It has also been suggested (Myrberg, 1981) that fishes are likely to "intercept" communication signals between other species, thereby enabling fishes that do not make sounds themselves to detect potential predators or prey acoustically.

What Do Fish Sven?

The mechanosensory lateral line functions remained puzzling for many decades. Recent biophysical techniques have elucidated the cellular responses of lateral line hair endorgans. Precisely controlled micromechanical stimuli and recent computer-controlled devices have permitted a highly quantitative analysis of the hydrodynamic stimuli that most effectively stimulate the hair cells in neuromasts of different kinds (Denton and Gray, 1989; Kalmijn, 1989; Van Netten and Kroese, 1989; Coombs and Janssen, 1990). These studies suggest "svenning" primarily involves detection of differential particle acceleration and/or velocity around the surface of the fish's body.

Acceleration, Velocity, and Displacement Sensitivity

The hair cells themselves respond to displacement of the bundle, but the structure of cupulae on neuromasts suggests that cupular deflection results from a frictional drag term, which depends on water velocity (Van Netten and Kroese, 1989). On the body surface, a free neuromast can detect the velocity component of the stimulus. However, in a rigid canal of diameter less than a few millimeters and over a range of frequencies, the velocity inside the canal is proportional to the acceleration outside, so a canal neuromast can be a detector of the acceleration component of an outside stimulus (see Denton and Gray, 1988, 1989). Canal neuromasts in general seem to respond best to the acceleration component of an oscillating stimulus probe near the fish, while free neuromasts generally respond best to the velocity component, in agreement with theoretical models dealing with canal morphology (Denton and Gray, 1988, 1989; see Coombs et al., 1992). The variety of canal sizes, rigidity, and pore dimensions suggest quite complex transformations may be available to some species.

Frequency Bandwidth

The frequency sensitivity for lateral line endorgans, determined by recording from afferent fibers, seems to be limited to values below about 200 Hz. Again, there is a distinction between the superficial neuromasts, generally having sensitivity at the low end of the range, from 10 to 60 Hz, while canal neuromasts usually respond best between 50 to 200 Hz (Münz, 1989).

COMPARATIVE STUDIES

The Auditory System — Comparative Structure

Retzius (1881) very much appreciated that there was broad diversity in the structure of the ear of fishes (see Figure 2). Clearly, the semicircular canal duct shapes show wide variation, as do the sizes and shapes of the endorgans, and particularly the saccular otolith. More recent investigators have noted additional diversity in other peripheral structures, particularly with the way the ear may be coupled to a pressure-detecting device (see Popper and Coombs, 1982; Popper, 1983; Schellart and Popper, 1992, for discussions of diversity).

As a result of our more recent analyses, particularly at the level of SEM and TEM, it has become apparent that the greatest diversity in the ear sensory epithelia is associated with those endorgans presumed to be involved with hearing. There are striking differences in the hair cell orientation patterns of the saccule among different species (Figure 5), and the most unusual patterns are consistently found in the hearing specialist species that have a coupling between a gas bubble and the inner ear. More specifically, the "standard" pattern (Figure 5) is generally found in species that do not have such coupling, while the other patterns (with the exception of the curved vertical) are found in species with such coupling. In fact, even within a single family, the Holocentridae, a hearing specialist (*Myripristis*) has an "opposing" pattern while the nonspecialist (*Adioryx*) has the "standard" pattern. The presence of a specialized saccular pattern is true for virtually all hearing "specialist" species studied to date (reviewed in Popper and Coombs, 1982; Popper, 1983; Schellart and Popper, 1992).

The presence of specializations in otic endorgans associated with audition extends beyond the saccule. In two cases where the utricle is known to be involved with audition, the clupeids (herring-like fishes) and a marine catfish (*Arius felis*), this endorgan has a sensory epithelium and hair cell orientation unlike those found in any other vertebrate utricle (see Blaxter et al., 1981; Popper and Tavolga, 1981; Platt, 1983).

A very important point to be made with regard to ear diversity is that similar saccular patterns are found in widely diverse taxonomic groups, and we suspect that these are multiple independent "inventions" of the same pattern. Perhaps the best-documented case is the standard pattern. This pattern is likely to have evolved from the curved vertical pattern encountered in nonteleosts (Popper and Northcutt, 1983) by the "addition" of a discrete group of cells at the rostral end of the epithelium. Teleost species ancestral to the otophysans apparently had the "standard" four-quadrant pattern (Popper, 1978; Popper and Platt, 1983) and the vertical pattern found in the otophysans is apparently derived by loss of the horizontally oriented cells. Similarly, the vertical pattern found in the mormyrids is likely derived from a four-quadrant pattern found in virtually all other osteoglossiform fishes (Popper, 1981). Other variously specialized patterns are found in taxonomically widely diverse fishes (e.g., Popper and Coombs, 1982; Schellart and Popper, 1992).

Another example of diversity is the striking interspecific differences encountered in the otoliths, and particularly the saccular otolith. The structure of the solid teleost otolith is species-specific, and saccular otoliths sometimes have elaborate shapes (see Popper, 1983). It has been proposed that the motion of the otolith may be a consequence of the shape, mass, and center of gravity of the stone; if so, we predict that the elaborated structures might affect the motions of the otolith relative to the sensory epithelium (reviewed in Schellart and Popper, 1992). However, almost nothing is known about such motions other than the results of a single study by Sand and Michelsen (1978), but it seems likely that the diversity in structure may result in variation in inner ear response, particularly to acoustic signals.

The presence of this structural diversity leads to a major interesting question about its functional significance in terms of hearing. One possible explanation is that the diversity reflects different mechanisms for the detection or processing of signals by different species. Along these same lines, the presence of similar structures in different taxa could mean that there has been parallel evolution among various species that results in evolution of similar structures. A second possible explanation for the diversity of structure is that different species have evolved different mechanisms for performing the same type(s) of analysis and/or detection. Whatever the ultimate explanation for the diversity found within auditory systems of fishes, it is clear that we have far too little data at this time to make reasonable suggestions as to its consequences in the ear and peripheral structures.

While very little is known about the CNS of fishes, it is worth noting that there seem to be some interspecific differences in the central projections of the ear in specialists as compared to nonspecialists (reviewed in McCormick, 1992). However, there are still far too few data upon which to make a reasonable assessment of this variation, and we also lack data on detailed projections from the inner ear endorgans and their regions to determine whether there are any patterns in projections that have any consequence with diversity in inner ear mechanisms.

Diversity in the Vestibular and Lateral Line Systems

Our knowledge of structural and functional diversity in parts of the octavolateralis system other than auditory is superficial and speculative by comparison. We are largely at the stage of still describing the differences in canal sizes or neuromast morphology, for example. The challenge here will be to unravel what kinds of differences are important to functional differences.

MAJOR QUESTIONS AND FUTURE DIRECTIONS

In this paper we have attempted to review much of what is known about the octavolateralis system of fishes. However, despite a significant increase in our knowledge over the past 20 years, particularly with regard to the auditory system, we still do not have a detailed understanding of how the parts of the octavolateralis system function in even a single species. One of the fascinations with fish is that even when we have such an understanding, the immense comparative issues that we have attempted to highlight in this paper will remain.

Many of the issues that are still open to question with regard to hearing are reviewed by Popper and Fay (1993) and we will not deal with them in any detail here. However, it is important to list at least a few of the issues that we feel are most important with regard to the major components of the octavolateralis system.

Auditory System

With regard to hearing, we still have little appreciation of the precise function of the swim bladder, other auxiliary gas bubbles, and the Weberian ossicles in hearing. We do not understand the motion of the otolith and the patterns of stimulation of the inner ear, nor do we have any understanding of why there is such striking diversity in the structure of the saccular otoliths in different species and whether this may be related to the relative motion in the endorgans. The functional significance of diversity in hair cell orientation patterns are not understood and we do not really have significant data on the consequences of having multiple types of hair cells or ciliary bundles. The whole question of the importance of cells with different channel properties is in need of study.

The innervation of the ear, the response of different epithelial regions, and how the neural projections "map" to the CNS are in need of study, as are associated comparative issues. Moreover, we know little about the responses of different levels of the CNS to auditory stimuli, although a reasonable body of data is accumulating for *Carassius*. We know almost nothing about the efferent system and how it might contribute to sound detection and processing in fish.

Finally, despite the great diversity in the auditory periphery of fishes, we need to ask questions about comparative function of the auditory system. In essence, are the differences we see peripherally associated with fishes doing different auditory tasks, or are the differences reflections of various ways to do a similar acoustic task?

Vestibular System

A fundamental question about the inner ear of fish is how the vestibular sense is separated from the auditory sense. In bony fish, unlike tetrapods having specialized separate auditory organs, the otolith organs handle acoustic input as well as gravistatic and locomotor accelerations. Anatomical tracing of fibers shows the projections from otolith organs have some overlap in the octaval nuclei. But such tracing does not resolve whether, for example, some utricular epithelial regions may be acoustically responsive, so the overlap would be to central auditory paths, or whether instead some saccular epithelial regions may be gravistatically responsive, so the overlap would be to vestibular paths.

The nearly complete lack of data on primary afferent responses to tilt from teleost otolith organs suggests that such physiology should be a priority in fish inner ear research. It would be extremely useful to have dye fills of characterized axons to determine the peripheral sites of different response properties as well.

Whether the differentiation of a striolar region is important in the same context remains unknown. In teleosts, the utricle (and possibly the lagena) always has a striola. But the saccule seems to lack a distinct striolar region. The apparent lack of striola in the organ considered the most important auditory organ may be significant. The recent demonstration in a teleost that gentamicin treatment produces a highly specific loss of ciliary bundles only in striolar regions of otolith organs (Yan et al., 1991) may provide a new approach to the function of regional distinctions in these maculae.

Lateral Line

The lateral line system in many fishes is a complex array over much of the body, as well as the elaboration of canals and free neuromasts on the head. Given the structural variety, it is somewhat surprising that the operation of the system, from hair cells to behavior, seems relatively consistent across different regions in terms of the rather narrow, low-frequency band of operation. However, the complex morphology suggests that this system may involve very complex spatial processing. Searches thus far have not yet revealed any mapping representations in the brain that might be analogous to those well known for vision, audition, and somatic sense. There simply are not yet sufficient data to know how the CNS extracts either timing or spatial cues from this complex periphery.

Biophysical studies of the hair cells and hydrodynamics continue to reveal how the accessory structures may affect the peripheral responsiveness to different kinds of stimuli. We need to know more about how the physiology and central projections correlate to determine how "svenning" is utilized by the animal for the various discriminations done by the lateral line system.

Interactions

Even beyond the interactions in the periphery of the vestibular and auditory systems, we know very little about how the auditory and lateral line systems may interact in the CNS. There have been long-standing arguments about the functional overlap between the two systems (see Dijkgraaf, 1963; van Bergeijk, 1967) that have been recently resolved, at least to some degree (Coombs et al., 1992). However, there are reasons to believe that the two systems may overlap in the CNS (Striedter, 1991) and the functional consequence of this overlap is not known, but is in need of study.

ACKNOWLEDGMENTS

We are grateful to our many colleagues and students who have, over the years, contributed to our research efforts and to our understanding of the octavolateralis system. In particular, we extend special thanks to Helen Popper and Carol Platt for putting up with over 15 years of our often-raucous collaboration. We are also grateful to the National Institutes of Health, Office of Naval Research, and the National Science Foundation for their support of our research efforts.

REFERENCES

Alexander, R. McN., The structure of the Weberian apparatus in the Cyprini, *Proc. Zool. Soc. London*, 139, 451, 1962.

Blaxter, J. H. S. and Fuiman, L. A., Function of the free neuromasts of marine teleost larvae, in *The Mechanosensory Lateral Line: Neurobiology and Evolution,* Coombs, S., Görner, P., and Münz, H., Eds., Springer-Verlag, New York, 1989, 481.

Blaxter, J. H. S., Denton, E. J., and Gray, J. A. B., Acoustico-lateralis systems in clupeid fishes, in *Hearing and Sound Communication in Fishes,* Tavolga, W. N., Popper, A. N., Fay, R. R., Eds., Springer-Verlag, New York, 1981, 39.

Bleckmann, H., Tittel, G., and Blubaum-Gronau, E., The lateral line system of surface feeding fish: anatomy, physiology and behavior, in *The Mechanosensory Lateral Line: Neurobiology and Evolution,* Coombs, S., Görner, P., and Münz, H., Eds., Springer-Verlag, New York, 1989, 501.

Boord, R. L. and Montgomery, J. C., Central mechanosensory lateral line centers and pathways among the elasmobranchs, in *The Mechanosensory Lateral Line: Neurobiology and Evolution,* Coombs, S., Görner, P., and Münz, H., Eds., Springer-Verlag, New York, 1989, 323.

Boyle, R. and Highstein, S. M., Resting discharge and response dynamics of horizontal semicircular canal afferents in the toadfish, *Opsanus tau, J. Neurosci.,* 10, 1557, 1990.

Braemer, W. and Braemer, H., Zur Gleichgewichtsorientierung schrägstehender Fische, *Z. Vgl. Physiol.,* 40, 529, 1958.

Budelli, R. and Macadar, O., Statoacoustic properties of utricular efferents, *J. Neurophysiol.,* 42, 1479, 1979.

Buwalda, R. J. A., Segregation of directional and non-directional acoustic information in the cod, in *Hearing and Sound Communication in Fishes,* Tavolga, W. N., Popper, A. N., and Fay, R. R., Eds., Springer-Verlag, New York, 1981, 139.

Carlström, D., A crystallographic study of vertebrate otoliths, *Biol. Bull.,* 125, 441, 1963.

Chang, J. Y. S., Popper, A. N., and Saidel, W. M., Heterogeneity of sensory hair cells in a fish ear, *J. Comp. Neurol.,* 324, 621, 1992.

Chapman, C. J. and Sand, O., Field studies of hearing in two species of flatfish, *Pleuronectes platessa* (L.) and *Limanda limanda* (L.) (Family Pleuronectidae), *Comp. Biochem. Physiol. A,* 47, 371, 1974.

Claas, B., Fritzsch, B., and Münz, H., Common efferents to lateral-line and labyrinthine hair cells in aquatic vertebrates, *Neurosci. Lett.,* 27, 231, 1981.

Coombs, S. and Janssen, J., Behavioral and neurophysiological assessment of lateral line sensitivity in the mottled sculpin, *Cottus bairdi, J. Comp. Physiol. A.,* 167, 557, 1990.

Coombs, S. and Popper, A. N., Hearing differences among Hawaiian squirrelfish (family Holocentridae) related to differences in the peripheral auditory system, *J. Comp. Physiol. A,* 132, 203, 1979.

Coombs, S., Janssen, J., and Webb, J. C., Diversity of lateral line systems, evolutionary and functional considerations, in *Sensory Biology of Aquatic Animals,* Atema, J., Fay, R. R., Popper, A. N., and Tavolga, W. N., Eds., Springer-Verlag, New York, 1988, 188.

Coombs, S., Görner, P., and Münz, H., Eds., *The Mechanosensory Lateral Line: Neurobiology and Evolution,* Springer-Verlag, New York, 1989.

Coombs, S., Janssen, J., and Montgomery, J. C., Functional and evolutionary implications of peripheral diversity in lateral line systems, in *The Evolutionary Biology of Hearing,* Webster, D. B., Fay, R. R., and Popper, A. N., Eds., Springer-Verlag, New York, 1992, 267.

Corwin, J. T., Functional anatomy of the auditory system in sharks and rays, *J. Exp. Zool. Suppl.,* 2, 62, 1989.

Crawford, J. D., Sex recognition by electric cues in a sound-producing mormyrid fish, *Pollimyrus isidori, Brain Behav. Evol.,* 38, 20, 1991.

Demski, L., Gerald, G. W., and Popper, A. N., Central and peripheral mechanisms in teleost sound production, *Am. Zool.,* 13, 1141, 1973.

Denton, E. J. and Gray, J. A. B., Mechanical factors in the excitation of the lateral line of fishes, in *Sensory Biology of Aquatic Animals,* Atema, J., Fay, R. R., Popper, A. N., and Tavolga, W. N., Eds., Springer-Verlag, New York, 1988, 598.

Denton, E. J. and Gray, J. A. B., Some observations on the forces acting on neuromasts in fish lateral line canals, in *The Mechanosensory Lateral Line: Neurobiology and Evolution,* Coombs, S., Görner, P., and Münz, H., Eds., Springer-Verlag, New York, 1989, 229.

Dijkgraaf, S., Hearing in bony fishes, *Proc. R. Soc. London Ser. B,* 152, 51, 1960.

Dijkgraaf, S., The functioning and significance of the lateral line organ, *Biol. Rev.,* 38, 51, 1963.

Enger, P. S., Kalmijn, A. J., and Sand, O., Behavioral investigations on the functions of the lateral line and inner ear in predation, in *The Mechanosensory Lateral Line: Neurobiology and Evolution,* Coombs, S., Görner, P., and Münz, H., Eds., Springer-Verlag, New York, 1989, 575.

Fay, R. R., The goldfish ear codes the axis of acoustic particle motion in three dimensions, *Science,* 225, 951, 1984.

Fay, R. R., *Hearing in Vertebrates, A Psychophysics Databook,* Hill-Fay Assoc., Winnetka, 1988.

Fay, R. R. and Popper, A. N., Modes of stimulation of the teleost ear, *J. Exp. Biol.,* 62, 379, 1975.

Fine, M., Winn, H., and Olla, B., Communication in fishes, in *How Animals Communicate,* Sebeok, T. A., Ed., Indiana University Press, Bloomingham, 1977, 472.

Flock, Å., Transducing mechanisms in lateral line canal organ receptors, *Cold Spring Harbor Symp.,* 30, 133, 1965.

Flock, Å, The lateral line organ mechanoreceptors, in *Physiology of Fishes,* Vol. 5, *Sensory Organs,* Hoar, W. S. and Randall, D. J., Eds., Academic Press, New York, 1971, 231.

Furukawa, T., Effects of efferent stimulation of the saccule of goldfish, *J. Physiol.,* 315, 203, 1981.

Furukawa, T. and Ishii, Y., Neurophysiological studies on hearing in goldfish, *J. Neurophysiol.,* 30, 1377, 1967.

Graf, W. and Baker, R., The vestibuloocular reflex of the adult flatfish. II. Vestibulooculomotor connectivity, *J. Neurophysiol.,* 54, 900, 1985

Graf, W. and Brunken, W. J., Elasmobranch oculomotor organization: anatomical and theoretical aspects of the phylogenetic development of vestibulo-oculomotor connectivity, *J. Comp. Neurol.,* 227, 569, 1984.

Hassan, E. S., Hydrodynamic imaging of the surroundings by the lateral line of the blind cave fish, *Anoptichthys jordani,* in *The Mechanosensory Lateral Line: Neurobiology and Evolution,* Coombs, S., Görner, P., and Münz, H., Eds., Springer-Verlag, New York, 1989, 217.

Hawkins, A. D. and Sand, O., Directional hearing in the median vertical plane by the cod, *J. Comp. Physiol. A,* 122, 1, 1977.

Highstein, S. M., The central nervous system efferent control of the organs of balance and equilibrium, *Neurosci. Res.,* 12, 13, 1991.

Highstein, S. M. and Baker, R., Action of the efferent vestibular system on primary afferents in the toadfish, *Opsanus tau, J. Neurophysiol.,* 54, 370, 1985.

Hudspeth, A. J., The cellular basis of hearing: the biophysics of hair cells, *Science,* 230, 745, 1985.

Iversen, R. T. B., Auditory thresholds of the scombrid fish *Euthynnus affinis,* with comments on the use of sound in tuna fishing, FAO Conference on Fish Behaviour in Relation to Fishing Techniques and Tactics, FAO Fisheries Rep. No. 62, Food and Agriculture Organization, Rome, 1969, p. 849.

Jacobs, D. W. and Tavolga, W. N., Acoustic intensity limens in the goldfish, *Anim. Behav.,* 15, 324, 1967.

Kalmijn, A. J., Hydrodynamic and acoustic field detection, in *Sensory Biology of Aquatic Animals,* Atema, J., Fay, R. R., Popper, A. N., and Tavolga, W. N., Eds., Springer-Verlag, New York, 1988, 1838.

Kalmijn, A. J., Functional evolution of lateral line and inner ear systems, in *The Mechanosensory Lateral Line: Neurobiology and Evolution,* Coombs, S., Görner, P., and Münz, H., Eds., Springer-Verlag, New York, 1989, 187.

Kleerekoper, H. and Roggenkamp, P. A., An experimental study on the effect of the swimbladder on hearing sensitivity in *Ameiurus nebulosus nebulosus* (Lesueur), *Can. J. Zool.,* 37, 1, 1959.

Lewis, T. N. and Rogers, P. H., Detection of scattered ambient noise by fish, *J. Acoust. Soc. Am.,* 91, 2435, 1992.

Lowenstein, O., The equilibrium function of the vertebrate labyrinth, *Biol. Rev.,* 11, 113, 1936.

Lowenstein, O., The concept of the acousticolateral system, in *Lateral Line Detectors,* Cahn, P. H., Ed., Indiana University Press, Bloomington, 1967.

Lowenstein, O., The labyrinth, in *Physiology of Fishes,* Vol. 5, *Sensory Organs,* Hoar, W. S. and Randall, D. J., Eds., Academic Press, New York, 1971, 207.

Mayne, R., A systems concept of the vestibular organs, in *Handbook of Sensory Physiology,* Vol. 6, Pt. 2, *Vestibular System: Psychophysics, Applied Aspects, and General Interpretations,* Kornhuber, H. H., Eds., New York, Springer, 1974, 493.

McCormick, C. A., Central projections of the lateral line and eighth nerves in the bowfin, *Amia calva, J. Comp. Neurol.,* 197, 1, 1981.

McCormick, C., Organization and evolution of the octavolateralis area in fishes, in *Fish Neurobiology,* Northcutt, R. G. and Davis, R. E., Ed., University of Michigan Press, Ann Arbor, 1983, 179.

McCormick, C. A., Central lateral line mechanosensory pathways in bony fish, in *The Mechanosensory Lateral Line: Neurobiology and Evolution,* Coombs, S., Görner, P., and Münz, H., Eds., Springer-Verlag, New York, 1989, 341.

McCormick, C. A., Evolution of central auditory pathways in anamniotes, in *The Evolutionary Biology of Hearing,* Webster, D. B., Fay, R. R., and Popper, A. N., Eds., Springer-Verlag, New York, 1992, 323.

McCormick, C. A. and Popper, A. N., Auditory sensitivity and psychophysical tuning curves in the elephant nose fish, *Gnathonemus petersii, J. Comp. Physiol.,* 55, 753, 1984.

McLaren, J. W. and Hillman, D. E., Displacement of htc semicircular canal cupula during sinusoidal rotation, *Neuroscience,* 4, 2001, 1979.

Meredith, G. E. and Roberts, B. L., Distribution and morphological characteristics of efferent neurons innervating end organs in the ear and lateral one of the European eel, *J. Comp. Neurol.,* 265, 494, 1987.

Meyer, D. L., Platt, C. J., and Distel, H. J., Postural control mechanisms in the upside-down catfish *(Synodontis nigriventris), J. Comp. Physiol. A,* 110, 323, 1976.

Montgomery, J. C., Lateral line detection of planktonic prey, in *The Mechanosensory Lateral Line: Neurobiology and Evolution,* Coombs, S., Görner, P., and Münz, H., Eds., Springer-Verlag, New York, 1989, 561.

Münz, H., Functional organization of the lateral line periphery, in *The Mechanosensory Lateral Line: Neurobiology and Evolution,* Coombs, S., Görner, P., Münz, H., Eds., Springer-Verlag, New York, 1989, 285.

Myrberg, A. A., Jr., Sound communication and interception in fishes, in *Hearing and Sound Communication in Fishes,* Tavolga, W. N., Popper, A. N., and Fay, R. R., Eds., Springer-Verlag, New York, 1981, 395.

Myrberg, A. A., Jr. and Spires, J. Y., Hearing in damselfishes: an analysis of signal detection among closely related species, *J. Comp. Physiol.,* 140, 135, 1980.

Nieuwenhuys, R., Comparative anatomy of the cerebellum, *Prog. Brain Res.,* 25, 1, 1967.

Northcutt, R. G., Central auditory pathways in anamniotic vertebrates, in *Comparative Studies of Hearing in Vertebrates,* Popper, A. N. and Fay, R. R., Eds., Springer-Verlag, New York, 1980, 79.

Northcutt, R. G., Audition in the central nervous system of fishes, in *Hearing and Sound Communication in Fishes,* Tavolga, W. N., Popper, A. N., and Fay, R. R., Eds., Springer-Verlag, New York, 1981, 331.

Ott, J. F. and Platt, C., Early abrupt recovery from ataxia during vestibular compensation in goldfish, *J. Exp. Biol.,* 138, 345, 1988a.

Ott, J. F. and Platt, C., Postural changes occurring during one month of vestibular compensation in goldfish, *J. Exp. Biol.,* 138, 359, 1988b.

Partridge, B. L., Lateral line function and the internal dynamics of fish schools, in *Hearing and Sound Communication in Fishes,* Tavolga, W. N., Popper, A. N., and Fay, R. R., Eds., Springer-Verlag, New York, 1981, 515.

Pfeiffer, W., Equilibrium orientation in fish, *Int. Rev. Gen. Exp. Zool.,* 1, 77, 1964.

Piddington, R. W., Central control of auditory input in the goldfish. I. Effect of shocks to the midbrain, *J. Exp. Biol.,* 55, 569, 1971.

Plassmann, W., Sensory modality interdependence in the octaval system of an elasmobranch, *Exp. Brain Res.,* 50, 283, 1983.

Platt, C., Central control of postural orientation in flatfish. I. Postural change dependence on central neural changes, *J. Exp. Biol.,* 59, 491, 1973.

Platt, C., Hair cell distribution and orientation in goldfish otolith organs, *J. Comp. Neurol.,* 172, 283, 1977.

Platt, C., The peripheral vestibular system in fishes, in *Fish Neurobiology,* Northcutt, R. G. and Davis, R. E., Eds., University of Michigan Press, Ann Arbor, 1983, 89.

Platt, C., Equilibrium in the vertebrates: signals, senses, and steering underwater, in *Sensory Biology of Aquatic Animals,* Atema, J., Fay, R. R., Popper, A. N., and Tavolga, W. N., Eds., Springer-Verlag, New York, 1988, 783.

Platt, C. and Popper, A. N., Fine structure and function of the ear, in *Hearing and Sound Communication in Fishes,* Tavolga, W. N., Popper, A. N., and Fay, R. R., Eds., Springer-Verlag, New York, 1981, 3.

Platt, C. and Popper, A. N., Variations in lengths of ciliary bundles on hair cells along the macula of the sacculus in two species of teleost fishes, *SEM* 4, 1915, 1984.

Platt, C., Popper, A. N., and Fay, R. R., The ear as part of the octavolateralis system , in *The Mechanosensory Lateral Line: Neurobiology and Evolution,* Coombs, S., Görner, P., and Münz, H., Eds., Springer-Verlag, New York, 1989, 663.

Poggendorf, D., Die absoluten Hörschwellen des Zwergwelses (*Amiurus nebulosus*) und Beitrage zur Physik des Weberischen Apparates der Ostariophysen, *Z. Vgl. Physiol.,* 34, 222, 1952.

Popper, A. N., A scanning electron microscopic study of the sacculus and lagena in the ears of fifteen species of teleost fishes, *J. Morphol.,* 153, 397, 1977.

Popper, A. N., Scanning electron microscopic study of the otolithic organs in the bichir (*Polypterus bichir*) and shovel-nose sturgeon (*Schaphirhynchus platorynchus*), *J. Comp. Neurol.,* 18, 117, 1978.

Popper, A. N., Comparative scanning electron microscopic investigations of the sensory epithelia in the teleost sacculus and lagena, *J. Comp. Neurol.,* 200, 357, 1981.

Popper, A. N., Organization of the ear and auditory processing, in *Fish Neurobiology,* Northcutt, R. G. and Davis, R. E., Eds., University of Michigan Press, Ann Arbor, 1983, 125.

Popper, A. N. and Coombs, S., The morphology and evolution of the ear in actinopterygian fishes, *Am. Zool.,* 22, 311, 1982.

Popper, A. N. and Fay, R. R., Sound detection and processing by fish: critical review and major research questions, *Brain Behav. Evol.,* 41, 14, 1993.

Popper, A. N. and Northcutt, R. G., Structure and innervation of the inner ear of the bowfin, *Amia calva, J. Comp. Neurol.,* 213, 279, 1983.

Popper, A. N. and Platt, C., Sensory surface of the saccule and lagena in the ears of ostariophysan fishes, *J. Morphol.,* 176, 121, 1983.

Popper, A. N. and Tavolga, W. N., Sound detection and inner ear structure in the marine catfish, *Arius felis, J. Comp. Physiol.,* 144, 27, 1981.

Popper, A. N., Platt, C., and Edds, P. L., Evolution of the vertebrate inner ear: an overview of ideas, in *The Evolutionary Biology of Hearing*, Webster, D. B., Fay, R. R., and Popper, A. N., Eds., Springer-Verlag, New York, 1992, 49.

Popper, A. N., Rogers, P. H., Saidel, W. M., and Cox, M., The role of the fish ear in sound processing, in *Sensory Biology of Aquatic Animals,* Atema, J., Fay, R. R., Popper, A. N., and Tavolga, W. N., Eds., Springer-Verlag, New York, 1988, 687.

Retzius, G., *Das Gehörorgan der Wirbelthiere,* Vol. 1, *Das Gehörogan der Fische und Amphibien,* Samson and Wallin, Stockholm, 1881.

Roberts, B. L. and Meredith, G. E., The efferent innervation of the ear: variations on an enigma, in *The Evolutionary Biology of Hearing,* Webster, D. B., Fay, R. R., and Popper, A. N., Eds., Springer-Verlag, New York, 1992, 185.

Rogers, P. H., What do fish listen to?, *J. Acoust. Soc. Am.,* 79, S22, 1986.

Rogers, P. H. and Cox, M., Underwater sound as a biological stimulus, in *Sensory Biology of Aquatic Animals,* Atema, J., Fay, R. R., Popper, A. N., and Tavolga, W. N., Eds., Springer-Verlag, New York, 1988, 131.

Rogers, P. H., Popper, A. N., Cox, M., and Saidel, W. M., Processing of acoustic signals in the auditory system of bony fish, *J. Acoust. Soc. Am.,* 83, 338, 1988.

Rovainen, C. M., Vestibulo-ocular reflexes in adult sea lamprey, *J. Comp. Physiol. A,* 112, 159, 1976.

Russell, I. J. and Roberts, B. L., Inhibition of spontaneous lateral-line activity by efferent nerve stimulation, *J. Exp. Biol.,* 57, 77, 1972.

Saidel, W. M. and Popper, A. N., Sound reception in two anabantid fishes, *Comp. Biochem. Physiol. A,* 88, 37, 1987.

Saidel, W. M., Presson, J. C., and Chang, J. S., S-100 immunoreactivity identifies a subset of hair cells in the utricle and saccule of a fish, *Hear. Res.,* 47, 139, 1990a.

Saidel, W. M., Popper, A. N., and Chang, J., Spatial and morphological differentiation of trigger zones in afferent fibers to the teleost utricle, *J. Comp. Neurol.,* 302, 620, 1990b.

Sand, O., Lateral-line systems, in *Comparative Physiology of Sensory Systems,* Bolis, L., Keynes, R. D., and Maddrell, S. H. P., Eds., Cambridge University Press, Cambridge, 1984, 3.

Sand, O. and Michelsen, A., Vibration measurements of the perch saccular otolith, *J. Comp. Physiol. A,* 123, 85, 1978.

Saunders, J. C. and Dear, S. P., Comparative morphology of stereocilia, in *Hearing and Other Senses: Presentations in Honor of E. G. Wever,* Fay, R. R. and Gourevitch, G., Eds., Amphora Press, Groton, CT, 1983, 175.

Schellart, N. A. M. and Popper, A. N., Functional aspects of the evolution of the auditory system of actinopterygian fish, in *The Evolutionary Biology of Hearing,* Webster, D. B., Fay, R. R., and Popper, A. N., Eds., Springer-Verlag, New York, 1992, 295.

Schneider, H., Morphology and physiology of sound-producing mechanisms in teleost fishes, in *Marine Bio-Acoustics, II,* Tavolga, W. N., Eds., Pergamon Press, Oxford, 1967, 135.

Schoen, L. and von Holst, E., Das Zusammenspiel von Lagena und Utriculus bei der Lageorientierung der Knochenfische, *Z. Vgl. Physiol.,* 39, 399, 1950.

Schöne, H., Über die Arbeitsweise der Statolithenapparate bei Plattfischen, *Biol. Jahresh.,* 4, 135, 1964.

Schuijf, A., Directional hearing of cod (*Gadus morhua*) under approximate free field conditions, *J. Comp. Physiol.,* 98, 307, 1975.

Schuijf, A., Models of acoustic localization, in *Hearing and Sound Communication in Fishes,* Tavolga, W. N., Popper, A. N., and Fay, R. R., Eds., Springer-Verlag, New York, 1981, 267.

Schuijf, A. and Hawkins, A. D., Eds., *Sound Reception in fish,* Elsevier, Amsterdam, 1976, 288.

Schuijf, A. and Hawkins, A. D., Acoustic distance discrimination by the cod, *Nature,* 302, 143, 1983.

Schuijf, A., Visser, C., Willers, A. F. M., and Buwalda, R. J. A., Acoustic localization in an ostariophysan fish, *Experientia,* 33, 1062, 1977.

Schultze, H. P., A new acanthodian from the Pennsylvanian of Utah, U.S.A., and the distribution of otoliths in gnathostomes, *J. Vertebr. Paleontol.,* 10, 49, 1990.

Spanier, E., Aspects of species recognition by sound in four species of damselfish, genus *Eupomacentrus* (Pisces: Pomacentridae), *Z. Tierpsychol.,* 51, 301, 1979.

Steinacker, A. and Romero, A., Voltage-gated potassium current and resonance in the toadfish saccular hair cells, *Brain Res.,* 574, 229, 1992.

Striedter, G. F., Auditory, electrosensory, and mechanosensory pathways through the diencephalon and telencephalon of channel catfishes, *Brain Behav. Evol.,* 312, 311, 1991.

Sugihara, I. and Furukawa, T., Morphological and functional aspects of two different types of hair cells in the goldfish sacculus, *J. Neurophysiol.,* 62, 1330, 1989.

Tavolga, W. N., Mechanisms of sound production in the ariid catfishes *Galeichthys* and *Bagre, Bull. Am. Mus. Nat. Hist.,* 124, 1, 1962.

Tavolga, W. N., Sound production and detection, in *Fish Physiology,* Vol. 5, Hoar, W. S. and Randall, D. J., Eds., Academic Press, New York, 1971, 135.

Tavolga, W. N., Sensory parameters in communication among coral reef fishes, *Mt. Sinai J. Med. N.Y.,* 41, 324, 1974.

Tavolga, W. N., Popper, A. N., and Fay, R. R., Eds., *Hearing and Sound Communication in Fishes,* Springer-Verlag, New York, 1981.

Tricas, T. C. and Highstein, S. M., Visually mediated inhibition of lateral line primary afferent activity by the octavolateralis efferent system during predation in the free-swimming toadfish, *Opsanus tau, Exp. Brain Res.,* 83, 233, 1990.

Tricas, T. C. and Highstein S. M., Action of the octavolateralis efferent system upon the lateral line of free-swimming toadfish, *Opsanus tau, J. Comp. Physiol.,* 169, 25, 1991.

van Bergeijk, W. A., The evolution of vertebrate hearing, in *Contributions to Sensory Physiology,* Neff, W. D., Ed., Academic Press, New York, 1967, 1.

van Netten, S. M. and Kroese, A. B. A., Dynamic behavior and micromechanical properties of the cupula in *The Mechanosensory Lateral Line: Neurobiology and Evolution,* Coombs, S., Görner, P., and Münz, H., Eds., Springer-Verlag, New York, 1989, 246.

von Holst, E., Uber den Lichtrückenreflex bei Fischen, *Publ. Stn. Zool. Napoli,* 15, 143, 1935.

von Holst, E., Die Arbeitsweise des Statolithenapparates bei Fischen, *Z. Vgl. Physiol.,* 32, 60, 1950.

Webster, D. B., Fay, R. R., and Popper, A. N., Eds., *The Evolutionary Biology of Hearing,* New York, Springer-Verlag, 1992, 859.

Wersäll, J., Vestibular receptor cells in fish and mammals, *Acta Oto-Laryngol. Suppl.,* 163, 25, 1961.

Wever, E. G., The evolution of vertebrate hearing, in *Handbook of Sensory Physiology,* Vol. V/l, *Auditory System,* Keidel, W. E. and Neff, W. D., Eds., Springer-Verlag, Berlin, 1974, 423.

Winn, H. E., Vocal facilitation and biological significance of toadfish sounds, in *Marine Bio-Acoustics II,* Tavolga, W. N., Ed., Pergamon Press, Oxford, 1967, 283.

Yan, H. Y., Saidel, W. M., Chang, J., Presson, J. C., and Popper, A. N., Sensory hair cells of the fish ear, evidence of multiple types based on ototoxicity sensitivity, *Proc. R. Soc. London Ser. B,* 145, 133, 1991.

5 Electrosensation

Walter Heiligenberg

INTRODUCTION

Although electrosensation appears to have evolved very early in vertebrates, it is now only found in a small set of aquatic organisms and has received much scientific attention only recently. The study of this modality has yielded a variety of powerful model systems for the exploration of sensory and neural phenomena at various levels, ranging from the behavioral to the cellular and synaptic. Particularly rewarding has been the close integration of behavioral, anatomical, and physiological approaches, and studies at a higher level of integration have generally identified the biological relevance of phenomena to be explored at the next lower level.

Electrosensory systems offer particular experimental advantages by appearing more transparent and less "cluttered" than more highly evolved systems, such as vision and audition in birds and mammals. Yet the basic neuronal designs found in electrosensation, appearing old and conservative in the evolutionary sense, are largely the same as those in higher and more derived systems. Most significantly, some behavioral responses of electric fish are so robust that they remain intact in physiological preparations, thus allowing simultaneous studies at the behavioral and cellular levels.

An extensive review of this field was edited by Bullock and Heiligenberg (1986), and more recent developments were reviewed by Hopkins (1988), Bell (1989), Kramer (1990), and Heiligenberg (1989, 1990, 1991a-d).

Since space limitations will not allow for an in-depth review of the current state of the field, this chapter will largely summarize findings that have been reviewed recently elsewhere. Original contributions will be quoted primarily to the extent that they have not been covered in recent reviews.

THE EVOLUTION OF ELECTROSENSATION

Two classes of electroreceptors are distinguished: ampullary and tuberous (Bennett and Obara, 1986; Zakon, 1986a). Whereas ampullary receptors sense low-frequency electric signals of no more than several tens of hertz, tuberous receptors are only found in fishes with electric organs and are largely tuned to the dominant spectral frequencies of the individual's own electric organ dis-

0-8493-8042-1/93/$0.00+$.50

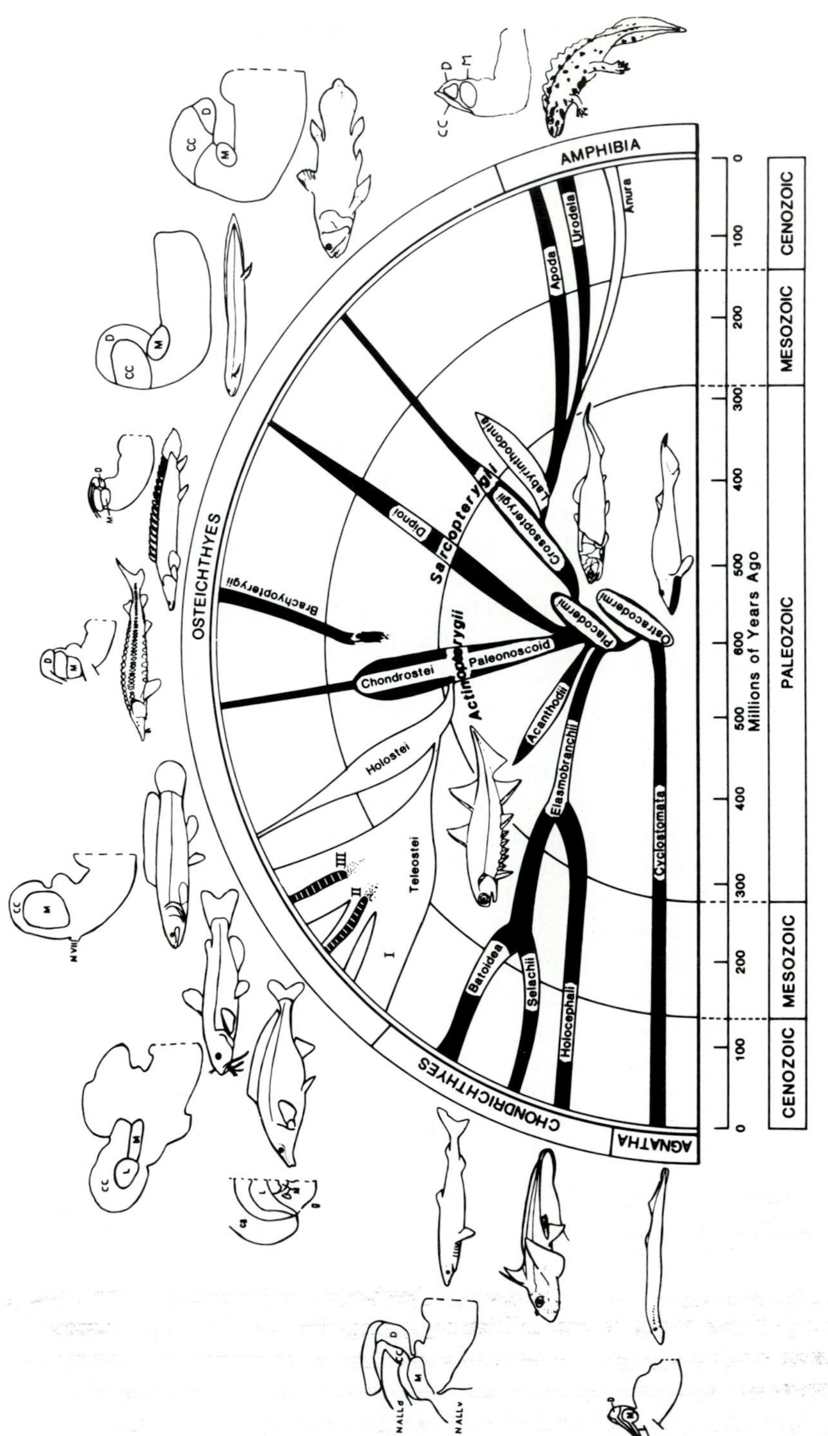

AMPHIBIA
OSTEICHTHYES
CHONDRICHTHYES
AGNATHA
Anura
Urodela
Apoda
Labyrinthodontia
Crossopterygii
Sarcopterygii
Dipnoi
Brachyopterygii
Actinopterygii
Paleonoscoid
Chondrostei
Holostei
Teleostei
I
II
III
Placodermi
Ostracodermi
Acanthodii
Elasmobranchii
Batoidea
Selachii
Holocephali
Cyclostomata
0
100
200
300
400
500
600
500
400
300
200
100
0
Millions of Years Ago
CENOZOIC
MESOZOIC
PALEOZOIC
MESOZOIC
CENOZOIC
CC
D
M
L
N VII
NALLd
NALLv

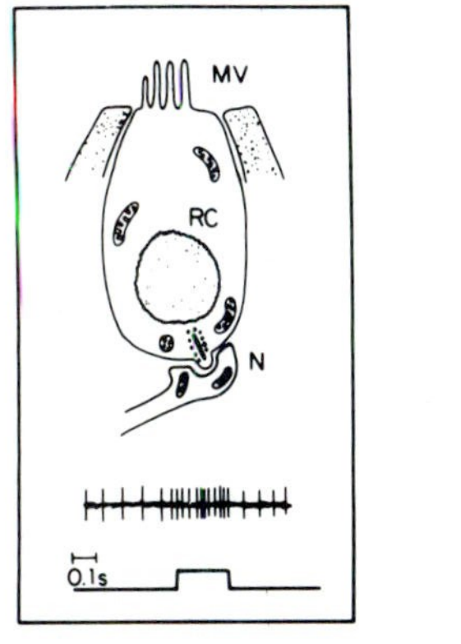

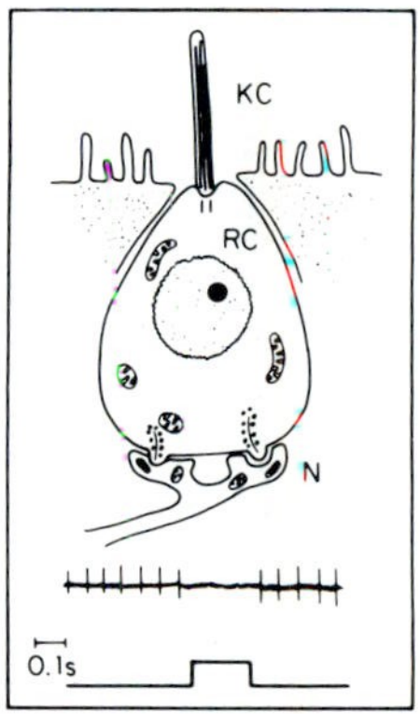

FIGURE 1. The evolution of the electric sense in fish and amphibia. Branches filled in with black represent types of fish and amphibia whose extant forms possess ampullary electroreceptor organs. (An exception are the marine Agnatha ["hagfish"], which apparently lost their electric sense together with other features.) The receptor cells (RC) of these organs (see schema on right) carry a kinocilium (KC). They are excited by outside-negative voltages and inhibited by outside-positive voltages. The primary afferent nerve (N) is excited by chemical synapses. Branches marked by hatching represent two lines of teleosts, the mormyriformes (II) on the one hand, and the gymnotiformes and catfish (III) on the other, that apparently "reinvented" the electric sense after it had been lost with the evolution of the holostean fish. The ampullary organs of these "modern" fish (see schema on left) have receptor cells (RC) covered with microvilli (MV) and respond oppositely to those of the "ancestral" fish, i.e., they are excited by outside-positive voltages and inhibited by outside-negative voltages. The schematic transverse section through one half of the hindbrain indicates the projections of electrosensitive afferent fibers, the dorsal (D) nucleus in ancestral fish, and the electrosensitive lateral line lobe (L) in modern fish. The medial nucleus (M) receives mechanosensory afferents. CC indicates the corpus cerebelli. (This diagram was kindly provided by Carl D. Hopkins, complemented by data kindly provided by Bernd Fritzsch and by additional drawings by Svenja Viete.)

charges (EODs), which lie in the range from tens to thousands of hertz. Ampullary receptors are found in almost all ancestral types of fish and aquatic forms of amphibians (Figure 1), and they may be as old as the origin of vertebrates (Figure 2). They are known as the ampullae of Lorenzini in elasmobranchs. Most curiously, electrosensation was lost with the evolution of holostean fish, but then reappeared in some isolated groups of more modern teleost fishes, such as the mormyriform fishes of Africa (Figure 3) and the gymnotiform fishes of Central and South America (Figure 4), and the catfish, which are related to the latter (Bodznick and Boord, 1986; Finger, 1986; Finger et al., 1986; Fritzsch and Münz, 1986; Northcutt, 1986; Ronan, 1986). These "modern" electrosensitive fish all have ampullary receptors, which differ, however, from those of the ancestral forms by being excited by the opposite current polarity. Mormyriform and gymnotiform fish also evolved electric organs (see review by Bass, 1986a) and, in addition, tuberous electroreceptors tuned to their EODs. These fish sense the EODs of neighbors either as isolated events or through interference with their own EODs. In addition to their role in social communication (Hagedorn, 1986; Hopkins, 1986, 1988), EODs also serve in the "electrolocation" of objects as the fish is able to assess the distortion of its own EOD field by the appearance of objects that differ electrically from water (Bastian, 1986a).

THE MORPHOLOGY, FUNCTION, AND PHYSIOLOGY OF ELECTRORECEPTORS

With regard to their morphology and physiology, electroreceptors are related to the hair cells of lateral line and vestibular organs. The most recent review of electroreceptor morphology and physiology was written by Zakon (1986a), and relatively few additional discoveries have been made since. Still relatively little is known about the ionic mechanisms and pharmacology of electroreceptors (Bennett and Obara 1986). As mentioned earlier, we discriminate between "ampullary" receptors, which are tuned to low frequencies, and "tuberous" receptors, which are tuned to high frequencies (Figures 5 and 6). Primary afferents from these two classes of receptors project to different targets in the central nervous system, forming the beginning of two separate electrosensory systems, which eventually converge at the level of the midbrain. Mormyriforms and gymnotiforms show significant similarities in the basic principles of their electrosensory systems, but differences in their respective implementations point to their convergent evolution.

Ampullary Receptors

Ampullary receptors are found in all types of electrosensitive fish. They are excited by outside-negative transepidermal fields in the "ancestral" types of fish (see Figure 1) and by outside-positive fields in the modern types of fish, the mormyriforms, catfish, and gymnotiforms (Figure 6). In marine fish, ampullary receptors are connected to the surface via a long, jelly-filled duct of low resistivity, whereas the receptors of all freshwater fish are characterized by short ducts. Kalmijn (1974 [Fig. 5]) has argued that the long duct of marine ampullary receptors enhances the sensitivity to voltage gradients in marine fish, which experience relatively small transepidermal voltage differences because of their relatively low skin resistance. The much higher relative skin resistance of freshwater fish causes much larger transepidermal voltage differences, and these can be detected even with short ampullary ducts. This interpretation is supported by the observation that marine catfish have long ampullary ducts, while their freshwater relatives have short ducts.

Ampullary receptors have become famous for their extreme sensitivity. Behavioral thresholds can be as low as several nanovolts per centi-

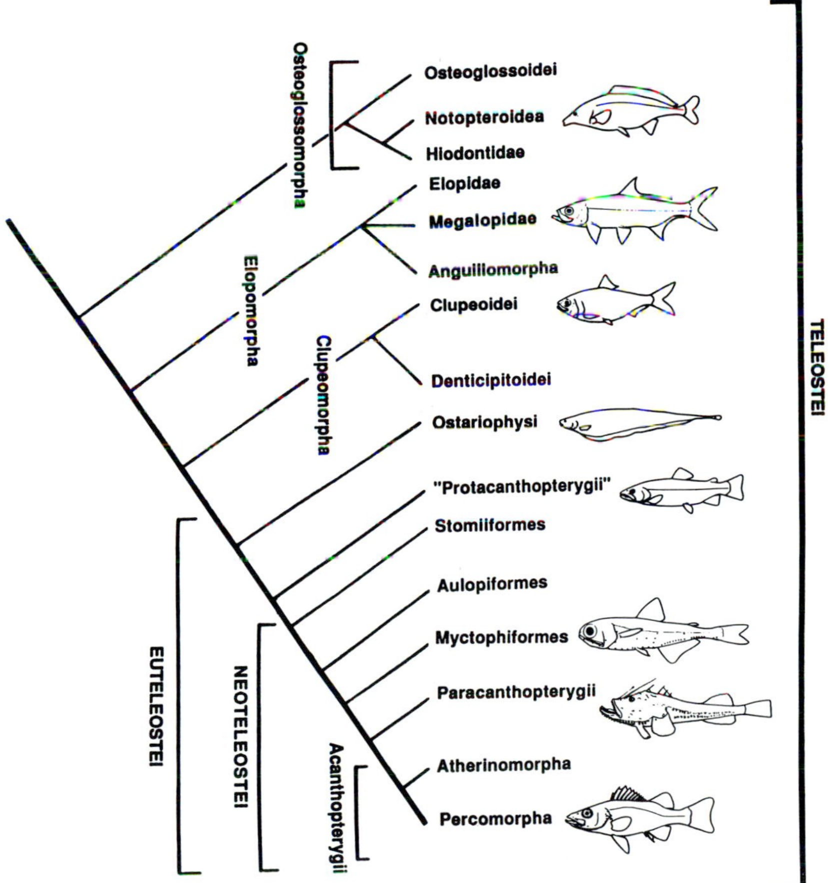

FIGURE 2. A cladogram of teleost fish evolution. A gymnotiform fish is shown as a representative of the Ostariophysi, which also comprise the catfish. A mormyrid fish is shown as a representative of the Notopteroidea. (This diagram was kindly provided by Carl D. Hopkins.)

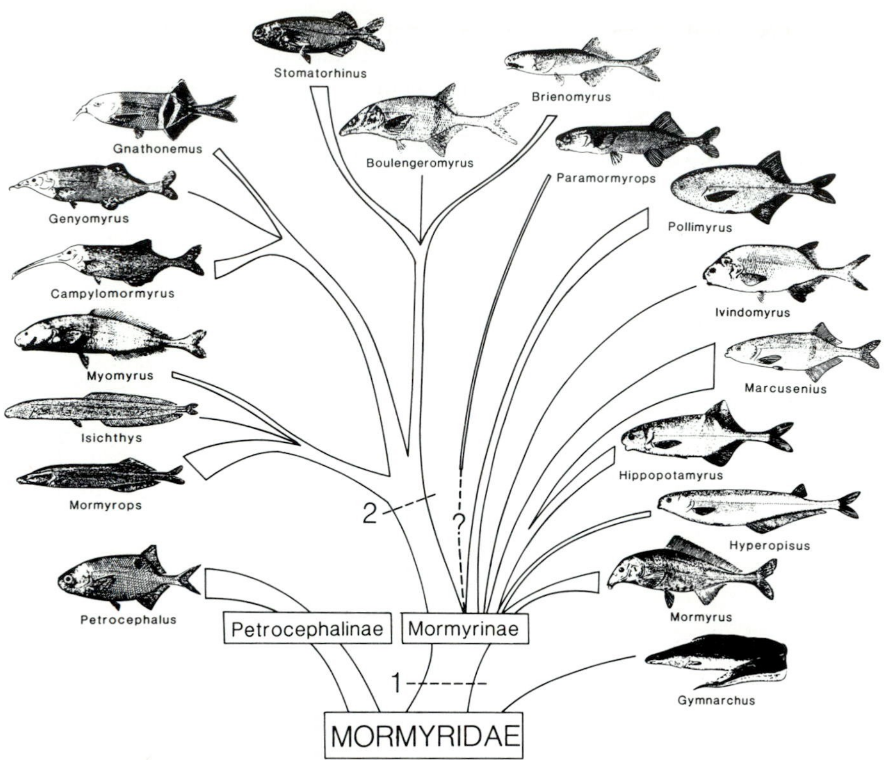

FIGURE 3. The African electric fish family, Mormyridae, comprises approximately 200 species. All known mormyrids, with the exception of the genus *Gymnarchus*, produce pulse-type discharges. The width of branches indicates the number of species in that genus. (This diagram was kindly provided by Carl D. Hopkins.)

meter (Kalmijn, 1987; Weaver and Astumian, 1990). This high sensitivity serves various functions in the life of elasmobranchs. It allows them to hunt prey by detecting, for example, the very faint potentials associated with the ionic leakage of the gills of a buried flounder. Further, the detection of electric gradients caused by various geophysical and geochemical processes may serve their orientation in space. Most remarkably, elasmobranchs are capable of determining magnetic vectors on the basis of the electric currents that are induced by their own motion across magnetic field lines. So far, elasmobranchs are the only animals with a robust behavioral discrimination of magnetic cues, but they should not require special magnetoreceptors for this task, since their electroreceptors are sufficiently sensitive (Kalmijn, 1984, 1987).

The ampullary receptors of mormyrids also respond to the fish's own EOD pulses. They project somatotopically to the ventrolateral zone of the electrosensory lateral line lobe (ELL). Higher-order neurons of the ELL, however, adapt to input associated with the fish's own EODs within minutes

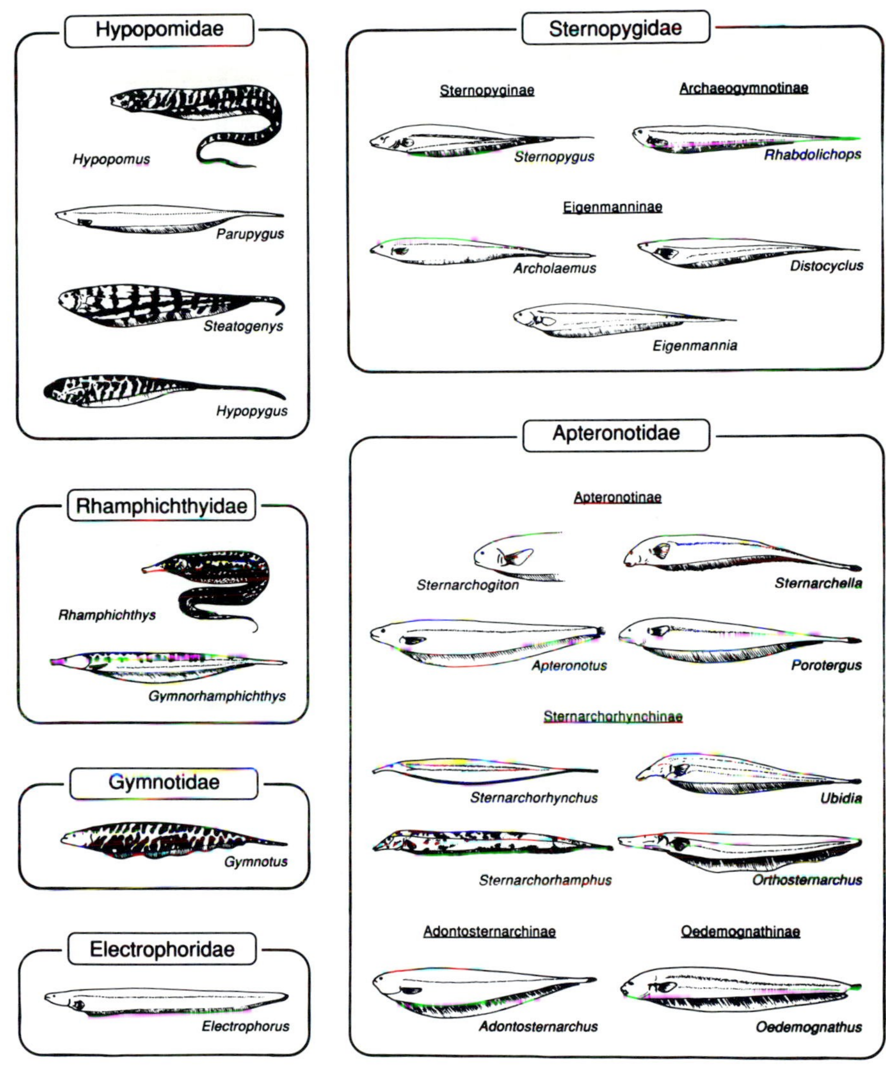

FIGURE 4. The South American electric fish order, Gymnotiformes, comprises six families (boxes). Whereas the Sternopygidae and Apteronotidae produce wave-type discharges, the remaining produce pulse-type discharges. The electric eel, *Electrophorus*, is the only known species that can produce high-voltage discharges in addition to its regular, weak discharges. (Diagram kindly provided by Carl D. Hopkins.)

and thus only respond to novel, low-frequency signals. This form of adaptation involves a modifiable central "template", which is updated continually and represents a negative image, or copy, of the expected sensory input that is due to the fish's own EOD activity. Updating of this template requires the continual occurrence of the corollary discharge signal associated with the electric organ pacemaker, at a rate of a few to several hertz, and this template serves to "null"

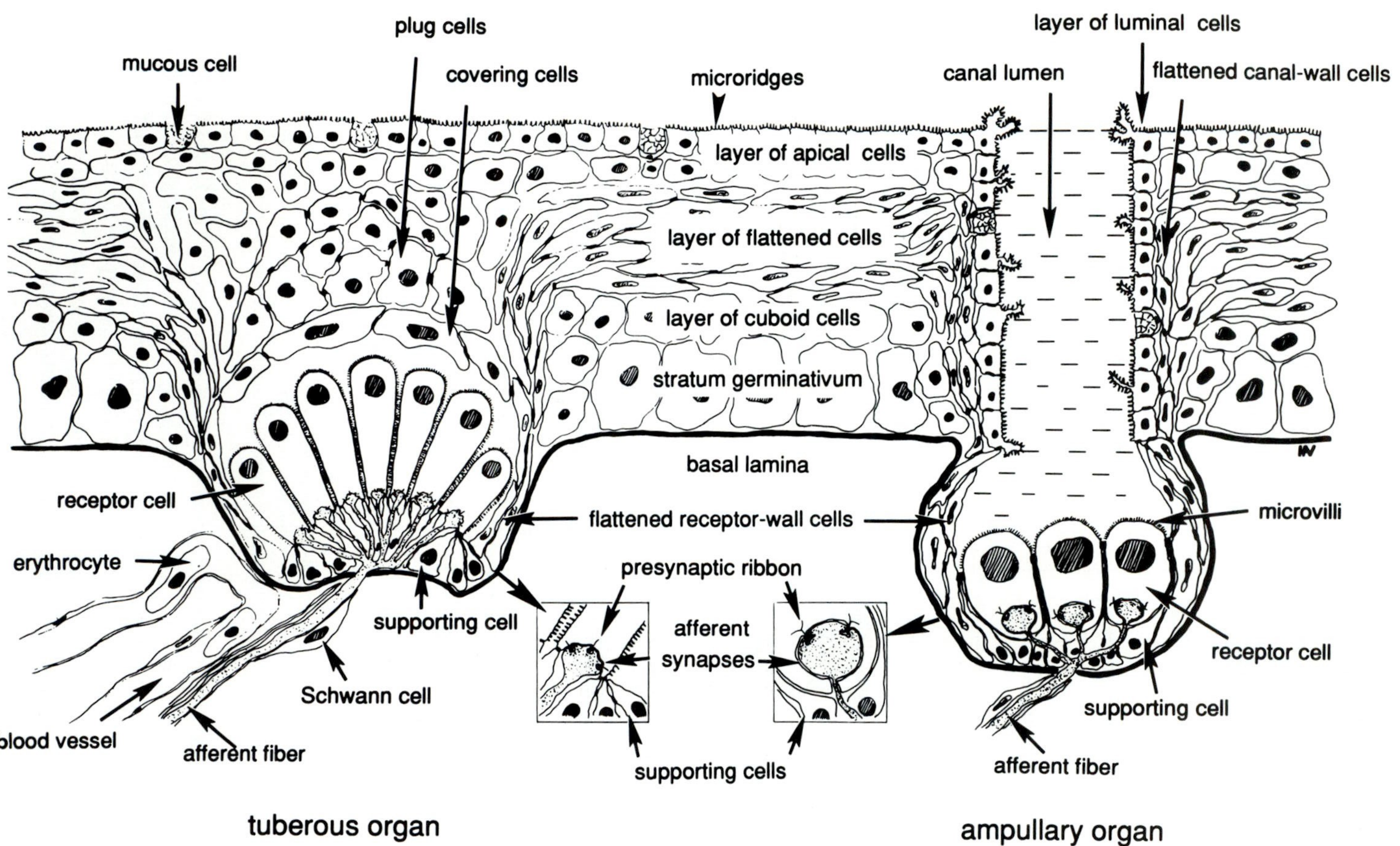
plug cells
mucous cell
covering cells
microridges
canal lumen
layer of luminal cells
flattened canal-wall cells
layer of apical cells
layer of flattened cells
layer of cuboid cells
stratum germinativum
basal lamina
receptor cell
flattened receptor-wall cells
microvilli
erythrocyte
presynaptic ribbon
supporting cell
afferent
synapses
receptor cell
Schwann cell
supporting cell
blood vessel
afferent fiber
supporting cells
afferent fiber
tuberous organ
ampullary organ

FIGURE 5. The structure of electroreceptor organs in the skin of the gymnotiform fish, *Eigenmannia*. A tuberous organ has been drawn on the left, an ampullary organ on the right, with insets showing synaptic details. The layer of flattened cells, interconnected by tight junctions (shown as thickened contact points between cells), provides a relatively high electrical resistance and wraps around the receptor organs, thus preventing current from bypassing these organs. Moreover, tight junctions between the receptor cells and supporting cells at their perimeter force current to flow through the synaptic region at the basis of the receptor cells. Afferent fibers (dotted profiles) form the postsynaptic sites and are excited by chemical transmission. As shown in the insets, presynaptic ribbons invade small protrusions of the receptor cell into the postsynaptic surface. Whereas the receptor cells of tuberous organs stand freely within the lumen of the organ, those of ampullary organs are linked together high up by supporting cells and their tight junctions. Ampullary organs are connected to the outside via a canal lumen filled with a jelly of high electrical conductivity. This jelly is produced at the tip of supporting cells. Tuberous organs are protected against a constant current flow by a layer of covering cells and an assembly of plug cells which form a capacitive link to the outside. The width of a receptor cell is approximately 10 µm. (This drawing is based upon ultrastructural studies and was kindly provided by Heinrich A. Vischer.)

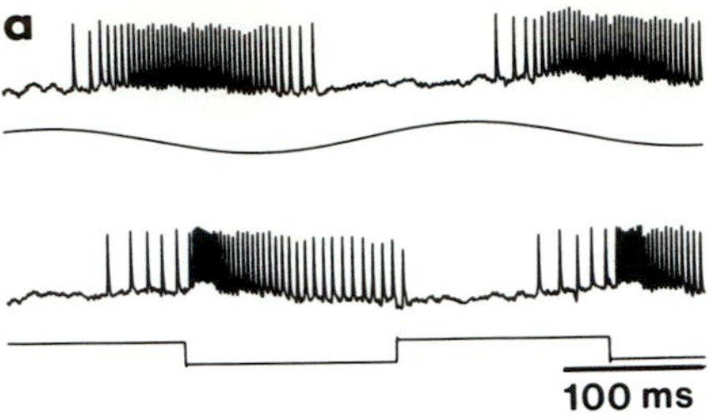

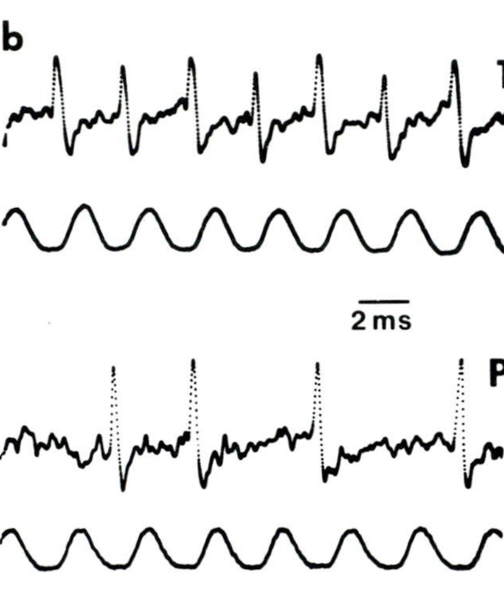

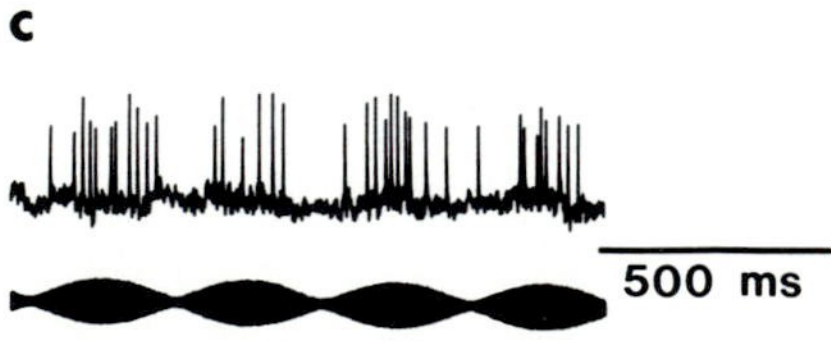

FIGURE 6. Responses of electroreceptor afferents in *Eigenmannia*. (a) Intracellular recording from an ampullary afferent fiber. The size of the action potentials is approximately 15 mV. The traces underneath show the voltage inside the fish's body with reference to the outside, with "positive" pointing up. The p-p amplitude of the stimulus, measured near and perpendicularly to the fish's head surface, was 1–2 mV/cm. Note that ampullary receptors are excited by the negative value of the sustained voltage as well as by changes of the voltage in the negative direction. (b) Extracellular recording from two tuberous electroreceptor afferents, a T-type fiber (top) and a P-type fiber (bottom), with the fish's natural EOD recorded underneath in each case. Note that the T-fiber fires on each EOD cycle, while the P-fiber fires intermittently. (These records were kindly provided by Joseph Bastian.) (c) Extracellular recording from a P-type afferent responding to a sinusoidal amplitude modulation of an EOD-like carrier signal (lower trace). P-type afferents code stimulus amplitude by the rate of their firing.

the predictable sensory input so that higher-order neurons no longer respond to the fish's own EOD signal (Bell, 1989 [Figs. 1 and 4]).

Tuberous Receptors

Tuberous receptors are found in the mormyriform and gymnotiform fish, both of which have developed electric organs for the rhythmic production of electric fields. The morphology and physiology of electric organs was reviewed by Bass (1986a). These organs are derived from segmental musculature, but have been replaced by a neurogenic organ in a single gymnotiform family, the apteronotids. This neurogenic organ is a modification of the efferent axons that innervate the myogenic organ of larval apteronotids (Kirschbaum, 1983).

Whereas the majority of electric fish produces discrete, pulse-like EODs, a few gymnotiform genera and the mormyriform genus *Gymnarchus* produce continual, wave-like EODs. The duration and shape of EOD pulses as well as the fundamental frequency, or repetition rate, and spectral content of wave-like EODs are typical for a given species and often are even sex-specific (Hopkins, 1988). Therefore, electric fish can identify conspecifics and potential mates on the basis of their EODs alone.

Tuberous receptors are generally most sensitive in the spectral range that dominates the individuals EODs (Zakon, 1986a). So-called "phase-coding" types of tuberous receptors code the timing or "phase" of an electric signal by marking its zero-crossing (the transition from a negative to a positive voltage) by a single spike. In addition, they may discriminate between signal forms that are identical in their amplitude spectra and only differ in their phase function (von der Emde and Bleckmann, 1992). As a consequence, electric fish can discriminate signals of identical power spectra but different phase functions. Audio-amplified versions of such signals cannot be distinguished by our own auditory system, which is known to have pure phase sensitivity. The response of some types of phase-coding receptors may vary little as a function of the amplitude of the signal. So called "amplitude-coding" types of tuberous receptors, on the other hand, alter the strength or latency of their response as a function of signal amplitude and normally are rather insensitive to the timing of the signal or its phase-function. In mormyriforms (Bell et al., 1989; Bell 1990a, 1990b), as well as in gymnotiforms (Carr and Maler, 1986), the phase-coding and amplitude-coding receptor inputs project to separate structures of the hindbrain and thus form separate pathways. In gymnotiforms, these two pathways remain distinct up the level of the midbrain. Phase and amplitude information are thus processed by separate neuronal substrates, much as in the auditory system of owls (Konishi, 1991) and other organisms. Since mormyriforms and gymnotiforms show some fundamental differences in the design of their tuberous receptors, they will be treated separately.

The Tuberous Receptors of Mormyriforms

There are two types of tuberous receptor organs in mormyrids, "Knollenorgans" and "mormyromasts". Knollenorgans mark the occurrence of an EOD pulse by firing a single spike, and their afferents project to the nucleus of the electrosensory lateral line lobe (nELL). The fish's own pacemaker system, which triggers each EOD pulse, generates a corollary discharge signal that inhibits the transmission of Knollenorgan inputs caused by the fish's own EOD pulses. From the level of the nELL onwards, a fish thus only receives information about EODs of its neighbors but not about its own. This allows for the detection and undisturbed analysis of specific EOD patterns that serve in social communication (Bell and Grant, 1989 [Fig. 5]). The nature of this inhibition appears to be GABA-ergic (Mugnaini and Maler, 1987).

Mormyromasts contain two types of receptor cells, A cells peripherally, and B cells deeper within the receptor organ. Whereas the primary afferents of A cells project somatotopically to the medial zone of the ELL, those of the B cells project to the dorsolateral zone of the ELL (Bell et al., 1989 [Fig. 2 and 7]). Both types of afferents code an increment of the local EOD amplitude by shortening the latency and increasing the number of spikes fired in response to the stimulus. They differ, however, in that A-afferents have a higher threshold and fire a smaller maximal number of spikes per burst (Bell, 1990a, 1990b). Moreover, von der Emde and Bleckmann (1992) have demonstrated that B-afferents, in contrast to A-afferents, are extremely sensitive to changes in the waveform, or "phase function", of the EOD. Changes in the phase function of the EOD are mainly caused by the capacitive load of objects in the environment. Through the joint evaluation of inputs from A-fibers and B-fibers, the fish should, therefore, be able to discriminate ohmic and capacitive aspects of an object, an ability that was demonstrated in behavioral experiments (von der Emde, 1990).

Mormyromasts have a higher threshold then Knollenorgans and are primarily driven by the fish's own EODs. A corollary discharge signal of the pacemaker system labels mormyromast input caused by the fish's own EODs within the ELL, and the latency between the corollary discharge signal and the first afferent spike appears to code the amplitude of the local EOD pulse (Bell, 1990a [Fig. 9]; Bell, 1990b [Fig. 3]).

The studies described here were performed almost exclusively on the genus *Gnathonemus* and a few additional genera, which all fire pulse-like EODs. A single genus, *Gymnarchus*, sometimes considered a member of a separate family, Gymnarchidae, fires continual, wave-like EODs. Comparatively little is known about the structure-function relation of its tuberous electroreceptors, which were last reviewed by Szabo (1974) and Szabo and Fessard (1974). Physiological studies by Bullock et al., (1975) revealed two types of tuberous afferents that can be classified as phase coders and amplitude coders, comparable to the T- and P-units in gymnotiforms with wave-type EODs (see below). This genus has attracted new attention after Kawasaki's (1991) discovery that its jamming avoidance response resembles that of the gymnotiform genus *Eigenmannia* by not requiring a corollary discharge signal of the pacemaker system. This is all the more surprising since this genus, much as the remaining mormyrids and in contrast to gymnotiforms, does have a corollary discharge signal of its pacemaker available.

The Tuberous Receptors of Gymnotiforms

Whereas the majority of gymnotiform genera produce pulse-like EODs, several genera, such as *Sternopygus, Eigenmannia,* and *Apteronotus*, generate continual, wave-like EODs. As reviewed by Zakon (1986a), pulse-type gymnotiforms have two types of tuberous receptor afferents, which appear to function as phase coders and amplitude coders, respectively. "Pulse-marker" afferents mark the occurrence of an EOD with a single spike, whereas "burst-duration" coders code the local EOD amplitude by the number of spikes within a burst of firing. Still, little is known about the central processing of information supplied by these two types of afferents. Behavioral experiments have shown, however, that the fish can discriminate EOD-like pulses with identical amplitude spectra but different phase functions, i.e., waveforms (Heiligenberg and Altes, 1978). Shumway and Zelick (1988) have argued that either type of afferents, pulse-markers or burst-duration coders, could be used to discriminate sex-related differences in EOD waveform.

Wave-type gymnotiforms have T-type and P-type afferents to code phase and amplitude, respectively. T-units fire a single spike per EOD cycle, phase-locked to the zero-crossing of the EOD waveform. For a given stimulus amplitude, the latency of firing may show a standard deviation of as little as 10 μsec. Individual T-units vary with regard to the extent that their latency shortens with increasing signal amplitude, with some units being almost immune to amplitude modulations. P-units, on the other hand, fire intermittently and with extremely variable latencies. With increasing amplitude, their latency of firing within the EOD cycle becomes less variable, and their probability of firing a spike within a given EOD cycle increases. By raising the stimulus amplitude sufficiently, one can induce some P-units to respond similarly to a T-unit, i.e., to fire on each EOD cycle and with a fairly constant latency. It thus appears that, under normal conditions, T- and P-units operate within different dynamic ranges but are otherwise rather similar. Sanchez and Zakon (1990) demonstrated that a T-afferent of *Sternopygus* receives inputs from several tuberous receptor organs, whereas a P-afferent receives input from a single pore. This structural organization would explain why T-afferents have lower thresholds than P-afferents and are normally being driven more closely to saturation.

Each tuberous electroreceptor afferent projects to three separate topographic maps within the ELL, and T-afferents and P-afferents, in turn, project to different layers and cell types within the ELL, forming the beginning of separate central nervous pathways for the processing of phase information and amplitude information, respectively. These pathways then merge within the torus semicircularis of the midbrain (Carr and Maler, 1986). Physiological and anatomical studies have shown that the three maps within the ELL differ inversely with regard to their temporal and spatial resolutions of electric images, with the most lateral map showing highest temporal and lowest spatial resolution (Shumway, 1989a, 1989b).

THE DEVELOPMENT OF ELECTRORECEPTORS AND LATERAL LINE ORGANS

Recent studies have focused on the development of electroreceptors, the differentiation of central nervous structures, and the emergence of the jamming avoidance response (JAR) in the gymnotiform fish *Eigenmannia*, with the ultimate goal of linking behavioral and neuronal development.

Vischer (1989a, 1989b) and Vischer et al. (1989) showed that mechanoreceptive and electroreceptive afferent fibers of the anterior lateral line nerve, which innervates all mechano- and electroreceptors on the head surface, are already present in the periphery before the respective lateral line receptors develop. Moreover, early ablation of such afferent fibers prevented the formation of mechanoreceptors and electroreceptors in those areas which would have been innervated by the ablated fibers (Vischer and Heiligenberg, 1989). This suggests that outgrowing fibers induce the development of mechanoreceptors and electroreceptors. These fibers, identified by the nature of their central projection, must either convey specific information to the periphery to induce the development of their respective kinds of receptors from general primordial cells, or they must recognize different types of primordial cells that give rise to their respective types of receptors. In a similar way, new receptors are induced to develop in the regenerating skin of adult fish after invasion by afferent fibers (Zakon, 1986b). The study of the development of mechanoreceptors and electroreceptors offers particular experimental advantages as processes of growth and migration can readily be followed and manipulated in the transparent skin of embryos and larvae.

Following the emergence of single electroreceptors, subsequent divisions of some of these receptors lead to the establishment of local clusters, or "receptor units" (Zakon, 1986a; Vischer, 1989b), and each receptor within a unit is innervated by collaterals of the same, original afferent fiber. Tuberous electrosensory afferents are either of the P type or of the T type, with the latter being more sensitive and more sharply tuned to the fish's fundamental EOD frequency. As mentioned above, Sanchez and Zakon (1990) discovered that T-type afferent fibers of *Sternopygus* innervate clusters of receptors, whereas P-type afferents innervate single, isolated receptors. P-type afferents and T-type afferents contact different targets in the ELL and thus can be distinguished on the basis of their central projections (Mathieson et al., 1987). Therefore, these two types of afferent fibers must also differ in their influence upon receptor formation in that only T-afferents induce the division of receptors, which leads to the establishment of clusters.

THE TUNING OF ELECTRORECEPTORS AND PACEMAKERS

Each EOD is triggered by a command pulse originating in the medullary pacemaker (see review by Dye and Meyer, 1986). The pacemaker of electric fish with continual, wave-type EODs triggers each discharge cycle and thus fires at a high, regular rate which may exceed 1000 Hz in some species. In an undisturbed individual, living at a stable temperature, the coefficient of variation of the pacemaker interval, which is its standard deviation divided by its mean, may be as small as 0.0001. For a pacemaker rate of 1000 Hz, this amounts to a jitter of less than 0.1 μsec (Bullock et al., 1975). Regular electronic function generators are not more stable.

As extensively reviewed by Zakon (1986a), tuberous electroreceptors of wave-type species are most sensitive at a stimulus frequency near the fundamental frequency, i.e., the repetition rate of the individual's EODs. Receptor tuning is measured by determining the minimal stimulus intensity required for receptor recruitment as a function of stimulus frequency, and the minimum of this tuning curve defines the "best frequency" of the receptor. Much as some hair cells in auditory systems (Crawford and Fettiplace, 1981), electroreceptors briefly oscillate, or "ring" at their best frequency when stimulated by a pulse. As the fish's EOD frequency changes with age, receptor tuning follows. In some species, the pacemaker frequency can be changed through administration of steroid hormones, with androgens shifting the frequency into a male-typical range. Such experimentally induced shifts in pacemaker frequency are followed within days by a corresponding shift in electroreceptor tuning, even if the electric organ has been silenced by high spinal transection so that the fish's electroreceptors are no longer driven (Keller et al., 1986; Meyer et al., 1987). Androgens affect the tuning of electroreceptors in the same manner even after lesions of the medullary pacemaker nucleus (Ferrari and Zakon, 1989).

Electroreceptors lack efferent innervation, and they are induced to develop in regenerating skin following innervation by primary afferents. Surprisingly, regenerated electroreceptors are also tuned to the fish's pacemaker frequency even if its EOD has been silent throughout the period of regeneration (Zakon, 1986b; Ferrari and Zakon, 1989). It thus appears that no form of electrical or neural communication between electroreceptors and pacemaker is required to achieve matching of their tuning frequencies. Which mechanism could then account for this phenomenon? Koch (1984) pointed out that neuronal membranes, due to their voltage-dependent resistances, can be seen as second-order systems and that their resonance frequency could be changed by altering their conductance. If age-related or hormone-induced changes in pacemaker frequency were accomplished in this manner, then similar conductance changes in electroreceptor membranes could achieve corresponding shifts in their tuning frequency. Yet, the accuracy of the match observed in these fish is surprising.

Steroid hormones also affect the cellular morphology of current-generating cells in electric organs, and the ensuing alterations of their electric properties lead to a change in the waveform of the EOD (Hagedorn and Carr, 1985; Bass and Volman, 1987). Most significantly, such changes so far could only be induced in species with sexually dimorphic EODs (Bass, 1986b).

ELECTROLOCATION: BEHAVIOR, NEUROPHYSIOLOGY, AND ANATOMY

Objects moving in the vicinity of the fish's body surface locally distort the amplitude and phase function, i.e., waveform, of the EOD, and these distortions represent the electric image of the object. The local EOD signal is monitored continually by electroreceptors. These are distributed all over the fish's body surface, though most densely at the rostral end, and primary afferents relay their information to somatotopically organized maps in the ELL of the hindbrain. Further processing of electric image information is performed in still higher-order electrosensory maps in the midbrain and cerebellar structures, and a topographic convergence with the retinotopic map in the tectum opticum yields a multimodal spatial representation of the world outside (Bastian, 1986a; Bell, 1986; Bell and Szabo, 1986; Carr and Maler, 1986).

Recent behavioral studies on mormyrids (von der Emde, 1990) have shown that electrolocating fish can discriminate between ohmic and capacitive properties of objects. The detection of capacitive features should be essential for the exploration and assessment of living tissues, and the behavioral discrimination of ohmic and capacitive loads is most likely achieved through a joint evaluation of signals provided by phase-coding and amplitude-coding tuberous electroreceptors (see section on tuberous receptors, above).

Most recent studies on central-nervous structures involved in electrolocation behavior have focused on the role of corollary discharges in mormyrids (see above) and upon a descending, recurrent pathway to the ELL of gymnotiforms.

The pyramidal cells of the ELL code amplitude modulations of EOD-like signals and project to the torus semicircularis of the midbrain. Collaterals of their axons also project topographically to the nucleus praeeminentialis which provides two kinds of recurrent, descending input to the ELL. Recent intracellular studies by Bastian and Bratton (1990 [Fig. 1]) and Bratton and Bastian (1990 [Fig. 10]) have shown that these two forms of inputs originate from different types of cells in the nucleus praeeminentialis. Multipolar cells project to the eminentia granularis pars posterior, which in turn sends axons of its granule cells to the ELL to form a parallel fiber system. This input appears to affect the ELL in its entirety by rendering the responses of pyramidal cells more phasic, i.e., more rapidly adapting to sustained inputs. This form of general inhibition functions as a gain control (Bastian, 1986b) by enabling the pyramidal cells of the ELL to code amplitude modulations over a wide range of mean amplitude levels, much as a visual system can operate under vastly different light levels.

A second form of descending input originates from the stellate cells of the nucleus praeeminentialis which form a topographically reciprocal projection to the ELL and generate another, apparently more localized, parallel fiber system. The function of this input is still unknown, but on the basis of its topographic restriction one may speculate that it forms the basis of a local attention mechanism. Unfortunately, selective elimination of this input is still not possible.

A further function of descending recurrent inputs may be the establishment of central representations of expected patterns of sensory inputs. Novelties would then be detected as contrasts against this background. Evidence for this function appears to emerge from more recent studies by Bastian (personal communication) on the electrosensory system of gymnotiform fish, and Bell's (1989) exploration of efference copy mechanisms in the electrosensory system of mormyrid fish demonstrate the existence of central representations of expected sensory information.

ELECTRIC COMMUNICATION: BEHAVIOR, NEUROPHYSIOLOGY, AND ANATOMY

Detailed reviews of this topic (Hopkins, 1988, Kramer, 1990) have described species- and gender-related differences in EODs and their significance in the context of social communication. Gender-related differences in EOD waveform can be traced back to sexual dimorphisms in electric organ morphology and physiology, which have been shown to be hormone dependent (see section above on tuning of electroreceptors). All electric fish appear to be capable of modulating the rate of their EODs in specific ways, and such modulations serve as signals in the context of aggression and courtship.

One of the most extensively studied forms of social communication is the JAR in gymnotiform electric fish of the genera *Eigenmannia* and *Apteronotus*. These fish produce nearly sinusoidal EODs of highly stable, though individually different, frequencies. When exposed to an interfering sinusoidal signal of a frequency close to that of its own EOD, *Eigenmannia* will lower its frequency in response to a slightly higher interfering frequency, and it will raise its frequency in response to a slightly lower frequency. Frequency differences of a magnitude between 2 and 6 Hz cause strongest responses. The behavioral, physiological, and anatomical aspects of the JAR have been reviewed (Heiligenberg, 1991d). A similar form of JAR is observed in the mormyriform fish, *Gymnarchus*, the only mormyrid with a wave-type EOD (Bullock et al., 1975). The behavioral rules and principles of neuronal implementation of the JAR in *Gymnarchus* are very similar to those of the JAR in *Eigenmannia*.

Recent studies have focused upon the organization of the medullary pacemaker in gymnotiforms and its modulation by diencephalic and mesencephalic inputs (see reviews in Heiligenberg, 1990, 1991d; Kawasaki and Heiligenberg, 1989, 1990; Keller et al., 1991). Fish of the genera *Eigenmannia* and *Apteronotus*, for example, may raise their pacemaker frequency either in a smooth manner or very abruptly. The latter form, called "chirping", may even lead to a brief interruption of their EODs. These two forms of modulation are induced by separate classes of cells in the diencephalic prepacemaker nucleus, which innervate the medullary pacemaker nucleus.

By applying glutamate receptor blockers to pacemaker nuclei *in vitro* and *in vivo*, Dye et al., (1989) discovered that the NMDA (*N*-methyl-D-aspartate) receptor blocker APV (D(–)-2-amino-5-phosphonovaleric acid) selectively and reversibly inhibited smooth frequency rises, while the AMPA ([±]α-amino-3-hydroxy-5-methylisoxazole-4-propionic acid) receptor blockers GAMS (γ-D-glutamylaminomethylsulfonic acid) and CNQX (6-cyano-7-nitroquinoxaline-2,3-dione) inhibited chirping. Through the use of different glutamate receptors, the same neuronal network can thus be induced to operate in strikingly different modes. Based upon preliminary anatomical and physiological studies, Dye et al., (1989) had assumed originally that the NMDA-mediated input originated from the medial portion of the diencephalic prepacemaker nucleus (PPn-G), and that the AMPA-mediated input originated from the lateral portion of this nucleus (PPn-C). Recent studies by Metzner and Heiligenberg (1991a) confirmed the findings made about the PPn-C but also discovered that the PPn-G of *Eigenmannia* generates an AMPA-mediated smooth frequency rise of the pacemaker frequency and that a newly discovered, additional prepacemaker nucleus, a sublemniscal nucleus (SPPn) of the midbrain (Keller et al., 1991), generates an NMDA-mediated tonic excitation of the pacemaker nucleus. This tonic excitation appears to be inhibited whenever the fish intends to lower its pacemaker frequency. In the related wave-species *Sternopygus*, the diencephalic PPn can generate an NMDA-mediated, smooth frequency rise of the pacemaker frequency, whereas the mesencephalic SPPn can shut off the EOD via an NMDA-mediated, selective, and sustained depolarization of the relay cells in the pacemaker nucleus (Keller et al., 1991).

An even larger variety of EOD modulations is seen in the gymnotiform genus *Hypopomus* which fires discrete, pulse-like EODs (Kawasaki and Heiligenberg, 1989, 1990). This genus has three functionally distinct subdivisions of its diencephalic prepacemaker nucleus, the PPn-I, PPn-G, and PPn-C, respectively. Activation of the PPn-I slows down, and ultimately stops, the rhythm of the pacemaker nucleus via GABA-mediated inhibition of the pacemaker cells. Activation of the PPn-C generates rapid EOD bursts, or "chirps", via AMPA-mediated excitation of the relay cells, and activation of the PPn-G causes a smooth rise of the EOD rate by NMDA-mediated excitation of the pacemaker cells. In addition, activation of the mesencephalic prepacemaker nucleus (SPPn) causes a sustained, NMDA-mediated depolarization of the relay cells and thus shuts off the EOD activity much as in the case of *Sternopygus*.

Gravid females of *Eigenmannia* hover in floating vegetation and spawn in response to repeated courtship signals of the resident male. These signals consist of short interruptions, or "chirps", in the otherwise continuous flow of the fish's EODs. In the absence of a male, a sufficiently gravid female may even deposit her eggs in the vicinity of an electrode from which a male's chirps are being broadcast (Hagedorn, 1986). More recent, unpublished studies show that even electronic mimics of chirps facilitate spawning. Since chirps contain both low-frequency and high-frequency spectral components, they are coded by the ampullary as well as by the tuberous electroreceptor system (Metzner and Heiligenberg, 1991b). A central nervous pathway formed by neurons that respond to chirps leads via the torus semicircularis of the midbrain and the nucleus electrosensorius of the diencephalon to the vicinity of the pituitary (Heiligenberg et al., 1991; Metzner and Heiligenberg, 1991b). Along this substrate, the perception of long barrages of a male's chirps may ultimately influence the state of the female's pituitary and initiate the necessary hormonal steps for the release of eggs. This system may be an excellent model system to study mechanisms through which social signals manipulate the hormonal state of a receiver.

Anatomical tracing studies and immunohistochemical investigations in the genus *Apteronotus* have revealed hypothalamic inputs to the pituitary that are linked to structures in the diencephalon and telencephalon (Johnston and Maler, 1992). These structures, in turn, could mediate effects upon the endocrine state that are caused by the perception of a variety of social and environmental signals. Conversely, extensive peptidergic projections from hypothalamic areas to diencephalic structures involved either in the detection of social signals or in their production could mediate pituitary influences upon the detection and generation of such signals (Weld and Maler, 1992; Yamamoto et al., 1992).

NEURAL MECHANISMS UNDERLYING SENSORY HYPERACUITY

The JAR of *Eigenmannia* requires the analysis of amplitude and phase modulations which characterize the signal resulting from the interference of a neighbor's EOD with the fish's own EOD. Weakest detectable jamming signals yield modulations in the phase, or the timing, of zero-crossings, as small as 0.3 μsec (see review by Heiligenberg, 1991d). So called T-type receptor afferents code the timing of zero-crossings with a mean standard deviation of

30 μsec, and the fish must, therefore, enhance its acuity in the detection of phase modulations by two orders of magnitude. Neurons in lamina 6 of the torus semicircularis code the timing of zero-crossings with a mean standard deviation already three times smaller than do T-type receptor afferents. Still higher-order neurons in deeper laminae of the torus respond to phase modulations on the order of a few microseconds, and, with very long recordings and subsequent averaging, one can demonstrate an even higher sensitivity. This suggests that parallel convergence of inputs from these neurons should yield the extreme sensitivity of the still higher-order neurons in the diencephalon which can detect still smaller phase modulations within the very short response latency of the JAR, which is 0.2 to 0.3 s. The necessity of massive spatial convergence of information becomes evident if one limits stimulus regimens to smaller areas of the body surface. The ensuing restriction in the number of contributing receptors significantly reduces the sensitivity of phase-coding neurons as well as that of the JAR itself (Kawasaki et al., 1988).

Network simulations of lamina 6 have explored the acuity of phase discrimination that individual neurons could achieve under very plausible biophysical conditions (Lytton, 1991). These neurons, having soma diameters of approximately 5 μm, are too small to be studied by classical physiological approaches. They respond to the "differential phase" between the arrival time of action potentials at their dendrites and the arrival time of potentials at their soma by being inhibited or excited if one input leads or lags, respectively, with reference to the other. On the basis of their ultrastructural organization they should be able to resolve temporal disparities as small as 1 μsec.

ACKNOWLEDGMENTS

I thank Carl D. Hopkins and Heinrich A. Vischer for kindly providing figures for this article, and John Spiro for his critical comments on the manuscript.

REFERENCES

Bass, A. H., Electric organs revisited: evolution of a vertebrate communication and orientation organ, in *Electroreception,* Bullock, T. H. and Heiligenberg, W., Eds., John Wiley & Sons, New York, 1986a, 13.

Bass, A. H., A hormone-sensitive communication system in an electric fish, *J. Neurobiol.,* 17, 131, 1986b.

Bass, A. H. and Volman, S. F., From behavior to membranes: testosterone-induced changes in action potential duration in electric organs, *Proc. Natl. Acad. Sci. U.S.A.,* 84, 9295, 1987.

Bastian, J., Electrolocation: behavior, anatomy, and physiology, in *Electroreception,* Bullock, T. H. and Heiligenberg, W., Eds., John Wiley & Sons, New York, 1986a, 577.

Bastian, J., Gain control in the electrosensory system: a role for the descending projections to the electrosensory lateral line lobe, *J. Comp. Physiol., A,* 158, 505, 1986b.

Bastian, J. and Bratton, B., Descending control of electroreception. I. Properties of nucleus praeeminentialis neurons projecting indirectly to the electrosensory lateral line lobe, *J. Neurosci.,* 10, 1226, 1990.

Bell, C. C., Electroreception in mormyrid fish: central physiology, in *Electroreception,* Bullock, T. H. and Heiligenberg, W., Eds., John Wiley, & Sons, New York, 1986, 423.

Bell, C. C., Sensory coding and corollary discharge effects in mormyrid electric fish, *J. Exp. Biol.,* 146, 229, 1989.

Bell, C. C., Mormyromast electroreceptor organs and their afferent fibers in mormyrid fish. II. Intra-axonal recordings show initial stages of central processing, *J. Neurophysiol.,* 63 (2), 303, 1990a.

Bell, C. C., Morymromast electroreceptor organs and their afferent fibers in mormyrid fish. III. Physiological differences between two morphological types of fibers, *J. Neurophysiol.,* 62 (2), 319, 1990b.

Bell, C. C. and Grant, K., Corollary discharge inhibition and preservation of temporal information in a sensory nucleus of mormyrid electric fish, *J. Neurosci.,* 9, 1029, 1989.

Bell, C. C. and Szabo, T., Electroreception in mormyrid fish: central anatomy, in *Electroreception,* Bullock, T. H. and Heiligenberg, W., Eds., John Wiley & Sons, New York, 1986, 375.

Bell, C. C., Zakon, H. H., and Finger, T. E., Mormyromast electroreceptor organs and their afferent fibers in mormyrid fish. I. Morphology, *J. Comp. Neurol.,* 286, 391, 1989.

Bennett, M. V. L. and Obara, S., Ionic mechanisms and pharmacology of electroreception, in *Electroreception,* Bullock, T. H. and Heiligenberg, W., Eds., John Wiley & Sons, New York, 1986, 157.

Bodznick, D. and Boord, R. L., Electroreception in chondrichthyes: central anatomy and physiology, in *Electroreception,* Bullock, T. H. and Heiligenberg, W., Eds., John Wiley & Sons, New York, 1986, 225.

Bratton, B. and Bastian, J., Descending control of electroreception. II. Properties of nucleus praeeminentialis neurons projecting directly to the electrosensory lateral line lobe, *J. Neurosci.,* 10, 1241, 1990.

Bullock, T. H. and Heiligenberg, W., Eds., *Electroreception,* John Wiley & Sons, New York, 1986.

Bullock, T. H., Behrend, K., and Heiligenberg, W., Comparison of jamming avoidance responses in gymnotoid and gymnarchid electric fish: a case of convergent evolution of behavior and its sensory basis, *J. Comp. Physiol.,* 103, 97, 1975.

Carr, C. E. and Maler, L., Electroreception in gymnotiform fish: central anatomy and physiology, in *Electroreception,* Bullock, T. H. and Heiligenberg, W., Eds., John Wiley & Sons, New York, 1986, 319.

Crawford, A. C. and Fettiplace, R., An electrical tuning mechanism in turtle cochlear hair cells, *J. Physiol.,* 312, 377, 1981.

Dye, J. C. and Meyer, J. H., Central control of the electric organ discharge in weakly electric fish, in *Electroreception,* Bullock, T. H. and Heiligenberg, W., Eds., John Wiley, & Sons, New York, 1986, 71.

Dye, J., Heiligenberg, W., Keller, C. H., and Kawasaki, M., Different classes of glutamate receptors mediate distinct behaviors in a single brainstem nucleus, *Proc. Natl. Acad. Sci. U.S.A.,* 86, 8993, 1989.

Ferrari, M. B. and Zakon, H. H., The medullary pacemaker nucleus is unnecessary for electroreceptor tuning plasticity in *Sternopygus, J. Neurosci.,* 9, 1354, 1989.

Finger, T. E., Electroreception in catfish: behavior, anatomy and electrophysiology, in *Electroreception,* Bullock, T. H. and Heiligenberg, W., Eds., John Wiley, & Sons, New York, 1986, 287.

Finger, T. E., Bell, C. C., and Carr, C. E., Comparisons among electroreceptive teleosts: why are electrosensory systems so similar?, in *Electroreception,* Bullock, T. H. and Heiligenberg, W., Eds., John Wiley & Sons, New York, 1986, 465.

Fritzsch, B. and M¨unz, H., Electroreception in amphibians, in *Electroreception,* Bullock, T. H. and Heiligenberg, W., Eds., John Wiley & Sons, 1986, 483.

Hagedorn, M., The ecology, courtship, and mating of gymnotiform electric fish, in *Electroreception,* Bullock, T. H. and Heiligenberg, W., Eds., John Wiley & Sons, New York, 1986, 497.

Hagedorn, M. and Carr, C. E., Single electrocytes produce a sexually dimorphic signal in South American electric fish, *Hypopomus occidentalis* (Gymnotiformes, Hypopomidae), *J. Comp. Physiol. A,* 156, 511, 1985.

Heiligenberg, W., Coding and processing of electrosensory information in gymnotiform fish, *J. Exp. Biol.,* 146, 255, 1989.

Heiligenberg, W., Electrosensory systems in fish, *Synapse*, 6, 196, 1990.

Heiligenberg, W., Recent advances in the study of electroreception, *Curr. Opinion Neurobiol.,* 1, 187, 1991a.

Heiligenberg, W., Sensory control of behavior in electric fish, *Curr. Opinion Neurobiol.,* 1, 633, 1991b.

Heiligenberg, W., The jamming avoidance response (JAR) of the electric fish, *Eigenmannia*: computational rules and their neuronal implementation, *Semin. Neurosci.,* 3 (1), 3, 1991c.

Heiligenberg, W., *Neural Nets in Electric Fish,* The Computational Neuroscience Ser., MIT Press, Cambridge, MA, 1991d.

Heiligenberg, W. and Altes, R. A., Phase sensitivity in electroreception, *Science,* 199, 1001, 1978.

Heiligenberg, W., Keller, C. H., Metzner, W., and Kawasaki, M., Structure and function of neurons in the complex of the nucleus electrosensorius of the gymnotiform fish *Eigenmannia*: detection and processing of electric signals in social communication, *J. Comp. Physiol. A,* 169, 151, 1991.

Hopkins, C. D., Behavior of mormyridae, in *Electroreception*, Bullock, T. H., and Heiligenberg, W., Eds., John Wiley & Sons, New York, 1986, 527.

Hopkins, C. D., Neuroethology of electric communication, *Annu. Rev. Neurosci.,* 11, 497, 1988.

Johnston, S. A. and Maler, L., Anatomical organization of the hypophysiotrophic systems in the electric fish, *Apteronotus leptorhynchus, J. Comp. Neurol.,* 317, 421, 1992.

Kalmijn, A., The detection of electric fields form inanimate and animate sources other than electric organs, in *Handbook of Sensory Physiology,* Vol. 3, Pt. 3, Fessard, A., Eds., Springer-Verlag, Berlin, 1974, 147.

Kalmijn, A., Theory of electromagnetic orientation: a further analysis, in *Comparative Physiology of Sensory Systems,* Bolis, L., Keynes, R. D., and Maddrell, S. H. P., Eds., Cambridge University Press, Cambridge, 1984, 525.

Kalmijn, A., Detection of weak electric fields, in *Sensory Biology of Aquatic Animals*, Atema, J., Fay, R. R., Popper, A. N., and Tavolga, W. N., Eds., Springer-Verlag, Berlin, 1987, 151.

Kawasaki, M., African electric fish, *Gymnarchus*, use the identical computational algorithm as South American electric fish for their jamming avoidance response, Neuroscience Meeting (New Orleans) Abstract, 650.11, 1991.

Kawasaki, M. and Heiligenberg, W., Distinct mechanisms of modulation in a neuronal oscillator generate different social signals in the electric fish *Hypopomus, J. Comp. Physiol. A,* 165, 731, 1989.

Kawasaki, M. and Heiligenberg, W., Different classes of glutamate receptors and GABA mediate distinct modulations of a neuronal oscillator, the medullary pacemaker of a gymnotiform electric fish, *J. Neurosci.,* 10, 3896, 1990.

Kawasaki, M., Rose, G. J., and Heiligenberg, W., Temporal hyperacuity in single neurons of electric fish, *Nature,* 336, 173, 1988.

Keller, C. H., Zakon, H. H., and Yialamas-Sanchez, D., Evidence for a direct effect of androgens upon electroreceptor tuning, *J. Comp. Physiol. A,* 158, 301, 1986.

Keller, C. H., Kawasaki, M., and Heiligenberg, W., The control of pacemaker modulations for social communication in the weakly electric fish *Sternopygus, J. Comp. Physiol. A,* 169, 441, 1991.

Kirschbaum, F., Myogenic electric organ precedes the neurogenic organ in apteronotid fish, *Naturwissenschaften*, 70, 205, 1983.

Koch, C., Cable theory in neurons with active linearized membranes, *Biol. Cybernetics*, 50, 15, 1984.

Konishi, M., Deciphering the brain's code, *Neural Computation*, 3(1), 1, 1991.

Kramer, B., *Electro-Communication in Teleost Fish: Behavior and Experiments*, Springer-Verlag, Berlin, 1990.

Lytton, W., Simulations of a phase-comparing neuron of he electric fish *Eigenmannia, J. Comp. Physiol. A,* 117, 1991.

Mathieson, W. B., Heiligenberg, W., and Maler, L., Ultrastructural studies of physiologically identified electrosensory afferent synapses in the gymnotiform fish *Eigenmannia, J. Comp. Neurol.,* 255, 526, 1987.

Metzner, W. and Heiligenberg, W., Midbrain and diencephalic links within the neuronal network underlying the jamming avoidance response in *Eigenmannia*, (Neurosciences Conf. Abstr.) 560.10, 1991a.

Metzner, M. and Heiligenberg, W., The coding of signals in the electric communication of the gymnotiform fish *Eigenmannia*: from electroreceptors to neurons in the torus semicircularis of the midbrain, *J. Comp. Physiol. A,* 169, 135, 1991b.

Meyer, J. H., Leong, M., and Keller, C. H., Hormone-induced and maturational changes in electric organ discharges and electroreceptor tuning in the weakly electric fish *Apteronotus, J. Comp. Physiol. A,* 160, 385, 1987.

Mugnaini, E. and Maler, L., Cytology and immunohistochemistry of the nucleus of the lateral line lobe in the electric fish *Gnathonemus petersii* (Mormyridae) brain. Evidence suggesting that GABA-ergic synapses mediate an inhibitory corollary discharge, *Synapse,* 1, 1, 1987.

Northcutt, R. G., Electroreception in non-teleost bony fishes, in *Electroreception*, Bullock, T. H. and Heiligenberg, W., Eds., John Wiley & Sons, New York, 1986, 257.

Ronan, M., Electroreception in cyclostomes, in *Electroreception*, Bullock, T. H. and Heiligenberg, W., Eds., John Wiley & Sons, New York, 1986, 209.

Sanchez, D. Y. and Zakon, H. H., The effects of postembryonic receptor cell addition on the response properties of electroreceptive afferents, *J. Neurosci.,* 10, 361, 1990.

Shumway, C. A., Multiple electrosensory maps in the medulla of weakly electric gymnotiform fish. I. Physiological differences, *J. Neurosci.*, 9, 4388, 1989a.

Shumway, C. A., Multiple electrosensory maps in the medulla of weakly electric gymnotiform fish. II. Anatomical differences, *J. Neurosci.*, 9, 4400, 1989b.

Shumway, C. A. and Zelick, R. D., Sex recognition and neuronal coding of electric organ discharge waveform in the pulse-type weakly electric fish, *Hypopomus occidentalis, J. Comp. Physiol. A,* 163, 465, 1988.

Szabo, T., Anatomy of the specialized lateral line organs of electroreception, in *Handbook of Sensory Physiology,* Vol. 3, Pt. 3, Fessard, A., Ed., Springer-Verlag, Berlin, 1974, 13.

Szabo, T. and Fessard, A., Physiology of electroreceptors, in *Handbook of Sensory Physiology,* Vol. 3, Pt. 3, Fessard, A., Ed., Springer-Verlag, Berlin, 1974, 59.

Vischer, H., The development of lateral line receptors in *Eigenmannia* (Teleostei, Gymnotiformes). I. The mechanoreceptive lateral line system, *Brain Behav. Evol.,* 33, 205, 19891.

Vischer, H., The development of lateral line receptors in *Eigenmannia* (Teleostei, Gymnotiformes). II. The electroreceptive system, *Brain Behav. Evol.,* 33, 223, 1989b.

Visher, H. A. and Heiligenberg, W., The development of electro- and mechanoreceptors in *Eigenmannia* is induced by innervation of epidermal cells, *Soc. Neurosci. Abstr.,* (15), 117, 10, 1989.

Vischer, H. A., Lannoo, M. J., and Heiligenberg, W., The development of the electrosensory nervous system in *Eigenmannia* (Gymnotiformes). I. The peripheral nervous system, *J. Comp. Neurol.,* 290, 16, 1989.

Von der Emde, G., Discrimination of objects through electrolocation in the weakly electric fish, *Gnathonemus petersii, J. Comp. Physiol. A,* 167, 413, 1990.

Von der Emde, G. and Bleckmann, H., Extreme phase sensitivity of afferents which innervate mormyromast electroreceptors, *Naturwissenschaften,* 79, 131, 1992.

Weaver, J. C. and Astumian, R. D., The response of living cells to very weak electric fields: the thermal noise limit, *Science*, 247, 459, 1990.

Weld, M. M., and Maler, L., Substance P-like immunoreactivity in the brain of the gymnotiform fish *Apteronotus leptorhynchus:* presence of sex differences, *J. Chem. Neuroanat.,* 5, 107, 1992.

Yamamoto, T., Maler, L., and Nagy, J. I., Organization of galanin-like immunoreactive neuronal systems in weakly electric fish *(Apteronotus leptorhynchus), J. Chem. Neuroanat.,* 5, 19, 1992.

Zakon, H. H., The electroreceptive periphery, in *Electroreception,* Bullock, T. H. and Heiligenberg, W., Eds., John Wiley & Sons, New York, 1986a, 103.

Zakon, H. H., The emergence of tuning in newly generated tuberous electroreceptors, *J. Neurosci.,* 6, 3297, 1986b.

6 Vision

Russell D. Fernald

INTRODUCTION

Fish vision has attracted the attention of scientists for a very long time, and inquiring minds might wonder why? There are many reasons, but most often cited are the enormous number of fish species (>26,000) and the fact that fish have succeeded in occupying virtually every living space in the hydrosphere. Because of this success, fish live in a remarkable variety of photic environments and their evolutionary histories reflect success in habitats with diverse levels of illumination and spectral range. Since the physical rules about how light acts in these different aquatic environments are known, this crucial selective force can be understood through its varied effects on the structure and function of the fish eye. Fish eyes are sufficiently similar to those of land-living vertebrates that discoveries in these animals also instruct us about retinal structure and function in other species.

There have been several recent comprehensive reviews of fish vision (e.g., Ali, 1974; Lythgoe, 1979; Nicol, 1989; Douglas and Djamgoz, 1990), so this chapter will identify major topics of interest in fish vision and assess recent progress towards understanding how and what fish see. The major theme will be that, through evolution, fish visual systems have become adapted to maximize detection of important stimuli in their photic environment. Understanding the visual system elements necessary for seeing to survive has led to important discoveries, specifically about fish vision and, more generally, about how vertebrate eyes work (Land and Fernald, 1992).

THE VISUAL ENVIRONMENT

LIGHT UNDERWATER

Since organic evolution began about 5 billion years ago, the earth has been exposed to about 10^{14} regular light/dark cycles (S. Hawking, personal communication) which have profoundly influenced the course of animal life. Consequently, light and its daily rhythm are certainly the most profound selective forces on earth. Fish visual systems beautifully illustrate abundant adaptations which maximize the visual detection and recognition of important objects. Such adaptations are perhaps more obvious in fish because the receptor and object both are in water, a well-defined medium whose physical properties can

0-8493-8042-1/93/$0.00+$.50

be described. The optical characteristics of water affect illumination intensity, spectral quality, and directional distribution. Water is also the transmission channel between stimulus and detector and provides the background for stimuli to be viewed (e.g., Muntz, 1990). Comparing species that live in water with distinctly different optical properties can reveal how mutual adaptations occur among stimulus, environment, and detector.

Light from the sun, moon, and stars is filtered by the atmosphere before entering the water column (Loew and McFarland, 1990). Photons will be reflected, refracted, and, to some extent, polarized at the water's surface before becoming available to fish below. These physical transformations, in turn, depend critically on the surface of the water since typically more light will enter through a still surface than a rough one. Generally, these transformations at the water's surface mean that the visual environment faced by fish depends on surface conditions, and it is quite different at dawn and dusk, when sunlight arrives at a lower angle via a longer atmospheric path, resulting in much lower intensities, shifted towards the shorter wavelengths (Munz and McFarland, 1977). Interestingly, even ultraviolet wavelengths penetrate water sufficiently to be used for fish vision (McFarland, 1986) and the recent discovery of ultraviolet (UV) detection systems in fish (Avery et al., 1983) supports this notion (see below). Irradiance in the water column can be expected to influence both the spectral sensitivity and the absolute sensitivity of the visual system of fish in each particular ecosystem.

The changes in lighting characteristics mentioned above are due to differential absorption or reflection of photons of different wavelengths, but once light enters water, scatter, or the redirection of photons due to reflection, refraction, and diffraction play an important role in determining how light forms images. Scatter decreases the penetration of light through water and diminishes the directionality of light, eliminating shadows and hence directionality cues. Light coming from an image is also scattered, reducing brightness and consequently reducing available information about the image (Lythgoe, 1972).

Fluctuations in time and space of the intensity of underwater light result from surface wave actions. Somewhat surprisingly, these effects can be seen at depths up to 70 m and it appears that the intensity fluctuations induced by surface wave actions differ at different depths as a function of the wavelength of the dominant waves (Loew and McFarland, 1990).

In addition to the illumination parameters familiar to human viewers, light from the sun is also polarized, which we cannot detect without instruments. This polarization can be measured underwater as well, where it can arise from three sources: transmission from polarized skylight, reflections at the air/water interface, and scatter by water and suspended particles underwater (Waterman, 1954). As a result, the distribution of polarized light underwater is very complex (e.g., Loew and McFarland, 1990). Despite the fact that polarized light can be detected by fish, it is not clear whether, or how, this information is used.

Fish viewing an object underwater need to make important judgments about what to do next, such as eat, mate, or flee. These judgments, which we subjectively know as seeing, correspond to integrating spectral irradiance over a two-dimensional space. In addition, polarization information might be important. To date, measurements of such irradiance plus polarization distributions have not been made underwater as comprehensively as they have, for example, in human psychophysical experiments (but see Loew and McFarland, 1990). Consequently, we do not know from measuring the stimulus in space directly what cues might be most salient for fish vision, so information about what might be important for fish vision comes from direct experiments designed to discover what, and how, fish see.

Objects Underwater

Seeing objects underwater requires that they differ sufficiently from the background along some visual dimension to be detectable. By viewing color and patterns in the objects fish look at, scientists have inferred what might constitute detectable differences and these have been tested. Numerous adaptations of the coloration of fish suggest that much can be done to reduce conspicuousness. For example, fish are typically dark on the dorsal surface and light on the ventral surface (countershading) because the downwelling light is much brighter so animals appear more uniform (Cott, 1940). Silvery sides are also common among fish because symmetrical underwater light distribution makes the light entering the eye the same whether the fish is present or not (Muntz, 1990). That these strategies have evolved to confuse predators is supported by countermeasures predators use to overcome such camouflage. Many fish capture prey by viewing them in silhouette, defeating both countershading and mirror strategies. In turn, some animals have evolved ventral photophores which generate light, obscuring their silhouette (Denton, 1970; Lawrey, 1974). Still other methods of concealment have been identified and used to suggest how the fish visual system might work (e.g., Muntz, 1990).

FISH OPTICAL SYSTEMS

The optical system of fish eyes has evolved to collect light and form a focused image for analysis by the retina. The challenge of seeing underwater has resulted in novel solutions to these fundamental problems. Understanding such solutions is one path towards learning how the visual system functions. This section will discuss how images are formed and how they are focused on the retina. Since fish eyes continue to grow throughout life (see below), the solutions must satisfy both immediate and future needs of vision.

Lenses obey fundamental physical laws about bending light, so constraints on their properties are known, providing a straightforward strategy for analysis. Despite this, details of exactly how the optics of the fish eye work have been assessed in only a handful of species.

Collecting Light

The sensitivity and acuity of an eye depend on the brightness of the image reaching the retina. In viewing a point of light, the brightness of the image increases in proportion to both the intensity of the source and area of the pupil, and decreases in proportion to the square of the distance from the source to the plane of the pupil (Fernald, 1988). For deep-sea fish, there are point sources of light, such as bioluminescent organisms or organs of other fish, so light-gathering ability has probably been the primary selective force leading to large eyes in these creatures.

However, fish living in higher levels of ambient daytime light near the surface of the water view scenes illuminated by extended sources for which the retinal illuminance is related both to pupillary area and the focal length of the eye (Fernald, 1990a). The focal length is defined as the distance behind the lens at which an object infinitely far away is focused. Because the focal length of the fish lens is proportional to the lens radius (Matthiessen's Ratio; Matthiessen, 1882) as is the pupillary area, retinal luminance depends only on the intensity of illumination for surface-dwelling fish (Fernald, 1990a). Consequently, proportionately larger eyes in this habitat do not collect proportionately more light, suggesting there must be other selective forces favoring larger eyes.

Focus of Light

The optics of animals living underwater are different from those of animals living in air in one fundamentally important respect: underwater there is no air-cornea interface which provides significant optical power to land-living vertebrates. In humans, for example, the air-cornea interface provides about 43 diopters of refractive power, with the lens responsible for an additional 13 diopters which is changed for focusing (Westheimer, 1968). Eyes underwater have no optical benefit from the cornea because its refractive index is nearly identical to that of water. Hence swimmers, deprived of 80% of their optical power, must strain to focus anything. Thus, in all underwater eyes, the refractive power of the cornea is neutralized and the lens must provide all the dioptric strength. In addition, lenses of fish must have a short focal length minimizing eye size to preserve their hydrodynamic profile. Together, these constraints have been met in fish through the evolution of a spherical lens with very high refractive power. Exceptions to this rule include skates and other rays, which have proportionally larger eyes and presumably, therefore, aspherical lenses (Cohen, 1990), as well as a few species of fish that live at the air-water interface, including the "four-eyed" minnow, *Anableps* sp. and the amphibious fish *Dialommus fuscus*.

Although the cornea does not contribute to the optical power of the aquatic eye, shallow-living diurnal fish often have yellow corneas and, occasionally, yellow lenses. These yellow optical filters are thought to serve at least two purposes (Walls, 1942; Tansley, 1965; Moreland and Lythgoe, 1968; Lythgoe,

1979). First, the yellow filter reduces the amount of shortwave light, which is most scattered and hence contains degraded information about the image. Second, the yellow filter can counter camouflage illumination of animals swimming above, making them appear as shadows. Since any filter necessarily reduces the sensitivity of the eye, some fish species have a differential distribution of corneal pigment (e.g., Moreland and Lythgoe, 1968), and others have corneal chromatophores, providing a filter in the day, but becoming virtually transparent at night (Orlov and Gamburtzeva, 1975).

Structurally, fish eyes are superficially quite similar to those of terrestrial vertebrates, although in fish the eye is typically nearly hemispheric and has a short axial length. The actual shape of the eye depends on the amount and nature of focusing (accommodation) in the species, since the lens moves within the globe to achieve different states of focus (see below). In fish that do not accommodate, the retinal surface is nearly hemispheric, whereas in fish that do accommodate, the axis along which accommodation occurs, typically the nasal-temporal axis, is considerably longer.

Deep-sea fishes have eyes specially adapted to allow greater photon capture or visual field size (see Locket, 1977). Many mesopelagic fish species have evolved asymmetrical tubular eyes, in which pupillary size increases as does the length of the tube because space does not allow enlargement of a spherical eye. This type of eye has been found in 30 of the 750 mesopelagic species, distributed among 11 families (Marshall, 1971), suggesting that it may have evolved several times. In these eyes, which appear like "sections" of a larger eye, the lens is spherical and often fills the dorsal part of the eye. The main retina lies at the end of the "tube" and satisfies Matthiessen's ratio, meaning that it is at a distance appropriate for a typical fish eye lens. This results in a larger photon capture and hence increased sensitivity along the main axis of the eye. There is an accessory retina, which extends up the medial wall of the eyecup, from the choroid fissure to the dorsal region of the eye. In some fish, this accessory retina has light directed onto it by a bundle of transparent fibers (Lockett, 1977). These tubular eyes, similar to those which have evolved in some species of owls, are largely immobile. The second adaptation to the deep-sea environment is the appearance of retinal diverticula presumed to increase the visual field (Munk, 1966; Frederiksen, 1973). The diverticula project from the eyecup laterally through a slitlike opening. They presumably receive light via reflecting crystals located in a well-developed argenteum located laterally near the primary lens. Retinas in these fish often contain only rod photoreceptors, arranged in layers and the rod-like photoreceptors are significantly longer than those in surface-dwelling fish (Lockett, 1977).

Lens Quality

As noted above, fish lenses are spherically shaped and have significant dioptric power. Spherical lenses have fascinated scientists since early attempts to understand how light is refracted by nearly spherical raindrops to produce rainbows. Spherical lenses could be subject to spherical aberration or any of five other primary aberrations characteristic of optical systems (Fincham, 1959). Of these, only chromatic aberration is a function of the material of the lens itself, while the rest depend on the structure of the lens and the shape of its focusing surfaces. Four of these aberration types represent failures of the lens to focus an image onto a flat plane. Since the retina is curved, this is not a constraint in physiological optics. However, spherical or chromatic aberration or both could adversely affect the image quality of spherical fish lenses. Spherical aberration occurs when light rays entering at different distances from the optic axis are focused at different distances from the lens and chromatic aberration occurs when light of different wavelengths focuses at different distances from a lens along its main axis (longitudinal chromatic aberration) or different distances off the main axis of the lens (lateral chromatic aberration).

As shown by direct measurement, fish lenses are of very high quality despite the fact that they are spherical (See Figure 1 and Fernald and Wright, 1983, 1985a). This quality is all the more needed because most fish eyes have no pupil to restrict the light path through the center as occurs in terrestrial vertebrates. It has long been posited that the spherical fish lens is of high quality because there exists a refractive index gradient within it, being high in the center of the lens and decreasing continuously and symmetrically with radius in all directions. Maxwell (1854) allegedly first postulated this gradient while contemplating his breakfast herring (cited in Pumphrey, 1961), though the idea arose independently in the writing of Young (1801), Maxwell (1854), and Matthiessen (1886). Brewster (1816) even described his attempts at producing a spherical lens with a refractive index gradient to test directly this hypothesis.

Matthiessen (1882) originally proposed that the refractive index (N) would vary with distance r from the lens center as $N^2 = a - br^2$. Only recently, Axelrod et al. (1988) devised a technique to measure the gradient within an intact lens. They measured the entrance and exit locations of thin laser beams passing through the meridional plane of a fish lens and then matched these to fitted curves. Although the function is not exactly that proposed by Matthiessen, the gradient is roughly as he predicted over 100 years ago!

The optical resolution of the fish lens is very good, and indeed it approaches the theoretical diffraction limit (Fernald, 1990a). The resolution of the lens can be compared with that of the retina which is a result of the spacing of photoreceptors, which sets an upper limit on the resolution possible given assumptions about their connections to the brain (Fernald, 1988). The fish lens resolution appears to be about 10 times better than the retinal resolution would predict (Northmore and Dvorak, 1979; Fernald and Wright, 1985a), which may give the animals increased contrast sensitivity. Only at very small lens sizes does the theoretical limit of the lens resolution match retinal resolution, suggesting that,

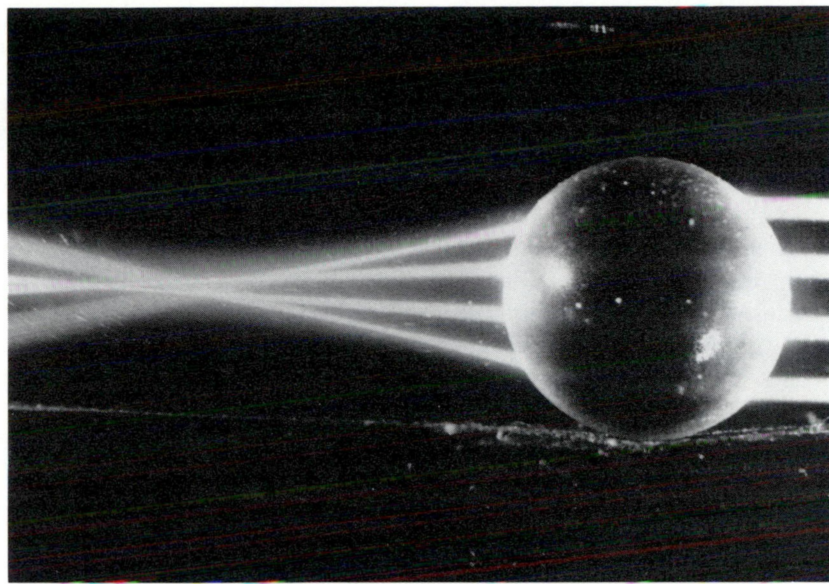

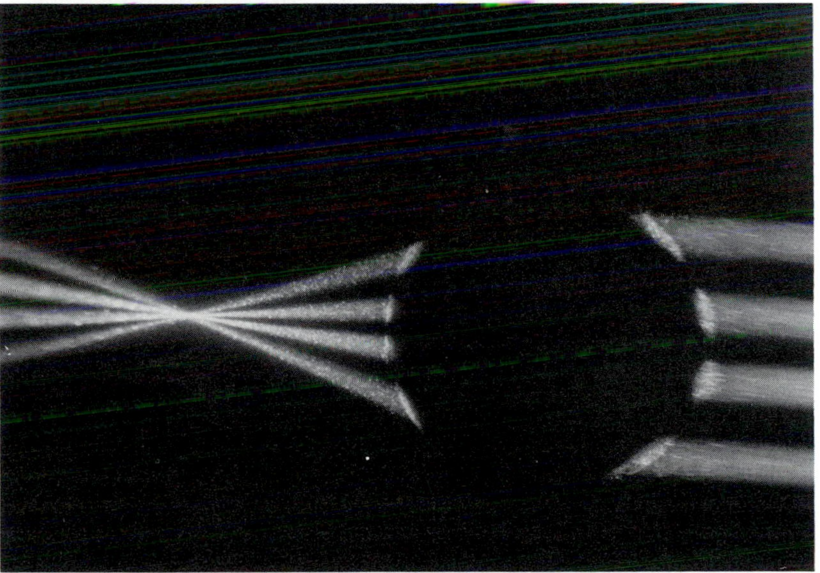

FIGURE 1. A comparison between (top) a glass sphere of uniform refractive index (n = 1.53) and (bottom) a fresh *Haplochromis burtoni* lens. In both cases, four argon laser beams (494 nm) are being focused. The glass bead shows significant longitudinal spherical aberration while the fish lens shows a fine focus, typical of lenses of this species.

for fish that do not live in the deep sea, growth of the eyes serves primarily to increase the optical resolution and perhaps to improve contrast detection. Ultimately, resolution available to the animal depends not only on lens quality but also on receptor size, receptor spacing, and the convergence of receptors on higher-order cells.

The chromatic aberration of teleost lenses which is the distance between focal planes of light of different wavelengths (see above), has been measured by numerous investigators and ranges from 2 to 5% of the focal length (reviewed in Fernald, 1990a). The important question is whether or not this amount of chromatic aberration is of any consequence to the animal. If we assume that a single wavelength is focused in the receptor layer of the retina, then the effect of any other wavelength would be to make a colored fringe around the images projected onto the retina. For these fringes to be visible, the width of the fringe would have to extend over the separation of at least two cones. For the cichlid, *Haplochromis burtoni*, with chromatic error of 1.9% of the focal length, the computed blur circle produced by a 650-nm light produced when a 2-mm lens is focused for 500-nm light would be too small to see (see Fernald and Wright, 1985a for details). Clearly, however, the potential problem is proportionately worse for fish with smaller lenses and may have been one of the selective forces for eye growth in the course of evolution.

Accommodation

The visual image available to the retina of the animal depends not only on the quality of the lens but also on the state of focus of the image. Accommodation, or focusing, in fishes, when it occurs, results from the movement of the spherically shaped lens within the globe, and not by changing the shape of the lens, as is the case in vertebrates that live in air. In most fish examined, the lens moves approximately parallel to the plane of the pupil along an axis of pupillary eccentricity (see Figure 2; e.g., Fernald, 1990a). This means that the different regions of the retina are focused at different distances of different parts of the visual world simultaneously. For example, in the relaxed accommodative state in fish, the lens lies nearer to the nasal pole of the eye. Thus, the temporal part of the retina (nasal visual field) is focused for near vision and the nasal pole retina (temporal visual field) for far vision (Fernald and Wright, 1985b).

Beginning in the last century (Beer, 1894), scientists have measured the refractive state of the fish. This is difficult because the state of accommodation is a function of the angle of view, since the lens moves within the globe. Thus, measures of accommodation may not be comparable among species and occasionally not even comparable within the same study. Since lens movement varies with fish size (Fernald and Wright, 1985b), the size of the specimen used will also affect the detectability and measurement of accommodative state.

Retinoscopic measurement of refractive state is difficult in fish eyes because they are small. Moreover, because the measurements are made underwater, retinoscopic reflection must be corrected for in the air-water interface,

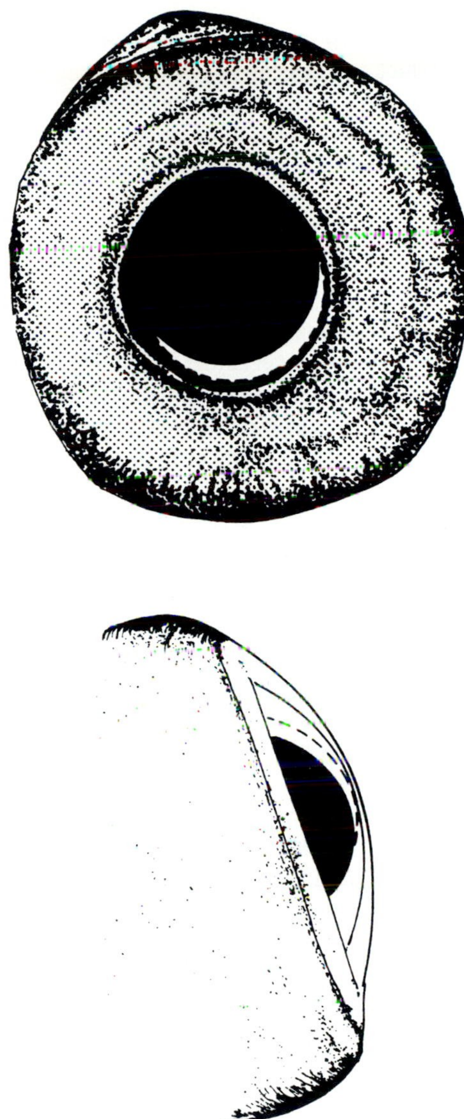

FIGURE 2. Top: Lateral view of the eye of the teleost, *Haplochromis burtoni,* showing the direction and amount of lens movement during maximal accommodation. Bottom: Dorsal view of the same eye. These show that the movement is essentially in the plane of the pupil. Both traced from a double exposure photograph. (Redrawn from Fernald and Wright, 1985b.)

or the point of focus will be estimated incorrectly (Hueter and Gruber, 1980). An additional difficulty is where to consider the source of the retinoscopic reflection in the eye. In many fish, there are no retinal blood vessels to use as landmarks, since the choroid rete provides oxygen to the eye. Assuming the wrong reflective surface can lead to sizable errors, particularly in small eyes, because the retinal thickness is nearly constant over a wide range of eye sizes. Finally, a significant error can be introduced because of the chromatic aberration of the optical system (Nuboer and Genderen-Takken, 1978).

Fernald and Wright (1985b) recently completed a thorough analysis of the lens movement and refractive state in the African cichlid fish, *H. burtoni*. This species is particularly suited for such studies because it grows throughout life and depends critically on vision for its social interactions. The results of these analyses reveal a number of general results about the teleost visual systems.

The magnitude of the lens movement responsible for accommodation increases with eye size, but the power of the lens decreases, so that the net effect is nearly neutralized. That is, as the animal gets larger, its accommodative mechanism is appropriately scaled to body size. The slight loss of accommodative amplitude with growth means that the distance to the nearest focal point increases from about 2 to 4.5 cm so that the visual near point maintains its same position in the front of the body.

Because accommodative amplitude depends strongly on retinal location, accommodative movement has substantially different effects at different retinal locations (Fernald, 1990a). The retinal poles have large potential accommodative changes of different sign while the central retina does not. Thus, images at a wide range of distances can be focused on either the temporal or nasal retina, in contrast to the central retina. Moreover, if the temporal pole of the retina is focused on a near object, the nasal pole must be focused on a distant object and vice versa.

Although there has been some dispute about the state of focus of the fish eye, much may be due to variability in the measurement techniques, suggesting that, as Nicol (1963) stated: "It still has to be demonstrated, convincingly, that the plane of focus in the fish eye at rest lies elsewhere than in the receptor layer; the so-called error may be one not of refraction but rather of method."

FISH RETINAL STRUCTURE, FUNCTION, AND DEVELOPMENT

The retinas of fishes do not differ fundamentally from those of other vertebrates. Images formed by the lens fall on the retina, a thin, transparent laminar structure comprised of seven major cell types (photoreceptors, bipolar, horizontal, interplexiform, amacrine, ganglion, and Müller) located at the back of the eye. It is sandwiched between the vitreous body towards the lens and the pigmented epithelium towards the sclera. Due to the enormous number of fish species and diversity of their habitats, fish retinas exhibit considerable structural variety.

Beginning with Cajal (1892), the many studies comparing retinas among fish species (e.g., Wunder, 1925; Verrier, 1928; Dathe, 1969; Bathelt, 1970;

Wagner, 1972), including a descriptive catalog (Ali and Anctil, 1976), reveal that behavioral and ecological constraints are the best predictors of retinal morphology. So, for example, it is not too surprising that nocturnal, solitary fish have radically different retinas than do closely related, day-active social fish. Over the course of evolution, retinal structure has been shaped by the selective forces of the light intensity and spectral content in the environment and the spatial resolution required for survival by the animal. Differences which have resulted from differential selective pressures can be found in (1) overall retinal thickness, (2) diversity of retinal cell subtypes, especially photoreceptors, and (3) regional specializations of retinal cells reflecting view-specific requirements.

Cell Phenotypes

Pigmented Epithelium

The pigmented epithelium (PE) is a layer of cuboidal cells joined along their common boundaries by complex junctions. Towards the retina, long processes of the PE interdigitate with the retina, surrounding the light-sensitive tips of the photoreceptors (Kuwabara, 1979; Bok, 1985). Within the cells are rod-shaped pigment granules which are highly light absorbent and optically isolate the cone outer segments and protect rod outer segments during bright illumination (see below). At low levels of illumination, these pigment granules migrate towards the back of the eye, into the cell bodies of the PE cells. Fish that have adapted to low light levels have fewer melanocytes and more reflecting particles, with the result being a retinal tapetum which reflects light (Wagner and Ali, 1978). These optical functions of the PE are only one part of its role in retinal function. Pigment epithelial cells phagocytize the tips of rod and cone outer segments, which are being continuously renewed (Young, 1970). Photoreceptor tip shedding and hence phagocytosis is highly rhythmic at both the cellular and molecular levels, having both circadian and diurnal features (Korenbrot and Fernald, 1989).

Photoreceptors

Photoreceptors absorb photons and transduce this light energy into electrical energy which can be interpreted by the nervous system. There are two principle types of photoreceptors, rods and cones. Müller (1851, 1857) first observed that of these two types of cells, cones were common in animals with diurnal activity cycles and rods common in animals active at night. From this, Schultze (1866) proposed a possible functional distinction, namely that cones were responsible for vision in bright light (photopic vision) and rods for vision in dim light (scotopic vision). In the subsequent 100 years, this has been shown to be true and is termed the duplicity theory of photoreceptor function. Generally, cones are responsible for vision during bright light and they are connected singly or in small groups to provide for high-resolution vision. In contrast, rods function best in dim light, and they are connected together to maximize sensitivity. The duplicity theory has proven to be an important cornerstone of our understanding of visual function.

The originally observed structural differences between rods and cones have been confirmed and extended to numerous ultrastructural features (e.g., Cohen, 1972). Recently, remarkably complete understanding of the biochemistry underlying the transduction of photons into neural signals has been elucidated (Yau and Baylor, 1989).

Differences between rod and cone photoreceptors are clear at the level of the light microscope. Rod photoreceptors have long, cylindrical outer segments while cones have shorter, conical outer segments. Within the rod outer segment, the light-absorbing opsin is localized on disks which are arranged in stacks contained within an outer membrane. Opsin in the cone outer segments is also located on disks, but these disks are contiguous with the outer membrane. Rod and cone inner segments also differ significantly. Rod inner segments are small and the cell body, located vitread of the inner limiting membrane contains darkly staining chromatin. Synaptic connections are spherical, and hence called spherules, which contain one or two ribbon-associated synaptic complexes. Cone inner segments are nearly as large as the outer segments, containing large mitochondria and endoplasmic reticulum; the cell nucleus is located at the outer limiting membrane and stains lightly. Synaptic connections to the cones (pedicles) are large, containing up to 100 dendrites from bipolar and horizontal cells. There is substantial evidence that the acidic amino acids L-glutamate or L-aspartate are the photoreceptor neurotransmitters (reviewed in Lasater, 1990). As in all vertebrates, cone photoreceptor subtypes differ in their spectral sensitivity, and this difference is reflected in the cone length in teleosts. The shortest cones are sensitive at the shortest wavelengths, with the very smallest sensitive in the UV range (Downing et al., 1986) and the longest cones are sensitive to the longest wavelengths (Levine and MacNichol, 1979). Although these differences in length were once postulated to compensate for chromatic error in the lens (Eberle, 1968), the size differences are too small to provide complete correction (Fernald, 1988). In teleost retinas, cones also appear in pairs which may or may not have similar spectral sensitivities. In fish that use vision extensively, the cone photoreceptors are arranged in highly ordered mosaics, as first noted by Ryder (1895). These mosaics appear organized in alternating rows of single and double cones ("row pattern") or in squares with four cone pairs set around a single, central cone ("square pattern") (e.g., Eigenmann and Shafer, 1900; Wagner, 1990). An example of a matrix is given in Figure 3. In animals with different cone types, the chromatic organization of the cone mosaic optimizes chromatic detection (Fernald, 1981b). Transitions from one mosaic type to another associated with a change in life habits have been reported in salmon and trout (Lyall, 1957a, 1957b) and in

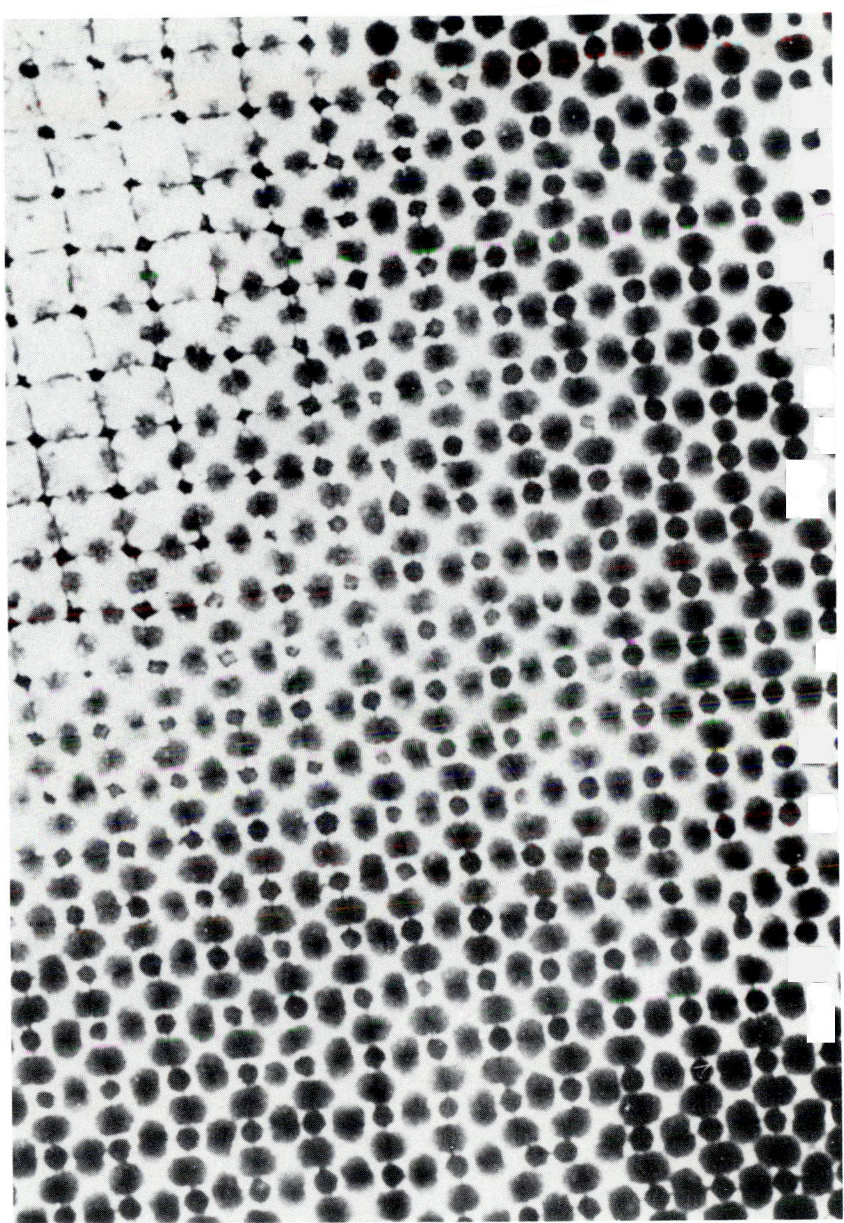

FIGURE 3. This photomicrograph illustrates the crystalline-like array of photoreceptors found in the photoreceptor layer of a teleost, *Haplochromis burtoni*. This histological section is 3 μm thick, cut at an oblique angle through the outer segments of the cone photoreceptors and stained with toluidine blue.

perch (Ahlbert, 1969). More recently, it has been reported that UV sensitive cones (Bowmaker et al., 1991) are lost in salmon when they migrate to the sea (e.g., Douglas and Hawryshyn, 1990). Changes in the cone phenotypes and mosaic also occur in fish that metamorphose (Evans and Fernald, 1990). In elasmobranchs, cone photoreceptors are rare and no such mosaics exist. Indeed, there is evidence that there is only one cone type in elasmobranchs (Cohen, 1990).

Photoreceptors and the pigmented epithelium exhibit some remarkable transformations in synchrony with the light/dark cycle. As mentioned above, phagocytosis and renewal of the photoreceptor outer segments occur in a regular rhythm (O'Day and Young, 1978). The disks containing opsin are produced every day, with a strong diurnal rhythm as reflected in opsin abundance and in opsin mRNA production (Korenbrot and Fernald, 1989). In contrast, phagocytosis of a single photoreceptor occurs only once every few days, but then is linked to the light/dark cycle as well.

Although photoreceptor outer segment renewal is found in all vertebrates, additional retinal transformations related to the light/dark cycle enhance retinal function in teleosts. The most dramatic of these are the retinomotor movements, which are the wholesale daily rearrangement of photoreceptor positions. In the light-adapted retina, cone photoreceptors move vitread so they are adjacent to the outer limiting membrane which lies between cone nuclei and the horizontal cells, pigment granules migrate vitread into the processes around the cone outer segments, forming a dense absorbing layer sclerad to the cone outer segment boundary, and the rod photoreceptors extend sclerad through elongation of the myoids beyond the pigment layer so they no longer receive photic input. Retinomotor movements occur in some other cold-blooded vertebrates as well as in some bird species (e.g., Walls, 1942). In teleosts, these movements can be as much as 40 to 80 μm and the pigment may migrate about 100 μm. The consequence of this positioning during the daytime is that only the cones receive light input, and the cone outer segments are optically isolated from each other. In darkness, these processes reverse so that the rod myoids contract bringing the rod photoreceptors adjacent to the outer limiting membrane, the pigment granules migrate to the perikarya of the pigmented epithelial cells, and the cone myoids elongate, placing the cones sclerad to the outer limiting membrane but vitread to the pigmented epithelial band so the rod photoreceptors are in the plane of focus. Consequently, the cone photoreceptors can receive light and may participate in vision at low light levels. These retinomotor movements exhibit a circadian rhythm in several species (John and Haut, 1964; John et al., 1967; John and Kaminester, 1969) The regulation of retinomotor responses and cellular events which underlie them have been reviewed (Burnside and Nagle, 1983) as has the distribution among teleosts (Ali and Wagner, 1975). It has been reported that the cone mosaic changes form during retinomotor movements (Kunz, 1980), although this does not appear to be the case (Fernald, 1982).

There are three generally accepted notions about the selective advantages of such movement. First, retinomotor movements result in essentially two separate functional retinas: cones for photopic vision and rods for scotopic vision (Herzog, 1905); second, the dispersed pigment granules contribute to photopic acuity by absorbing scattered rays refracted out of the cone outer segments (Garten, 1907); third, the pigmented epithelium shields rod outer segments from bleaching during bright light (Garten, 1907). A fourth advantage is to increase retinal acuity, especially in animals with small eyes (Fernald, 1988). In small fish, the retinal acuity of an all-cone retina resulting from retinomotor movements is twice what it would be if rod photoreceptors were also in place (Fernald, 1988). Conversely, the uniform field of rod photoreceptors during periods of low light intensity would increase threshold detection since no photons fall on low-sensitivity cones. Thus, retinomotor movements offer a neat solution to maximal packing of cone photoreceptors during daylight.

Three other dynamic morphological changes related to the light/dark cycle have been documented in teleost retinas. In some species, the cone synaptic connections decrease in the dark relative to the light, as measured by the length of the synaptic ribbons (Wagner, 1973a and b). This effect has not been seen in rod photoreceptor connections. Although functional measurements which define the possible role of these changes have not been made, the strong inference is that the cone input is strongly diminished in the dark. A second change in the strength of the connection between cones and other cells during the light/dark cycle are the finger-like extensions arising from the lateral horizontal cell dendrites which contact the cone plasmalemma (Wagner, 1980). These "spinules", which only occur on lateral dendrites of two types of horizontal cells (H1 and H2), are almost completely absent in dark-adapted retina whereas as many as 29 per dendrite may be present in the light (Wagner and Speck, 1983). As a result, contact area can increase by as much as 25% during bright light illumination (Kriete et al., 1984). There is now some evidence that these spinules mediate feedback information between cones and horizontal cells (Weiler and Wagner, 1984) which may be responsible for the change in receptive fields in dark-adapted retinas. The area surrounding the receptive field typically has response properties which are antagonistic to those of the center (antagonistic surround) and these disappear in dark-adapted retinas (Raynauld et al., 1979). The third change occurs in the coupling between horizontal cells (Kaneko et al., 1981; Teranishi et al., 1983). The coupling is achieved via gap junctions which increase in density and organization during the dark and rapidly (about 5 min) become loosely arranged in the light (Kurz-Isler and Wolburg, 1986, 1988). This increase in the effective receptive field size appears to increase the capture area and hence sensitivity of the rod receptors in dim light.

Bipolar Cells

Information about photon capture in photoreceptors is transmitted to the brain via bipolar and then ganglion cells in the retina. Bipolar cells in the inner nuclear layer which serve as the main conduit from photoreceptors to ganglion cells also have synapses from horizontal cells which regulate the lateral interactions across the retina and from dopaminergic interplexiform cells. In a pioneering work, Scholes (1975) examined teleost bipolar cell connections, identifying at least 10 types. He found that selective bipolar cells connect only to various spectral cone types (cone bipolar cells), that some bipolar cells connect predominantly to cones but also receive some rod input (mixed bipolar cells), and that still other bipolar cells have massive input, primarily from rods with a few red-sensitive cones included. Scholes' (1975) identification of different bipolar pathways suggests that separate processing of visual information which begins with the photoreceptors is continued in the next level of cells. The linking of the scotopic/rod pathway to the red cone bipolars in fish is different from mammalian retinas which have an exclusive cone (scotopic) pathway (e.g., Sterling et al., 1986). The bipolar cells are organized with antagonistic receptive fields with the "OFF" center cells located distal to the "ON" center cells in the inner plexiform layer. Although serotonin has been indicated as a neurotransmitter in bipolar cells of elasmobranchs (Brunn et al., 1984), in teleosts it is unknown what bipolar cells use to communicate with their partners.

Interplexiform Cells

Interplexiform cells were first described by Cajal (*Box salpa*, 1892) but only recently understood to be dopaminergic (Dowling and Ehinger, 1975, 1978) or glycine accumulating (Marc et al., 1979). The dopaminergic interplexiform cells receive input from horizontal cell axons (Marshak and Dowling, 1987), from amacrine dendrites (Dowling and Ehinger, 1978), and from the nervus terminalis via peptidergic transmitters (Zucker and Dowling, 1987). These terminate primarily on horizontal cells but also on some bipolar cells (Wulle and Wagner, 1990). The glycine-accumulating cells receive input from H1 horizontal cells (Marc and Liu, 1984) and from horizontal cell axons (Marshak and Dowling, 1987). These cells appear to have light-evoked slow responses similar to those of amacrine and bipolar cells (Hashimoto et al., 1980), suggesting that they might comprise a separate functional cell class. The dopaminergic interplexiform cells can be considered a final link in the efferent control pathway, acting via dopamine release at the level of the outer plexiform layer. Although a specific role for these cells is not yet certain, their characteristics suggest that they may modulate retinal responses.

Horizontal Cells

Horizontal cells comprise a large, distinct cell class located just vitread of the external limiting membrane. These cells are also organized in a mosaic pattern, particularly visible in young animals (Hagedorn and Fernald, 1992). There are three horizontal cells with long axons (about 500 μm) connected to cones (H1, H2, H3) and one connected to rods (H4) which has no axon (Stell and Lightfoot, 1975; Weiler, 1978; Downing, 1983). Following the initial proposals of Stell et al. (1975), four interrelated roles for horizontal cells in transforming photic information can be stated (Wagner, 1990): first they mediate chromatic interactions between different spectral cone types; second, they generate the antagonistic surround of bipolar cell receptive fields; third, they modulate spatially summed signals in the outer plexiform layer via gap junctions; and fourth, they constitute an additional pathway between the outer plexiform and inner nuclear layers. The H1 cells use gamma-aminobutyric acid (GABA) as a neurotransmitter (Lam and Steinman, 1971) but the neurotransmitters in the remaining horizontal cell types are unknown (e.g., Lasater, 1990).

Amacrine Cells

Amacrine cells are local circuit neurons which contact with every type of retinal cell except photoreceptors (Dowling, 1979). These cells may be the most diverse cell class in the central nervous system, with a record number of 48 distinct subtypes identified based on structure in one teleost, the roach, *Rutilus rutilus* (Wagner and Wagner, 1988). Matching this diversity is that found in amacrine neurotransmitter types (Lasater, 1990). GABA, glycine, and acetylcholine have been found in amacrine cells, as have numerous neuropeptides. Given the enormous diversity, amacrine cell function is difficult to characterize. Wagner (1990) suggests three possible principles of operation: first, populations of amacrine cells may act as multicellular aggregates via gap junction coupling; second, individual amacrine cells might function alone under certain conditions (Miller, 1979); third, parts of the dendritic field of amacrine cells may be functional microcircuits under certain conditions (Bloomfield and Miller, 1982).

Ganglion Cells

Ganglion cells integrate visual information and send it to the brain, not unlike the bipolar cells but in ganglion cells action potentials serve to transmit the signals. Beginning with Cajal (1892), ganglion cells have been morphologically characterized in many teleosts (e.g., Kock and Reuter, 1978; Stell and Witkowski, 1973; cichlids, Wagner, 1973b) and classified into two major groups: large (G) and small (S). The functional significance of these findings is not clear and it also is not understood whether the dendritic field size and shape have a bearing on the function of the ganglion cell. Ganglion cells may interact with each other (Sakai et al., 1986), but this also is speculative. The functional organization of their receptive fields is antagonistic center-surround organization, reminiscent of bipolar cells. Considering the chromatic information, Daw (1968) described four types of cells: (1) nonselective units which respond with a center and surround organization to all wavelengths; (2) P units which have color opponency only in the center of the receptive field; (3) O units which have color opponency throughout the receptive field; and (4) Q units which have double color opponency and are driven by the green cones. Subsequent work has confirmed and expanded this general scheme (e.g., Spekreijse et al., 1972).

Variability in Fish Retinas

Retinal thickness measured from the optic fiber layer to the outer plexiform layer is typically about 150 μm in most teleost retinas. The relative thickness of the photoreceptor layer is diagnostic for the light intensity of the habitat. For example, in diurnal species which live in relatively bright habitats, the ratio of photoreceptor thickness to neural retina is 1:1, whereas fish in low light level habitats have a ratio of 1.3:1, with the highest values found in deep-sea fish (Wagner, 1990). Animals that live in low light levels often have strikingly different photoreceptors, including arrangements of bundles of photoreceptors (Wagner and Ali, 1978).

The variation in the density of different cell types across the retina also predicts habitat and habits of the fish. Most commonly found are increases in the photoreceptor density, often in the form of a fovea-like concentration. These regions are typically in the temporal retina which registers scenes from the front of the animal (Verrier, 1928; Kahmann, 1936; Schwassmann, 1974; Wagner, 1972; Fernald, 1983; Collin and Pettigrew, 1989) and is on the axis along which focusing (accommodation) occurs (Fernald and Wright, 1985b). In regions of specialization, the ratio of photoreceptors to other neural cell types may remain constant so there is a general increase in cell density (Fernald, 1983). Although most of the regions of specialization are related to improved scotopic vision and correspond to increased cone densities, a few cases of rod foveas have been found (Munk, 1966; Locket, 1977).

Another variation in teleost retinas which has been of great interest is the relationship of the cone visual pigment spectral absorbances to the wavelength distributions characteristic of the habitats. To maximize quantum catch, the visual pigment spectral characteristics should overlap with the spectral irradiance in the water. For deep-living animals, this is certainly the case (Crescitelli et al., 1985; Bowmaker et al., 1988; Partridge et al., 1988). For many surface-living fish, this proved not to be the case, leading Lythgoe (1972) to propose that the absorbance properties of visual pigments may provide needed contrast when they are offset from the peak ambient illumination. This hypothesis has been confirmed and extended (McFarland and Munz, 1975; Munz and McFarland, 1977; Crescitelli et al., 1985; Crescitelli, 1991). Generally, it was demonstrated that the peak absorbances of photoreceptor pigments were offset as predicted and appeared to provide considerably enhanced contrast for the animals.

In some animals where color vision is used in behavioral interactions (Fernald, 1990b), the cone spectral distribution is greater than expected for simple contrast detection (Fernald and Liebman, 1980). Additional data and discussion can be found in recent reviews (Levine and MacNichol, 1979, 1982; Lythgoe, 1984; Bowmaker, 1990).

CENTRAL VISUAL PROJECTIONS

Information about the visual world leaves the retina and travels via the optic nerve to several sites in the brain for processing. The vast majority of the fibers terminate in the optic tectum, although several other regions have been found to receive visual input using morphological or physiological criteria (e.g., Fernald, 1981a; Schellart, 1990). Since most of the results from experiments aimed at understanding visual processing have been done in the tectum, they will be reviewed here.

VISUAL MAP

Some of the earliest measurements of electrical activity demonstrated that there is a one-to-one correspondence between points in visual space and points on the tectal surface (termed a retinotopic map) (Jacobson and Gaze, 1964; Schwassmann and Kruger, 1965). This organized representation of visual space also exists in the optic nerve (Scholes, 1979). The point-to-point correspondence between stimulus location and tectal response has been found in every case and is consistent with similar maps in other vertebrate visual systems. In animals with a foveal specialization consisting of increased density of retinal cells, that region is appropriately over-represented in the tectum (e.g., Schwassman, 1968), also in congruence with data from other species.

Although this point-to-point mapping appears to be a universal principle of visual systems, it does little to explain how seeing happens. In teleosts, as in other species, there has been a search for other principles of image processing, but not much is known yet about how visual information is processed in fish brains. Many investigators have measured receptive field sizes and shapes, discovering that the smallest-sized receptive fields (<1 degree) are in regions of high receptor concentration (reviewed in Guthrie, 1990). These results have all been obtained without use of chromatic stimuli. Since there is considerable behavioral evidence that many species of fish use chromatic information (see above), it will be instructive in the future to understand how colors are represented in the brain.

Visual pathways to other nontectal visual areas have been mapped in several fish (reviewed by Schellart, 1990), but there is very little physiological or behavioral evidence to correlate with this information. This is an area which needs some experiments combining physiological recordings with more naturalistic stimuli to understand the role of nontectal visual areas in seeing.

DEVELOPMENT OF THE VISUAL SYSTEM

Unlike warm-blooded vertebrates, the central nervous system of fish grows by adding neurons throughout life. This is not a small effect, since in rapidly growing animals as much as 90% of the retinal cells present at one year will have been added after hatching (Fernald, 1984). Growth continues while the animal behaves normally, feeding and breeding without interruption. Because of this capability, fish have become model systems for the analysis of growth and development, spawning numerous recent reviews (Fernald, 1989, 1991; Powers and Raymond, 1990). Consequently, the discussion here will summarize the current understanding and present some unsolved issues.

As the eye grows, one immediate question is whether the field of view changes. It does not. Several studies have shown that the field of view is independent of eye size (Easter et al., 1977; Fernald, 1984). In addition, neither the lens quality (Fernald and Wright, 1983, 1985a) nor the accommodative mechanism (Fernald and Wright, 1985b) is compromised during growth. As the animal grows, the image delivered to the retina actually improves in quality (Fernald, 1990a). Regulation of the growth of fish is complex, but in some species this is socially regulated (Fraley and Fernald, 1982). How these social signals result in metabolic changes is still unknown, but social effects on some physiological systems are beginning to be understood (Davis and Fernald, 1990).

Retinal development has been studied in several teleost species (Müller, 1952; Hollyfield, 1972; Grün, 1975; Sandy and Blaxter, 1980; Sharma and Ungar, 1980; Hagedorn and Fernald, 1992) and all of these studies confirm two general principles. First, there is a sequential production of retinal cell phenotypes which is essentially the same in all vertebrates, beginning with ganglion cells and ending with rod photoreceptors. Second, there are two distinct developmental gradients: one which extends vitread to sclerad and the other which extends from the center to the periphery of the developing eye. Fish that metamorphose exhibit an extreme form of the first developmental principle, adding rod photoreceptors at metamorphosis which can be as much as 4 months after the animal is living independently (review in Evans and Fernald, 1990).

Although the phenomenology of retinal development is known, the controlling factors are not. Recent work in other vertebrate species has shown that the final cell phenotype is not related to cell lineage (e.g., Turner and Cepko, 1987; Holt et al., 1988), but rather that unspecified local factors govern cell fate. An interesting difference between fish retinal development and that of mammals is that although cell death is an important factor in shaping mammalian retinas it is not in teleosts. Cell specification in teleosts appears to be accurate enough during development to not require subsequent adjustment through pruning.

Perhaps the most unusual feature of teleost retinal development is that it does not stop at hatching. Instead, there is continued growth of the retina from a germinal zone located at the margin of the eye. In addition to the cell addition, retinal stretching also occurs (e.g., Fernald, 1984). First described by Müller (1952), this germinal zone is a compressed version of the central-to-peripheral gradient present in embryogenesis. At the outermost edge, a single layer of pseudo-stratified epithelial cells represents the beginning of cell addition and about 150 μm centralwards a fully differentiated, laminated retinal structure exists. Consistent with the late addition of rod photoreceptors in embryogenesis (e.g., Raymond, 1985; Hagedorn and Fernald, 1992), there is a secondary zone of rod neurogenesis, central to the marginal germinal zone (Fernald, 1989). Even more remarkable, the teleost retina adds rod photoreceptors throughout the central retina (Fernald and Johns, 1980; Johns and Fernald, 1981). An example of new cell addition is given in Figure 4. These cells arise from a set of precursor cells of unknown origin which divide in the outer nuclear layer amid the differentiated layer of rod nuclei. Discovery and details of this phenomenon has recently been reviewed (Fernald, 1989). The role of these new rods appears to be to preserve the retinal sensitivity in the face of massive stretching of the extant retinal tissue (Fernald, 1985; Powers and Raymond, 1990).

As the retina grows, many more ganglion cells are added which must be integrated into the primary projection to the optic tectum. The cells are added to the optic nerve in an unusual fashion first recognized and described by Scholes (1979). The problem is that the teleost optic nerve in visually oriented animals has a retinotopic order at all ages, despite the fact that it contains an increasing number of optic fibers. Scholes (1979) found that in the ribbon-shaped optic nerve, the new fibers are all added at one edge, preserving retinotopy in the nerve but allowing for the dramatic increase in cell fiber number.

New axons reaching the optic tectum pose another topological problem during adult growth because the tectum adds cells in a pattern different from that of the retina. Whereas the retinal growth is symmetrical around the margin of the eye (Müller, 1952), the tectal growth pattern is asymmetric, appearing roughly in a horseshoe shape with the major growth zone at the rear (caudal) margin (Meyer, 1978). Despite the noncongruence of retinal and tectal growth patterns, the visual field is represented on the surface of the tectum in a consistent fashion in all sizes of animals. This requires the continuous remodeling of the terminations of optic fibers during growth.

The dynamic nature of visual system growth in adulthood is all the more remarkable because animals remain socially interactive without any signs that the nervous system is being enlarged and rearranged. This suggests that neural remodeling could be a very rapid process, possibly in phase with the light/dark cycle similar to that seen in dynamic changes of retinal cell connectivity (see above). Because of this continued visual system modification, teleosts have become favored models for studies of nervous system development.

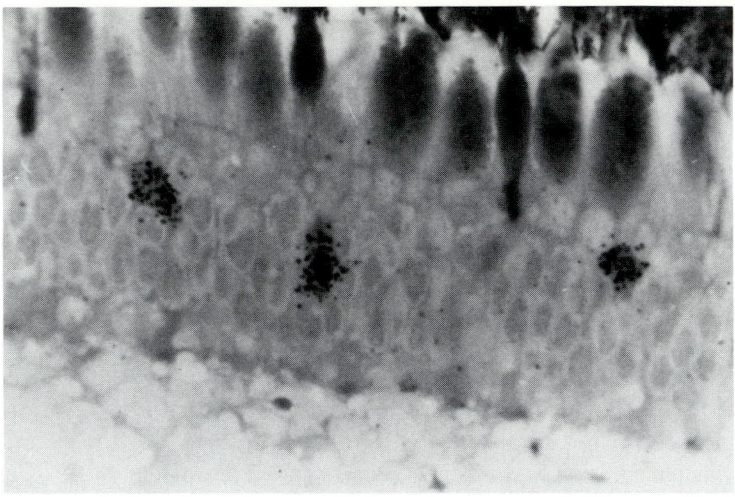

FIGURE 4. Three labeled nuclei in the outer nuclear layer of an adult *H. burtoni* injected 24 h previously with ^{3}[H]-thymidine. Such labeled nuclei are found exclusively in the outer nuclear layer and these newly divided cells differentiate into rod photoreceptors.

SUMMARY

Fish eyes are adapted to the aquatic environment in several striking ways. The lens, which collects light and focuses images, is spherical but has excellent optical properties so it can be used across its entire extent. This spherical lens focuses images on the retina by changing position within the globe since its shape must remain constant to provide the optical power needed for underwater vision. To provide effective vision in both bright and dim light, the fish retina has two remarkably different states resulting from the photoreceptors moving each day. During the day, there is an all-cone retina, the rods residing behind a thick veil of pigmented epithelium. At night, there is a reversal and the rods occupy the plane of focus while the cones recede to the rear of the eye. In this way, the photoreceptors most suited for the level of illumination are in position at the appropriate time. The cone photoreceptors have spectral sensitivities appropriate to optimize contrast in water, where the animal lives. Finally, the fish eye grows throughout life so that all the specialized adaptations required for life in water are generated anew as cells are added and nervous tissue of the visual system expanded. Taken together, these adaptations represent some of the most direct examples of natural selection at work.

ACKNOWLEDGMENTS

I thank the members of my laboratory for useful discussions and D.H. Evans for comments on an earlier draft of the manuscript. This work was supported in part by National Institutes of Health Grants EY 05051 and EY 8306.

REFERENCES

Albert, L.-B., The organization of the cone cells in the retinae of four teleosts with different feeding habits *(Perca fluviatilis* L., *Lucioperca lucioperca* L., *Acerin cernua* L. and *Coregonus albula* L.), *Ark. Zool. Ser. 2,* 22, 445, 1969.

Ali, M. A., Ed., *Vision in Fishes,* Plenum Press, New York, 1974.

Ali, M. A. and Anctil, M., *Retinas of Fishes: An Atlas,* Springer-Verlag, Berlin, 1976.

Ali, M. A. and Wagner, H.-J., Distribution and development of retinomotor responses, in *Vision in Fishes,* Plenum Press, New York, Ali, M. A., Ed., 1975, 369.

Allen, E. E. and Fernald, R. D., Scotopic visual threshold in the African cichlid fish, *Haplochromis burtoni, J. Comp. Physiol.,* 157, 247, 1985.

Avery, J. A., Bowmaker, J. K., Djamgoz, M. B. A., and Downing, J. E. G., Ultraviolet sensitive receptor in freshwater fish, *J. Physiol. London.,* 334, 23, 1983.

Axelrod, D., Lerner, D., and Sands, P. J., Refractive index within the lens of a goldfish eye, determined from the paths of thin laser beams, *Vision Res.,* 28, 57, 1988.

Bathelt, D., Experimentelle und Vergleichend morphologische untersuchungen am Visuellen System von Teleostiern, *Zool. Jahrb. Abt. Anat.,* 87, 402, 1970.

Beer, T., Die Accommodation des Fischauges, *Pflugers Arch. Gesamte Physiol.,* 58, 523, 1894.

Bloomfield, S. A. and Miller, R. F., A physiological and morphological study of the horizontal cell types of the rabbit retina, *J. Comp. Neurol.,* 208, 288, 1982.

Bok, D., Retinal photoreceptor — pigment epithelium interactions, Friedenwald Lecture, *Invest. Ophthalmol. Vis. Sci.,* 26, 1659, 1985.

Bowmaker, J. K., Visual pigments of fishes, in *The visual System of Fish,* Douglas, R. H. and Djamgoz, M. B. A., Eds., Chapman and Hall, New York, 1990, 81.

Bowmaker, J. K., Dartnall, H. J. A., and Herring, P. J., Longwave-sensitive visual pigments in some deep-sea fishes: segregation of 'paired' rhodopsins and porphyropsins, *J. Comp. Physiol. A,* 163, 685, 1988.

Bowmaker, J. K., Thorpe, A., and Douglas, R. H., Ultraviolet-sensitive cones in the goldfish, *Vision Res.,* 31, 349, 1991.

Brewster, D., On the structure of the crystalline lens in fishes and quadrupeds, as ascertained by its action on polarised light, *Philos. Trans. R. Soc. London.,* 311, 1816.

Bruun, A., Ehinger, B., and Sytsma, V. M., Neurotransmitter localization in the skate retina, *Brain Res.,* 295, 233, 1984.

Burnside, B. and Nagle, W., Retinomotor movements of photoreceptor and retinal pigment epithelium: mechanism and regulation, in *Progress in Retinal Research,* Osborne, N. and Chader, G., Eds., Pergamon Press, Oxford, 1983, 67.

Cajal, S. R., La retine des vertebres, *Cellule,* 9, 121, 1892.

Cohen, A. I., Rods and cones, in *Physiology of Photoreceptor Organs,* Fuortes, M. G. F., Ed., Springer-Verlag, New York, 1972, 63.

Cohen, J. L., Vision in elasmobranchs, in *The Visual System of Fish,* Douglas, R. H. and Djamgoz, M. B. A., Eds., Chapman and Hall, New York, 1990, 465.

Cott, H. B., *Adaptive Coloration in Animals,* Methuen, London, 1940.

Collin, S. P. and Pettigrew, J. D., Quantitative comparison of the limits on visual spatial resolution set by the ganglion cell layer in twelve species of reef teleosts, *Brain Behav. Evol.,* 34, 184, 1989.

Crescitelli, F., Visual pigments. The scotopic photoreceptors and their visual pigments of fishes: functions and adaptations, *Vision Res.,* 31, 339, 1991.

Crescitelli, F., McFall-Ngai, M., and Horowitz, J., The visual pigment sensitivity hypothesis: further evidence from fishes of varying habitats, *J. Comp. Physiol. A,* 157, 323, 1985.

Dathe, H. H., Vergleichende Untersuchungen an der Retina Mitteleuropaischer Sußwasserfische, *Z. Mikrosk. Anat. Forsch.,* 80, 269, 1969.

Davis, M. R. and Fernald, R. D., Social control of neuronal soma size, *J. Neurobiol.,* 21, 1180, 1990.

Daw, N. W., Colour-coded ganglion cells in the goldfish retina: extension of their receptive fields by means of new stimuli, *J. Physiol. London,* 197, 567, 1968.

Denton, E. J., On the organization of reflecting surfaces in some marine animals, *Philos. Trans. R. Soc. London Ser. B,* 258, 285, 1970.

Douglas, R. H. and Djamgoz, M. B. A., Eds., *The Visual System of Fish,* Chapman and Hall, London, 1990.

Douglas, R. H. and Hawryshyn, C. W., Behavioral studies of fish vision: an analysis of visual capabilities, in *The Visual System of Fish,* Douglas, R. H. and Djamgoz, M. B. A., Eds., Chapman and Hall, New York, 1990, 373.

Dowling, J. E., Information processing by local circuits: the vertebrate retina as a model system in *The Neurosciences. Fourth Study Program* Schmitt, F. O. and Warden, F. G., Eds., MIT Press, Cambridge, MA, 1979, 163.

Dowling, J. E. and Ehinger, B., Synaptic organization of the amine-containing interplexiform cells of the goldfish and cebus monkey retinas, *Science,* 188, 270, 1975.

Dowling, J. E. and Ehinger, B., The interplexiform cell system. I. Synapses of the dopaminergic neurons of the goldfish retina, *Proc. R. Soc. London Ser. B,* 201, 7, 1978.

Downing, J. E. G., Functionally Identified Interneurones in the Vertebrate (Fish) Retina: Electrophysiological, Ultrastructural and Pharmacological Studies, Ph.D. thesis, University of London, 1983.

Downing, J. E. G., Djamgoz, M. B. A., and Bowmaker, J. K., Photoreceptors of a cyprinid fish, the roach: morphological and spectral characteristics, *J. Comp. Physiol. A,* 159, 859, 1986.

Easter, S. S., Jr., Johns, P. A., and Baumann, L. R., Growth of the adult goldfish eye. I. Optics, *Vision Res.,* 17, 469, 1977.

Eberle, H., Zapfenbau, Zapfenläenge und Chromatische Aberration im Auge von *Lebistes reticulatus* (Peters Guppy), *Zool. Jahrb. Abt. Allg. Zool. Physiol. Tiere,* 74, 121, 1968.

Eigenmann, C. H. and Shafer, G. E., The mosaic of single and twin cones in the retinas of fishes, *Am. Nat.,* 34, 109, 1900.

Evans, B. I. and Fernald, R. D., Metamorphosis and fish vision, *J. Neurobiol.,* 21, 1037, 1990.

Fernald, R. D., Visual field and retinal projections in the African cichlid fish, *Haplochromis burtoni, Neurosci. Abstr.,* 7, 844, 1981a.

Fernald, R. D., Chromatic organization of a chichlid fish retina, *Vision Res.,* 21, 1749, 1981b.

Fernald, R. D., Cone mosaic in a teleost retina: no difference between light and dark adapted states, *Experentia,* 38, 1337, 1982.

Fernald, R. D., Neural basis of visual pattern recognition in fish, in *Advances in Vertebrate Neuroethology,* Ewert, J.-P., Capranica, R. R., and Ingle, D. J., Eds., Plenum Press, New York, 1983, 569.

Fernald, R. D., Vision and behaviour in an African cichlid fish, *Am. Sci.,* 72, 58, 1984.

Fernald, R. D., Growth of the teleost eye: novel solutions to complex constraints, *Environ. Biol. Fishes,* 13, 113, 1985.

Fernald, R. D., Aquatic adaptations in fish eyes, in *Sensory Biology of Aquatic Animals,* Atema, J., Fay, R. R., Popper, A. N., and Tavolga, W. N., Eds., Springer-Verlag, Berlin, 1988, 435.

Fernald, R. D., Retinal rod neurogenesis, in *Development of the Vertebrate Retina,* Finlay, B. L. and Sengelaub, D. R., Eds., Plenum Press, New York, 1989, 31.

Fernald, R. D., The optical system of fishes, in *The Visual System of Fish*, Douglas, R. H. and Djamgoz, M. B. A., Eds., Chapman and Hall, New York, 1990a, 45.

Fernald, R. D., *Haplochromis burtoni*: a case study, in *The Visual System of Fish,* Douglas, R. H. and Djamgoz, M. B. A., Eds., Chapman and Hall, New York, 1990b, 443.

Fernald, R. D., Teleost vision: seeing while growing, *J. Exp. Zool.*, 5, 167, 1991.

Fernald, R. D. and Johns, P., Retinal structure and growth in the cichlid fish, *Haplochromis burtoni, Invest. Ophthalmol. Vis. Sci.*, 69, 1980.

Fernald, R. D. and Liebman, P., Visual receptor pigments in the African cichlid fish, *Haplochromis burtoni, Vision Res.*, 20, 857, 1980.

Fernald, R. D. and Wright, S. E., Maintenance of optical quality during crystalline lens growth, *Nature (London)*, 301, 618, 1983.

Fernald, R. D. and Wright, S. E., Growth of the visual system of the African cichlid fish, *Haplochromis burtoni*: optics, *Vision Res.*, 25, 155, 1985a.

Fernald, R. D. and Wright, S. E., Growth of the visual system of the African cichlid fish *Haplochromis burtoni:* accommodation, *Vision Res.*, 25, 163, 1985b.

Fincham, W. H. A., *Optics,* Hatton Press, London, 1959.

Fraley, N. B. and Fernald, R. D., Social control of development rate in the African cichlid fish, *Haplochromis burtoni, Z. Tierpsychol.*, 60, 66, 1982.

Frederikson, R. D., On the retinal diverticula in the tubular-eyed opisthoproctid deep-sea fishes *Macropinna microstoma* and *Dolichopteryx longipes, Vidensk. Medd. Dan. Naturhist. Foren. Khobenhavn,* 136, 233, 1973.

Garten, S., Die Veranderungen der Netzhaut durch Licht, *Graefe-Saemisch Handbuch der gesamten Augenheilkunde,* Leipzig, 1907.

Grün, G., Structural basis of the functional development of the retina in the cichlid *Tilapia leucostica, J. Embryol. Exp. Morphol.*, 33, 243, 1975.

Guthrie, S. D. M., The physiology of the telostean optic tectum, in *The Visual System of Fish,* Douglas, R. H. and Djamgoz, M. B. A., Eds., Chapman and Hall, London, 1990, 279.

Hagedorn, M. and Fernald, R. D., Retinal growth and cell addition during embryogenesis in the teleost, *Haplochromis burtoni, J. Comp. Neurol.*, 321, 193, 1992.

Hashimoto, Y., Abe, M., and Inokuchi, M., Identification of the interplexiform cell in the dace retina by dye injection method, *Brain Res.*, 197, 331, 1980.

Herzog, H., Experimentelle Untersuchungen zur Physiologie der Bewegungsorgange in der Netzhaut, *Arch. Anat. Physiol. (Physiol. Abstr.)*, 516, 413, 1905.

Hollyfield, J. G., Histogenesis of the retina of the killifish *Fundulus heteroclitus, J. Comp. Neurol.*, 144, 373, 1972.

Holt, C. E., Bertsch, T. W., Ellis, H. M., and Harris, W. A., Cellular determination in the *Xenopus* retina is independent of lineage and birth date, *Neuron,* 1, 15, 1988.

Hueter, R. E. and Gruber, S. H., Retinoscopy of the aquatic eye, *Vision Res.*, 20, 197, 1980.

Jacobson, M. and Gaze, R. M., Types of visual response from single units in the optic tectum and the optic nerve of the goldfish, *Q. J. Exp. Physiol.*, 49, 199, 1964.

John, K. R. and Haut, M., Retinomotor cycles and correlated behaviour in the teleost *Astyanax mexicanus* (Fillipi), *J. Fish. Res. Board Can.*, 21, 591, 1964.

John, K. R. and Kaminester, L. H., Further studies on retinomotor rhythms in the teleost *Astyanax mexicanus, Physiol. Zool.*, 42, 60, 1969.

John, K. R., Segall, M., and Zawatzky, L., Retinomotor rhythms in the goldfish *Carassius auratus, Biol. Bull. Mar. Biol. Lab. Woods Hole,* 132, 200, 1967.

Johns, P. R. and Fernald, R. D., Genesis of rods in teleost fish retina, *Nature (London)*, 293, 141, 1981.

Kahmann, H., Uber das Foveale Sehen der Wirbeltiere. I. Uber die Fovea Centralis und die Fovea Latralis bei einigen Wirbeltieren, *Albrecht v. Graefes Arch. Ophththalmol.*, 135, 265, 1936.

Kaneko, A., Nishimura, M., Techibana, M., Tauchi, M., and Shimai, K., Physiological and morphological studies of single pathways in the carp retina, *Vision Res.,* 21, 1519, 1981.

Kirsch, M., Wagner, H., and Djamgoz, M. B. A., Dopamine and plasticity of horizontal cell function in the teleost retina: regulation of a spectral mechanism through D_1-receptors, *Vision Res.,* 31, 401, 1991.

Kock, J-H. and Reuter, T., Retinal ganglion cells in the crucian carp (Carassius carassius). I. Size and number of somata in eyes of different size, *J. Comp. Neurol.,* 179, 535, 1978.

Korenbrot, J. I. and Fernald, R. D., Circadian rhythm and light regulate opsin mRNA in rod photoreceptors, *Nature,* 337, 454, 1989.

Kriete, A., Wagner, H.-J., Haucke, M., Gerlach, B., Harms, H., and Anus, H. M., Dreidimensionale Rekonstruktion Elekronenmikroskopischer Serienschnitte zur Erfassung Synaptischer Plastizitat, *Mikroskopie (Wien),* 41, 192, 1984.

Kunz, Y. W., Cone mosaics in a teleost retina: changes during light and dark adaptation, *Experientia,* 36, 1371, 1980.

Kurz-Isler, G. and Wolburg, H., Gap-junctions between horizontal cells in the cyprinid fish alter rapidly their structure during light and dark adaptation, *Neurosci. Lett.,* 67, 7, 1986.

Kurz-Isler, G. and Wolburg, H., Light-dependent dynamics of gap junctions between horizontal cells in the retina of the crucian carp, *Cell Tissue Res.,* 251, 641, 1988.

Kuwabara, T., Species differences in the retinal pigment epithelium in *The Retinal Pigment Epithelium* Zinn, K. M. and Marmor, M. F., Eds., Harvard University Press, London, 1979, 58.

Lam, D. M. K. and Steinman, L., The uptake of gamma-^{3}H-aminobutyric acid in the goldfish retina, *Proc. Natl. Acad. Sci. U.S.A.,* 68, 2777, 1971.

Land, M. F. and Fernald, R. D., The evolution of eyes, *Annu. Rev. Neurosci.,* 15, 1, 1992.

Lasater, E. M., Neurotransmitters and neuromodulators of the fish retina, in *The Visual System of Fish,* Douglas, R. H. and Djamgoz, M. B. A., Eds., Chapman and Hall, New York, 1990, 211.

Lawrey, J. V., Lantern fish compare downwelling light and bioluminescence, *Nature,* 247, 155, 1974.

Levine, J. S. and MacNichol, E. F., Jr., Visual pigments in teleost fishes: effects of habitat microhabitat and behavior on visual system evolution, *Sens. Process,* 3, 95, 1979.

Levine, J. S. and McNichol, E. F., Jr., Color vision in fishes, *Sci. Am.,* 246, 108, 1982.

Locket, N. A., Adaptations to the deep-sea environment, in *Handbook of Sensory Physiology* Crescitelli, F., Ed., Springer-Verlag, Berlin, 1977, 67.

Loew, E. R. and McFarland, W. N., The underwater visual environment, in *The Visual System of Fish,* Douglas, R. H. and Djamgoz, M. B. A., Eds., Chapman and Hall, New York, 1990, 1.

Lyall, A. H., The growth of the trout retina, *Q. J. Micros. Sci.,* 98, 101, 1957a.

Lyall, A. H., Cone arrangement in teleost retinae, *Q. J. Microsc. Sci.,* 98, 189, 1957b.

Lythgoe, J. N., The adaptation of visual pigments to the photic environment, in *The Handbook of Sensory Physiology,* Dartnall, H. J. A., Ed., Springer-Verlag, Berlin, 1972, 566.

Lythgoe, J. N., *The Ecology of Vision,* Clarendon Press, Oxford, 1979.

Lythgoe, J. N., Visual pigments and environmental light, *Vision Res.,* 24, 1539, 1984.

Marc, R. E. and Liu, W.-L. S., Horizontal cell synapses onto glycine-accumulating interplexiform cells, *Nature, (London),* 312, 266, 1984.

Marc, R. E., Lam, D. M. K., and Stell, W. K., Glycinergic pathways in the goldfish retina, *Invest. Ophthalmol. Vis. Sci.,* 18, 34, 1979.

Marschak, D. W. and Dowling, J. E., Synapses of cone horizontal cell axons in goldfish retina, *J. Comp. Neurol.,* 256, 430, 1987.

Marshall, N. B., *Explorations in the Life of Fishes,* Harvard University Press, Cambridge, MA 1971, 185.

Matthiessen, L., Über die Beziehungen, Welche zwischen dem Brechungsindex des Kernzentrums der Krystallinse und den Dimensionen des Auges bestehen, *Pflugers Arch. Gesamte Physiol.,* 27, 510, 1882.

Matthiessen, L., Über den Physikalisch-optischen Bau des Auges der Cetacean und der Fische, *Pflugers Archiv. Gesamte Physiol. Menschen Tiere,* 38, 521, 1886.

Maxwell, J. C., Some solutions of problems, *Cambridge & Dublin Mathematical Journal,* 1, 76, 1854.

McFarland, W. N., Light in the sea — correlations with behaviors of fishes and invertebrates, *Am. Zool.,* 26, 389, 1986.

McFarland, W. N. and Munz, F. W., Part II: The photic environment of clear tropical seas during the day, *Vision Res.,* 15, 1063, 1975.

Meyer, R. L., Evidence from thymidine labelling for continuing growth of retina and tectum in juvenile goldfish, *Exp. Neurol.,* 59, 99, 1978.

Miller, R. F., The neuronal basis of ganglion cell receptive field organization and the physiology of amacrine cells, in *The Neurosciences: Fourth Study Program,* Schmitt, F. O. and Warden, F. G., Eds., MIT Press. Cambridge, MA, 1979, 227.

Moreland, J. D. and Lythgoe, J. N., Yellow corneas in fishes, *Vision Res.,* 8, 1377, 1968.

Müller, H., Zue Histologie der Netzhaut, *Z. Wiss. Zool.,* 3, 234, 1851.

Müller, H., Anatomisch-physiologische Untersuchungen uber die Netzhaut bei Menschen und Wirbelthieren, *Z. Wiss. Zool.,* 8, 1, 1857.

Müller, H., Bau und Wachstum der Netzhaut des Guppy *(Lebistes reticulatus), Zool. Jahrb. Abt. Allg. Zool. Physiol.,* 63, 275, 1952.

Munk, O., Ocular anatomy of some deep-sea teleosts, *Dana Rep.,* 70, 1, 1966.

Muntz, W. R. A., Stimulus, environment and vision in fishes, in *The Visual System of Fish,* Douglas, R. H. and Djamgoz, M. B. A., Eds., Chapman, and Hall, New York, 1990, 491.

Munz, F. W. and McFarland, W. N., Evolutionary adaptations of fishes to the photic environment, in *The Handbook of Sensory Physiology,* Crescitelli, F., Ed., Springer-Verlag, Berlin, 1977, 193.

Nicol, J. A. C., Some aspects of photoreception and vision in fishes, *Adv. Mar. Biol.,* 1, 171, 1963.

Nicol, J. A. C., *The eyes of Fishes,* Clarendon Press, Oxford, 1989.

Northmore, D. P. M. and Dvorak, C. A., Contrast sensitivity and acuity of the goldfish, *Vision Res.,* 19, 225, 1979.

Nuboer, J. F. W. and Genderen-Takken, H. V., The artefact of retinoscopy, *Vision Res.,* 18, 1091, 1978.

O'Day, W. T. and Young, R. W., Rhythmic shedding of outer segment membranes by visual cells in goldfish, *J. Cell Biol.,* 76, 593, 1978.

Orlov, O. Y. and Gamburtzeva, A. G., Dynamics of corneal colorations in fish, *Hexagrammos octagrammus, Biofizika,* 21, 362, 1975.

Partridge, J. C., Archer, S. N., and Lythgoe, J. N., Visual pigments in the individual rods of deep-sea fishes, *J. Comp. Physiol. A,* 162, 543, 1988.

Powers, M. K. and Raymond, P. A., Development of the visual system, in *The Visual System of Fish,* Douglas, R. H. and Djamgoz, M. B. A., Eds., Chapman and Hall, New York, 1990, 419.

Pumphrey, R. J., Concerning vision, in *The Cell and the Organism,* Ramsay, J. A., Eds., Cambridge University Press, Cambridge, MA, 1961, 193.

Raymond, P. A., Cytodifferentiation of photoreceptors in larval goldfish: delayed maturation of rods, *J. Comp. Neurol.,* 236, 90, 1985.

Raynauld, J. P. Laviolette, J. R., and Wagner, H.-J., Goldfish retina: a correlate between cone activity and morphology of the horizontal cell in cone pedicles, *Science,* 204, 1436, 1979.

Ryder, J. A., An arrangement of retinal cells in the eyes of fishes partially stimulating compound eyes, *Proc. Natl. Acad. Sci. U.S.A.,* 161, 1895.

Sakai, H. M., Naka, K. I., and Dowling, J. E., Ganglion cell dendrites are presynaptic in catfish retina, *Nature, (London),* 319, 495, 1986.

Sandy, J. M. and Blaxter, J. H. S., A study of retinal development in larval herring and sole, *J. Mar. Biol. Assoc. U.K.,* 60, 59, 1980.

Schellart, N. A. M., The visual pathways and central non-tectal processing, in *The Visual System of Fish,* Douglas, R. H. and Djamogz, M. B. A., Eds., Chapman and Hall, London, 1990, 345.

Scholes, J. H., Colour receptors, and their synaptic connexions, in the retina of a cyprinid fish, *Philos. Trans. R. Soc. London Ser. B,* 270, 61, 1975.

Scholes, J. H., Nerve fibre topography in the retinal projection to the tectum, *Nature (London),* 278, 620, 1979.

Schultze, M., Anatomie und Physiologie der Netzhaut, *Arch. Mikrosk. Anat. Entwicklungs Mech.,* 2, 175, 1866.

Schwassmann, H. O., Visual projections upon the tectum in foveate marine teleosts, *Vision Res.,* 8, 1337, 1968.

Schwassmann, H. O., Central projections of the retina and vision, in *Vision in Fishes,* Ali, M. A., Ed., Plenum Press, New York, 1974, 113.

Schwassmann, H. O. and Kruger, L., Organization of the visual projection upon the optic tectum of some freshwater fish, *J. Comp. Neurol.,* 124, 113, 1965.

Sharma, S. C. and Ungar, F., Histogenesis of the goldfish retina, *J. Comp. Neurol.,* 191, 373, 1980.

Spekreijse, H., Wagner, H. G., and Wohlbarsht, M. L., Spectral and spatial coding of ganglion cell responses in goldfish retina, *J. Neurophysiol.,* 35, 73, 1972.

Stell, W. K. and Lightfoot, D. O., Color-specific interconnections of cones and horizontal cells in the retina of the goldfish, *J. Compr. Neurol.,* 159, 473, 1975.

Stell, W. K. and Witkowsky, P., Retinal structure in the smooth dogfish, *Mustelus canis:* general description and light microscopy of giant ganglion cells, *J. Comp. Neurol.,* 148, 1, 1973.

Stell, W. K., Lightfoot, D., Wheeler, T., and Leeper, H., Goldfish retina: functional polarization of cone horizontal cell dendrites and synapses, *Science, N.Y.,* 190, 989, 1975.

Sterling, P., Freed, M., and Smith, R. G., Microcircuitry and functional architecture of the cat retina, *Trends Neurosci.,* 9, 186, 1986.

Tansley, K., *Vision in Vertebrates,* Chapman and Hall, London, 1965.

Teranishi, T. and Negishi, K., Dendritic morphology of a class of interstitial and normally placed amacrine cells revealed by intracellular lucifer yellow injection in carp retina, *Vision Res.,* 31, 463, 1991.

Teranishi, T., Negishi, K., and Kato, S., Dopamine modulates S-potential amplitude and dye-coupling between external horizontal cells in carp retina, *Nature (London),* 301, 243, 1983.

Turner, D. L. and Cepko, C. L., A common progenitor for neurons and glia persists in rat retina late in development, *Nature,* 328, 131, 1987.

Verrier, M. L., Recherches sur les yeux et la vision des poissons, *Bull. Biol. Fr. Belg. Suppl.,* 11, 1, 1928.

Wagner, H.-J., Vergleichende Untersuchungen uber das Muster der Sehzellan und Horizontalen in der Teleostier-Retina (Pisces), *Z. Morphol. Tiere,* 72, 77, 1972.

Wagner, H.-J., Die Nervosen Netzhautelemente von *Nannacara anomala* (Cichlidae, Teleostei) I. Darstellung nach Silberimpragnation, *Z. Zellforsch. Mikrosk. Anat.,* 137, 63, 1973a.

Wagner, H.-J., Darkness-induced reduction of the number of synaptic ribbons in fish retina, *Nature (London) New Biol.,* 246, 53, 1973b.

Wagner, H.-J., Light dependent plasticity of the morphology of horizontal cell terminals in cone pedicles of fish retinas, *J. Neurocytol.,* 9, 573, 1980.

Wagner, H.-J. and Ali, M. A., Retinal organization in goldeye and mooneye (Teleostei: Hiodontidae), *Rev. Can. Biol.,* 37, 65, 1978.

Wagner, H., Retinal structure of fishes, in *The Visual System of Fish,* Douglas, R. H. and Djamgoz, M. B. A., Eds., Chapman and Hall, New York, 1990, 109.

Wagner, H.-J. and Speck, P. T., Ein Mikroncomputergestutztes System zur Graphischen Rekonstruktion von Serienschnitten, *Verh. Anat. Ges.,* 77, 187, 1983.

Wagner, H.-J. and Wagner, E., Amacrine cells in the retina of a teleost fish, the roach (*Rutilus rutilus*). A golgi study on differentiation and layering, *Philos. Trans. R. Soc. London Ser. B,* 321, 263, 1988.

Walls, G. L., *The Vertebrate Eye and its Adaptive Radiation,* Hafner, New York, 1942.

Waterman, T. H., Polarization patterns in submarine illumination, *Science, N.Y.,* 120, 927, 1954.

Weiler, R., Horizontal cells of the carp retina. Golgi impregnation and Procion Yellow injection, *Cell Tissue Res.,* 195, 515, 1978.

Weiler, R. and Wagner, H.-J., Light-dependent change of cone-horizontal cell interactions in carp retina, *Brain Res.,* 298, 1, 1984.

Westheimer, G., The eye, in *Medical Physiology,* Mountcastle, V. B., Ed., C. V. Mosby, St. Louis, MO, 1968, 1532.

Wulle, I. and Wagner, H.-J., Dopaminergic Neurone in der Fischretina: Immunhistochemische Untersuchungen der Morphologie, Verteilung und Synaptologie, *Verh. Anat. Ges.,* 74, 21, 1990.

Wunder, W., Physiologische und Vergleichend-anatomische Untersuchungen an der Knochenfischnetzhaut, *Z. Vgl. Physiol.,* 3, 1, 1925.

Yau, K.-W. and Baylor, D. W., Cyclic GMP-activated conductance of retinal photoreceptor cells, *Annu. Rev. Neurosci.,* 12, 289, 1989.

Young, T., On the mechanism of the eye, *Philos. Trans. Royal Soc. London,* 92, 23, 1801.

Young, R. W., Visual cells, *Sci. Am.,* 223, 81, 1970.

Zucker, C. L. and Dowling, J. E., Centrifugal fibres synapse on dopaminergic interplexiform cells in the teleost retina, *Nature, (London),* 330, 166, 1987.

7 Chemoreception

Toshiaki J. Hara

INTRODUCTION

Fish are immersed in their physical and chemical environment, and their sensory systems are in continuous interactions with environmental perturbations. The aquatic environment is similar to the terrestrial environment in that it contains a multitude of chemical mixtures. However, the aquatic environment differs from the terrestrial environment in the ways in which chemical substances can be distributed: (1) molecules need to be in solution rather than in the gaseous phase to be transported, and (2) water is a slower carrier medium, both in terms of diffusion and current. Thus, solubility, rather than volatility, of compounds determines their capacity as chemical signals. Consequently, nonvolatile compounds with relatively small molecular weights are prominent substances of fish chemoreception and have been implicated in various behavioral roles.

The chemical senses are the most ancient of sensory systems, having evolved some 500 million years ago. They mediate the functions most basic to the survival of the individual and the species: feeding and reproduction. Fish detect chemical stimuli through two major channels of chemoreception: olfaction (smell) and gustation (taste). Solitary chemosensory cells, though less defined, provide additional chemoreceptor function (Kotrschal, 1991; Whitear, 1992). In terrestrial animals, olfaction is normally distinguished as a distance chemical sense with high sensitivity and specificity, and gustation is primarily a contact or close-range sense with moderate sensitivity. The aquatic environment surrounding fish makes their chemical senses unique and the distinction between these two sensory modalities is not always as clear as in terrestrial organisms. In air or water, however, all chemoreception can be considered an aquatic phenomenon, because the receptor is covered with fluid materials where the initial sensory transduction process takes place.

This chapter describes fundamental features of organization and function of the chemosensory systems and provides up-to-date information on their roles in fish behavior. The reader is referred to previous work (Hara, 1982b, 1992a) for detailed discussion on the same topics.

0-8493-8042-1/93/$0.00+$.50

STRUCTURE OF CHEMOSENSORY ORGANS

Olfactory System

Peripheral Olfactory Organ

Gross Anatomy

The olfactory organ of fishes shows a considerable diversity, reflecting the degree of development and ecological habitats. The cyclostomes are monorhinal, i.e., having a single olfactory organ with a single nostril. In elasmobranchs the paired olfactory pits or sacs are usually situated on the ventral side of the snout. The opening of each pit is divided into two parts by a fold of skin: anterior inlet (naris) and posterior outlet. In the teleost fishes the paired olfactory organs are usually located on the dorsal side of the head. Unlike terrestrial vertebrates, there is no direct contact between the olfactory and respiratory systems in any teleost species. Water enters the anterior and leaves through the posterior naris as the fish swims.

The floor of the nasal cavity is lined with the olfactory epithelium or mucosa, which is raised from the floor into a series of folds or lamellae to form a rosette (Figure 1). The arrangement, shape, and degree of development of the lamellae vary considerably from species to species. In the majority of species, however, the lamellae radiate from a central ridge (raphe) arising rostrocaudally from the floor of the cavity. The number of olfactory lamellae varies among species; from a few in sticklebacks (*Gasteorsteus*) to as many as 120 in eels (*Anguilla*) and morays (*Conger*). A sexual dimorphism of the olfactory organ exists in some fish species (Marshall, 1967). The number of lamellae increases to some extent with growth of an individual, but remains relatively constant after the fish reaches a certain stage in development. Secondary folding of the lamellae occurs in salmoniformes, sharks, and dipnoans. There are 5 to 10 secondary foldings per lamella in adult salmoniformes, but none in the parr and the young.

The olfactory lamella is composed of two layers of epithelium enclosing a thin stromal sheet. The epithelium, columnar pseudostratified, is separated into two regions: sensory and nonsensory (indifferent). The sensory epithelium shows various distribution patterns, but may fall into one of the following four types: (1) continuous except for the lamella margin (*Anguilla*, *Ictalurus*), (2) separated regularly by the nonsensory epithelium (*Salmoniformes*), (3) interspread irregularly with the nonsensory epithelium (*Gasterosteus*, *Thunnus*), and (4) scattered in islets (*Phoxinus*, *Cyprinus*, *Carassius*). The sensory epithelium consists of three main cell types: (1) receptor cells (olfactory neurons), (2) supporting (sustentacular) cells, and (3) basal cells.

Fine Structure

Two morphologically distinct receptor cell types, ciliated and microvillar, generally exist in teleosts (Figure 1; Zielinski and Hara, 1988; Zeiske et al., 1992). Only microvillar receptor cells are found in elasmobranchs and Australian lung fish (*Neoceratodus forsteri*) while African lung fish (*Protopterus annectens*) and lamprey (*Lampetra fluviatilis*) show only ciliated receptor cells (Zeiske *et al.*, 1992). Approximately 5 to 10 million olfactory receptor cells comprise the sensory epithelium on each side of the nasal cavity in an average teleost (Yamamoto, 1982).

The receptor cell is a bipolar primary neuron with a cylindrical dendrite (1.5 to 2.5 μm in diameter) which terminates at the free surface of the epithelium, i.e., directly exposed to the external environment. The distal end of the dendrite forms a swelling (olfactory knob) which protrudes slightly above the epithelial surface. The proximal part of the perikaryon tapers off to form an axon. The axons pass through the basement membrane, become grouped in the submucosa, and form the olfactory nerve fascicles, which run posteriorly to end in the olfactory bulb. The ciliated receptor cell has 4 to 8 cilia (2 to 7 μm long, 0.2 to 0.3 μm wide) radiating from an olfactory knob (Figure 1). Usually, the cilia show the 9 + 2 arrangement of microtubules identical to that of common

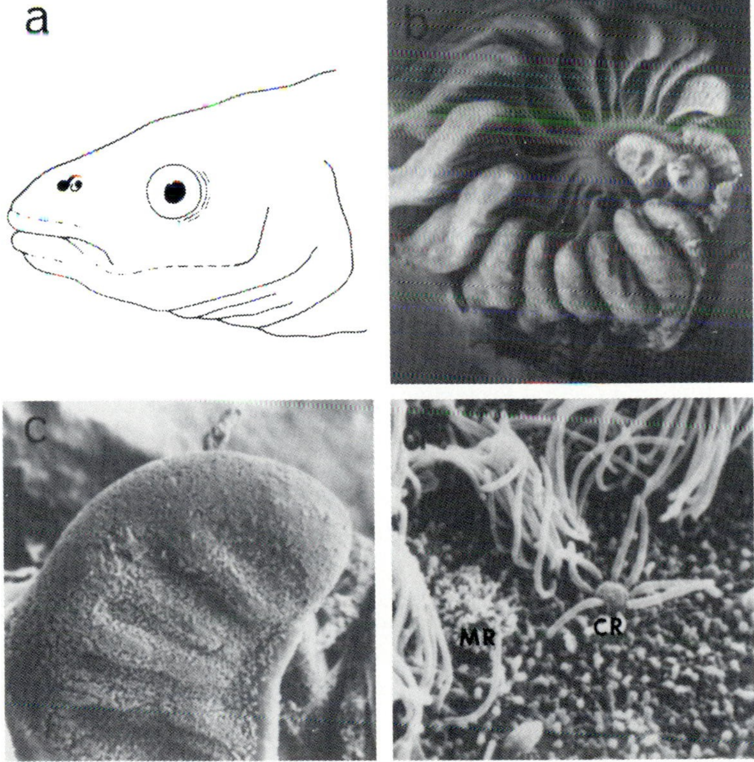

FIGURE 1. Position of the nose (a) in a teleost, and scanning electron micrographs of an olfactory rosette (b), lamella (c), and a surface view of the sensory epithelium (d). CR, ciliated receptor cell; MR, microvillar receptor cell. (From Hara, T. J., in *The Behaviour of Teleost Fishes,* Pitcher, T. J., Ed., Croom Helm, London, 1986, 152. With permission.)

kinocilia. The microvilli (2 to 5 μm long, 0.1 μm wide) number from 30 to 80, depending upon the species. The cytoplasm of microvillar receptor cells exhibit similar ultrastructural features to that of the ciliated receptors.

The receptor cells are continually renewed in normal adults. Experimental severance of the olfactory nerve or treatment of the olfactory organ with toxicants causes degeneration of the olfactory neurons followed by reconstitution of a new population of functional neurons (Evans *et al.*, 1982; Evans and Hara, 1985). The uniqueness of the renewal process of the olfactory receptor cells may be an adaption of the system to impairment due to environmental hazards during the life cycle of the animal.

The supporting cells are columnar epithelial cells extending vertically from the epithelial surface to the basal lamina, forming a mosaic interspersed with receptor cells. The free surface is flat, with relatively few irregular microvilli. Supporting cells adjoin receptor cells, ciliated non-sensory cells, and other supporting cells at the free surface by an apical tight junction. The function of the supporting cells in olfaction is not defined, but a significance beyond mere mechanical support has been suggested in higher vertebrates (Getchell et al., 1988). Ciliated nonsensory cells are columnar epithelial cells with a flat surface from which a number of long kinocilia (20 to 30 μm long) extend. The beating of these cilia creates weak currents over the olfactory lamellae, presumably assisting in the transport of stimulant molecules in the olfactory organ (Døving, 1986).

The basal cells are small and undifferentiated cells lying adjacent to the basal lamina and having no cytoplasmic processes reaching the free surface. The basal cells in the sensory epithelium are assumed to be the progenitors of the receptor or supporting cells. Increased mitotic figures are seen in the basal region in a reconstituting olfactory epithelium (Evans et al., 1982). In addition to the cell types above, mucus (goblet) cells are also abundant in the nonsensory epithelium.

Olfactory Bulb and Central Projections

The olfactory nerve fibers, unmyelinated axons of the receptor cells, course to the first relay station, the olfactory bulb. The olfactory nerve endings synapse with the second-order bulbar neurons, mitral cells, as glomeruli. In some fish species, e.g., *Cyprinus carpio*, the olfactory nerve consists of medial and lateral bundles; the former derive from the more rostral lamellae and the latter from the more caudal (Sheldon, 1912). The fibers do not branch until they terminate. Fish may have either short olfactory nerves and a long olfactory tract, pedunculated (e.g., *Carassius*, *Ictalurus*) or long olfactory nerves and a short olfactory tract, sessile (e.g., *Anguilla*, salmoniformes, and the majority of teleosts).

The olfactory bulb is remarkably consistent in morphology throughout the vertebrates. However, the fish olfactory bulb is poorly differentiated and the lamination is not distinct as in higher vertebrates (Allison, 1953). The mitral cell, the major bulbar neuron, has a large cell body and more than one dendrite ending in different glomeruli. This contrasts to mammalian mitral cells, in which only a single main dendrite ends in each glomerulus.

The axons of the mitral cells form the olfactory tract through which information from the olfactory bulb is conveyed to the telencephalic hemispheres. The olfactory tract consists of two main bundles, lateral and medial. Both bundles are further subdivided into several small bundles. Some fibers run directly to the hypothalamus, and some cross in the anterior commissure. The centrifugal fibers originating in the telencephalon run posteriorly through the medial portion of the olfactory tract and terminate in the bulb (Oka et al., 1982; Satou, 1992). The olfactory bulb is thus under constant influence of higher centers through the centrifugal fibers. The olfactory tract fibers are estimated to be 10^4 in number (Døving, 1986). The convergence of the primary olfactory neurons (about 5 to 10 million) upon secondary neurons (mitral cells) would therefore be about 1000:1. This high convergence ratio is approximately the order that is estimated for the mammalian olfactory system (Allison, 1953). The glomerulus is thus suitably designed for effective spatial summation and this factor must be at least responsible for the extremely high sensitivity of the system.

The olfactory system is separable into two subsystems: (1) the lateral olfactory system, in which the lateral part of the olfactory bulb receives input signals mainly from the lateral bundle of the olfactory nerve and sends output information to the lateral olfactory tract; and (2) the medial olfactory system, in which the medial part of the olfactory bulb receives input signals mainly from the medial bundle and sends outputs to the medial tract (Satou, 1990, 1992). The olfactory tract fibers terminate bilaterally in the following telencephalic hemispheres: (1) medial terminal field in the area ventralis telencephali, (2) lateral terminal field in the ventrolateral part of the area dorsalis telencephali, and (3) posterior terminal field in the central part of the area dorsalis telencephali (Oka et al., 1982; Satou, 1990, 1992).

Gustatory System

Taste Buds

The taste bud (also termed terminal or end bud) constitutes the structural basis of the gustatory organ (Figure 2). In teleosts, taste buds are located on the gill rakers and arches, on appendages such as barbels and fins, as well as within the oral cavity and pharynx. In some species (silurids, ictalurids), taste buds are distributed over the whole body surface. Although most species, fresh- or seawater, have well-developed taste buds in branchial regions (Iwai, 1964), no taste buds are found on gill rakers of rainbow trout (Ezeasor, 1982). Taste buds of elasmobranchs are restricted to the mouth and pharynx. Numbers and densities of taste buds vary in different locations and species. For instance, in rainbow trout (20 cm body length), taste buds number as many as 30 per mm^2 in some areas of the palate, totaling 3000 to 4000 taste buds on the whole palate

(Figure 2). A minnow, *Pseudorasbora parva* (6 cm body length), has over 140 taste buds per mm^2 on the lip and palatal organ (Kiyohara et al., 1980). Taste buds totaling more than 175,000 are estimated on the entire body surface of the yellow bullhead, *Ictalurus natalis* (25 cm body length; Atema, 1971).

The taste buds are usually bulbiform in shape and extend perpendicularly through the entire thickness of the epidermis. They vary in size, ranging from 45 to 75 μm in height and from 30 to 50 μm in width in the majority of species (Iwai, 1964). The apex of the taste bud (taste bud pore) usually protrudes, or sometimes is retracted, from the surface of the surrounding epidermis. Gustatory cells, supporting cells, and basal cells constitute the taste bud. Based primarily on the electron density of the cell cytoplasm, they are sometimes

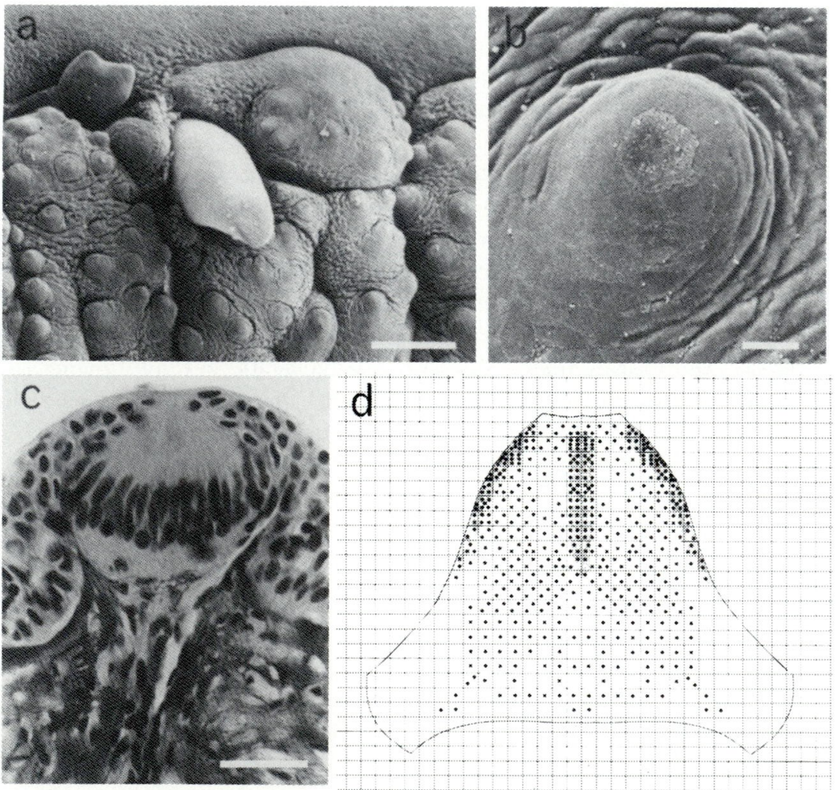

FIGURE 2. Morphology and distribution of taste buds on the rainbow trout palate. (a) Scanning electron micrograph of the inside upper lip showing distribution of taste buds; (b) scanning electron micrograph of a taste bud located on an elevated papilla; (c) light micrograph of a longitudinal section through a taste bud; (d) distribution of the palatal taste bud. Each dot represents five taste buds and each square corresponds to 1 mm^2. Scale bars = 250, 20, and 30 μm in a, b, and c, respectively. (From Marui, T. et al., *J. Comp. Physiol. A*, 153, 423, 1983b. With permission.)

distinguished as light and intermediate (gustatory), dark (supporting), and basal cells (see Ezeasor, 1982; Roper, 1989; Reutter, 1992).

The gustatory (receptor) cells have a single or occasionally two apical processes (0.5 μm thick and 1.5 to 3 μm long). The number of gustatory cells in a taste bud varies considerably; e.g., from 5 in *Pomatoschistus* to 67 in *Corydoras* (Jakubowski and Whitear, 1990). The elongated gustatory cell is cylindrical to spindle shaped. The supranuclear cytoplasm contains numerous electron-dense tubules (0.4 to 0.6 μm) and the infranuclear portion is characterized by the presence of abundant vesicles of 50 to 70 nm diameter.

The supporting cell is slender, fusiform in shape, with a smooth apical contour. The supporting cells are provided with a number of small microvilli (0.1 to 0.2 μm thick and 0.5 to 1 μm long) at their apical surfaces. The distal cytoplasm contains secretory vesicles, which vary in number, size, and appearance in different species. The supporting cell vesicles open at the surface to secrete the cap of mucus that covers the taste bud pore. In the neck of the bud the supporting cells surround and separate the distal processes of the gustatory cells. There are desmosomal connections to adjacent epithelial cells, other supporting cells, gustatory cells, and basal cells.

The basal cells, typically one to five per taste bud, are situated at the bottom of the taste bud, directly on the basement membrane. Some gadid species lack basal cells in external taste buds (Jakubowski and Whitear, 1990). Basal cells, flattened or oblate in shape, are attached to the gustatory and supporting cells through desmosomes. Unmyelinated nerve fibers, packed closely together, are intermingled with the processes of gustatory and basal cells, forming the nerve plexus. This is the terminal structure of the gustatory nerve. Taste bud cells, too, are capable of renewal; their life span in fish are between 12 and 42 d, depending on temperature (Raderman-Little, 1979). The basal cells of teleostean taste buds are different from those of mammalians which are considered to be stem cells or regenerative cells (Reutter, 1992). By their resemblance to Merkel cells, the basal cells are suggested to function as interneurons or mechanoreceptors (Reutter, 1992).

Cell Contacts, Gustatory Nerves and Central Projections

The gustatory cells synapse with afferent gustatory nerve fibers. In catfish, *Ictarulus*, synaptic contacts exist between basal cells and other gustatory cells (Reutter, 1982). In higher vertebrates, the synaptic organization of the taste bud is very complicated and there is evidence indicating the existence of peripheral integrative mechanisms including cross-talk between taste cells, summation of chemoreceptor responses by interneurons (basal cells), and centrifugal control of taste buds via efferent inputs from the central nervous system (CNS) (Roper, 1989). Gustatory synapses are marked by membrane densities, but not always by transmitter vesicles (Jakubowski and Whitear, 1990). Thus, the unequivocal identification of neurotransmitters at synapses in taste buds has not yet been achieved.

Peripheral gustatory inputs are transmitted to the central nervous system through three separate cranial nerves: facial (VIIth), glossopharyngeal (IXth), and vagal (Xth). Normally, the facial nerve transmits gustatory information from the extraoral surface, the glossopharyngeal from the anterior part of the oral cavity, and the vagal nerve from the oropharynx.

The facial and vagal nerves, respectively, terminate in the facial and vagal lobes in the medulla. The glossopharyngeal nerve terminates in a dorsal medullary region between the facial and vagal lobes. In fish, neither the facial nor the vagal lobe is well developed. In bullhead catfish, the taste system is segregated into two distinct, though interrelated, subsystems: facial and oral or vagal subsystems (Kanwal and Finger, 1992). The former detects chemical stimuli at a distance and can therefore subserve a food-localization function during food search, while the latter performs a discriminative function leading to the selective ingestion of food. The facial and vagal lobes are further elaborated into several longitudinal columns or lobules. Each lobule receives segregated inputs from discrete portions of the body. Such well-defined mapping of the peripheral gustatory apparatus onto the gustatory sensory column is considered to be a feature common to all teleosts (Kanwal and Finger, 1992).

Solitary Chemosensory Cells

In addition to the olfactory and gustatory systems, there exists a differentiated epithelial sensory cell system composed of solitary chemosensory cells (SCCs), which closely resemble the gustatory cell but are not organized into discrete taste buds (Whitear, 1992). SCCs are found in a variety of actinopterygian fishes, including sturgeons and lungfish, but also in lampreys (oligovillous cells) and in ranid tadpoles (Stifchenzellen) (Kotrschal, 1991). In teleosts, they are widely distributed in the external skin and oropharyngeal epithelium, including the gills. The SCCs are bipolar, with apical process(es) at the surface of the epidermis, and the proximal region associated with a neurite profile. The number of SCCs varies greatly in different species. As many as 3 to 6 million SCCs have been estimated in the anterior dorsal fin of rockling, *Gaidropsarus mediterraneus*. In cyprinid species, the density of SCCs in external skin approximates 100 per mm^2 (Kotrschal, 1991; Whitear, 1992). Slender nerve fibers make synaptic contact with the SCCs, usually near the base of the cell. SCCs on the head and oropharyngeal epithelium could share innervation with the taste buds, from the facial or vagal component.

FUNCTIONS OF CHEMORECEPTORS

Signal Transduction in Olfaction and Gustation

Olfactory and gustatory receptors detect chemical stimuli, transduce them into electrical signals, and transmit this information to the central nervous system. It is generally accepted that the initial event in chemoreception is reversible binding of a stimulus molecule (ligand-binding) to a specific membrane receptor located in the cilia or microvilli of the receptor cell. Recent studies of molecular mechanisms of chemosensory transduction indicate that multiple mechanisms exist for the transduction of chemosensory information in teleosts and, in some cases, multiple transduction pathways can be stimulated by a single receptor class (Brand and Bruch, 1992). In olfaction, ligand binding activates G-proteins located in receptor membranes. The activated G-proteins then activate adenylyl cyclase or phospholipase C to produce second messengers (cyclic AMP [cAMP], inositol triphosphate [IP_3], diacylglycerol [DAG], etc.). The resulting increase in second messenger concentrations opens ion channels, causing membrane depolarization, leading to the generation of nerve impulses. At some receptors, ligand binding translocates ions directly without involving the activation of second messengers. In gustation, two transduction sequences, one involving the second mesengers (IP_3 or DAG) and the other direct ion translocation, operate in the channel catfish gustatory system (Brand and Bruch, 1992).

Electrophysiologists are able to tap, amplify, and display electrical signals thus transduced at various levels of the olfactory and gustatory systems. In olfaction, the electroolfactogram (EOG), summated generator potentials of the olfactory neurons, is the method most widely used to monitor olfactory responses to chemical stimuli (Figure 3). Electrical responses are also obtained from the olfactory bulb (electroencephalogram, EEG), olfactory nerve, and tract. In gustation, recordings are usually made from branches of the facial nerve innervating barbels (catfish) or the palate (salmonids and others) (Figure 3).

Sensitivity and Specificity

Olfactory System

Adrian and Ludwig (1938) were the first to record electrical activities of the olfactory tract in response to natural chemical stimulation of the olfactory organ in catfish, carp, and tench, *Tinca tinca*. This marked the first study of the vertebrate olfactory system using an electrophysiological technique. Despite this, most of the following studies of olfaction utilized vertebrates other than fish. However, since Sutterlin and Sutterlin's (1971) and Suzuki and Tucker's (1971) works on amino acid stimulation of the olfactory receptors in Atlantic salmon, *Salmo salar*, and white catfish, *Ictalurus catus*, respectively, research on fish olfaction has centered on response characteristics to amino acids in a variety of fish species (see reviews in Hara, 1982a, 1986, 1992b; Caprio, 1984, 1988). Earlier studies on fish olfaction almost exclusively employed chemicals primarily odorous to humans (Hara, 1971, 1975). Many of the classical data on detection thresholds reported for chemical stimuli such as β-phenethyl alcohol, butyl alcohol, and morpholine are not supported by studies using modern electrophysiological techniques. Recently, new groups of chemostimulatory substances have been identified and their stimulatory characteristics for fish investigated. These include steroid hormones, bile salts, and prostaglandins.

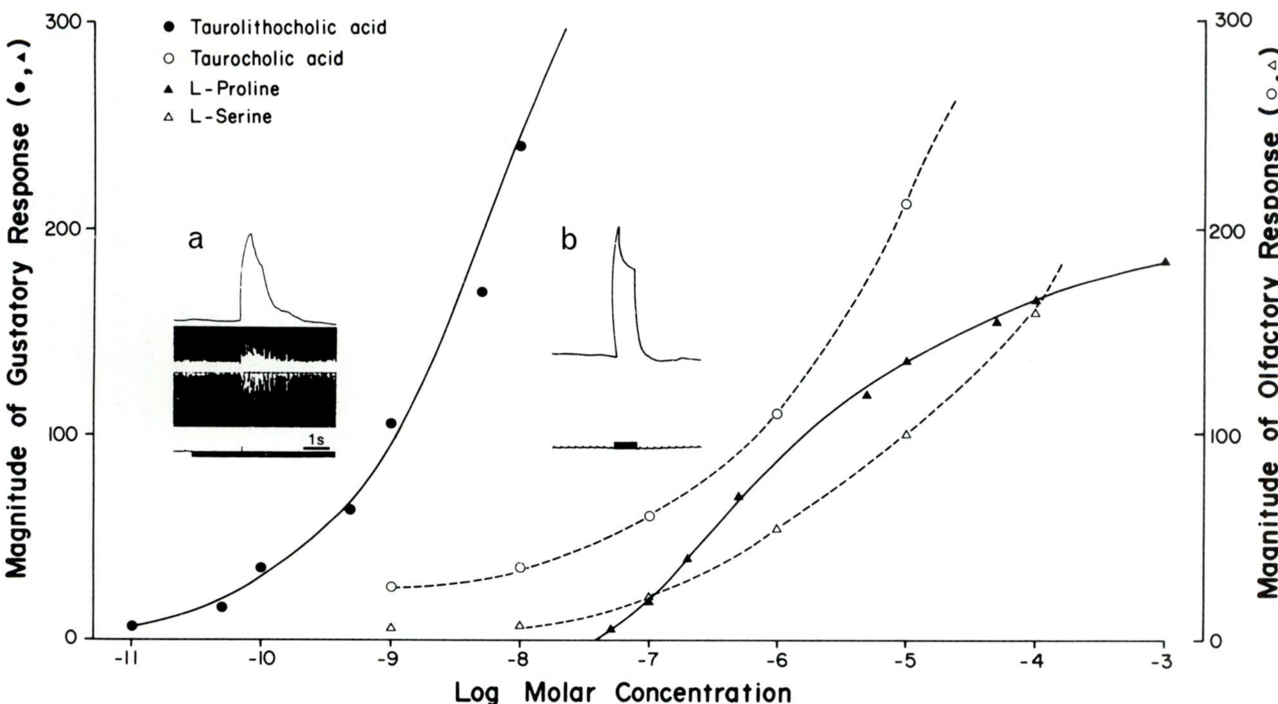

FIGURE 3. Comparison of the olfactory (EOG) and gustatory (palatine nerve) sensitivity to representative chemical stimuli examined electrophysiologically in rainbow trout. Inserts, examples of palatine nerve (a) and EOG (b) response recordings.

Amino Acids

The fish olfactory system detects low levels of naturally occurring amino acids. Thresholds obtained electrophysiologically from over 20 species lie between 10^{-7} and 10^{-9} *M*, which approximate levels of free amino acids found in natural waters (Table 1; Hara, 1982a, 1992b). Generally, unsaturated L-α-amino acids containing unbranched and uncharged side chains are the most effective olfactory stimuli. Ionically charged α-amino and α-carboxyl groups are essential; acylation of the former or esterification of the latter results in reduced activity. The amino acids effective as olfactory stimuli are thus characterized by being simple, short, and straight-chained, with only certain attached groups (Hara, 1975, 1982a). The responses to amino acids are pH dependent, and most of the highly stimulatory amino acids have their maximal activities near their isoelectric points (Hara, 1976). Although some variabilities exist, the olfactory spectrum of amino acids is generally similar across species.

Peptides are normally nonstimulatory, however, salmon gonadotropin-releasing hormone (GnRH) and its analogs have been reported recently to be potent olfactory stimulants for rainbow trout, with threshold concentrations of 10^{-16} to 10^{-14} *M* (Andersen and Døving, 1991).

Steroids

Water-borne 17α,20β-dihydroxy-4-pregnen-3-one (17,20P) stimulates the endocrine system of mature male goldfish via their olfactory system to increase milt production at the time of spawning (for details, see below). Among gonadal steroids tested, 17,20P has the lowest detection threshold of 10^{-13} to 10^{-12} *M* (Table 1; Sorensen et al., 1987, 1990). EOG response magnitude for 17,20P increases sharply with an increase in concentration, and at 10^{-8} *M*, it reaches 2 to 3 times that evoked by 10^{-5} *M* L-serine. Some steroid glucuronides are stimulatory for the olfactory system of female African catfish, *Clarias gariepinus* (Resink et al., 1989). The threshold concentration for 5β-pregnen-3α,17α-diol-20-one-3α-glucuronide, the most stimulatory steroid glucuronide tested, is approximately 10^{-11} *M*. 17,20P is not stimulatory for African catfish, while steroid glucuronides are not effective for goldfish. None of the sex steroids tested above are stimulatory for salmonids, suggesting high species specificity. Testosterone, however, is a potent odorant in precocious male Atlantic salmon parr (Moore and Scott, 1991). Interestingly, the olfactory epithelia of these fish only appear to be responsive for a limited period during the year (October). Immature fish do not respond at any time. Unlike amino acid receptors, olfactory receptors for sex pheromones exhibit high stereospecificity. Of the 24 closely related steroid hormones tested, for instance, only 17,20P and 17α,20β,21P evoke EOG responses at 10^{-8} *M* in goldfish (Sorensen et al., 1990).

TABLE 1
Olfactory Thresholds for Three Major Groups of Aquatic Odorants for Fish, Determined Electrophysiologically

Species	Threshold (M)			Source
	Amino Acids	Steroids	Prostaglandins	
Myxine glutinosa	10^{-6}-10^{-5}	10^{-6}-10^{-5}		Døving and Holmberg (1974)
Negaprion brevirostris	10^{-8}-10^{-7}			Zeiske et al. (1986)
Dasyatis sabina	10^{-8}-10^{-6}			Silver (1979)
Salmo salar	10^{-9}-10^{-5}	10^{-14}		Sutterlin and Sutterlin (1971), Moore and Scott (1991)
Salvelinus fontinalis	10^{-8}-10^{-7}			Hara et al. (1973)
Salvelinus alpinus	10^{-9}-10^{-7}	10^{-9}-10^{-8}	10^{-11}-10^{-9}	Belghaug and Døving (1977), Døving et al. (1980), Sveinsson (unpublished)
Salvelinus namaycush	10^{-8}-10^{-7}	10^{-9}-10^{-8}		Hara et al. (1989), Hara (unpublished)
Oncorhynchus mykiss (*Salmo gairdneri*)	10^{-8}-10^{-7}	10^{-10}-10^{-9} 10^{-16}[a]		Hara (1973), Hara et al. (1984), Anderson and Døving (1991)
Oncorhynchus nerka	10^{-7}-10^{-6}			Hara (1972)
Oncorhynchus kisutch	10^{-7}-10^{-6}			Hara (1972)
Coregonus clupeaformis	10^{-7}-10^{-6}			Hara et al. (1973)
Ictalurus catus	10^{-9}-10^{-7}			Suzuki and Tucker (1971), Caprio (1980)
Ictalurus punctatus	10^{-9}-10^{-7}	10^{-5}-10^{-4}		Caprio (1978), Erickson and Caprio (1984)
Ictalurus serracanthus	10^{-9}-10^{-7}			Caprio (1980)
Clarias gariepinus		10^{-11}-10^{-10}		Resink et al. (1989)
Cyprinus carpio	10^{-9}-10^{-7}			Goh and Tamura (1978), Ohno et al. (1984)
Carassius auratus	10^{-9}-10^{-8}	10^{-13}-10^{-11}	10^{-12}-10^{-10}	Sorensen et al. (1987, 1988)
Anguilla rostrata	10^{-9}-10^{-7}			Silver (1982)
Misgurnus anguillicaudatus			10^{-13}-10^{-10}	Kitamura and Ogata (1990)
Chrysophyrys major	10^{-7}			Goh et al. (1979) Goh and Tamura (1980)
Mugil cephalus	10^{-7}			Goh and Tamura (1980)
Seriola quinqueradiata	10^{-7}			Kobayashi and Fujiwara (1987)
Pagrus major	10^{-7}-10^{-6}			Kobayashi and Goh (1985)
Conger myriaster	10^{-9}-10^{-8}			Goh et al. (1979)

[a] Gonadotropin-releasing hormone (GnRH)

Modified from Hara (1992b).

Gallbladder bile and some of its major constituent bile salts are potent olfactory stimulants for some fish species, especially salmonids (Døving et al., 1980; Hara et al., 1984; Zhang and Hara, unpublished). The bile is stimulatory at a dilution of 10 nl/l in rainbow trout and char. Detection thresholds for bile salts range between 10^{-11} and 10^{-9} *M*. The most stimulatory bile salts tested are sulfotaurolithocholate, taurolithocholate, and taurodeoxycholate. Bile salts are stimulatory for the goldfish olfactory system (Sorensen et al., 1987), but in channel catfish they are only stimulatory at high concentrations, 10^{-4} *M* and higher (Erickson and Caprio, 1984).

Prostaglandins

The olfactory system of goldfish is sensitive to F-series prostaglandins (PGFs), postovulatory pheromones that stimulate male sexual arousal (see below). $PGF_{2\alpha}$ and its metabolite 15-keto-$PGF_{2\alpha}$ are detected electrophysiologically at a threshold of approximately 10^{-10} *M* (Table 1; Sorensen et al., 1988). At 10^{-6} *M*, $PGF_{2\alpha}$ elicits EOG responses three times those elicited by 10^{-5} *M* L-serine. $PGF_{1\alpha}$ (having an extra double bond), $PGF_{3\alpha}$ (lacking a double bond), and $PGE_{2\alpha}$ (having a ketone group, instead of -OH) show concentration-response curves similar to that for $PGF_{2\alpha}$ but shifted to the right by about one log unit. High sensitivity of olfactory receptors to PGs is also demonstrated in Arctic char, *Salvelinus alpinus*, and lake char, *S. namaycush*, but not other salmonids (Sveinsson and Hara, unpublished). The structure-activity relationship for PGs obtained in Arctic char is almost identical to that for goldfish.

Amino acids, sex steroids, bile salts, and prostaglandins are detected by separate receptor mechanisms, and experimental evidence further suggests that multiple receptor mechanisms exist within each chemical group.

Gustatory System

The first electrophysiological responses of the fish gustatory system were obtained from the facial/trigeminal nerve complex innervating the barbel of the bullhead catfish, *Ictalurus nebulosus* (Hoagland, 1933). In this study, he tested various acids, salts, and natural substances, as well as mechanical stimulation. Since this pioneer work, a number of reports dealing with physiological responses of the fish gustatory system have appeared (see reviews in Bardach and Atema, 1971; Caprio, 1982, 1984, 1988). Like the olfactory studies, early fish gustatory studies utilized as stimuli the four classical taste substances (salt, sugar, acid, and bitter substance) used on humans (Hara, 1971). Sutterlin and Sutterlin's (1971) and Suzuki and Tucker's (1971) works on amino acid stimulation of the olfactory system again opened an avenue for new gustatory research that resulted in the finding that the facial nerves innervating the maxillary barbel of the channel catfish are also highly sensitive to amino acids (Caprio, 1975, 1978). Subsequently, amino acids have been shown to be potent gustatory stimuli in various fish species.

Besides amino acids, the fish gustatory system detects various chemicals, including aliphatic acids, nucleotides, bile salts, carbon dioxide, and some marine toxins.

Amino Acids

The palatine nerve response to an amino acid is fast-adapting, returning to the baseline within a few seconds (see Figure 3). Generally, the L-isomer of an amino acid is more stimulatory than its D-enantiomer in all species examined, except the sea catfish, *Arius felis*, in which D-alanine is more stimulatory than the L-form (Marui and Caprio, 1992). Thresholds for the more stimulatory amino acids range from 10^{-9} to 10^{-7} *M* in most species, while that of the channel catfish barbel for L-alanine averages 5×10^{-11} *M* (Caprio, 1978).

Unlike the olfactory system, gustatory response spectra for amino acids vary considerably. Among fish species whose gustatory responses to amino acids have been examined systematically (23 species), two groups can be identified: (1) those that respond to many amino acids, i.e., "wide response range", and (2) those that respond to few amino acids, i.e., "limited response range" (Table 2; Hara and Zielinski, 1989). The channel catfish may represent the extreme of group 1. Chemoreceptors on their maxillary barbels are extremely sensitive to most of the natural amino acids (Caprio, 1975, 1978). Similar to its olfactory counterpart, the concentration-response relationship exhibits a broad dynamic range, covering over 9 log units. Molecular requirements of the amino acids for stimulation of the olfactory and gustatory systems in the channel catfish are also similar, i.e., amino acids containing three to six carbon atoms having unbranched and unchanged side chains are all stimulatory (Caprio, 1978). Among other group 1 species, topmouth minnow, red sea bream, and mullet have broad amino acid spectra similar to that of channel catfish, though some differences exist in the effectiveness of individual amino acids (Goh and Tamura, 1980; Kiyohara et al., 1981).

Group 2 species are characterized by their specific responses to L-proline, L-alanine and a few related amino acids. Detection thresholds for L-proline lie between 10^{-8} and 10^{-7} for most species. Typical and probably prototypic of this group are the salmonids. In rainbow trout, of natural amino acids, L-proline, hydroxyl-L-proline, L-alanine, L-leucine, L-phenylalanine, L-α-aminoguanidino-propionic acid (L-AGPA), and betaine are stimulatory under natural pH (Marui et al., 1983b). These amino acids appear to be detected by at least three independent, though not exclusive, palatal chemoreceptors: (1) proline receptor (proline, hydroxyproline, alanine), (2) betaine-receptor (betaine, AGPA), and (3) leucine-receptor (leucine, phenylalanine). The Pacific salmon (kokanee) *Oncorhynchus nerka*, shows the same spectral pattern as that of rainbow trout. Members of the genus *Salvelinus* (Arctic char, brook char, lake char) respond only to three amino acids: L-proline, hydroxy-L-proline, and L-alanine (Hara et al., 1989). *Salmo* (Atlantic salmon, brown trout) respond strongly to

L-AGPA, in addition to the above three amino acids (Hara et al., unpublished). In both *Salvelinus* and *Salmo*, the activity of betaine is totally lacking. Lake whitefish, *Coregonus clupeaformis,* and Arctic grayling, *Thymallus arcticus,* do not respond to hydroxy-L-proline and L-alanine, but are highly responsive to L-arginine. Phylogenetic relationships are thus clear in the gustatory amino acid spectral patterns of salmonids. Among nonsalmonid species, carp (Marui et al., 1983a) and goldfish (Sorensen and Hara, unpublished) may be included in this group. Their spectral patterns are slightly wider, responding additionally to L-cysteine, L-glutamate, and glycine. In puffer, *Fugu pardalis*, three receptor groups have been identified: (1) proline (proline, dimethylglycine), (2) alanine (alanine, glycine, sarcosine), and (3) betaine (betaine, dimethylglycine) (Kiyohara and Hidaka, 1991).

The wide response range group is thus typified by their general responsiveness to short-chained, unbranched α-amino acids, and has so far been found more among marine species. The limited response range group is of a proline receptor type, and strongly represented by salmonids.

Steroids

Some of the bile salts identified as olfactory stimulants also stimulate the gustatory receptors of salmonids. Rainbow trout detect taurolithocholate at concentrations of 10^{-12} to 10^{-11} *M*, when electrical responses are recorded from the palatine nerve. The threshold is approximately 4 log units lower than that for L-proline, the most potent gustatory amino acid for this species. Taurocholate, taurodeoxycholate, and cholate are also stimulatory to lesser degrees, but still exceed the potency of the amino acids. Gustatory receptors of Arctic char, brook char, and lake char are also sensitive to these bile salts, with threshold concentrations of 10^{-9} to 10^{-8} *M*. So far no gustatory sensitivity has been reported in species other than salmonids.

Glucuronide conjugates of certain sex steroids are stimulatory for the gustatory receptors of some salmonids. In rainbow trout, 17β-estradiol-3 (β-D-glucuronide) has a detection threshold of approximately 5×10^{-6} *M*, and at 10^{-4} *M* it evokes a response three times the magnitude of that for 10^{-3} *M* L-proline (Hara, unpublished). Etiocholanolon-3α-ol-17-one glucuronide is the only other stimulatory sex steroid examined. None of these steroids are stimulatory for the Arctic char gustatory system. High detection thresholds obtained, however, make their behavioral implication difficult.

Aliphatic Acids and Nucleotides

Gustatory stimulation by aliphatic acids has been examined in Atlantic salmon (Sutterlin and Sutterlin, 1970), Japanese eel (Yoshii et al., 1979), and carp (Marui and Caprio, 1992). Detection thresholds for carboxylic acids range between 10^{-7} and 10^{-4} *M*. The stimulatory effectiveness generally increases with increasing carbon chain length of the aliphatic acids in the Atlantic salmon and

TABLE 2
Grouping of Fish According to the Specificity of Gustatory Responses to Amino Acids

Species	Stimulatory Amino Acids	Source
	(Wide Response Range)	
Channel catfish, *Ictalurus punctatus*	Neutral, basic, and acidic	Caprio (1975, 1978), Davenport and Caprio (1982), Kanwal and Caprio (1983)
Topmouth minnow, *Pseudorasbora parva*	Neutral, basic, and acidic	Kiyohara et al. (1981)
Tiger fish, *Therapon oxyrhynchus*	Neutral, basic, and acidic	Hidaka and Ishida (1985)
Rabbitfish, *Siganus fuscescens*	Neutral, basic, and acidic	Ishida and Hidaka (1987)
Red sea bream, *Chrysophrys major*	Neutral and basic	Goh and Tamura (1980)
Mullet, *Mugil cephalus*	Neutral and basic	Goh and Tamura (1980)
Isaki grunt, *Parapristipoma trilineatum*	Neutral and basic	Ishida and Hidaka (1987)
Jack mackerel, *Trachurus japonicus*	Neutral and basic	Ishida and Hidaka (1987)
Chub mackerel, *Scomber japonicus*	Neutral and basic	Ishida and Hidaka (1987)
	(Limited Response Range)	
Japanese eel, *Anguilla japonica*	Arg, Gly, Ala, Pro, Lys, Ser	Yoshii et al. (1979)
Rainbow trout, *Oncorhynchus mykiss*	AGPA, Pro, Hpr, Bet, Leu, Ala, Phe	Marui et al. (1983b)
Kokanee, *Oncorhynchus nerka*	AGPA, Pro, Hpr, Bet, Leu, Ala, Phe, Arg	Hara et al. (unpublished)
Common carp, *Cyprinus carpio*	Pro, Ala, CysH, Glu, Bet, Gly	Marui et al. (1983a)
Arctic charr, *Salvelinus alpinus*	Pro, Hpr, Ala, AGPA	Hara et al. (unpublished)
Brook trout, *Salvelinus fontinalis*	Pro, Hpr, Ala, Phe, AGPA	Hara et al. (unpublished)
Lake trout, *Salvelinus namaycush*	Pro, Hpr, Ala, Arg	Hara et al. (unpublished)
Atlantic salmon, *Salmo salar*	AGPA, Pro, Ala, Hpr	Hara et al. (unpublished)
Brown trout, *Salmo trutta*	AGPA, Pro, Ala, Hpr	Hara et al. (unpublished)
Lake whitefish, *Coregonus clupeaformis*	AGPA, Pro, Arg	Hara et al. (unpublished)

TABLE 2 (continued)

Species	Stimulatory Amino Acids (Limited Response Range)	Source
Arctic grayling, *Thymallus arcticus*	Pro, Arg, AGPA, Ala	Hara et al. (unpublished)
Puffer, *Fugu pardalis*	Pro, Ala, Gly, Bet	Kiyohara et al. (1975)
Yellowtail, *Seriola quinqueradiata*	Pro, Bet, Try, Val, Ala	Hidaka et al. (1985)
Amberjack, *Seriola dumerili*	Pro, Try, Bet, Ala	Ishida and Hidaka (1987)

Modified from Hara and Zielinski (1989)

Japanese eel. In carp, aromatic carboxylic acids are also stimulatory, and for aliphatic acids dicarboxylic acids are more stimulatory than monocarboxylic acids.

Nucleotides including adenosine 5′-monophosphate, inosine 5′-monophosphate, uridine 5′-monophosphate, and adenosine 5′-diphosphate, are stimulatory for gustatory receptors in many species (for review see Marui and Caprio, 1992). Detection thresholds for nucleotides are generally high, ranging between 10^{-6} and 10^{-4} *M*. In turbot, *Scophthalmus maximus*, of the 47 nucleotides and nucleosides tested, only inosine and guanosine 5′-monophosphates have feeding stimulant activities (Mackie and Adron, 1978).

Other Gustatory Chemicals

The palatal chemoreceptors of carp (Hidaka, 1970), Japanese eel (Yoshii et al., 1980), and rainbow trout (Yamashita et al., 1989) are sensitive to CO_2 and capable of distinguishing between CO_2 and H^+. The threshold concentration estimated for the rainbow trout palatal organ is 4×10^{-5} *M*, which is slightly above the level of CO_2 found in natural waters when equilibrated with air (1.8×10^{-5}*M*).

Behavioral observations show that some fish species reject pieces of liver from toxic puffer or artificial food pellets containing tetrodotoxin (TTX). When tested for gustatory stimulation, rainbow trout and Arctic char detect TTX at a concentration of 2×10^{-7} *M* (Yamamori *et al.*, 1988). A paralytic shellfish toxin, saxitoxin (STX), has a similar gustatory effect on both species. STX is identical in its pharmacological effect, in which the voltage-sensitive sodium channels in nerve and muscle tissues are specifically blocked.

Solitary Chemoreceptor Cells

Reflecting high variabilities in their structure and distribution, solitary chemoreceptor cells (SCCs) display a wide variety of chemical sensitivity.

Oligovillous cells of brook lamprey, *Lampetra planeri*, show exceptional sensitivity to sialic acid, with threshold concentrations of 10^{-9} to 10^{-8} *M* (Baatrup and Døving, 1985). Olfactory and gustatory amino acids have no effects, suggesting a separate functional role from the pharyngeal taste buds. The free fin rays of the searobin, *Prionotus carolinus*, detect several amino acids at concentrations of 10^{-6} to 10^{-4} *M* (Silver and Finger, 1984). These include betaine, dimethylglycine, sarcosine, alanine, and glycine. The amino acid spectral pattern seen in the searobin is similar to those of the gustatory system of some species, especially marine species (see Kiyohara and Hidaka, 1991). The searobin fin rays are also extremely sensitive to mechanical stimulation. SCCs of the anterior dorsal fin of rockling, *Ciliata mustela*, are sensitive to skin mucus of various fish species, but not to amino acids (Peters et al., 1991). The fish skin mucus is a potent olfactory stimulus for some species, and free amino acids present in the mucus are primarily responsible for olfactory stimulation (Hara et al., 1984).

CHEMORECEPTION AND FISH BEHAVIOR

The encoded chemosensory information is transmitted and integrated into behavioral patterns through spatially separated neuroanatomical substrates within the central nervous system. The following four principal areas of chemosensory-behavior interactions deserve special attention: (1) homing migration in salmon, (2) hormones, pheromones, and reproduction, (3) intra- and interspecific interactions, and (4) feeding.

Homing Migration of Salmon

Homing migrations of salmon is well documented (Hasler, 1966; Harden Jones, 1968; Thorpe, 1988), and a number of studies have demonstrated the importance of chemical information in home-stream discrimination (see reviews in Hara, 1970; Cooper and Hirsch, 1982; Stabell, 1984, 1992). Despite considerable research effort, however, many aspects of homing migration are still poorly understood and remain largely controversial (Quinn, 1990). The imprinting hypothesis, i.e., learning of environmental chemical cues during a sensitive period of early development, and the pheromone hypothesis, i.e., innate responses to pheromones specific for local fish populations, are the two prevailing hypotheses.

Imprinting Hypothesis

The imprinting hypothesis proposes that juvenile salmon "imprint" to certain distinctive odors of the home stream during the early period of residence, and as adults they use this information to locate the home stream (Hasler and Wisby, 1951). Harden Jones (1968) further elaborates on this hypothesis, proposing that salmon are imprinted to a sequence of odors and retrace them as adults. The odors are learned at a sensitive period during the smolt stage. The physiological mechanisms by which the information is retained are not understood, but greater olfactory learning activity concomitant with elevated levels of blood thyroid hormones during smolting has been implicated in olfactory imprinting (Morin et al., 1989). There is also evidence that the imprinting process is rapid. Less than ten days, or perhaps only several hours, appears sufficient for imprinting to take place (Hasler and Scholz, 1983). The nature of the olfactory cues in home streams has been variously characterized as volatile, nonvolatile, organic, and inorganic, however no specific home-stream odor has been identified.

A series of artificial imprinting experiments on Lake Michigan coho salmon using the synthetic chemical morpholine attracted much public attention (for review see Hasler and Scholz, 1983). The basic principle was to expose salmon to low levels of either morpholine or phenethyl alcohol in place of, or in addition to, natural home-stream odors, to determine whether as adults they could be attracted to a stream scented with that chemical. The general conclusions drawn were that the fish are lured to a stream by the synthetic chemicals, able to learn or imprint to them during a brief period of the smolt stage, and retain these cues without being again exposed to the chemicals.

Pheromone Hypothesis

The pheromone hypothesis states that (1) populations or races of salmon in different streams emit pheromones that serve to identify fish distinctly from each other, (2) the memory of this population-specific pheromone is inherited, and (3) homing adults follow pheromone trails released by juveniles residing in the stream, i.e., the juveniles provide a constant source of population odors (Nordeng, 1971, 1977). The pheromones are suggested to be released from the skin mucus (Nordeng, 1971), or produced in the liver and released into the environment via the bile and the feces (Stabell, 1992). The olfactory bulbar neurons of Arctic char respond differentially to mucus emanating from different populations of the same species (Døving et al., 1974). The skin mucus of salmonids contains species-specific compositions of amino acids that are primarily responsible for olfactory simulation (Hara et al., 1984). As discussed above, the bile and bile salts are potent olfactory and gustatory stimulants for salmonids. Laboratory tests with juvenile salmonids have shown that they can distinguish the chemical traces of conspecifics of one population from another, but field studies have failed to demonstrate that homing is based on the presence of members of the population on or near the spawning grounds (Quinn, 1990).

Specific homing to a native spawning site is believed to be under genetic influence. Thus, a possible genetic contamination of pheromones resulting from hatchery escapes or random stocking program might seriously interfere with homing performance and population structure (Stabell, 1992).

Hormones, Pheromones, and Reproduction

Olfaction exerts a functional role in every aspect of the reproductive process, and considerable work on the role of chemical signals, pheromones, in reproductive behavior of fishes have been conducted. It has long been known that some substances originating from the gonad have pheromonal activities in eliciting courtship behavior in fish (Tavolga, 1956; Colombo et al., 1982; van den Hurk and Lambert, 1983). However, the sensory system involved and the nature of pheromones have not been rigorously investigated (for reviews see Liley, 1982; Liley and Stacey, 1983). It is only recently that Sorensen et al. (1987, 1988) have demonstrated that hormones and their metabolites may commonly serve as reproductive pheromones in fishes.

Preovulatory female goldfish sequentially release two hormonal pheromones, 17,20P and PGFs. Environmental cues trigger an ovulatory surge in gonadotropic hormones (GtH) in a vitellogenic female, which subsequentially stimulates 17,20P synthesis by the ovary. 17,20P induces female oocyte maturation and is released to the water, where it functions as a preovulatory priming pheromone. This pheromone evokes a surge in circulating GtH in males (Kobayashi *et al.*, 1986), which stimulates the synthesis of testicular 17,20P (Dulka et al., 1987). This in turn evokes an increase in milt production by the time of ovulation and spawning (Sorensen et al., 1987). At the time of ovulation, females produce PGFs to mediate follicular rupture and to trigger female spawning behavior. Circulating PGFs are subsequently metabolized and released into the water, where they function as a postovulatory pheromone that stimulates male sexual arousal, thus resulting in spawning synchrony (Sorensen et al., 1988).

In African catfish, steroid glucuronides such as 5β-pregnane-3α,17α-diol-20-one-3α glucuronide originating in male seminal vesicles play a pheromonal role in attracting ovulated females (Resink et al., 1989). Ovulated females also induce ovulation in other females only in the presence of males, suggesting the existence of bisexual primer pheromones (van Weerd and Richter, 1991). $PGF_{2\alpha}$, when injected intramuscularly, induces the release of a female-specific chemical in fathead minnows that triggers courtship behavior in conspecific males (Cole and Smith, 1987). In Arctic char, males release $PGF_{2\alpha}$ into water to attract females and stimulate their spawning behavior (Sveinsson and Hara, unpublished).

As discussed earlier, the fish olfactory system consists of medial and lateral subsystems. Selective stimulation of the lateral olfactory tract of cod, *Gadus morhua*, induces feeding behavior, while stimulation of the medial tract induces behavior patterns involved in spawning (Døving and Selset, 1980). In goldfish, stimulation by 17,20P and $PGF_{2\alpha}$ is mediated primarily by the medial olfactory tract, while stimulation by amino acids is mediated largely by the lateral tract (Sorensen et al., 1991). These two subsystems may be correlated to the main and accessory olfactory systems of higher vertebrates.

Intra- and Interspecific Interactions

Fright Reaction

When the skin of a fish is damaged (e.g., by a predator), alarm substance cells are broken and release alarm substance (Schreckstoff). Nearby conspecifics smell the alarm substance and show a fright reaction (Schreckreaktion). The fright reaction may then be treated as a visual signal by other conspecifics, leading to rapid transmission of the signal through a group of fish. This system, with its distinctive cell type, occurs only in the superorder Ostariophysi. Since von Frisch's (1938) discovery, a substantial amount of work has been reported on the alarm substance-fright reaction system (reviewed in Pfeiffer, 1974, 1982; Smith, 1982, 1986), In his recent review, Smith (1992) described other forms of "alarm pheromone systems" in the percid darters, sculpin, Cyprinodontiform, and other fish groups, in addition to the original alarm substance-fright reaction system in the Ostariophysi. All known fish alarm substance-fright reaction systems are based on release of chemicals by mechanical damage.

Kin Recognition

Juvenile coho salmon distinguish water conditioned by either familiar or unfamiliar siblings from water conditioned by nonsiblings of the same population, but if the two sibling groups were reared together, the test fish show preference for neither group (Quinn and Hara, 1986). Juvenile Arctic char are also able to discriminate between siblings and nonsiblings from the same population (Olsén, 1992). Sibling recognition might be important during shoaling, and that information about kinship can later reduce the risk of inbreeding (Quinn and Busack, 1985). Amino acids, ammonia, steroids, and other substances released from fish along with urine, feces, mucus, and through the gills have been suggested as possible chemical cues (Olsén, 1992).

Feeding

Feeding behavior is a stereotyped sequence of behavioral components which can normally be differentiated into several phases: (1) arousal, (2) search, and (3) uptake and ingestion (consummatory). Fish rely upon information received through all sensory channels. The relative importance of individual sensory systems differs in different species and is determined by their ecological niches, feeding strategy, motivation, and other biotic and abiotic environmental factors (Pavlov and Kasumyan, 1990). A wide variety of fish species utilize chemical signals in search, location, and ingestion of food. Compounds inducing feeding behavior are widely distributed throughout nature. However, fish do not respond to all food flavors; each species seems to have its own preference.

Feeding Stimulants

To date, all feeding stimulants identified for fish are (1) of low molecular weight (<1000), (2) nonvolatile, (3) nitrogenous, and (4) amphoteric. This generalization applies to most of the cases in which amino acids, betaine, other amino acid-like substances, and nucleotides have been implicated. A number of studies have shown conclusively that amino acids acting singly and in combination stimulate feeding behavior in fishes (see Carr, 1982; Hidaka, 1982; Mackie, 1982; Marui and Caprio, 1992; Takeda and Takii, 1992; Jones, 1992). High correlations exist between the potent gustatory amino acids determined electrophysiologically and those determined behaviorally in many species. However, for visual feeders such as rainbow trout, potent gustatory agents do not necessarily represent potent feeding stimulants. L-Proline, for example, is one of the most effective gustatory stimulants and in highest concentration in food extracts, but is inactive as a feeding stimulus (Adron and Mackie, 1978; Mackie, 1982; Marui et al., 1983b). The high sensitivity to a relatively broad spectrum of amino acids and other chemicals, and apparent species-specific array of relative acuities suggest that a biologically meaningful feeding stimulant may consist of a fingerprint-like mixture or chemical images (Atema, 1980). The fact that mixtures are almost always more potent than any individual component further indicates that the active compounds somehow interact additively or synergistically.

Detection of noxious or toxic substances is as important as detection of palatable materials. Some salmonids have developed high gustatory sensitivity to marine toxins and carbon dioxide (Yamamori et al., 1988; Yamashita et al., 1989). Some fish species may have evolved a mechanism to avoid poisonous prey organisms.

Feeding Stimulants, Nutrition, and Growth

Feeding activities change with dietary acceptability and palatability associated with chemical and physical properties as well as surrounding environmental conditions. The incorporation of feeding stimulants increases the palatability of formula diets for fish, particularly where alternative proteins are substituted for widely used fish meal. A diet supplemented with feeding stimulant amino acids and nucleotides stimulates feeding activity, survival rate, and feeding efficiency of larval and juvenile fish, leading to enhanced growth performance (Takeda and Takii, 1992). These effects are attributable primarily to improved food intake, digestion, and absorption. Chemosensory stimulation by dietary feeding stimulants induces the cephalic reflex to enhance secretion of gastric juices in the postprandial period and maintain increased protein and carbohydrate digestion. Analyses of hepatic enzymes further suggest that dietary feeding stimulants seem to enhance the utilization of dietary carbohydrate as an energy source through glycolysis, and the synthesis of body protein and fat by supplying nucleic acid and nicotinamide adenine dinucleotide phosphate (NADP) through the hexose monophosphate shunt (Takeda and Takii, 1992). Thus, feeding stimulant properties of glycogenic amino acids such as L-alanine, L-proline, and glycine might be related to the preferential utilization for energy production in fish.

ACKNOWLEDGMENTS

I thank Karla DeCaigny for her excellent secretarial assistance. Research discussed here was supported by Natural Science and Engineering Research Council of Canada grants (A 7576 and OGP 0007576).

REFERENCES

Adrian, E. D. and Ludwig, C., Nervous discharges from the olfactory organs of fish, *J. Physiol. (London)*, 94, 44, 1938.

Adron, J. W. and Mackie, A. M., Studies on the chemical nature of feeding stimulants for rainbow trout, *Salmo gairdneri* Richardson, *J. Fish Biol.*, 12, 303, 1978.

Allison, A. C., The morphology of the olfactory system in the vertebrates, *Biol. Rev.*, 28, 195, 1953.

Andersen, Ø. and Døving, K. B., Gonadotropin releasing hormone (GnRH) — a novel olfactory stimulant in fish, *NeuroReport*, 2, 458, 1991.

Atema, J., Structures and functions of the sense of taste in the catfish (*Ictalurus natalis*), *Brain Behav. Evol.*, 4, 273, 1971.

Atema, J., Chemical senses, chemical signals, and feeding behaviour in fishes, in *Fish Behaviour and Its Use in the Capture and Culture of Fishes*, Bardach, J. E., Magnuson, J. J., May, R. C., and Reinhart, J. M., Eds., ICLARM, Manila, 1980, 57.

Baatrup, E. and Døving, K. B., Physiological studies on solitary receptors of the oral disc papillae in the adult brook lamprey, *Lampetra planeri*, *Chem. Senses*, 10, 559, 1985.

Bardach, J. E. and Atema, J., The sense of taste in fishes, in *Handbook of Sensory Physiology*, Vol. 4, Beidler, L.M., Ed., Springer-Verlag, Berlin, 1971, 293.

Belghaug, R. and Døving, K. B., Odour threshold determined by studies of the induced waves in the olfactory bulb of the char *(Salmo alpinus)*, *Comp. Biochem. Physiol. A*, 57, 327, 1977.

Brand, J. G. and Bruch, R. C., Molecular mechanisms of chemosensory transduction: gustation and olfaction, in *Fish Chemoreception*, Hara, T. J., Ed., Chapman and Hall, London, 1992, 126.

Caprio, J., High sensitivity of catfish taste receptors to amino acids, *Comp. Biochem. Physiol. A*, 52, 247, 1975.

Caprio, J., Olfaction and taste in the channel catfish: an electrophysiological study of the responses to amino acids and derivatives, *J. Comp. Physiol. A*, 123, 357, 1978.

Caprio, J., Similarity of olfactory receptor responses (EOG) of freshwater and marine catfish to amino acids, *Can. J. Zool.*, 58, 1778, 1980.

Caprio, J., High sensitivity and specificity of olfactory and gustatory receptors of catfish to amino acids, in *Chemoreception in Fishes*, Hara, T. J., Ed., Elsevier, Amsterdam, 1982, 109.

Caprio, J., Olfaction and taste in fish, in *Comparative Physiology of Sensory Systems*, Bolis, L., Keynes, R. D., and Madrell, S.H.P., Eds., Cambridge University Press, Cambridge, 1984, 257.

Caprio, J., Peripheral filters and chemoreceptor cells in fishes, in *Sensory Biology of Aquatic Animals*, Atema, J., Fay, R.R., Popper, A. N., and Tavolga, W. N., Eds., Springer-Verlag, Berlin, 1988, 313.

Carr, W.E.S., Chemical stimulation of feeding behavior, in Chemoreception in Fishes, Hara, T. J., Ed., Elsevier, Amsterdam, 1982, 259.

Cole, K. S. and Smith, R. J. F., Release of chemicals by prostaglandin-treated female fathead minnows, *Pimephales promelas,* that stimulate male courtship, *Horm. Behav.,* 21, 440, 1987.

Colombo, L., Belvedere, P. C., Marconato, A., and Bentivegna, F., Pheromones in teleost fish, in *Reproductive Physiology in Fish,* Richter, C. J. J. and Goos, H. J. Th., Eds., Pudoc, Wageningen, 1982, 84.

Cooper, J. C. and Hirsch, P. J., The role of chemoreception in salmonid homing, in *Chemoreception in Fishes,* Hara, T. J., Ed., Elsevier, Amsterdam, 1982, 343.

Davenport, C. J. and Caprio, J., Taste and tactile recordings from the ramus recurrens facialis innervating flank taste buds in the catfish, *J. Comp. Physiol. A,* 147, 217, 1982.

Døving, K. B., Functional properties of the fish olfactory system, in *Progress in Sensory Physiology,* Vol. 6, Ottoson, D., Ed., Springer-Verlag, Berlin, 1986, 39.

Døving, K.B. and Holmberg, K., A note on the function of the olfactory organ of the hagfish *Myxine glutinosa, Acta Physiol. Scand.,* 91, 430, 1974.

Døving, K.B. and Selset, R., Behavior patterns in cod released by electrical stimulation of olfactory tract bundlets, *Science,* (Wash., D.C.), 207, 559, 1980.

Døving, K. B., Nordeng, H., and Oakley, B., Single unit discrimination of fish odours released by char (*Salmo alpinus* L.) populations, *Comp. Biochem. Physiol. A,* 47, 1051, 1974.

Døving, K. B., Selset, R., and Thommesen, G., Olfactory sensitivity to bile acids in salmonid fishes, *Acta Physiol. Scand.,* 108, 123, 1980.

Dulka, J. G., Stacey, N. E. Sorensen, P. W., and Van Der Kraak, G. J., A steroid sex pheromone synchronizes male-female spawning readiness in goldfish, *Nature (London),* 325, 251, 1987.

Erickson, J. R. and Caprio, J., The spatial distribution of ciliated and microvillus olfactory receptor neurons in the channel catfish is not matched by a differential specificity to amino acid and bile salt stimuli, *Chem. Senses,* 9, 127, 1984.

Evans, R. E. and Hara, T. J., The characteristics of the electro-olfactogram (EOG): its recovery following olfactory nerve section in rainbow trout (*Salmo gairdneri), Brain Res.,* 330, 65, 1985.

Evans, R. E., Zielinski, B., and Hara, T. J., Development and regeneration of the olfactory organ in rainbow trout, in *Chemoreception in Fishes,* Hara, T. J., Ed., Elsevier, Amsterdam, 1982, 15.

Ezeasor, D. N., Distribution and ultrastructure of taste buds in the oropharyngeal cavity of the rainbow trout, *Salmo gairdneri* Richardson, *J.Fish Biol.,* 20, 53, 1982.

Getchell, M. L., Zielinski, B., and Getchell, T. V., Odorant and autonomic regulation of secretion in the olfactory muscosa, in *Molecular Neurobiology of the Olfactory System,* Margolis, F.L. and Getchell, T. V., Eds., Plenum Press, New York, 1988, 71.

Goh, Y. and Tamura, T., The electrical responses of the olfactory tract to amino acids in carp, *Bull. Jpn. Soc. Sci. Fish.,* 44, 341, 1978.

Goh, Y. and Tamura, T., Olfactory and gustatory responses to amino acids in two marine teleosts - red sea bream and mullet, *Comp. Biochem. Physiol. C,* 66, 217, 1980.

Goh, Y., Tamura, T., and Kobayashi, H., Olfactory responses to amino acids in marine teleosts, *Comp. Biochem. Physiol. A,* 62, 863, 1979.

Hara, T. J., An electrophysiological basis for olfactory discrimination in homing salmon: a review, *J. Fish. Res. Board Can.,* 29, 569, 1970.

Hara, T. J., Chemoreception, in *Fish Physiology,* Vol. 5, Hoar, W. S. and Randall, D. J., Eds., Academic Press, New York, 1971, 79.

Hara, T. J., Electrical responses of the olfactory bulb of Pacific salmon *Oncorhynchus nerka* and *Oncorhynchus kisutch, J. Fish. Res. Board Can.,* 29, 1351, 1972.

Hara, T. J., Olfactory responses to amino acids in rainbow trout, *Salmo gairdneri, Comp. Biochem. Physiol. A,* 44, 407, 1973.

Hara, T. J., Olfaction in fish, *Progr. Neurobiol.,* 5, 271, 1975.

Hara, T. J., Effects of pH on the olfactory responses to amino acids in rainbow trout, *Salmo gairdneri, Comp. Biochem. Physiol. A,* 54, 37, 1976.

Hara, T. J., Structure-activity relationships of amino acids as olfactory stimuli, in *Chemoreception in Fishes*, Hara, T. J., Ed., Elsevier, Amsterdam, 1982a, 135.

Hara, T. J., Ed., *Chemoreception in Fishes*, Elsevier, Amsterdam, 1982b.

Hara, T. J., Role of olfaction in fish behaviour, in *The Behaviour of Teleost Fishes*, Pitcher, T. J., Ed., Croom Helm, London, 1986, 152.

Hara, T. J., Ed., *Fish Chemoreception*, Chapman and Hall, London, 1992a, 373.

Hara, T. J., Mechanisms of olfaction, in *Fish Chemoreception*, Hara, T. J., Ed., Chapman and Hall, London, 1992b, 150.

Hara, T.J. and Zielinski, B., Structural and functional development of the olfactory organ in teleosts, *Trans. Am. Fish. Soc.*, 118, 183, 1989.

Hara, T. J., Law, Y. M. C., and Hobden, B. R., Comparison of the olfactory response to amino acids in rainbow trout, brook trout, and whitefish, *Comp. Biochem. Physiol. A*, 45, 969, 1973.

Hara, T. J., Macdonald, S., Evans, R. E., Marui, T., and Arai, S., Morpholine, bile acids and skin mucus as possible chemical cues in salmonid homing: electrophysiological re-evaluation, in *Mechanisms of Migration in Fishes*, McCleave, J. D., Arnold, G. P., Dodson, J. J., and Neill, W. H., Eds., Plenum Press, New York, 1984, 363.

Hara, T. J., Sveinsson, T., Evans, R. E., and Klaprat, D. A., Morphological and functional characteristics of the chemosensory organs of Canadian char species, *Physiol. Ecol. Jpn. Spec.*, 1, 506, 1989.

Harden Jones, F. R., *Fish Migration,* Edward Arnold, London, 1968.

Hasler, A. D., *Underwater Guideposts*, University of Wisconsin Press, Madison, 1966.

Hasler, A. D. and Scholz, A. T., *Olfactory Imprinting and Homing in Salmon*, Springer-Verlag, Berlin, 1983.

Hasler, A. D. and Wisby, W. J., Discrimination of stream odors by fishes and relation to parent stream behavior, *Am. Nat.*, 85, 223, 1951.

Hidaka, I., The effect of carbon dioxide on the carp palatal chemoreceptors, *Bull. Jpn. Soc. Sci. Fish.*, 36, 1034, 1970.

Hidaka, I., Taste receptor stimulation and feeding behavior in the puffer, in *Chemoreception in Fishes*, Hara, T. J., Ed., Elsevier, Amsterdam, 1982, 243.

Hidaka, I. and Ishida, Y., Gustatory response in the shimaisaki (tigerfish) *Therapon oxyrhynchus, Bull. Jpn. Soc. Sci. Fish.*, 51, 387, 1985.

Hidaka, I., Ohsugi, T., and Yamamoto, Y., Gustatory response in the young yellowtail *Seriola quinqueradiata, Bull. Jpn. Soc. Sci. Fish.*, 51, 21, 1985.

Hoagland, H., Specific nerve impulses from gustatory and tactile receptors in catfish. *J. Gen. Physiol.*, 16, 685, 1933.

Ishida, Y. and Hidaka, I., Gustatory response profiles for amino acids, glycinebetaine, and nucleotides in several marine teleosts, *Nippon Suisan Gakkaishi*, 53, 1391, 1987.

Iwai, T., A comparative study of the taste buds in gill rakers and gill arches of teleostean fishes, *Bull. Misaki Mar. Biol. Inst. Kyoto Univ.*, 7, 19, 1964.

Jakubowski, M. and Whitear, M., Comparative morphology and cytology of taste buds in teleosts, *Z. Mikrosk. Anat. Forsch.*, 104, 539, 1990.

Jones, K.A., Food search behaviour in fish and the use of chemical lures in commercial and sports fishing, in *Fish Chemoreception,* Hara, T. J., Ed., Chapman and Hall, London, 1992, 288.

Kanwal, J. S. and Caprio, J., An electrophysiological investigation of the oro-pharyngeal (IX-X) taste system in the channel catfish, *Ictalurus punctatus, J. Comp. Physiol. A*, 150, 345, 1983.

Kanwal, J. S. and Finger, T. E., Central representation and projections of gustatory systems, in *Fish Chemoreception*, Hara, T. J., Ed., Chapman and Hall, London, 1992, 79.

Kitamura, S. and Ogata, H., Olfactory response of male loach, *Misgurnus anguillicaudatus,* to F-type prostaglandins, *Taste* and *Smell*, 24, 163, 1990.

Kiyohara, S. and Hidaka, I., Receptor sites for alanine, proline, and betaine in the palatal taste system of the puffer, *Fugu pardalis, J. Comp. Physiol. A*, 169, 523, 1991.

Kiyohara, S., Hidaka, E., and Tamura, T., Gustatory response in the puffer - II. Single fiber analyses, *Bull. Jpn. Soc. Sci. Fish.*, 41, 383, 1975.

Kiyohara, S., Yamashita, S., and Kitoh, J., Distribution of taste buds on the lips and inside the mouth in the minnow, *Pseudorasbora parva, Physiol. Behav.*, 24, 1143, 1980.

Kiyohara, S., Yamashita, S., and Harada, S., High sensitivity of minnow gustatory receptors to amino acids, *Physiol. Behav.*, 26, 1103, 1981.

Kobayashi, H. and Fujiwara, K., Olfactory response in the yellowtail *Seriola quinqueradiata, Nippon Suisan Gakkaishi*, 53, 1717, 1987.

Kobayashi, H. and Goh, Y., Comparison of the olfactory responses to amino acids obtained from receptor and bulbar levels in a marine teleost, *Exp. Biol.*, 44, 199, 1985.

Kobayashi, M., Aida, K., and Hanyu, I., Pheromone from ovulatory female goldfish induces gonadotropin surge in males, *Gen. Comp. Endocrinol.*, 63, 451, 1986.

Kotrschal, K., Solitary chemosensory cells — taste, common chemical sense or what?, *Rev. Fish Biol. Fish.*, 1, 3, 1991.

Liley, N. R., Chemical communication in fish, *Can. J. Fish. Aquat. Sci.*, 39, 22, 1982.

Liley, N. R. and Stacey, N. E., Hormones, pheromones and reproductive behavior, in *Fish Physiology*, Vol. 9, Hoar, W. S., Randall, D. G., and Donaldson, E. M., Eds., Academic Press, New York, 1983, 1.

Mackie, A. M., Identification of the gustatory feeding stimulants, in *Chemoreception in Fishes*, Hara, T. J., Ed., Elsevier, Amsterdam, 1982, 275.

Mackie, A. M. and Adron, J. W., Identification of inosine and inosine-5'-monophosphate as the gustatory feeding stimulants for the burbot, *Scophthalmus maximus, Comp. Biochem. Physiol. A*, 60, 79, 1978.

Marshall, N. B., The olfactory organs of bathypelagic fishes, *Symp. Zool. Soc. London*, 19, 57, 1967.

Marui, T. and Caprio, J., Teleost gustation, in *Fish Chemoreception*, Hara, T. J., Ed., Chapman and Hall, London, 1992, 171.

Marui, T., Harada, S., and Kasahara, Y., Gustatory specificity for amino acids in the facial taste system of the carp, *Cyprinus carpio* L., *J. Comp. Physiol. A*, 153, 299, 1983a.

Marui, T., Evans, R. E., Zielinski, B., and Hara, T. J., Gustatory responses of the rainbow trout (*Salmo gairdneri*) palate to amino acids and derivatives, *J. Comp. Physiol. A*, 153, 423, 1983b.

Moore, A. and Scott, A. P., Testosterone is a potent odorant in precocious male Atlantic salmon (*Salmo salar* L.) parr, *Philos. Trans. R. Soc. London Ser. B*, 332, 241, 1991.

Morin, P.-P., Dodson, J. J., and Doré, F. Y., Thyroid activity concomitant with olfactory learning and heart rate changes in Atlantic salmon, *Salmo salar*, during smoltification, *Can. J. Fish. Aquat. Sci.*, 46, 131, 1989.

Nordeng, H., Is the local orientation of anadromous fishes determined by pheromones? *Nature (London)*, 233, 411, 1971.

Nordeng, H., A pheromone hypothesis for homeward migration in anadromous salmonids, *Oikos*, 28, 155, 1977.

Ohno, T., Yoshii, K., and Kurihara, K., Multiple receptor type for amino acids in the carp olfactory cells revealed by quantitative cross-adaptation method, *Brain Res.*, 310, 13, 1984.

Oka, Y., Ichikawa, M., and Ueda, K., Synaptic organization of the olfactory bulb and central projection of the olfactory tract, in *Chemoreception in Fishes*, Hara, T.J., Ed., Elsevier, Amsterdam, 1982, 61.

Olsén, K. H., Kin recognition in fish mediated by chemical cues, in *Fish Chemoreception*, Hara, T. J., Ed., Chapman and Hall, London, 1992, 229.

Pavlov, D. S. and Kasumyan, A. O., Sensory principles of the feeding behaviour of fishes, *J. Ichthyol.*, 30, 77, 1990.

Peters, R. C., Kotrschal, K., and Krautgartner, W.-D., Solitary chemoreceptor cells of *Ciliata mustela* (Gadidae, Teleostei) are tuned to mucoid stimuli, *Chem. Senses*, 16, 31, 1991.

Pfeiffer, W., Pheromones in fish and amphibia, in *Pheromones*, Birch, M. C., Ed., North-Holland, Amsterdam, 1974, 269.

Pfeiffer, W., Chemical signals in communication, in *Chemoreception in Fishes*, Hara, T. J., Ed., Elsevier, Amsterdam, 1982, 307.

Quinn, T. P., Current controversies in the study of salmon homing, *Ethol. Ecol. Evol.*, 2, 49, 1990.

Quinn, T. P. and Busack, C. A., Chemosensory recognition of siblings in juvenile coho salmon *(Oncorhynchus kisutch), Anim. Behav.,* 33, 51, 1985.

Quinn, T. P. and Hara, T. J., Sibling recognition and olfactory sensitivity in juvenile coho salmon *(Oncorhynchus kisutch), Can. J. Zool.*, 64, 921, 1986.

Raderman-Little, R., The effect of temperature on the turnover of taste bud cells in catfish, *Cell. Tissue Kinet.*, 12, 269, 1979.

Resink, J. W., Voorthuis, P. K., van den Hurk, R., Peters, R. C., and van Oordt, P. G. W. J., Steroid glucuronides of the seminal vesicle as olfactory stimuli in African catfish, *Clarias gariepinus, Aquaculture*, 83, 153, 1989.

Reutter, K., Taste organ in the barbel of the bullhead, in *Chemoreception in Fishes*, Hara, T. J., Ed., Elsevier, Amsterdam, 1982, 77.

Reutter, K., Structure of the peripheral gustatory organ, represented by the siluroid fish *Plotosus lineatus* (Thunberg), in *Fish Chemoreception*, Hara, T. J., Ed., Chapman and Hall, London, 1992, 60.

Roper, S. D., The cell biology of vertebrate taste receptors, *Annu. Rev. Neurosci.*, 12, 329, 1989.

Satou, M., Synaptic organization, local neuronal circuitry, and functional segregation of the teleost olfactory bulb, *Prog. Neurobiol.*, 34, 115, 1990.

Satou, M., Synaptic organization of the olfactory bulb and its central projection, in *Fish Chemoreception*, Hara, T. J., Ed., Chapman and Hall, London, 1992, 40.

Sheldon, R. E., The olfactory tracts and centers in teleosts, *J. Comp. Neurol.*, 22, 177, 1912.

Silver, W. L., Olfactory responses from a marine elasmobranch, the Atlantic stingray, *Dasyatis sabina, Mar. Behav. Physiol. A*, 6, 297, 1979.

Silver, W. L., Electrophysiological responses from the peripheral olfactory system of the American eel, *Anguilla rostrata, J. Comp. Physiol.*, 148, 379, 1982.

Silver, W. L. and Finger, T. E., Electrophysiological examination of a non-olfactory, non-gustatory chemosense in the sea robin, *Prionotus carolinus, J. Comp. Physiol. A*, 154, 167, 1984.

Smith, R. J. F., The adaptive significance of the alarm substance-fright reaction system, in *Chemoreception in Fishes*, Hara, T. J., Ed., Elsevier, Amsterdam, 1982, 327.

Smith, R. J. F., The evolution of chemical alarm signals in fishes, in *Chemical Signals in Vertebrates*, Vol. 4, Duvall, D., Müller-Schwarze, D., and Silverstein, R. M., Eds., Plenum Press, New York, 1986, 99.

Smith, R. J. F., Alarm signals in fishes, *Rev. Fish Biol. Fish.*, 2, 33, 1992.

Sorensen, P. W., Hormones, pheromones and chemoreception, in *Fish Chemoreception*, Hara, T. J., Ed., Chapman and Hall, London, 1992, 199.

Sorensen, P. W., Hara, T. J., and Stacey, N. E., Extreme olfactory sensitivity of mature and gonadally-regressed goldfish to a potent steroidal pheromone, 17α,20β-dihydroxy-4-pregnen-3-one, *J. Comp. Physiol. A*, 160, 305, 1987.

Sorensen, P. W., Hara, T. J., Stacey, N. E., and Goetz, F. Wm., F Prostaglandins function as potent olfactory stimulants that comprise the postovulatory female sex pheromone in goldfish, *Biol. Reprod.*, 39, 1039, 1988.

Sorensen, P. W., Hara, T. J., Stacey, N. E., and Dulka, J. G., Extreme olfactory specificity of male goldfish to the preovulatory steroidal pheromone 17α,20β-dihydroxy-4-pregnen-3-one, *J. Comp. Physiol. A*, 166, 373, 1990.

Sorensen, P. W., Hara, T. J., and Stacey, N. E., Sex pheromones selectively stimulate the medial olfactory tracts of male goldfish, *Brain Res.*, 558, 343, 1991.

Stabell, O. B., Homing and olfaction in salmonids: a critical review with special reference to the Atlantic salmon, *Biol. Rev.*, 59, 333, 1984.

Stabell, O. B., Olfactory control of homing behaviour in salmonids, in *Fish Chemoreception*, Hara, T. J., Ed., Chapman and Hall, London, 1992, 249.

Sutterlin, A. M. and Sutterlin, N., Taste responses in Atlantic salmon (*Salmo salar*) parr, *J. Fish Res. Bd. Can.*, 27, 1927, 1970.

Sutterlin, A. M. and Sutterlin, N., Electrical responses of the olfactory epithelium of Atlantic salmon *(Salmo salar), J. Fish. Res. Board Can.*, 28, 565, 1971.

Suzuki, N. and Tucker, D., Amino acids as olfactory stimuli in freshwater catfish, *Ictalurus catus* (Linn.), *Comp. Biochem. Physiol. A*, 40, 399, 1971.

Takeda, M. and Takii, K., Gustation and nutrition in fishes: application to aquaculture, in *Fish Chemoreception*, Hara, T. J., Ed., Chapman and Hall, London, 1992, 271.

Tavolga, W. N., Visual, chemical and sound stimuli as cues in sex discriminatory behavior of the gobiid fish, *Bathygobius soporator, Zoologica*, 41, 49, 1956.

Thorpe, J. E., Salmon migration, *Sci. Prog. Oxf.*, 72, 346, 1988.

van den Hurk, R. and Lambert, J. G. D., Ovarian steroid glucuronides function as sex pheromones for male zebrafish, *Brachydanio rerio, Can. J. Zool.*, 61, 2381, 1983.

van Weerd, J. H. and Richter, C. J. J., Sex pheromones and ovarian development in teleost fish, *Comp. Biochem. Physiol. A*, 100, 517, 1991.

von Frisch, K., Zur Psychologie des Fisch-Schwarmes, *Naturwissenschaften*, 26, 601, 1938.

Whitear, M., Solitary chemosensory cells, in *Fish Chemoreception*, Hara, T. J., Ed., Chapman and Hall, London, 1992, 103.

Yamamori, K., Nakamura, M., Matsui, T., and Hara, T. J., Gustatory responses to tetrodotoxin and saxitoxin in fish: a possible mechanism for avoiding marine toxins, *Can. J. Fish. Aquat. Sci.*, 45, 2182, 1988.

Yamamoto, M., Comparative morphology of the peripheral olfactory organ in teleosts, in *Chemoreception in Fishes*, Hara, T. J., Ed., Elsevier, Amsterdam, 1982, 39.

Yamashita, S., Evans, R. E., and Hara, T. J., Specificity of the gustatory chemoreceptors for CO_2 and H^+ in rainbow trout *(Oncorhynchus mykiss), Can. J. Fish. Aquat. Sci.*, 46, 1730, 1989.

Yoshii, K., Kamo, N., Kurihara, K., and Kobatake, Y., Gustatory responses of eel palatine receptors to amino acids and carboxylic acids, *J. Gen. Physiol.*, 74, 301, 1979.

Yoshii, K., Kashiwayanagi, M., Kurihara, K., and Kobatake, Y., High sensitivity of the eel palatine receptors to carbon dioxide, *Comp. Biochem. Physiol. A*, 66, 327, 1980.

Zeiske, E., Caprio, J., and Gruber, S. H., Morphological and electrophysiological studies on the olfactory organ of the lemon shark, *Negaprion brevirostris* (Poey), in *Indo-Pacific Fish Biology*, Uyeno, T., Arai, R., Taniuchi, T., and Mutsuura, K., Eds., Ichthyological Society of Japan, Tokyo, 1986, 381.

Zeiske, E., Theisen, B., and Breucker, H., Structure, development, and evolutionary aspects of the peripheral olfactory system, in *Fish Chemoreception*, Hara, T. J., Ed., Chapman and Hall, London, 1992, 13.

Zielinski, B. and Hara, T. J., Morphological and physiological development of olfactory receptor cells in rainbow trout (*Salmo gairdneri*) embryos, *J. Comp. Neurol.*, 271, 300, 1988.

8 Cardiovascular System

Anthony P. Farrell

INTRODUCTION

Fish are the most successful vertebrate group in terms of biomass and number of species. However, prior to the reviews by Randall (1970) and Satchell (1971), early descriptions of fish cardiovascular physiology were based largely, and sometimes inappropriately, on mammalian cardiovascular systems. Our understanding of the cardiovascular physiology specific to fishes has improved considerably and much of this information has been presented already in a number of informative reviews, monographs, and perspectives (Jones and Randall, 1978; Johansen and Burggren, 1980; Laurent et al., 1983; Farrell, 1984, 1985; Santer, 1985; Butler, 1986; Nilsson and Axelsson, 1987; Butler and Metcalfe, 1988; Farrell, 1991a; Satchell, 1991; Farrell and Jones, 1992; Bushnell et al., 1992). This chapter focuses on features which are common to water breathing fishes. The extent of diversity between fish groups, especially between the cyclostomes, elasmobranchs, and teleosts is also characterized where it is known. The first section provides a brief overview of the general organization of the fish cardiovascular system, followed by a discussion of cardiac anatomy. Next, the regulation of cardiac output ($\dot{Q}$), and vascular patterns and the distribution of blood flow are presented.

GENERAL OVERVIEW OF THE CARDIOVASCULAR SYSTEM

The following description of the general organization of the fish cardiovascular system serves two purposes: (1) it highlights some of the differences with mammalian cardiovascular systems, and (2) it provides a framework for the remainder of the chapter. The branchial heart, which is contained within a pericardial sac, consists of four chambers, the sinus venosus, the atrium, the ventricle, and either the bulbus arteriosus (in teleosts) or the conus arteriosus (in cyclostomes and elasmobranchs). Venous blood returning to the heart is first collected by the sinus venosus and then pumped sequentially by the atrium and the ventricle into the conus or bulbus. The four chambers are anatomically distinct, unlike the mammalian heart. The sinus venosus is the site of the pacemaker tissue that initiates the heart beat. Atrial contraction is the main means for filling the ventricle. The ventricle is the largest and main pressure-

0-8493-8042-1/93/$0.00+$.50

generating chamber. There are large interspecific differences in ventral aortic blood pressure — lowest in cyclostomes and highest in very active teleosts — and to some extent stroke volume, and these are reflected in the considerable interspecific variability in ventricular morphology, histology, and vascular supply. The bulbus arteriosus of teleosts and the conus arteriosus of cyclostomes are elastic chambers which serve to depulse blood flow ejected from the ventricle, thereby creating a continuous flow of blood in the major arteries. The conus arteriosus of elasmobranchs contains cardiac muscle, is contractile, and has two to six sets of valves.

The entire cardiac output ($\dot{Q}$) passes through the conus or bulbus and the ventral aorta and is distributed to the respiratory (gill) circulation via afferent branchial arteries (four to seven bilateral pairs). In contrast with other vertebrates, blood goes directly to the systemic circulation and does not return to the heart after passing through the respiratory circuit; the respiratory and systemic circulations are arranged in series. As a result, systemic dorsal aortic blood pressure (P_{da}) is around one-third lower than ventral aortic blood pressure (P_{va}).

The fish systemic circulation is unique in that it is divided into primary and secondary circulations; apparently there is no lymphatic system in fish. The ventral and dorsal aortas are the main distribution vessels for the primary circulation. The secondary circulation arises from primary arteries at numerous gill and systemic locations as narrow, convoluted arterial vessels. The secondary circulation is a low pressure and low hematocrit system which generally perfuses surface structures that exchange gases directly with the water (gills, scales, and skin). Thus, the secondary circulation is predominantly nutritive, in contrast with the largely respiratory role of the primary circulation. The secondary circulation drains into the veins of the primary circulation. Blood pressures in veins are generally low and sometimes subambient in fishes. Thus, accessory hearts are found which aid in the return of venous blood to the branchial heart. In addition, venous blood can be aspirated toward the branchial heart in certain fishes as a result of a *vis-a-fronte* cardiac filling mechanism.

Regulation of $\dot{Q}$ is achieved by changes in both heart rate (f_H) and stroke volume (SV_H) which are controlled by intrinsic, neural, and humoral control mechanisms. Up to a threefold increase in $\dot{Q}$ is observed in fish. Changes in the distribution of blood flow between the various in-parallel vascular circuits are in turn brought about through changes in vascular resistance.

CARDIAC ANATOMY

This section focuses on the principal features of the anatomy, fine structure and morphometrics of the cardiac chambers.

Sinus Venosus

The thin-walled sinus venosus receives venous blood primarily via paired Cuverian ducts, hepatic veins, and anterior jugular veins. The openings of the

Cuverian ducts into the sinus venosus are not valved, but the smaller hepatic veins are guarded by muscular sphincters. The sinus wall can have either a complete or partial layer of cardiac muscle, or no cardiac muscle (Yamauchi, 1980). The major functional role of the sinus venosus is related to the initiation and control of the heart beat. The sinus venosus is the site of the cardiac pacemaker tissue, a ring of specialized myocardial cells (nodal tissue) usually located at the base of the sino-atrial ostium. Myogenic activity of atrial and ventricular tissues and particularly the atrioventricular region can also initiate a slower rate of heart beat.

Atrium

The main functional role of the atrium is to act as a contractile chamber to fill the ventricle. The single atrium has thin muscular trabeculae in its walls. The trabeculae (19 to 35 μm in diameter) form a mesh-like network, which in benthopelagic teleosts has a more spacious arrangement, giving the atrium a frail appearance (Santer, 1985). Atrial mass is generally 8 to 25% of ventricular mass and constitutes 0.01 to 0.03% of body mass (Table 1). However, the atria of the New Zealand hagfish (*Eptatretus cirrhatus*), tunas, and two species of red-blooded Antarctic fishes (*Pagothenia bernacchii* and *P. borchgrevinki*) are exceptionally large.

Ventricle

The single ventricle in fishes shows considerable species variability with neither a "typical" shape (Santer, 1985), nor a "typical" ventricular mass, histology, or vascular network (Farrell and Jones, 1992) . Santer (1985) suggested three categories of ventricular shape:

1. A *sac-like* ventricle, which is common in elasmobranchs and marine teleosts
2. A *tubular* ventricle, which is found only in fish with an elongate body shape
3. A *pyramidal* ventricle which is restricted to teleost species with an active life style.

Fish ventricles have two basic types of myocardium: the spongiosa which forms a sponge-like network of muscular trabeculae and the compacta which can be an outer layer enclosing the inner spongiosa. The spongiosa accounts for the greater proportion of ventricular mass in almost all fishes (Table 1). Most fish species have a type I (subtype Ia) ventricle with only spongiosa myocardium (Table 2). There are no capillaries per se in the myocardium of any type I hearts. In type I hearts venous blood contained in the lumen and intertrabecular spaces of the ventricle (luminal blood) provides the only blood supply (hence the terms venous, lacunary, or avascular hearts). Given that ventricular (and atrial) trabeculae of most fish derive O_2 from luminal blood, the diameter of the trabeculae is likely a compromise between minimizing the distance for O_2 diffusion and maximizing the cross-sectional area for tension development (Davie and Farrell, 1991a).

TABLE 1
Cardiac Morphometrics in Selected Fishes

	Relative Ventricular Mass (%)	*Compacta* (%)	Relative Atrial Mass (%)
	Teleosts with Compacta		
Katsuwonus pelamis (skipjack tuna)	0.38	65.6	0.061
Thunnus albacares (yellowfin tuna)	0.29	55.4	0.056
Makaira nigricans (Pacific blue marlin)	0.087	48.0	0.013
Clupea harengus (herring)	—	24.8	—
Salmo gairdneri (rainbow trout)	0.08–0.13	30–39	0.017
Anguilla dieffenbachii (longfinned eel)	0.03–0.10	40.9	0.007–0.013
	Elasmobranchs		
Isurus oxyrinchus (shortfin mako shark)	0.14	41.5	0.028
Squalus acanthias (dogfish)	0.086	22.1	0.017
Chimaera monstrosa	—	5.0	—
	Species with Only Spongiosa		
Eptatretus cirrhatus (New Zealand hagfish)	0.096	None	0.039
Chionodraco hamatus (hemoglobin-free icefish)	0.39	None	—
Pagothenia borchgrevinki (red-blooded Antarctic fish)	0.16	None	0.040
Pleuronectes platessa (flounder)	0.035	None	—

Adapted from Farrell and Jones (1992).

The three other categories of ventricles all have capillaries in the compacta. Type II ventricles have coronary capillaries only in the compacta, e.g., most teleosts possessing compacta. The coronary circulation in ventricle types III and IV reaches the spongiosa as well as the compacta. Most elasmobranchs have type III ventricles. Endothermic sharks and active teleosts have type IV ventricles (Tota, 1989).

Relative ventricular mass (ventricular mass divided by body mass) varies 10-fold in fish (Table 1). The following generalizations were made by Farrell

TABLE 2.
Ventricle Types Based on the Myocardial Tissue Type in the Ventricle and on the Pattern of the Coronary Circulation

	Ventricle Tissue Type		Coronary Vessels				
Type	*Spongiosa*	*Compacta*	Cranial Circulation	Pectoral Circulation	*Compacta* Capillaries	*Spongiosa* Capillaries	Atrial Capillaries
Ia	Yes	No	No	No	No	No	No
b	Yes	No	Superficial	No	No	No	No
c	Yes	No	Superficial	No	No	No	No
II	Yes	Yes	Yes	In some species	Yes	No	No
III	Yes	Yes (<40%)	Yes	In some species	Yes	Yes	Yes

Adapted from Tota (1989).

and Jones (1992) concerning ventricular mass: active species have larger ventricles; endothermic sharks have larger ventricles than ectothermic sharks; cold-acclimation increases relative ventricular mass in rainbow trout, carp, and goldfish; and Antarctic fishes have relatively large ventricles. These generalizations implicate a large relative ventricular mass as being important in (1) developing higher blood pressure in active fishes, (2) compensating for the negative inotropic effect of low temperature, and (3) accommodating large cardiac stroke volumes.

Myocytes

A number of features distinguish fish cardiac muscle cells (myocytes) from those in mammals. These include: a small size (1 to 12.5 μm) compared with 10 to 25 μm in mammals; a limited anatomical development of the sarcoplasmic reticulum (SR); a peripheral arrangement of myofibrils; a variable amount of intracellular myoglobin.

Myocyte diameters are most often between 2.5 and 6.0 μm. A narrow myocyte provides for a shorter diffusion distance from the outside to the center of the cell, and a higher ratio of sarcolemmal area to intracellular volume. The effect of cardiac growth on myocyte size has been assessed for perch (*Perca perca*; Karttunen and Tirri, 1986) and rainbow trout (Farrell et al., 1988a) and in both species hyperplastic growth predominates. Hyperplastic growth of fish myocytes contrasts with the situation in mammals where myocyte growth is essentially hypertrophic, except during the embryonic period and possibly the first few weeks of neonatal life. The "high-speed" cardiac pump of small mammals is characterized by small myocytes. The small myocytes of the "low-speed" fish heart clearly contrast with this dogma (Farrell, 1991a).

Unlike skeletal muscle and mammalian cardiac muscle, the SR and T-tubule systems are rudimentary in fish. The limited anatomical development of the SR with respect to excitation-contraction coupling is discussed later. The fine structure of myocardial cells is described in considerable detail by Yamauchi (1980), Helle (1983), and Santer (1985).

Innervation Patterns

Vagal cardiac innervation is found in all fishes except the hagfishes. The vagus contains cholinergic efferent fibers which provide a rich innervation of the sino-atrial region. The atrium and atrioventricular region are innervated to a lesser extent whereas the ventricle is only sparsely and partially innervated by cholinergic fibers. These fibers stimulate inhibitory muscarinic cholinoceptors in all species except lampetroids, which have excitatory nicotinic cholinoceptors. Adrenergic cardiac innervation is found in the majority of teleosts studied, but not in cyclostomes or elasmobranchs. Typically, the sino-atrial and atrioventricular regions, are well innervated. The compacta of the ventricle, especially near the atrioventricular region, the coronary vessels, and the bulbus arteriosus also receive adrenergic fibers. The literature on the innervation patterns of fish hearts and pacemaker tissue is comprehensively reviewed by Nilsson (1983), Laurent et al. (1983), and Santer (1985).

Coronary Circulation

All of the elasmobranchs and about one quarter of the teleosts have coronaries and no single factor adequately accounts for the presence of a coronary circulation (Davie and Farrell 1991a). Typically, teleosts that either tolerate environmental hypoxia or are active swimmers have a coronary circulation as well as a high myoglobin level. The coronary circulation has two sites of origin: a cranial (cephalad) circulation and, in some fish, an additional pectoral (caudal) circulation (e.g., eel, skate, and swordfish). The anatomical origins of both the cranial and pectoral coronary circuits are such that oxygenated blood is delivered directly from the gills to the heart at the highest possible postbranchial blood pressure. The cranial circulation arises as paired branches off the efferent branchial arch arteries and reaches the ventricle across the surface of the bulbus arteriosus or conus arteriosus. The pectoral coronary circulation arises from the first branch of the dorsal aorta, the coracoid artery, and vessels penetrate the pericardium, reaching the ventricle via pericardial ligaments. The cranial and pectoral circulations are functionally interconnected at the arterial level (Davie and Farrell, 1991b).

In elasmobranchs and teleosts the compacta is well vascularized with capillaries connecting to veins that drain into the atrial chamber close to the atrioventricular region. However, the trabeculae of the spongiosa in types III and IV ventricles have centrally located arteries, terminal arterioles, a few capillaries, and no venous system. Terminal arterioles and capillaries open directly to the lumen of the heart, in a manner analogous to the Thebesian system of the mammalian heart (Tota, 1989). On the conus arteriosus, numerous very small, perpendicular branches from the main coronary artery circle and supply the conal cardiac muscle and large veins run parallel to the main coronary arteries More thorough descriptions of the coronary circulations in fishes are presented by Tota et al. (1983), Tota (1989), and Davie and Farrell (1991a).

Arteriosclerotic lesions are abundant in the coronary vessels of mature, migratory Pacific salmon. Typically, coronary lesions are found in more than 95% and often 100% of a sample population, occur along 66 to 80% of the length of the main coronary artery, and often obstruct the arterial lumen by 10 to 30% (see Farrell et al., 1986, 1990). These lesions consist of extensive intimal proliferations of vascular smooth muscle without the calcium and lipid inclusions typically found in mature mammalian lesions (Moore et al., 1976). The etiology of coronary lesions in migratory salmonids is unclear. Lesion growth is stimulated by maturation and by injection of sex hormones (House et al., 1979) but lesions are well developed at least one year before maturation in Atlantic salmon (Farrell et al., 1986). Factors related to growth are also implicated in lesion development because lesions are fewer and less severe in the slower-growing and land-locked varieties of salmonids (Kubasch and Rourke, 1990; Saunders et al., 1992). Preliminary studies show no lesions in the main coronary artery that lies on the conus arteriosus of elasmobranchs (Farrell et al., 1992).

REGULATION OF CARDIAC OUTPUT

Cardiac Cycle

The synchronized contractions and relaxations of the cardiac chambers are referred to as the cardiac cycle. The electrical events (waves of muscle depolarization and repolarization) are recorded as an electrocardiogram (ECG). Although the ECG terminology is the same in fishes as in mammals, there are aspects of the cardiac cycle that are unique to fishes. As in mammals, the P-wave represents atrial contraction, the QRS-complex ventricular contraction, and the T-wave ventricular relaxation. In fishes such as eels and hagfishes, contraction of the sinus venosus appears as a V-wave during the P-T interval. An additional B-wave appears during the S-T interval, marking the contraction of conal muscle in elasmobranchs. The atrioventricular delay, as indicated by the P-Q interval, ensures that atrial contraction provides complete and proper ventricular filling. While contraction of the ventricle is a coordinated and synchronized event, the fish ventricle lacks the specialized conducting fibers found in mammalian hearts (Farrell and Jones, 1992).

The contraction/relaxation cycle of either the atrium or ventricle can be separated into systolic and diastolic periods. Systole refers to the initial isometric contraction phase (when blood pressure is raised but there is no ejection of blood from the cardiac chamber) and the subsequent isotonic contraction phase (when blood is ejected from the chamber). Diastole refers to the relaxation of cardiac muscle during which cardiac filling occurs.

Cardiac Output ($\dot{Q}$)

$\dot{Q}$ is the product of SV_H and f_H. This section focuses on the overall changes in $\dot{Q}$ in response to experimental or environmental perturbations. The mechanisms controlling SV_H and f_H are described in subsequent sections.

Less than 100 studies have measured $\dot{Q}$ in fish and the majority of these have employed the Fick principle. Unfortunately, the Fick estimates in fish are unreliable because the method can introduce substantial errors due to gill O_2 consumption and O_2 uptake across the skin among other factors (see Randall, 1985; Butler and Metcalfe, 1988). Thus, statements about $\dot{Q}$ in fishes are made with the understanding that, first, there is a limited data base and second, many of these values may be inaccurate. These considerations are amplified by Farrell and Jones (1992).

Activity

The routine activity state of fishes is highly species specific. Accordingly, fish that display high levels of activity often have a higher resting $\dot{Q}$ than sluggish forms (Table 3). In fish species ranging from hagfish to tuna, resting $\dot{Q}$ varies by 15-fold (8.7 to 132 ml/min/kg). About half of this range is accounted for by the 7-fold difference in f_H.

Fish exercise anaerobically (burst exercise) for short periods. Burst exercise is accompanied by decreases in f_H, $\dot{Q}$ and arterial blood pressure. During the recovery from burst exercise $\dot{Q}$, f_H and arterial blood pressure all increase. Burst exercise can reduce O_2 supply to the heart (a decrease in venous PO_2) and expose the heart to a marked extracellular acidosis (a decrease in blood pH from 7.9 to 7.4) for a considerable portion of the recovery period. Both hypoxia and acidosis directly affect the heart (see below).

In response to prolonged swimming, $\dot{Q}$ increases from 47 to 200% depending on the species (Table 3). It seems likely that the $\dot{Q}$ value observed with maximum exercise *in vivo* is probably close to the maximum $\dot{Q}$ possible for that species (see Farrell and Jones, 1992).

Farrell (1991a) noted that the increase in $\dot{Q}$ associated with aerobic swimming in fishes is largely effected by increases in SV_H and that this pattern is different to that in other vertebrates where increases in f_H are considerably larger than those of SV_H. During prolonged swimming the percentage change in SV_H is 25 to 200% vs. 7 to 50% for f_H (Table 3). This generalization may be limited, however, to temperate water species since the Antarctic notothenid *P. borchgrevinki* (Table 3) and possibly the tropical tuna increase f_H rather than SV_H during exercise. Also, the European pike, *Esox lucia*, shows large changes in f_H (from 30 to 80 bpm) between active and inactive states and this type of control may be characteristic of ambush-predators (Lucas et al., 1991).

Temperature

Temperature has a profound effect on $\dot{Q}$ and its regulation. So far all studies indicate that $\dot{Q}$ increases with increases in water temperature. For an acute temperature change, the Q_{10} values for resting $\dot{Q}$ range between 1.56 and 2.40 and following acclimation to different water temperatures, the "chronic Q_{10}" values for resting Q is 2.6 to 2.7. These data suggest that resting $\dot{Q}$ conforms to a given temperature change (Graham and Farrell, 1989; Farrell and Jones, 1992).

Maximum aerobic swimming performance in rainbow trout is rather similar over a reasonably broad range of water temperatures (Randall and Brauner, 1991), i.e., a Q_{10} value nearer 1.0 than 2.0. At first sight, this finding is inconsistent with the important role of $\dot{Q}$ in the O_2 supply to skeletal muscle and the observation that resting $\dot{Q}$ conforms to a temperature change. However, this discrepancy may be explained if resting $\dot{Q}$ and maximum $\dot{Q}$ are affected differently by temperature acclimation.

TABLE 3
Cardiovascular Data Measured in Fish at Rest (Bold) and During Either Exercise (Swimming Speed Given in cm/s or % of U_{crit}) or Hypoxia (PO_2 Given in mmHg)

Species (Mass, kg)	Condition	$\dot{Q}$ (ml/min/kg)	f_H (beats/min)	Stroke Volume (ml/kg)	P_{va} (mmHg)	P_{da} (mmHg)	Vascular Resistance (mmHg/ml/min/kg) Branchial (Rg)	Systemic (Rs)	Ref.
Myxine glutinosa	**Rest**	**8.7**	**22**	**0.41**	**7.8**	**5.8**	**0.23**	**0.66**	a
(0.55–0.91)									
Scyliorhinus stellaris	**Rest**	**52.5**	**43**	**1.21**	**25.2**	**18.5**	**0.13**	**0.35**	b
(2.8)	26 cm/s	89.2	46	1.94	25.5	17.5	0.09	0.20	
S. canicula	**Rest**	**23.2**	**30**	**0.81**	**31.8**	**24.8**	**0.30**	**1.07**	c
(0.56–1.05)	40 mmHg	27.9	20	1.43	27.5	18.8	0.31	0.67	
Triakis semifasciata	**Rest**	**33.1**	**51**	**0.77**	**47.7**	**32.3**	**0.46**	**0.98**	d
(1.93)	32–43 cm/s	56.2	55	1.02	58.1	36.2	0.39	0.64	
	Postexercise	60.4	50	1.22	55.0	35.4	0.33	0.58	
Anguilla anguilla	**Rest**	**11.5**	**37**	**0.29**	**38.0**	**23.3**	**1.28**	**2.02**	e
(0.51)	40 mm Hg	7.2	23	0.31	29.6	15.7	1.87	2.01	
A. australis	**Rest**	**11.3**	**54**	**0.21**	**38.6**	**23.5**	**1.34**	**2.08**	f
(0.62)	15 cm/s	11.3	54	0.21	39.0	23.6	1.36	2.09	
A. japonica	**Rest**	**10.0**	**30**	**0.33**	**28.4**	**18.9**	**0.96**	**1.89**	g
(0.3–0.6)	70 mmHg	10.0	27	0.37	27.9	25.0	0.29	2.50	
	40 mmHg	6.9	10	0.69	23.5	19.4	0.60	2.81	
Gadus morhua	**Rest**	**17.3**	**43**	**0.39**	**36.8**	**24.0**	**0.74**	**1.39**	h
(0.4–0.8)	26 cm/s	25.4	51	0.49	46.5	30.0	0.65	1.18	
G. morhua	**Rest**	**19.2**	**41**	**0.51**	**36.8**	**23.3**	**0.70**	**1.21**	i
(0.4–1.3)	35 mmHg	17.0	21	0.86	36.0	23.3	0.74	1.37	
Hemitripterus americanus	**Rest**	**18.8**	**37**	**0.51**	**28.5**	**23.3**	**0.28**	**1.24**	j
(0.67–1.4)	30 cm/s	30.9	49	0.64	35.3	26.3	0.29	0.85	
Ophiodon elongatus	**Rest**	**11.2**	**29**	**0.38**	**38.0**	**28.3**	**0.87**	**2.53**	k

(3.8–8.5)	75 cm/s	9.9	28	0.37	39.2	30.2	0.91	3.05	
	35 cm/s	7.7	12	0.70	33.8	23.9	1.29	3.10	
Oncorhynchus mykiss	**Rest**	**17.6**	**38**	**0.46**	**39.0**	**31.0**	**0.46**	**1.76**	l
(0.7–1.5)	51% U_{crit}	28.4	43	0.62	40.1	28.7	0.40	1.01	
	75% U_{crit}	34.8	49	0.70	44.7	32.7	0.34	0.94	
	80% U_{crit}	42.9	51	0.80	61.7	37.0	0.58	0.86	
	U_{crit}	52.6	51	1.00	—	—	—	—	
O. mykiss	**Rest**	**36.7**	**79**	**0.46**	**31.2**	**25.3**	**0.25**	**0.52**	m
(0.1–0.7)									
Pagothenia borchgrevinki	**Rest**	**29.6**	**11.3**	**2.16**	**28**	**—**	**(0.95)**	**—**	n
(0.064)	20 cm/s	51.8	21.0	2.16	28	—	(0.54)	—	
Thunnus albacares	**Spinalized**	**115.0**	**97**	**1.30**	**89.7**	**32.6**	**0.49**	**0.28**	o
(1.4)	130 cm/s	115.0	97	1.30	89.7	32.6	0.50	0.28	
	90 cm/s	115.0	97	1.30	89.7	32.6	0.50	0.28	
	50 cm/s	74.1	71	1.00	89.7	32.6	0.77	0.44	
T. Alalunga	**Spinalized**	**29.4**	**90**	**0.33**	**82.4**	**47.1**	**1.20**	**1.60**	p
(7.8–10.7)									
Katsuwonus pelamis	**Spinalized**	**132.0**	**126**	**1.08**	**87.3**	**40.2**	**0.36**	**0.30**	o
(1.6)	130 cm/s	132.0	108	1.22	87.3	40.2	0.36	0.30	
	90 cm/s	105.0	86.2	1.22	87.3	36.2	0.49	0.34	
	50 cm/s	75.0	65.5	1.15	87.3	40.2	0.63	0.54	

Note: Central venous pressure was assumed to be zero in all conditions. Vascular resistances shown in parentheses are values for total peripheral resistance (Rp), the sum of Rg and Rs.

References: a, Axelsson et al. (1990); b, Piiper et al. (1977); c, Butler and Taylor (1975); d, Lai et al. (1989a, 1990a); e, Peyraud-Waitznegger and Soulier (1989); f, Davie and Forster (1980); g, Chan (1986); h, Axelsson and Nilsson (1986); i, Fritsche and Nilsson, 1989; j, Axelsson et al. (1989); k, Farrell (1982); l, Kiceniuk and Jones (1977); m, Wood and Shelton (1980a, 1980b); n, Axelsson et al. (1992); o, Bushnell (1988); p, Lai et al. (1987).

Environmental Hypoxia

When water is made moderately hypoxic (water $PO_2 > 70$ torr), fish generally maintain $\dot{Q}$. A reflex bradycardia develops at a lower water PO_2 (40 to 70 torr). In some fishes, there is an additional compensatory increase in SV_H so that $\dot{Q}$ is unchanged (e.g., rainbow trout and *Scyliorhinus canicula*), whereas in other fish there is a smaller increase in SV_H and $\dot{Q}$ decreases (e.g., lingcod and eel). Tuna are unusually hypoxia-sensitive, decreasing $\dot{Q}$ at a water PO_2 between 85 and 104 mmHg as a result of reflex bradycardia and no change in SV_H (Bushnell, 1988; Brill and Bushnell, 1991). With severe hypoxia, $\dot{Q}$ decreases in all fish studied except for hagfishes (Axelsson et al., 1990; Forster et al., 1992).

Control of f_H

The intrinsic rhythm of the sino-atrial pacemaker is modulated by stretch of the pacemaker cells, cholinergic nerve fibers, adrenergic nerve fibers, and hormones. Since the exact interplay of the various effectors is not completely understood, the following describes the known mechanisms and identifies under what conditions and in which species they play a major role. Significant changes in f_H occur with changes in water temperature, during burst swimming, on exposure to environmental hypoxia, with aerobic swimming, and with feeding.

Intrinsic f_H, Resting f_H, and Maximum f_H

Intrinsic f_H is both species-specific and directly affected by temperature. However, f_H *in vivo* is rarely the intrinsic f_H. Adrenergic and (or) cholinergic controls exert the greatest influence on f_H in fish. Therefore, good estimates of intrinsic f_H *in vivo* are possible after pharmacological blockade of cardiac β-adrenoceptor and muscarinic receptors with injections of suitable antagonist drugs. Based on this type of information, Farrell and Jones (1992) suggested that intrinsic f_H is lowest in cyclostomes, higher in elasmobranchs, and highest in certain teleosts, with a sevenfold difference in f_H existing between hagfish and tuna. Also active fish tend to have a higher intrinsic f_H within a phylogenetic grouping.

The cardiac pacemaker rate varies directly with temperature. With acute changes in temperature, the Q_{10} for intrinsic f_H is 2.0 or greater. Cold acclimation, however, results in a higher pacemaker rate at a given test temperature. As a result, a Q_{10} value derived from the f_H at the two acclimation temperatures is typically less than 2.0. The compensatory mechanism underlying the adjustment of pacemaker rate in response to temperature acclimation is not understood.

All species examined so far, except for cyclostomes, have an inhibitory resting cholinergic tone to the heart. In addition, all species have an excitatory resting adrenergic tone mediated by β-adrenoceptors. Therefore, resting heart rate in most fish species is determined by a "push-pull" type modulation as set by the relative levels of adrenergic and (or) cholinergic tone on the heart

(Farrell, 1991a). In fact, resting f_H can be slower than, faster than, or the same as the intrinsic pacemaker frequency depending on the relative contributions of adrenergic and cholinergic tone.

Maximum f_H in fish is generally achieved by removing the cholinergic inhibitory tone and possibly maximizing the adrenergic excitation. Maximum f_H is rarely more than twice the resting f_H, being highly species-specific and temperature dependent (see Table 3). The red-blooded Antarctic fish *P. borchgrevinki* is exceptional; the very low resting f_H at 0°C increases by 300% after injection of atropine to remove the inhibitory cholinergic tone (Axelsson *et al.*, 1992).

A curious and unexplained observation is that maximum f_H rarely exceeds values around 120 bpm (Farrell, 1991a). This is the case even in small, developing fish larvae and embryos (Burggren and Pinder, 1991). An exception to this generalization are the tunas, in which f_H can be 180 to 240 bpm.

Cholinergic Control of f_H

Cholinergic fibers carried in the vagus nerve are responsible for bradycardia, probably through stimulation of muscarinic receptors in pacemaker cells (Saito, 1973). The resting cholinergic tone which exists in most fish increases (1) during exposure to environmental hypoxia, (2) at the initiation of burst swimming, and (3) with visual and olfactory stimuli. The level of cholinergic inhibition of resting f_H is greater in cold-acclimated compared with warm-acclimated *Scyliorhinus canicula* and rainbow trout (Taylor et al., 1977; Wood et al., 1979), but is lower in cold-acclimated vs. warm-acclimated *Anguilla anguilla* (Siebert, 1979). Release of this cholinergic inhibition (vagal release) results in tachycardia, as observed in exercise. Obviously, the greater the resting cholinergic tone, the greater the potential for tachycardia.

Adrenergic Control of f_H

Evidence is accumulating that the adrenoceptor pool of various tissues, including cardiac tissue, varies between fish species and with temperature and possibly other environmental factors (Peyraud Waitzenegger et al., 1980; Keen, 1992). Most fish hearts have the β-adrenoceptors which, when stimulated, produce positive chronotropy (see Laurent et al., 1983; Nilsson, 1983; Farrell, 1984; Butler and Metcalfe, 1988). Chronotropic actions are likely at the level of the pacemaker and A-V conduction. Some species have additional α-adrenoceptors. Stimulation of α-adrenoceptors produces negative chronotropy. In those fish species possessing both α- and β-adrenoceptors, the overall effect of adrenergic stimulation reflects the relative membrane populations of the two receptor types and the relative ratios of adrenaline and noradrenaline. In fish, unlike mammals, adrenaline is more potent at stimulating β-adrenoceptors and noradrenaline is more potent at stimulating α-adrenoceptors.

Both adrenaline and noradrenaline can stimulate cardiac adrenoceptors following their release in varying ratios from (1) chromaffin tissue outside the heart and into the blood (possibly in all fish), (2) chromaffin tissue within the heart (in cyclostomes and dipnoans), and (3) adrenergic nerves (in teleosts possessing adrenergic cardiac innervation). The blood levels of catecholamines found in resting fishes and/or endogenous cardiac catecholamines may provide the resting excitatory adrenergic tone. The substantial rise in circulating levels of adrenaline and noradrenaline during and following stressful events (e.g., hypoxia, burst exercise, physical disturbance, and near the maximum prolonged swimming speed) probably has chronotropic as well as inotropic effects.

Control of f_H by Mechanic Factors

Mechanical modulation of heart rate by stretching pacemaker cells with increased cardiac filling (venous blood pressure) may be important for increasing f_H by 10 to 20% in hagfishes. In other, more phylogenetically advanced fishes (elasmobranchs and teleosts) this mechanical modulation has been superseded by neurohumoral modulation of heart rate (Farrell, 1991a; Farrell and Jones, 1992).

Barostatic Reflex Control of f_H

Even though baroreceptors have not been localized precisely in fish, evidence for a putative barostatic reflex exists (Jones and Milsom, 1982). Increases in arterial blood pressure typically produce bradycardia coincident with rising or peak blood pressures (see Wood and Shelton, 1980b; Farrell, 1986). This reflex bradycardia is abolished by vagotomy and atropine, indicating the efferent output is via cholinergic fibers in the cardiac vagus nerve. Tachycardia associated with hypotension does not exceed the f_H achieved with zero vagal tone, suggesting a sole involvement of cholinoceptors in the efferent output. The control of f_H in fishes is more fully discussed by Nilsson (1983) and Farrell and Jones (1992).

Control of SV_H

As noted above, an important difference between the control of $\dot{Q}$ in fishes and in mammals is the much larger increase in SV_H associated with exercise in fishes. Based on *in vivo* and *in vitro* studies, two- to threefold increases in SV_H are possible in fishes (Table 3). Clearly, then, the control of SV_H is very important in the regulation of $\dot{Q}$ in fishes.

SV_H is set by ventricular end-diastolic volume (EDV) minus ventricular end-systolic volume (ESV). Ventricular EDV is determined by the volume of blood delivered from the atrium, and ventricular volume and distensibility. Ventricular ESV is determined primarily by the force of contraction and the diastolic blood pressure in the ventral aorta. In fish, unlike in mammals, ventricular ESV normally approaches zero and so atrial stroke volume is approximately the same as ventricular EDV.

Mechanical Properties of Cardiac Muscle

Force of contraction is altered either through the Frank-Starling mechanism, or by factors that modulate cardiac contractility. The Frank-Starling mechanism of the heart refers to an intrinsic property of cardiac muscle and, as such, is applicable to both the atrium and the ventricle. In essence, increased EDV results in a greater force of contraction and a greater SV_H. Since an increase in cardiac filling pressure increases EDV, a relationship exists between cardiac filling pressure and SV_H, which is referred to as a Starling curve. Starling curves are available for perfused hearts from a variety of teleosts, the elasmobranch, *Squalus acanthias*, and the cyclostome, *Eptatretus cirrhatus* (Figure 1A). Fish hearts are much more sensitive to filling pressure than mammalian hearts in that an increase of only 1 to 2 mmHg in cardiac filling pressure is needed to move along the Starling curve from resting SV_H to near maximum SV_H (Farrell, 1991a).

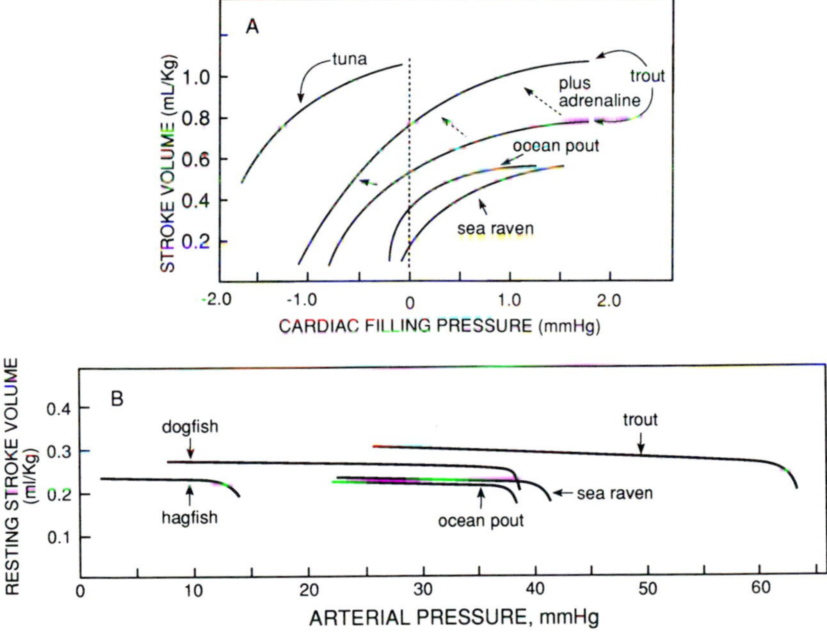

FIGURE 1. (A) A comparison of Starling curves obtained for perfused hearts from several teleost species. Note that, through *vis-a-fronte* filling, a large portion of the curve for rainbow trout is at subambient pressure. The positive inotropic effect of adrenaline is to shift the Starling curve for rainbow trout upwards and to the left (adapted from Farrell and Jones, 1992). (B) The effect of increased output pressure on the ability of perfused fish hearts to maintain constant SV_H (homeometric regulation). At resting $\dot{Q}$ levels, increases in output pressure have little effect on resting SV_H over a physiological pressure range. The maximum pressure is species specific, being higher in active fish (adapted from Farrell and Jones, 1992).

Neurohumoral factors that alter cardiac contractility make the heart more or less sensitive to filling pressure and create a family of Starling curves. The positive inotropic effect of adrenaline, for example, shifts the Starling curve to the left and upwards (Figure 1A). Thus, increased cardiac performance due to the intrinsic Frank-Starling mechanism causes a shift along a given curve, whereas altered contractility causes a shift to another curve. What is not clear is whether the regulation of SV_H *in vivo* is through changes in venous filling pressure per se, changes in cardiac contractility at a constant filling pressure, or some combination of both.

Another intrinsic property of cardiac muscle is the ability to maintain $\dot{Q}$ independent of diastolic output pressure (homeometric regulation) (Farrell, 1984). Thus, fish hearts independently maintain SV_H over a broad range for P_{va}. (Most man-made mechanical pumps are pressure dependent; the higher one raises the end of the hose of a water pump, the lower the volume output.) This range is species specific, being greater in more active fishes that normally have higher P_{va} (Figure 1B). Positive inotropic agents such as adrenaline can extend the homeometric range.

Cardiac Filling

Because atrial contraction is the primary determinant of ventricular filling, an appreciation of the two types of atrial filling is central to an understanding of the control of SV_H in fishes. *Vis-a-tergo* (force from behind) atrial filling utilizes potential and kinetic energy that either remains in the venous circulation or that is generated by sinus contraction. In this case, central venous blood pressure is the most important determinant of atrial filling. *Vis-a-fronte* (force from in front) atrial filling uses some of the energy of ventricular contraction directly to distend the atrium and so central venous blood pressures can be subambient. Of the two mechanisms for atrial filling, *vis-a-tergo* filling is probably used by cyclostomes and most teleosts. Elasmobranchs and active teleosts apparently use both *vis-a-fronte* and *vis-a-tergo* filling. Even though a low SV_H is generated with subambient filling pressures (*vis-a-fronte* filling), maximum SV_H requires a positive filling pressure (*vis-a-tergo* filling). The potential advantages and disadvantages of these two mechanisms of atrial filling, and the requirement for, and limitations of, a relatively rigid pericardial cavity are discussed in detail elsewhere (Shabetai et al., 1985; Farrell et al., 1988b; Farrell, 1991a).

It has long been held that ventricular filling occurs solely as a result of atrial contraction based on blood pressure measurements and angiographs (see Randall, 1970; Satchell, 1971; Johansen and Burrgren, 1980). However, Lai et al. (1990b), working with the leopard shark, provided evidence for biphasic ventricular filling. Further studies are needed, therefore, to provide a clear picture of the relative importance of atrial contraction and direct ventricular filling from the venous system.

Cardiac Contractility

The force with which the atrium and ventricle contract is a major determinant of ESV and hence cardiac SV_H. Force of contraction is augmented by an increase in muscle fiber length (see Frank-Starling mechanism). A change in cardiac contractility, an expression of the vigor of contraction, is another mechanism by which the force of contraction is modulated. Changes in contractility, as noted above, have important consequences to the Frank-Starling mechanism and homeometric regulation. Contractility, defined as the change in developed force at a given resting fiber length, is usually measured directly *in vitro* by peak isometric tension developed at a fixed initial fiber length and indirectly *in vivo* by the ventricular dP/dt during the isovolumic phase of the cardiac cycle (isometric contraction). There are clear phylogenetic differences in the *in vivo* values for ventricular dP/dt for fishes. Values for tuna are similar to those in mammalian heart, whereas other teleost values are five times lower, and those for elasmobranchs and hagfish are 50 times lower (Farrell and Jones, 1992). The basis for these differences is not clear at this time.

Cardiac contractility is modulated neurally (through vagal and adrenergic nerve fibers), humorally, and locally. Farrell and Jones (1992) list the following factors known to alter contractility (inotropy) of the atrium and/or ventricle in fishes. Positive inotropic effects are produced by the following: increased temperature, β-adrenergic stimulation, extracellular calcium, the peptides arginine, vasotocin, and oxytocin, adenosine, prostacyclin, and histamine. Negative inotropic effects are known for the following: hypoxia and acidosis, acetylcholine, α-adrenergic stimulation in some species, purinergic agents, and adrenaline in combination with adenosine. It is important to note that atrial and ventricular sensitivity to inotropic agents may be quite different. Also, these inotropic actions can be interactive. For example, adrenergic stimulation can offset the effect of extracellular acidosis (see Farrell, 1984)

Contractility *in vitro* is also dependent on the duration of the active state (the period of contraction and relaxation) and its intensity (rate of contraction). In hagfish and various teleosts, an inverse relationship exists between maximum isometric tension and contraction frequency which is referred to as a negative staircase. However, cardiac tissues from elasmobranchs (Driedzic and Gesser, 1988), skipjack tuna (Keen et al., 1992), and mammals have a positive force-frequency relationship at low frequencies and a negative one at higher frequencies. Interestingly the apices of the force-frequency relationships in elasmobranchs and skipjack tuna occur at contraction frequencies (0.3–0.4 Hz and 1.4–1.6 Hz, respectively), which correspond to f_H *in vivo*.

Cardiac Power Output

Farrell and Jones (1992) suggested that myocardial power output, the product of Q and pressure development by the heart and expressed per gram ventricular mass, is a useful measure of the integrated performance of the heart for comparative purposes. Even though most species have resting power output values of 0.9 to 1.8 mW/g ventricular mass, there is a sixfold range for the highest and lowest resting values (Table 4). Power output increases two- to fourfold with exercise or adrenergic stimulation such that maximum power output values for elasmobranchs and sluggish species of teleosts are about half those values for the more active salmonids (3.2 mW/g *vs.* 6 to 7 mW/g).

A larger ventricular mass means a greater potential for power development. By expressing myocardial power output "per kilogram body mass" rather than per gram ventricular mass, one can assess the total power output of the heart relative to the whole animal. In this way, it is possible to appreciate how the relatively large ventricles of the very active and tropical tunas and much less

TABLE 4
Myocardial Power Output in Selected Fishes at Rest and While Swimming

	Power output (mW/g)				
	Rest	Exercise	Temperature (°C)	Body Mass (kg)	Source
Temperate-water fishes					
Myxine glutinosa+	0.08	0.27	11	0.08	a
Triakis semifasciata	1.71	3.30	14–24	1.93	b
Scyliorhinus stellaris	1.43	2.46	19	2.6	c
Gadus morhua	1.77	3.29	10.5	0.4–0.8	d
Ophiodon elongatus	1.18	3.21	10	4.2	e
Hemitripterus americanus	1.16	3.13	10	1.2	f
Anguilla australis	1.08	2.19	16–20	0.9–1.1	g
Oncorhynchus mykiss	1.53	7.03	11	1.0	h
Oncorhynchus kisutch	1.22	5.97	5	1.4	i
Tropical fishes					
Katsuwanus pelamis	4.70	—	26	1–2	j
Thunnus albacares	5.60	—	26	1–2	j
Antarctic fishes					
Chaenocephalus aceratus	0.98	—	0.5–2	1.0	k
*Chionodraco hamatus**	—	1.6–3.4	0–2	0.29–0.47	l
Pagothenia borkgrevinki	1.05	2.00	0	0.06	m

Note: Where ventricle mass is not known values of 0.2 g/kg body and 0.08 g/kg body were assumed for elasmobranch and teleost species. Plus (+) denotes maximum value for postadrenaline infusion. Asterisk (*) denotes maximum value in a perfused heart preparation.

References: a, Axelsson et al. (1990) and Forster et al. (1991); b, Lai et al. (1989a, 1989b, 1990a); c, Piiper et al. (1977); d, Axelsson and Nilsson (1986); e, Farrell (1982); f, Axelsson et al. (1989) and Farrell et al. (1985); g, Hipkins (1985); h, Kiceniuk and Jones (1977); i, Davis (1966); j, Bushnell (1988); k, Hemmingsen and Douglas (1977); l, Tota et al. (1991); m, Axelsson et al. (1992).

active Antarctic fishes result in an increase in *total* myocardial power output. For example, the total resting myocardial power output for skipjack tuna (26 mW/kg body mass) is more than 10 times that for rainbow trout (1.5 mW/kg body mass) and even the total myocardial power output of the Antarctic fish *Chionodraco hamatus* is impressive (2.8 mW/kg body mass).

Myocardial power output has also proven useful in estimating the O_2 cost of cardiac pumping since myocardial $\dot{V}_{O_2}$ is known to be around 0.3 ml/s per mW of power output in various species (Davie and Farrell, 1991a). In general, the O_2 cost of cardiac pumping at rest is relatively small in terms of the overall O_2 requirements of the fish, being between 0.6 and 4.6% of resting $\dot{V}_{O_2}$ (Farrell and Jones, 1992). With swimming, even though the total myocardial $\dot{O}_2$ consumption increases, the O_2 cost of cardiac pumping relative to total $\dot{V}_{O_2}$ may decrease slightly. However, the hemoglobin-free Antarctic fish *Chaenocephalus aceratus* may be unique in having an extremely high O_2 cost of cardiac pumping (23% of resting $\dot{V}_{O_2}$; Farrell and Jones, 1992). Hagfishes are also unique in that they can generate a substantial portion, if not all, of their ATP requirement for resting power output via anaerobic metabolism (see Forster et al., 1991; Farrell, 1991a). Thus, while anoxia generally kills fish hearts, the normally very low myocardial power output of the hagfish is maintained during severe hypoxia.

VASCULAR PATTERNS AND BLOOD FLOW DISTRIBUTION

VASCULAR PATTERNS

Gill Circulation

The general gill anatomy (Hughes, 1984) and its internal vascular pathways (Laurent, 1984) have been comprehensively described. In view of this, only the general pattern of the gill circulation is presented here. The entire $\dot{Q}$ enters the afferent branchial arteries (*af.BA*) in all water breathing fish. From the *af.*BA, blood is distributed along the length of each gill filament by an afferent filamental artery (*af.FA*). After passing through the respiratory exchange site, the secondary lamellae, blood passes on to efferent filamental arteries (*ef.FA*) and efferent branchial arteries (*ef.BA*). Blood is then distributed either anteriorly by the carotid arteries or posteriorly by the dorsal aorta. In some fishes, branches from the *ef.BA* create the mandibular artery, which supplies the pseudobranch and choroid gland, and the hypobranchial artery, which supplies some of the pectoral muscles and the coronary circulations (Figure 2).

Lamellae are arranged like rungs of a ladder along the length of each gill filament, connecting the *af.FA* and *ef.FA* (Figure 3). Each secondary lamella typically consists of a thin vascular sheet of lamellar capillaries connected by a long afferent lamellar arteriole (*af.LA*) and a short efferent lamellar arteriole (*ef.LA*) to the filamental arteries. Lamellae have a variety of shapes, with the

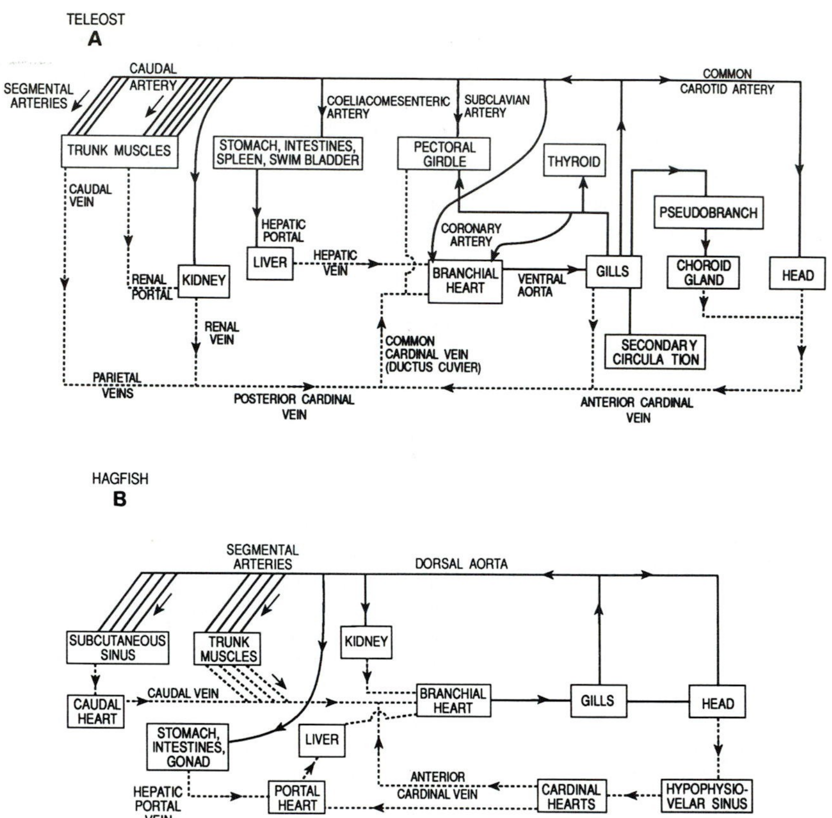

FIGURE 2. Schematic representations of the primary arterial (solid lines) and venous (broken lines) circulations in (A) a salmonid, as a representative of a teleost fish (adapted from Smith and Bell, 1976), and (B) a hagfish.

blood vessels occupying more than 80% of the lamellar surface area (Farrell *et al.*, 1980). Pillar cells account for the remaining lamellar surface area and span the width of the vascular sheet, presumably preventing it from ballooning under the stress of internal blood pressure. A somewhat larger-diameter marginal vessel is found around the periphery of the lamellae. The marginal vessel may be involved in blood distribution to, collection from, and shunting within the main lamellar vascular sheet. A variable portion of the lamellar vascular sheet lies below the epithelial surface of the gill filament. The additional distance between blood and water that results represents a significant diffusion barrier compared with the remainder of the lamellar vascular sheet, which lies above the filament surface. In elasmobranchs and lungfish, a sinus-like corpus cavernosum lies between and connected to the *af.*FA and *af.*La. For most of the lamellae, blood must first pass through the corpus cavernosum before reaching the *af.*La. In hagfish, the gill has a basket rather

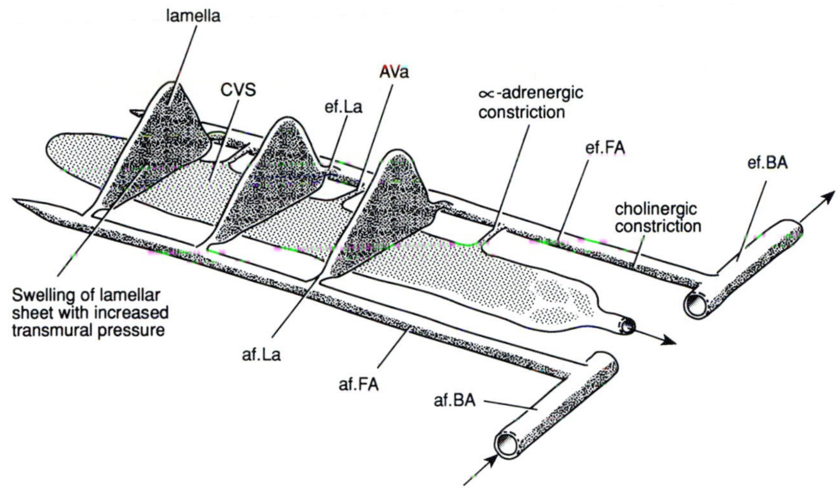

FIGURE 3. A schematic representation of the major vascular pathways in the gill filament of a teleost fish. Known sites for changes in vascular resistance or dimensions are indicated. The marked decrease in gill vascular resistance associated with β-adrenergic vasodilation may result from relaxation of *af.La* and (or) *ef.La* (af, afferent; ef, efferent; BA, branchial artery; FA, filament artery; La, lamellar arteriole; AVa, arteriovenous anastomoses; CVS, central venous sinus; lamella, secondary lamella).

than ladder-like arrangement with radial arteries feeding a discrete array of arching vessels within the lamellae.

An additional, highly variable network of vessels exist in the gills (see Laurent, 1984). This complex network, or parts of it, has been incorrectly referred to as lymphatics and venolymphatics in the past. Instead, this network is part of the secondary circulation and probably serves a nutritive function. A feature common in most fish gills is a central venous sinus (*CVS*) which lies underneath the lamellae and extends along the filament length (Figure 3). The *CVS* has narrow arteriolar connections, efferent arteriovenous anastomoses (ef.AVas), with the *ef.FA*. The *ef.AVa*s can allow a significant and variable diversion of blood from the *ef.FA* into the *CVS*. Sometimes nutritive arteries (*NA*) which lie parallel to the *ef.FA* are connected to the main secondary circulation (e.g., rainbow trout). The *CVS* also connects to the branchial vein via numerous, meandering vessels which run across and along the afferent side of the filament towards the gill arch. Afferent arteriovenous anastomoses (af.AVas) are found in species such as the eel and these connect the af.FA with the CVS providing for a potential arteriovenous shunt.

Systemic Circulation

The dorsal aorta is the main distribution artery for the primary circulation. The general distribution pattern for most fish is similar to that presented in Figure 2A. It is important to note that the coeliacomesenteric artery (*CMA*) is a major branch and in some fishes the arrangement of the dorsal aorta and the CMA is such that blood from the posterior *ef.BA*s may be preferentially directed to the *CMA* (Farrell, 1980).

Veins of the primary circulation return blood from the head and gills directly to the heart. However, venous return from the trunk muscles and gastrointestinal tract passes first through the kidney and liver, respectively (Figure 2). In certain teleosts and elasmobranchs, caudal hearts that are located near the tail and powered by skeletal muscle assist venous return toward the branchial heart (See Satchell, 1991, 1992). The vascular pattern of hagfishes (Figure 2B) has a number of different features. There are more accessory hearts, no renal portal vein, and a large subcutaneous sinus. While the cardinal and caudal hearts are powered by skeletal muscle, the portal heart is composed of true cardiac muscle. The blood volume of hagfishes is approximately twice that of teleosts and elasmobranchs, but the majority of this difference can be accounted for by the subcutaneous sinus which contains low hematocrit blood (Hct = 4 to 5%) and has a slow turnover rate. Even though hagfishes and lampreys are both classified as cyclostomes, lampreys have no caudal heart, portal heart, or subcutaneous sinus.

Distinctions between the primary and secondary circulations and the misconceptions regarding lymphatics in fishes are well described by Vogel (1985), Satchell (1991), and Steffensen and Lomholt (1992). The relationship of the primary and secondary circulations is illustrated in Figure 4. It is important to note that the connections between the primary and secondary circulations are

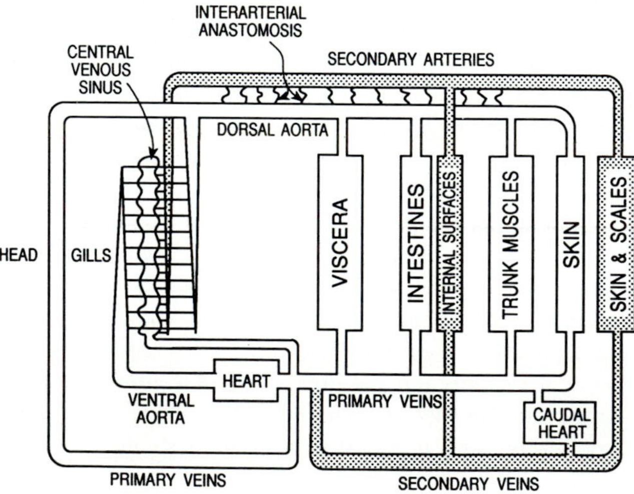

FIGURE 4. The general distribution pattern of the secondary circulation in teleost fish and its relationship to the primary circulation (adapted from Vogel, 1985).

such that they reduce blood pressure considerably and exclude the majority of the red blood cells. Consequently, because the volume of the secondary circulation is at least as large as, if not larger than, the primary circulation but the blood pressure is low, the circulation times of the secondary circulation is probably of the order of hours rather than minutes. Whereas the role of the secondary circulation to the gills, skin, and scales is likely to be largely nutritive (these organs can exchange gases directly with the water), the role of the secondary circulation to the intestines is so far unresolved.

Blood Flow Distribution

Vascular Resistance.

An increase in $\dot{Q}$ obviously increases blood flow to all vascular beds if vascular resistance does not change. Since most vascular beds are arranged in parallel, the relative vascular resistance of a vascular bed determines its relative blood flow. Thus, redistribution of blood flow requires a change in the vascular resistance of one or more of the vascular beds. Changes in vascular resistance, either accompanying or independent of changes in $\dot{Q}$, can be brought about by neural, humoral, or local regulation of smooth muscle activity in resistance vessels. Because the gill and systemic circuits are arranged in series, a change in systemic vascular resistance, in addition to effecting a redistribution of blood flow within the systemic circulation, is likely to affect the pattern of gill blood flow pattern (see below).

Bushnell et al. (1992) tabulated data on vascular resistance in fishes and from their table a number of points are apparent. With the exception of Antarctic icefishes where vascular resistance is unusually low, total peripheral vascular resistance (*Rp*), the arithmetic sum of the gill and systemic resistances, ranges between 0.6 to 3.4 mmHg/ml·min·kg (Table 3). With the exception of tunas, systemic vascular resistance (*Rs*) accounts for the majority (61 to 82%) of *Rp* in cyclostomes, elasmobranchs, and teleosts. Neither hypoxia nor swimming are reportedly associated with more than a twofold change in either gill vascular resistance (*Rg*) or *Rs*.

Information on the extent of and mechanisms for blood flow redistribution in fish is very limited. Thus, the following descriptions of gill, coronary, skeletal muscle, and gut blood flow are incomplete and should be considered tentative. Of these circulations, gill blood flow patterns have been studied far more extensively than the others combined.

Gill Blood Flow

There are many studies describing the pharmacology and innervation of fish gills (see Nilsson, 1983), but few of these provide conclusive information on the exact pattern of blood flow through the gills and how it is controlled (see Randall and Daxboeck, 1982; Randall, 1985). Gill blood flow can change in three major ways: (1) in the number of lamellae perfused (this affects respiratory surface area), (2) in the evenness with which blood perfuses individual lamellae, and (3) in the amount of blood diverted from the primary (arterio-arterial) circulation via the *AVa*s to the secondary circulation. Several sites of vasoactivity have been established, so an integrated and complex picture of the involvement of vasoactivity in controlling gill blood flow is beginning to emerge. The mechanisms involved in controlling gill blood flow are humoral, neural, and physical (changes in blood pressure). In cyclostomes, elasmobranchs, and teleosts, excitatory cholinergic, excitatory adrenergic, and inhibitory adrenergic responses are all implicated in site-specific gill vasoactivity to alter the pressure profile within the gill. In teleosts and cyclostomes, additional excitatory responses to 5-hydroxytryptamine and purinergic agents are possible. In terms of blood pressure affecting gill blood flow, an increase in $\dot{Q}$ usually increases input pressure, and an increase in *Rs* increase P_{da} and all blood pressures within the gills.

Lamellar Recruitment

In rainbow trout and lingcod, it appears that only approximately two thirds of the lamellae are normally perfused at one time (Booth, 1978; Farrell et al., 1979). More lamellae are perfused after adrenaline injections *in vivo* (Booth, 1978), perhaps as a result of β-adrenergic mediated dilation of *af.Las* (the major site of gill resistance; Farrell, 1980) and *ef.La*s (where adrenergic nerve fibers are located). In addition, lamellar recruitment occurs when mean and pulse perfusion pressures are increased *in vitro,* possibly because increases in input pressure “pop” open unperfused lamellae once the critical opening pressure of the *af.LA* is overcome (Farrell *et al.*, 1979). Similarly, α-adrenergic constriction of the efferent gill vessels (and perhaps systemic vessels) may recruit lamellae by raising input pressure.

Intralamellar Flow Pattern

Because the lamellar vascular sheet is highly compliant (Farrell *et al.*, 1980), an increase in lamellar transmural pressure alters the perfusion pattern within an individual lamellae. With increased blood pressure, blood flow shifts from the larger marginal vessels and creates a more even flow across the main vascular sheet. An increase in $\dot{Q}$ (with no change in vascular resistance), branchial vasoconstriction distal to the secondary lamellae, and systemic vasoconstriction would all increase lamellar transmural pressure. Acetylcholine injections, which probably constrict of the muscular sphincter at the base of the *ef.FA* (Figure 3) that is innervated by cholinergic fibers (Nilsson, 1983), increase the lamellar vascular space but do not recruit lamellae (Booth, 1979; Holbert et al., 1979).

Pillar cells contain myosin. However, whether pillar cells are contractile and actively modify the lamellar sheet thickness, perhaps in response to either hormones or autocoids, or by being myogenic, is not known.

Arterio Venous Flow

Clear evidence exists for the control of blood flow through the *ef.AVa*s. The *ef.AVa*s are innervated with adrenergic fibers (Nilsson, 1983) and are constricted by α-adrenergic agonists (Payan and Girard, 1977). Thus, the actions of circulating catecholamines and stimulation of spinal sympathetic nerves to the gills reduce arteriovenous flow while increasing arterio-arterial outflow from the gills (Nilsson, 1983). Similar changes in flow would be produced when vessels downstream of the *ef.AVa* sites (e.g., the *ef.FA* sphincter and systemic vessels) were constricted (Figure 3).

Systemic Blood Flow

The resistance to systemic blood flow (*Rs*) is approximately three to four times greater than *Rg* (Table 3). Most studies have considered *Rs* as a lumped parameter. Thus, observed changes in *Rs* induced by drug injections provide only a conceptual framework on which to examine the specific, and presumably independent, mechanisms that control gut, skeletal muscle, and other vascular resistances. In hagfish, *Myxine glutinosa*, *Rs* is reduced by 50% with adenosine and by 10 to 30% with adrenaline and acetylcholine (Axelsson *et al.*, 1990). In elasmobranchs, *Rs* is controlled at rest mainly via circulating catecholamines; injection of catecholamines produces a large α-adrenoceptor-mediated vasoconstriction as well as a smaller β-adrenoceptor-mediated vasodilation (see Butler and Metcalfe, 1988). Injections of catecholamines have similar effects in teleosts, but a systemic adrenergic tone is apparently either absent (e.g., Atlantic cod, Fritsche and Nilsson, 1990; *Anguilla australis*, Hipkins et al., 1986) or maintained neurally (e.g., rainbow trout, see Bushnell et al., 1992).

The distribution pattern of blood in the visceral and somatic vascular beds has been studied in a few species. Studies using microspheres (Figure 5A) indicate that the greater proportion of $\dot{Q}$ goes to the skeletal muscle (red = 10 to 36%; white = 28 to 49%), but because of the large mass of white muscle, the actual perfusion rate of white muscle is quite low. Blood flow to the gut (stomach plus intestine) is 3 to 9% of $\dot{Q}$ measured with the microsphere technique (Figure 5A). However, direct measurements of visceral blood flow in sea raven and Atlantic cod reveal a somewhat higher proportion of $\dot{Q}$ is directed to the gut. Flow in the coeliac artery (approximately half of the gut blood flow) is normally 2.9 ml/min/kg (15% of $\dot{Q}$) in unfed sea raven, but this flow rate doubles after feeding, probably through a release of tonically active α-adrenergic vasoconstriction of the visceral vasculature (Axelsson et al., 1989). In Atlantic cod, gastrointestinal (stomach, pyloric ceca, liver, and intestine) blood flow is 7.5 ml/min/kg (40% of $\dot{Q}$); it also increases with feeding (52% of $\dot{Q}$), but is unchanged after α-adrenoceptor blockade, unlike the sea raven (Axelsson and Fritsche, 1991).

Generally, swimming causes an increase in $\dot{Q}$, a decrease in *Rs*, and increases in P_{va} and P_{da} (Table 3). Blood flow is redistributed to the swimming

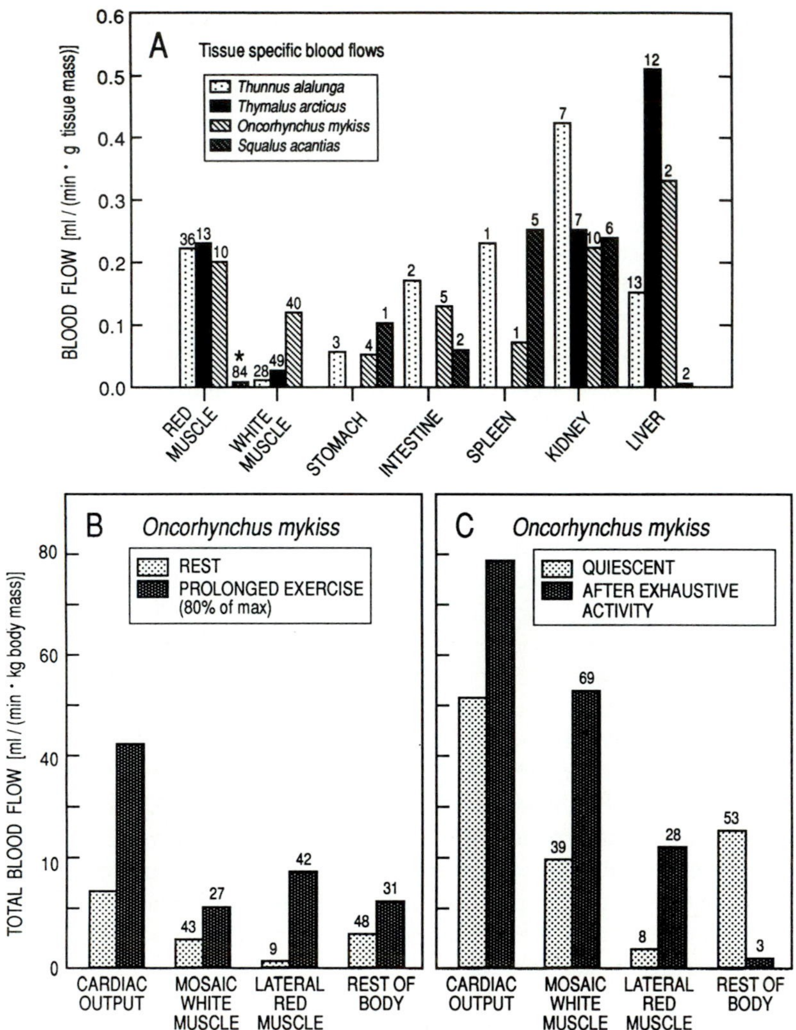

FIGURE 5. Distribution of blood flow and cardiac output as determined by studies with microspheres. (A) Tissue-specific blood flows and the percentage of cardiac output (numbers on top of each bar) in four species of fish at rest (data obtained for *Thunnus alalunga* from White et al., 1988; for *Thymalus articus* from Cameron, 1975; for *Oncorhynchus mykiss* from Barron et al., 1987; and for *Squalus acanthias* from Kent et al., 1971). The red and white muscle was not separated for *Squalus acanthias* and so the value presented is for the combined red and white muscle mass which accounted for 84% of body mass. Panels B and C show the effect of prolonged (B) and exhaustive (C) exercise on cardiac output and its distribution in *Oncorhynchus mykiss* (bars = ml/min·kg body mass; numbers = percentage of cardiac output). (Prolonged exercise data from Randall and Daxboeck, 1982; exhaustive activity data from Neumann et al., 1983). The mosaic white muscle of rainbow trout, which accounts for approximately 66% of body mass, is comprised of approximately 2% red fibers and 98% white fibers.

muscles at the expense of the viscera (Figures 5B and 5C) with lateral red muscle experiencing the greatest increase in blood flow, more than quadrupling in rainbow trout. This redistribution of blood flow most likely occurs as a result of vasoconstriction in the visceral circulation. For example, in exercising Atlantic cod, coeliaco-mesenteric vascular resistance almost doubles, reducing blood flow in the CMA by about one third (Axelsson et al., 1989). However, the extent of the involvement of vasoactive mechanisms within the skeletal muscle circulation is unknown.

The effects of environmental hypoxia on $\dot{Q}$, P_{va}, P_{da} and Rs are quite variable between species (Table 3; see also Farrell and Jones, 1992, and Bushnell et al., 1992). For example, in Arctic grayling, blood flow distribution is unchanged at a water PO_2 of 55 mmHg (Cameron, 1975). In contrast, blood flow in the CMA of Atlantic cod is halved at a water PO_2 of 30 to 40 mmHg as a result of an α-adrenergic vasoconstriction (Axelsson and Fritsche, 1991).

Coronary Blood Flow

A coronary circulation, when present, supplements the venous O_2 supply to the myocardium and as such appears to be central to proper cardiac function. Nonetheless, the extent to which cardiac function depends on coronary circulation varies between species. Rainbow trout survive even though the coronary artery has been surgically tied off and can even swim up to 70% of their maximum prolonged swimming speed. In contrast, the suggestion has been made that the very large ventricle of skipjack tuna may be obligately dependent on a continuous coronary blood flow (Farrell et al., 1992).

Most fish probably have a maintenance level of coronary blood flow. Various measurements and estimates of coronary blood flow in several fish species indicate that this is only a small percentage (1 to 2%) of $\dot{Q}$ (see Davie and Farrell, 1991a; Farrell and Jones, 1992). Coronary blood flow more than doubles with hypoxia in coho salmon (Axelsson and Farrell, in press), probably as a result of a large vasodilatory reserve. With exercise, coronary blood flow in rainbow trout may have to increase fourfold to match the fourfold increase in myocardial O_2 demand if O_2 extraction from the coronary circulation does not change significantly (Farrell, 1987).

Coronary vascular resistance is altered through α-adrenergic constriction, β-adrenergic relaxation, cholinergic constriction, and physical factors related to the compression of vessels by the contraction of cardiac muscle (see Davie and Farrell, 1991a). Coronary vasoactivity is also temperature dependent (Farrell, 1987). Even though most circulatory control is normally exerted at the arteriolar level, the main coronary artery is also extremely vasoactive (Davie and Farrell, 1991a).

ACKNOWLEDGMENT

The author acknowledges the financial support provided by the Natural Sciences and Engineering Research Council Canada and the British Columbia and Yukon Heart Foundation.

REFERENCES

Axelsson, M. and Farrell, A. P., Coronary blood flow *in vivo* in the coho salmon (*Oncorhynchus kisutch*), *Am. J. Physiol.,* in press.

Axelsson, M. and Fritsche, R., Effects of exercise, hypoxia and feeding on the gastrointestinal blood flow in the Atlantic cod (*Gadus morhua), J. Exp. Biol.*, 158, 181, 1991.

Axelsson, M. and Nilsson, S., Blood pressure control during exercise in the Atlantic cod, *Gadus morhua, J. Exp. Biol.*, 126, 225, 1986.

Axelsson, M., Driedzic, W. R., Farrell, A. P., and Nilsson, S., Regulation of cardiac output and blood flow in the sea raven, *Hemitripterus americanus, Fish Physiol. Biochem.*, 196, 2, 1989.

Axelsson, M., Farrell, A. P., and Nilsson, S., Effects of hypoxia and drugs on the cardiovascular dynamics of the Atlantic hagfish *Myxine glutinosa, J. Exp. Biol.*, 151, 297, 1990.

Axelsson, M., Davison, W., Forster, M. E., and Farrell, A. P., Cardiovascular responses of the red-blooded Antarctic fishes, *Pagothenia bernacchii* and *P. borchgrevinki, J. Exp. Biol.*, 167, 179, 1992.

Barron, M. G., Tarr, B. D., and Hayton, W. L., Temperature-dependence of cardiac output and regional blood flow in rainbow trout, *Salmo gairdneri* Richardson, *J. Fish. Biol.*, 31, 735, 1987.

Booth, J. H., The distribution of blood flow in the gills of fish: application of a new techniques to rainbow trout (*Salmo gairdneri), J. Exp. Biol.*, 73, 119, 1978.

Booth, J. H., Circulation in trout gills: the relationship between branchial perfusion and the width of the lamellar blood space, *Can. J. Zool.*, 57, 2183, 1979.

Brill, R. W. and Bushnell, P. G., Metabolic and cardiac scope of high energy demand teleosts — the tunas, *Can. J. Zool.*, 69, 2002, 1991.

Burggren, W. W. and Pinder, A. W., Ontogeny of cardiovascular and respiratory physiology of lower vertebrates, *Annu. Rev. Physiol.*, 53, 107, 1991.

Bushnell, P. G., Cardiovascular and respiratory responses to hypoxia in three species of Obligate Ram Ventilating Fishes, Skipjack Tuna, *Katsuwonus pelamis*, Yellowfin Tuna, *Thunnus albacares*, and Bigeye Tuna, *Thunnus obesus*, Ph.D. thesis, University of Hawaii, Honolulu, 1988.

Bushnell, P. G., Jones, D. R., and Farrell, A. P., The Arterial System, in *Fish Physiology: Cardiovascular Systems*, Vol. 12, Hoar, W. S., Randall, D. J., and Farrell, A. P., Eds., Academic Press, New York, 1992, 89.

Butler, P. J., Exercise, in *Fish Physiology: Recent Advances*, Nilsson, S. and Holmgren, S., Eds., Croom Helm, London, 1986, 102.

Butler, P. J. and Metcalfe, J. D., Cardiovascular and respiratory systems, in *Physiology of Elasmobranch Fishes*, Shuttleworth, T. V., Ed., Springer-Verlag, New York, 1988, 1.

Butler, P. J. and Taylor, E. W., The effect of progressive hypoxia on respiration in the dogfish (*Scyliorhinus canicula), J. Exp.Biol.*, 63, 117, 1975.

Cameron, J. N., Morphometric and flow indicator studies of the teleost heart, *Can. J. Zool.*, 53, 691, 1975.

Chan, D. K. O., Cardiovascular respiratory and blood adjustments to hypoxia in the Japanese eel, *Anguilla japonica, Fish Physiol. Biochem.*, 2, 179, 1986.

Davie, P. S. and Farrell, A. P., The coronary and luminal circulations of the myocardium of fishes, *Can. J. Zool.*, 69, 1993, 1991a.

Davie, P. S. and Farrell, A. P., Cardiac performance of an isolated heart preparation from the dogfish (*Squalus acanthias*): the effects of hypoxia and coronary artery perfusion, *Can. J. Zool.*, 69, 1822, 1991b.

Davie, P. S. and Forster, M. E., Cardiovascular responses to swimming in eels, *Comp. Biochem. Physiol. A*, 67, 367, 1980.

Davis, J. C., The Influence of Water and Blood Flow on Gas Exchange at the Gills of Rainbow Trout, *Salmo gairdneri*, M.Sc. Thesis, University of British Columbia, Vancouver, Canada, 1966.

Driedzic, W. R. and Gesser, H., Differences in force-frequency relationships and calcium dependency between elasmobranch and teleost hearts, *J. Exp. Biol.*, 140, 227, 1988.

Farrell, A. P., Gill morphometrics, vessel dimensions and vascular resistance in lingcod, *Ophiodon elongatus, Can. J. Zool.*, 58, 807, 1980.

Farrell, A. P., Cardiovascular changes in the unanaesthetized lingcod (*Ophiodon elongatus*) during short-term progressive hypoxia and spontaneous activity, *Can. J. Zool.*, 60, 933, 1982.

Farrell, A. P., A review of cardiac performance in the teleost heart: intrinsic and humoral regulation, *Can. J. Zool.*, 62, 523, 1984.

Farrell, A. P., Cardiovascular and hemodynamic energetics of fishes, in *Circulation, Respiration and Metabolism*, Gilles, R., Ed., 1985, 377.

Farrell, A. P., Cardiovascular responses in the sea raven, *Hemitripterus americanus*, elicited by vascular compression, *J. Exp. Biol.*, 122, 65, 1986.

Farrell, A. P., Coronary flow in a perfused rainbow trout heart, *J. Exp. Biol.*, 129, 107, 1987.

Farrell, A. P., From hagfish to tuna — a perspective on cardiac function, *Physiol. Zool.*, 64, 1137, 1991a.

Farrell, A. P., Circulation of body fluids, in *Environmental and Metabolic Animal Physiology*, Prosser, C. L., Ed., Wiley-Liss, New York, 1991b, 509.

Farrell, A. P., Davie, P. S., Franklin, C. E., Johansen, J. A., and Brill, R. W., Cardiac physiology in tunas. I. *In vitro* perfused heart preparations from yellowfin and skipjack tunas, *Can. J. Zool.*, 70, 1200, 1992.

Farrell, A. P., Davie, P. S., and Sparksman, R., The absence of coronary arterial lesions in five species of elasmobranches, *Raja nasuta, Squalus acanthias* L, *Isurus oxyrinchus* Rafinesque, *Prionace glauca* L, and *Lamna nasus* Bonnaterre, *J. Fish. Diseases,* 15, 537, 1992.

Farrell, A. P., Daxboeck, C., and Randall, D. J., The effect of input pressure and flow on the pattern and resistance to flow in isolated perfused gills, *J. Comp. Phys.*, 133, 233, 1979.

Farrell, A. P., Hammons, A. M., Graham, M. S., and Tibbits, G. F., Cardiac growth in rainbow trout, *Salmo gairdneri, Can. J. Zool.*,66, 2368, 1988a.

Farrell, A. P., Johansen, J. A., and Saunders, R. L., Coronary lesions in Pacific salmonids, *J. Fish Dis.*, 13, 97, 1990.

Farrell, A.P., Johansen, J.A., and Graham, M.S., The role of the pericardium in cardiac performance of the trout *(Salmo gairdneri), Physiol. Zool.*, 61, 213, 1988b.

Farrell, A. P. and Jones, D. R., The Heart, in *Fish Physiology: Cardiovascular Systems*, Vol. 12, Hoar, W. S., Randall, D. J., and Farrell, A. P., Eds., Academic Press, New York, 1992, 1.

Farrell, A. P., Saunders, R. L., Freeman, H. C., and Mommsen, T. P., Arteriosclerosis in Atlantic salmon: effects of dietary cholesterol and maturation, *Arteriosclerosis,* 6, 453, 1986.

Farrell, A. P., Sobin, S. S., Randall, D. J., and Crosby, S., Sheet blood flow in the secondary lamellae in teleost gills, *Am. J. Physiol.*, 239, R428, 1980.

Forster, M. E., Axelsson, M., Farrell, A. P., and Nilsson, S., Cardiac function and circulation in hagfishes, *Can. J. Zool.*, 69, 1985, 1991.

Forster, M. E., Axelsson, M., Davison, W., and Farrell, A. P., Hypoxia in the New Zealand hagfish, *Eptatretus cirrhatus, Respir. Physiol.*, 88, 373, 1992.

Fritsche, R. and Nilsson, S., Cardiovascular responses to hypoxia in the Atlantic cod, *Gadus morhua, Exp. Biol.*, 48, 153, 1989.

Fritsche, R. and Nilsson, S., Autonomic nervous control of blood pressure and heart rate during hypoxia in the cod, *Gadus morhua, J. Comp. Physiol. B*, 160, 287, 1990.

Graham, M. S. and Farrell, A. P., The effect of temperature acclimation and adrenaline on the performance of a perfused trout heart, *Physiol. Zool.*, 62, 38, 1989.

Helle, K. B., Structures of functional interest in the myocardium of lower vertebrates, *Comp. Biochem. Physiol. A*, 76, 447, 1983.

Hemmingsen, E. A. and Douglas, E. L., Respiratory and circulatory adaptations to the absence of hemoglobin in Chaenichthyid fishes, in *Adaptations within Antarctic Ecosystems*, Llano, G. A., Ed., Smithsonian Institute, Washington, D.C., 1977, 479.

Hipkins, S. F., Adrenergic responses of the cardiovascular system of the eel, *Anguilla australis, in vivo, J. Exp. Zool.*, 235, 7, 1985.

Hipkins, S. F., Smith, D. G., and Evans, B. K., Lack of adrenergic control of dorsal aortic blood pressure in the resting eel, *Anguilla australis, J. Exp. Zool.*, 238, 155, 1986.

Holbert, P. W., Boland, E. J., and Olson, K. R., The effect of epinephrine and acetylcholine on the distribution of red cells within the gills of the channel catfish *(Ictalurus punctatus), J. Exp. Biol.*, 79, 135, 1979.

House, E. W., Dornauer, R. J., and Van Lenten, B. J., Production of coronary arteriosclerosis with sex hormones and human chorionic gonadotrophin (HCG) in juvenile steelhead and rainbow trout, *Salmo gairdneri, Atherosclerosis*, 34, 197, 1979.

Hughes, G. M., General anatomy of the gills, in *Fish Physiology,* Vol. 10A, Hoar, W. S. and Randall, D. J., Eds., Academic Press, New York, 1984, 1.

Johansen, K. and Burggren, W. W., Cardiovascular function in the lower vertebrates, in *Heart and Heart-Like Organs*, Bourne, G. H.,Ed., Academic Press, New York, 1980, 61.

Jones, D. R. and Milsom, W. K., Peripheral receptors affecting breathing and cardiovascular function in non-mammalian vertebrates, *J. Exp. Biol.*, 100, 59, 1982.

Jones, D. R. and Randall, D. J., The respiratory and circulatory systems during exercise, in *Fish Physiology*, Hoar, W. S. and Randall, D. J., Eds., Academic Press, New York, 1978, 425.

Karttunen, P. and Tirri, R., Isolation and characterization of single myocardial cells from the perch, *Perca fluviatilis, Comp. Biochem. Physiol. A,* 84, 181, 1986.

Keen, J. E., Thermal Acclimation, Cardiac Performance and Adrenergic Sensitivity in Rainbow Trout, Ph.D. thesis, Simon Fraser University, Burnaby, British Colombia, 70, 1211, 1992.

Keen, J. E., Farrell, A. P., Tibbits, G. F., and Brill, R. W., Cardiac dynamics in tunas. II. Effect of ryanodine, calcium and adrenaline on force-frequency relationships in arterial strips from skipjack tuna, *Katsuwonus pelamis, Can. J. Zool.*, 70, 1211, 1992.

Kent, B. B., Peirce, E. C., II, and Bever, C. T., Distribution of blood flow in *S. acanthias*: a preliminary study, *Bull. Mt. Desert Isl. Biol. Lab.*, 11, 53, 1971.

Kiceniuk, J. W. and Jones, D. R., The oxygen transport system in trout (*Salmo gairdneri*) during sustained exercise, *J. Exp. Biol.*, 69, 247, 1977.

Kubasch, A. and Rourke, A. W., Arteriosclerosis in steelhead trout, *Oncorhynchus mykiss* (Walbaum): a developmental analysis, *J. Fish Biol.*, 37, 65, 1990.

Lai, N. C., Graham, J. B., Lowell, W. R., and Laurs, R. M., Pericardial and vascular pressures and blood flow in the albacore tuna, *Thunnus alalunga, J. Exp. Biol.*, 146, 187, 1987.

Lai, N. C., Graham, J. B., and Lowell, W. R., Elevated pericardial pressure and cardiac output in the leopard shark *Triakis semifasciata* during exercise: the role of the pericardioperitoneal canal, *J. Exp. Biol.*, 147, 263, 1989a.

Lai, N. C., Graham, J. B., Bhargava, V., Lowell, W. R., and Shabetai, R., Branchial blood flow distribution in the blue shark (*Prionace glauca*) and the leopard shark (*Triakis semifasciata), Exp. Biol.*, 48, 273, 1989b.

Lai, N. C., Graham, J. B., and Burnett, L., Blood respiratory properties and the effect of swimming on blood gas transport in the leopard shark *Triakis semifasciata, J. Exp. Biol.*, 151, 161, 1990a.

Lai, N. C., Shabetai, R., Graham, J. B., Hoit, B. D., Sunnerhagen, K. S., and Bhargava, V., Cardiac function of the leopard shark, *Triakis semifaciata, J. Comp. Physiol. B*, 160, 259, 1990b.

Laurent, P., Gill internal morphology, in *Fish Physiology*, Vol. 10A, Hoar, W. S. and Randall, D. J., Eds., Academic Press, New York, 1984, 73.

Laurent, P., Holmgren, S., and Nilsson, S., Nervous and humoral control of the fish heart: structure and function, *Comp. Biochem. Physiol. A*, 76, 525, 1983.

Lucas, M. C., Priede, I. G., Armstrong, J. D., Gindy, A. N. Z., and De Vera, L., Direct measurements of metabolism, activity and feeding behaviour of pike, *Esox lucius* L., in the wild, by the use of heart rate telemetry, *J. Fish Biology*, 39, 325, 1991.

Moore, J. F., Mayr, W., and Hougie, C., Ultrastructure of coronary arterial changes in spawning Pacific salmon genus *Oncorhynchus* and steelhead trout *Salmo gairdneri, J. Comp. Pathol.*, 86, 259, 1976.

Neumann, P., Holeton, G. F., and Heisler, N., Cardiac output and regional blood flow in gills and muscles after strenuous exercise in rainbow trout *(Salmo gairdneri), J. Exp. Biol.*, 105, 1, 1983.

Nilsson, S., Autonomic Nerve Function in the Vertebrates, Springer-Verlag, Berlin, 1983, 1.

Nilsson, S. and Axelsson, M., Cardiovascular control systems in fish, in *Neurobiology of the Cardio-Respiratory System*, Taylor, E. W., Ed., Manchester University Press, Manchester, 1987, 115.

Payan, P. and Girard, J. P., Adrenergic receptors regulating patterns of blood flow through the gills of trout, *Am. J. Physiol.*, 232, H18, 1977.

Peyraud-Waitzenegger, M., Barthelemy, L., and Peyraud, C., Cardiovascular and ventilatory effects of catecholamines in unrestrained eels *(Anguilla anguilla L.), J. Comp. Physiol. B*, 138, 367, 1980.

Peyraud-Waitzenegger, M. and Soulier, P., Ventilatory and circulatory adjustments in the European eel *(Anguilla anguilla* L.) exposed to short term hypoxia, *Exp. Biol.*, 48, 107, 1989.

Piiper, J., Meyer, M., Worth, H., and Willmer, H., Respiration and circulation during swimming activity in the dogfish *Scyliorhinus stellaris, Respir. Physiol.*, 30, 221, 1977.

Randall, D.J., The circulatory system, in *Fish Physiology*, Vol. 4., Hoar, W. S. and Randall, D. J., Eds., Academic Press, New York, 1970, 133.

Randall, D., Shunts in fish gills, in *Cardiovascular shunts — Phylogenetic, Ontogenetic and Clinical aspects,* Alfred Benson Symp. XXI, Johansen, K. and Burggren, W., Eds., Munksgaard, Copenhagen, 1985, 71.

Randall, D. J. and Brauner, C., Effects of environmental factors on exercise in fish, *J. Exp. Biol.*, 160, 113, 1991.

Randall, D. J. and Daxboeck, C., Cardiovascular changes in the rainbow trout *(Salmo gairdneri* Richardson) during exercise, *Can. J. Zool.*, 60, 1135, 1982.

Saito, T., Effects of vagal stimulation on the pacemaker action potentials of carp heart, *Comp. Biochem. Physiol. A*, 44, 191, 1973.

Santer, R., *Morphology and Innervation of the Fish Heart*, Springer-Verlag, Berlin, 1985, 102.

Satchell, G. H., *Circulation in Fishes,* Cambridge University Press, Cambridge, 1971, 1.

Satchell, G. H., *Physiology and Form of Fish Circulation,* Cambridge University Press, Cambridge, 1991, 1.

Satchell, G. H., The venous system, in *Fish Physiology*, Vol. 12, Hoar, W. S., Randall, D. J., and Farrell, A. P., Eds., Academic Press, New York, 1992, 141.

Saunders, R. L., Farrell, A. P., and Knox, D. E., Progression of coronary arterial lesions in Atlantic salmon as a function of growth rate, *Can. J. Aquat. Fish. Sci.*, 49, 878, 1992.

Shabetai, R., Abel, D. C., Graham, J. B., Bhargava, V., Keyes, R. S., and Witztum, K., Function of the pericardium and pericardioperitoneal canal in elasmobranch fishes, *Am. J. Physiol.*, 248, H198, 1985.

Siebert, H., Thermal adaptation of heart rate and its parasympathetic control in the European eel *Anguilla* (L), *Comp. Biochem. Physiol. C*, 64, 275, 1979.

Smith, L. S. and Bell, G. R., A Practical Guide to the Anatomy and Physiology of Pacific Salmon, Miscellaneous Special Publication 27, Department of the Environment, Fisheries and Marine Service, Ministry of Supply and Services, Ottawa, Canada, 1976.

Steffensen, J. F. and Lomholt, J. P., The secondary circulation in *Fish Physiology: Cardiovascular Systems*, Vol. 12, Hoar, W. S., Randall, D. J., and Farrell, A. P., Eds., Academic Press, New York, 1992, 185.

Taylor, E. W., Short, S., and Butler, P. J., The role of the cardiac vagus in the response of the dogfish *Scyliorhinus canicula* to hypoxia, *J. Exp. Biol.*, 70, 57, 1977.

Tota, B., Myoarchitecture and vascularization of the elasmobranch heart ventricle, *J. Exp. Zool. Suppl.*, 2, 122, 1989.

Tota, B., Acierno, R., and Agnisola, C., Mechanical performance of the isolated and perfused heart of the hemoglobinless anarctic icefish *Chionodraco humatas* (Lönnberg): effects of loading conditions and temperature, *Phil. Trans. R. Soc. Lond. Ser. B*, 332, 191, 1991.

Tota, B., Cimini, V., Salvatore, G., and Zummo, G., Comparative study of the arterial and lacunary systems of the ventricular myocardium of elasmobranch and teleost fishes, *Am. J. Anat.*, 167, 15, 1983.

Vogel, W. O. P., Systemic vascular anastomoses, primary and secondary vessels in fish, and the phylogeny of lymphatics, in *Cardiovascular Shunts*, Munksgaard, Copenhagen, 1985, 143.

White, F. C., Kelly, R., Kemper, S., Schumacker, P. T., Gallagher, K. R., and Laurs, R. M., Organ blood flow haemodynamics and metabolism of the albacore tuna *Thunnus alalunga* (Bonnaterre), *Exp. Biol.*, 47, 161, 1988.

Wood, C. M. and Shelton, G., Cardiovascular dynamics and adrenergic responses of the rainbow trout *in vivo, J. Exp. Biol.*, 87, 247, 1980a.

Wood, C. M. and Shelton, G., The reflex control of heart rate and cardiac output in the rainbow trout: interactive influences of hypoxia, haemorrhage and systemic vasomotor tone, *J. Exp. Biol.*, 87, 271, 1980b.

Wood, C. M., Peiprzak, P., and Trott, J. N., The influence of temperature and anaemia on the adrenergic and cholinergic mechanisms controlling heart rate in the rainbow trout, *Can. J. Zool.*, 57, 2440, 1979.

Yamauchi, A., Fine structure of the fish heart, in *Hearts and Heart-Like Organs*, Bourne, G. H., Ed., Academic Press, New York, 1980, 119.

9 Gas Exchange

Steve F. Perry and Gordon McDonald

INTRODUCTION

Some basic features of the gas exchange apparatus of fish are summarized in Figure 1, which shows: (A) a horizontal section of the head of a generalized teleost with the four pairs of gill arches, and indicates the direction of water flow through the branchial and opercular cavities; (B) part of a single gill arch with two rows of gill filaments. The basic units of gas exchange are the lamellae, which arise perpendicularly from the upper and lower surfaces of the filaments. Each lamella (C) is comprised of two epithelia that are kept separated by a series of pillar cells (circles) between which blood can flow. Blood flow is in a direction opposite to that of water flow. Lamellar spacing along the filament ranges anywhere from 10 to 60 per millimeter depending on species and is correlated with level of activity. Total lamellar numbers vary greatly, from as few as 0.5 million in an inactive bottom dweller like the toadfish (1 kg body mass) to over 6 million in a very active species such as a 1-kg tuna (Hughes, 1984). Lamellar thicknesses range from 10 to 25 μm, and spacing between lamellae is typically larger, from 20 to 100 μm (Piiper and Scheid, 1982).

The transfer of oxygen from the environment to the cells can be visualized as a series of steps: gill ventilation, branchial diffusion, blood oxygen transport, and tissue diffusion. Each step can be described quantitatively (Figure 1D); a useful approach for identifying factors responsible for determining and limiting O_2 delivery. In this chapter we briefly review the first three steps (there are insufficient data concerning tissue diffusion to warrant discussion), focusing on O_2 delivery. We also give attention to the transfer of CO_2 from the cells to the environment, especially for those circumstances where the principles might differ from those governing O_2 delivery. Several aspects of gas exchange in fishes have been reviewed in recent years (e.g., Jones and Randall, 1978; Randall, 1982, 1990; Randall et al. 1982; Randall and Daxboeck, 1984; Perry, 1986; Butler and Metcalfe, 1988; Cameron, 1989; Heisler, 1989; Perry and Wood, 1989; Piiper, 1989; Jensen, 1991; Wood, 1991). The reader is referred to this literature for more detail than is provided here. In this chapter, the classical aspects of blood gas transport are reiterated, with emphasis being placed on the nature of inter- and intraspecific differences. Particular attention is given to recent developments in the area of physiological adaptation to stress.

0-8493-8042-1/93/$0.00+$.50

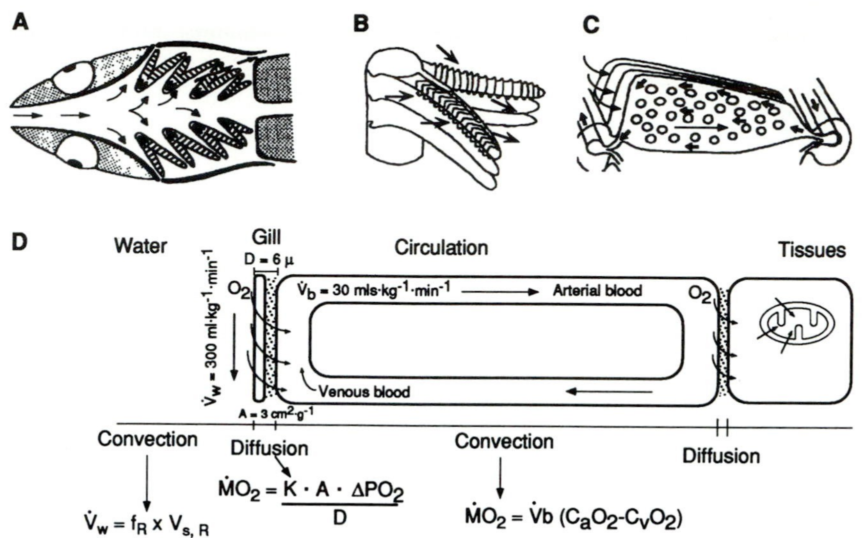

FIGURE 1. Anatomic and physiological model for gas exchange in fish. (A) Horizontal section through the head of a generalized teleost showing four pairs of gill arches. Arrows indicate direction of water flow through branchial cavity. (Redrawn from McDonald, 1983.) (B) A part of a single gill arch with two rows of gill filaments. Gas exchange units are the lamellae projecting above and below the surfaces of the filaments. (Redrawn from Satchell, 1984.) (C) Transverse section of a filament showing a lamella in side view. Circles represent pillar cells. (Redrawn from Satchell, 1984.) (D) Quantitative model for gas transfer in fish. $\dot{V}_W$, ventilatory water flow; f_R, = breathing frequency; $V_{s,R}$, ventilatory stroke volume; $\dot{M}O_2$, oxygen uptake; K, Krogh's permeation coefficient of diffusion; A, gill functional surface area; ΔPO_2, mean water-to-blood O_2 partial pressure gradient; D, diffusion distance; $\dot{V}b$, cardiac output; C_aO_2, arterial blood oxygen content; C_vO_2, venous (prebranchial) blood oxygen content). See text for further details.

BRANCHIAL GAS TRANSFER

VENTILATION

The basic mechanism of gill ventilation is essentially the same in all fish although details vary greatly among species (see Milsom, 1989 for review). Ventilatory flow is typically generated by the combined action of a buccal force pump and an opercular suction pump operating out of phase so that a more or less continuous flow of water through the gill curtain is produced (Figure 1A). In addition to the action of the buccal and opercular pumps, there are rhythmic contractions of the intrinsic musculature of the gill arches and filaments during the ventilatory cycle. These dynamically adjust the filaments and lamellae to the changing volume of the branchial cavity and to the flow stream, and thereby maintain continuity of the gill sieve and regulate the resistance to water flow (Hughes and Morgan, 1973; Hughes, 1984).

Interspecific differences in the gills and their ventilatory apparatus include variations in the number and relative sizes of inhalant and exhalant openings, in the relative contributions of buccal and opercular pumps to the generation of flow, variations in the development of gill support structures, and in the degree of fusion of filaments and lamellae. Much of this variation can be attributed to variation in life-style and habitat of fish. For example, tuna gills show fusion of adjacent secondary lamellae at their tips, making a sieve-like structure which is thought to eliminate dead space that would arise from separation of the filament tips at high rates of swimming (Stevens, 1972; Hughes, 1984). A contrasting example is the bottom-dwelling flatfish, in which a well-developed branchiostegal apparatus (the ventral surfaces of the opercular cavity), a dominant opercular pump, a reduction in the size of the exhalent openings, and their closure at specific points in the ventilatory cycle act to keep the branchial cavities clear of debris when the fishes are buried in sand or silt (Hughes, 1960, 1984).

Ventilatory flow ($\dot{V}_W$) is determined by both the frequency (f_R) and depth (i.e. stroke volume, $V_{s,R}$) of breathing (Figure 1D) and is typically increased by small increases in f_R and large increases in $V_{s,R}$ (Jones and Randall, 1978). Routine breathing frequencies in juvenile and adult fish are in the range of 30 to 70 min^{-1} (Roberts, 1975) while in recently hatched larval fishes, with rudimentary ventilation mechanisms and as yet poorly developed gills, f_R can be as low as 10 min^{-1} (McDonald and McMahon, 1977). Also, in truly resting fish, ventilation can be intermittent. Some benthic species, such as the bullhead catfish (*Ictalurus nebulosus*), can cease breathing for as long as 1 min, whereas others alternate between periods of very shallow breathing (branchiostegal movements only) and brief periods of strong ventilatory movements of the opercular flaps (Roberts and Rowell, 1988).

In juvenile to adult fish (body weights ≥100 g) routine ventilatory flows typically fall within the range of 100 to 300 $ml \cdot kg^{-1} \cdot min^{-1}$ (Wood et al, 1979; Johansen, 1982). Under circumstances of routine activity in normoxic water the gill convection requirement, the $\dot{V}_W$ required per unit O_2 uptake, is 200 to 400 ml $H_2O \cdot ml\ O_2^{-1}$ (Johansen, 1982). Although the countercurrent arrangement of water and blood flow (Figure 1C) is a more efficient gas exchange mechanism than the cross-current or tidal ventilation of other vertebrates, convection requirements for fish are 4- to 8-fold higher than those typical for terrestrial ectotherms of similar metabolic demands (Milsom, 1989). The higher convection requirement is largely related to the fact that air-saturated water has only about 1/30 of the O_2 content of air.

Ventilatory flow does not appear to be directly limiting to oxygen uptake ($\dot{M}O_2$) for the former can typically be increased much more than the latter; up to 10- to 15-fold in active species (e.g., salmonid species) compared to $\dot{M}O_2$ increases of rarely more than 5-fold (Brett, 1972; Priede, 1985). However, the increasing cost of breathing at higher ventilation rates may limit any gains in O_2 delivery. A number of estimates of cost of breathing have been published and these are quite variable, in part because of species differences, but more importantly due to differences in methodology. Nonetheless, estimates of costs for routine ventilation of around 10% of routine $\dot{M}O_2$ have become generally accepted (e.g., Jones and Randall, 1978; Milsom, 1989). The cost, however, may increase to as high as 70% with ventilation volumes three times routine (Schumann and Piiper, 1966). By way of comparison, in humans the resting cost of breathing is estimated to be about 2%. A higher cost of breathing in aquatic vertebrates is to be expected because of the higher viscosity (840 × air) and density (60 × air) of the medium.

In actively swimming species the problem of increasing ventilatory cost is circumvented by transferring the work of ventilation from the respiratory muscles to the muscles of locomotion (ram ventilation), a strategy which all but eliminates the cost of ventilation. At swimming speeds ranging from 20 to 60 $cm \cdot sec^{-1}$, active ventilation ceases and fish swim with an open mouth. The ram transition velocity varies with species, environmental PO_2, and temperature, but not apparently with body size (Roberts, 1975; Roberts and Rowell, 1988). The consequence of ram ventilation is that the cost of breathing remains constant and independent of the exercise level of the fish (Jones and Randall, 1978; Randall and Daxboeck, 1984). Thus, fish probably avoid substantial increases in ventilation wherever possible; reports of very high ventilation rates are from the somewhat unusual circumstance of animals confined in ventilation chambers and exposed to hypoxic conditions. Saunders (1962), for example, reported $\dot{V}_W$s as high as 7 and 12 $l \cdot kg^{-1} \cdot min^{-1}$ in *I. nebulosus* and sucker (*Catastomus commersoni*), respectively, at low PO_2. Under normal circumstances, fish encountering hypoxia either attempt to escape (Dandy, 1970) or reduce their $\dot{M}O_2$ to conform to the environmental O_2 supply (Hughes, 1981).

Branchial Diffusion

In any particular species the total surface area for gas diffusion, the lamellar surface area (LSA), scales with body mass according to the allometric relationship: area = $a \cdot weight^b$. The exponent, b, varies from 0.5 to 1.0, but a value of 0.8 is quite common (Hughes, 1984). A similar value for the exponent is also usually found for the scaling of MO_2 to body mass. Although LSA scales with body mass, diffusion distance remains fairly constant for each species.

Gill surface area and diffusion distance also tend to be highly correlated with life-style and habitat of fish. The range of variation in both parameters is shown in Figure 2. At one extreme are the air-breathing fishes which have thick lamellae and very small lamellar surface areas, in some cases so small that the gills are almost functionless in terms of O_2 uptake (Hughes and Morgan, 1973), while on the other extreme are the scombrid family, e.g., tuna and mackerel, which have very thin lamellae and large surfaces, comparable, in fact, in dimensions to the lungs of mammals. Most fish species, however, are found in a relatively narrow range between these two extremes (indicated by the shaded area in Figure 2). Nonetheless, within this range, free-living and more active fishes tend to have relatively larger areas than bottom-dwelling forms. In

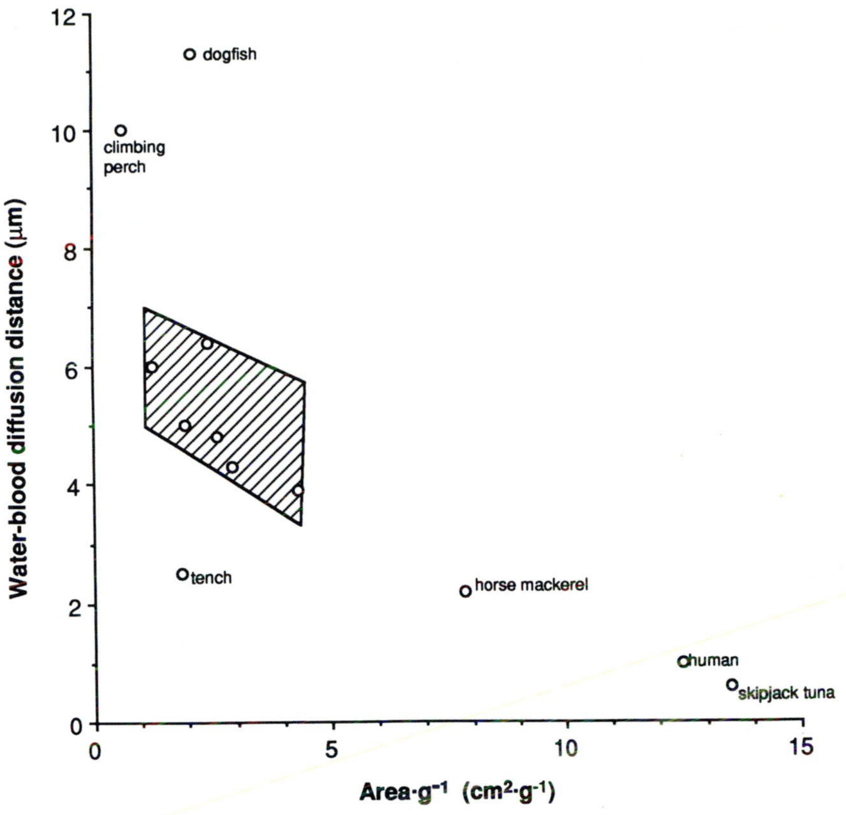

FIGURE 2. Total gill area per unit body weight ($cm^2\ g^{-1}$) and estimated diffusion distance between water and blood for fish of differing life-styles and habitats. Cross-hatched area is the range within which most teleost species fall. Dogfish, *Scyliorhinus canicula* (Hughes, 1972); climbing perch, *Anabas testudineus*, (Hughes et al., 1973); tench, *Tinca tinca* (Hughes and Morgan, 1973); horse mackerel, *Trachurus trachurus* (Hughes, 1966); skipjack tuna, *Katsuwonus pelamis* (Muir and Hughes, 1969). (Adapted from Roberts, 1975.)

addition to variations in surface area and thickness, the size of the lamella is larger and the interlamellar spacing smaller in fast-swimming fish compared to more sluggish forms (Hughes, 1966a and b).

The dimensional variations among fish species (13-fold range in surface area; 10-fold range in thickness) are related to a >10 fold variation in maximum O_2 uptake rate (MO_{2max}), from <100 $nmol \cdot g^{-1}\ min^{-1}$ (plaice) to >1000 $nmol \cdot g^{-1}\ min^{-1}$ (tuna) (Stevens and Neill, 1978; Priede and Holliday, 1980). The quantitative relationship between these morphometric features and branchial O_2 transfer (MO_2) is expressed by the modified Fick equation (Figure 1D):

$$MO_2 = K^*A^*\Delta PO_2/D \qquad (1)$$

where K is the oxygen permeation coefficient in $\mu mol \cdot \mu m^{-1} cm^{2-1} kPa^{-1}$, A is the functional surface area of the gills (surface area available for gas exchange), in $cm^2 g^{-1}$, ΔPO_2 is the mean O_2 partial pressure gradient between the blood and the water in kPa, and D is the mean blood-to-water diffusion distance in µm.

Equation 1 provides a useful context for discussing the mechanisms fish employ to increase branchial O_2 diffusion because each variable, potentially at least, is available for adjustment.

The Permeation Coefficient, K

The value of K has not been determined on fish gill epithelia. Indeed, there have been relatively few measurements of this value in fish tissues other than an estimate for eel (*Anguilla anguilla*) skin (Kirsch and Nonnette, 1977) which is thought to be too high to describe the gill (Randall and Daxboeck, 1984). It is clear, however, that O_2 diffuses more slowly through tissues than water, tissue Ks from 1/2 to 1/10 of K_{water} have been reported; a value of 1/3 K_{water} is considered to be a reasonable estimate for the fish gill (Randall and Daxboeck, 1984). Although it is assumed to be constant for different parts of the barrier, this is uncertain; also, it is uncertain the extent to which it varies with temperature; Krogh (1941) established that the permeation coefficient for frog connective tissue declined about 20% from 20 to 10°C. This would be important, of course, for those fish species which occupy a wide thermal range.

Diffusion Distance, D

Diffusion distance, as noted above, varies considerably with species. It also varies within the gill apparatus of a single animal (Hughes and Morgan, 1973). What is less clear is the extent to which, if any, D is dynamically regulated in relation to activity in individuals. There is some evidence, however, that contraction of the myosin filaments within the pillar cells and perhaps also increased gill water pressures with increased ventilation may serve to counteract the tendency for the thickness of the blood sheet to increase in response to increasing intralamellar blood pressure (Randall and Daxboeck, 1984).

Functional Surface Area, FSA

It has long been recognized that the entire gill is not equally ventilated or perfused at any one time. This has led to a distinction being drawn between the total surface area, estimated from morphometric studies, and the functional surface area (FSA), the surface area available for gas transfer. During quiet ventilation the overall resistance to flow is relatively large towards the tips of the filaments so that water flow mainly occurs through the basal and middle lamellar channels. Similarly, there is preferential perfusion of proximal lamellae. Indeed, according to studies by Booth (1978) on rainbow trout (*Oncorhynchus mykiss*), only about 60% of all lamellae are perfused at rest. Increasing ventilation leads to the recruitment of water channels toward the distal ends of the filaments. At the same time, increases in blood pressure and pulsatility, along with active dilation of afferent lamellar arterioles, leads to opening or recruitment of distal lamellae and, within the open lamellae, shunting of blood from basal to more central channels (Nilsson, 1986).

The Blood-to-Water Oxygen Partial Pressure Gradient (ΔPO_2)

The O_2 diffusion gradient continually changes, of course, as the blood flows through the gill vasculature and the calculation of the mean gradient is complicated by the fact that the relationship between blood PO_2 and O_2 content is nonlinear (Figure 3). However, the following equation is considered a reasonable approximation (Wood and Perry, 1985): $\frac{1}{2}(P_IO_2 + P_EO_2) - \frac{1}{2}(P_aO_2 + P_vO_2)$ where P_I and P_E = inspired and expired O_2 tensions and P_a and P_v = arterial and venous tensions. The first term is dependent on ventilation while the second term is dependent on tissue O_2 demand, gill blood flow, and the O_2 transport properties of the blood (Johansen, 1982) (see below).

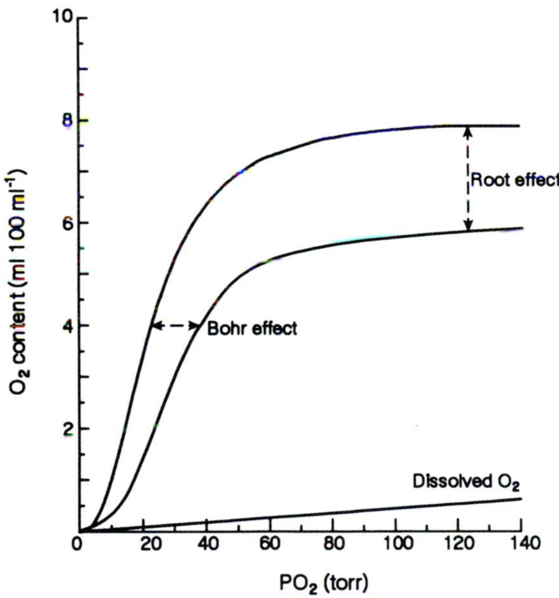

FIGURE 3. Typical sigmoidal oxygen dissociation curves for tetrameric teleost hemoglobins showing the effects of blood acid-base status on the affinity of hemoglobin-O_2 binding (the Bohr effect) and on the capacity of binding (the Root effect). Intracellular acidosis shifts the curve downward (lowered capacity) and to the right (lowered affinity) and vice versa. The fraction of oxygen carried in the plasma as dissolved O_2 generally constitutes less than 5% of the total O_2 content at physiological PO_2 levels.

Compromises Between Branchial Gas Transfer and Osmoregulation

The two main variables available for increasing branchial O_2 diffusion are FSA and ΔPO_2. The choice between the two is complicated by the fact that the gills must serve not only for respiratory gas exchange but also for osmo- and iono-regulation. Fish must balance the advantage of maximizing O_2 diffusion at the gills with the osmoregulatory disadvantage resulting from the accompanying increases in ionic loss/water influx (freshwater teleosts) or water loss/ionic gain (marine teleosts). Although this phenomenon has not been widely studied, for the freshwater rainbow trout at least, there is a clear increase in Na^+ diffusive efflux whenever $\dot{M}O_2$ increases (Randall et al, 1972; Wood and Randall, 1973; Gonzalez and McDonald, 1992) which would necessitate additional energy expenditure for the regulation of ion balance. It has long been thought that fish managed the osmorespiratory compromise by limiting their functional surface area (Satchell, 1984; Nilsson, 1986) implying that given a choice between FSA and ΔPO_2, adjustments to ΔPO_2 would be preferred. However, it is now apparent that the limiting factors are more complex. According to Equation 1, any increase in FSA, providing there are no accompanying changes in other variables, would mean that the diffusive efflux of Na^+ ($J_{out}Na^+$) would increase no more than the influx of O_2; i.e., the ratio of the two fluxes, the ion/gas ratio (IGR), would remain constant. However, recent studies on rainbow trout (Gonzalez and McDonald, 1992) found substantial increases in $J_{out}Na^+$ relative to increases in $\dot{M}O_2$ under a number of different circumstances. For example, intravascular infusion of adrenaline (epinephrine) to yield a plasma concentration of ~300 nmol l^{-1}, elevated MO_2 by about 3-fold but $J_{out}Na^+$ by over 20-fold; i.e., an increase in the IGR of about 10-fold. The authors concluded that the key event responsible for the disproportionate increase in $J_{out}Na^+$ was an increase in intralamellar blood pressure resulting from one or more of the following effects of adrenaline: increased cardiac output, dilation of afferent lamellar arterioles (Farrell, 1980), and/or constriction of efferent lamellar arterioles (Petterson, 1983). The increased pressure is thought to increase ionic diffusion by pressure distortion of paracellular channels, an event which would not similarly increase transcellular O_2 diffusion, i.e., it would lead to an increase in IGR. Injection of adrenaline undoubtedly exaggerates the hemodynamic changes in the gills compared to those occurring with normal aerobic exercise. Nonetheless, these findings suggest that it is more important for fish to limit increases in intralamellar blood pressure rather than increases in functional surface area when managing the osmorespiratory compromise.

Gonzalez and McDonald (manuscript submitted) have now tested this hypothesis by examining adjustments to gas exchange in several fish species that differ in habitat and life-style; ranging from still-water residents (e.g., sunfish species) that lead largely inactive life-styles, to stream dwellers (e.g., rainbow trout and shiner species) who spend a large part of their time expending energy holding station against current. For all species the IGR under routine conditions averaged about 60 pmol $Na^+ \cdot$nmol O_2^{-1}. When, however, the species were exercised, the inactive species showed a much larger increase in IGR compared to the active species. Indeed, a very close inverse correlation exists between postexercise $\dot{M}O_2$ and IGR (Figure 4). The interpretation of this finding is that

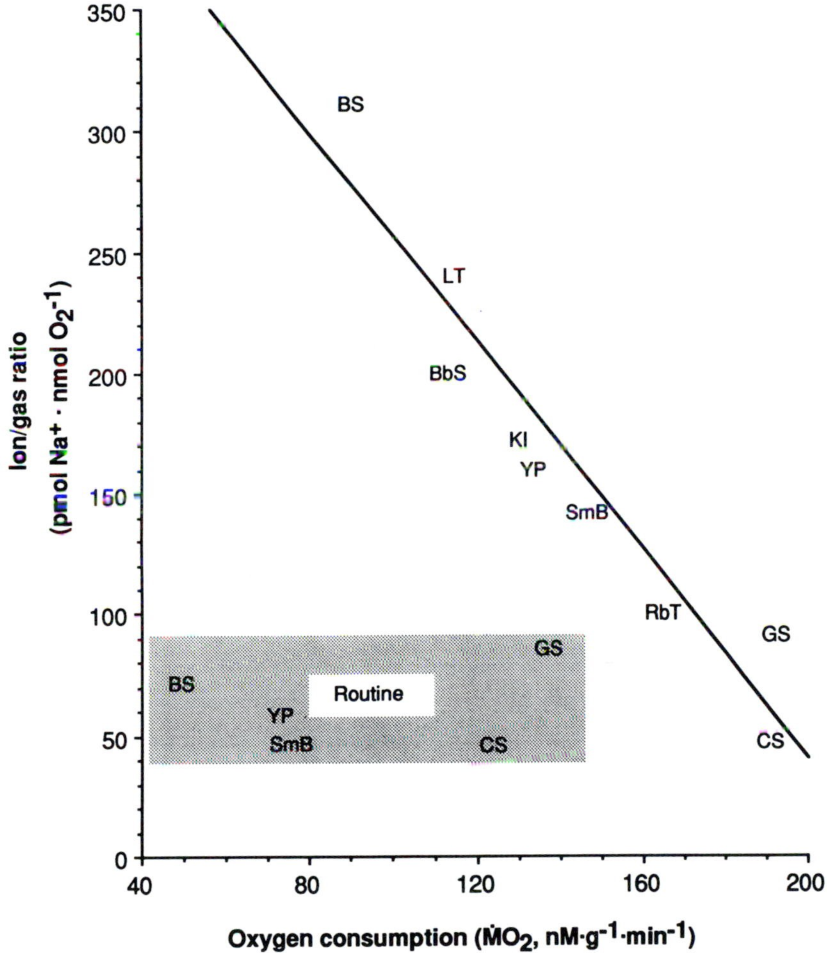

FIGURE 4. Correlation between postexercise oxygen consumption ($\dot{M}O_2$) and postexercise ion/gas ratio (IGR) for nine species of fish. Each species was exercised for 5 min by manual chasing. BS, banded sunfish; BbS, black banded sunfish; YP, yellow perch; SmB, smallmouth bass; GS, golden shiner; CS, common shiner; Ki, killifish; LT, lake trout; RbT, rainbow trout. All measurements made at 18°C, except for LT and RbT at 10°C. Shaded area indicates range of routine and IGR for five of the species. Routine measurements were made on well-rested animals. (Redrawn from Gonzalez and McDonald, submitted for publication.)

the more active a species normally is, the more it relies on increases in FSA to meet increased O_2 demand. Less active species, which rarely exploit the upper limits of their metabolic scope, rely more on hemodynamic adjustments which, while increasing blood flow through the gills, also increase intralamellar blood pressure. They are better able to tolerate the resulting ion losses because they happen so rarely.

BLOOD GAS TRANSPORT

OXYGEN

Basic Principles

Oxygen is transported in the blood in two forms, as physically dissolved O_2 and as O_2 chemically bound to the respiratory pigment hemoglobin. The amount of physically dissolved O_2 is determined essentially by the prevailing PO_2 and the solubility coefficient of O_2 in the plasma. Owing to the low solubility of O_2 in fish plasma (Boutilier et al., 1984), the physically dissolved O_2 usually constitutes only a small fraction (less than 5%) of the total O_2 carried in the blood (see Figure 3). A striking exception is the Antarctic teleost, *Chaenocephalus aceratus*, which lacks hemoglobin (Holeton, 1970). Thus, in this species, O_2 is transported in the blood exclusively in physical solution. The relatively high arterial PO_2 in *C. aceratus* (e.g., 120 torr; Holeton, 1970), coupled with an unusually high O_2 solubility (a consequence of the low temperature of the plasma), allow greater quantities of dissolved O_2 to be transported in comparison to other teleosts. In addition, an unusually high cardiac output in *C. aceratus* assures that sufficient quantities of O_2 can be taken up at the gill and delivered to the tissues despite a low arterial-venous O_2 content difference. The high cardiac output is achieved at a relatively low heart rate (14 beats per minute). Furthermore, the unusually low peripheral resistance to blood flow ensures that the work performed by the heart is kept low despite the large volume of blood being pumped. Thus, the low metabolic rate of *C. aceratus*, resulting from its low temperature and sluggish life-style, is sustainable without the need for a respiratory pigment.

In all other fishes the presence of the respiratory pigment hemoglobin within the red blood cells (rbcs) serves to increase the oxygen-carrying capacity of the blood sufficiently to meet the metabolic requirements without necessitating unduly high cardiac outputs. The concentration of hemoglobin within the rbcs (mean cellular hemoglobin concentration, MCHC) appears to be relatively constant (approximately 30 $g \cdot 100\ ml^{-1}$) among species, although relatively few species have been examined. On the other hand, the concentration of rbcs within the plasma (hematocrit) is highly variable, ranging between about 15 and 40%. Thus, the hemoglobin levels in the blood and the blood oxygen-carrying capacity are primarily related to the hematocrit. Although there are several exceptions, it is generally apparent that blood hemoglobin levels in the more inactive species are lower than in the more active species. The oxygen-carrying capacity of the blood can be increased acutely by liberation of sequestered rbcs from the spleen (see below) or chronically by enhancement of erythropoiesis.

The relationship between blood PO_2 and hemoglobin-O_2 (Hb-O_2) saturation is described by the oxygen equilibrium curve (see Figure 3). With the exception of monomeric oxygenated hemoglobin in the agnathans, piscine hemoglobins are tetrameric and exhibit cooperativity of binding. Thus, the oxygen equilibrium curves generally are sigmoidal in their shape (Hill number; $N_{HILL} > 1$). The oxygen equilibrium curve of agnathan hemoglobin is hyperbolic ($N_{HILL} = 1$).

The dynamic nature of Hb-O_2 binding is affected by several intracellular allosteric modulators, including nucleoside triphosphates (NTPs) and H^+ ions (see reviews by Weber and Jensen, 1988; Boutilier and Ferguson, 1989). The predominant NTPs in the nucleated rbcs of fish are adenosine triphosphate (ATP) and guanosine triphosphate (GTP) although their relative concentrations within the rbc are highly species dependent. ATP is the principal intracellular NTP in salmonids and elasmobranchs, whereas GTP predominates in carp, eel, goldfish, and tench (Weber and Jensen, 1988). Decreased NTP levels cause an increase in the affinity of Hb-O_2 binding as indicated by a lowering of the P_{50} value (the PO_2 at which Hb is 50% saturated with O_2). GTP is a more potent allosteric modifier of P_{50} than ATP (Weber and Jensen, 1988). The effect of altered pH on Hb-O_2 affinity is termed the Bohr effect. Reductions in pH owing to increased production of metabolic acids or CO_2 reduce the affinity of Hb-O_2 binding (increase in the P_{50}) and vice versa. In fish, the Bohr effect is mediated almost exclusively by H^+ ions (arising from dissociation of metabolic acids or CO_2) and is termed the "fixed acid Bohr effect". The "specific CO_2 Bohr effect" is physiologically less important in fish than in other vertebrates because acetylation of the α-amino groups of the α-chains limits the CO_2 binding to the β-chains of hemoglobin where there is considerable competition with the NTPs for the binding sites. Owing to the Bohr effect, there is a transient increase in Hb-O_2 affinity as blood pH rises in the branchial vasculature during the process of CO_2 excretion. This assists in the loading of O_2 at the gill. Conversely, the reduction of blood pH at the tissue level assists in the unloading of O_2 from Hb as the P_{50} is increased.

In some fish, changes in rbc pH not only affect the affinity of Hb-O_2 binding (the Bohr effect) but also the maximal O_2 binding capacity such that at reduced pH complete saturation of hemoglobin cannot be achieved even at supraphysiological PO_2 levels. The effects of acidosis on lowering blood oxygen content was first described by Root (1931), and is now known commonly as the Root effect (see Brittain, 1987; Riggs, 1988; Pelster and Weber, 1991 for reviews). The Root effect is physiologically important in the establishment of extremely high oxygen partial pressures in the swim bladder or the vitreous humour of the eye (Pelster and Weber, 1991) via acidification of the blood and consequent liberation of O_2 from hemoglobin in the rete mirabile and the choroid rete, respectively. The existence of Root effect hemoglobins is thus well correlated with the existence of a choroid rete and/or swim bladder. The elasmobranchs, which lack both structures, do not generally display a pronounced Root effect. The Root effect, while beneficial for elevating oxygen partial pressures in the swim bladder and eye, may contribute to a lowering of blood oxygen content during periods of internal acidosis. Generally, however, such acid-base disturbances are accompanied by the release of catecholamines into the circulation which serve to minimize changes in rbc pH (see below).

Certain adult teleost species possess multiple hemoglobins (isoforms) with differing functional characteristics. Generally, anodal hemoglobins are characterized by a low affinity and a marked sensitivity to pH and organic phosphates (NTPs), whereas the cathodal isoforms are characterized by a high affinity and relative insensitivity to pH and organic phosphates. The presence of multiple hemoglobin isoforms is physiologically advantageous, especially in those fish displaying a Root effect, because blood oxygen transport is less likely to be impaired during acid-base disturbances. On the other hand, some species (e.g., carp) possess multiple hemoglobins, all of which display a pronounced Root effect (Brittain, 1987).

In summary, the quantity of oxygen taken up at the gills and delivered to the tissues is affected by several variables, including the oxygen partial pressure, the shape of the oxygen equilibrium curve, the blood oxygen-carrying capacity, and cardiac output (see Figure 1D). Thus, different strategies can be utilized by different species (interspecific differences) to achieve their oxygen transport requirements. Furthermore, each of these variables can be modified within a single species (intraspecific differences) to optimize oxygen transport during physical or environmental disturbances.

Interspecific Differences

Table 1 summarizes several important blood respiratory/cardiovascular variables in four teleost species. These data effectively illustrate two widely different strategies that are utilized by fish to transport oxygen within their blood (see Figure 5). The apparent strategy employed by the more active species, such as rainbow trout, is to utilize a relatively low-affinity hemoglobin (high P_{50}) in conjunction with high arterial PO_2 values. On the other hand, the more sluggish species, such as carp (*Cyprinus carpio*), flounder (*Platichthys stellatus*), and eel (*Anguilla* sp.), appear to utilize a relatively high-affinity hemoglobin (low P_{50}) in conjunction with low arterial PO_2 values. In each case the hemoglobin in the arterial blood is approximately 90 to 100% saturated with O_2 and arterial O_2 content is maintained relatively constant despite the very different arterial PO_2 values. Thus, the differences in arterial blood O_2 content between species is primarily determined by the blood O_2 carrying capacity. For example, in flounder (Table 1) the low arterial O_2 content is a function of the low hematocrit (approximately 14%) in this species (Wood et al., 1979).

TABLE 1
A Summary of Selected Cardiorespiratory Variables in Four Teleost Species; Rainbow Trout (*Oncorhynchus mykiss*), Carp (*Cyprinus carpio*), Flounder (*Platichthys stellatus*), and European Eel (*Anguilla anguilla*)

	Trout[a]	Carp[b]	Flounder[c]	Eel[d]
PaO_2 (torr)	133.2	23.2	34.9	49.1
PvO_2 (torr)	31.9	9.0	13.4	10.8
CaO_2 (ml 100 ml^{-1})	8.2	7.4	4.6	6.5
CvO_2 (ml 100 ml^{-1})	5.0	4.4	3.3	2.6
CaO_2-CvO_2 (ml 100 ml^{-1})	3.2	3.0	1.3	3.9
$\dot{V}b$ (ml min^{-1} kg^{-1})	18.3	24.6	39.2	11.5
P_{50} (torr)	22.9	7.3	8.6	11.1

PaO_2, Arterial oxygen tension; PvO_2, venous oxygen tension; CaO_2, arterial oxygen content; CvO_2, venous oxygen content; $\dot{V}b$, cardiac output; P_{50}, the PO_2 at 50% hemoglobin saturation.

[a] Data from Cameron and Davis (1970), except the P_{50} (Perry and Reid, 1992a).
[b] Data from Takeda (1990).
[c] Data from Wood et al. (1979).
[d] Data from Peyreaud-Waitzenegger and Soulier (1989), except the P_{50} (Perry and Reid, 1992a).

The amount of oxygen transported by the blood to the tissues is relatively constant at rest (assuming constant temperature) and is determined by the product of the arterial-venous O_2 content difference (C_aO_2 - C_vO_2) and the cardiac output ($\dot{V}b$) (see Figure 1D). In the flounder, the relatively low arterial-venous O_2 content difference is offset by an unusually high $\dot{V}b$ (Table 1). The latter strategy permits the flounder to deliver adequate oxygen to the tissues at high cardiac efficiency but an obvious consequence of such a strategy is that the ability to increase $\dot{V}b$ further will be compromised. The incapacity to substantially elevate $\dot{V}b$ may contribute to the relative inability of the flounder to perform exercise. Rainbow trout, on the other hand, operate at a relatively low $\dot{V}b$ and thus have a marked capacity to elevate blood flow further in the event of increased O_2 delivery requirements.

There are obvious advantages and disadvantages to the two basic strategies of blood oxygen transport described above (see Malte and Weber, 1987). High-affinity hemoglobin in combination with low arterial PO_2 is clearly advantageous for optimizing oxygen extraction efficiency from the water, because it assures that a large water-to-blood oxygen partial pressure gradient is sustained as blood flows through the gill vasculature. In other words, the lower the arterial PO_2, the greater will be the mean water-to-blood O_2 diffusion gradient (PO_2). This strategy, therefore, may allow a considerable reduction in the ventilatory requirement with an associated energetic benefit. A disadvantage of low arterial PO_2 is the associated reduction in the O_2 diffusion gradient

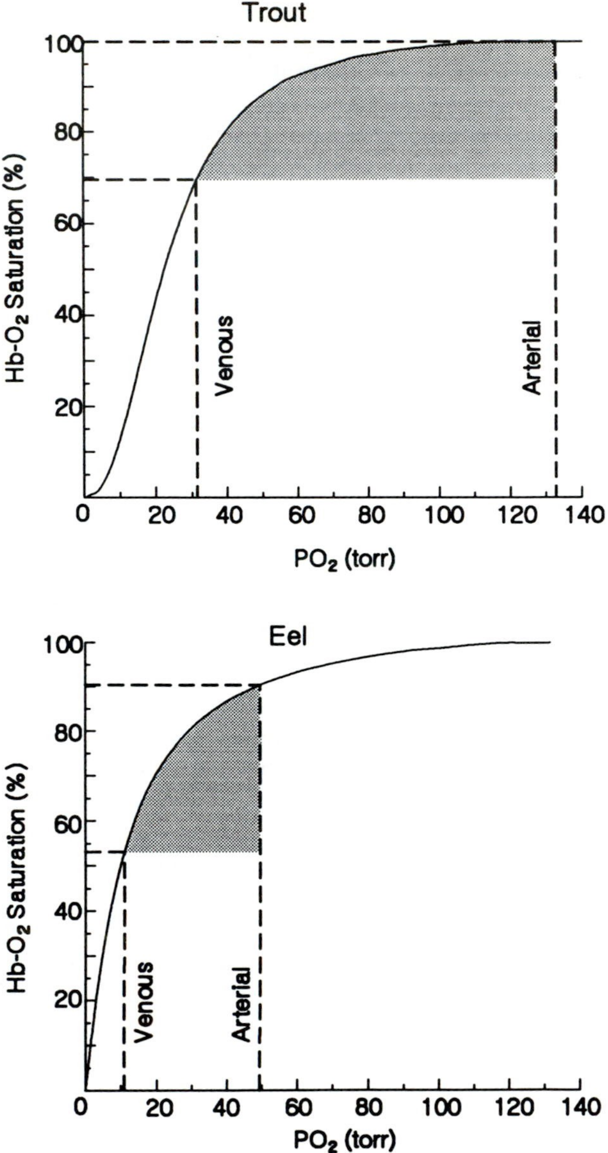

FIGURE 5. Representative oxygen dissociation curves for trout and eel that illustrate two different strategies for blood oxygen transport in active (e.g., trout) and sluggish (e.g., eel) species. The trout utilizes a high-affinity hemoglobin and consequently arterial PO_2 is high and arterial-venous O_2 content differences are reflected by large arterial-venous PO_2 differences. The eel utilizes a low-affinity hemoglobin and consequently arterial PO_2 is low and arterial-venous O_2 content differences are reflected by relatively small arterial-venous PO_2 differences. See text for further details.

between the blood and tissues. This strategy, therefore, is appropriate for the more sluggish fish that rarely perform exercise. In an active species such as rainbow trout, high arterial PO_2 values in concert with low-affinity hemoglobin permit a large O_2 diffusion gradient between the blood and tissues; this strategy is designed to optimize O_2 unloading. The disadvantage of such a strategy is the resultant low efficiency of O_2 extraction from the water and the accompanying high cost of ventilation. This is partially compensated for by the high degree of cooperativity (large N_{HILL}) of O_2 binding normally associated with low-affinity hemoglobins (Malte and Weber, 1987). In addition, rainbow trout and other active species possessing low-affinity hemoglobins normally inhabit well-aerated waters, which serves to enhance the initial water-to-blood O_2 diffusion gradient. Clearly, the particular strategy utilized by a given species will reflect, in part, its habitat and life-style. There is an obvious advantage to high-affinity hemoglobin and low arterial PO_2 so long as the O_2 diffusion gradients between the blood and tissues are not compromised.

Intraspecific Differences

Fish routinely encounter stressful situations, including a variety of environmental disturbances (e.g., hypoxia, hypercapnia) and elevation of physical activity. Such stresses often require that blood oxygen transport either be enhanced or optimized. At any particular PO_2 (set by gas transfer at the gill; see above), blood O_2 transport is a function of cardiac output, blood O_2 carrying capacity, and O_2-hemoglobin binding affinity/capacity (see Figure 1D). Changes in cardiac output that accompany physical/environmental perturbations have been dealt with elsewhere (see Chapter 8 by Farrell, this volume), and will not be considered further.

Modification of Blood Oxygen Carrying Capacity

In many fish species, the spleen acts as a storage reservoir for rbcs. Under appropriate conditions, including hypoxia (Yamamoto et al., 1985; Wells and Weber, 1990), hypercapnia (Perry and Kinkead, 1989), and exercise (Yamamoto et al., 1980, Yamamoto, 1987, 1988; Yamamoto and Itazawa, 1989; Wells and Weber, 1990; Pearson and Stevens, 1991) a contraction of the smooth muscle associated with the spleen can liberate the sequestered rbcs into the circulation and, thus, acutely elevate the blood O_2 carrying capacity (see also reviews by Wood and Perry, 1985; Perry and Wood, 1989; Wood, 1992). The mechanism underlying the contraction of the spleen is thought to involve stimulation of splenic α-adrenoceptors (Nilsson and Grove, 1974) either by direct innervation from the sympathetic nervous system or by circulating catecholamines (Perry and Kinkead, 1989). The importance of rbc release from the spleen during impairment of blood O_2 transport induced by blood acidosis (a function of the Root effect) was demonstrated using pharmacological techniques (Vermette and Perry, 1988a; Perry and Kinkead, 1989). In those studies, treatment of rainbow trout with the α-adrenoceptor antagonist phentolamine totally prevented (Perry and Kinkead, 1989) or significantly impaired (Vermette and Perry, 1988a) the regulation of blood O_2 transport during hypercapnic acidosis. The liberation of rbcs from the spleen can be induced by intra-arterial injections of catecholamines or synthetic α-adrenoceptor agonists (Vermette and Perry, 1988b; Perry and Kinkead, 1989). Splenectomy prevents the usual rise in blood hemoglobin levels after injection of catecholamines, indicating that the spleen is indeed the only significant source of additional rbcs during adrenergic stimulation. An alternate strategy for acutely increasing the O_2 carrying capacity of the blood is via hemoconcentration (Wood and Randall, 1973; Milligan and Wood, 1982, 1986; Yamamoto et al., 1985) owing to fluid shifts between the vascular and intracellular or external compartments. In either case, a consequence of the increased concentration of rbcs in the vascular fluid is an increase in the viscosity of the blood. Depending on the magnitude of the viscosity increase, there could be an associated detrimental impact on the cardiovascular system because of the increased cardiac work required to pump the more viscous blood. Interestingly, it was shown recently (Wells et al., 1991) that adrenergic swelling of the rbc (see below) may reduce the blood viscosity and thereby possibly offset increases in viscosity that otherwise would occur owing to the increased numbers.

During chronic stresses the blood O_2 carrying capacity additionally could be increased by a stimulation of erythropoiesis. Few studies have addressed this possibility, although Davie et al. (1986) demonstrated that long-term swimming (200 d) did not affect blood O_2 capacity in rainbow trout.

Modification of O_2-Hemoglobin Affinity/ Capacity

Under resting conditions H^+ ions are distributed passively across the rbc membrane according to a Donnan equilibrium. Thus, *in vitro*, rbc intracellular pH (pHi) varies as a linear function of extracellular pH (pHe; see Figure 6). The slope of this relationship varies among species and among the various studies that have been performed, with reported values ranging between approximately 0.3 and 0.9. Regardless, in the absence of regulatory processes, rbc pHi would be expected to reflect changes in pHe. Owing to this relationship as well as to the relationship between rbc pH and Hb-O_2 affinity (the Bohr effect) or capacity (the Root effect) it is clear that blood O_2 transport would be compromised during periods of extracellular acidosis in the absence of any cellular regulation. Several species of teleost, however, are capable of minimizing or totally preventing changes in rbc pHi during extracellular acidosis *in vivo* (e.g., Boutilier et al., 1986; Primmett et al. 1986; Milligan and Wood, 1987; Perry and Kinkead, 1989) by effectively uncoupling the usual pHe-pHi relationship (see Figure 6). Consequently, the potentially deleterious effects of acid-base disturbances on Hb-O_2 binding are minimized (e.g., see reviews by Perry and Wood, 1989; Randall, 1990; Jensen, 1991; Pelster and Weber, 1991).

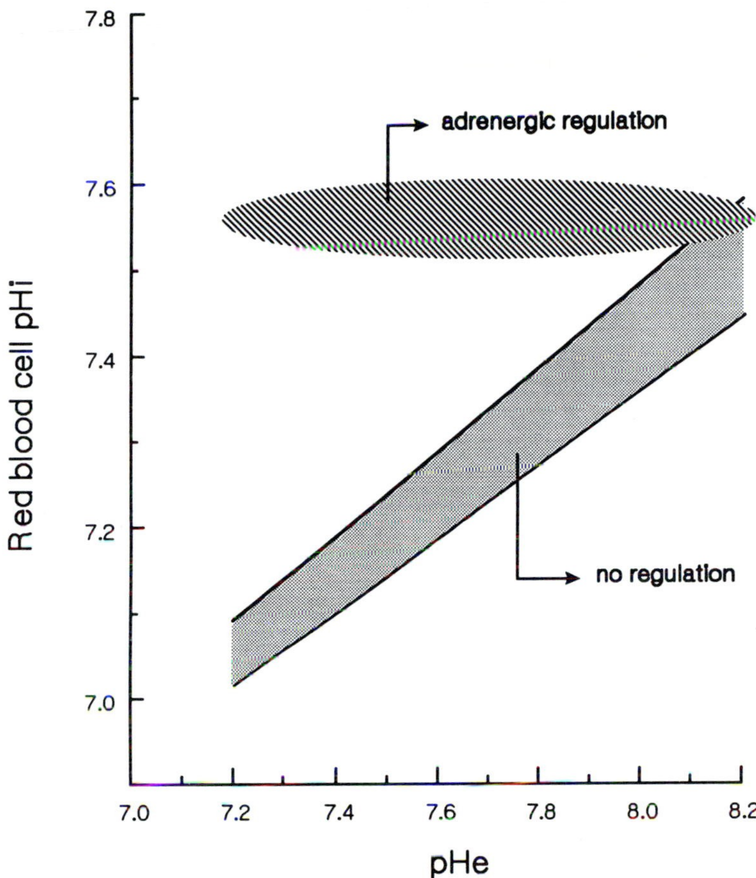

FIGURE 6. The relationship between whole blood extracellular pH (pHe) and red blood cell intracellular pH (pHi). *In vitro* or in the absence of adrenergic regulation *in vivo,* red blood cell pHi varies as a linear function of pHe (stippled area). In many teleost species, depression of pHe is accompanied by release of catecholamine into the circulation. The resultant stimulation of red blood cell Na^+/H^+ exchange can maintain rbc pHi relatively constant (adrenergic regulation) in the face of extracellular acidosis (hatched area).

The underlying mechanism of the apparent regulation of rbc pHi during extracellular acidosis is the adrenergic activation of rbc Na^+/H^+ exchange activity by acute elevation of circulating catecholamines (Baroin et al., 1984; Nikinmaa and Huestis, 1984; Cossins and Richardson, 1985; see also reviews by Boutilier and Ferguson, 1989; Nikinmaa and Tufts, 1989; Motais et al., 1990; Thomas and Motais, 1990; Fievet and Motais, 1991; Thomas and Perry, 1992). The series of events culminating in the extrusion of H^+ ions from the rbc is summarized in Figure 7. The elevation of circulating catecholamines originates from their release from chromaffin tissue. In teleosts the chromaffin tissue is most highly concentrated in the walls of the posterior cardinal vein at the level of the head kidney (see review by Randall and Perry, 1992). In addition, chromaffin cells are scattered throughout the kidney tissue, with the concentration of these cells often being highest in the anterior regions of the kidney. In elasmobranchs, the chromaffin tissue is located predominantly in the axillary bodies, while in cyclostomes and dipnoans, the heart is an important site of chromaffin tissue (see Nilsson, 1983). In teleosts the chromaffin tissue is innervated by sympathetic preganglionic cholinergic nerve fibers. Stimulation of these nerve fibers during periods of stress probably is the most significant mechanism contributing to the elevation of circulating catecholamines at such times (Randall and Perry, 1992), although the intermediary pathways are unknown. Indeed, the proximate stimulus for the release of catecholamines from the chromaffin tissue of the head kidney is also unclear, although both extracellular acidosis (Tang and Boutilier, 1988) and/or blood hypoxemia (Perry et al., 1989) are likely candidates. In addition to the established neural control of catecholamine release, it is also apparent that changes in the chemical status of the blood perfusing the chromaffin tissue can independently modulate the release of catecholamines. For example, elevation of plasma potassium levels (Opdyke et al., 1983) or regional hypoxemia (Perry et al., 1991b) can stimulate catecholamine release directly (see review by Randall and Perry, 1992).

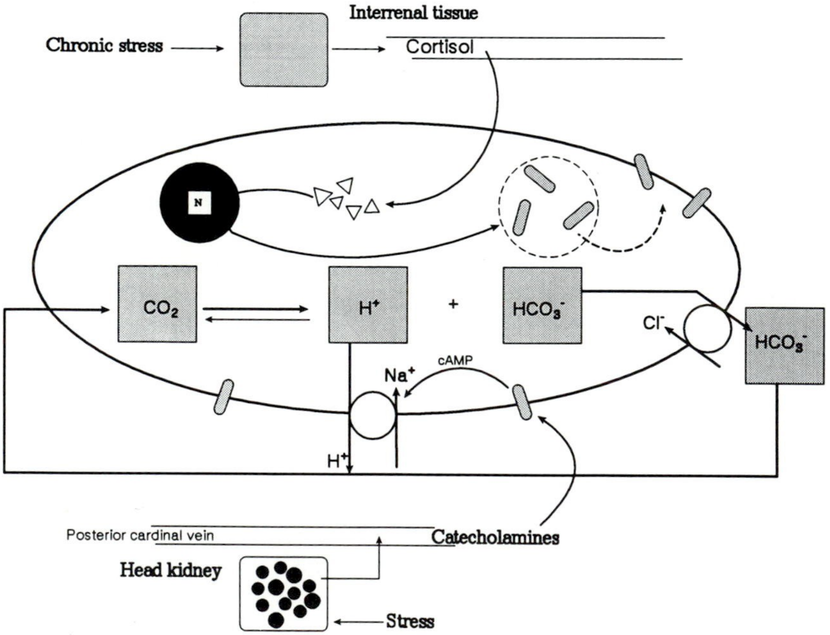

FIGURE 7. A model showing the effects of catecholamines on trout red blood cell Na^+/H^+ exchange. In response to acute stress, catecholamines are released from chromaffin tissue within the head kidney. The catecholamines interact with β_1-receptors (shaded ellipses) on the cell surface to initiate cAMP-dependent Na^+/H^+ exchange activity. Nascent cytosolic β-receptors can be recruited rapidly to the cell surface under periods of acute stress to enhance the adrenergic responsiveness of the red blood cell. N, Nucleus.

Once released into the circulation, the catecholamines stimulate the rbc Na^+/H^+ exchanger by binding to β-adrenoceptors on the plasma membrane. The catecholamine binding stimulates the production of intracellular cyclic AMP, a prerequisite for activation of Na^+/H^+ exchange. Upon activation, the exchanger extrudes H^+ ions into the plasma in exchange for Na^+ ions with a 1:1 stoichiometry; consequently, rbc pHi rises. Ultimately, the rise in rbc pHi is achieved despite the presence of a rapid Cl^-/HCO_3^- exchanger on the plasma membrane, because the rate of the Cl^-/HCO_3^- exchange at such times does not match the rate of the Na^+/H^+ exchange. This in turn reflects the slow uncatalyzed dehydration reaction of H^+ and HCO_3^- in the plasma which limits the inward flux of H^+ ions into the rbc via the Jacobs-Stewart cycle (see Motais et al., 1989; Nikinmaa et al., 1990).

In addition to the alkalinization of the rbc, adrenergic activation of the Na^+/H^+ exchanger promotes several other responses that are also known to enhance Hb-O_2 binding, including cell swelling and depletion of NTPs (Nikinmaa, 1986; Boutilier and Ferguson, 1989; Nikinmaa and Tufts, 1989). Adrenergic swelling of the rbc arises from the osmotic entry of water owing to the accumulation of Na^+ (and to a lesser extent Cl^-) within the cell after activation of Na^+/H^+ exchange. Cell swelling, per se, can elevate rbc pHi, simply by altering the Donnan distribution of H^+ across the rbc membrane, as the fixed negative charges on impermeable proteins (predominantly organic phosphates and hemoglobin) are diluted. A rapid and important consequence of adrenergic activation of red blood cell Na^+/H^+ exchange is depletion of NTPs. The decrease in NTP levels is thought to result from the additional energetic demands of the Na^+/K^+ pump as Na^+ levels in the cell rise (Ferguson and Boutilier, 1989). Thus, Hb-O_2 affinity/capacity are enhanced after adrenergic activation of rbc Na^+/H^+ exchange by the combined effects of alkalinization and reduced NTP levels.

Exposure of several fish species to hypoxia has been shown to cause adaptive increases in Hb-O_2 affinity (reduction of P_{50}; Wood and Johansen, 1973; Greaney and Powers, 1978; Soivio et al., 1980; Nikinmaa and Weber, 1984; Tetens and Lykkeboe, 1985; Tetens and Christensen, 1987; Claireaux et al., 1988). Several mechanisms contribute to the increased Hb-O_2 affinity although not all fish appear to use the same mechanisms. During acute hypoxia, the respiratory alkalosis caused by hyperventilation contributes to an elevation of rbc pHi and hence a decrease in the P_{50}. Depending on the severity of the hypoxia, elevations of plasma catecholamine levels can also contribute to the decrease in P_{50} (Tetens and Christensen, 1987) by its concerted effects on rbc pHi and NTP levels (see above). In hypoxic rainbow trout, Tetens and Christensen (1987) showed that the dominant effect of catecholamines on reducing the P_{50} was by raising rbc pHi rather than by altering NTP levels. On the other hand, the study of Boutilier et al., (1988) demonstrated a marked reduction in rbc NTP levels during hypoxia without a concomitant change in the P_{50}. Addition of catecholamines to blood, *in vitro*, also is known to reduce the P_{50}. In trout, the response appears to be mediated by changes in both pHi and NTP levels (Nikinmaa, 1983; Milligan and Wood, 1987) while in the mummichog (*Fundulus heteroclitus*) the response appears to be mediated solely by modulation of NTP levels (Dalessio et al., 1991). Clearly, further studies are required before any general conclusions can be formulated. During chronic hypoxia the reduction in P_{50} is often associated with reduced levels of NTP in the rbc, perhaps as a consequence of the reduced oxidative metabolism (Greaney and Powers, 1978). In the lamprey (*Lampetra fluviatilis*), the hypoxia-mediated increase in $Hb-O_2$ affinity is not associated with any changes in NTP levels but instead results from cell swelling and changes in pHi (Nikinmaa and Weber, 1984). The deoxygenation of hemoglobin, itself, during hypoxia also may contribute to the alkalinization of the rbc via the Haldane effect (i.e., hemoglobin behaves as a weaker acid when deoxygenated; Jensen, 1986).

The importance of catecholamine-mediated effects on rbc physiology are dependent upon species and the acid-base/respiratory status of the blood at the moment of catecholamine mobilization (see review by Thomas and Perry, 1992). Elasmobranch rbcs appear to lack appreciable β-adrenergic Na^+/H^+ exchange activity. It has been suggested that the absence of adrenergic Na^+/H^+ exchange in the elasmobranchs may reflect the absence of a significant Root effect in this group (Tufts and Randall, 1989). Relatively few species have been examined but, nevertheless, it is apparent that within the teleosts, salmonid rbcs display the greatest responsiveness to catecholamines. The rbcs of several species, including tench (*Tinca tinca*), carp (*C. carpio*), and eel (*A. rostrata*) display attenuated, if not entirely absent, adrenergic Na^+/H^+ exchange activity. These interspecific differences may reflect differences in one or more of the many steps in the β-adrenergic signal transduction pathway. Interestingly, a recent study (Perry and Reid, 1992b) demonstrated that the attenuated response of American eel rbcs to catecholamines was not related to insufficient cell surface β-adrenoceptors or uncoupling of these receptors from cyclic AMP production. Instead, it was suggested that the numbers of Na^+/H^+ exchangers, themselves, on the plasma membrane may be the limiting factor. Clearly, this too is an area that warrants further research.

Extracellular acidosis markedly enhances the activity of adrenergic Na^+/H^+ exchange with optimal activity occurring at an extracellular pH of approximately 7.3 (Borgese et al., 1987; Nikinmaa et al., 1987; Cossins and Kilbey, 1989). Blood hypoxemia also markedly enhances adrenergic Na^+/H^+ exchange activity (Motais et al., 1987; Salama and Nikinmaa, 1988, 1990; Reid and Perry, 1991; Perry and Reid, 1992b). The mechanisms underlying the greater responsiveness of hypoxic blood have not been established unequivocally, although it would appear that an important mechanism is the rapid mobilization of nascent cytosolic β-adrenoceptors to the cell surface, where they become functionally coupled to adenylate cyclase (Marttila and Nikinmaa, 1988; Reid and Perry, 1991; Perry and Reid, 1992b; Reid et al., 1993).

Carbon Dioxide

Basic Principles

Carbon dioxide transport in fish blood was reviewed most recently by Perry (1986) and Perry and Laurent (1990). The total amount of CO_2 transported in the blood is the sum of physically dissolved CO_2 and "chemically bound CO_2". Physically dissolved CO_2 is simply proportional to the PCO_2 and the solubility coefficient of CO_2 in the plasma (see Boutilier et al., 1984). At physiological PCO_2 levels, physically dissolved CO_2 constitutes only a minor fraction (generally less than 5%) of the total carbon dioxide carried in the blood. Chemically bound CO_2 is the difference between total CO_2 and physically dissolved CO_2 and is equivalent to the HCO_3^- concentration (CO_3^{2-} is negligible at physiological pH) plus any CO_2 bound to plasma proteins or hemoglobin (termed carbamino CO_2). Carbamino CO_2 is relatively unimportant in fish, perhaps owing to the acetylation of the terminal α-amino groups on the alpha chains of hemoglobin. Thus, the largest fraction of total CO_2 is carried in the form of HCO_3^- and generally constitutes about 90 to 95% of the total CO_2 in the blood. HCO_3^- is carried both in the plasma and the rbc, with the largest fraction residing in the plasma.

Carbon dioxide produced by aerobic metabolism diffuses from the tissues into the plasma. During the period of transit within the venous system the CO_2 is converted to HCO_3^- within the rbc, a reaction which is catalyzed by the enzyme carbonic anhydrase. The HCO_3^- thus formed exits the rbc in exchange for plasma Cl^- (the chloride shift). As blood flows through the gill vasculature plasma HCO_3^- enters the rbc in exchange for intracellular Cl^- and is rapidly dehydrated to CO_2 in the presence of carbonic anhydrase. The H^+ ions for the dehydration reaction are supplied by buffer groups on hemoglobin during oxygenation. The CO_2 thus formed diffuses into the plasma and then into the ventilatory water after traversing the gill epithelium. Owing to the integral role of the rbc in CO_2 transport and excretion, a reduction in the hematocrit can impair CO_2 excretion. In the absence of changes in cardiac output, CO_2 excretion is linearly related to hematocrit in rainbow trout (Perry et al., 1982). Even though the rbc Cl^-/HCO_3^- exchanger is rapid, it is nevertheless considered to be the rate-limiting step in carbon dioxide excretion (Perry, 1986).

Unlike the mammalian lung, the fish gill endothelium lacks carbonic anhydrase (see review by Perry and Laurent, 1990), thus the rbc is the sole significant site of HCO_3^- dehydration. An additional site of HCO_3^- dehydration may not be required in fish owing to the relatively long transit time of blood in the gill vasculature (0.5 to 3.0 s), the high capacitance of the water for CO_2, and the efficient countercurrent system of gas transfer across the gill (see above). Alternatively, it is quite conceivable that CO_2 is *not* in equilibrium in the arterial blood leaving the gills as a result of incomplete HCO_3^- dehydration during gill transit (D.J. Randall, personal communication). Theoretically, an extracellular site of HCO_3^- dehydration could be disadvantageous because it would prevent elevation of rbc pHi after adrenergic stimulation of the rbc Na^+/H^+ exchanger (see Nikinmaa et al., 1990) and might also impede the adaptive accumulation of plasma HCO_3^- during internal acidosis.

Adrenergic Effects on Blood CO_2 Transport

The adrenergic activation of rbc Na^+/H^+ exchange causes a large extracellular acidosis owing to the extrusion of H^+ ions. The titration of plasma HCO_3^- by H^+ at the uncatalyzed rate in the closed venous system, by theory, will elevate venous blood PCO_2. This may partially explain the elevation of arterial blood PCO_2 after release of endogenous or after injection of exogenous catecholamines in rainbow trout (Perry and Thomas, 1991). In addition, adrenergic activation of rbc Na^+/H^+ exchange would be expected to impair CO_2 excretion transiently during the initial disequilibrium phase of the response (see review by Thomas and Perry, 1992) when the normal outwardly directed PCO_2 gradient between the red blood cell and plasma is temporarily reversed. Furthermore, an elevation of intracellular HCO_3^- associated with the alkalinization would be expected to retard the inward movement of HCO_3^- and thus impede overall CO_2 excretion. There is substantial evidence that rbc CO_2 production from plasma HCO_3^- is reduced *in vitro* in the presence of catecholamines (Perry et al., 1991b; Wood and Perry, 1991), although supportive evidence *in vivo* is still lacking (Steffensen et al., 1987; Playle et al., 1990).

REFERENCES

Baroin, A., Garcia-Romeu, F., Lamarre, T., and Motais, R., A transient sodium-hydrogen exchange system induced by catecholamines in erythrocytes of rainbow trout, *Salmo gairdneri, J. Physiol. (London)*, 356, 21, 1984.

Borgese, F., Garcia-Romeu, F., and Motais, R., Ion movements and volume changes induced by catecholamines in erythrocytes of rainbow trout: effect of pH, *J. Physiol. (London)*, 382, 145, 1987.

Booth, J. L., The distribution of blood flow in gills of fish: application of a new technique to rainbow trout *(Salmo gairdneri), J. Exp. Biol.*, 73, 119, 1978.

Boutilier, R. G. and Ferguson, R. A., Nucleated red cell function: metabolism and pH regulation, *Can. J. Zool.*, 67, 2986, 1989.

Boutilier, R. G., Heming, T. A. and Iwama, G. K., Physiochemical parameters for use in fish respiratory physiology, in *Fish Physiology*, Vol. 10A, Hoar, W. S. and Randall, D. J., Eds., Academic Press, New York, 1984, 403.

Boutilier, R. G., Iwama, G. K., and Randall, D. J., The promotion of catecholamine release in rainbow trout, *Salmo gairdneri*, by acute acidosis: interaction between red cell pH and haemoglobin oxygen carrying-capacity, *J. Exp. Biol.*, 123, 145, 1986.

Boutilier, R. G., Dobson, G. P., Hoeger, U., and Randall, D. J., Acute exposure to graded levels of hypoxia in rainbow trout (Salmo gairdneri): metabolic and respiratory adaptations, *Respir. Physiol.*, 71, 69, 1988.

Brett, J. R., The metabolic demand for oxygen in fish, particularly salmonids, and a comparison with other vertebrates, *Respir. Physiol.*, 14, 151, 1972.

Brittain, T., The root effect, *Comp. Biochem. Physiol. B*, 86, 473, 1987.

Butler, P. J. and Metcalfe, J. D., *Physiology of Elasmobranch Fishes*, Springer-Verlag, Berlin, 1988.

Cameron, J. N., *The Respiratory Physiology of Animals*, Oxford University Press, New York, 1989.

Cameron, J. N. and Davis, J. C., Gas exchange in rainbow trout *(Salmo gairdneri)* with varying blood oxygen capacity, *J. Fish Res. Board Can.*, 27, 1069, 1970.

Claireaux, G., Thomas, S., Fievet, B. and Motais, R., Adaptive respiratory responses of trout to acute hypoxia. II. Blood oxygen carrying properties during hypoxia, *Respir. Physiol.*, 74, 91, 1988.

Cossins, A. R. and Kilbey, V. K., The seasonal modulation of Na^+/H^+ exchanger activity in trout erythrocytes, *J. Exp. Biol.*, 144, 463, 1989.

Cossins, A. R. and Richardson, P. A., Adrenalin-induced Na^+/H^+ exchange in trout erythrocytes and its effects upon oxygen-carrying capacity, *J. Exp. Biol.*, 88, 229, 1985.

Dandy, J. W. T., Activity response to oxygen in brook trout, *Salvelinus fontinalis* (Mitchill), *Can. J. Zool.*, 48, 1067, 1970.

Dalessio, P. M., DiMichele, L., and Powers, D. A., Adrenergic regulation of erythrocyte oxygen affinity, pH, and nucleoside triphosphate/hemoglobin ratio in the mummichog, *Fundulus heteroclitus*, *Physiol. Zool.*, 64, 1391, 1991.

Davie, P. S., Wells, R. M. G., and Tetens, V., Effects of sustained swimming on rainbow trout muscle structure, blood oxygen transport, and lactate dehydrogenase isozymes: evidence for increased aerobic capacity of white muscle, *J. Exp. Zool.*, 237, 159, 1986.

Farrell, A. P., Gill morphometrics, vessel dimensions, and vascular resistance in ling cod, *Ophiodon elongatus*, *Can. J. Zool.*, 58, 807, 1980.

Ferguson, R. A. and Boutilier, R. G., Metabolic-membrane coupling in red blood cells of trout: the effects of anoxia and adrenergic stimulation, *J. Exp. Biol.*, 143, 149, 1989.

Fievet, B. and Motais, R., Na^+/H^+ exchanges and red blood cell functions in fish, in *Advances in Comparative and Environmental Physiology*, Vol. 8, Gilles, R. et al., Eds., Springer-Verlag, Berlin, 1991.

Gonzalez, R. J. and McDonald, D. G., The relationship between oxygen consumption and ion loss in a freshwater fish, *J. Exp. Biol.*, 163, 317, 1992.

Gonzalez, R. J. and McDonald, D. G., Differences in the relationship between ion loss and oxygen uptake among fish from diverse habitats, *J. Exp. Biol.*, submitted.

Greaney, G. S. and Powers, D. A., Allosteric modifiers of fish hemoglobins: in vitro and in vivo studies of the effect of ambient oxygen and pH on erythrocyte ATP concentrations, *J. Exp. Zool.*, 203, 339, 1978.

Heisler, N., Interactions between gas exchange, metabolism, and ion transport in animals: an overview, *Can. J. Zool.*, 67, 2923, 1989.

Holeton, G. F., Oxygen uptake and circulation by a hemoglobinless Antarctic fish *(Chaenocephalus aceratus* Lonnberg) compared with three red-blooded Antarctic fish, *Comp. Biochem. Physiol.*, 34, 457, 1970.

Hughes, G. M., A comparative study of gill ventilation in marine teleosts, *J. Exp. Biol.*, 37, 28, 1960.

Hughes, G. M., The dimensions of fish gills in relation to their function, *J. Exp. Biol.*, 45, 177, 1966a.

Hughes, G. M., Morphometrics of fish gills, *Respir. Physiol.*, 14, 1, 1966b.

Hughes, G. M., Effects of low oxygen and pollution on the respiratory systems of fish, in *Stress and Fish*, Pickering, A. D., Ed., Academic Press, New York, 1981, 121.

Hughes, G. M., General anatomy of the gills, in *Fish Physiology*, Vol. 10A, Hoar, W. S. and Randall, D. J., Eds., Academic Press, New York, 1984, 1.

Hughes, G. M. and Morgan, M., The structure of fish gills in relation to their respiratory function, *Biol. Rev.*, 48, 419, 1973.

Hughes, G. M., Dube, S. C. and Datta Munshi, J. S., Surface area of the respiratory organs of the climbing perch, *Anabas testudineus* (Pisces: Anabantidae), *J. Zool (London)*, 170, 227, 1973.

Jensen, F. B., Pronounced influence of Hb-O_2 saturation on red cell pH in tench blood *in vivo* and *in vitro, J. Exp. Zool.*, 238, 119, 1986.

Jensen, F. B., Multiple strategies in oxygen and carbon dioxide transport by haemoglobin, in *Physiological Strategies for Gas Exchange and Metabolism*, Society for Experimental Biology Seminar Ser., Woakes, A. J., Grieshaber, M. K., and Bridges, C. R., Eds., Cambridge University Press, Cambridge, 1991, 55.

Johansen, K., Respiratory gas exchange of vertebrate gills, in *Gills,* Vol. 16, Society for Experimental Biology Seminar Ser., Houlihan, D. F., Rankin, J. C. and Shuttleworth, T. J., Eds., Cambridge University Press, Cambridge, 1982, 99.

Jones, D. R. and Randall, D. J., The respiratory and circulatory systems during exercise in fish, in *Fish Physiology,* Vol. 7, Hoar, W. S. and Randall, D. J., Eds., Academic Press, New York, 1978, 425.

Kirsch, R. and Nonnote, G., Cutaneous respiration in three freshwater teleosts, *Respir. Physiol.*, 29, 339, 1977.

Krogh, A., *The Comparative Physiology of Respiratory Mechanisms*, University of Pennsylvania Press, Philadelphia, 1941.

Malte, H. and Weber, R. E., The effect of shape and position of the oxygen equilibrium curve on extraction and ventilation requirement in fishes, *Respir. Physiol.*, 70, 221, 1987.

McDonald, D. G., The effects of H^+ upon the gills of freshwater fish, *Can. J. Zool.*, 63, 691, 1983.

McDonald, D. G. and McMahon, B. R., Respiratory development in Arctic char *Salvelinus alpinus* under conditions of normoxia and chronic hypoxia, *Can. J. Zool.*, 55, 1461, 1977.

Marttila, O. N. T. and Nikinmaa, M., Binding of β-adrenergic agonists ^{3}H-DHA and ^{3}H-CGP 12177 to intact rainbow trout *(Salmo gairdneri)* and carp *(Cyprinus carpio)* red blood cells, *Gen. Comp. Endocrinol.*, 70, 429, 1988.

Milligan, C. L. and Wood, C. M., Disturbances in haematology, fluid volume distribution and circulatory function associated with low environmental pH in the rainbow trout, *Salmo gairdneri, J. Exp. Biol.*, 99, 397, 1982.

Milligan, C. L. and Wood, C. M., Tissue intracellular acid-base status and the fate of lactate after exhaustive exercise in the rainbow trout, *J. Exp. Biol.*, 123, 123, 1986.

Milligan, C. L. and Wood, C. M., Regulation of blood oxygen transport and red cell pHi after exhaustive exercise in rainbow trout *(Salmo gairdneri)* and starry flounder *(Platichthys stellatus), J. Exp. Biol.*, 133, 263, 1987.

Milsom, W. K., Mechanisms of ventilation in lower vertebrates: adaptations to respiratory and non-respiratory constraints, *Can. J. Zool.*, 67, 2943, 1989.

Motais, R., Garcia-Romeu, F. and Borgese, F., The control of Na^+/H^+ exchange by molecular oxygen in trout erythrocytes, *J. Gen. Physiol.*, 90, 197, 1987.

Motais, R., Fievet, B., Garcia-Romeu, F. and Thomas, S., Na^+-H^+ exchange and pH regulation in red blood cells: role of uncatalyzed H_2CO_3 dehydration, *Am. J. Physiol.*, 256, C728, 1989.

Motais, R., Scheuring, U., Borgese, F. and Garcia-Romeu, F., Characteristics of β-adrenergic-activated Na-proton transport in red blood cells, in *Progress in Cell Research,* Vol. 1, Ritchie, J. M., Magistretti, P. J., and Bolis, L., Eds., Elsevier, Amsterdam, 1990, 179.

Muir, B. S. and Hughes, G. M., Gill dimensions for three species of tuna, *J. Exp. Biol.*, 51, 271, 1969.

Nikinmaa, M., Adrenergic regulation of haemoglobin oxygen affinity in rainbow trout red cells, *J. Comp. Physiol. B*, 152, 67, 1983.

Nikinmaa, M., Control of red cell pH in teleost fishes, *Ann. Zool. Fenn.*, 23, 223, 1986.

Nikinmaa, M. and Huestis, W. H., Adrenergic swelling of nucleated erythrocytes: cellular mechanisms in a bird, domestic goose, and two teleosts, striped bass and rainbow trout, *J. Exp. Biol.*, 113, 215, 1984.

Nikinmaa, M. and Tufts, B. L., Regulation of acid and ion transfer across the membrane of nucleated erythrocytes, *Can. J. Zool.*, 67, 3039, 1989.

Nikinmaa, M. and Weber, R. E., Hypoxic acclimation in the lamprey, *Lampetra fluviatilis*: organismic and erythrocytic responses, *J. Exp. Biol.*, 109, 109, 1984.

Nikinmaa, M., Steffensen, J. F., Tufts, B. L., and Randall, D. J., Control of red cell volume and pH in trout: effects of isoproterenol, transport inhibitors, and extracellular pH in bicarbonate/carbon dioxide-buffered media, *J. Exp. Zool.*, 242, 273, 1987.

Nikinmaa, M., Tiihonen, K., and Paajaste, M., Adrenergic control of red cell pH in salmonid fish: roles of the sodium/proton exchange, Jacobs-Stewart cycle and membrane potential, *J. Exp. Biol.*, 154, 257, 1990.

Nilsson, S., *Zoophysiology*, Vol. 13, Autonomic Nerve Function in the Vertebrates, Springer-Verlag, New York, 1983.

Nilsson, S., Control of gill blood flow, in *Fish Physiology: Recent Advances*, Nilsson, S. and Holmgren, S., Eds., Croom Helm, London, 1986, 87.

Nilsson, S. and Grove, D. J., Adrenergic and cholinergic innervation of the spleen of the cod, *Gadus morhua, Eur. J. Pharmacol.*, 28, 135, 1974.

Opdyke, D. F., Bullock, J., Keller, N. E. and Holmes, K., Dual mechanism for catecholamine secretion in the dogfish shark Squalus acanthias, *Am. J. Physiol.*, 244, R641, 1983.

Pearson, M. P. and Stevens, E. D., Size and hematological impact of the splenic erythrocyte reservoir in rainbow trout, *Oncorhynchus mykiss, Fish Physiol. Biochem.*, 9, 39, 1991.

Pelster, B. and Weber, R. E., The physiology of the Root effect, in *Advances in Comparative and Environmental Physiology,* Vol. 8, Gilles, R. et al., Eds., Springer-Verlag, Berlin, 1991, 51.

Perry, S. F., Carbon dioxide excretion in fish, *Can. J. Zool.*, 64, 565, 1986.

Perry, S. F. and Kinkead, R., The role of catecholamines in regulating arterial oxygen content during hypercapnic acidosis in rainbow trout *(Salmo gairdneri), Respir. Physiol.*, 77, 365, 1989.

Perry, S. F. and Laurent, P., The role of carbonic anhydrase in carbon dioxide excretion, acid-base balance and ionic regulation in aquatic gill breathers, in *Transport, Respiration and Excretion: Comparative and Environmental Aspects*, Truchot, J.-P. and Lahlou, B., Eds., S. Karger, Basel, 1990, 39.

Perry, S. F. and Reid, S. D., Relationships between blood oxygen content and catecholamine levels during hypoxia in rainbow trout and American eel, *Am. J. Physiol.*, 263, R240, 1992a.

Perry, S. F. and Reid, S. D., The relationships between β-adrenoceptors and adrenergic responsiveness in trout (*Oncorhynchus mykiss*) and eel *(Anguilla rostrata)* erythrocytes, *J. Exp. Biol.*, 167, 235, 1992b.

Perry, S. F. and Thomas, S., The effects of endogenous or exogenous catecholamines on blood respiratory status during acute hypoxia in rainbow trout *(Oncorhynchus mykiss), J. Comp. Physiol. B.*, 161, 489, 1991.

Perry, S. F. and Wood, C. M., Control and coordination of gas transfer in fishes, *Can. J. Zool.*, 67, 2961, 1989.

Perry, S. F., Davie, P. S., Daxboeck, C., and Randall, D. J., A comparison of CO_2 excretion in a spontaneously ventilating blood-perfused trout preparation and saline-perfused gill preparations: contribution of the branchial epithelium and red blood cell, *J. Exp. Biol.*, 101, 47, 1982.

Perry, S. F., Kinkead, R., Gallaugher, P., and Randall, D. J., Evidence that hypoxemia promotes catecholamine release during hypercapnic acidosis in rainbow trout *(Salmo gairdneri), Respir. Physiol.*, 77, 351, 1989.

Perry, S. F., Fritsche, R., Kinkead, R., and Nilsson, S., Control of catecholamine release in vivo and in situ in the Atlantic cod (Gadus morhua) during hypoxia, *J. Exp. Biol.*, 155, 549, 1991a.

Perry, S. F., Wood, C. M., Thomas, S. and Walsh, P. J., Adrenergic inhibition of bicarbonate dehydration through trout erythrocytes is mediated by activation of Na^+/H^+ exchange, *J. Exp. Biol.*, 157, 367, 1991b.

Petterson, K., Adrenergic control of oxygen transfer in perfused gills of the cod, *Gadus morhua, J. Exp. Biol.*, 102, 327, 1983.

Peyreaud-Waitzenegger, M. and Soulier, P., Ventilatory and cardiovascular adjustments in the European eel *(Anguilla anguilla)* exposed to short term hypoxia, *Exp. Biol.*, 48, 107, 1989.

Piiper, J., Factors affecting gas transfer in respiratory organs of vertebrates, *Can. J. Zool.*, 67, 2956, 1989.

Piiper, J. and Scheid, P., Physical principles of respiratory gas exchange in fish gills, in *Gills*, Vol. 16, Society for Experimental Biology Seminar Ser., Houlihan, D. F., Rankin, J. C., and Shuttleworth, T. J., Eds., Cambridge University Press, Cambridge, 1982, 45.

Playle, R. C., Munger, R. S., and Wood, C. M., Effects of catecholamines on gas exchange and ventilation in rainbow trout *(Salmo gairdneri), J. Exp. Biol.*, 152, 353, 1990.

Priede, I. G., Metabolic scope in fishes, in *Fish Energetics: New Perspectives*, Tytler, P. and Calow, P., Eds., The Johns Hopkins University Press, Baltimore, 1985, 33.

Priede, I. G. and Holliday, F. G. T., The use of a new tilting respirometer to investigate some aspects of metabolism and swimming activity of the plaice *(Pleuronectes platessa L.), J. Exp. Biol.*, 85, 295, 1980.

Primmett, D. R. N., Randall, D. J., Mazeaud, M., and Boutilier, R. G., The role of catecholamines in erythrocyte pH regulation and oxygen transport in rainbow trout *(Salmo gairdneri)* during exercise, *J. Exp. Biol.*, 122, 139, 1986.

Randall, D. J., The control of respiration and circulation in fish during exercise and hypoxia, *J. Exp.Biol.*, 100, 275, 1982.

Randall, D. J., Control and co-ordination of gas exchange in water breathers, in *Advances in Comparative and Environmental Physiology*, Boutilier, R. G., Ed., Springer-Verlag, Berlin, 1990, 253.

Randall, D. J. and Daxboeck, C., Oxygen and carbon dioxide transfer across fish gills, in *Fish Physiology*, Hoar, W. S. and Randall, D. J., Eds., Academic Press, New York, 1984, 263.

Randall, D. J. and Perry, S. F., Catecholamines, in *Fish Physiology*, Vol. 12, *The Cardiovascular System*, Randall, D. J. and Hoar, W. S., Eds., Academic Press, New York, in press.

Randall, D. J., Baumgarten, D., and Malyusz, M., The relationship between gas and ion transfer across the gills of fishes, *Comp. Biochem. Physiol. A*, 41, 629, 1972.

Randall, D. J., Perry, S. F., and Heming, T. A., Gas transfer and acid-base regulation in salmonids, *Comp. Biochem. Physiol. B*, 73, 93, 1982.

Reid, S. D. and Perry, S. F., The effects and physiological consequences of elevated cortisol on rainbow trout *(Oncorhynchus mykiss)* β-adrenoceptors, *J. Exp. Biol.*, 158, 217, 1991.

Reid, S. D., LeBras, Y. M., and Perry, S. F., The *in vivo* effects of hypoxia on the trout *(Oncorhynchus mykiss)* erythrocyte β-adrenergic signal transduction system, *J. Exp. Biol.*, 176, 103, 1993.

Riggs, A. F., The Bohr effect, *Annu. Rev. Physiol.*, 50, 181, 1988.

Roberts, J. L., Respiratory adaptations of aquatic animals, in *Physiological Adaptation to the Environment*, Vernberg, F. J., Ed., Intext Educational Publishers, New York, 1975, 395.

Roberts, J. L. and Rowell, D. M., Periodic respiration of gill breathing fishes, *Can. J. Zool.*, 66, 182, 1988.

Root, R. W., The respiratory function of the blood of marine fishes, *Biol. Bull. Mar. Biol. Lab. Woods Hole*, 61, 427, 1931.

Salama, A. and Nikinmaa, M., The adrenergic responses of carp *(Cyprinus carpio)* red cells: effects of PO_2 and pH, *J. Exp. Biol.*, 136, 405, 1988.

Salama, A. and Nikinmaa, M., Effect of oxygen tension on catecholamine-induced formation of cAMP and on swelling of carp red blood cells, *Am. J. Physiol.*, 259, C723, 1990.

Satchell, G. H., Respiratory toxicology of fishes, in *Aquatic Toxicology*, Vol. 2, Weber, L. J., Ed., Raven Press, New York, 1984, 1.

Saunders, R. L., The irrigation of the gills of fishes. II. Efficiency of O_2 uptake in relation to respiratory flow, activity and concentrations of O_2 and CO_2, *Can. J. Zool.*, 40, 817, 1962.

Schumann, D. and Piiper, J., Der sauerstoffbedarf der atmung bei fischen nach messungen au dernarkotisierten schleie *(Tinca tinca), Arch. Gesamte. Physiol.*, 288, 14, 1966.

Stevens, E. D., Some aspects of gas exchange in tuna, *J. Exp. Biol.*, 56, 809, 1972.

Stevens, E. D. and Neill, W. H., Body temperature relations of tunas, especially skipjack, in *Fish Physiology*, Vol. 7, Hoar, W. S. and Randall, D. J., Eds., Academic Press, New York, 1978, 316.

Soivio, A., Nikinmaa, M., and Westman, K., The blood oxygen binding properties of hypoxic *Salmo gairdneri, J. Comp. Physiol.*, 136, 83, 1980.

Steffensen, J. F., Tufts, B. L. and Randall, D. J., Effect of burst swimming and adrenaline infusion on O_2 consumption and CO_2 excretion in rainbow trout, *Salmo gairdneri, J. Exp. Biol.*, 131, 427, 1987.

Takeda, T., Ventilation, cardiac output and blood respiratory parameters in the carp, *Cyprinus carpio*, during hyperoxia, *Respir. Physiol.*, 81, 227, 1990.

Tang, Y. and Boutilier, R. G., Correlation between catecholamine release and degree of acidotic stress in rainbow trout, *Salmo gairdneri, Am. J. Physiol.*, 255, R395, 1988.

Tetens, V. and Christensen, N. J., Beta-adrenergic control of blood oxygen affinity in acutely hypoxia exposed rainbow trout, *J. Comp. Physiol. B*, 157, 667, 1987.

Tetens, V. and Lykkeboe, G., Acute exposure of rainbow trout to mild and deep hypoxia: O_2 affinity and O_2 capacitance of arterial blood, *Respir. Physiol.*, 61, 221, 1985.

Thomas, S. and Motais, R., Acid-base balance and oxygen transport during acute hypoxia in fish, *Comp. Physiol.*, 6, 76, 1990.

Thomas, S. and Perry, S. F., Control and consequences of adrenergic activation of red blood cell Na^+/H^+ exchange on blood oxygen and carbon dioxide transport, *J. Exp. Zool.*, 263, 160, 1992.

Tufts, B.L. and Randall, D. J., The functional significance of adrenergic pH regulation in fish erythrocytes, *Can. J. Zool.*, 67, 235, 1989.

Vermette, M. G. and Perry, S. F., Adrenergic involvement in blood oxygen transport and acid-base balance during hypercapnic acidosis in the rainbow trout, *Salmo gairdneri, J. Comp. Physiol. B*, 158, 107, 1988a.

Vermette, M. G. and Perry, S. F., Effects of prolonged epinephrine infusion on blood respiratory and acid-base states in the rainbow trout: alpha and beta effects, *Fish Physiol. Biochem.*, 4, 189, 1988b.

Weber, R. E. and Jensen, F. B., Functional adaptations in hemoglobins from ectothermic vertebrates, *Annu. Rev. Physiol.*, 50, 161, 1988.

Wells, R. M. G. and Weber, R. E., The spleen in hypoxic and exercised rainbow trout, *J. Exp. Biol.*, 150, 461, 1990.

Wells, R. M. G., Davie, P. S., and Weber, R. E., Effect of β-adrenergic stimulation of trout erythrocytes on blood viscosity, *Comp. Biochem. Physiol. C*, 100, 653, 1991.

Wood, C. M., Acid-base and ion balance, metabolism, and their interactions after exhaustive exercise in fish, *J. Exp. Biol.*, 160, 285, 1991.

Wood, S. C. and Johansen, K., Blood oxygen transport and acid-base balance in eels during hypoxia, *Am. J. Physiol.*, 225, 849, 1973.

Wood, C. M. and Perry, S. F., Respiratory, circulatory, and metabolic adjustments to exercise in fish, in *Circulation, Respiration, Metabolism,* Gilles, R., Ed., Springer-Verlag, Berlin, 1985, 2.

Wood, C. M. and Perry, S. F., A new *in vitro* assay for CO_2 excretion by trout red blood cells: effects of catecholamines, *J. Exp. Biol.*, 157, 349, 1991.

Wood, C. M. and Randall, D. J., The influence of swimming activity on sodium balance in the rainbow trout *(Salmo gairdneri), J. Comp. Physiol.*, 82, 207, 1973.

Wood, C. M., McMahon, B. R. and McDonald, D. G., Respiratory gas exchange in the resting starry flounder, *Platichthys stellatus*: a comparison with other teleosts, *J. Exp. Biol.*, 78, 167, 1979.

Yamamoto, K., Contraction of spleen in exercised cyprinid, *Comp. Biochem. Physiol. A*, 87, 1083, 1987.

Yamamoto, K., Contraction of spleen in exercised freshwater teleost, *Comp. Biochem. Physiol. A*, 89, 65, 1988.

Yamamoto, K. and Itazawa, Y., Erythrocyte supply from the spleen of exercised carp, *Comp. Biochem. Physiol. A*, 92, 139, 1989.

Yamamoto, K., Itazawa, Y. and Kobayashi, H., Supply of erythrocytes into the circulating blood from the spleen of exercised fish, *Comp. Biochem. Physiol. A*, 65, 5, 1980.

Yamamoto, K., Itazawa, Y. and Kobayashi, H., Direct observation of fish spleen by an abdominal window method and its application to exercised and hypoxic yellowtail, *Jpn. J. Ichthyol.*, 31, 427, 1985.

10 Autonomic Nerve Functions

Stefan Nilsson and Susanne Holmgren

INTRODUCTION

Our knowledge of the organization and function of the autonomic nervous system in fish stems from detailed anatomical investigations in the 19th century (e.g., Stannius, 1849; Chevrel, 1889). Functional studies, using pharmacologically active agonists and antagonists were pioneered by John Z. Young (Young, 1931, 1933c, 1936), and have been continued in several laboratories. The findings from studies of fish, especially the teleosts, are remarkably consistent with observations of autonomic nerve functions in mammals. Some organs that do not occur in mammals (e.g., gills and swim bladders) offer an insight into the diversity of autonomic innervation patterns that aid the understanding of the evolution of autonomic nerve function. Furthermore, the number of neurons, and probably also the number of transmitter types, is smaller in fish than in the more advanced tetrapods. These facts provide new angles for the study of autonomic nerve functions.

ANATOMICAL CONSIDERATIONS

In 1898, J.N. Langley proposed the term "autonomic nervous system" for *"the sympathetic system and the allied nervous system of the cranial and sacral nerves, and for the local nervous system of the gut"* (Langley, 1898). An anatomical subdivision into a sympathetic, a parasympathetic, and an enteric nervous system is still valid for mammalian systems. However, in the nonmammalian vertebrates a subdivision of the nervous outflow from the spinal cord into an anterior "sympathetic" (corresponding to the thoracicolumbar outflow in mammals) and a posterior "parasympathetic" (corresponding to the sacral outflow in mammals) outflow is generally difficult to make. A terminology that simply differentiates between a cranial and a spinal autonomic system has been proposed (Nilsson, 1983). The term "enteric nervous system" is retained as originally proposed by Langley (1898). For full discussions on anatomical classification of vertebrate autonomic nervous systems, see Nilsson (1983) and Gibbins (1993). Additional reviews and descriptions of the autonomic nervous anatomy and functions are those by Nicol (1952), Burnstock (1969), and Pick (1970) (vertebrates generally); Stannius (1849), Chevrel (1889), and Campbell (1970) (fish generally); Rovainen (1979) (cy-

0-8493-8042-1/93/$0.00+$.50

clostomes); Nilsson and Holmgren (1988) (elasmobranchs); Nilsson and Holmgren (1992a) (lungfish).

An anatomical classification of the divisions of the autonomic nervous system certainly aids the discussion of autonomic nerve functions. It must, however, be emphasized that the anatomical terminology "spinal autonomic" or "sympathetic" is not equivalent to the functional terminology that describes the neurotransmitter, such as "adrenergic" or "peptidergic". Many neurotransmitters have been identified in all anatomical divisions of the autonomic nervous system, and it becomes increasingly important to separate the anatomical and functional terms (see below in section on Neurotransmitters of the Autonomic Nervous System).

As in the tetrapods, spinal and cranial autonomic pathways in fish generally comprise two neurons. A preganglionic neuron leaves the central nervous system to run to an autonomic ganglion, and a postganglionic neuron runs from the ganglion to the effector tissue. The autonomic ganglion may be situated close to the spinal column, as in the paired sympathetic chains of teleosts (and tetrapods), or close to (or within) the effector organ itself, as in the cranial autonomic pathways (see Nilsson, 1983).

It should be pointed out that the simple view of a "two-neuron pathway" may not be valid in certain regions. For instance, the enteric nervous system receives an input from both cranial and spinal autonomic sources, and in the innervation of the gut the number of connected neurons in a certain pathway may be very hard to establish.

Cyclostomes

The anatomy of the autonomic nervous system in fish shows great variation, from the very simple arrangement in cyclostomes to the elaborate and well-developed system in teleosts. In the cyclostomes, particularly in the myxinoids (hagfish), the autonomic nervous system is poorly developed, and it is often impossible to decide whether peripheral neurons are autonomic or sensory. There are no sympathetic chains, but nerve branches that can be regarded as corresponding to a spinal autonomic outflow leave in the ventral (myxinoids) or dorsal and ventral (lampetroids; lampreys) spinal roots. There are ganglionic clusters along the dorsal aorta and elsewhere and fibers are also found in blood vessels (Fänge et al., 1963; Leont'eva, 1966; Nakao and Ishizawa, 1982). The nature and function of a spinal autonomic innervation of the viscera in cyclostomes clearly require much further attention.

Cranial autonomic fibers are present in the vagus (X) in both cyclostome groups, and in the lampetroids autonomic pathways may also run in the facial (VII) and glossopharyngeal (IX) nerves. The facial and glossopharyngeal pathways probably innervate the gills (Iijima and Wasano, 1980; Nakao, 1981). Cardiac vagal fibers occur in lampetroids, but there is no extrinsic innervation of the heart in myxinoids (Nicol, 1952; Nakao et al., 1981; Nilsson, 1983). However, a vagal innervation of the gallbladder has been shown in *Myxine glutinosa* (Augustinsson et al., 1956; Fänge and Johnels, 1958).

Ganglionated nerves form a plexus in the cyclostome gut, but the nature and function of these nerves are not well understood (Fänge et al., 1963).

Elasmobranchs

Sympathetic chains of the type found in the teleosts (see later) and tetrapods are absent in elasmobranchs, but segmental paravertebral ganglia are irregularly connected longitudinally and by transverse commisures. The ganglia receive their input from the spinal nerves via white rami communicantes, but recurrent grey rami are absent in elasmobranchs (Young, 1933c).

The most anterior pair of paravertebral ganglia, and clusters of chromaffin cells (often known as the suprarenal tissue), form the axillary bodies (Figure 1). Posterior to these, the paravertebral ganglia are smaller and lie dorsally in the posterior cardinal sinus. Also, these ganglia are often found in association with the suprarenal tissue (Young, 1933c; Nilsson and Holmgren, 1988). Nerves from the paravertebral ganglia, the splanchnic nerves, are made up of postganglionic fibers that run to the viscera (Figure 1).

Cranial autonomic fibers have been postulated in the oculomotor (III), facial (VII), glossopharyngeal (IX), and vagus (X) nerves (Young, 1933c; Nicol, 1952). A ciliary ganglion is present, and fibers from this ganglion innervate the eye (Young, 1933c). The function of the facial and glossopharyngeal fibers is not clear, and a vasomotor innervation of the gill blood vessels has been denied (Metcalfe and Butler, 1984). Vagal fibers innervate the heart, where a ganglion is found in the wall of the sinus venosus. The fibers extend posteriorly as far as the pylorus and anterior part of the spiral intestine (Young, 1933c; Nicol, 1952).

Myenteric and submucous plexuses are present in the elasmobranch gut (Kirtisinghe, 1940; Nilsson and Holmgren, 1988).

Teleosts

The sympathetic chains in teleosts and ganoid fish are generally well developed, and the arrangement resembles that found in the tetrapods. A unique feature in teleosts and holosteans is the cephalic portions of the sympathetic chains: the chains continue into the head, bearing ganglia in connection with the trigeminal/facial, glossopharyngeal, and vagus nerves (Figure 2). The connections between these ganglia and the cranial nerves are composed solely of postganglionic fibers (i.e., gray *rami communicantes*), and the preganglionic fibers of these pathways are all of spinal autonomic origin. Conclusions about the anatomical origin of neurons based solely on the neurotransmitter content may be premature. However, it is tempting to regard the presence of adrenergic ganglion cells in the ciliary ganglion as an indication that this ganglion, in part, represents the most anterior end of the teleost sympathetic chains (Young, 1931; Nicol, 1952; Nilsson,1983).

At the level of the third to fourth spinal nerves on the right side, there is a substantial sympathetic chain ganglion. From this ganglion (the coeliac ganglion) the anterior splanchnic nerves run along the coeliac and mesenteric

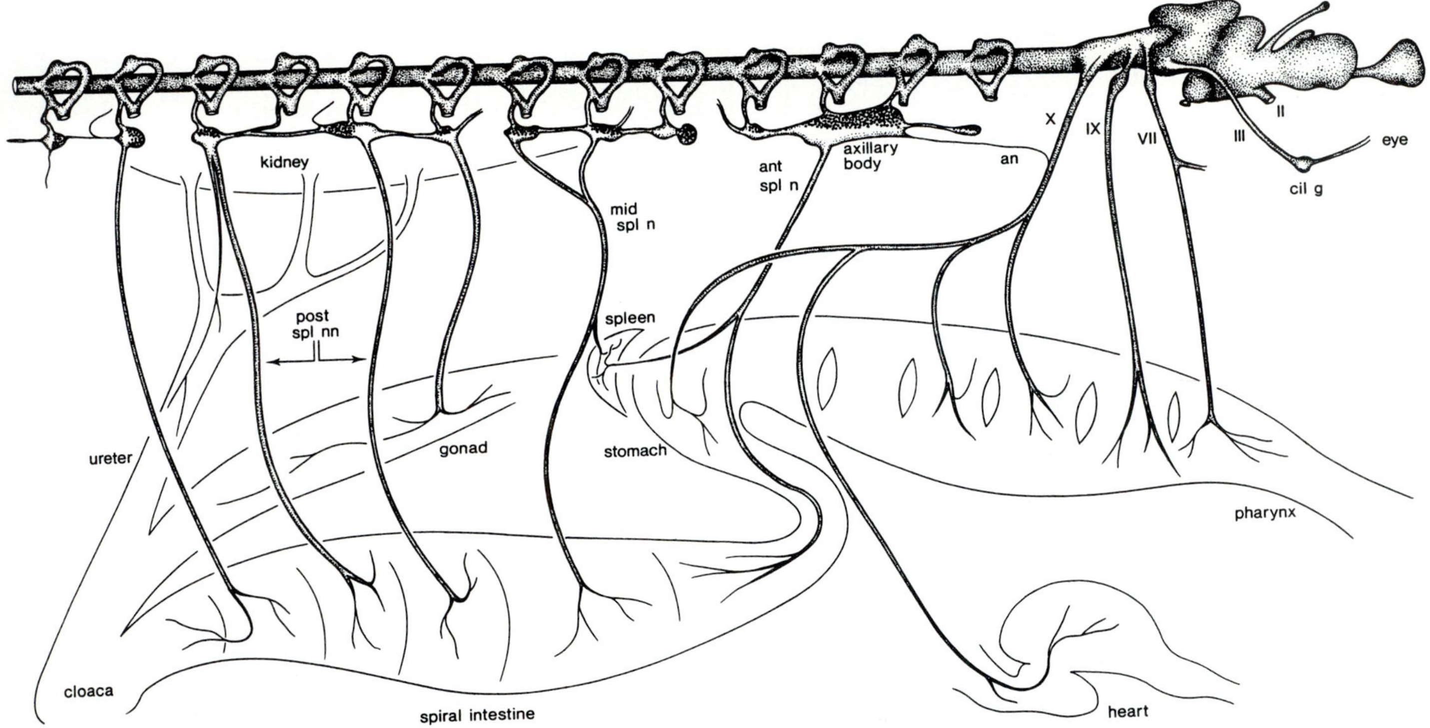

FIGURE 1. Generalized view of the arrangement of the autonomic nervous system in elasmobranchs. Note the chromaffin tissue which is associated with the paravertebral ganglia (e.g., the axillary body). Abbreviations: an, anastomosis connecting the anterior paravertebral ganglia to the vagus nerve; ant spl n, anterior splanchnic nerve; cil g, ciliary ganglion; mid spl n, middle splanchnic nerve; post spl nn, posterior splanchnic nerves. Roman numbers refer to their respective cranial nerves. (Reproduced from Nilsson, S., *Autonomic Nerve Function in the Vertebrates,* Springer-Verlag, Berlin, 1983, 18. With permission.)

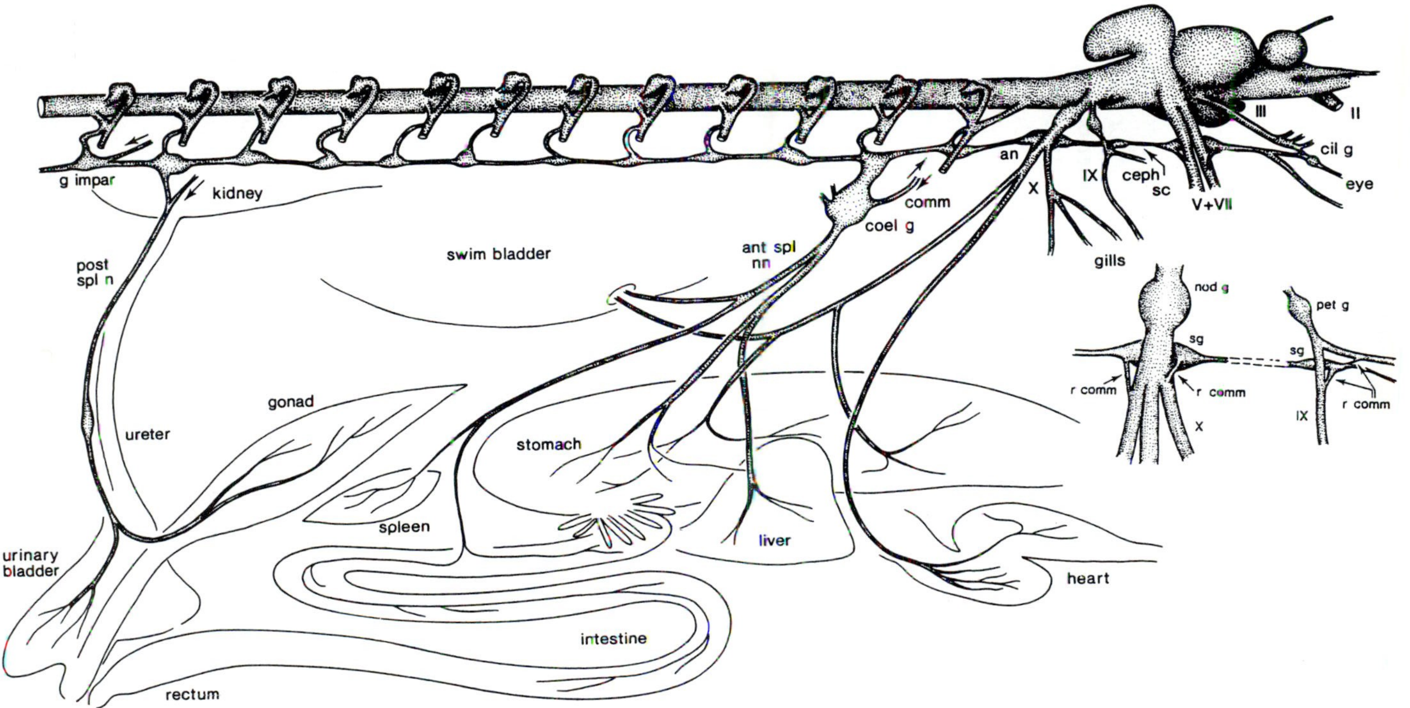

FIGURE 2. Generalized view of the arrangement of the autonomic nervous system in teleosts. Note the cephalic sympathetic chain which carries ganglia in contact with the cranial nerves. Abbreviations as in Figure 1 and: ceph sc, cephalic sympathetic chain; coel g, coeliac ganglion; comm, commissure between left and right sympathetic chain; g impar, ganglion impar (=posterior fusion of left and right sympathetic chains); nod g, nodose ganglion; pet g, petrous ganglion; r comm, ramus communicans; sg, sympathetic ganglion. (Reproduced from Nilsson, S., *Autonomic Nerve Function in the Vertebrates*, Springer-Verlag, Berlin, 1983, 21. With permission.)

arteries to the gut, swim bladder, spleen, and liver. The coeliac ganglion in teleosts is part of the right sympathetic chain and not, as in the tetrapods, a prevertebral or collateral ganglion situated away from the sympathetic chains (Figure 2). Thus, as in the elasmobranchs, the splanchnic nerves are composed largely of postganglionic fibers (cf. Nilsson, 1983). However, in some species preganglionic neurons that innervate small groups of ganglion cells are also present in the splanchnic nerve (cf. Nilsson, 1983; Gibbins, 1992)

The sympathetic chains fuse to a single strand in the posterior part of the abdomen (Chevrel, 1889; Young, 1931). In this region a posterior splanchnic nerve (vesicular nerve) leaves the chains to innervate the posterior part of the gut and the urinogenital organs (Young, 1931; Nilsson, 1976, 1983; Uematsu, 1985, 1986; Lundin and Holmgren, 1986; Uematsu *et al.*, 1989). In *Gadus morhua, Uranoscopus scaber* and some other teleost species, a substantial ganglion is present in the nerve running along the ureters (Young, 1931; Nilsson, 1976), which suggests an arrangement of the posterior spinal autonomic ganglia similar to the "sacral parasympathetic" system in the terminology of Langley (1898, 1921).

The presence of cranial autonomic ("parasympathetic") fibers has been shown unequivocally only in the oculomotor (III) and vagus (X) nerves of teleosts. In view of the extent of the cephalic sympathetic chains, innervation of the vasculature of the head and gills is probably of spinal autonomic origin (Nilsson, 1984b). A pair of large vagus nerves run to the gut and swim bladder. There are substantial contributions of spinal autonomic neurons to the vagi, which may therefore be best described as "vago-sympathetic nerve trunks" (Young, 1931; Nilsson, 1983).

As in the elasmobranchs, both myenteric and submucous plexuses are present in the enteric nervous system of teleosts (Kirtisinghe, 1940; Burnstock, 1959; Holmgren and Nilsson, 1983a; Holmgren, 1985; see below in the section on Autonomic Nerve Control, The Gut.)

Dipnoans

The sympathetic chains in lungfish are extremely fine strands, and were overlooked by early anatomists (Parker, 1892). The strands form loops (annuli) around the segmental (intercostal) arteries, and fibers run to the chromaffin tissue that is embedded in the wall of these arteries (Giacomini, 1906; Jenkin, 1928; Holmes, 1950; Abrahamsson et al., 1979a and b). The sympathetic chain ganglia are minute, and adrenergic cell bodies were not observed in these ganglia in *Protopterus aethiopicus* or *Lepidosiren paradoxa*. The nature and function of the spinal autonomic pathways (apart from a probable control of chromaffin tissue) remain to be clarified (Abrahamsson et al., 1979a and b; Axelsson et al., 1989b).

Little is known about the autonomic component in the cranial nerves of lungfish, but a cardiac, pulmonary, and probably gastrointestinal innervation by vagal fibers has been postulated (Abrahamsson et al., 1979a and b; Axelsson *et al.*, 1989b). In addition, autonomic fibers to the eye may run in the oculomotor nerve of *Neoceratodus forsteri* (Nicol, 1952).

NEUROTRANSMITTERS OF THE AUTONOMIC NERVOUS SYSTEM

Autonomic neurons in mammals usually contain more than one neurotransmitter. This is often one or more neuropeptides in combination with each other or with the classical transmitters acetylcholine or adrenaline/noradrenaline (Furness and Costa, 1987; Gibbins, 1989). Therefore, we can no longer truly separate, e.g., adrenergic or cholinergic neurons from peptidergic neurons; the use of this denomination merely characterizes one of the transmitters common to a group of neurons. Among fish, only few cases of coexistence are established at the time being, but with the increasing use of more and more specific methods it is highly likely that similar cases of coexistence, as in mammals, will be found.

Catecholamines

Adrenergic neurons in fish possess the enzyme phenylethanolamine-*N*-methyltransferase (PNMT), and thus have the ability to convert noradrenaline into adrenaline. Measurements of catecholamine levels show that most adrenergically innervated organs of fish contain a mixture of adrenaline and noradrenaline; usually adrenaline is the dominant of the two (Holtzbauer and Sharman, 1972). The two catecholamines are presumably stored and released together in the adrenergic terminals (Nilsson, 1983).

Acetylcholine

Acetylcholine was for long considered the sole transmitter of all preganglionic neurons and all postganglionic, parasympathetic neurons of the autonomic nervous system in all vertebrate groups (Nilsson, 1983). This view has now been modified to include several other transmitters in coexistence with acetylcholine in these neurons (Furness and Costa, 1987; Gibbins, 1989).

The presence of acetylcholine in a neuron is difficult to establish histochemically. However, acetylcholine is formed from acetyl-coenzyme A and choline by the enzyme choline acetyltransferase (ChAT; Nachmansohn and Machado, 1943). ChAT is restricted to cholinergic neurons and can be identified by several biochemical methods. The presence of ChAT has been demonstrated in the spleen, the heart, and the swim bladder in some teleost species (Holmgren, 1981; Winberg et al., 1981; Fänge and Holmgren, 1982).

Nothing is known so far about the identity of possible peptidergic neurotransmitters, which, judging from the situation in other vertebrates, could coexist with acetylcholine in fish. However, acetylcholine and adrenaline might coexist in autonomic neurons innervating the spleen of the cod, *G.*

morhua. This suggests an interesting evolutionary stage, in parallel with what is often seen in developing nerve cells in culture (Holmgren and Nilsson, 1976; Burnstock, 1978; Winberg et al., 1981).

5-Hydroxytryptamine

Serotonergic (5-HT-containing) nerve fibers have been demonstrated in several fish species, including cyclostomes (e.g., Baumgarten et al., 1973), elasmobranchs (Holmgren and Nilsson, 1983a), and teleosts (Watson, 1979; Goodrich et al., 1980; Anderson, 1983). The nerves predominantly innervate the gut, but vagal neurons innervating the gills of the rainbow trout (*Oncorhynchus mykiss*) have also been reported (Dunel-Erb et al., 1989).

5-HT occurs in several types of cells, including neurons, paracrine cells, endocrine cells, polymorphous granular cells (PGCs), and neuroepithelial cells (NECs). Using the formaldehyde-induced fluorescence histochemical technique (Falck-Hillarp technique), Anderson (1983) demonstrated 5-HT neurons in some teleost species but not in others. However, by the use of the more sensitive immunohistochemical method, 5-HT neurons could be demonstrated in this latter group, which possibly also indicates differences in transmitter storing capacity. Interestingly, 5-HT-storing enterochromaffin cells were common in species with a low neuronal content of 5-HT, but were absent in fishes with high content of neuronal 5-HT (Anderson and Campbell, 1988).

Purines

In several cases, ATP (adenosine triphosphate) has been proposed to be a transmitter or cotransmitter in mammalian autonomic neurons (e.g., Burnstock, 1986). An effect of ATP or related adenyl compounds has been reported in a number of autonomically innervated fish tissues (Young, 1980a and b, 1983; Meghji and Burnstock, 1984a and b; Small et al., 1990; Farrell and Davie, 1991), but not as many criteria for a transmitter action as in mammals have been fulfilled, and at this stage it is premature to conclude that purinergic nerves exist in the autonomic nervous system of fish.

Neuropeptides

A considerable number of neuropeptides have now been identified by the use of immunohistochemistry (often in combination with radioimmunoassay) in autonomic nerves of representatives of all groups of fish (cyclostomes, elasmobranchs, holocephalans, holosteans, dipnoans, brachiopterygians, and teleosts). (Figure 3; Nilsson and Holmgren, 1992b; Jensen and Holmgren, 1993). The immunochemical methods do not reveal the exact identity of the peptide, but close relationships to the mammalian counterpart (which was usually used for the preparation of the antisera) can be established. In the instances where the sequence of the fish peptide has been established, a close similarity to the equivalent mammalian peptide is often found, particularly in the biologically active part of the molecule.

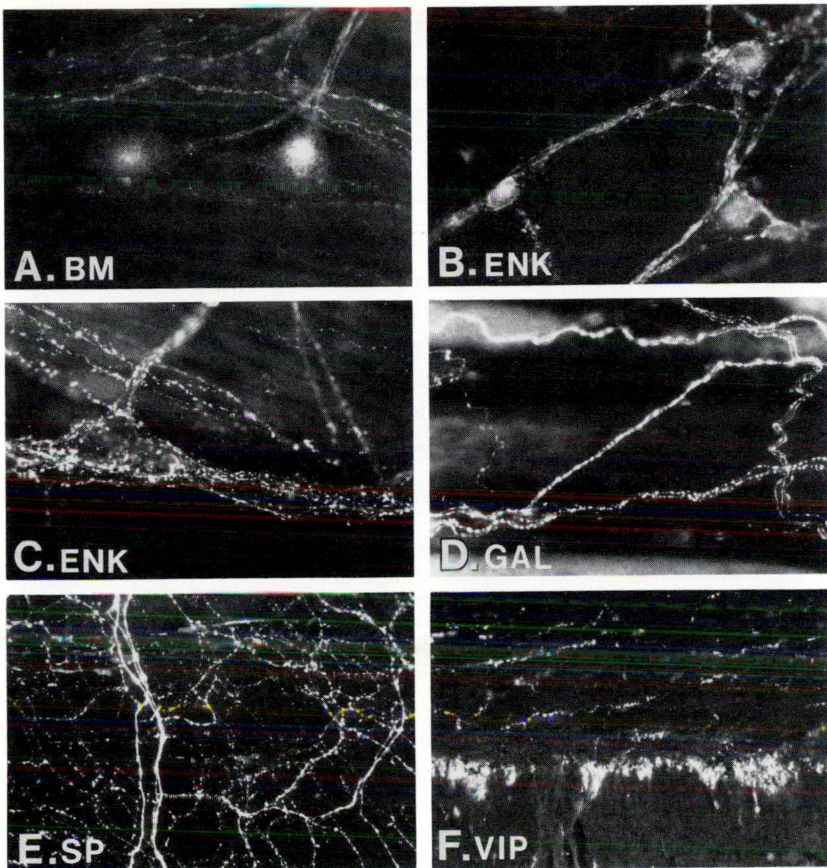

FIGURE 3. Neuropeptide immunoreactivity (IR) in autonomic nerve of fish, shown by the use of immunohistochemistry. (A) Bombesin-IR (BM) in the myenteric plexus of the stomach of *Myoxocephalus scorpius*, ×300. (B) Met-enkephalin-IR (ENK) in the myenteric plexus of the intestine of the eel, *Anguilla anguilla*, ×300. (C) Met-enkephlin-IR (ENK) in the myenteric plexus of the rectum of the lungfish, *Lepidosiren paradoxa*, ×300. (D) Galanin-IR (GAL) in perivascular nerves of a mesenterial artery of the cod, *Gadus morhua*, ×120. (E) Substance P-IR (SP) in the myenteric plexus of the stomach of the cod, ×100. (F) VIP-IR in the myenteric plexus of the intestine of the cod, ×240.

Bombesin

Bombesin is a 14-amino acid peptide, which was originally isolated from the skin of amphibians (Anastasi et al., 1971). The related *gastrin releasing peptides* (GRPs) were first found in mammals and birds, and are neuropeptides of 27, 23, or 10 amino acids, with the C-terminal 9 or 10 amino acids identical to bombesin (e.g., McDonald et al., 1979, 1980). Immunohistochemistry has revealed the presence of bombesin/GRP-like peptides in autonomic nerves of the gut and circulatory system in several fish species. A shark GRP and a cod

GRP have been sequenced, and show similarities to mammalian GRP in the C-terminal region (Conlon et al., 1987; Jensen and Conlon, 1992b). Radioimmunoassays, immunohistochemistry, and pharmacological studies in the Atlantic cod, *G. morhua,* indicate that several bombesin/GRP-like peptides are present in the cod gut (Holmgren and Jönsson, 1988).

Neuropeptide Y (NPY)

NPY appears to be one of the best-preserved neuroendocrine peptides, maybe even the most well conserved, considering its length of 36 amino acids (Larhammar et al., 1992). NPYs from goldfish (*Carassius auratus*), ray (*Torpedo marmorata*), and dogfish (*Scyliorhinus canicula*) have been sequenced (Blomqvist et al., 1992; Conlon et al., 1992). In addition, several peptides isolated from the pancreas of fish show large similarities to NPY (see Conlon et al., 1991); some of these peptides occur in pancreatic nerves (Noe et al., 1986).

Somatostatin

A number of fish somatostatins have been sequenced, and several forms comprising 14-, 25-, or 28-amino acid residues are reported (see, e.g., Holmgren and Jensen, 1993). Somatostatin-14 is extremely well preserved among vertebrates; identical forms are reported from elasmobranch, avian, and mammalian species. It is, however, not yet clear which forms occur in fish neurons, although there are indications from, e.g., the toad, *Bufo marinus,* that somatostatin-14 is the neuronally contained variant (Campbell et al., 1982).

Tachykinins

The tachykinins form a large family with the C-terminal pentapeptide Phe-X-Gly-Leu-Met-NH2 in common, and several subgroups have been identified. To date, two different tachykinins from elasmobranchs have been sequenced, scyliorhinin I, which is related to phyasalaemin and substance P, and scyliorhinin II which shows cross-reactions with neurokinin A (NKA) (Conlon et al., 1986). A substance P-related and a NKA-related peptide have recently been isolated from the cod and rainbow trout brains; these peptides are also present in the intestine of the same species (Jensen and Conlon, 1992a; see also Holmgren and Jensen, 1993).

Vasoactive Intestinal Polypeptide (VIP)

In contrast to most other neuropeptide families, the VIP-like peptides share the N-terminal, and the N-terminal is also the biologically most active part of the peptide. Dogfish and cod VIP have been sequenced, and like their mammalian counterpart, both consist of 28 amino acids. They differ from each other in four positions, and both differ from mammalian VIP in five positions (Dimaline et al., 1987; Thwaites et al., 1989).

Other Neuropeptides

Gastrin/caerulein/CCK-like peptides from fish have so far not been sequenced, but radioimmunological studies in Atlantic cod, rainbow trout, and spiny dogfish indicate the presence of several forms of these peptides with different regional distributions in the gut (Vigna et al., 1985; Jönsson et al., 1987; Aldman et al., 1989). Similarly, neurotensin-like, opioid-like, calcitonin gene-related peptide (CGRP)-like, and galanin-like immunoreactivities have been demonstrated in peripheral nerves of fish, but the exact sequences are so far unknown (e.g., Bjenning and Holmgren, 1988; Burkhardt-Holm and Holmgren, 1992; Karila et al., 1993; Nilsson and Holmgren, 1992b).

AUTONOMIC NERVE CONTROL OF DIFFERENT ORGANS

The Chromaffin System

Chromaffin cells show many similarities with the postganglionic adrenergic neurons (e.g., catecholamines synthesis, storage, and release, and control by preganglionic cholinergic fibers). Although they are obviously not neurons, these cells may provide supplementary adrenergic control of many systems in fish (Nilsson, 1984a).

Fixation of adrenal glands using dichromate solution was shown by Henle (1865) to cause a darkening of the adrenal medulla, while the cortex remained unstained. Later, Kohn (1902) coined the term "chromaffin" to describe these cells and it is now known that the reaction depends on the oxidation of stored catecholamines to their respective adrenochromes (Coupland, 1965, 1972).

With the introduction of histochemical techniques that produce fluorescent reaction products from intracellular catecholamines (Eränkö, 1952, 1955; Falck et al., 1962), it is possible to detect the catecholamine-storing cells even if the catecholamine content is too low to produce a clear chromaffin reaction.

Chromaffin cells are present in all fish groups, and occur either in distinct chromaffin tissue or in scattered clusters of intensely fluorescent cells (small intensely fluorescent cells, SIF cells, Eränkö and Härkönen, 1963, 1965). These scattered clusters are sometimes known as "para-ganglia" or, more often, "extra-adrenal chromaffin tissue" in mammals. In fish, where adrenals are absent, the term "extra-adrenal" loses its meaning, and the catecholamine-storing tissue is simply called "chromaffin tissue" (Santer, 1993).

Cyclostomes

In the cyclostomes, chromaffin tissue is found in scattered cell clusters in the walls of arteries and veins, and in large quantities within both the atrium and ventricle of the heart (Table 1, Figure 4; Augustinsson et al., 1956; Bloom et al., 1961; Shibata and Yamamoto, 1976). The control of the

TABLE 1
Catecholamine Content in Tissues Known to Contain Chromaffin Cells

	Adrenaline	Noradrenaline	Ref.
Cyclostomes			
Atrium			
Myxine glutinosa	8.1	18	Euler and Fänge, 1961
Lampetra fluviatilis	127.1	16	Stabrovskii, 1967
Ventricle			
Myxine glutinosa	59	6.5	Euler and Fänge, 1961
Lampetra fluviatilis	81	11.6	Stabrovskii, 1967
Elasmobranchs			
Axillary body			
Squalus acanthias	445	2139	Abrahamsson, 1979b
Scyliorhinus canicula	14670	20410	Mazeaud, 1971
Teleosts and ganoids			
Posterior cardinal vein			
Gadus morhua	38.2	14.3	Abrahamsson and Nilsson, 1976
Huso huso	19.8	4.8	Balashov et al., 1981
Lepisosteus platyrhincus	47.5	21.5	Nilsson, 1981
Head kidney			
Salmo gairdneri (=Oncorhynchus mykiss)	4.7	4.5	Nakano and Tomlinson, 1968
Dipnoans			
Whole heart			
Protopterus aethiopicus	4.2	70.8	Abrahamsson et al., 1979a
Intercostal arteries			
Protopterus aethiopicus	216	94	Abrahamsson et al., 1979a

Note: The levels are expressed in μg/g tissue, but the ratio, of chromaffin tissue to other tissues in the samples, varies.

chromaffin cells in the myxinoid heart, which receives no extrinsic innervation, is not clear. It might be speculated that the chromaffin cells of the cyclostome heart represent an original form of chemoreceptor, directly sensitive to chemical stimuli such as hypoxia. This hypothesis does, however, require confirmation.

Elasmobranchs

In the elasmobranchs, chromaffin tissue is found in segmental bodies, the "suprarenal bodies", together with the paravertebral ganglia. The largest clusters of chromaffin cells form the axillary bodies together with the gastric ganglia (see earlier). Noradrenaline is the predominant catecholamine in the axillary bodies of those elasmobranchs investigated (Table 1)(see Nilsson, 1983; Santer, 1993).

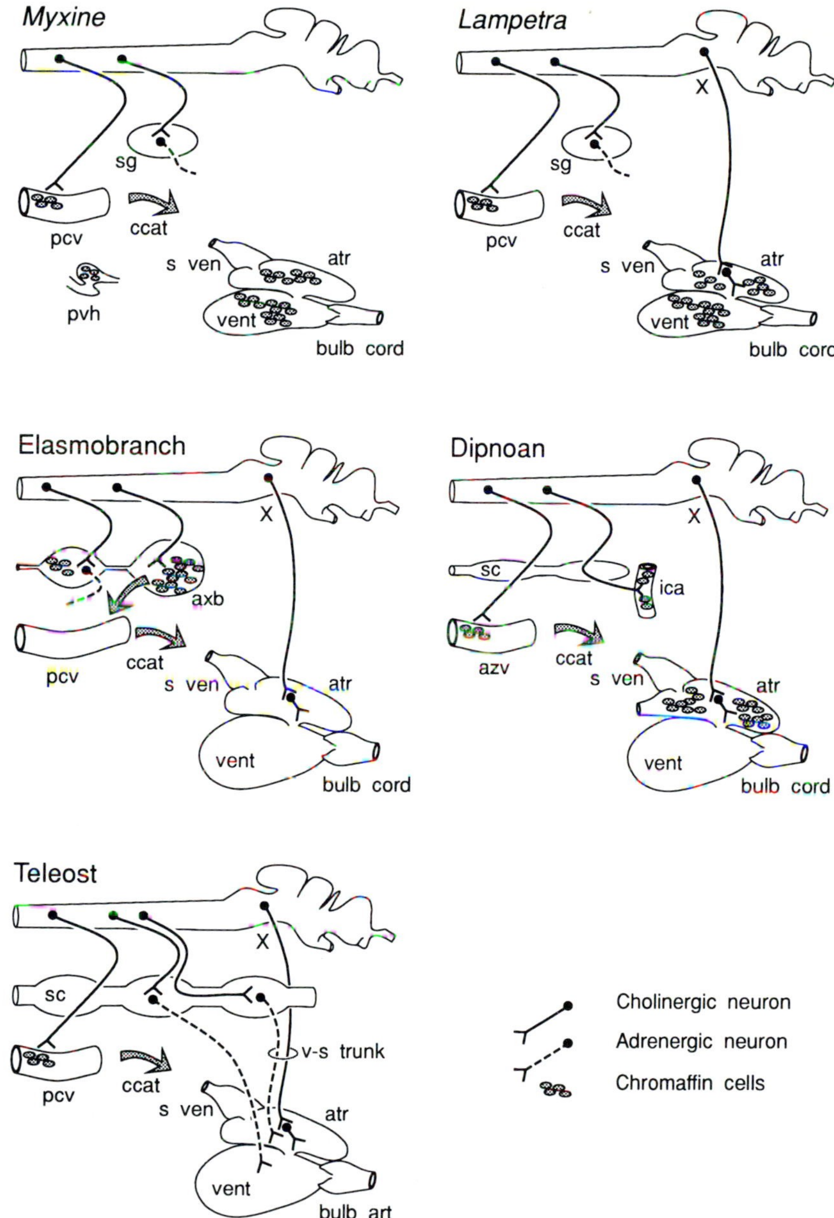

FIGURE 4. Summary of the patterns of autonomic innervation of the heart of cyclostomes (*Myxine* and *Lampetra*), elasmobranchs, dipnoans, and teleosts. Also shown is the distribution of chromaffin cells in different parts of the circulatory system. Abbreviations: atr, atrium; axb, axillary body; azv, azygos vein (=left posterior cardinal vein); bulb art, bulbus arteriosus; bulb cord, bulbus cordis (=conus arteriosus);ccat, circulating catecholamines; ica, intercostal artery; pcv, posterior cardinal vein; pvh, portal vein heart; sc, sympathatic chain; sg, sympathatic ganglion; s ven, sinus venosus; vent, ventricle; v-s trunk, "vago-sympathetic trunk".

Teleosts

The chromaffin tissue of teleosts and the ganoids, *Huso huso* and *Lepisosteus platyrhinchus,* occurs mainly in the wall of the posterior cardinal veins within the head kidney (Giacomini, 1902; Mazeaud, 1971; Nilsson, 1976, 1981; Nilsson et al., 1976; Balashov et al., 1981). A functional innervation of the chromaffin tissue by preganglionic fibers of spinal autonomic origin has been shown in the cod, *G. morhua.* Electrical stimulation of the nerve supply produces release of catecholamines (Nilsson et al., 1976). Under experimental conditions, this release is substantial enough to affect both the heart (Holmgren, 1977) and gill vasculature (Wahlqvist, 1981; Nilsson, 1984b). During near-exhaustive swimming, a release of catecholamines from the chromaffin tissue aids swimming performance in the cod (Butler et al., 1989). Catecholamine release as a direct effect of hypoxemia has also been shown in the cod (Perry et al., 1991).

Dipnoans

The spinal autonomic nervous system of dipnoans is poorly developed, and adrenergic neurons may be lacking altogether in this group of fish (Nilsson and Holmgren, 1992a). There is, however, a well-developed chromaffin system. In *P. aethiopicus* and *L. paradoxa*, chromaffin tissue is found in three major locations: (1) in the walls of the proximal portion of the intercostal arteries, (2) in the wall of the left posterior cardinal vein ("azygos vein") and (3) within the atrium (Holmes, 1950; Abrahamsson et al., 1979a; Scheuermann, 1979; Axelsson et al., 1989b). The atrial chromaffin tissue has also been shown histochemically in *N. forsteri*, and is thus a feature in all the lungfish genera (S. Holmgren and R. Fritsche, unpublished).

An innervation of the chromaffin tissue of the azygos vein and intercostal arteries has been described and it seems likely that there is also an innervation of the chromaffin tissue of the atrium (Giacomini, 1906; Jenkin, 1928; Holmes, 1950; Scheuermann, 1979; Nilsson and Holmgren, 1992a).

The Circulatory System

The need for an efficient system for oxygen transport from the surrounding medium (water or air) to the tissues and organs of the organism has been a strong selective element in the evolution of circulatory and respiratory systems in the vertebrates. The role of the autonomic nervous system in the regulation of the circulatory system has been the subject of interest for more than a century (see Morris and Nilsson, 1993), and there are numerous studies of the distribution and function of the autonomic innervation of the cardiovascular system. The information includes anatomical and histochemical investigations of the distribution and nature of the autonomic nerves, via physiological and pharmacological studies of isolated tissues and organs to the integrated mechanisms *in vivo.*

Innervation of the Heart

The fish heart is a convoluted tube of three or four chambers connected in series: the sinus venosus, the atrium, the ventricle, and, in elasmobranchs, the bulbus cordis. In teleosts, the bulbus cordis is replaced by an elastic segment of the ventral aorta, the bulbus arteriosus (Satchell, 1991).

Cyclostomes

Blood pressure in the cyclostomes is generally low, and blood circulation is aided by the work of auxiliary "hearts" at the gills, in the caudal venous sinus, and a pulsatile segment in the hepatic portal vein "portal heart" (Fänge et al., 1963; Satchell, 1984, 1986; Davie et al., 1987; Forster et al., 1991).

The systemic heart of lampetroids is innervated by vagal cholinergic fibers. Contrary to the situation in the other vertebrates, this innervation is *excitatory*, acting via nicotinic cholinoceptors. The heart of myxinoids lacks an extrinsic innervation by autonomic fibers, and is insensitive to exogenously applied acetylcholine and catecholamines (Fänge and Östlund, 1954; Augustinsson et al., 1956; Hirsch et al., 1964; Beringer and Hadek, 1972; Lukomskaya and Michelson, 1972). A regulation of cardiac function via intrinsic catecholamines from the chromaffin stores has been documented (Bloom et al., 1961; Axelsson et al., 1990; Figure 4).

Elasmobranchs

The elasmobranch heart is innervated by inhibitory cholinergic vagal fibers, acting via muscarinic cholinoceptors as in the other vertebrates (except cyclostomes) (Laurent et al., 1983; Nilsson, 1983; Nilsson and Holmgren, 1988). The possible role of neuropeptides in the autonomic innervation of the elasmobranch heart is not yet clear. Nerve fibers showing bombesin-immunoreactivity have been shown in the heart and coronary vessels of *Raja erinacea*, and in the sinus venosus and the atrium of *R. erinacea* and *R. radiata* a sparse innervation by neuropeptide Y-immunoreactive nerve fibers has been shown (Bjenning et al., 1989, 1991).

There appears to be no functional adrenergic innervation of the elasmobranch heart (Laurent et al., 1983; Nilsson and Holmgren, 1988), although a sparse innervation of the sinus venosus by adrenergic fibers has been described in the Port Jackson shark, *Heterodontus portusjacksoni* (Gannon et al., 1972). Catecholamines stimulate the elasmobranch heart via β-adrenoceptors, but unequivocal evidence for an adrenergic control of the elasmobranch heart is lacking. A control via circulating catecholamines from chromaffin stores within the posterior cardinal sinuses is possible (Satchell, 1971; Gannon et al., 1972; Abrahamsson, 1979b; Capra and Satchell, 1977; Nilsson and Holmgren, 1988; Figure 4).

Teleosts

The heart of teleosts receives one pair of vagal branches which enter the sinus venosus. The cell bodies of the postganglionic neurons are collected to one or more ganglia situated near the sino-atrial border, and postganglionic fibers run to the sinus and atrium (Laurent, 1962; Yamauchi and Burnstock, 1968; Gannon and Burnstock, 1969; Santer, 1972; Holmgren, 1977, 1981).

As in the elasmobranchs, the vagal fibers are cholinergic and inhibitory, their effects mediated by muscarinic cholinoceptors of the pacemaker and atrium (Cameron, 1979; Laurent et al., 1983; Nilsson, 1983). The vagal fibers do not reach the ventricle, and inotropic cholinergic effects are restricted to the atrium. This control is sufficient to modify ventricular filling and thus cardiac stroke volume (Johansen and Burggren, 1980).

The possible role, or even existence, of a peptidergic innervation of the teleost heart is not clear. Injected porcine vasoactive intestinal polypeptide (VIP) produces an increased stroke volume in the cod, *G. morhua*, *in vivo* (Jensen et al., 1991), but the function of the intrinsic peptide is not clear.

There is an adrenergic innervation of the heart in most teleost species, except the pleuronectids (Govyrin and Leont'eva, 1965; Gannon and Burnstock, 1969; Gannon, 1971; Holmgren, 1977; Cameron, 1979; Donald and Campbell, 1982). The adrenergic fibers reach the heart in the vagi (*via* the connections from the cephalic sympathetic chains), along the most anterior spinal nerves, or along the coronary arteries (Gannon and Burnstock, 1969; Holmgren, 1977). The adrenergic innervation, which is sometimes augmented by circulating catecholamines, is tonically active and produces positive inotropic and chronotropic effects via β-adrenoceptors (Fänge and Östlund, 1954; Forster, 1976; Holmgren, 1977; Cameron, 1979; Axelsson, 1988; Axelsson et al., 1989a; Figure 4).

Dipnoans

There is a high degree of separation between oxygenated and deoxygenated blood in the dipnoan heart, despite the anatomically single ventricle (Johansen and Burggren, 1980; Nilsson and Holmgren, 1991). An inhibitory cholinergic vagal innervation is present, but an innervation by fibers of spinal autonomic origin is most unlikely (Jenkin, 1928; Abrahamsson et al., 1979b). The vagal cholinergic cardiac tonus varies with the breathing cycles in *L. paradoxa*, and there is an additional continuous adrenergic excitatory tonus (Axelsson et al., 1989b). A small cholinergic vagal tonus occurs in the resting Australian lungfish, *N. forsteri*, and there is a similarly small adrenergic tonus (R. Fritsche, M. Axelsson, G.C. Grigg, C.E. Franklin, and S. Nilsson, unpublished).

As in the cyclostomes, there are large quantities of catecholamines in the atrium of *P. aethiopicus, L. paradoxa, N. forsteri* (Abrahamsson et al., 1979a; Scheuermann, 1979; Axelsson et al., 1989b; S. Holmgren and R. Fritsche, unpublished; Figure 4).

Innervation of the Vasculature

A vasoconstrictor innervation by adrenergic neurons is present in practically all vertebrates. In the cyclostomes, adrenergic nerve fibers have been shown by histochemical studies in *Lampetra fluviatilis* (Leont'eva, 1966). Catecholamines or acetylcholine increase systemic as well as branchial vascular resistance in *M. glutinosa*, but little is known about the existence and function of vasomotor nerves (Reite, 1969; Axelsson et al., 1990).

A branchial vasomotor innervation is probably absent in elasmobranchs, (Nicol, 1952; Metcalfe and Butler, 1984). A vasomotor control of gill blood vessels may occur due to release of catecholamines from chromaffin tissue and the catecholamine concentration in blood plasma from the dogfish, *S. canicula*, was shown to be high enough to affect the branchial vasculature (Davies and Rankin, 1973). In teleosts, branchial vasomotor fibers run in the branchial branches of the vagus. Control of the teleost branchial vasculature is highly complex, and both acetylcholine and 5-hydroxytryptamine have been postulated in cranial autonomic vasomotor fibers of the teleost gill vasculature (Smith, 1977; Pettersson and Nilsson, 1979; Nilsson, 1984b; Bailly and Dunel-Erb, 1986; Bailly et al., 1989). In the cod, CCK causes a marked branchial vasoconstriction (Sundin and Nilsson, 1992).

Spinal autonomic fibers control the branchial and systemic vascular beds in teleosts and other actinopterygians. Vasomotor fibers to the viscera run in the anterior and posterior splanchnic nerves (Nilsson, 1976; Uematsu et al., 1989). Gray *rami communicantes* are present in teleosts, and postganglionic spinal autonomic fibers may join the spinal nerves to reach the somatic vasculature. Adrenergic nerves regulate blood pressure of teleosts both at rest and during exercise or hypoxia (Wood and Shelton, 1975; Smith, 1978; Axelsson and Nilsson, 1986; Axelsson, 1988; Axelsson et al., 1989a; Fritsche and Nilsson, 1989, 1990, 1993). There is a complex pattern of adrenergic nerves within the gill filaments (Donald, 1984, 1987; Dunel-Erb et al., 1989), and this innervation appears to affect primarily the arterio-venous vascular pathways, including the nutritive vasculature (Nilsson and Pettersson, 1981; Nilsson, 1984b; Sundin and Nilsson, 1992).

The peptidergic cardiovascular innervation has been studied mainly in the gut. In the cyclostome, *Lampetra fluviatilis*, there is histochemical evidence of bombesin-like and CGRP-like peptides in perivascular fibers of gut vessels (Nilsson and Holmgren, 1992b), and in the holostean fish, *Lepisosteus platyrhinchus*, a VIP-like peptide innervates gut vessels (Holmgren and Nilsson, 1983b). In teleost and/or elasmobranch species, bombesin-, galanin-, NPY-, somatostatin-, substance P-, and VIP-like immunoreactivities have been reported in fibers innervating the coeliac and mesenteric arteries and their branches (see Nilsson and Holmgren, 1992b).

Feeding produces an increased blood flow to the gut in the sea raven, *Hemitripterus americanus*, and the cod (Axelsson et al., 1989a; Axelsson and Fritsche, 1991). Exercise, on the other hand, reduces the blood flow to the gut.

Both adrenergic nerves and circulating catecholamines contract the gut vessels, increasing the gastrointestinal vascular resistance (Axelsson et al., 1989a; Axelsson and Fritsche, 1991; Holmgren et al., 1992a), while the responses to neuropeptides are less clearcut (see below).

Physiological/pharmacological experiments on gut blood flow involving neuropeptides have been performed in elasmobranchs and teleosts only. In general, substance P increases flow to the gut (Jensen et al., 1991; Holmgren et al., 1992a). VIP is vasodilatory in most species (Holder et al., 1983; Jensen et al., 1991), but increases gut vascular resistance in the spiny dogfish (Holmgren et al., 1992a). Bombesin shows variable responses in the spiny dogfish, while NPY increases the flow to the gut (Bjenning et al., 1990; Holmgren et al., 1992b). Galanin contracts isolated gut arteries from the cod (Karila et al., 1993).

THE GUT

Morphology, Extrinsic and Intrinsic Nerves

The fish gut can, in most species, be divided into an esophagus, a stomach, an intestine, and sometimes a rectum. However, there are several species, such as the carp, which lack a stomach, and the food enters the intestine directly.

The morphology of the gut wall in fish follows the common vertebrate pattern of a thin outer, longitudinal muscle layer, and a thicker inner, circular muscle layer. The inner lining of the gut is formed by a mucosa, with an underlying submucosa of variable thickness. Ganglionated plexuses occur between the muscle layers (the myenteric plexus) and in the submucosa (the submucous plexus); nerve fibers from these plexuses penetrate all layers of the gut. Bundles of nerve fibers run between the bundles of circular smooth muscle, and single varicose fibers or small fiber bundles run along the smooth muscle cells. The innervation within the longitudinal muscle layer is sparse. Single varicose fibers from the submucous plexus innervate the mucosa (Holmgren and Nilsson, 1991). In comparison to mammals, the number of enteric ganglion cells in fish are few. On the other hand, it has been demonstrated in the cod, that a large number of ganglion cells are present in microganglia along the length of both the splanchnic nerves and the vagi (Abrahamsson, 1979a; Karila et al., 1993).

The extrinsic control of the fish gut is performed by nerve fibers running in the vagi and the splanchnic nerves. Adrenergic neurons innervating the gut are of spinal autonomic origin. Stimulation of these nerves causes excitation of gut smooth muscle in some species, and an inhibition or mixed effects in other species, partly due to an effect of the transmitter on different subtypes of adrenergic receptors (α- and β-adrenoceptors) (Nilsson, 1983). Similarly, stimulation of the vagal innervation of the gut has different effects in different species. When present, an excitatory effect can be attributed to stimulation of cholinergic nerves, while the nature of the inhibitory vagal neurons present in some species still remains to be elucidated (Nilsson, 1983).

Enteric neurons may release acetylcholine, 5-hydroxytryptamine, purines (?), and neuropeptides in various combinations. A number of reports are available on the effects of the putative transmitters in several elasmobranch and teleost fish species (see Holmgren and Nilsson, 1991, Jensen and Holmgren, 1993). In general, bombesin, gastrin/CCK, and tachykinins are excitatory on gut smooth muscle, but exceptions occur. Bombesin and substance P increase blood flow to the fish gut, while opioids and neurotensin show varying effects, depending on tissue and species. VIP is usually inhibitory on gut smooth muscle, and increases blood flow to the gut.

The same transmitter may have different effects on stomach and intestine. This is the case with the bombesin-related peptide, litorin, which is excitatory on stomach preparations, but inhibitory on the intestine in the cod gut (Holmgren and Jönsson, 1988). So far there are few cases of interactions between neurotransmitters established in fish. One of these is the potentiation of the response to acetylcholine caused by bombesin in the gut of the rainbow trout and cod (Thorndyke and Holmgren, 1990).

Neuropeptides and, in some cases, serotonergic neurons occur in the gut of the holocephalan *Chimaera monstrosa* (Yui et al., 1990), the holosteans *Lepisosteus platyrhinchus* (Holmgren and Nilsson, 1983b) and *Amia calva* (Rajjo et al., 1989), the dipnoans *Lepidosiren paradoxa* and *Protopterus annectens* (Nilsson and Holmgren, 1992a), and the brachiopterygian fish *Polypterus senegalensis* (Burkhardt-Holm and Holmgren, 1992), but so far there are no studies on the function of these putative transmitters.

Development and Evolution of the Gut Innervation in Fish

There are but few studies of the development of the gut innervation in individual fish. These investigations, using embryos from egg-laying dogfish species, show that most types of nerves develop during early embryonal stages, while a few (bombesin, endorphin) appear around, or shortly after, hatching. Nerves (and endocrine cells) of the intestine develop earlier than those of the stomach. There are fewer nerve fibers present in the embryonal or newly hatched fish than in adults, and the distribution is slightly different. It has been suggested that the nerves that appear before hatching are involved in the control of cell growth and proliferation, but after hatching their role changes to the control of food processing (El-Salhy, 1984; Tagliafierro et al., 1989).

Another aspect of the evolution of the gut innervation is the development of increasingly sophisticated control mechanisms. A group of peptides with an important role in the control of gut motility is the tachykinin group. When the presence and the effect of substance P-like peptides in the gut of different fish groups are compared, it is found that some species, such as cyclostomes, rays (e.g., *Raja radiata*) and the bichir (*P. senegalensis*), store tachykinins in endocrine cells only, while in other species they occur in both nerve fibers and endocrine cells. It is possible that the presence in endocrine cells is a more primitive feature. In concert with the different anatomical features, the mechanism of action appears to have undergone an evolution from weak effects, if any, in cyclostomes, *via* an effect directly on the muscle cells only in the "older" fish groups, to more complicated patterns involving cholinergic and serotonergic neurons in teleosts (see Jensen and Holmgren, 1993).

The involvement of autonomic neurons in some integrated functions of the fish gut will be described below. In these examples, the function of the autonomic nerves have been singled out, but it is imperative to keep in mind that the final effect is also dependent on local and circulating hormones, and on the intrinsic properties of the smooth muscles and secretory cells themselves.

Gut Secretion

Gut secretion in fish, *in vivo*, has been studied mainly in the cod intestine, continuously infused with 33% seawater to mimic a natural situation. In this state the fish maintains water balance and stops drinking, which allows sampling of the undiluted secretory product of the oxynticopeptic cells of the stomach mucosa (Holstein, 1979).

Acid secretion is stimulated by histamine from mast cells. In addition to the secretomotor control exerted by a number of hormones released from local endocrine cells (substance P, bombesin, somatostatin, gastrin), autonomic nerves are involved in the control of the secretion: VIP-nerves are inhibitory, while cholinergic (vagal) nerves are excitatory. 5-HT, tachykinin, and bombesin neurons may be involved in the excitatory control of acid secretion (see Jönsson and Holmgren, 1989).

The endocrine pancreas has an extensive innervation by fibers containing galanin-like and VIP-like material, and there are also reports on the presence of CCK-, enkephalin-, NPY-, oxytocin-, and substance P-like immunoreactivity in nerve fibers surrounding the endocrine cells, and bombesin-, CGRP-, and VIP-like immunoreactivity in perivascular nerves in the pancreas (e.g., Van Noorden and Patent, 1980; Noe et al., 1986; McDonald et al., 1987; Yui et al., 1988; Jönsson, 1991). There are, however, no studies on the function of these nerves in fish.

Little or nothing is known of the autonomic control of pepsin secretion (from oxynticopeptic cells), bicarbonate secretion (from intestinal mucosa and exocrine pancreas), and mucous secretion in the fish gut.

A special case of "gut secretion" is the salt (Na^+ and Cl^-)-secreting function of the rectal gland of elasmobranch fish. As the name implies, this gland is situated near the rectum, and the glandular duct empties in the hindgut posterior to the spiral valves. Autonomic neurons containing VIP-, GRP-, gastrin/CCK, and bombesin innervate the capsule or run along the secretory ducts (Holmgren and Nilsson, 1983a; Chipkin et al., 1988; Yui et al., 1990). Porcine VIP stimulates the Cl^- secretion in *Squalus acanthias* (Stoff et al., 1979), but not in *Scyliorhinus canicula* or *Raja radiata* (Shuttleworth and Thorndyke, 1984; Thorndyke and Shuttleworth, 1985). Instead, an endogenous peptide, rectin, has been shown to act as a potential stimulant in *S. canicula* (Thorndyke and Shuttleworth, 1985). Bombesin and somatostatin both inhibit salt secretion (Stoff et al., 1979; Silva et al., 1988, 1990), but clearly the function of the peptidergic nerves in the control of the rectal gland deserves further attention.

Gut Motility

Studies on gut motility in fish have been using isolated strip preparations, isolated vascularly perfused stomach or intestine, or the whole fish *in vivo*. A fluid-filled balloon may be inserted into the stomach or intestine, and either the pressure or the volume changes within the balloon recorded (Holmgren and Nilsson, 1981; Holmgren et al., 1985). The gut can perform several types of movements: reception of food, mixing of the food, emptying of the food from the stomach to the intestine, peristalsis, and defecation are some examples (Holmgren, 1989).

Reception of Food

Feeding patterns vary considerably between fish species. Some fish can instantly swallow big prey, while others only nibble their food and will vomit if forced to eat bigger prey. Studies of distension of the stomach of the rainbow trout, cod, and some species of flatfish have demonstrated a similar mechanical response to gut filling in all species. During a first phase the stomach relaxes rapidly, then a rhythmic activity is started and the relaxation proceeds to its maximum at a slower rate. However, the nervous and hormonal control may vary between species. In the rainbow trout, the initiation of activity depends on cholinergic and serotonergic nerves. The cholinergic nerves are inhibited by somatostatin, the serotonergic nerves apparently by the neuropeptide VIP. In the cod, the serotonin/VIP control system is absent (Grove and Holmgren, 1992a and b).

Gallbladder Motility

The gallbladder in fish, as in mammals, is contracted by cholecystokinin and related peptides. In *in vivo* experiments, the fish are kept swimming in a flume, and the pressure inside the gallbladder is measured *via* a catheter inserted into the bladder. Injection of food (moistened pellets) or acid into the proximal intestine of the rainbow trout causes contraction of the gallbladder. Similar to the situation in mammals, this effect may be mediated by a gastrin/CCK-related peptide. Gastrin/CCK-containing cells are present in the intestinal mucosa, and CCK injected to the blood stream mimics the effect of food or acid inserted into the intestine. The effect of CCK may be modulated by VIP-neurons present in the gallbladder wall (Aldman et al., 1992; Aldman and Holmgren, 1993).

Gastric Emptying

Another aspect of gut motility is gastric emptying. If fish are fed food pellets prepared with electron-dense beads, the passage of food through the gut may be followed by X-ray photography. Preliminary results in rainbow trout indicate a clear delay in gastric emptying if the fish are treated with CCK (S. Holmgren, G. Aldman, and D.J. Grove, in preparation). CCK is normally released when food enters the intestine, which facilitates the processing of the food, e.g., by release of bile (see above). A delay in the emptying of more food into the intestine before this first portion is processed may be an important coordination that may thus be controlled by the release of CCK.

Urinogenital Organs

Cyclostomes, female elasmobranchs, and many teleosts have a urinary bladder (Romer, 1971). The elasmobranch bladder is innervated by posterior splanchnic nerves (Young, 1933c). Similarly, the teleost bladder and gonads receive excitatory cholinergic and inhibitory adrenergic fibers from the posterior splanchnic nerve (Nilsson, 1983). The excitatory response of the urinary bladder in *Lophius piscatorius*, and the inhibitory response in the cod, to splanchnic nerve stimulation could not be completely blocked by atropine and adrenergic antagonists, respectively, suggesting the presence of a nonadrenergic, noncholinergic innervation (Young, 1936; Lundin and Holmgren, 1986).

Immunohistochemistry has shown serotonergic neurons and a peptidergic innervation by VIP-fibers only in both the urinary bladder and the gonads of the cod. Porcine VIP has no effects on the gonads, but inhibits rhythmic contractions of the wall of the urinary bladder (while the inhibitory response of the urinary bladder to nerve stimulation is a decrease in basic tonus) (Lundin and Holmgren, 1986; Uematsu et al., 1989).

Swim Bladders and Lungs

Swim Bladders

Nerves innervating the swim bladder are involved in the control of gas secretion, blood flow through the different parts of the swim bladder, and resorption of gas. Vagal cholinergic fibers innervate the gas gland and cause inflation of the swim bladder by an increased secretion from the gas gland cells. Gas resorption is controlled by adrenergic nerves, which constrict the gas gland vasculature via α-adrenoceptors, reduce the area of the secretory mucosa via α-adrenoceptors, and increase the resorptive area by an action on β-adrenoceptors (Fänge, 1953; Nilsson, 1983).

VIP-immunoreactive nerves are present in both secretory and resorptive areas of all teleost swim bladders that have been investigated. VIP reduces the tension of swimbladder wall muscle, increases the flow through the swim bladder, and increases gas secretion in atropinized cod. However, VIP decreases gas secretion in untreated cod, and a dual effect of VIP has been postulated: a direct stimulatory effect on the gas gland cells and an indirect effect via inhibition of cholinergic neurons (Lundin and Holmgren, 1984, 1991; Lundin, 1991; Figure 5).

Substance P-, neurotensin-, and enkephalin- fibers occur in some species (Lundin and Holmgren, 1989). Substance P is excitatory on swimbladder wall muscle, and increases the rate of gas filling in atropinized cod. The effects may be hormonal, or indirect on secretory neurons (Lundin, 1991; Lundin and Holmgren, 1991). No effects of neurotensin have been found, while enkephalin is excitatory on strip preparations of the eel (*Anguilla anguilla*) swimbladder wall (Lundin, 1991).

Lungs

Dipnoans use lungs for air breathing, sometimes as their main means of oxygen uptake. Some other fish species use their swim bladders for air breathing. These "lungs" are essentially simple sacs with different degrees of folding

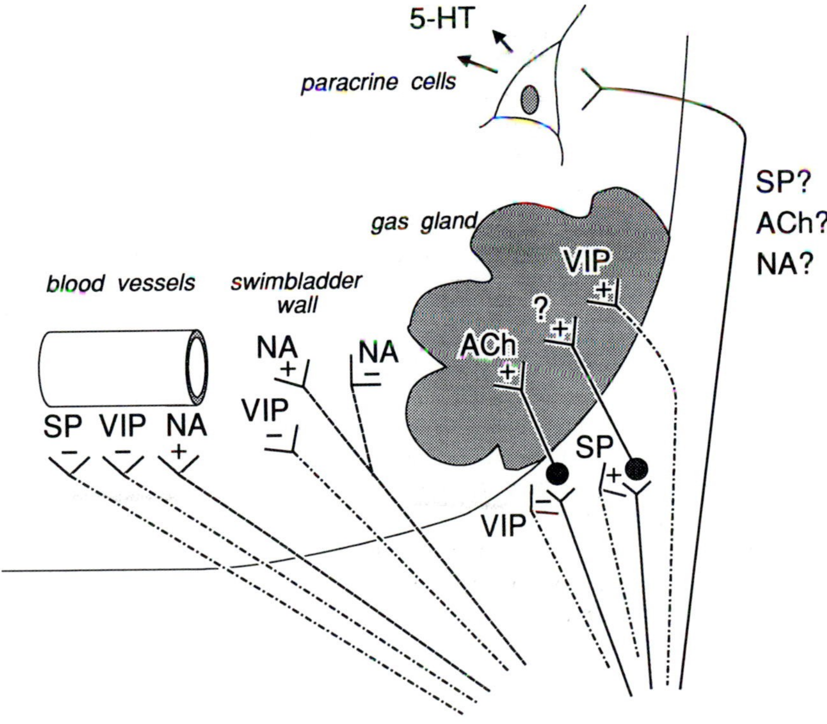

FIGURE 5. Summary of the autonomic innervation of the swim bladder in the cod, *Gadus morhua*. Abbreviations refer to the neurotransmitters/hormones released: 5-HT, 5-hydroxytryptamine; ACh, acetylcholine; NA, noradrenaline; SP, substance P; VIP, vasoactive intestinal polypeptide, and + and – indicate whether the effects are excitatory or inhibitory, respectively. (Based primarily on Lundin, 1991.)

of the inner epithelium into septa for surface enlargement. In dipnoans and holosteans, there is an excitatory vagal cholinergic innervation of the lung. Adrenergic nerves are present in the holostean lung, but not in the dipnoans (Abrahamsson et al., 1979a and b; Nilsson, 1981; Axelsson et al., 1989b). A moderate innervation by VIP and enkephalin-containing fibers is present in the dipnoan lung wall (Nilsson and Holmgren, 1992a).

The Iris

The autonomic pathways to the vertebrate iris show remarkable uniformity. Cranial autonomic nerve fibers run to the ciliary ganglion in the oculomotor nerve and short ciliary root, and postganglionic neurons to the iris run in the ciliary nerves. The anterior sympathetic chain ganglia send fibers to the iris via the ciliary nerves. However, the anatomical arrangement of the oculomotor nerve and ciliary roots and nerves may be similar in the different vertebrate groups, but the same cannot be said about the nature of the autonomic neurons that control the smooth muscles of the iris. Even in related species, the nature of the iris innervation can be quite different (Nilsson, 1983).

The iris is regulated much like the aperture of a camera, and the pupillary diameter determines the amount of light reaching the retina. In most tetrapods, the change in pupillary aperture depends on both circular smooth muscles at the edge of the iris (the *sphincter pupillae*) and radial muscles (the *dilator pupillae*). In many fish, the dilator muscle is absent. An interesting feature of the fish (and amphibian and reptilian) iris is an additional regulatory mechanism, a light-sensitive pigment within the iris sphincter that produces a direct pupilloconstriction in response to light (see Nilsson, 1993).

Elasmobranchs

Young (1933a) investigated the autonomic innervation of the iris in several elasmobranch species and showed the direct effects of light on the iris sphincter. There appears to be no autonomic innervation of the sphincter, but electrical stimulation of oculomotor nerve fibers contracts the dilator muscles. The fibers are probably cholinergic, since atropine reduced the pupillary diameter. There is no contribution to the iris innervation by fibers of spinal autonomic origin (Young, 1933a). Pupillary diameter in elasmobranchs is regulated by the antagonistic actions of the light-sensitive sphincter and the cholinergically innervated radial dilator muscle (Young, 1933a; Nilsson, 1993).

Teleosts

Several teleosts show a pupilloconstriction as a direct effect of light on the sphincter (Nilsson, 1980, 1992; Rubin and Nolte, 1981). The nature of the autonomic innervation varies between species. Cranial autonomic cholinergic excitatory fibers innervate the dilator muscle (*Uranoscopus scaber*) and there is a spinal autonomic cholinergic excitatory innervation of the sphincter (*U. scaber* and *L. piscatorius*) (Young, 1933b). In the cod, *G. morhua*, dilator muscles are absent but the sphincter contracts in response to light and in response to acetylcholine, adrenaline, and tyramine. Pharmacological studies confirm an excitatory adrenergic innervation of the sphincter, mediated by α-adrenoceptors (Nilsson, 1980). As mentioned earlier, histochemical evidence suggests that adrenergic ganglion cells lie within the ciliary ganglion (Nilsson, 1980, 1992).

Melanophores and Color Change

The skin of fish, like that in amphibians and some reptiles, can change hue due to redistribution of pigments within chromatophores, which are specialized flattened epidermal or dermal cells, often showing extensive branching. Contrary to the slow *morphological color change* that occurs due to new synthesis of pigment or proliferation of pigment cells (e.g., suntan in humans), the *physiological* color change in fish is often quite rapid (minutes). The brown/black chromatophores, melanophores, that are common in the skin of many vertebrates contain granules of the pigment melanin. The granules can move to occupy a smaller or larger part of the cell, thus modifying the shade of the skin surface. Pigment movement depends on intracellular transport operated by the microtubule system (for details, see Bagnara and Hadley, 1973; Fujii et al., 1991; Rodionov et al., 1991).

In teleosts, the movement of pigment (dispersion or aggregation) is controlled both by hormones and by the autonomic nervous system. The chief hormone involved in this control is an MSH (melanocyte-stimulating hormone), released from the pituitary gland. The effect of this hormone varies between species: in *A. anguilla* pigment dispersion is induced, while in other species, e.g., *Pleuronectes platessa* and *Phoxinus phoxinus*, there is, instead, a pigment aggregation (Parker, 1934; Bagnara and Hadley, 1973).

The autonomic nerve control is exerted by spinal autonomic pathways forming "chromatic tracts", with the postganglionic neurons distributed to the skin along the spinal nerves (von Frisch, 1911; Grove, 1969a and b; Fernando and Grove, 1974a and b; Fernando, 1989). Histochemical studies have revealed adrenergic nerves running to the melanophores (Falck et al., 1969). Suggestions that the melanophores receive a double antagonistic innervation remain to be confirmed, and the postganglionic fibers are, with few known exceptions, adrenergic (for a full discussion, see Grove, 1993). Pigment aggregation occurs via activation of α-adrenoceptors of the melanophores, and pharmacological evidence suggests that these adrenoceptors resemble α_2-adrenoceptors (Andersson et al., 1984).

REFERENCES

Abrahamsson, T., Axonal transport of adrenaline, noradrenaline and phenylethanolamine-N-methyl transferase (PNMT) in sympathetic neurons of the cod, *Gadus morhua, Acta Physiol. Scand.*, 105, 316, 1979a.

Abrahamsson, T., Phenylethanolamine-N-methyl transferase (PNMT) activity and catecholamine storage and release from chromaffin tissue of the spiny dogfish, *Squalus acanthias, Comp. Biochem. Physiol. C*, 64, 169, 1979b.

Abrahamsson, T. and Nilsson, S., Phenylethanolamine-N-methyl transferase (PNMT) activity and catecholamine content in chromaffin tissue and sympathetic neurons in the cod, *Gadus morhua, Acta Physiol. Scand.*, 96, 94, 1976.

Abrahamsson, T., Holmgren, S., Nilsson, S., and Pettersson, K., On the chromaffin system of the African lungfish, *Protopterus aethiopicus, Acta Physiol. Scand.*, 107, 135, 1979a.

Abrahamsson, T., Holmgren, S., Nilsson, S., and Pettersson, K., Adrenergic and cholinergic effects on the heart, the lung and the spleen of the African lungfish, *Protopterus aethiopicus, Acta Physiol. Scand.*, 107, 141, 1979b.

Aldman, G. and Holmgren, S., VIP inhibits CCK-induced gallbladder contarction involving a beta-adrenoceptor mediated pathway in the rainbow trout, *Oncorhynchus mykiss, in vivo, Gen. Comp. Endocrinol.*, 88, 351, 1992.

Aldman, G., Jönsson, A.C., Jensen, J., and Holmgren, S., Gastrin/CCK-like peptides in the spiny dogfish, *Squalus acanthias*; concentrations and actions in the gut, *Comp. Biochem. Physiol. C*, 92, 103, 1989.

Aldman, G., Grove, D.J., and Holmgren, S., Duodenal acidification and intraarterial injection of CCK8 increase gallbladder motility in the rainbow trout *Oncorhynchus mykiss, Gen. Comp. Endocrinol.*, 86, 20, 1992.

Anastasi, A., Erspamer, V., and Bucci, M., Isolation and structure of bombesin and alytesin, two analogous active peptides from the skin of the European amphibians *Bombina* and *Alytes, Experentia,* 27, 166, 1971.

Anderson, C., Evidence for 5-HT-containing intrinsic neurons in the teleost intestine, *Cell Tissue Res.*, 230, 377, 1983.

Anderson, C. and Campbell, G., Immunohistochemical study of 5- HT-containing neurons in the teleost intestine: relationship to the presence of enterochromaffin cells, *Cell Tissue Res.*, 254, 553, 1988.

Andersson, R. G. G., Karlsson, J. O., and Grundström, N., Adrenergic nerves and alpha-2 adrenoceptor system regulating melanosome aggregation within fish, *Labus ossifragus* melanophores, *Acta Physiol. Scand.*, 121, 173, 1984.

Augustinsson, K.-B., Fänge, R., Johnels, A., and Östlund, E., Histological, physiological and biochemical studies on the heart of two cyclostomes, hagfish *(Myxine)* and lamprey *(Lampetra), J. Physiol.*, 131, 257, 1956.

Axelsson, M., The importance of nervous and humoral mechanisms in the control of cardiac performance in the Atlantic cod, *Gadus morhua*, at rest and during non-exhaustive exercise, *J. Exp. Biol.*, 137, 287, 1988.

Axelsson, M. and Fritsche, R., Effects of exercise, hypoxia and feeding on the gastrointestinal blood flow in the Atlantic cod, *Gadus morhua, J. Exp. Biol.*, 158, 181, 1991.

Axelsson, M. and Nilsson, S., Blood pressure control during exercise in the Atlantic cod *Gadus morhua, J. Exp. Biol.*, 126, 225, 1986.

Axelsson, M., Driedzic, W. R., Farrell, A.P., and Nilsson, S., Regulation of cardiac output and gut blood flow in the sea raven, *Hemitripterus americanus, Fish Physiol. Biochem.*, 6, 315, 1989a.

Axelsson, M., Abe, A. S., Bicudo, J. E. P. W., and Nilsson, S., On the cardiac control in the South American lungfish, *Lepidosiren paradoxa, Comp. Biochem. Physiol.*, 93A, 561, 1989b.

Axelsson, M., Farrell, A. P., and Nilsson, S., Effects of hypoxia and drugs on the cardiovascular dynamics of the Atlantic hagfish, *Myxine glutinosa, J. Exp. Biol.*, 151, 297, 1990.

Bagnara, J. T. and Hadley, M. E., *Chromatophors and Color Change*, Prentice Hall, Englewood Cliffs, NJ, 1973.

Bailly, Y. and Dunel-Erb, S., The sphincter of the efferent filament artery in teleost gills. I. Structure and parasympathetic innervation, *J. Morphol.*, 187, 219, 1986.

Bailly, Y., Dunel-Erb, S., Geffard, M., and Laurent, P., The vascular and epithelial serotonergic innervation of the actinopterygian gill filament with special reference to the trout, *Salmo gairdneri, Cell Tissue Res.*, 258, 349, 1989.

Balashov, N. V., Fänge, R., Govyrin, V. A., Leont'eva, G. R., Nilsson, S., and Prozorovskaya, M. P., On the adrenergic system of ganoid fish: the beluga, *Huso huso* (Chondrostei), *Acta Physiol. Scand.*, 111, 435, 1981.

Baumgarten, H.G., Björklund, A., Lachenmayer, L., Nobin, A., and Rosengren, E., Evidence for the existence of serotonin-, dopamine- and noradrenaline-containing neurons in the gut of *Lampetra fluviatilis, Z. Zellforsch.*, 141, 33, 1973.

Beringer, T. and Hadek, R., Cardiac innervation in the lamprey, *Petromyzon marinus, Anat. Rec.*, 172, 269, 1972.

Bjenning, C. and Holmgren, S., Neuropeptides in the fish gut. A study of evolutionary trends, *Histochemistry*, 88, 155, 1988.

Bjenning, C., Driedzic, W., and Holmgren, S., Neuropeptide Y-like immunoreactivity in the cardiovascular nerve plexus of the elasmobranchs Raja erinacea and Raja radiata, *Cell Tissue Res.*, 255, 481, 1989.

Bjenning, C., Jönsson, A.C., and Holmgren, S., Bombesin-like immunoreactive material in the gut, and the effect of bombesin on the stomach circulatory system of an elasmobranch fish, *Squalus acanthias, Regul. Pept.*, 28, 57, 1990.

Bjenning, C., Farrell, A. P., and Holmgren, S., Bombesin-like immunoreactivity in the gut and heart region of elasmobranch fish (skates), and the in vitro effect of bombesin on coronary vessels from the longnose skate, *Raja rhina, Regulatory Pept.*, 35, 207, 1991.

Blomqvist, A.G., Söderberg, C., Lundell, I., Milner, R.J., and Larhammar, D., Strong evolutionary conservation of neuropeptide Y: sequences of chicken, goldfish, and *Torpedo marmorata* DNA clones, *Proc. Natl. Acad. Sci. U.S.A.*, 89, 2350, 1992.

Bloom, G., Östlund, E., Euler, U. S. V., Lishajko, F., Ritzén, M., and Adams-Ray, J., Studies on catecholamine-containing granules of specific cells in cyclostome hearts, *Acta Physiol. Scand.*, 53 (Suppl. 185), 1, 1961.

Burkhardt-Holm, P. and Holmgren, S., A study of the alimentary canal of the brachiopterygian fish *Polypterus senegalensis* with electron microscopy and immunohistochemistry, *Acta Zool. (Stockholm)*, 73, 85, 1992.

Burnstock, G., The innervation of the gut of the brown trout *(Salmo trutta), Q. J. Microsc. Sci.*, 100, 199, 1959.

Burnstock, G., Evolution of the autonomic innervation of visceral and cardiovascular systems in vertebrates, *Pharmacol. Rev.*, 21, 247, 1969.

Burnstock, G., Do some sympathetic neurones synthesize and release both noradrenaline and acetylcholine?, *Progr. Neurobiol.*, 11, 205, 1978.

Burnstock, G., The changing face of autonomic neurotransmission, *Acta Physiol. Scand.*, 126, 67, 1986.

Butler, P. J., Axelsson, M., Ehrenström, F., Metcalfe, J. D., and Nilsson, S., Circulating catecholamines and swimming performance in the Atlantic cod, *Gadus morhua, J. Exp. Biol.*, 141, 377, 1989.

Cameron, J. S., Autonomic nervous tone and regulation of heart rate in the gold fish, *Carassius auratus, Comp. Biochem. Physiol.*, 63C, 341, 1979.

Campbell, G., Autonomic nervous system, in *Fish Physiology* Vol. 4, Hoar, W. S. and Randall, D. J., Eds., Academic Press, New York, London, 1970, 109.

Campbell, G., Gibbins, I. L., Morris, J. L., Furness, J. B., Costa, M., Oliver, J. R., Beardsley, A. M., and Murphy, R., Somatostatin is contained in and released from cholinergic nerves in the heart of the toad *Bufo marinus, Neuroscience*, 7, 2013, 1982.

Capra, M. F. and Satchell, G. H., Adrenergic and cholinergic responses of the isolated saline-perfused heart of the elasmobranch fish *Squalus acanthias, Gen. Pharmacol.*, 8, 59, 1977.

Chevrel, R., Système nerveux grand-sympathique des elasmobranchs et des poissons osseux, *Arch. Zool. Exp. Gen.*, 5, V, 1887–1890.

Chipkin, S. R., Stoff, J. S., and Aronin, N., Immunohistochemical evidence for neural mediation of VIP activity in the dogfish rectal gland, *Peptides*, 9, 119, 1988.

Conlon, J. M., Deacon, C. F., O'Toole, L., and Thim, L., Scyliorhinin I and II: two novel tachykinins from the dogfish gut, *FEBS Letters*, 200, 111, 1986.

Conlon, J. M., Henderson, I. W., and Thim, L., Gastrin-releasing peptide from the intestine of the elasmobranch fish, *Scyliorhinus canicula* (common dogfish), *Gen. Comp. Endocrinol.*, 68, 415, 1987.

Conlon, J. M., Bjenning, C., Moon, T. W., Youson, J. H., and Thim, L., Neropeptide Y-related peptides from the pancreas of teleostean (eel), holostean (bowfin) and elasmobranch (skate) fish, *Peptides*, 12, 221, 1991.

Conlon, J. M., Bjenning, C., and Hazon, N., Structural characterization of neuropeptide Y from the brain of the dogfish, *Scyliorhinus canicula, Peptides*, 13, 493, 1992.

Coupland, R. E., *The Natural History of the Chromaffin Cell*, Longmans, London, 1965.

Coupland, R. E., The chromaffin system, in Handbook of Experimental Pharmacology, Vol. 33. *Catecholamines*, Blaschko, M., and Muscholl, E., Eds., Springer, Berlin, 1972, 16.

Davie, P. S., Forster, M. E., Davison, B., and Satchell, G. H., Cardiac function in the New Zealand hagfish, *Eptatretus cirrhatus, Physiol. Zool.*, 60, 233, 1987.

Davies, D.T. and Rankin, J.C., Adrenergic receptors and vascular responses to catecholamines of perfused dogfish gills, *Comp. Gen. Pharmacol.*, 4, 139, 1973.

Dimaline, R., Young, J., Thwaites, D. T., Lee, C. M., Shuttleworth, T. J., and Thorndyke, M. C., A novel vasoactive intestinal peptide (VIP) from elasmobranch intestine has full affinity for mammalian pancreatic VIP receptors, *Biochim. Biophys. Acta*, 930, 97, 1987.

Donald J., Adrenergic innervation of the gills of brown and rainbow trout, *Salmo trutta* and *S. gairdneri, J. Morph.*, 182, 307, 1984.

Donald, J. A., Comparative study of the adrenergic innervation of the teleost gill, *J. Morphol.*, 193, 63, 1987.

Donald, J. and Campbell, G., A comparative study of the adrenergic innervation of the teleost heart, *J. Comp. Physiol.*, 147, 85, 1982.

Dunel-Erb, S., Bailly, Y., and Laurent, P., Neurons controlling the gill vasculature in five species of teleosts, *Cell Tissue Res.*, 255, 567, 1989.

El-Salhy, M., Immunocytochemical investigation of the gastro- entero-pancreatic (GEP) neuro-hormonal peptides in the pancreas and gastrointestinal tract of the dogfish *Squalus acanthias, Histochemistry*, 80, 193, 1984.

Eränkö, O., On the histochemistry of the adrenal medulla of the rat, with special reference to acid phosphatase, *Acta Anat.*, 16 (Suppl. 17), 308, 1952.

Eränkö, O., Distribution of adrenaline and noradrenaline in the adrenal medulla, *Nature (London)*, 8, 88, 1955.

Eränkö, O. and Härkönen, M., Histochemical demonstration of fluorogenic amines in the cytoplasm of sympathetic ganglion cells of the rat, *Acta Physiol. Scand.*, 58, 285, 1963.

Eränkö, O. and Härkönen, M., Monoamine-containing small cells in the superior cervical ganglion of the rat and an organ composed of them, *Acta Physiol. Scand.*, 63, 511, 1965.

Euler, U. S. V. and Fänge, R., Catecholamines in nerves and organs of *Myxine glutinosa, Squalus acanthias*, and *Gadus callarias, Gen. Comp. Endocrinol.*, 1, 191, 1961.

Falck, B., Hillarp, N. Å., Thieme, G., and Torp, A., Fluorescence of catecholamines and related compounds condensed with formaldehyde, *J. Histochem. Cytochem.*, 10, 348, 1962.

Falck, B., Müntzing, J., and Rosengren, A.M., Adrenergic nerves to the dermal melanophores of the rainbow trout, *Salmo gairdneri, Z. Zellforsch.*, 99, 430, 1969.

Fänge, R., The mechanisms of gas transport in the euphysoclist swimbladder, *Acta Physiol. Scand.*, 30, 1, 1953.

Fänge, R. and Holmgren, S., Choline acetyltransferase activity in the fish swimbladder, *J. Comp. Physiol. B*, 146, 57, 1982.

Fänge, R. and Johnels, A. G., An autonomic nerve plexus control of the gall bladder in *Myxine, Acta Zool.*, 39, 1, 1958.

Fänge, R. and Östlund, E., The effects of adrenaline, noradrenaline, tyramine and other drugs on the isolated heart from marine vertebrates and a cephalopod *(Eledone cirrosa), Acta Zool. (Stockholm)*, 35, 289, 1954.

Fänge, R., Johnels, A. G., and Enger, P. S., The autonomic nervous system, in *The biology of Myxine*, Brodal, A., and Fänge R., Eds., Univ. Forlaget, Oslo, 1963, 124.

Farrell, A. P. and Davie, P. S., Coronary vascular reactivity in the skate, *Raja nasuta, Comp. Biochem. Physiol. C*, 99, 555, 1991.

Fernando, M. M., Monoaminergic nerves in the skin of plaice, *Pleuronectes platessa* (L.), *Comp. Biochem. Physiol. C*, 92, 1, 1989.

Fernando, M. M. and Grove, D. J., Melanophore aggregation in the plaice *(Pleuronectes platessa* L.). I. Changes in in vivo sensitivity to sympathomimetic amines, *Comp. Biochem. Physiol. A*, 48, 711, 1974a.

Fernando, M. M. and Grove, D. J., Melanophore aggregation in the plaice, *Pleuronectes platessa*. II. *In vitro* effects of adrenergic drugs, *Comp. Biochem. Physiol. A*, 48, 723, 1974b.

Forster, M. E., Effects of adrenergic blocking drugs on the cardiovascular system of the eel, *Anguilla anguilla* (L.), *Comp. Biochem. Physiol. C*, 55, 33, 1976.

Forster, M. E., Axelsson, M., Farrell, A. P., and Nilsson, S., Cardiac function and circulation in hagfishes, *Can. J. Zool.*, 69, 1985, 1991.

Fritsche, R. and Nilsson, S., Cardiovascular responses to hypoxia in the Atlantic cod, *Gadus morhua, Exp. Biol.*, 48, 153, 1989.

Fritsche, R. and Nilsson, S., Autonomic nervous control of blood pressure and heart rate during hypoxia in the cod, *Gadus morhua, J. Comp. Physiol.*, 160B, 287, 1990.

Fritsche, R. and Nilsson, S., Cardiovascular and respiratory control during hypoxia, in *Fish Ecophysiology*, Jensen, F. B. and Rankin, J. C., Eds., Chapman and Hall, London, New York, 1993, 180.

Fujii, R., Wakatabi, H., and Oshima, N., Inositol 1,4,5- triphosphate signals the motile response of fish chromatophores. I. Aggregation of pigment in the *Tilapia* melanophore, *J. Exp. Zool.*, 259, 9, 1991.

Furness, J. B. and Costa, M., *The Enteric Nervous System,* Churchill Livingstone, Edinburgh, 1987.

Gannon, B. J., A study of the dual innervation of teleost heart by a field stimulation technique, *Comp. Gen. Pharmacol.*, 2, 175, 1971.

Gannon, B. J. and Burnstock, G., Excitatory adrenergic innervation of the fish heart, *Comp. Biochem. Physiol.*, 29, 765, 1969.

Gannon, B.J., Campbell, G.D., and Satchell, G.H., Monoamine storage in relation to cardiac regulation in the Port Jackson shark *Heterodontus portusjacksoni, Z. Zellforsch.*, 131, 437, 1972.

Giacomini, E., Sulla esistenza della sostanza midollare nelle capsule surrenale del Teleostei, *Monit. Zool. Ital.*, 13, 183, 1902.

Giacomini, E., Sulle capsule surrenali e sul simpatico dei Dipnoi. Ricerche in *Protopterus annectens, R. C. Acad. Lincei*, 15, 394, 1906.

Gibbins, I. L., Co-existence and co-function, in *The Comparative Physiology of Regulatory Peptides*, Holmgren, S., Ed., Chapman and Hall, London, 1989, 308.

Gibbins, I. L., Comparative anatomy and evolution of the autonomic nervous system, in *Comparative Physiology and Evolution of the Autonomic Nervous System*, Nilsson, S. and Holmgren, S., Eds.; Burnstock, G., Series Ed., Harwood Academic, Chur, 1993.

Goodrich, J. T., Bernd, P., Sherman, D. L., and Gershon, M. D., Phylogeny of enteric serotonergic neurons, *J. Comp. Neurol.*, 190, 15, 1980.

Govyrin, V. A. and Leont'eva, G. R., Distribution of catecholamines in myocardium of vertebrates, *J. Evol. Biochem. Physiol.*, 1, 38, 1965.

Grove, D. J., The effects of adrenergic drugs on melanophores of the minnow, *Phoxinus phoxinus* (L.), *Comp. Biochem. Physiol.*, 28, 37, 1969a.

Grove, D. J., Melanophore dispersion in the minnow, *Phoxinus phoxinus* (L.), *Comp. Biochem. Physiol.*, 28, 55, 1969b.

Grove, D. J., Chromatophores, in *Comparative Physiology and Evolution of the Autonomic Nervous System*, Nilsson, S. and Holmgren, S., Eds.; Burnstock G., Series Ed., Harwood Academic, Chur, 1993.

Grove, D. J. and Holmgren, S., Intrinsic mechanisms controlling cardiac stomach volume of the rainbow trout *(Oncorhynchus mykiss)* following gastric distension, *J. Exp. Biol.*, 163, 33, 1992a.

Grove, D. J. and Holmgren, S., Mechanisms controlling stomach volume of the Atlantic cod *(Gadus morhua)* following gastric distension, *J. Exp. Biol.*, 163, 49, 1992b.

Henle, J., Über das Gewebe der Nebenniere und der Hypophysis, *Z. Rat. Mat.*, 24, 143, 1865.

Hirsch, E. F., Jellinek, M., and Cooper, T., Innervation of the hagfish heart, *Circ. Res.*, 14, 212, 1964.

Holder, F.C., Vincent, B., Ristori, M.T., and Laurent, P., Vascular perfusion of an intestinal loop in the catfish *Ictalurus melas*. Demonstration of the vasoactive effects of the mammalian vasoactive intestinal peptide and gastro-intestinal extracts from teleost fish, *C. R. Seances Acad. Sci. Ser. III Sci. Vie*, 296, 783, 1983.

Holmes, W., The adrenal homologues in the lungfish *Protopterus, Proc. R. Soc. London Ser. B*, 137, 549, 1950.

Holmgren, S., Regulation of the heart of a teleost, *Gadus morhua*, by autonomic nerves and circulating catecholamines, *Acta Physiol. Scand.*, 99, 62, 1977.

Holmgren, S., Choline acetyltransferase activity in the heart from two teleosts, *Gadus morhua* and *Salmo gairdneri, Comp. Biochem. Physiol.*, 69C, 403, 1981.

Holmgren, S., The effect of putative non-adrenergic, non- cholinergic autonomic transmitters on isolated strips from the stomach of the rainbow trout, *Salmo gairdneri, Comp. Biochem. Physiol. C*, 74, 229, 1983.

Holmgren, S., Substance P in the gastrointestinal tract of *Squalus acanthias, Molec. Physiol.*, 8, 119, 1985.

Holmgren, S., Ed., *The Comparative Physiology of Regulatory Peptides*, Chapman and Hall, London, 1989.

Holmgren, S. and Jensen, J., Comparative aspects on the biochemical identity of neurotransmitters of autonomic neurons, in *Comparative Physiology and Evolution of the Autonomic Nervous System*, Nilsson, S. and Holmgren, S., Eds., Burnstock, G., Series Ed., Harwood Academic, Chur, in press.

Holmgren, S. and Jönsson, A.C., Occurrence and effects on motility of bombesin-related peptides in the gastrointestinal tract of the Atlantic cod, *Gadus morhua, Comp. Biochem. Physiol. C*, 89, 249, 1988.

Holmgren, S. and Nilsson, S., Effects of denervation, 6- hydroxydopamine and reserpine on the cholinergic and adrenergic responses of the spleen of the cod, *Gadus morhua, Eur. J. Pharmacol.*, 39, 53, 1976.

Holmgren, S. and Nilsson, S., On the non-adrenergic, non- cholinergic innervation of the rainbow trout stomach, *Comp. Biochem. Physiol.*, 70, 65, 1981.

Holmgren, S. and Nilsson, S., Bombesin-, gastrin/CCK-, 5- hydoxytryptamine-, neurotensin-, somatostatin-, and VIP-like immunoreactivity and catecholamine fluorescence in the gut of the elasmobranch, *Squalus acanthias, Cell Tissue Res.*, 234, 595, 1983a.

Holmgren, S. and Nilsson, S., VIP-, bombesin-, and neurotensin-like immunoreactivity in neurons of the gut of the holostean fish, *Lepisosteus platyrhincus, Acta Zool. (Stockholm)*, 64, 25, 1983b.

Holmgren, S. and Nilsson, S., Novel neurotransmitters in the autonomic nervous systems of non-mammalian vertebrates, in *Novel Peripheral Neurotransmitters.* Sect. 135 of *Int. Encycl. Pharmacol. Ther.*, Bell, C., Series Ed., Pergamon Press, New York, 1991, 293.

Holmgren, S., Grove, D.J., and Nilsson, S., Substance P acts by releasing 5-hydroxytryptamine from enteric neurons in the stomach of the rainbow trout, *Salmo gairdneri, Neuroscience*, 14, 683, 1985.

Holmgren, S., Axelsson, M., and Farrell, A. P., The effect of catecholamines, substance P and vasoactive intestinal polypeptide (VIP) on blood flow to the gut in the dogfish *Squalus acanthias, J. Exp. Biol.*, 168, 161, 1992a.

Holmgren, S., Axelsson, M., and Farrell, A. P., The effects of neuropeptide Y and bombesin on blood flow to the gut in the dogfish *Squalus acanthias*, 9th Int. Symp. Gastrointest. Hormones, *Regul. Peptides*, 40, 169, 1992b.

Holstein, B., Gastric acid secretion and water balance in the marine teleost "*Gadus morhua', Acta Physiol. Scand.*, 105, 93, 1979.

Holtzbauer, M. and Sharman, D. F., The distribution of catecholamines in vertebrates, in *Catecholamines. Handbuch der Experimentellen Pharmakologie*, Vol. 33, Blaschko, H. and Muscholl, E., Eds., Springer, Berlin, 1972, 110.

Iijima, T. and Wasano, T., A histochemical and ultrastructural study of serotonin-containing nerves in cerebral blood vessels of the lamprey, *Anat. Rec.*, 198, 671, 1980.

Jenkin, P. M., Note on the nervous system of *Lepidosiren paradoxa, Proc. R. Soc. Edinburgh*, 48, 55, 1928.

Jensen, J. and Conlon, J.M., Substance P-related and neurokinin A-related peptides from the brain of the cod and trout, *Eur. J. Biochem.*, 206, 659, 1992a.

Jensen, J. and Conlon, J.M., Isolation and primary structure of gastrin-releasing peptide from a teleost fish, the trout *(Oncorhynchus mykiss), Peptides*, 13, 995, 1992b.

Jensen, J. and Holmgren, S., Neurotransmitters in the intestine of the Atlantic cod, *Gadus morhua, Comp. Biochem. Physiol.*, 82C, 81, 1985.

Jensen, J. and Holmgren, S., The gastrointestinal canal, in *Comparative Physiology and Evolution of the Autonomic Nervous System,* Nilsson, S. and Holmgren, S., Eds.; Burnstock, G., Series Ed., Harwood Academic, Chur, 1993.

Jensen, J., Axelsson, M., and Holmgren, S., Effects of substance P and vasoactive intestinal polypeptide on gastrointestinal blood flow in the Atlantic cod *Gadus morhua, J. Exp. Biol.*, 156, 361, 1991.

Johansen, K. and Burggren, W., Cardiovascular function in the lower vertebrates, in *Hearts and Heart-Like Organs*, Bourne, G. H., Ed., Academic Press, London, 1980, 61.

Jönsson, A. C., Regulatory peptides in the pancreas of two species of elasmobranchs and in the Brockmann bodies of four teleost species, *Cell Tissue Res*, 266, 163, 1991.

Jönsson, A. C. and Holmgren, S., Gut secretion, in *The Comparative Physiology of Regulatory Peptides*, Holmgren, S., Ed., Chapman and Hall, London, 1989, 256.

Jönsson, A. C., Holmgren, S., and Holstein, B., Gastrin/CCK-like immunoreactivity in endocrine cells and nerves in the gastrointestinal tract of the cod, *Gadus morhua*, and the effect of peptides of the gastrin/CCK family on cod gastrointestinal smooth muscle, *Gen. Comp. Endocrinol.*, 66, 190, 1987.

Karila, P., Jönsson, A.C., Jensen, J., and Holmgren, S., Galanin- like immunoreactivity in extrinsic and intrinsic nerves to the gut of the Atlantic cod, *Gadus morhua*, and the effect of galanin on gut smooth muscle, *Cell Tissue Res.*, 271, 537, 1993.

Kirtisinghe, P., The myenteric nerve-plexus in some lower chordates, *Q. J. Microsc. Sci.*, 81, 521, 1940.

Kohn, A., Das chromaffin Gewebe, *Ergeb. Anat. Entwicklungsgesch.*, 12, 253, 1902.

Langley, J. N., On the union of cranial autonomic (visceral) fibres with the nerve cells of the superior cervical ganglion, *J. Physiol. (London)*, 23, 240, 1898.

Langley, J.N., *The Autonomic Nervous System,* Part I, Heffer, Cambridge, 1921.

Larhammar, D., Söderberg, C., and Blomqvist, A. G., Evolution of the neuropeptide Y family of peptides, in *Neuropeptide Y and Related Peptides*, Wahlestedt, C. and Colmers, W. F., Eds., Humana Press, Clifton, NJ, 1993.

Laurent, P., Contribution a l'étude morphologique et physiologique de l'innervation du coeur des téléosténs, *Anat. Microsc. Morphol. Exp.*, 51, 339, 1962.

Laurent, P., Holmgren, S., and Nilsson, S., Nervous and humoral control of the fish heart: structure and function, *Comp. Biochem. Physiol.*, 76A, 525, 1983.

Leont'eva, G. R., Distribution of catecholamines in blood vessel walls of cyclostomes, fishes, amphibians and reptiles, *J. Evol. Biochem. Physiol.*, 2, 31, 1966.

Lukomskaya, N.J. and Michelson, M.J., Pharmacology of the isolated heart of the lamprey, *Lampetra fluviatilis, Comp. Gen. Pharmacol.*, 3, 213, 1972.

Lundin, K., The Teleost Swimbladder. A Study of the Non- Adrenergic, Non-Cholinergic Innervation, thesis, University of Göteborg, Göteborg, Sweden, 1991.

Lundin, K. and Holmgren, S., Vasointestinal polypeptide-like immunoreactivity and effects of VIP in the swimbladder of the cod, *Gadus morhua, Comp. Physiol.*, 154B, 627, 1984.

Lundin, K. and Holmgren, S., Non-adrenergic, non-cholinergic innervation of the urinary bladder of the Atlantic cod, *Gadus morhua, Comp. Biochem. Physiol.*, 84C, 315, 1986.

Lundin, K. and Holmgren, S., The occurrence and distribution of peptide- or 5-HT-containing nerves in the swimbladder of four different species of teleosts, (*Gadus morhua, Ctenolabrus rupestris, Anguilla anguilla, Salmo gairdneri), Cell Tissue Res.*, 257, 641, 1989.

Lundin, K. and Holmgren, S., An X-ray study of the influence of vasoactive intestinal polypeptide and substance P on the secretion of gas into the swimbladder of a teleost, *Gadus morhua, J. Exp. Biol.*, 157, 286, 1991.

Lundin, K., Holmgren, S., and Nilsson, S., Peptidergic functions in the dogfish rectum, *Acta Physiol. Scand.*, 121, 46A, 1984.

Mazeaud, M. M., Recherches sur la Biosynthése, la Sécrétion et le Catabolisme de l'Adrénaline et de la Noradrénaline chez quelques Espèces de cyclostomes et de Poissons, Ph.D thesis, Université de Paris, Paris, 1971.

McDonald, T. J., Jörnvall, H., Nilsson, G., Vagne, M., Ghatei, M., Bloom, S. R., and Mutt, V., Characterization of a gastrin releasing peptide from porcine non-antral gastric tissue, *Biochem. Biophys. Res. Commun.*, 90, 227, 1979.

McDonald, T. J., Jörnvall, H., Ghatei, M., Bloom, S.R., and Mutt, V., Characterization of an avian gastric (proventricular) peptide having sequence homology with porcine gastrin-releasing peptide and the amphibian peptides, bombesin and alytesin, *FEBS Lett.*, 122, 45, 1980.

McDonald, J., Greiner, F., Wood, J. G., and Noe, B. D., Oxytocin- like immunoreactive nerves are associated with insulin-containing cells in pancreatic islets of anglerfish *(Lophius americanus), Cell Tissue Res.*, 249, 7, 1987.

Meghji, P. and Burnstock, G., The effect of adenyl compounds on the heart of the dogfish *Scyliorhinus canicula, Comp. Biochem. Physiol.*, 77, 295, 1984a.

Meghji, P. and Burnstock, G., Actions of some autonomic agents on the heart of the trout *Salmo gairdneri* with emphasis on the effects of adenyl compounds, *Comp. Biochem. Physiol.*, 78, 69, 1984b.

Metcalfe, J. D. and Butler, P. J., On the nervous regulation of gill blood flow in the dogfish *Scyliorhinus canicula, J. Exp. Biol.*, 113, 253, 1984.

Morris, J. L. and Nilsson, S., The circulatory system, in *Comparative Physiology and Evolution of the Autonomic Nervous System*, Nilsson, S. and Holmgren, S., Eds.; Burnstock, G., Series Ed., Harwood Academic, Chur, 1993.

Nachmansohn, D. and Machado, A. L., The formation of acetylcholine. A new enzyme: choline acetylase, *J. Neurophysiol.*, 6, 397, 1943.

Nakano, T. and Tomlinson, N., Addendum: Catecholamine and carbohydrate concentrations in rainbow trout (*Salmo gairdneri*) in relation to physical disturbance, *J. Fish. Res. Bd. Can.*, 25, 603, 1968.

Nakao, T., An electron microscopic study on the innervation of the gill filaments of a lamprey, *Lampetra japonica, J. Morphol.*, 169, 325, 1981.

Nakao, T. and Ishizawa, A., An electron microscopic study of autonomic nerve cells in the cloacal region of the lamprey, *Lampetra japonica, J. Neurocytol.*, 11, 517, 1982.

Nakao, T., Susuki, S., and Saito, M., An electron microscopic study of the cardiac innervation in larval lamprey, *Anat. Rec.*, 199, 555, 1981.

Nicol, J. A. C., Autonomic nervous systems in lower chordates, *Biol. Rev.*, 27, 1, 1952.

Nilsson, S., Fluorescent histochemistry and cholinesterase staining of sympathetic ganglia in a teleost, *Gadus morhua, Acta Zool. (Stockholm)*, 57, 69, 1976.

Nilsson, S., Sympathetic nervous control of the iris sphincter of the Atlantic cod, *Gadus morhua, J. Comp. Physiol.*, 138, 149, 1980.

Nilsson, S., On the adrenergic system of ganoid fish: the Florida spotted gar, *Lepisosteus platyrhincus, Acta Physiol. Scand.*, 111, 447, 1981.

Nilsson, S., *Autonomic Nerve Function in the Vertebrates*, Springer-Verlag, Berlin, 1983.

Nilsson, S., Adrenergic control systems in fish, *Mar. Biol. Lett.*, 5, 127, 1984a.

Nilsson, S., Innervation and pharmacology of the gills, in *Fish Physiology,* Vol. 10A, Hoar, W. S. and Randall, D. J., Eds., Academic Press, Orlando, 1984b, 185.

Nilsson, S., The iris, in *Comparative Physiology and Evolution of the Autonomic Nervous System*, Nilsson, S. and Holmgren, S., Eds.; Burnstock, G., Series Ed., Harwood Academic, Chur, 1993.

Nilsson, S. and Holmgren, S., The autonomic nervous system of Elasmobranchs: structure and function, in *Physiology of Elasmobranch Fishes*, Shuttleworth, T. J., Ed., Springer, Berlin, Tokyo, 1988, 143.

Nilsson, S. and Holmgren, S., Autonomic nerve function and cardiovascular control in lungfish, in *Physiological Adaptations in Vertebrates*, Wood, S. C., Weber, R. E., Hargens, A. R., and Millard, R. W., Eds., Marcel Dekker, Inc., New York, 1992a, 377.

Nilsson, S. and Holmgren, S., Cardiovascular control by purines, 5-hydroxytryptamine and neuropeptides, in *Fish Physiology,* Vol 12B, *The Cardiovascular System*, Randall, D. J., and Farrell, A. P., Eds., Academic Press, New York, 1992b, 301.

Nilsson, S. and Pettersson, K., Sympathetic nervous control of blood flow in the gills of the Atlantic cod, *Gadus morhua, J. Comp. Physiol.*, 144, 157, 1981.

Nilsson, S., Abrahamsson, T., and Grove, D. J., Sympathetic nervous control of adrenaline release from the head kidney of the cod, *Gadus morhua, Comp. Biochem. Physiol.*, 55C, 123, 1976.

Noe, B. D., McDonald, J. K., Greiner, F., Wood, J. G., and Andrews, P. C., Anglerfish islets contain NPY immunoreactive nerves and produce the NPY analog aPY, *Peptides*, 7, 147, 1986.

Parker, G. H., Color changes in the catfish, *Ameiurus* in relation to neurohumors, *J. Exp. Zool.*, 69, 199, 1934.

Parker, W. N., On the anatomy and physiology of *Protopterus annectens, Trans. R. Irish Acad.*, 30, 109, 1892.

Perry, S. F., Fritsche, R., Kinkead, R., and Nilsson, S., Control of catecholamine release *in vivo* and *in situ* in the Atlantic cod (*Gadus morhua*) during hypoxia, *J. Exp. Biol.*, 155, 549, 1991.

Pettersson, K. and Nilsson, S., Nervous control of the branchial vascular resistance of the Atlantic cod, *Gadus morhua, J. Comp. Physiol.*, 129, 179, 1979.

Pick, J., *The Autonomic Nervous System. Morphological, Comparative, Clinical and Surgical Aspects*, Lippincott, Philadelphia, 1970.

Rajjo, I. M., Vigna, S. R., and Crim, J. W., Immunocytochemical localization of vasoactive intestinal polypeptide in the digestive tracts of a holostean and a teleostean fish, *Comp. Biochem. Physiol.*, 94C, 411, 1989.

Reite, O. B., The evolution of vascular smooth muscle responses to histamine and 5-hydroxytryptamine. I. Occurrence of stimulatory actions in fish, *Acta Physiol. Scand.*, 75, 221, 1969.

Rodionov, V. I., Gyoeva, F. K., and Gelfand, V. I., Kinesin is responsible for centrifugal movement of pigment granules in melanophores, *Proc. Natl. Acad. Sci. U.S.A.*, 88, 4956, 1991.

Romer, A. S., *The Vertebrate Body,* W. B. Saunders, Philadelphia, 1971.

Rovainen, C. M., Neurobiology of lampreys, *Physiol. Rev.*, 59, 1007, 1979.

Rubin, L. R. and Nolte, J., Autonomic innervation and photo sensitivity of the sphincter pupillae muscle of 2 teleosts *Lophius piscatorius* and *Opsanus tau, Curr. Eye Res.*, 1, 543, 1981.

Santer, R. M., Ultrastructural and histochemical studies on the innervation of the heart of a teleost *(Pleuronectes platessa* L.), *Z. Zellforsch.*, 131, 519, 1972.

Santer, R. M., Chromaffin systems, in *Comparative Physiology and Evolution of the Autonomic Nervous System,* Nilsson, S. and Holmgren, S., Eds.; Burnstock, G., Series Ed., Harwood Academic, Chur, 1993.

Satchell, G. H., *Circulation in Fishes*, Cambridge University Press, Cambridge, 1971.

Satchell, G. H., On the caudal heart of *Myxine* (Myxinoidea: Cyclostomata), *Acta Zool. (Stockholm)*, 65, 125, 1984.

Satchell, G. H., Cardiac function in the hagfish, *Myxine* (Myxinoidea:Cyclostomata), *Acta Zool. (Stockholm)*, 67, 115, 1986.

Satchell, G. H., *Physiology and Form of Fish Circulation,* Cambridge University Press, Cambridge, 1991.

Scheuermann, D. W., Untersuchungen hinsichtlich der Innervation des Sinus venosus und des Aurikels von *Protopterus annectens, Acta Morphol. Neerl. Scand.*, 17, 231, 1979.

Shibata, Y. and Yamamoto, T., Fine structure and cytochemistry of specific granules in the lamprey atrium, *Cell Tissue Res.*, 172, 487, 1976.

Shuttleworth, T. J. and Thorndyke, M., An endogenous peptide stimulates secretory activity in the elasmobranch rectal gland. *Science,* 225, 319, 1984.

Silva, P., Lear, S., and Epstein, F. H., Bombesin inhibits stimulated chloride secretion by the rectal gland of *Squalus acanthias. Clin. Res.,* 36, 598A, 1988.

Silva, P., Lear, S., Reichlin, S., and Epstein, F. H., Somatostatin mediates bombesin inhibition of chloride secretion by the rectal gland, *Am. J. Physiol.*, 258, R1459, 1990.

Small, S. A., MacDonald, C., and Farrell, A. P., Vascular reactivity of the coronary artery in rainbow trout *Oncorhynchus mykiss, Am. J. Physiol.*, 258, R1402, 1990.

Smith, D. G., Sites of cholinergic vasoconstriction in trout gills, *Am. J. Physiol.*, 233, R222, 1977.

Smith, D. G., Neural regulation of blood pressure in rainbow trout *(Salmo gairdneri), Can. J. Zool.*, 56, 1678, 1978.

Stabrovskii, E. M., The distribution of adrenaline and noradrenaline in the organs of the baltic lamprey *Lampetra fluviatilis* at rest and during various functional stresses, *J. Evol. Biochem. Physiol.*, 3, 216, 1967.

Stannius, H., *Das peripherische Nervernsystem der Fische*, Stiller, Rostock, 1849.

Stoff, J. S., Rosa, R., Hallac, R., Silva, P., and Epstein, F. H., Hormonal regulation of active chloride transport in the dogfish rectal gland, *Am. J. Physiol.*, 237, F138, 1979.

Sundin, L., and Nilsson, S., Arterio-venous branchial blood flow in the Atlantic cod *Gadus morhua, J. Exp. Biol.*, 165, 73, 1992.

Tagliafierro, G., Rossi, G. G., Bonini, E., Faraldi, G., and Farina, L., Ontogeny and differentiation of regulatory peptide and serotonin-immunoreactivity in the gastrointestinal tract of an elasmobranch, *J. Exp. Zool, Suppl.,* 2, 165, 1989.

Thorndyke, M. and Holmgren, S., Bombesin potentiates the effect of acetylcholine on isolated strips of fish stomach, *Regul. Pept.*, 30, 125, 1990.

Thorndyke, M. and Shuttleworth, T. J., Biochemical and physiological studies on peptides from the elasmobranch gut, *Peptides*, 6 (Suppl. 3), 369, 1985.

Thwaites, D. T., Young, J., Thorndyke, M. C., and Dimaline, R., The isolation and chemical characterization of a novel vasoactive intestinal peptide-related peptide from a teleost fish, the cod, Gadus morhua, *Biochim. Biophys. Acta*, 999, 217, 1989.

Uematsu, K., Effects of drugs on the responses of the ovary to field and nerve stimulation in a tilapia *Sarotherodon niloticus, Bull. Jpn. Soc. Sci. Fish.*, 51, 47, 1985.

Uematsu, K., The autonomic innervation of the ovary of the dab, *Limanda yokohamae, Jpn. J. Ichthyol.*, 33, 293, 1986.

Uematsu, K., Holmgren, S., and Nilsson, S., Autonomic innervation of the ovary in the Atlantic cod, *Gadus morhua, Fish Physiol. Biochem.*, 6, 213, 1989.

Van Noorden, S. and Patent, G. J., Vasoactive intestinal polypeptide-like immunoreactivity in nerves of the pancreatic islet of teleost fish, *Gillichtys mirabilis, Cell Tissue Res.*, 212, 139, 1980.

Vigna, S. R., Fisher, B. L., Morgan, J. L. M., and Rosenquist, G. L., Distribution and molecular heterogeneity of cholecystokinin-like immunoreactive peptides in the brain and gut of the rainbow trout, *Salmo gairdneri, Comp. Biochem. Physiol. C.*, 82, 143, 1985.

von Frisch, K., Beiträge zur Physiologie der Pigmentzellen in der Fischhaut, *Pfluegers Arch. Gesamte. Physiol.*, 138, 319, 1911.

Wahlqvist, I., Branchial vascular effects of catecholamines released from the head kidney of the Atlantic cod, *Gadus morhua, Mol. Physiol.*, 1, 235, 1981.

Watson, A. H. D., Fluorescent histochemistry of the teleost gut: evidence for the presence of serotonergic neurones, *Cell Tissue Res.*, 197, 155, 1979.

Winberg, M., Holmgren, S., and Nilsson, S., Effects of denervation and 6-hydroxydopamine on the activity of choline acetyltransferase in the spleen of the cod, *Gadus morhua, Comp. Biochem. Physiol. C*, 69, 141, 1981.

Wood, C. M. and Shelton, G., Physical and adrenergic factors affecting systemic vascular resistance in the rainbow trout: a comparison with branchial vascular resistance, *J. Exp. Biol.*, 63, 505, 1975.

Yamauchi, A. and Burnstock, G., An electronmicroscope study of the innervation of the trout heart, *J. Comp. Neurol.*, 132, 567, 1968.

Young, J. Z., On the autonomic nervous system of the teleostean fish *Uranoscopus scaber, Q. J. Microsc. Sci.*, 74, 491, 1931.

Young, J. Z., Comparative studies on the physiology of the iris. I. Selachians., *Proc. R. Soc. Ser. B. Biol. Sci.*, 112, 228, 1933a.

Young, J. Z., Comparative studies on the physiology of the iris. II. *Uranoscopus and Lophius, Proc. R. Soc. Ser. B Biol. Sci.*, 112, 242, 1933b.

Young, J. Z., The autonomic nervous system of selachians, *Q. J. Micr. Sci.*, 75, 571, 1933c.

Young, J. Z., The innervation and reactions to drugs of the viscera of teleostean fish, *Proc. R. Soc.*, 120, 303, 1936.

Young, J. Z., Nervous control of stomach movements in dogfishes and rays, *J. Mar. Biol. Assoc. U.K.*, 60, 1, 1980a.

Young, J. Z., Nervous control of gut movements in *Lophius, J. Mar. Biol. Assoc. U.K.*, 60, 19, 1980b.

Young, J. Z., Control of movements of the stomach and spiral intestine of *Raja* and *Scyliorhinus, J. Mar. Biol. Assoc. U.K.*, 63, 557, 1983.

Yui, R., Nagata, Y., and Fujita, T., Immunohistochemical studies on the islet and the gut of the Arctic lamprey, *Lampetra japonica, Arch. Histol. Cytol.*, 51, 109, 1988.

Yui, R., Shimada, M., and Fujita, T., Immunohistochemical studies of peptide- and amine-containing endocrine cells and nerves in the gut and the rectal gland of the ratfish *Chimaera monstrosa, Cell Tissue Res.*, 260, 193, 1990.

11 Osmotic and Ionic Regulation

David H. Evans

INTRODUCTION

The fossil record supports the hypothesis that the origin and early evolution of the vertebrates was in the marine environment but that some agnathan fish groups invaded fresh water in the upper Silurian, some 420 million years ago (Halstead, 1985). It also suggests that lampreys and the subsequent osteichthyan fishes went through an evolutionary period in fresh water before reinvasion of the marine environment by some groups. The fact that all vertebrates, except for the hagfishes, display a plasma salt concentration much less than seawater levels supports the idea of an early vertebrate sojourn in fresh water since life in fresh water would presumably select for a reduction in the salt gradients across the permeable, branchial epithelium to reduce the net efflux of salt into the very dilute medium (see below). However, chondrichthyan fishes also have reduced plasma salt concentrations, despite no paleontological evidence for early evolution in fresh water (Halstead, 1985), and Bray (1985) has suggested that there is a physiological advantage for reduction in plasma salt concentrations per se, even in the marine environment. Nevertheless, the extant fishes, with the exception of the hagfishes, have plasma salt concentrations distinctly different from either their current freshwater or marine habitats (Table 1). Therefore, these aquatic vertebrates must osmo- and ionoregulate in order to maintain their plasma salt content relatively consistent, which is required to regulate cell volume and intracellular salt content.

The vast majority of the literature in fish osmotic and ionic regulation deals with teleosts, probably because they are so common, relatively easy to collect, and have species that are easily maintained in the laboratory. Nevertheless, to keep an evolutionary perspective, this chapter will examine what is currently known about water and salt regulation in the agnatha, chondrichthyes, and osteichthyes, in that order. Because of space limitations, in most cases, only very recent and review publications have been cited.

0-8493-8042-1/93/$0.00+$.50

TABLE 1
Plasma Ionic Concentrations of Various Fishes*

	Na	Cl	K	Mg	Ca	SO_4	Urea	TMAO	Total
Seawater	439	513	9.3	50	9.6	26	0	0	1050
Hagfish[1]	486	508	8.2	12	5.1	3	—	—	1035
Lamprey[2]	156	159	32	7.0	3.5	—	—	—	333
Shark[3]	255	241	6.0*	3.0*	5*	0.5*	441	72*	1118
Teleost[4]	180	196	5.1	2.5	2.8	2.7	—	—	452
Euryhaline teleost[5]	142	168	3.4	—	3.3	—	—	—	297
Freshwater (soft)	0.25	0.23	0.005	0.04	0.07	0.05	—	—	1
Lamprey[6]	120	104	3.9	2.0	2.5	—	—	—	272
Stingray[7]	150	150	—	—	—	—	1.3	—	308
Teleost[8]	130	125	2.9	1.2	2.1	—	—	—	274
Euryhaline teleost[5]	124	132	2.9	—	2.7	—	—	—	240

[1]*Myxine glutinosa;* [2]*Petromyzon marinus*; [3]*Scyliorhinus canicula*;* [4]*Lophius piscatorius;* [5]*Pleuronectes flesus;* [6]*Lampetra fluviatilis;* [7]*Potamotrygon sp.;* [8]*Cyprinus carpio.*

*Data for *Squalus acanthias* from Forster et al. (1972); other data from Evans (1979).

AGNATHA

Hagfishes

The total osmolarity of hagfish plasma is approximately equivalent (isoosmotic) to that of seawater, so this group is generally considered to be the only vertebrate group that, like the invertebrates, is potentially free from any need for osmoregulation (e.g., Hardisty, 1979). However, some studies have determined that ionic gradients may exist (Table 1 and Evans, 1979), suggesting that some mechanisms for ionoregulation are present. The lack of any osmotic gradients is associated with an extremely high epithelial permeability to water (see Evans, 1979). This is to be contrasted with what appears to be an extremely low epithelial Na^+ permeability (Evans and Hooks, 1983), suggesting that salt and water must traverse the gill (or skin) epithelium via separate, or electrically charged, pathways.

Hagfishes maintain their plasma K^+, Ca^{2+}, Mg^{2+}, and SO_4^{2-} concentrations below that of seawater apparently by renal excretion and mucus gland secretion (Evans, 1979; Hardisty, 1979), although extant data are somewhat conflicting. The hagfish kidney is quite simple (Figure 1), with segmentally arranged glomeruli draining into short neck segments and thence into paired, common archinephric ducts that have some of the structural and functional attributes of the proximal tubule of the other fishes and tetrapods (e.g., Fels et al., 1989). Despite the presence of distinct glomeruli, it is unclear how primary urine is

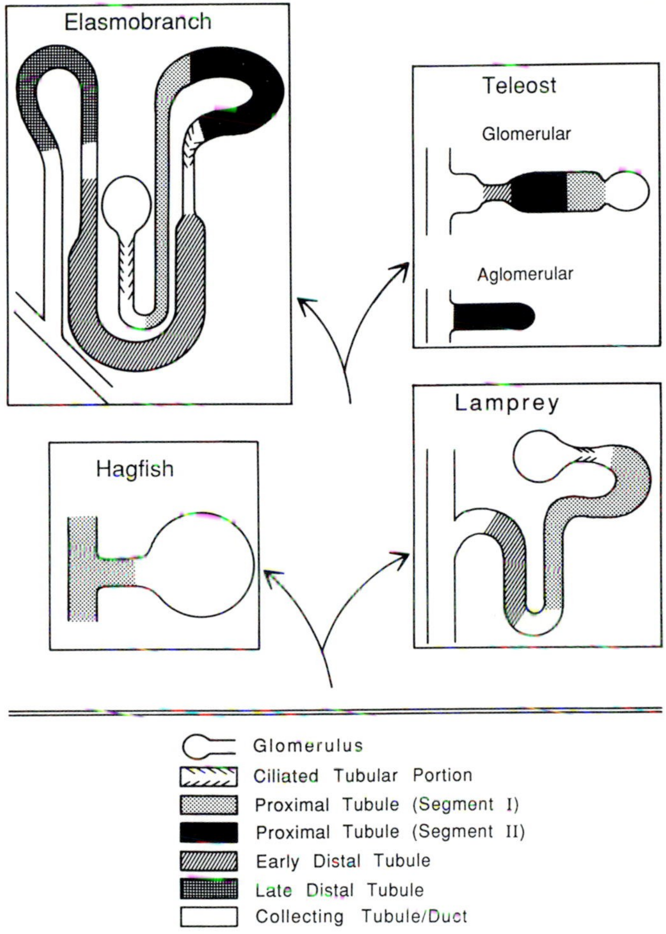

FIGURE 1. Evolution of the piscine kidney tubules. See text for details for individual taxa. (Redrawn from Hentschel and Elger, 1989.)

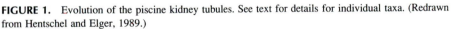

formed, since the hydrostatic pressure in the glomerular capillary is actually below the blood colloidal osmotic pressure, suggesting that other mechanisms of filtration or secretion must be operative (Riegel, 1986). Nevertheless, the glomerular filtration rate in *Eptatretus stouti* and *Myxine glutinosa* averages about 25 $\mu l \cdot 100g^{-1}\ h^{-1}$; urine flows are equivalent, indicating that there is no tubular reabsorption of water (see Evans, 1979), as might be expected in an animal in an iso-osmotic solution. The source of the water necessary for urine formation is unknown, but may be hypo-osmotic body fluids ingested during feeding on teleost prey. In an early study of *E. stouti* (Munz and McFarland, 1964), urine to plasma (U/P) ratios were shown to be greater than unity for K^+, Mg^{2+}, and SO_4^{2-}, suggesting that secretion of these ions takes place in the archinephric duct. However, a more recent study of *M. glutinosa* found a U/P ratio slightly less than unity for Mg^{2+} and only 1.15 for SO_4^{2-} (Alt et al., 1981), although the U/P ratio for K^+ was 1.5 as in the earlier study. Thus the role of the kidney in maintenance of hagfish plasma Mg^{2+} and SO_4^{2-}concentrations below that of seawater remains unclear. Hagfish plasma Ca^{2+} concentrations are 50 to 75% those in seawater; however, U/P ratios for this divalent ion are 0.5 to 0.8 (Munz and McFarland, 1964; Alt et al., 1981), indicating that extrarenal sites for Ca^{2+} secretion must be present. Since hagfish slime contains some 10.7 m$M{\cdot}l^{-1}$ Ca^{2+}, 207 m$M{\cdot}l^{-1}$ K^+, and 38.5 m$M{\cdot}l^{-1}$ Mg^{2+}, this is presumably the major extrarenal pathway for all of these ions (Munz and McFarland, 1964).

Several analyses of hagfish plasma Na^+ concentrations have demonstrated that plasma levels are maintained distinctly above those in seawater (see Evans, 1979), but a more recent study found no Na^+ gradient between *M. glutinosa* plasma and seawater (Alt et al., 1981), indicating that more studies are necessary. In addition, no measurements of transepithelial electrical potential (TEP) between the plasma and the external medium have been made, so it is unknown if Na^+ is actually kept out of electrochemical equilibrium or maintained higher in the plasma by a slight plasma-negative TEP. If indeed, plasma Na^+ is regulated above seawater levels by hagfishes, the kidney is not involved because U/P ratios are near unity (Evans, 1979; Alt et al., 1981). This putative regulation may be via branchial (skin?) Na^+ uptake, since we found that H^+ extrusion from *M. glutinosa* is dependent upon external Na^+ (Evans, 1984a), suggesting the presence of the Na/H exchange mechanism that is associated with Na^+ balance in fresh water fishes (see below and Evans, 1979). Our data also suggested the presence of a parallel Cl/HCO_3 exchange system, somewhat surprising since plasma Cl^- concentrations are at or below seawater levels in hagfishes (e.g., Evans, 1979; Alt et al., 1981). Since it has been shown that these parallel ionic exchangers are fundamental to acid-base regulation in other fishes (e.g., Heisler, 1989, and this volume), one might argue that the apparent unlinking of Na^+ and Cl^- regulation in hagfishes is secondary to the need to excrete excess H^+ (or retain HCO_3^-) in this group, so that plasma Na^+ concentrations are increased and plasma Cl^- concentrations are reduced secondary to stimulated Na^+ influx via Na/H exchange and inhibited Cl^- uptake via Cl/HCO_3 exchange. Acid-base regulation via these ionic exchange systems may be crucial for these burrowing, largely anaerobic fishes (Mallatt et al., 1987). The actual site of these ionic exchanges is yet to be determined, but mitochondrion-rich cells (Figure 2A), similar to the "chloride cells" of lampreys, sharks, and teleosts (which have been demonstrated to be the site of active ionic fluxes in seawater, and possibly fresh water; see below), have been described recently in the gill epithelium of both *M. glutinosa* (Bartels, 1985; Bartels and Welsch, 1986; Elger, 1987) and *E. stoutii* (Mallatt et al., 1987), with densities in the gill epithelium of the latter species approaching those found in the teleost gill (Mallatt and Paulsen, 1986). It has been suggested that these cells function in acid-base regulation in hagfishes (Mallatt et al., 1987); this proposition has been supported by the histochemical localization in these cells of both Na^+/K^+-

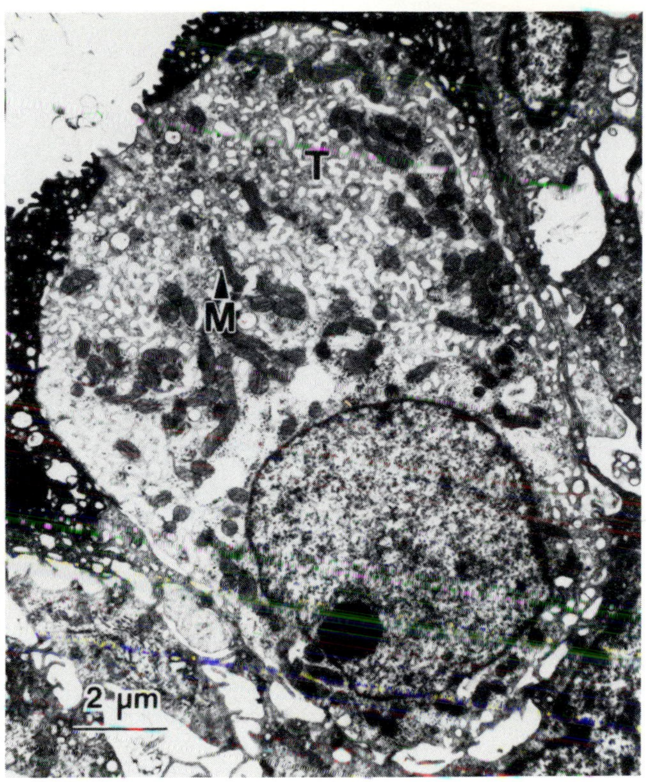

FIGURE 2A–D. Electron micrographs of mitochondrion-rich cells in the branchial epithelium of various fishes. See text for details for individual species. (A) Single "ionocyte" in the gill of the marine hagfish, *Eptatretus stouti*: M, mitochondrion; T, cytoplasmic tubule system. (From Mallatt, J. et al., *Can. J. Zool.*, 65, 1956, 1987. With permission.)

ATPase and carbonic anhydrase, two enzymes probably involved in acid-base regulation (Conley and Mallatt, 1988). The hypothesis that specific, mitochondrion-rich cells in the hagfish gill are involved in acid-base regulation needs to be tested. However, it is important to note that the ionic exchange systems necessary for NaCl uptake in fresh water are apparently present in modern hagfishes as well as sharks and teleosts (see below), and were presumably present in early, marine agnathan fishes for acid-base regulation. Thus, the ionic exchanges that were critical for ionoregulation as the vertebrates entered fresh water may have evolved for acid-base regulation before the vertebrates entered fresh water (Evans, 1984a).

Given their deep-water habitat, it is unlikely that hagfishes ever are exposed to varying salinities; however, at least two studies have demonstrated some limited tolerance of salinities below and above seawater. *M. glutinosa* was shown to tolerate salinities ranging from 150 to 60% (1540 to 600 $mOsm \cdot l^{-1}$) seawater after a gradual concentration change of 15 mOsm/d (Cholette et al., 1970). The

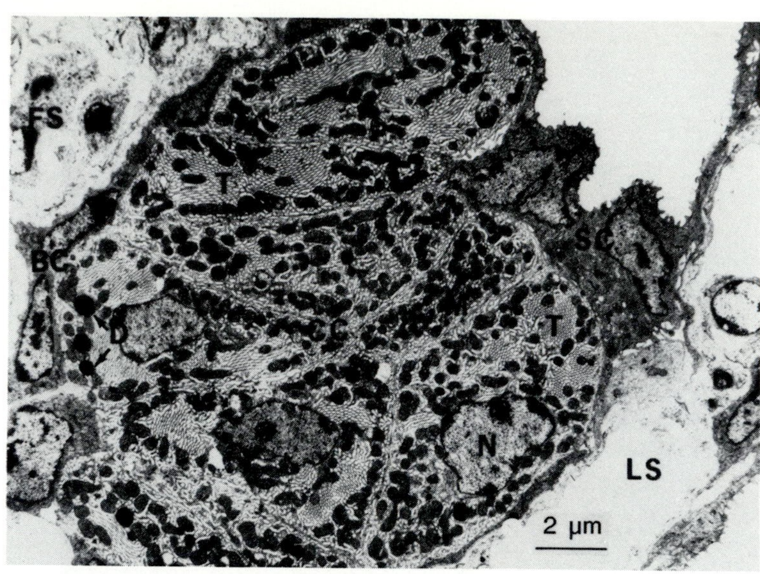

FIGURE 2B. Multiple "chloride cells" (CC) in the gill of the freshwater-adapted lamprey, *Petromyzon marinus*: M, mitochondria; T, cytoplasmic tubule system; N, nucleus; LS, lamellar blood space; FS, filamental blood space; SC, superficial cells; BC, basal cells; D, dense bodies. (From Peek, W. D. and Youson, J. H., *J. Morphol.*, 160, 143, 1979. With permission of copyright owner, ©Wiley-Liss, a division of John Wiley & Sons.)

animals were osmoconformers over the entire range. Another study, using *E. stouti* (McFarland and Munz, 1965), found that this species is able to regulate its body weight after transfer to 80% seawater, but body weights remained elevated after transfer to 60 to 40% seawater. The body weight decreased and was not regulated after transfer to 122% seawater. These data indicate that *E. stouti* has mechanisms for ridding the body of excess water in at least slightly hypo-osmotic salinities, but apparently cannot balance the osmotic withdrawal of water in hyper-osmotic salinities, presumably because it either cannot ingest the medium or retrieve needed water from gut contents after ingestion (McFarland and Munz, 1965). Maintenance of body weight in hypo-osmotic salinities is probably associated with an increase in the number of functioning glomeruli (glomerular recruitment) because single glomerular filtration rates (SGFR) have been measured in seawater-acclimated hagfish (see Hardisty, 1979; Alt et al., 1981) and they are higher than expected given the known number of glomeruli and total urinary outflow, suggesting that not all glomeruli are functioning under normal conditions. Recruitment of nonfunctioning glomeruli may be merely secondary to increased blood flow to the kidney since various studies have shown that SGFR is proportional to perfusion pressure (e.g., Riegel, 1978), despite the uncertainty of the actual mechanism of urine formation (Riegel, 1986).

The fact that hagfishes are osmoconformers in dilute salinities indicates that mechanisms for substantial salt uptake are not present, despite the fact that Na/H

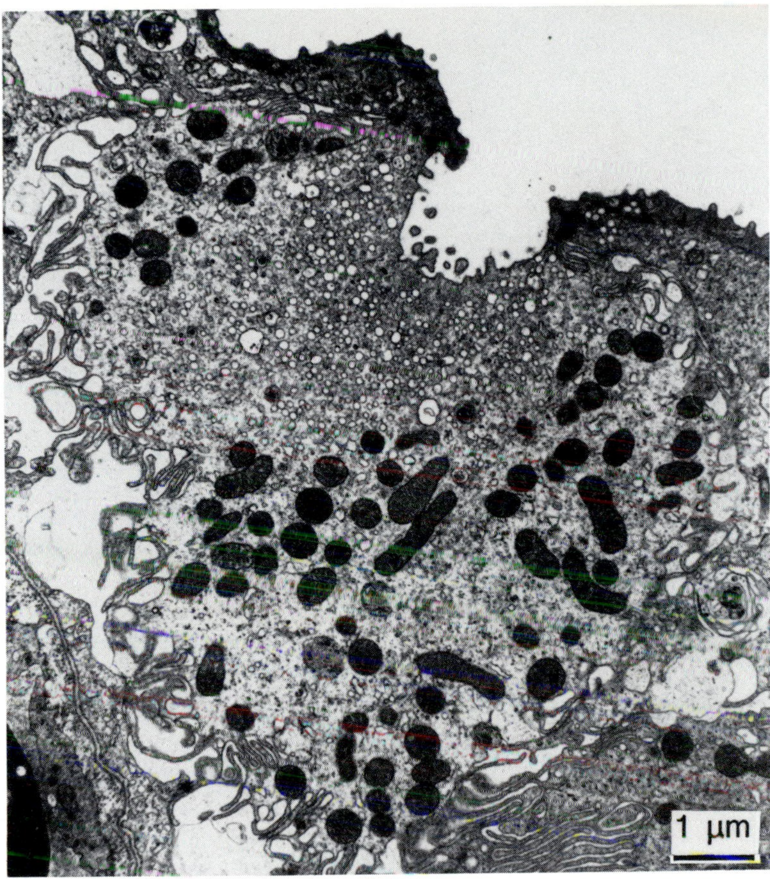

FIGURE 2C. Single CC in the gill of the skate, *Raja* sp., note characteristic multiple mitochondria and cytoplasmic vesicles and tubules. (Kindly supplied by Drs. S. Dunel Erb and Pierre Laurent.)

and Cl/HCO_3 exchange mechanisms are apparently present in the gill or skin (see above). In addition, it appears that reabsorption of ions in the kidney tubules is slight at best (McInerney, 1974), so renal losses of ions would increase in dilute salinities due to increased GFR. Thus, despite relatively low epithelial, ionic permeabilities, the hagfish cannot balance the net loss of ions. Whether this imbalance is secondary to a limitation of the kinetics of the gill ionic uptake systems (see Evans, 1979; 1984b) and a limitation or lack of ionic reabsorptive mechanisms in the kidney, or a lack of stimulatory hormones, remains to be determined.

In conclusion, despite osmoconformity, relative stenohalinity, and limited ionoregulation in seawater, modern hagfishes appear to have the branchial and at least some of the renal mechanisms that theoretically would allow osmoregulation in reduced salinities. It is unclear what restricts modern hagfishes to

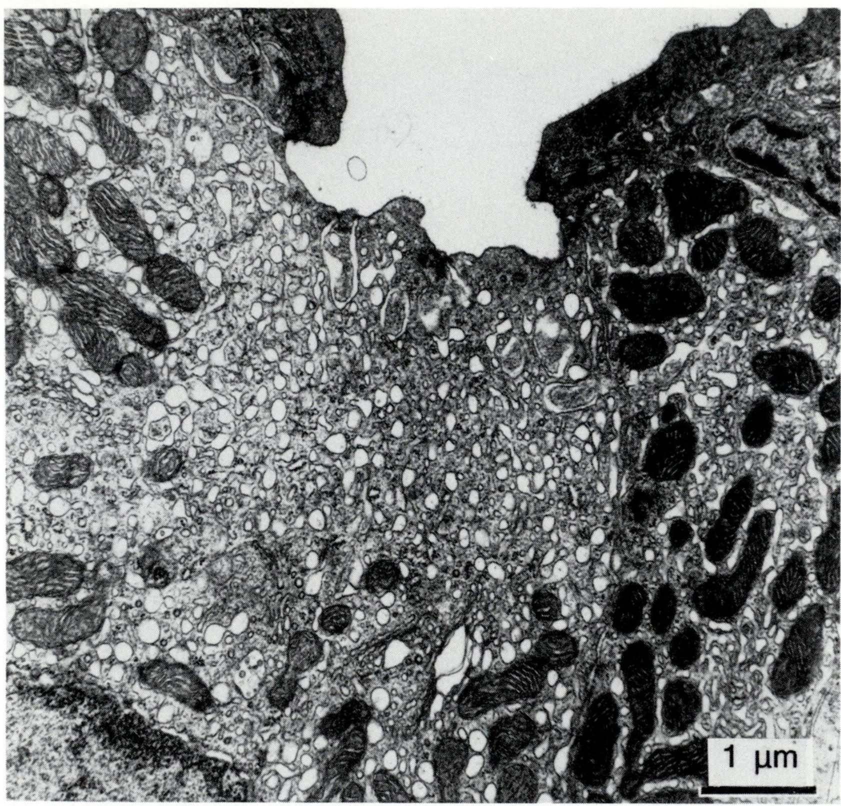

FIGURE 2D. Single CC in the gill of the sole, *Solea solea*; note characteristic multiple mitochondria and cytoplasmic vesicles and tubules. (Kindly supplied by Drs. S. Dunel Erb and Pierre Laurent.)

the marine environment, but it is apparent that the renal glomerulus and branchial ionic regulatory mechanisms necessary for survival in dilute salinities were probably present in ancestral agnathan fishes, thus allowing the entry into fresh water for the stem group that gave rise to lampreys, as well as subsequent, jawed vertebrates.

Lampreys

This other extant group of agnatha fishes are osmoregulators (Table 1), often euryhaline, and have apparently developed the same suite of osmoregulatory mechanisms that have been much better characterized in the teleosts. All lampreys develop in fresh water, pass through a larval stage (termed the ammocoete), and then either remain in fresh water or migrate to the sea before returning to fresh water for spawning. The few studies measuring either net ionic losses or isotopic fluxes suggest that lampreys have retained the low epithelial ionic permeability characteristic of hagfishes, and therefore the presumed ancestral, marine agnathan. No direct measurements of tl osmotic permeability of lampreys have been reported, but their rel urine flows in fresh water (see below) are presumably associated with a substantial gill or skin osmotic permeability, again, similar to the modern hagfishes, and presumed to be the ancestral condition (see Evans, 1979). The strategies of lamprey osmotic and ionic regulation have been reviewed most recently by Evans (1979) and Hardisty (1979).

Fresh Water

Since lampreys are hyperosmotic to fresh water, they must balance the net osmotic influx of water, as well as the net ionic loss, in order to maintain a consistent plasma (and cellular) osmolarity and ionic content. No measurements of the transepithelial electrical potential have been made in lampreys, but it is assumed that both Na^+ and Cl^- are out of electrochemical equilibrium and therefore must be maintained by active transport by either the kidney or gill epithelium. The freshwater lamprey kidney (Figure 1) is much more complicated than that of the hagfish, with distinct glomeruli, proximal and distal tubules, an intermediate thin segment, and a collecting duct (Hentschel and Elger, 1989). The proximal and distal tubules are arranged in a loop (Hardisty, 1979), remniscent of the loop of Henle of the mammalian kidney, as well as the elasmobranch kidney (see below and Henderson et al., 1988). Total and single glomerular filtration rates are substantial in *Lampetra fluviatilis* (2800 $\mu l \cdot 100g^{-1}\ h^{-1}$ and 7 $nl\ min^{-1}$, respectively), 5 to 7 times as great as those parameters in freshwater teleosts (compare Moriarty et al., 1978 with Jackson et al., 1977 and Evans, 1979). Urinary salt loss is minimized by tubular reabsorption of both Na^+ and Cl^-, mostly in the collecting duct (Logan et al., 1980), producing urine concentrations of approxmately 10 $mM\ l^{-1}$ for both ions and an osmolarity of only 20 to 30 mOsm l^{-1} (Moriarty et al., 1978; Logan et al., 1980). Urine flows are approximately 50% of GFRs, indicating that tubular reabsorption of water is also present (Moriarty et al., 1978). The bulk (90%) of this urine reabsorption is in the collecting duct, rather than the proximal tubule as is found in tetrapods (Logan et al., 1980). Knowing the urine flow and concentration, one can calculate the renal loss of Na^+ or Cl^- to be some 15 $\mu M \cdot 100\ g^{-1}\ h^{-1}$, of the same order as the unidirectional ion efflux from the whole animal (Evans, 1979), suggesting extremely low branchial or skin ion permeability. A recent study that measured branchial ionic fluxes across *Petromyzon marinus* supported this conclusion (Stinson and Mallatt, 1989).

Despite extremely low epithelial Na^+ and Cl^- permeability and urinary salt output, one must assume that the net loss of these ions from lampreys into fresh water must be balanced by some uptake, presumably across the gill epithelium. There have been only a few data published addressing this putative uptake directly (see Evans, 1979), but they have demonstrated that the kinetics of Na^+ uptake from dilute solutions are similar to those published for freshwater teleosts. A recent study demonstrated that branchial influxes were equivalent to branchial effluxes, suggesting operative uptake mechanisms (Stinson and Mallatt, 1989). Histological data (e.g., Morris and Pickering, 1976; Peek and Youson, 1979) have demonstrated that the interlamellar gill epithelium of both *L. fluviatilis* and *P. marinus* in fresh water contains mitochondrion-rich cells (Figure 2b) morphologically similar to the "chloride cells" found in teleosts, as well as the mitochondrion-rich cells of the hagfish gill, which are assumed to be involved in acid-base balance (see above). Assuming functional as well as morphological homology, one can hypothesize that these cells are the site of ionic uptake by freshwater-acclimated lampreys via parallel Na/H and Cl/ HCO_3 exchange, which also functions in acid-base regulation. Physiological data are necessary to test this hypothesis; however, it appears that the full complement of the renal and branchial mechanisms necessary for osmotic and ionic regulation in fresh water are present in the modern-day lampreys, and therefore presumably were present in the earliest freshwater vertebrates. Other than presumed adjustments of the kinetics of ionic uptake across the branchial epithelium, to facilitate NaCl uptake from dilute solutions, the only major evolutionary addition (from the presumed ancestral marine agnathan) was the distal renal tubule and collecting duct, the sites of ion retrieval from the filtered urine (Logan et al., 1980).

Seawater

Some species of lampreys are euryhaline and anadromous, entering the sea after metamorphosis and returning to fresh water to spawn. Determinations of plasma solute concentrations are relatively rare, and are often from "fresh run" individuals caught in rivers after upstream migration from the sea. Plasma from such individuals, or those acclimated to high salinities (e.g., Table 1), is hypo-osmotic to seawater with concentrations of major ions, especially Na^+ and Cl^-, maintained significantly below seawater levels. Thus, osmotic and ionic gradients are reversed from the condition in fresh water, dictating that marine lampreys must balance the osmotic loss of water and diffusional gain of ions. No measurements of transepithelial electrical potentials across the gills of "fresh run" lampreys have been published, so we can only assume that both Na^+ and Cl^- are out of electrochemical equilibrium, and hence must be actively controlled.

Acclimation of "fresh run" *L. fluviatilis* to 50% seawater is associated with a 80% reduction in both single glomerular filtration and total GFR (Rankin et al., 1980), demonstrating that the reduced GFR is secondary to reduction in filtration across individual glomeruli, rather than reduction in the number of glomeruli filtering, as is found in teleosts and amphibians (McVicar and Rankin, 1985). In addition, tubular water reabsorption increased from 40% in fresh water to 80% in 50% seawater. Because of reduced GFR and increased tubular water reabsorption, the final urine flow is also reduced by some 80% (Rankin et al., 1980). Despite this substantial reduction in urinary output, the hyporegulating lamprey still must balance the sum of this relatively small water loss and the osmotic water loss across permeable epithelial surfaces, presumably the gills and skin. To gain needed water, both *L. fluviatilis* and *P. marinus* ingest the medium in 50% seawater, and absorb needed water secondary to intestinal ionic uptake of Na^+ and Cl^- (see Evans, 1979). Ingested Na^+ and Cl^- are not excreted by the kidney; U/P ratios for both ions are at or below unity (Rankin et al., 1980). Ingested divalent ions largely remain in the intestine. However, some divalent ion uptake from the intestine must take place because the urine of lampreys in 50% seawater has U/P ratios of 2.5, 11.7, and 12.9 for Ca^{2+}, Mg^{2+} and SO_4^{2-}, respectively (Pickering and Morris, 1970).

The net diffusional and intestinal uptake of Na^+ and Cl^- by hyporegulating lampreys is balanced by extrarenal excretion, presumably across the gills, although no studies have been published addressing this proposition directly. The mitochondrion-rich "chloride cells" in the gill epithelium of both *P. marinus*_(Peek and Youson, 1979) and *L. fluviatilis* (Morris and Pickering, 1976; Bartels and Welsch, 1986; Bartels et al., 1987) lose their characteristic basolateral invaginations when these lampreys are acclimated to fresh water, suggesting that these cells play a role in salt secretion, as they do in teleosts (see below). Moreover, the levels of gill Na^+/K^+ATPase, an enzyme known to be involved in salt extrusion in teleosts (see below), also declines when *P. marinus* is acclimated to fresh water (see Evans, 1979).

In summary, contrary to hagfishes, lampreys have developed a suite of osmotic and ionic regulatory mechanisms that enable some members to migrate between the freshwater and marine environments. What is especially intriguing is that these mechanisms display an amazing similarity to the pathways of osmotic and ionic regulation in teleosts (see below). Lampreys and teleosts certainly have been phylogenetically separate for at least 350 million years, so this is a fine example of convergent evolution between two vertebrate lineages.

CHONDRICHTHYES

The sharks, skates, and rays evolved some 350 million years ago, from a different stock than the bony fishes (see Gilbert, this volume). It is unclear whether these cartilaginous fishes evolved in fresh or seawater (Bray, 1985), but extant species are predominantly marine, with rare freshwater immigrants, and a single family of stenohaline, freshwater stingrays (Potamotrygonidae). Whatever the osmolarity of the environment during their early evolution, chondrichthyan fishes, like all other vertebrates (except hagfishes), have evolved a blood NaCl concentration less than that of seawater, but above that of fresh water. However, the concentrations of both ions are significantly above those in the plasma of lampreys or teleosts (Table 1). Despite reduced plasma salt levels, marine chondrichthyan fishes are actually slightly hyperosmotic to seawater because of the addition of two organic solutes, urea and trimethylamine oxide (TMAO), to their plasma and cells (Table 1). They therefore face a net osmotic influx of water. Since their plasma NaCl concentration is below that of seawater and the few measurements of transepithelial electrical potentials indicate that both Na^+ and Cl^- are out of electrochemical equilibrium, chondrichthyan fishes also face a net influx of NaCl (Evans, 1980). Isotopic flux studies have determined that, like hagfishes and lampreys, the gill epithelium of chondricthyan fishes is extremely permeable to water, but relatively impermeable to NaCl. Water fluxes as high as 150% of the body water per hour have been reported, but salt fluxes are only 50 to 100 $\mu M \cdot 100\ g^{-1}\ h^{-1}$, in the same range as those described for hagfishes, lampreys, and freshwater teleosts (Evans, 1979). The data base on the mechanisms of chondrichthyan osmoregulation deals almost exclusively with the largest subclass containing the sharks, skates, and rays (elasmobranchs), and has been most recently reviewed by Pang et al. (1977), Evans (1979), and Shuttleworth (1988).

Retention of Urea and Trimethylamine Oxide

The urea concentration of elasmobranch blood would be fatal to most other vertebrates because of the destabilizing effects of urea on most proteins, including enzymes. These fishes are able to tolerate this uremia because TMAO is able to counteract the toxic effects of urea when present at approximately 50% the concentration of urea (Yancey and Somero, 1980), which is the case in elasmobranch cells. The extremely high concentrations of urea and TMAO in elasmobranch plasma are maintained by a combination of low branchial permeability and reabsorption in the kidney tubules (Shuttleworth, 1988). The actual tubular site of renal reabsorption is unknown, but Boylan (1972) suggested that it may be passive, via countercurrent exchange. Recent anatomical and physiological data (Friedman and Hebert, 1990; Lacy and Reale, 1991a, 1991b) support this conclusion.

Regulatory Mechanisms in Seawater

The Kidney

The kidney of elasmobranchs has an extremely complex architecture, probably the most complex of any vertebrate (Figure 1). The substantial osmotic influx of water across the elasmobranch gill is balanced by a GFR (350 $\mu l \cdot 100\ g^{-1}\ h^{-1}$) that approaches that of freshwater teleosts, and is substantially greater than that of marine teleosts (Evans, 1979). Na^+ and Cl^- U/P ratios may approach unity (Hickman, 1969), so the kidney probably is not the site of net salt efflux to balance the net influx of salt across the gills, even when the rectal gland has been removed (e.g., Burger, 1965). Despite no apparent net alteration of urine NaCl concentration in the kidney, it is now clear that both active secretion and active reabsorption of salt takes place, in different segments. Recent studies have demonstrated that the distal region of the isolated, perfused, proximal tubule (segment II, Figure 1) of the shark, *Squalus acanthias,* is able to secrete NaCl into the lumen via a coupled Na+Cl cotransport system (Beyenbach and Baustian, 1989), similar to that described for salt secretion by the shark rectal gland and teleost gill (see below). This secretion is associated with substantial fluid movement into the lumen of the proximal tubule, so it presumably does not generate net secretion of salt, but may add to the net output of urine (Beyenbach and Baustian, 1989). Interestingly, the same Na^+Cl cotransport system has been shown to reabsorb salt and fluid further down the renal tubule (early distal), thereby setting up osmotic gradients that allow for the tubular countercurrent system to facilitate the passive reabsorption of urea (Friedman and Hebert, 1990). Net tubular reabsorption of urine must be greater than filtration plus secretion because urine flows (approximately 100 $\mu l \cdot 100\ g^{-1}\ h^{-1}$; Evans, 1979) are distinctly below the GFRs, but still relatively substantial compared to those of marine teleost fishes (see below).

The Rectal Gland

The digitiform rectal gland is the dominant, and possibly only, site of extrarenal secretion of excess NaCl by marine elasmobranchs (see Greger et al., 1986; Shuttleworth, 1988, for recent reviews). It produces a solution that contains even more NaCl than seawater but is iso-osmotic to the plasma (e.g., Evans, 1979). The gland is composed of thousands of secretory tubules which drain into a central canal, emptying into a duct which opens into the posterior intestine just proximal to the cloaca. The secretion is produced by a Na+Cl cotransport system, which is, in turn, dependent upon Na/K exchange generated by the extremely high concentrations of Na^+/K^+-ATPase, which characterize the rectal gland (Shuttleworth, 1988). In recent years, the steps of this NaCl secretory system in the rectal gland (Figure 3) have been characterized by the use of isolated, perfused glands (facilitated by the presence of a single afferent artery, efferent vein, and secretory duct), as well as by biophysical dissection of isolated, perfused tubules (e.g., Greger et al., 1986; Silva et al., 1990). In fact, tissue-cultured sheets of rectal gland tubular epithelium are also now being studied (Karnaky et al., 1991; Valentich and Forrest, 1991). The interest in rectal gland transport is primarily generated by the similarities between this NaCl transport system and that found in the thick ascending limb of the loop of Henle of mammals, the site of action of a variety of antihypertensive, diuretic drugs (Silva et al., 1990). In addition, recent studies have demonstrated that the apical Cl^- channels of the shark rectal gland secretory cell share a variety of characteristics with the human cystic fibrosis transmembrane conductance regulator that appears to be the site of malfunction in this important disease (Marshall et al., 1991). This is an excellent example of the importance of comparative studies, and their important potential relationship with human physiology and disease.

A priori, one might expect that an increase in plasma salt concentration would be the major stimulus of rectal gland secretion, but it has been demonstrated that an increase in volume, rather than salt content, stimulates rectal gland flow in intact *S. acanthias* (Solomon et al., 1985). Since elasmobranchs

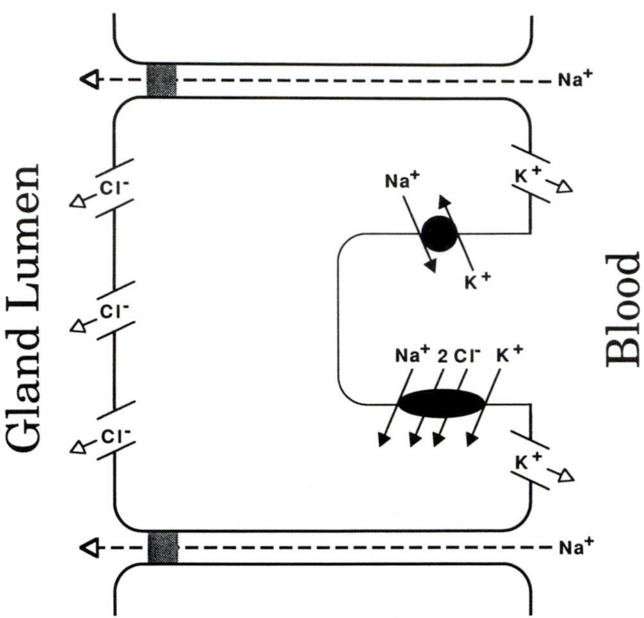

FIGURE 3. Current model for NaCl secretion by the shark rectal gland. The basolateral (serosal) membrane displays extensive infoldings (see Ernst et al., 1981) into the cytoplasm, here depicted as a single crypt. Basolateral Na^+/K^+-ATPase generates an electrochemical gradient for Na^+, which provides the driving force for Na+K+2Cl transport into the cytoplasm across the basolateral membrane. Na^+ is recycled via Na/K exchange and K^+ exits the basolateral membrane via K^+ channels, which balances the uptake of K^+ via both Na/K exchange and Na+K+2Cl cotransport. Intracellular Cl^- exits the cell via apical (mucosal) Cl^- channels, providing the electrical gradient (lumen negative) which drives Na^+ into the lumen across leaky, paracellular junctions. See Greger et al. (1986) and Silva et al. (1990) for more details.

face constant influx of water, this would provide for a tonic stimulation of rectal gland function, necessary for salt balance. However, the rare elasmobranch that enters brackish water would face a stimulation of rectal gland secretion, counter to its needs to maintain constant plasma NaCl concentrations in the face of a more dilute environment. In fact, early studies on rectal gland function did find that the gland was very active in *S. acanthias* exposed to dilute salinities, but that the plasma Cl^- levels were maintained at only slightly reduced levels (Burger, 1965), probably not too surprisingly, since the external salinity was approximately 80% seawater, which was probably still hyperchloric to the plasma. Nevertheless, it is clear that the rectal gland responses to dilute salinities need to be studied further.

The Gills

There are no direct data supporting a role for the elasmobranch gill in ion regulation, but some interesting indirect data do exist suggesting that sharks, like marine lampreys and teleosts, may secrete excess NaCl across the gill epithelium. Anatomical studies have identified cells (Figure 2c) in the epithelium that are similar to the mitochondrion-rich cells of hagfishes, lampreys, and teleosts (e.g., Laurent, 1982), although it is possible that these cells are functioning in acid-base balance, rather than NaCl extrusion. Our recent study of the pathways for ammonia transport across the perfused dogfish shark gill demonstrated that a small component was probably via Na+NH_4+2Cl transport, suggesting that some net extrusion of NaCl was possible via this pathway (Evans and More, 1988), in a manner similar to that which has been very well characterized for the teleost gill epithelium (see below). Finally, various studies have demonstrated that extirpation of the rectal gland does not seriously impair the salt balance of sharks (e.g., Evans, 1979; Evans et al., 1982), suggesting that other pathways for salt excretion exist. Since it appears that the kidney does not start to excrete hyperosmotic urine after rectal gland removal (e.g., Burger, 1965), one might suppose that gill extrusion mechanisms exist. This interesting possibility of extrusion of salt across the elasmobranch gill epithelium needs to be explored.

Euryhaline and Freshwater Elasmobranchs

The data base on the physiological changes associated with entry into lowered salinities by euryhaline elasmobranchs remains relatively small and largely that analyzed in previous reviews (Evans, 1979; Shuttleworth, 1988). Sharks such as *Carcharhinus leucas* (e.g., Thomerson and Thorson, 1977), sawfishes in the genus *Pristis*, and the stingray *Dasyatis* are relatively common invaders of brackish and even fresh water, but they are apparently immigrants, rather than "land-locked". The few analyses of plasma solute concentrations have demonstrated that these species reduce both NaCl and urea concentrations substantially in lowered salinities. Urea concentration diminution is apparently the result of reductions in both synthetic rate and renal reabsorption. As might be expected in this distinctly hypo-osmotic medium, urine flows increase (secondary to both an increase in GFR and decrease in tubular reabsorption), but urine salt concentrations decline. However, the urinary salt efflux actually increases because of the substantial increase in urine flow, presumably exacerbating the plasma salt decline, produced by diffusional loss down the electrochemical gradient. Nevertheless, in low salinities *C. leucas* is able to maintain plasma Na^+ and urea concentrations of approximately 200 and 130 m*M*, respectively (Shuttleworth, 1988), indicating that ionic regulation is substantial and that plasma urea levels remain relatively high, despite the obvious need to reduce plasma solute concentrations in the reduced salinity.

Apparently, the only truly freshwater elasmobranchs are members of the stingray family Potamotrygonidae in the Amazon and Orinoco river systems, some 4000 km from the sea. *Potamotrygon* plasma Na^+ concentrations are only about 150 m*M* and the urea concentrations are vanishingly small (about 1 m*M*; Evans, 1979). In fact, *Potamotrygon* has apparently lost the ability to maintain high plasma urea levels; acclimation to 50% seawater is associated with an increase in plasma NaCl concentrations, but plasma urea levels rise to only 6 m*M* (Bitter and Lang, 1980). Interestingly, the kidney of *Potamotrygon* lacks the complicated tubular architecture (see above) which is apparently associated with urea reabsorption in other elasmobranchs (Hentschel and Elger, 1987). One might expect that evolution in fresh water might be associated with a reduction in gill water permeability, but *Potamotrygon* maintains the same extraordinarily high water permeability that has been described for marine elasmobranchs (Evans, 1979).

Both elasmobranch freshwater invaders and land-locked stingrays presumably have gill mechanisms for the necessary uptake of both Na^+ and Cl^-; however, no direct evidence for their presence has been published. Pang et al. (1977) described unpublished results demonstrating that Na^+ influx from dilute media into *Potamotrygon* was saturable, suggesting a carrier-mediated system, but they did not demonstrate that any net uptake took place. In addition, it has been demonstrated that H^+ extrusion by the marine dogfish, *S. acanthias*, is linked to external Na^+ (Evans, 1982), suggesting the presence of Na/H exchange, which could be used in Na^+ regulation in fresh water. These data corroborated an earlier study indicating that *S. canicula* in seawater had a component of Na^+ influx that was coupled to plasma H^+ or NH_4^+ (Payan and Maetz, 1973). These data suggest that the usual ionoregulatory exchange systems are present even in the marine shark gill, presumably for acid-base regulation, and are therefore presumably present in euryhaline and fresh water species. However, critical data are lacking.

As might be expected, since freshwater *Charcharhinus* and *Potamotrygon* do not need to excrete excess salt, the rectal gland in these individuals is reduced in size, and presumably in function (Shuttleworth, 1988). Functional studies of this atrophied gland would be interesting.

TELEOSTS

It is generally assumed that the teleosts evolved from freshwater predecessors; both the fossil record and the fact that teleostean plasma NaCl concentration is some 30% that of seawater (Table 1) support this proposition. Since teleost plasma does not contain significant concentrations of organic solutes, these fishes are hyperosmotic to fresh water but hypo-osmotic to seawater. Thus, like lampreys, freshwater teleosts face a volume load and salt loss, while marine teleosts are potentially volume depleted and salt loaded. Euryhaline species may face varying, and opposing, gradients and therefore a complex suite of osmoregulatory problems over the time frame of hours or days. The data base of teleost osmoregulation has been reviewed extensively (e.g., Evans, 1979, 1981, 1984b; Kirschner, 1979), with specific, branchial mechanisms discussed by various authors in a symposium volume (*American Journal of Physiology*, 238: R139–R276, 1980) for Jean Maetz, an early leader in this field, and in Hoar and Randall (1984). More recent, and specific, reviews will be referenced below.

Regulatory Mechanisms in Fresh Water

Modern, freshwater teleosts have retained the primitive, low branchial ionic permeability characteristic of all the earlier groups (see above), as indicated by isotopically measured, unidirectional Na^+or Cl^- fluxes generally less than 100 $\mu M \cdot 100\ g^{-1}\ h^{-1}$ (Evans, 1979), in the same range as that described for hagfishes, lampreys, and elasmobranchs (see above). This rather low ionic flux, despite substantial electrochemical gradients,* is associated with numerous and significant intercellular junctions between adjacent, mitochondrion-rich "chloride cells" and accessory cells in the gill epithelium of fresh water teleosts (e.g., Pisam and Rambourg, 1991; Karnaky, 1992). Interestingly, the diffusional water permeability of fresh water teleosts is far below that described for hagfishes, lampreys, and elasmobranchs; isotopically measured, unidirectional water fluxes average some 50% of the body water per hour (Evans, 1979).

The Kidney and Urinary Bladder

The teleostean kidney is morphologically much simpler than the kidney of the elasmobranchs (Figure 1). Freshwater teleosts display well-developed glomeruli, proximal and distal tubules, as well as collecting tubules and ducts. The proximal tubule actually has two subunits: segment I, which is homologous to the proximal tubule of the tetrapod vertebrates, and segment II, which is found only in elasmobranchs and teleosts (and other actinopterygians) and has no homologue in the tetrapods (see Hentschel and Elger, 1989, for an excellent review of vertebrate kidney morphology).

Despite a somewhat reduced water permeability compared with agnatha and elasmobranchs (see above), freshwater teleosts maintain a relatively high GFR, offsetting the substantial osmotic uptake of water. GFRs approximate 400 $\mu l \cdot 100\ g^{-1}\ h^{-1}$ and urine flows are usually only slightly less (e.g., Evans, 1979). However, NaCl is conserved by reabsorption in the distal segment and collecting tubule/duct, resulting in a urine concentration of some 20 mOsm l^{-1}, less than 10% of the plasma concentration. Recent studies have shown that the proximal tubule epithelium (probably PII) actually secretes NaCl (possibly via Na+Cl cotransport) into the urine, drawing water into the lumen osmotically, and presumably aiding in the formation of urine. In fact, urine flows exceed the GFR in at least the eel, *Anguilla rostrata* (Beyenbach and Baustian, 1989).

* Diffusive movements of ions are down electrochemical gradients, the sum of ionic and electrical potential differentials. To account for the ionic gradients between freshwater teleost plasma and the medium, the gills would have to maintain electrical potentials of some 180 mV (inside negative for the observed Na^+ distribution; inside positive for (Cl^-), far greater than the electrical potentials of a few millivolts that have been measured (see Evans, 1979; Potts, 1984 for a more complete discussion).

Many species of teleosts have a urinary bladder that is not homologous to that found in the tetrapods because of its mesodermal rather than endodermal origin. Final processing of urine can take place in the bladder by the degree of coupling between NaCl uptake and water reabsorption. Ionic uptake is predominantly a cotransport of NaCl (Dawson and Frizzell, 1989), but some evidence suggests that parallel exchanges of Na/H and Cl/HCO_3 may play a role (Demarest, 1984). In freshwater fishes, ionic uptakes are relatively high, but the bladder epithelium is relatively impermeable to water so that the resorbate is hyperosmotic to the urine, aiding in the retention of salt and elimination of water to balance the osmotic uptake across the gills (e.g., Evans, 1979; Demarest, 1984). Recent *in vivo* studies in the rainbow trout, *Oncorhynchus mykiss*, suggest that the role of the urinary bladder in the final reabsorption of salt and water may be much more substantial than formerly believed because previous cannulation studies had often drained urine from the bladder before final processing could occur (Curtis and Wood, 1991). In fact, this study determined that urine flows were actually intermittent and at least 20% less than previously published values, with salt loss at least 40% lower, when an external catheterization technique was used. This suggests that the role of the urinary bladder in controlling final urinary salt loss may be substantial.

Branchial Salt Uptake Mechanisms

The mechanisms of NaCl uptake by the gill epithelium of freshwater teleosts have been studied since the pioneering work of August Krogh (reviewed in Krogh, 1939). He was the first to suggest that ionic influxes of Na^+ and Cl^- were separate, and probably in exchange for plasma ions such as NH_4^+ and HCO_3^-, respectively. More recent evidence suggests that Na/H exchange may be more common than Na/NH_4 exchange (e.g., Heisler, 1989; Heisler, this volume). In addition, there are emerging data suggesting that the coupling between Na^+ and H^+ may be electrical, rather than chemical (Avella and Bornancin, 1989; Lin and Randall, 1991), and it appears that most ammonia excretion may be via both nonionic and ionic diffusion (Evans and Cameron, 1986; Wood, this volume). Nevertheless, the net result is still uptake of necessary NaCl and extrusion of acid/base regulating ions (e.g., Wood, 1991). In fact, somewhat surprisingly, current data indicate that these ionic exchange systems are present in the gills of marine fishes, including elasmobranchs and hagfishes, suggesting that the coupled transport (chemical or electrical) arose for acid-base regulation before the vertebrates entered fresh water (Evans, 1984a), an interesting case of "preadaptation".

Unfortunately, it is still not clear which cell type in the freshwater teleost gill epithelium is responsible for these transfer processes (Karnaky, 1986, 1992). A recent study suggested that the mitochondrion-rich, "chloride cell" of the branchial epithelium may be the site of these exchanges because acclimation of *O. mykiss* to artificial fresh water, with reduced NaCl content, was followed by a stimulation of ionic uptake and a proliferation of gill "chloride cells" (Perry and Laurent, 1989). However, the data are correlative rather than direct. The presence of similar cells in the gill epithelia of hagfishes and lampreys, presumably involved in acid-base regulation (see above), also supports the hypothesis that "chloride cells" are the site of ionic uptake in freshwater teleosts, but direct data are lacking. Therefore, at the present time, the specific cellular site of gill ionic and/or acid-base regulation involving the chemical or electrical Na/H (NH_4) and Cl/HCO_3 exchanges is undetermined.

Regulatory Mechanisms in Seawater

Marine teleosts display a much higher branchial permeability to salts than any other fish group, generally associated with reduced intercellular junctions between adjacent chloride and accessory cells (e.g., Pisam and Rambourg, 1991; Karnaky, 1992). Unidirectional Na^+ and Cl^- fluxes are 10 to 50 times greater than those measured in freshwater fishes, marine hagfishes, or elasmobranchs. The ionic gradients across the gill epithelium are of the same order as those in fresh water (~200 m*M*), but, of course, reversed. In addition, the transepithelial electrical potential (TEP) measured across marine teleosts is often of the order of +25 mV (plasma relative to seawater), suggesting that Na^+ is nearly in electrochemical equilibrium, while Cl^- is distinctly out of electrochemical equilibrium (e.g., Evans, 1980; Potts, 1984; Potts and Hedges, 1991), and therefore must be excreted. However, the TEP across some species of teleosts is significantly less positive, and even slightly negative (e.g., Evans, 1980; Potts and Hedges, 1991), suggesting that these species probably have to actively excrete both Na^+ and Cl^- into the seawater. Interestingly, the current model for salt extrusion by the marine teleost gill epithelium (see below) assumes that Na^+ is in electrochemical equilibrium, which is apparently not the case in some species of teleosts. The diffusional water permeability of marine teleosts is somewhat below that of their freshwater relatives, presumably secondary to the higher concentration of Ca^{2+} in the marine environment (Evans, 1969), although this proposition has not been studied extensively.

Oral Ingestion

Marine teleosts must ingest the surrounding seawater to balance the osmotic loss of water across the gills (and skin?), just as fresh water teleosts urinate large volumes of urine to balance the osmotic uptake of water. Rates of oral ingestion vary considerably, correlated with osmotic water permeability, but generally are approximately 1 ml·100 g^{-1} h^{-1}, providing a theoretical net Na^+ and Cl^- influx of some 500 μM·100 g^{-1} h^{-1}. This approaches 25 to 50% of the unidirectional salt fluxes that have been measured (Evans, 1979), and presumably a significantly greater percentage of the net ionic influx. Indeed, if Na^+ is actually in electrochemical equilibrium across the gill epithelium, the oral ingestion of Na^+ may be the only net influx of this ion. Thus, regulation of the body water content adds to, or may even provide, the net salt influx which must be excreted.

The ingested seawater is diluted by some 50% by diffusional uptake of salts from the esophagus (and to a lesser extent the stomach), but the bulk of water and additional salt uptake (in this case active) takes place in the small intestine (Evans, 1979). NaCl uptake is via a Na+Cl cotransport, with water following osmotically (Figure 4; e.g., Groot and Bakker, 1988; Musch et al., 1990). A classic study by Hickman (1968) demonstrated that, in the southern flounder, *Paralichthyes lethostigma*, over 95% of the ingested Na^+, K^+, and Cl^-, 69% of the Ca^{2+}, but only <15% of the Mg^{2+} and SO_4^{2+} are absorbed in the intestine. The bulk of the Mg^{2+} and SO_4^{2-} are therefore excreted rectally; all of the absorbed Mg^{2+} and SO_4^{2-} and 30% of the absorbed Ca^{2+} are excreted renally. The remaining Ca^{2+}, and all of the absorbed Na^+, K^+, and Cl^- are excreted extrarenally. Similar studies of other species would be quite informative, but it is apparent that divalent

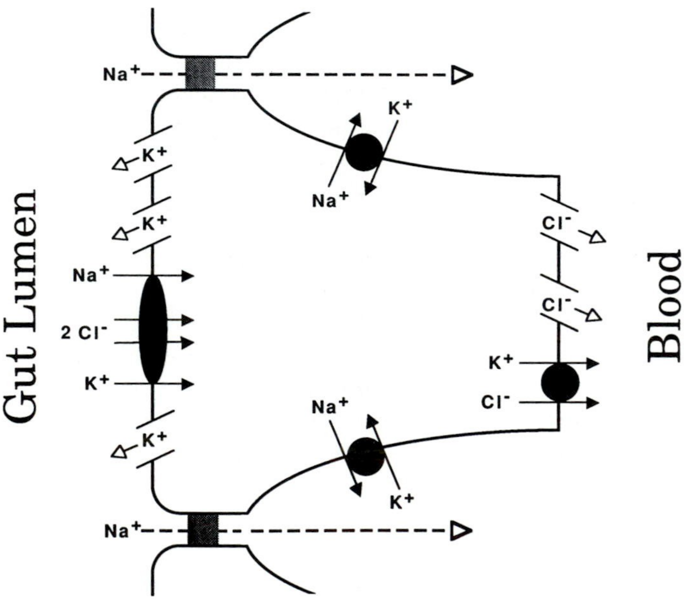

FIGURE 4. Current model for NaCl uptake by the marine teleost intestine. Extensive lateral channels, rather than basolateral infoldings, are present (e.g., Groot and Bakker, 1988). Basolateral (serosal) Na^+/K^+-ATPase generates an electrochemical gradient for Na^+, which provides the driving force for Na+K+2Cl transport into the cytoplasm, across the apical (mucosal) membrane. Net Na^+ uptake is via basolateral Na/K exchange, as well as paracellular diffusion down an electrochemical gradient generated by basolateral Cl^- channels. Basolateral KCl secretion also provides for net uptake of Cl^-. *In vivo*, apical K^+ channels provide for net secretion of K^+ into the gut lumen. See Groot and Bakker (1988) and Musch et al. (1990) for more details.

and monovalent ions are processed quite differently after the oral ingestion of seawater.

The Kidney and Urinary Bladder

The kidney of marine teleosts is generally somewhat morphologically reduced when compared to that of freshwater species. Gomeruli are often reduced in number, and glomeruli and proximal tubule I are actually missing in some 30 species. The distal tubule is usually lacking in both aglomerular and glomerular species of marine teleosts (Hentschel and Elger, 1987). In the latter, GFRs are of the order of 50 μl·100 g^{-1} h^{-1}, only 10% of those in fresh water. From studies of euryhaline species, it is apparent that this reduction in GFR is produced by a reduction in the number of filtering glomeruli, rather than a fall in single nephron filtration rates (e.g., Brown et al., 1983). Somewhat surprisingly, it has been shown recently that the proximal tubules in both the flounder, *Pseudopleuronectes americanus*, and the killifish, *Fundulus heteroclitus*, in seawater secrete fluid, secondary to NaCl transport, into the lumen; however, it is unclear if this salt transport is via the same Na+Cl cotransport system that secretes NaCl in the proximal tubule of elasmobranchs and freshwater teleosts (Beyenbach and Baustian, 1989; Cliff and Beyenbach, 1992). Urine flows approximate 30 μl·100 g^{-1} h^{-1} in glomerular species, indicating that some reabsorption of urine does occur, presumably in the collecting duct or tubule, or the urinary bladder where water permeability is increased over the freshwater condition, so that the resorbate is iso-osmotic rather than hyperosmotic (e.g., Evans, 1979; Demarest, 1984). Aglomerular species have urine flows equivalent to those of glomerular species (e.g., Evans, 1979), produced by secretion of divalent salts in the PII, as well as possibly by secretion of NaCl (Beyenbach and Baustian, 1989). Given the apparent ubiquity of proximal tubular secretion of salt, it is not clear why, with the apparent exception of *F. heteroclitus*, marine teleosts have not been demonstrated to produce a urine that is hyperosmotic to the plasma (Beyenbach and Baustian, 1989).

Branchial Salt Extrusion Mechanisms

Since the marine teleost kidney apparently is unable to produce a urine hyperosmotic to the plasma, extrarenal salt secretion must exist. Early experiments determined that the branchial epithelium was probably the site of this necessary salt secretion, and mitochondrion-rich cells (Figure 2D) were described by Keys and Willmer (1932) as "chloride cells" because of silver staining. Morphologically similar cells have now been described in the branchial epithelium of hagfishes, lampreys, and elasmobranchs (see above and Figure 2). The fact that the electrical current characteristic of salt transport can be localized over the "chloride cells" (CC) in the killifish opercular epithelium (which contains ~60% CC) provides direct evidence that the CC is the site of branchial salt extrusion (Foskett, 1987). In fact, this preparation (and other similar skin epithelia) provides a model system that has allowed the biophysical dissection of the transport of NaCl across the teleost gill epithelium (Zadunaisky, 1984).

The currently accepted model (e.g., Zadunaisky, 1984; Karnaky, 1986; Pisam and Rambourg, 1991) for NaCl extrusion by CC is the same as that proposed for the shark rectal gland (Figure 3, substitute "seawater" for "gland lumen"); however, some aspects, such as apical Cl^- conductance, K^+ sensitivity of the basolateral carrier, and basolateral K^+ conductance, have not been demonstrated definitively (e.g., Marshall, 1988). Moreover, it is important to note that this model assumes that Na^+ is in electrochemical equilibrium across the gill (i.e., that the electropositivity of the plasma relative to the medium is sufficient to offset the diffusional uptake of Na^+ across the gill epithelium and the uptake of Na^+ across the intestine). Several species of marine teleosts have now been found to maintain TEPs different from the Na^+ equilibrium potential (see above), so, at least in these species, other mechanisms for Na^+ extrusion must be involved. These mechanisms are still unstudied.

SUMMARY AND CONCLUSIONS

It is clear that the evolution of fishes from marine ancestors has also involved the evolution of suites of osmotic and ionic regulatory mechanisms that enable these aquatic vertebrates to maintain relatively consistent plasma salt content, cellular volume, and intracellular salt content. Such control is crucial for the movement of the vertebrates into fresh water, and thence back into the marine environment. A relatively few number of species of teleosts (and also some lampreys, and the very rare elasmobranch) are able to actually regulate against varying, and reversing, osmotic and ionic gradients and are therefore euryhaline. In contrast to the domination of the kidney in osmotic and ionic regulation in terrestrial vertebrates, fishes use such structures as the intestine, gill, and rectal gland, as well as the kidney, to regulate cellular volume and salt content. To a large extent, the basic models which describe the organ-systems, and even transport steps, involved in fish osmotic and ionic regulation were elucidated in the 1930s to 1970s (see Evans, 1979); more recent publications have dealt in large measure with the finer points of ionic transport, and the hormonal control (see Wendelaar Bonga, this volume) of this system. Because of the difficulty of studying the physiology of whole animals, our knowledge of many aspects of fish osmotic and ionic regulation is still relatively rudimentary. For example, the extremely complex structure of the fish gill precludes many of the perfusion and isolated-membrane techniques that are necessary for more in-depth dissection of critical salt transport steps. The fact that most of the questions about specific points in fish osmotic and ionic regulation that were posed in a review in 1984 (Evans, 1984b) still are unanswered is an indication of the relatively slow progress in this field in the last decade. Thus, basic questions about the mechanisms of osmotic and ionic regulation in fishes remain for future research.

ACKNOWLEDGMENTS

The author's research reported in this chapter was supported by various grants from the National Science Foundation, most recently DCB 8916413. In addition, some of our most recent work has also been supported by NIEHS grant EHS-1 P30 ESO3128 to the Center for Membrane Toxicity Studies at the Mt. Desert Island Biological Laboratory.

REFERENCES

Alt, J. M., Stolte, H., Eisenbach, G. M., and Walvig, F., Renal electrolyte and fluid excretion in the Atlantic hagfish *Myxine glutinosa, J. Exp. Biol.*, 91, 323, 1981.

Avella, M. and Bornancin, M., A new anaylsis of ammonia and sodium transport through the gills of the freshwater rainbow trout *(Salmo gairdneri), J. Exp. Biol.*, 142, 155, 1989.

Bartels, H., Assemblies of linear arrays of particles in the apical plasma membrane of mitochondria-rich cells in the gill epithelium of the Atlantic hagfish *(Myxine glutinosa), Anat. Rec.*, 211, 229, 1985.

Bartels, H. and Welsch, U., Mitochondria-rich cells in the gill epithelium of cyclostomes. A thin section and freeze fracture study, in *Indo-Pacific Fish Biology: Proc. of the Second Int. Conf. on Indo-Pacific Fishes,* Uyeno, T., Arai, T., Taniuchi, T., and Matsuura, K., Eds., Ichthyological Society of Japan, Tokyo, 1986, 58.

Bartels, H., Hilliard, R. W., and Potter, I. C., Structural changes in the plasma membrane of chloride cells in the gills of the sea water- and fresh water-adapted lampreys, *J. Cell Biol.*, 105, 305a, 1987.

Beyenbach, K. W. and Baustian, M. D., Comparative physiology of the proximal tubule, in *Structure and Function of the Kidney*, Kinne, R. K. H., Eds., S. Karger, Basel, 1989, 103.

Bitter, A. and Lang, S., Some aspects of the osmoregulation of Amazonia freshwater stingrays *(Potamotrygon hystrix).* I. Serum osmolality, sodium and chloride content, water content, hematocrit and urea level, *Comp. Biochem. Physiol.*, 67A, 9, 1980.

Boylan, J. W., A model for passive urea absorption in the elasmobranch kidney, *Comp. Biochem. Physiol.*, 42, 27, 1972.

Bray, A. A., The evolution of the terrestrial vertebrates: environmental and physiological considerations, *Philos. Trans. R. Soc. London Ser. B*, 309, 289, 1985.

Brown, J. A., Taylor, S. M., and Gray, C. J., Glomerular ultrastructure of the trout, *Salmo gairdneri, Cell Tissue Res.*, 230, 205, 1983.

Burger, J. W., Roles of the rectal gland and kidneys in salt and water excretion in the spiny dogfish, *Physiol. Zool.*, 38, 191, 1965.

Cholette, C., Gagnon, A., and Germain, P., Isosmotic adaptation in *Myxine glutinosa* L. I. Variations of some parameters and role of the amino acid pool of the muscle cells, *Comp. Biochem. Physiol.*, 33, 333, 1970.

Cliff, W. H. and Beyenbach, K. W., Secretory renal proximal tubules in seawater- and freshwater-adapted killifish, *Am. J. Physiol.*, 262, F108, 1992.

Conley, D. M. and Mallatt, J., Histochemical localization of Na^+K-ATPase and carbonic anhydrase activity in the gills of 17 fish species, *Can. J. Zool.*, 66, 2398, 1988.

Curtis, B. J. and Wood, C. M., The function of the urinary bladder in vivo in the freshwater rainbow trout, *J. Exp. Biol.*, 155, 567, 1991.

Dawson, D. C. and Frizzell, R. A., Mechanism of active K^+ secretion by flounder urinary bladder, *Pfluegers Arch.*, 414, 393, 1989.

Demarest, J. R., Ion and water transport by the flounder urinary bladder: salinity dependence, *Am. J. Physiol.*, 246, F395, 1984.

Elger, M., The branchial circulation and the gill epithelia in the Atlantic hagfish, *Myxine glutinosa* L., *Anat. Embryol.*, 175, 489, 1987.

Ernst, S. A., Hootman, S. R., Schreiber, J. H., and Riddle, C. V., Freeze-fracture and morphometric analysis of occluding junctions in rectal glands of elasmobranch fish, *J. Membr. Biol.*, 58, 101, 1981.

Evans, D. H., Studies on the permeability to water of selected marine, freshwater and euryhaline teleosts, *J. Exp. Biol.*, 50, 689, 1969.

Evans, D. H., Fish, in *Comparative Physiology of Osmoregulation in Animals,* Maloiy, G. M. O., Ed., Academic Press, Orlando, 1979, 305.

Evans, D. H., Kinetic studies of ion transport by fish gill epithelium, *Am. J. Physiol.*, 238, R224, 1980.

Evans, D. H., Osmotic and ionic regulation by freshwater and marine fish, in *Environmental Physiology of Fishes*, Ali, M. A., Ed., Plenum Press, New York, 1981, 93.

Evans, D. H., Mechanisms of acid extrusion by two marine fishes; the teleost, *Opsanus beta*, and the elasmobranch, *Squalus acanthias, J. Exp. Biol.*, 97, 289, 1982.

Evans, D. H., Gill Na/H and Cl/HCO_3 exchange systems evolved before the vertebrates entered fresh water, *J. Exp. Biol.*, 113, 464, 1984a.

Evans, D. H., The roles of gill permeability and transport mechanisms in euryhalinity, in *Fish Physiology*, Hoar, W. S. and Randall, D. J., Eds., Academic Press, Orlando, 1984b, 239.

Evans, D. H. and Cameron, J. N., Gill ammonia transport, *J. Exp. Zool.*, 239, 17, 1986.

Evans, D. H. and Hooks, C., Sodium fluxes across the hagfish, *Myxine glutinosa, Bull. Mt. Desert Isl. Biol. Lab.*, 23, 61, 1983.

Evans, D. H. and More, K., Modes of ammonia transport across the gill epithelium of the dogfish pup (*Squalus acanthias), J. Exp. Biol.*, 138, 375, 1988.

Evans, D. H., Oikari, A., Kormanik, G. A., and Mansberger, L., Osmoregulation by the prenatal spiny dogfish, *Squalus acanthias, J. Exp. Biol.*, 101, 295, 1982.

Fels, L. M., Raguse-Degener, G., and Stolte, H., The archinephron of *Myxine glutinosa* L. (Cyclostomata), in *Structure and Function of the Kidney*, Kinne, R. K. H., Ed., Forster, 1979. S. Karger, Basel, 1989, 73.

Forster, R. P., Goldstein, L., and Rosen, J. K., Intrarenal control of urea reabsorption by renal tubules of the marine elasmobranch, *Squalus acanthias., Comp. Biochem. Physiol.*, 42A, 3, 1972.

Foskett, J. K., The chloride cell, in *Comparative Physiology of Environmental Adaptations*, Kirsch, R. and Lahlou, B., Eds., S. Karger, Basel, 1987, 83.

Friedman, P. A. and Hebert, S. C., Diluting segment in kidney of dogfish shark. I. Localization and characterization of chloride transport, *Am. J. Physiol.*, 258, R398, 1990.

Greger, R., Schlatter, E., and Gogelein, H., Sodium chloride secretion in rectal gland of dogfish, Squalus acanthias, *News Physiol. Sci.*, 1, 134, 1986.

Groot, J. A. and Bakker, R., NaCl transport in the vertebrate intestine, in *Comparative and Environmental Physiology. NaCl Transport in Epithelia*, Gregor, R., Ed., Springer-Verlag, Berlin, 1988, 103.

Halstead, L. B., The vertebrate invasion of freshwater, *Philos. Trans. R. Soc. London Ser. B.*, 309, 243, 1985.

Hardisty, M. W., *Biology of the Cyclostomes*, Chapman and Hall, London, 1979.

Heisler, N., Interactions between gas exchange, metabolism, and ionic transport in animals: an overview, *Can. J. Zool.*, 67, 2923, 1989.

Henderson, I. W., O'Toole, L. B., and Hazon, N., Kidney function, in *Physiology of Elasmobranch Fishes*, Shuttleworth, T. J., Ed., Springer-Verlag, Berlin, 1988, 201.

Hentschel, H. and Elger, M., The distal nephron in the kidney of fishes, *Adv. Anat. Embryol. Cell Biol.*, 108, 1, 1987.

Hentschel, H. and Elger, M., Morphology of glomerular and aglomerular kidneys, in *Structure and Function of the Kidney*, Kinne, R. K. H., Ed., S. Karger, Basel, 1989, 1.

Hickman, C. P., Jr., Ingestion, intestinal absorption and elimination of sea water and salts in the southern flounder, *Paralichthyes lethostigma, Can. J. Zool.*, 46, 457, 1968.

Hickman, C. P., Jr., The kidney, in *Fish Physiology*, Hoar, W. S. and Randall, D. J., Eds., Academic Press, New York, 1969, 91.

Hoar, W. S. and Randall, D. J., Eds., *Fish Physiology*, Vol. 10B, Academic Press, Orlando, 1984.

Jackson, B. A., Brown, J. A., Oliver J.A., and Henderson, I.W., Actions of angiotensin on single nephron filtration rates of trout, *Salmo gairdneri*, adapted to fresh- and sea-water environments, *J. Endocrinol.*, 75, 32P, 1977.

Karnaky, K. J., Jr., Structure and function of the chloride cell of *Fundulus heteroclitus* and other teleosts, *Am. Zool.*, 26, 209, 1986.

Karnaky, K. J., Jr., Teleost osmoregulation: changes in the tight junction in response to the salinity of the environment, in *Tight Junctions*, Cereijido, M., Ed., CRC Press, Boca Raton, FL, 1992, 175.

Karnaky, K. J., Jr., Valentich, J. D., Currie, M. G., Oehlenschlager, W. F., and Kennedy, M. P., Atriopeptin stimulates chloride secretion in cultured shark rectal gland cells, *Am. J. Physiol.*, 260, C1125, 1991.

Keys, A. B. and Willmer, E. N., "Chloride-secreting cells" in the gills of fishes with special reference to the common eel, *J. Physiol. (London)*, 76, 368, 1932.

Kirschner, L. B., Control mechanisms in crustaceans and fishes, in *Mechanisms of Osmoregulation in Animals. Maintenance of Cell Volume*, Gilles, R., Ed., John Wiley & Sons, Chichester, 1979, 157.

Krogh, A., *Osmotic Regulation in Aquatic Animals*, Cambridge University Press, Cambridge, 1939.

Lacy, E. R. and Reale, E., Fine structure of the elasmobranch renal tubule: neck and proximal segments of the little skate, *Am. J. Anat.*, 190, 118, 1991a.

Lacy, E. R. and Reale, E., Fine structure of the elasmobranch renal tubule: intermediate, distal, and collecting duct segments of the little skate, *Am. J. Anat.*, 192, 478, 1991b.

Laurent, P., Structure of vertebrate gills, in *Gills*, Houlihan, D. F., Rankin, J. C., and Shuttleworth, T. J., Eds., Cambridge University Press, Cambridge, 1982, 25.

Lin, H. and Randall, D., Evidence for the presence of an electrogenic proton pump on the trout gill epithelium, *J. Exp. Biol.*, 161, 119, 1991.

Logan, A. G., Moriarty, R. J., and Rankin, J. C., A micropuncture study of kidney function in the river lamprey, *Lampetra fluviatilis*, adapted to fresh water, *J. Exp. Biol.*, 85, 137, 1980.

Mallatt, J. and Paulsen, C., Gill ultrastructure of the Pacific hagfish *Eptatretus stouti, Am. J. Anat.*, 177, 243, 1986.

Mallatt, J., Conley, D. M., and Ridgway, R. L., Why do hagfish have gill "chloride cells" when they need not regulate plasma NaCl concentration?, *Can. J. Zool.*, 65, 1956, 1987.

Marshall, J., Martin, K. A., Picciotto, M., Hockfield, S., Nairn, A. C., and Kaczmarek, L. K., Identification and localization of a dogfish homolog of human cystic fibrosis transmembrane conductance regulator, *J. Biol. Chem.*, 266, 22749, 1991.

Marshall, W. S., NaCl transport in gills and related structures-vertebrates, in *Comparative and Environmental Physiology. NaCl Transport in Epithelia*, Gregor, R., Ed., Springer-Verlag, Berlin, 1988, 48.

McFarland, W. N. and Munz, F. W., Regulation of body weight and serum composition by hagfish in various media, *Comp. Biochem. Physiol.*, 14, 393, 1965.

McInerney, J. E., Renal sodium reabsorption in the hagfish, *Eptatretus stouti, Comp. Biochem. Physiol.*, 49, 273, 1974.

McVicar, A. J. and Rankin, J. C., Dynamics of glomerular filtration in the river lamprey, *Lampetra fluviatilis* L., *Am. J. Physiol.*, 249, F132, 1985.

Moriarty, R. J., Logan, A. G., and Rankin, J. C., Measurement of single nephron filtration rate in the kidney of the river lamprey, *Lampetra fluviatilis* L., *J. Exp. Biol.*, 77, 57, 1978.

Morris, R. and Pickering, A. D., Changes in the ultrastructure of the gills of the river lamprey, *Lampetra fluviatilis* (L.), during the anadromous spawning migration, *Cell Tissue Res.*, 173, 271, 1976.

Munz, F. W. and McFarland, W. N., Regulatory function of a primitive vertebrate kidney, *Comp. Biochem. Physiol.*, 13, 381, 1964.

Musch, M. W., O'Grady, S. M., and Field, M., Ion transport of marine teleost intestine, *Methods Enzymol.*, 192, 746, 1990.

Pang, P. K. T., Griffith, R. W., and Atz, J. W., Osmoregulation in elasmobranchs, *Am. Zool.*, 17, 365, 1977.

Payan, P. and Maetz, J., Branchial sodium transport mechanisms in *Scyliorhinus canicula*: evidence for Na^+/NH_4^+ and Na^+/H^+ exchanges and for a role of carbonic anhydrase, *J. Exp. Biol.,* 58, 487, 1973.

Peek, W. D. and Youson, J. H., Ultrastructure of chloride cells in young adults of the anadromous sea lamprey, *Petromyzon marinus* L., in fresh water and during adaptation to sea water, *J. Morphol.*, 160, 143, 1979.

Perry, S. F. and Laurent, P., Adaptational responses of rainbow trout to lowered external NaCl concentration: contribution of the branchial chloride cell, *J. Exp. Biol.*, 147, 147, 1989.

Pickering, A. D. and Morris, R., Osmoregulation of *Lampetra fluviatilis* L. and *Petromyzon marinus* (Cyclostomata) in hyperosmotic solutions, *J. Exp. Biol.*, 53, 231, 1970.

Pisam, M. and Rambourg, A., Mitochondria-rich cells in the gill epithelium of teleost fishes: an ultrastructural approach, *Int. Rev. Cytol.*, 130, 191, 1991.

Potts, W. T. W., Transepithelial potentials in fish gills, in *Fish Physiology*, Hoar, W. S. and Randall, D. J., Eds., Academic Press, Orlando, 1984, 105.

Potts, W. T. W. and Hedges, A. J., Gill potentials in marine teleosts, *J. Comp. Physiol.*, 161, 401, 1991.

Rankin, J. C., Logan, A. G., and Moriarty, R. J., Changes in kidney function in the river lamprey, *Lampetra fluviatilis* L., in response to changes in external salinity, in *Epithelial Transport in the Lower Vertebrates*, Lahlou, B., Ed., Cambridge University Press, Cambridge, 1980, 171.

Riegel, J. A., Factors affecting glomerular function in the Pacific hagfish *Eptatretus stouti* (Lockington), *J. Exp. Biol.*, 73, 261, 1978.

Riegel, J. A., The absence of an arterial pressure effect on filtration by perfused glomeruli of the hagfish, *Eptatretus stouti* (Lockington), *J. Exp. Biol.*, 126, 361, 1986.

Shuttleworth, T. J., Salt and water balance—extrarenal mechanisms, in *Physiology of Elasmobranch Fishes*, Shuttleworth, T. J., Ed., Springer-Verlag, Berlin, 1988, 171.

Silva, P., Solomon, R. J., and Epstein, F. H., Shark rectal gland, *Methods Enzymol.,* 192, 754, 1990.

Solomon, R., Taylor, M., Sheth, S., Silva, P., and Epstein, F. H., Primary role of volume expansion in stimulation of rectal gland function, *Am. J. Physiol.*, 248, R638, 1985.

Stinson, C. M. and Mallatt, J., Branchial ion fluxes and toxicant extraction efficiency in lamprey *(Petromyzon marinus)* exposed to methylmercury, *Aquat. Tox.*, 15, 237, 1989.

Thomerson, J. E. and Thorson, T. B., The bull shark, *Carcharhinus leucas*, from the upper Mississippi river near Atlon, Illinois, *Copeia*, 1977, 166, 1977.

Valentich, J. D. and Forrest, J. N., Jr., Cl^- secretion by cultured shark rectal gland cells. I. Transepithelial transport, *Am. J. Physiol.*, 260, C813, 1991.

Wood, C. M., Branchial ion and acid-base transfer in freshwater teleost fish: environmental hyperoxia as a probe, *Physiol. Zool.*, 64, 68, 1991.

Yancey, P. H. and Somero, G. N., Methylamine osmoregulatory solutes of elasmobranch fishes counteract urea inhibition of enzymes, *J. Exp. Zool.*, 212, 205, 1980.

Zadunaisky, J., The chloride cell: the active transport of chloride and the paracellular pathways, in *Fish Physiology*, Hoar, W. S. and Randall, D. J., Eds., Academic Press, Orlando, 1984, 129.

12 Acid-Base Regulation

Norbert Heisler

INTRODUCTION

pH is a central parameter for the function of metabolism and, accordingly, for the production of energy, which is essential for maintenance of life. Certain key enzymes (such as phosphofructokinase, PFK) exhibit pronounced pH activity optima (Heisler, 1990b), such that deviations by more than a few tenths of a pH unit may considerably reduce metabolic flux, or even completely inhibit further energy production. Regulation of pH within narrow limits is accordingly an indispensable prerequisite for maintenance of homeostasis.

Acid-base regulation in fish is challenged more than in terrestrial animals by a number of environmental and endogenous factors. This is mainly related to utilization of water rather than air as the respiratory gas exchange medium. The composition of water varies to a much larger extent than that of air, due to a number of natural or contamination-related factors. Fish frequently encounter large and rapid changes in water oxygen and CO_2 concentrations, as well as changes in temperature and environmental electrolyte concentrations.

The typical countercurrent water/blood flow orientation (Heisler, 1989) at the large gill surface area provides a very efficient interface for respiratory gas exchange between environment and body fluids. This system satisfies the oxygen demand of metabolism in spite of the much lower oxygen content of water as compared to air. This can be accomplished only by a large specific ventilation (volume per unit oxygen consumption) of the viscous breathing medium water, which, however, requires considerably more energy for convective gas exchange than in terrestrial animals. The resulting close contact of the gill epithelium with a large amount of water (of considerable thermal capacity) per unit time (in the range of 0.1 to 0.5 of the body volume/minute) subjects animals to frequent and often large changes in body temperature.

The gills of fish are the combined site of respiratory gas exchange and main ionic regulation, in contrast to air-breathing terrestrial animals with specifically adapted gas exchange structures. This system allows for easy ionic exchange, facilitated by the large rate of water flow past the gill structures with hardly any inspired-expired differences in ionic composition. Ions are readily transferred across the large surface area without the necessity for establishment of large gradients between "internal" (plasma) and "external" (tubular urine) fluids by "multiplying" concentration mechanisms like in the mammalian kidney.

The "open" epithelial surface of the gill system is, however, subject to changes of the water ionic composition which will directly affect ionic transfer

0-8493-8042-1/93/$0.00+$.50

processes for osmo- and acid-base regulation, and often will also have considerable impact on respiratory gas exchange. In comparison, the "closed, one way" kidney system carries the advantage of being mechanically protected against changes of the environment and is accordingly less sensitive to environmental disturbances.

This chapter briefly describes characteristics of fish acid-base regulation in response to exogenous and endogenous disturbances, and delineates features of principal mechanisms available to fish for maintenance of homeostasis. The description concentrates on common regulatory attributes in selected species, and depicts factors known to limit the effective capacity of involved regulatory systems. The finite amount of available print space requires the reader to be referred to review articles wherever possible.

MECHANISMS AVAILABLE FOR FISH ACID-BASE REGULATION

The regulatory mechanisms available for fish are, in principle, the same as for terrestrial animals, but are utilized to a quite different extent. The situation is characterized for fish by intimate contact with the aqueous environment, generally including continuous immersion and utilization of water as the gas exchange medium. Water immersion favors ion transfer mechanisms supporting acid-base regulation, but provides severe restrictions for the regulation of body fluid P_{CO_2}.

ADJUSTMENT OF P_{CO_2}

The much lower oxygen content of water as compared to air (~0.03, depending on temperature) is a most unfavorable factor for fish homeostatic control. The large gill surface area generally facilitates uptake of oxygen satisfying metabolic demand, but the required large specific ventilation (volume per unit oxygen consumption) results in much lower P_{CO_2} values than in terrestrial animals. This is related to the much larger capacitance of pure water for CO_2 than for O_2, being in the same range as that of air. The effective capacitance in normal aquatic habitats with water P_{CO_2} of less than 1 mmHg is even larger, due to carbonate and other nonbicarbonate buffering of CO_2. The high relative rate of gill ventilation (about 5 ml/μmol O_2 consumption as compared to 0.6 ml pulmonary ventilation/μmol O_2 in man) results accordingly in much lower CO_2 partial pressure differences between arterial blood and water in fish (typically 1 to 4 mmHg) than in terrestrial vertebrates (typically 30 to 45 mmHg) (Heisler, 1989).

Small environmental-arterial P_{CO_2} differences (ΔP_{CO_2}), however, limit the scope for compensatory modulation of P_{CO_2} by hyper- or hypoventilation, a mechanism frequently exploited in mammals (Woodbury, 1965). A further complicating fact is that ΔP_{CO_2} in fish is not only related to the rate of gill ventilation, but is also a function of other factors like blood or water shunting past the gas exchange surface (Heisler, 1989), or incomplete CO_2 hydration and disequilibrium as a result of insufficient carbonic anhydrase activity on the mucosal side of the gill epithelium (e.g., Henry et al., 1988; Heisler, 1990a). Arterial P_{CO_2} can accordingly be reduced by hyperventilation to a significant extent only in fish with relatively large ΔP_{CO_2}, like carp (3 to 4 mmHg; Claiborne and Heisler, 1984, 1986), provided that the environmental P_{CO_2} background is close to zero (see below). Carp are capable of reducing arterial P_{CO_2} during hypoxia by a factor of two, when the rate of ventilation is elevated by a factor of ~3 (Glass et al., 1990). At least a fraction of ~1 mmHg of P_{CO_2} in this species is related, therefore, to factors like incomplete CO_2 hydration and/or blood or water shunting (Heisler, 1989).

Buffering

Surplus H^+ ions are transferred into nondissociated state and masked by association with buffer bases, induced by changes in pH. This mechanism is capable of reducing pH changes to values still compatible with life functions until the cause of disturbance is eliminated (or pH is compensated by other means), but is incapable of restoring the original pH. Buffering serves valuable functions during the time period between introduction of surplus H^+ and OH^- ions into the body fluids by metabolism or other sources, and their final elimination, but, due to the limited capacity of the buffer systems and to the associated shift in pH, is generally a mechanism for only transient acid-base regulation (for details: Heisler, 1986a). Because of their largely different characteristics, two types of buffers are distinguished.

Nonbicarbonate Buffers

In biological fluids, nonbicarbonate buffers are mainly protein residues (histidine, cysteine, and terminal NH_2 groups), characterized by pK′ values close to physiological pH values. Nonbicarbonate buffering takes place generally in a system closed for acid and base forms of the buffer (definitions: Brønsted, 1923; Heisler, 1986a), at constant total buffer concentration. The buffer value (β) as a measure for the buffering capability is defined as the change in base or acid form of the buffer system per change in pH (Van Slyke, 1922), being highest at the center of the buffer curve (pK′ = pH) and falling in a bell-shaped manner towards higher and lower pH values (for details see Heisler, 1986a, 1989).

In fish, nonbicarbonate buffer values of blood and intracellular tissue compartments are generally lower than in terrestrial vertebrates (factors of 1.5 to 4; Heisler, 1986a), while the general relationships of intracellular tissue buffer values, being much higher than those of blood and extracellular space, are similar. The largest proportion of the buffer capacity ($\kappa = \beta \cdot V$; V = volume) is accordingly located in the intracellular body compartments.

Bicarbonate Buffer System

In contrast to nonbicarbonate buffers, the bicarbonate buffer system (consisting of CO_2, H_2CO_3, HCO_3^- and CO_3^{2-}) is an open buffer system, with the volatile acid-anhydride form (CO_2) being affected by respiratory gas exchange (for details see Heisler, 1986a, 1989, 1990a). While in terrestrial animals this allows for an improvement of the effectiveness of the buffer by reducing the concentration of CO_2, this is generally not possible in fish (see above section; also Heisler, 1986a, 1989, 1990a). Since the buffer value of bicarbonate is a direct function of its concentration, the effectiveness of bicarbonate in reducing pH shifts is further limited in fish by much lower control [HCO_3^-] as compared to terrestrial animals (Heisler, 1984b, 1986b, 1986c). A severe limitation of the bicarbonate buffer system is related to the fact that in fish the difference between extracellular pH and pK_1' is generally larger than 1.5 pH units, and therefore buffering is weak as long as the advantage of open system buffering cannot be utilized. This advantage can only be employed when blood passes the gas exchange structures, whereas during tissue passage buffering takes place in a closed system. Under such conditions the increase in P_{CO_2} from combination of H^+ ions with bicarbonate may be considerable, and the fall in perfusate pH induced by addition of H^+ ions during tissue passage may become the limiting factor for fast washout of H^+ ions from intracellular compartments (for details see Heisler, 1986c). According to these factors the contribution of the bicarbonate buffer system is small in fish.

Adjustment of Bicarbonate by Ionic Transfer

P_{CO_2} and bicarbonate concentration are the two variables of the Henderson-Hasselbalch equation for the CO_2/HCO_3^- buffer system that can be modified *in vivo* (Heisler, 1986c, 1989, 1990a). As outlined above, changes in P_{CO_2} attained within the limitations provided by water as the gas exchange medium are too small for any significant contribution with respect to acid-base regulation. Fish, therefore, rely much more than terrestrial animals on mechanisms affecting the bicarbonate concentration in the body fluids. This can be achieved by transfer of acid-base relevant ions across the epithelial surface of the animal. Transfer of H^+, OH^-, and bicarbonate directly affect [HCO_3^-] (H^+ and OH^- via relevant buffer equilibria; Heisler, 1986c, 1989, 1990a), whereas other acid-base relevant ions only indirectly influence the acid-base status. When acid or base forms (Brønsted, 1923; Heisler, 1986c) of a nonbicarbonate buffer system are transferred between fluid compartments of different pH then titration of the buffer results in release or removal of H^+ ions to or from the free water phase of the respective fluid compartment. Transepithelial transfer of acid-base relevant ions is much more efficient in fish than in terrestrial animals and compensates for the poor efficiency of respiratory adjustment. The involved mechanisms generally include transfer of co- and counterions in order to maintain electroneutrality (see below).

CHARACTERISTIC FEATURES OF ACID-BASE REGULATION

Acid-base regulation in fish is continuously challenged by various endogenous and exogenous factors. Metabolism provides a continuous load of CO_2 and nonvolatile acid-base relevant end products, which have to be eliminated in order not to accumulate. The rate of endogenous production of acid-base relevant ions may rise tremendously during extreme muscular activity and during extreme hypoxia, when lactic acid is released by anaerobic metabolism. Exogenous disturbances of the acid-base regulation are induced by changes in environmental temperature, oxygen, and CO_2 concentrations, as well as by low water pH and associated effects of toxic substances such as aluminum on ion transfer epithelia. Typical patterns of the regulatory response to changes in temperature, lactacidosis, and several types of hypercapnia are described in the following sections. For the effects of low water pH and associated toxicity, the reader is referred to a number of excellent and extensive reviews (e.g., Fromm, 1980; Spry et al., 1981; Wood and McDonald, 1982, 1987; Wood and Soivio, 1991; Heisler 1984b).

CHANGES IN ENVIRONMENTAL TEMPERATURE

Steady state plasma pH, P_{CO_2}, and $[HCO_3^-]$ in heterothermal animals are a function of body temperature (Figure 1; for review, see Heisler, 1984b, 1986b). This fact was first recognized by Winterstein (1954) and Robin (1962) and later confirmed by numerous further studies (for review see Heisler, 1986c). Those authors independently expressed changes in pH by the $[OH^-]/[H^+]$ ratio (constant relative alkalinity), which features a negative correlation of pH with temperature. The imidazole alphastat hypothesis, a model of pH/temperature regulation based on adjustment of pH via ventilation related to the ionization of biological histidine-imidazole moieties (Reeves 1972; Reeves and Malan, 1976), failed to be confirmed by the majority of available experimental data. Plasma pH in fish generally changes much less (average $\Delta pH/\Delta t$: -0.011 U/°C; Heisler, 1986c) than expected from $\Delta pK'/\Delta t$ of imidazole compounds (-0.018 to 0.024 U/°C, Edsall and Wyman, 1958). Also, pH in intracellular compartments (Figure 2) and the magnitude of temperature-induced changes in ventilation are generally in contrast with the above hypotheses (Heisler, 1986c).

The most important regulatory measure for acid-base regulation in fish, ionic transfer of acid-base relevant ions, is explicitly excluded by the alphastat hypothesis (Reeves, 1972; Reeves and Malan, 1976). However, the experimental data available to this point do not support this proposition. Although direction and magnitude are diverse, all three fish species studied to date exhibit reversible transfer of acid-base relevant ions (*Scyliorhinus stellaris*, Heisler, 1978; *Ictalurus punctatus*, Cameron and Kormanik, 1982; *Synbranchus marmoratus*, Heisler, 1984a). In *Scyliorhinus* and *Ictalurus* bicarbonate equivalents are translocated from intracellular to extracellular space and even further

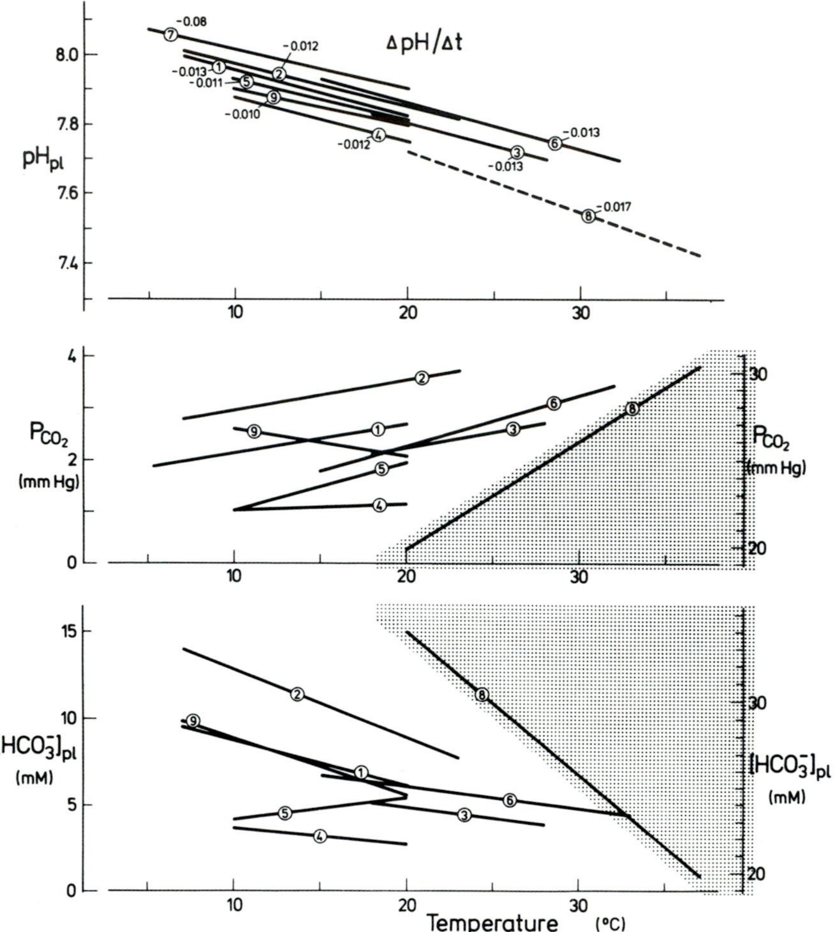

FIGURE 1. Plasma pH, P_{CO_2} and [HCO_3^-] as a function of environmental temperature in selected fish species. (1) *Salmo gairdneri* (Randall and Cameron, 1973); (2) *Cyprinus carpio* (N. Heisler and P. Neumann, unpublished); (3) *Cynoscion arenarius* (Cameron, 1978); (4) *Scyliorhinus stellaris* (juvenile; Heisler et al., 1980); (5) *Scyliorhinus stellaris* (adult; Heisler et al., 1980); (6) *Ictalurus punctatus* (Cameron and Kormanik, 1982); (7) *Anguilla rostrata* (Walsh and Moon, 1982); (8) *Synbranchus marmoratus* (Heisler, 1984a); (9) *Scyliorhinus stellaris* ($\Delta P_{CO_2}/\Delta t$ reversed; Heisler et al., 1980). Solid lines indicate water-breathers, the dotted line the only air-breather. Note: different scales of P_{CO_2} and [HCO_3^-] in *Synbranchus marmoratus* indicated by shaded areas.

released to the environment, whereas in *Synbranchus* transmembrane transfer is much larger and in the opposite direction, with generally little transfer occurring at the epithelial site (Figure 3; Heisler 1982a, 1984a). Differences in temperature-dependent intracellular buffering characteristics and pH changes are responsible for these completely different patterns (Heisler and Neumann,

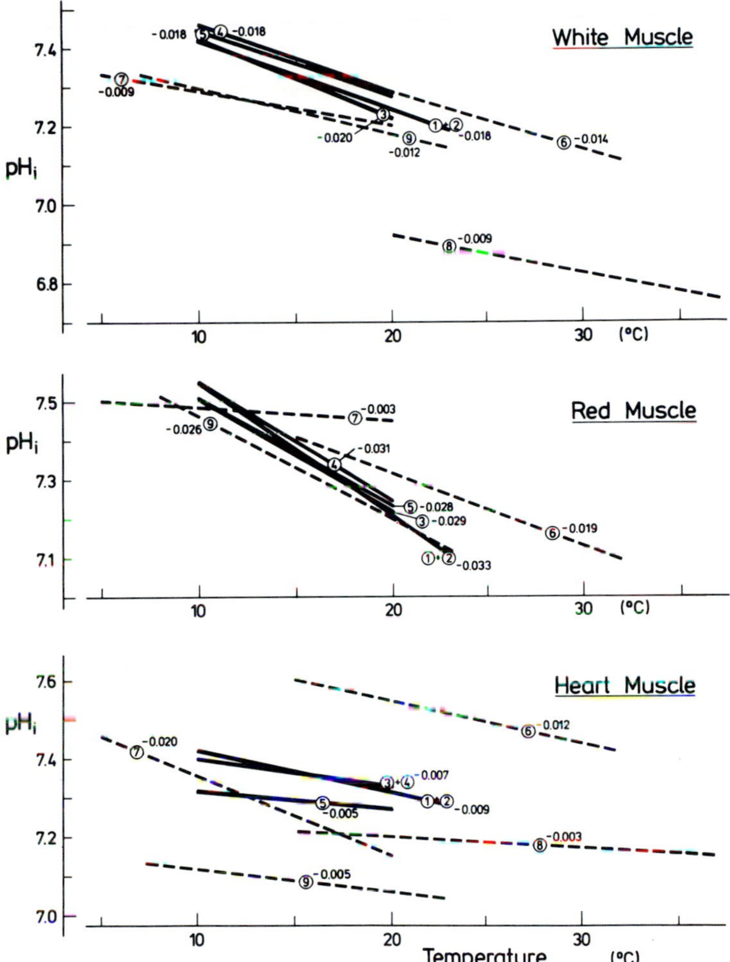

FIGURE 2. Intracellular pH as a function of environmental temperature in muscle tissues of selected fish species. (1) *Scyliorhinus stellaris* (juvenile; Heisler et al., 1976b); (2) *S. stellaris* (adult; Heisler et al., 1976b); (3) *S. stellaris* (juvenile, acclimated for 3 weeks; Heisler et al., 1980); (4) *S. stellaris* (adult, acclimated for 3 weeks; Heisler et al., 1980); (5) *S. stellaris* (adult, acclimated for 3 weeks, $\Delta P_{CO_2}/\Delta t$ reversed; Heisler et al., 1980); (6) *Ictalurus punctatus* (Cameron and Kormanik, 1982); (7) *Anguilla rostrata* (Walsh and Moon, 1982); (8) *Synbranchus marmoratus* (Heisler, 1984a); (9) *Cyprinus carpio* (N. Heisler and P. Neumann, unpublished). Note: solid lines indicate marine elasmobranchs, dashed lines freshwater teleost fish.

1980; Heisler, 1984a), indicating that extrapolation towards unstudied species is difficult.

The necessity of adjusting the bicarbonate concentration renders acid-base adjustment after changes in temperature a slow process, which is also effected

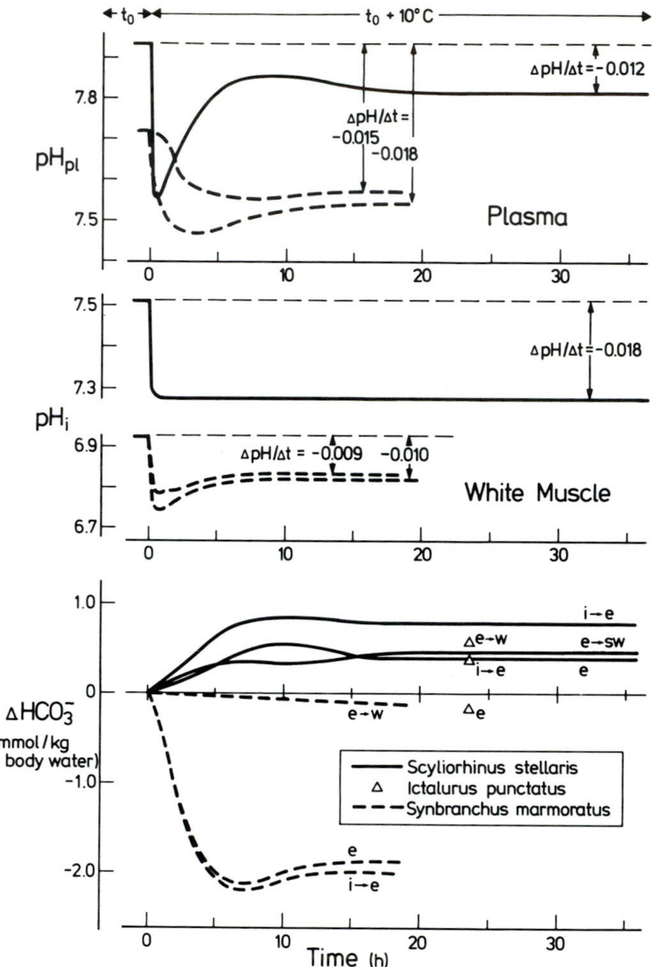

FIGURE 3. Kinetics of pH adjustment and bicarbonate equivalent ion transfer (i→e, intra- to extracellular; e→w, e→sw, extracellular space to water or seawater, respectively) in *Scyliorhinus stellaris* (Heisler, 1978) and *Synbranchus marmoratus* (Heisler, 1984a) upon temperature rising by 10°C. ΔpH/Δt values designate steady-state temperature coefficients of pH (U/°C). See also text.

by initial overshoot phenomena in P_{CO_2} (Figure 3). In *Scyliorhinus stellaris* a 10°C change in temperature (10 to 20°C) effects plasma pH to shift more than threefold (ΔpH/Δt ~ –0.039 U/°C) of the finally attained steady state change (ΔpH/Δt ~ –0.012 U/°C) (Heisler, 1978, 1984a). A new steady state is attained only after more than 12 h (Figure 3). *Synbranchus* exhibits similarly slow acid-base adjustment kinetics in response to an analogous temperature step (20 to 30°C)(Figure 3; Heisler 1984a).

Direct determination of intracellular pH kinetics is technically difficult and has been impossible under physiological conditions. Some information to this point can be obtained by appropriate model calculations based on the rate of transmembrane bicarbonate-equivalent transfer, and tissue-specific buffer characteristics (Heisler and Neumann, 1980; Heisler, 1984a, 1986a). The uncertainties of such estimates, related to assumptions with respect to homogeneity of transmembrane transfer (Heisler, 1984a, 1986b), are small in animals for which the fractional contribution of the modeled tissue to the body weight is large. Intracellular pH kinetics are accordingly modeled with rather good confidence for white muscle of *Scyliorhinus* (~54%; Heisler, 1978) and *Synbranchus* (~75% of body weight; Heisler, 1982a). In *Scyliorhinus* white muscle adjustment of intracellular pH is almost complete 30 min after changing temperature (Figure 3), whereas the process takes much longer in heart muscle (2 h) and red muscle (8 h), provided the same specific transfer rate is valid for white, red, and heart muscle. According to the much larger amounts of bicarbonate net transferred, the process of intracellular pH adjustment takes much longer in white muscle of *Synbranchus* (7 to 8 h; Figure 3; Heisler, 1984a).

The rate of transfer of bicarbonate equivalent ions after changes in temperature is at least one order of magnitude smaller than those achieved during peak perturbations of other acid-base stress conditions (see below). Similarly low transfer rates are attained only during mild disturbances or during approach to control values, suggesting that temperature-induced pH shifts do not affect cell function to a great extent. This notion is supported by the wide pH optima (at least compared with temperature-induced pH changes) of energy-producing enzymes (Heisler, 1986c, 1990b).

Tissue Anoxia and Lactacidosis

When tissue oxygen supply falls below metabolic demand, energy is supplied by anaerobic glycolysis. Such events of anaerobiosis are frequently encountered by fish, induced by environmental hypoxia or anoxia, or by failure of the oxygen transport chain (external respiration – diffusional gas exchange – blood gas transport – tissue diffusion) to supply sufficient oxygen, for instance during high levels of muscular activity. The main anaerobic intermediary product in vertebrates is lactic acid, which, according to its low pK′ value (~3.9), dissociates an equimolar amount of H^+ ions (only few species such as *Carassius auratus* and *C. carassius* have advantageously adapted to long-term anoxia by switching to the acid-base neutral intermediary product ethanol; see Shoubridge and Hochachka, 1980; van den Thillart et al., 1983; van den Thillart and van Waarde, 1991). Released H^+ ions are buffered by nonbicarbonate buffers or combine with bicarbonate before being eliminated from the body fluids by aerobic processing of lactic acid or by transepithelial ion transfer.

CO_2 Production and Gill Ventilation during Lactacidosis

Environmental hypoxia as one of the most common factors responsible for tissue anoxia results *per se* in largely elevated gill ventilation and excess elimination of CO_2 (Glass et al., 1990). This generally does not lead, however, to any reduction of P_{CO_2} for pH compensation. During this stage of lactacidosis the amount of CO_2 produced largely exceeds the amount that can be eliminated. In the dogfish *Scyliorhinus stellaris*, for instance, the amount of bicarbonate combined with H^+ ions (forming CO_2) during a short period of anaerobic muscular activity is equivalent to the amount of CO_2 produced during 3.5 h of resting steady-state aerobic metabolism (Holeton and Heisler, 1983; Randall et al., 1976). Accordingly, arterial P_{CO_2} rises from about 2 to more than 5 mmHg for 3 h, and remains above controls for more than 10 h (Figure 4; Holeton and Heisler, 1983). This long-term elevation certainly cannot be attributed to CO_2 produced from combination of H^+ and bicarbonate, but the rate of aerobic CO_2 production is elevated during recovery from lactacidosis due to payment of the oxygen dept by reestablishment of ATP and creatine-phosphate stores, aerobic processing of lactic acid, and certainly also due to the elevated energy requirements for ionic transfer (see below). Factors like incomplete CO_2 hydration in the gill effluent water, or larger than normal heterogeneities between blood and water flow in the gill epithelium may also play a role in producing an elevated ΔP_{CO_2} between arterial blood and expired water during such periods of enhanced CO_2 elimination. Accordingly, the regulation is taxed by a mixed metabolic and respiratory acidosis, not allowing pH to be restored by modulation of P_{CO_2}.

Ion Transfer Processes

There are generally three mechanisms to restore pH during lactacidosis: adjustment of P_{CO_2}, which is quite commonly utilized in terrestrial animals, but is not available for fish due to the limitations of aqueous gas exchange (see above), aerobic processing of lactic acid by breakdown to CO_2 or resynthesis to glycogen, and the elimination of surplus H^+ ions from the body fluids. Aerobic processing of lactic acid is a slow process, whereas transepithelial transfer of H^+ is apparently the mechanism of choice for fish.

Except for some details, the general regulatory patterns are quite similar among different species. After production and dissociation of lactic acid during extensive muscular activity, the dissociation products H^+ and lactate$^-$ are gradually eliminated from the intracellular space (Figure 4; Heisler, 1984b, 1986a). The time course of elimination is different for the two dissociation products and variable among fish species. Generally, the extracellular acid-base deflections peak much earlier than extracellular lactate and the acid-base variables are also restored long before plasma lactate (Figure 4). This is achieved by transient net transfer of H^+ ions to the ambient water, while lactate is never released to the environment (Holeton and Heisler, 1983; Holeton et al., 1983).

The dynamics of acid-base relevant ionic transfer reflect a complicated pattern of interaction of different mechanisms. In the first period after muscular activity intracellular pH in *Scyliorhinus stellaris* is generally less affected than extracellular pH. This is due to elimination of H^+ ions from the cells at an enormous rate (>250 μmol · min^{-1} · kg^{-1}) during the initial efflux time (Figure

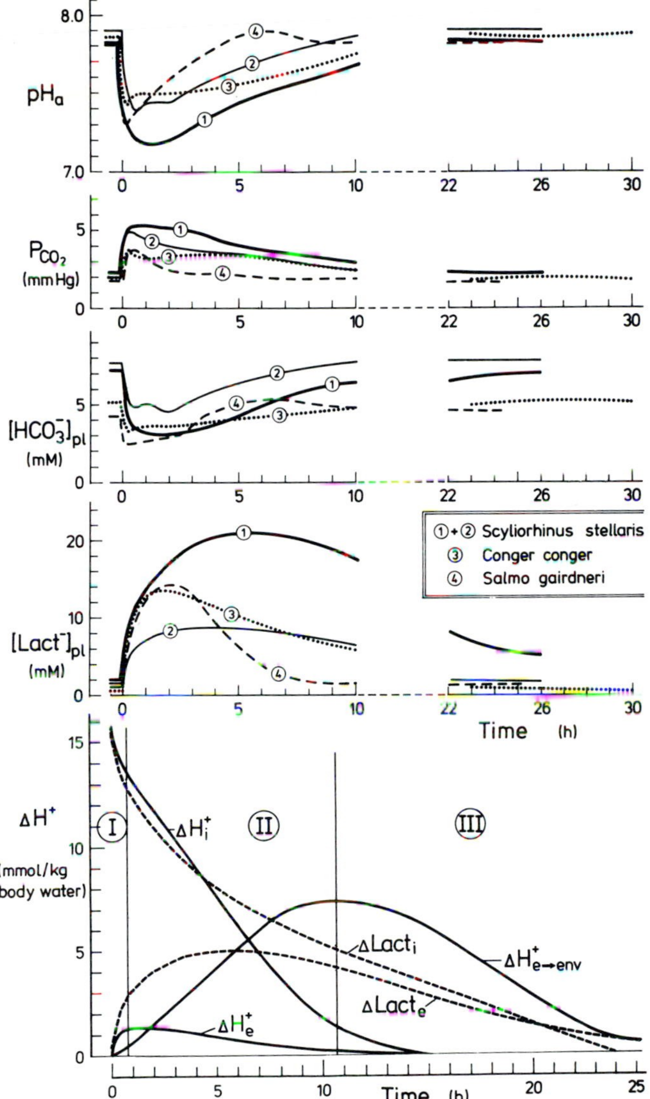

FIGURE 4. *Upper four panels*: Changes in plasma pH, P_{CO_2}, $[HCO_3^-]$, and lactate concentration during lactacidosis induced by exhausting muscular activity in three selected fish species (*Scyliorhinus stellaris*, Holeton and Heisler, 1983; *Salmo gairdneri*, Holeton et al., 1983; *Conger conger*, Toews et al., 1983). Note: elevation of P_{CO_2} and the normalization of pH before lactate concentration. See also text. *Lower panel*: Dynamics of surplus H^+ and lactate ions in intracellular (subscript i) and extracellular body fluid compartments (subscript e), and transferred to the environmental water (subscript i→e) of *Scyliorhinus stellaris* during recovery from severe muscular lactacidosis. After initially very fast elimination of H^+ ions from muscle cells (Phase I; ΔH^+_i)) the transmembrane H^+ elimination kinetics are determined by transfer to the environment (Phase II; $\Delta H^+_{e\to env}$) and the oxidative metabolization of lactic acid (see text). After normalization of extracellar and intracellular pH, H^+ ions are net recovered from the environment for further metabolism of lactic acid (Phase III; $\Delta H^+_{e\to env}$ and $\Delta Lact_i$ and $\Delta Lact_e$). Based on data of Holeton and Heisler (1983) using model calculations of Heisler (1984b, 1986a).

4, phase I). After this initial period the rate is retarded, due to the establishment of a transient equilibrium between intracellular and extracellular pH, eliminating the driving force for net H^+ transfer across the cell membrane (equilibrium limitation; Holeton and Heisler, 1983; Figure 4, phase II). H^+ ions are then further released from the intracellular space only at the rate of transfer to the environment. In the case of *Scyliorhinus stellaris*, less than 8% of surplus H^+ ions are buffered in blood and interstitial space. The whole process is attenuated by the limited net H^+ transfer rate to the environment (~15 $\mu mol \cdot min^{-1} \cdot$ kg body water^{-1}), similar to the maximal rate observed during hypercapnia). When the extracellular acid-base status is normalized (at still high levels of plasma lactate) H^+ ions required for further aerobic processing of lactate are returned from the environment at the rate of lactic acid metabolization (Figure 4, phase III). The transient storage of surplus H^+ ions in the ambient water is common among all studied species (e.g., *Conger conger*, Toews et al., 1983; *Salmo gairdneri*, Holeton et al., 1983).

Hypercapnia

Compensatory Effect of Hyperventilation

As described above, the environmental-arterial P_{CO_2} difference (ΔP_{CO_2}) is generally small in water-breathing fish, such that the effect of hyperventilation is small even at low environmental P_{CO_2}. It certainly is negligible on the background of the levels of environmental and endogenous hypercapnia fish may encounter. Although surface seawater P_{CO_2} usually reflects equilibrium with air (0.15 to 0.3 mmHg), P_{CO_2} at 200 to 500 m depths may rise to 5 to 10 mmHg due to anaerobic metabolism of microorganisms. Natural freshwater P_{CO_2} may range from extremely low, photosynthesis-related values (less than atmospheric, $<\sim 0.26$ mmHg) to as high as 60 mmHg (Heisler et al., 1982) as a result of insufficient surface gas exchange, and CO_2 release from bicarbonate induced by anaerobic microbial metabolism. Any adjustment of ΔP_{CO_2} brought about by changes in gill ventilation will have little effect on the level of plasma P_{CO_2}. A rise of inspired P_{CO_2} by 10 mmHg will result in fish in a rise of arterial P_{CO_2} by a factor of 3 to 6 with deflections of arterial pH by ~0.4 to 0.5 pH units (Heisler, 1984b, 1986b). In contrast, the conditions in mammals allow the acid-base status to be maintained normal during a similar increase in inspired P_{CO_2}: slight hyperventilation (+30%) will reduce the inspired/alveolar P_{CO_2} difference from 40 to 30 mmHg. Since its effect is modest, hyperventilation is hardly exploited during environmental hypercapnia in fish: the response is generally small and transient without being related to the actual time course of pH normalization (e.g., Randall et al., 1976; Dejours, 1975, 1981).

Regulatory Pattern during Environmental Hypercapnia

Environmental hypercapnia is almost immediately transmitted to the arterial blood by the large gill surface area, generally resulting in a comparable rise of plasma P_{CO_2} and a correlated decrease in plasma (Figure 5) and intracellular pH values. However, with some notable exceptions (see below), pH recovers towards control values soon after initiation of hypercapnia by elevation of the bicarbonate concentration, partially or even completely compensating the elevation in P_{CO_2}. During the initial phase with large deflections of pH, the rise in extracellular bicarbonate is partially due to nonbicarbonate buffering of CO_2, particularly in well-buffered intracellular muscle compartments (Figure 5). With increasing compensation and return of pH towards control values, the fraction of bicarbonate supplied by this mechanism falls, approaching zero when pH compensation is complete. In fish, mobilization of carbonate from the osteal structures is generally not involved during short term hypercapnia (directly measured: Cameron, 1985; estimated from changes in plasma [Ca^{2+}] and [Mg^{2+}]: N. Heisler and N.A. Andersen, unpublished data).

While the amount of bicarbonate supplied by nonbicarbonate buffering is relatively large during the first stages of hypercapnia, the main fraction of finally accumulated bicarbonate is derived from acid-base relevant transepithelial ion transfer mechanisms (Figure 5). The amount of bicarbonate supplied by the environment during hypercapnia (1% CO_2) is in the range of 3 to 6 mmol/(kg body weight) (e.g., Heisler et al., 1976a; Claiborne and Heisler, 1984, 1986; Heisler, 1986b). This is much more than the total bicarbonate pool at control conditions.

The activation of transepithelial transfer upon a step change in environmental hypercapnia is often delayed by a short time (~15 to 30 min; Figure 5; e.g., *Scyliorhinus stellaris*, Heisler et al., 1976a; *Conger conger*, Toews et al., 1983). During this period the plasma concentration is elevated by transmembrane transfer of acid-base relevant ions from the well-buffered intracellular fluid compartments. Simultaneously, bicarbonate is often lost to the environment, a maladaptive response indicating still incomplete adjustment of ion retaining and resorbing structures to the elevated plasma bicarbonate levels. The initial release of bicarbonate equivalents from intracellular to extracellular space may represent an analogously inadequate adjustment at the cellular level. The bicarbonate lost from intracellular and extracellular fluid compartments during this time is returned when transepithelial transfer is fully activated (Figure 5). Further transfer of bicarbonate equivalents to both extracellular and intracellular compartments generally results in almost complete pH compensation in both compartments (Figure 6).

The extracellular space receives a much larger fraction of the accumulated bicarbonate than the intracellular space, in spite of its small relative volume (20 to 25% of the body water; e.g., Heisler, 1978, 1982a). The relatively small amount of bicarbonate transferred to the large intracellular space (80 to 75% of body water) nevertheless results in almost complete pH compensation and even closer restoration of pH to normocapnic controls than in the extracellular fluid (Heisler, 1984b, 1986a, 1988a). This is due to extracellular control bicarbonate levels generally larger than intracellular bicarbonate by factors of 4 to 8, and to the supportive effect of the high capacity of intracellular nonbicarbonate buffers (Heisler, 1984b, 1986a, 1986b).

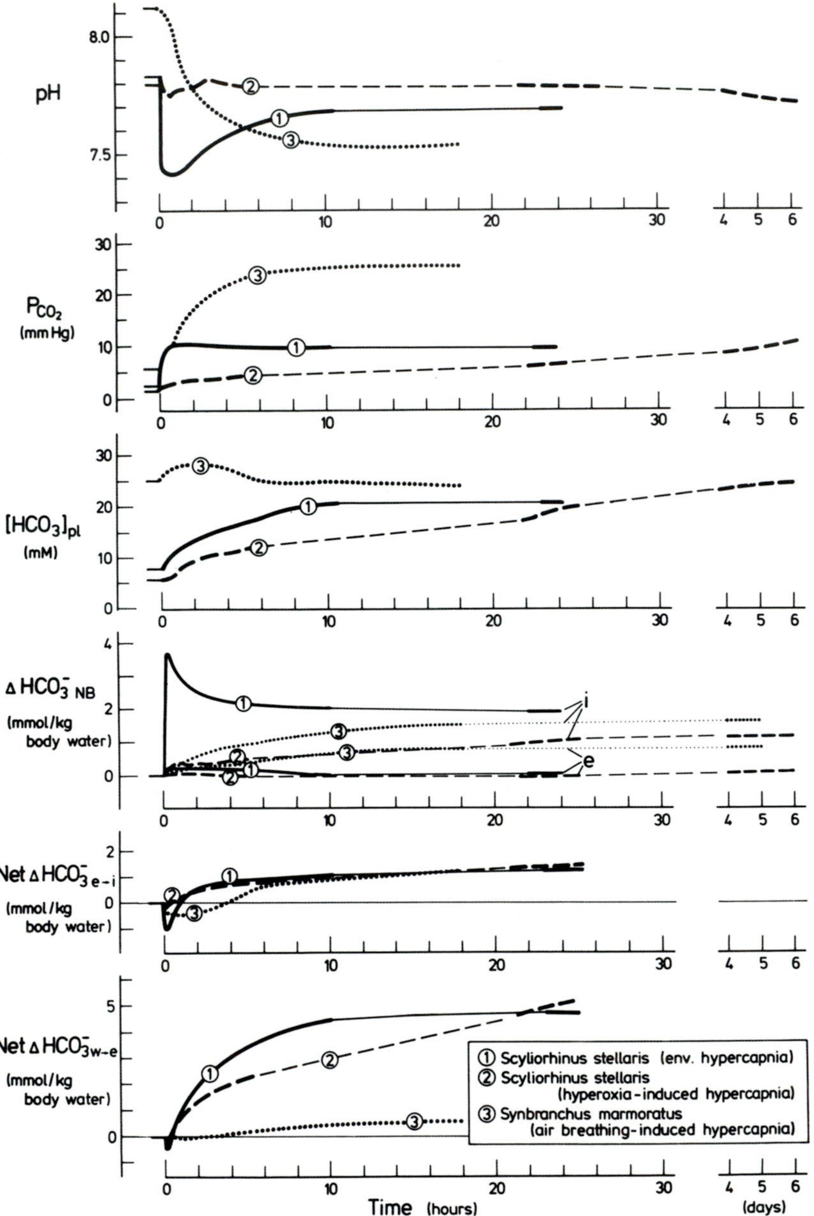

FIGURE 5. Plasma pH, P_{CO_2}, and $[HCO_3^-]$ in response to hypercapnia of various origin (environmental, hyperoxia-induced, or induced by transition from water to air breathing), as well as the contributions of nonbicarbonate buffering ($\Delta HCO^-_{3\,NB}$, e, extracellular, i, intracellular) and bicarbonate-equivalent ion transfer ($\Delta HCO^-_{3\,e\rightarrow i}$: extra- to intracellular; $\Delta HCO^-_{3\,w\rightarrow e}$: environmental water to extracellular space) to pH compensation. Note: differences in the time course of P_{CO_2} elevation; differences in the amount of bicarbonate gained from the environment (see also text). Based on data of Heisler et al. (1976a, 1980, 1988, *Scyliorhinus stellaris*) and Heisler (1982a; *Synbranchus marmoratus*). See text.

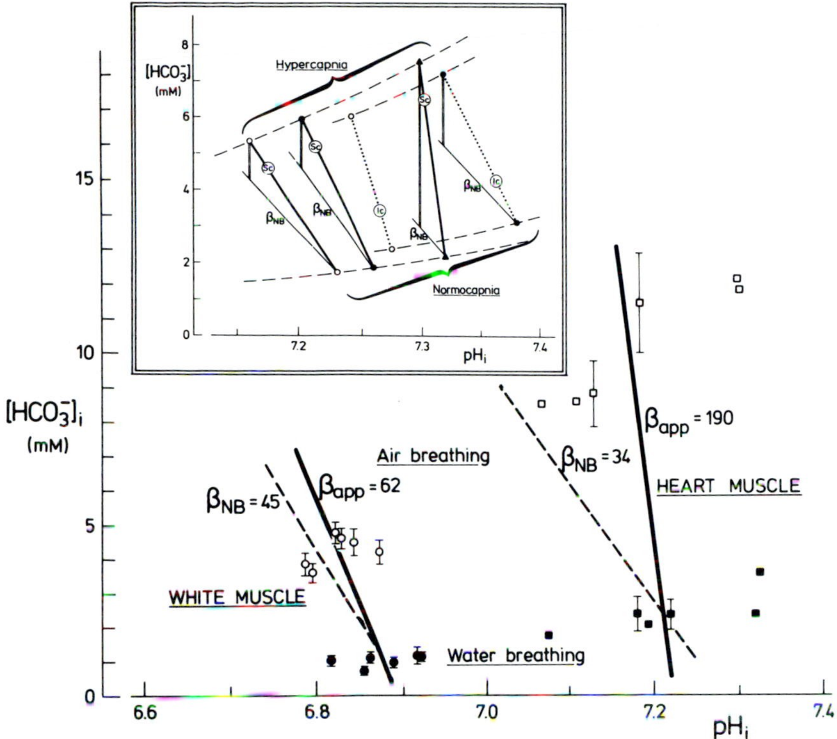

FIGURE 6. Compensation of intracellular pH in response to air-breathing-induced (*Synbranchus marmoratus*, Heisler, 1982a) and environmental hypercapnia (insert: Sc = *Scyliorhinus stellaris*, Heisler et al., 1980; Ic = *Ictalurus punctatus*; Cameron, 1980). The bicarbonate concentration is elevated above the nonbicarbonate buffer lines (β_{NB}), indicating translocation of bicarbonate equivalents into the intracellular fluid compartments. See also text.

Hyperoxia-Induced Hypercapnia

Regulation of ventilation in fish is primarily oxygen-oriented (Dejours, 1975, 1981), characterized by adjustment in order to provide sufficient amounts of oxygen for metabolism (in contrast, terrestrial animals regulate ventilation aiming for constant arterial P_{CO_2}). On the background of sparse availability of oxygen in the water, this type of regulation is certainly advantageous for tissue oxygen supply, but carries the inherent disadvantage of endogenous respiratory acid-base disturbances. Even moderate hypoxia results in considerably elevated gill ventilation, supporting the animal's oxygen supply by reducing the inspired-arterial oxygen partial pressure difference (ΔP_{O_2}), but simultaneously may effect changes in arterial P_{CO_2} and pH, at least in animals with relatively high plasma bicarbonate (e.g., Glass et al., 1990). Conversely, environmental hyperoxia results in substantially reduced ventilation with simultaneously increased ΔP_{O_2} and ΔP_{CO_2} (e.g., Dejours, 1973; Heisler et al., 1981, 1988; Wilkes et al., 1981; Hōbe et al., 1984; for review: Heisler, 1986b).

Po_2 may be extremely variable in freshwater habitats, with low values related to oxygen consumption by animals, plants, and microorganisms during dark periods, and high values induced by photosynthesis during daytime (for reference, see Heisler, 1984b, 1986b). In seawater, the variability of Po_2 is smaller, ranging from 20 to 60 mmHg at water depths of 400 to 500 m of some particular ocean areas up to values of 150 to 190 mmHg in near-surface water layers (Harvey, 1974).

The reduction of ventilation upon exposure to environmental hyperoxia results in an increase in Pco_2, which is, similar to environmental hypercapnia, compensated for by equivalent changes in bicarbonate concentration. The kinetics of the regulatory process, however, are very much different, owing to the fact that hyperoxia-induced hypercapnia develops slowly by accumulation of endogenously produced CO_2 (Heisler et al., 1988), whereas CO_2 is actually taken up at a high rate from the ambient water via the gill gas exchange surface during environmental hypercapnia. The slower time course of hyperoxia-induced hypercapnia reveals the high degree of interrelationship between respiratory and ionic regulation in fish.

In *Scyliorhinus stellaris* exposed to hyperoxia the time course of Pco_2 rise is much slower than expected on the basis of the water oxygen content (Figure 5). This is partially related to conversion of endogenously produced CO_2 to HCO_3^- by nonbicarbonate buffering, but the main factor is that the reduction of ventilation is less than equivalent to the elevated oxygen availability. Pco_2 attains the final value of 11 mmHg only after 5 days (d) of hyperoxia, and the rise is well matched to the accumulation of plasma bicarbonate: pH is little affected during adaptation to the new environment (>–0.08 units; Heisler et al., 1988). This indicates a switch from primarily oxygen-oriented towards pH-stat regulation (Heisler et al., 1988) during hyperoxia, with the adjustment of gill ventilation tracking (with some regulatory oscillations) the accumulation of bicarbonate (Figure 5). The resulting disparity between oxygen demand and ventilation becomes obvious from the elevation of arterial Po_2 to initially more than 250 mmHg, returning to normoxic control values below 100 mmHg only after 5 days of hyperoxia (Heisler et al., 1988). During this time plasma [HCO_3^-] rises steadily to values of almost 25 m*M*, concurrent with, and almost completely compensating, the elevation of Pco_2 (Figure 5; Heisler et al., 1988). When the primary stimulus for ventilatory drive, the low availability of oxygen, is eliminated, ventilation is apparently very closely related to plasma pH. Only when the reduction in ventilation reapproaches levels just sufficient to satisfy the metabolic oxygen demand, is ventilatory control returned to the oxygen-related drive, which is indicated by deviations of pH from the controls becoming larger than during initial phases of hyperoxia-induced hypercapnia (Heisler et al., 1988).

Although the general pattern after establishment of a new steady state is similar in hyperoxia-induced and environmental hypercapnia, the bicarbonate dynamics are considerably different (Figure 5). In contrast to the response during environmental hypercapnia with almost immediately complete elevation of arterial P_{CO_2}, the rise in P_{CO_2} during hyperoxia is a slow process with only small deflections of pH from the control values. Accordingly, there is little production of bicarbonate from nonbicarbonate buffering in hyperoxia-induced hypercapnia (Figure 5). Also, the rate of bicarbonate equivalent gain is much lower during hyperoxia. A high transepithelial transfer rate (close to the maximal capacity of 15 $\mu mol \cdot min^{-1} \cdot kg^{-1}$) of bicarbonate-equivalent resorption for this species (Heisler et al., 1976a; Heisler and Neumann, 1977; Holeton and Heisler, 1983; Heisler, 1988b) is attained only during the first hours of hyperoxia, but is reduced to ~5$\mu mol \cdot min^{-1} \cdot kg^{-1}$ with plasma pH recovering to less than –0.03 after about 3 to 5 h, and falls to about 2 $\mu mol \cdot min^{-1} \cdot kg^{-1}$ after 21 to 25 h of hyperoxia. Accordingly, it takes much longer to accumulate the same quantity of bicarbonate during hyperoxia-induced than during environmental hypercapnia. During this time ventilation is much higher than required on the basis of the oxygen demand, although the capacity of branchial ion transfer mechanisms would allow much faster compensation of pH and reduction of ventilation to the final low rate. The reason for this delayed regulation is unknown, but may be associated with the involved energetics (the additional energetic cost for ventilation may be less than that of rapid accumulation of bicarbonate), to the kinetics of readjustment of the respiratory center, or to differential sensitivity of sensors for respiratory adjustment and ionic transfer.

Air Breathing-Induced Hypercapnia

The oxygen content of air (20.8 vol%) is even higher than that of water equilibrated with pure oxygen (in the range of 3 vol% at 20°C). Facultative air-breathing fish switching from water breathing to air breathing are therefore expected to reduce ventilation of their gas exchange organ similar to water-breathers in hyperoxic water: in both cases the concentration of oxygen in the breathing medium is largely enhanced as compared to normoxic water breathing. Upon air breathing, the rate of "gill" ventilation in the tropical teleost *Synbranchus marmoratus* is actually reduced considerably. P_{CO_2} rises from about 6 to 25 mmHg within 10 to 72 h, effecting plasma pH drop from 8.15 to 7.5 (Figure 5; Heisler, 1982a). In contrast to the typical response of fish to hypercapnia, however, pH in this species remains deflected, with no sign of compensation. Bicarbonate is adjusted, after some initial elevations due to blood nonbicarbonate buffering, at essentially the same level as during water breathing. *Synbranchus* does not gain significant amounts of bicarbonate from the environment during this stage of air-breathing-induced hypercapnia (Figure 5; Heisler, 1982a). A similar lack of extracellular compensation is observed also in other air-breathers like *Channa argus* (Ishimatsu and Itazawa, 1983) and *Amia calva* (Daxboeck et al., 1981). Only dehydration leads to a slight rise in bicarbonate in *Protopterus* as indicated by overproportionate concentration increases of the other ions (DeLaney et al., 1974, 1977).

In contrast to the complete lack of extracellular compensation, intracellular pH of white and heart muscle of *Synbranchus* is almost completely restored (Figure 6; Heisler, 1982a). The elevated intracellular bicarbonate concentration is effected by transfer of bicarbonate to the intracellular space, mainly produced by blood nonbicarbonate buffering. Under the prevalent conditions of only a limited amount of additional bicarbonate being available, regulation of intracellular homeostasis evidently gains higher priority. This type of regulation is efficient: the small amount of bicarbonate produced by extracellular buffering would not effect any significant extracellular compensation on the background of very high control levels of bicarbonate (24 m*M*), but serves for almost complete intracellular pH compensation on the basis of much lower intracellular bicarbonate levels (Heisler, 1982a).

With respect to extracellular compensation of hypercapnia the conditions are quite adverse for *Synbranchus*. The observed fivefold rise in P_{CO_2} would imply an equivalent rise in bicarbonate in order to achieve complete compensation (Heisler, 1986c). On the basis of the bicarbonate level during water breathing (24 m*M*), however, this would require an equimolar reduction in extracellular [Cl^-] in order to maintain electroneutrality; even a fraction of this could hardly be tolerated (Heisler, 1982a). Independent of this factor, the animals will hardly be able to gain sufficient amounts of bicarbonate from the environmental water anyway. Filling the buccal cavity with air for respiratory gas exchange prevents irrigation with water, reducing the contact time of branchial ion-transporting epithelia with water to short flush periods every 5 to 35 min (Heisler, 1982a). Also, the environmental conditions for these fish are extremely unfavorable with respect to ion exchange with the environment (Heisler, 1982a; see also below).

LIMITING FACTORS FOR ACID-BASE REGULATION

Evidently, transepithelial acid-base relevant ion transfer is the main mechanism for fish acid-base regulation. This transfer is generally performed at a rate much higher than that of comparable mechanisms in terrestrial animals, with typical overall transfer rates per unit standard metabolic rate in the range of 0.13 to 0.47 (μmol ion transfer per μmol O_2 consumption) for fish as compared to 0.02 to 0.03 in typical mammals like dog and man (Heisler, 1988b). An important factor is the direct contact of the large surface area of the main transfer site, the gill epithelium, with a large volume of environmental water (high specific ventilation, see above). Resting water flow past the gill epithelium in fish is in the range of 5000 to 20000 $ml \cdot kg^{-1} \cdot h^{-1}$ (e.g., Randall et al., 1976) and larger than the typical urine flow rate in fish (1 to 10 $ml \cdot kg^{-1} \cdot h^{-1}$; Hunn, 1982) and mammals (e.g., man, 1.1 $ml \cdot kg^{-1} \cdot h^{-1}$) by factors of 10^3 to 10^4. Avoiding the necessity to establish large ionic gradients, like in the mammalian renal tubular system, the high water flow facilitates ion transfer processes to a much greater extent, provided the ionic composition of the environmental water is favorable.

Branchial Ion Transfer and Water Ionic Composition

Due to their exposed location the branchial ion transfer mechanisms are directly affected by changes of the environment. Comparison of literature data suggested first that the ion composition of the environmental water was an important factor for fish in dealing with environmental hypercapnia (Heisler, 1982b). Trout in dilute water ($[Na^+]$ <0.1 m*M*; $[HCO_3^-]$ 0.050 m*M*) required much longer to achieve about the same level of pH compensation (72 h, Janssen and Randall, 1975) than trout in water of higher ionic concentration ($[Na^+]$ 0.5 m*M*; $[HCO_3^-]$ 3 m*M*, 24 h, Eddy et al., 1977) and *Conger* in seawater ($[Na^+]$ 480 m*M*; $[HCO_3^-]$ 2.5 m*M*, 10 h, Toews et al., 1983). Closer analysis in *Scyliorhinus stellaris* indicated that bicarbonate was net gained from the environment at a constant rate in a range of $[HCO_3^-]_{pl}[HCO_3^-]_{sw} = 0.3$ to 4, but the rate fell in an apparently linear fashion at higher $[HCO_3^-]_{pl}[HCO_3^-]_{sw}$ ratios, attaining zero at $[HCO_3^-]_{pl}/[HCO_3^-]_{sw} = 13$. At even higher ratios, bicarbonate was lost to the water (Figure 7). Since all other ions remained constant in the environmental water these data suggest water $[HCO_3^-]$ as a limiting factor for acid-base relevant ion transfer in this species (Heisler and Neumann, 1977; Heisler, 1988b). Also in trout at different environmental salinities and in carp at different environmental bicarbonate concentrations, environmental hypercapnia was compensated more rapidly at higher salinities (Iwama and Heisler, 1991) or bicarbonate concentrations (N.A. Andersen and N. Heisler, unpublished data), respectively. After lactic acid infusion (Tang and Boutilier, 1988) or exhausting exercise (Tang et al., 1989) genetically identical trout populations achieved much larger branchial bicarbonate-equivalent flux rates in seawater than in freshwater.

Limits of Plasma Bicarbonate Concentration

In spite of large differences in time course, the extent of compensation achieved during hypercapnia is rather independent of the environmental ionic composition. This is illustrated by the regulatory pattern observed during hyperoxia-induced hypercapnia in *Scyliorhinus stellaris*. After some initial regulatory oscillations, extracellular pH is kept compensated to within 0.03 pH below the normoxic controls, until plasma bicarbonate exceeds 20 m*M*. With still higher P_{CO_2} values, bicarbonate is not elevated equivalently, such that the compensation is increasingly incomplete (Figure 5). Even after 6 d of hyperoxia, plasma bicarbonate does not rise above 25 m*M* (Heisler et al., 1988). A similar value is attained also during environmental hypercapnia (Heisler et al., 1976a).

Literature data on the compensation of hypercapnia indicate that quite a number of aquatic lower vertebrates never exceed bicarbonate concentrations of 23 to 33 m*M* (Heisler, 1988b). With such a limit or threshold for plasma bicarbonate, the degree of compensation that can be achieved must generally

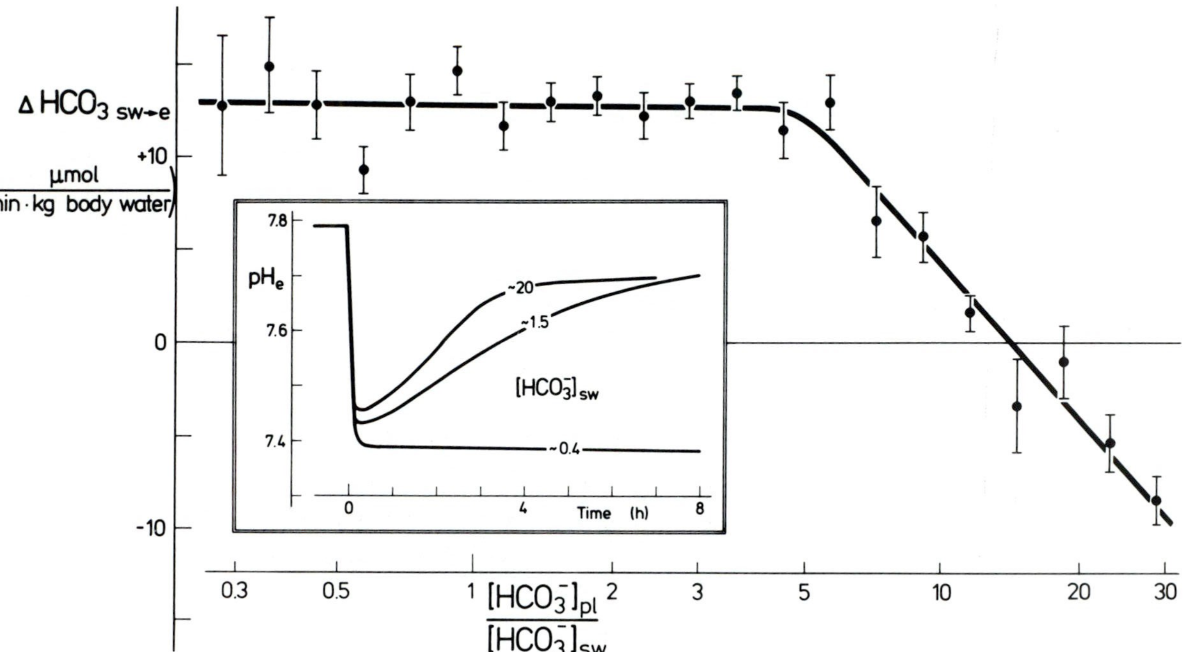

FIGURE 7. Transepithelial bicarbonate-equivalent transfer rate in *Scyliorhinus stellaris* as a function of the ratio between plasma and environmental bicarbonate concentration ($[HCO_3^-]_{pl}/[HCO_3^-]_{sw}$). Insert: kinetics of pH recovery at different water $[HCO_3^-]$. (Based on data of Heisler and Neumann, 1977). See text.

be lower in animals with high control bicarbonate concentration, due to the inability to elevate bicarbonate by the same factor as P_{CO_2} (Heisler, 1986c). Due to the high arterial P_{CO_2} of about 4 mmHg, *Cyprinus carpio* maintains a comparatively high control bicarbonate concentration of about 13 m*M*, much higher than other water-breathers (trout ~4 m*M*, Holeton et al., 1983; *Conger* ~5 m*M*, Toews et al., 1983; *Scyliorhinus* ~7 m*M*, Heisler et al., 1976a). Carp did not raise plasma bicarbonate above 25 m*M* during exposure to hypercapnia of 1 or 5% CO_2, in spite of incomplete pH compensation (80 or 45%) even after exposure for up to 3 weeks (Claiborne and Heisler, 1984, 1986). Also, the lack of any compensation in the tropical air-breather *Synbranchus* (Heisler, 1982a) may, at least partially, be related to the fact that the control bicarbonate concentration in this species is as high as 24 m*M*, which may represent, or be close to, a limiting bicarbonate threshold. In this species, however,alternative factors such as the limited exposure time of the ion-transporting epithelia to water during air breathing, or adversely low environmental ion concentrations may also be involved (Heisler, 1982a). Interestingly, the limit does not seem to be directly related to problems in gaining bicarbonate from the water, but may rather represent the inability to maintain a high plasma [HCO_3^-]. In at least two species of aquatic animals, *Cyprinus carpio* and the urodele amphibian *Siren lacertina*, bicarbonate infused into the blood stream during environmental hypercapnia is not retained, but quantitatively released to the environment (Claiborne and Heisler, 1986; Heisler et al., 1982). These data, in concert with the bulk of literature reports, support the notion of an epithelial threshold for retention and net resorption of bicarbonate equivalents. However, a few experiments in three fish species have indicated that much higher bicarbonate concentrations can actually be attained during compensation of hypercapnia (Børjeson, 1976, 1977; Jensen and Weber, 1982; Dimberg, 1988; Cameron and Iwama, 1987).

As evident from bicarbonate-equivalent resorption as a function of water bicarbonate in *Scyliorhinus* (see above), the water [HCO_3^-] is an important factor in this matter, at least for this species. Preliminary data indicate that also in carp bicarbonate-equivalent resorption is enhanced with rising water [HCO_3^-] (N.A. Andersen and N. Heisler, unpublished), but even at 50 m*M* (Figure 8), carp fail to exhibit any degree of compensation comparable to the phenomenal compensatory capacity of Uppsala trout (Børjeson, 1976, 1977; Dimberg, 1988), Odense tench (Jensen and Weber, 1982), and Port Aransas channel catfish (Cameron and Iwama, 1987), yielding up to 50 m*M* plasma [HCO_3^-]. Long-term acclimation may seem to be an obvious factor, particularly for the Uppsala trout, which were adapted to elevated levels of environmental P_{CO_2} for several months before the experiments; much longer than the 3 weeks of adaptation during which carp did not attain higher levels of plasma bicarbonate than 26 m*M* (Claiborne and Heisler, 1986). In catfish, however, plasma [HCO_3^-]s of 50 m*M* were attained after 5 d of hypercapnia (Cameron and Iwama, 1987).

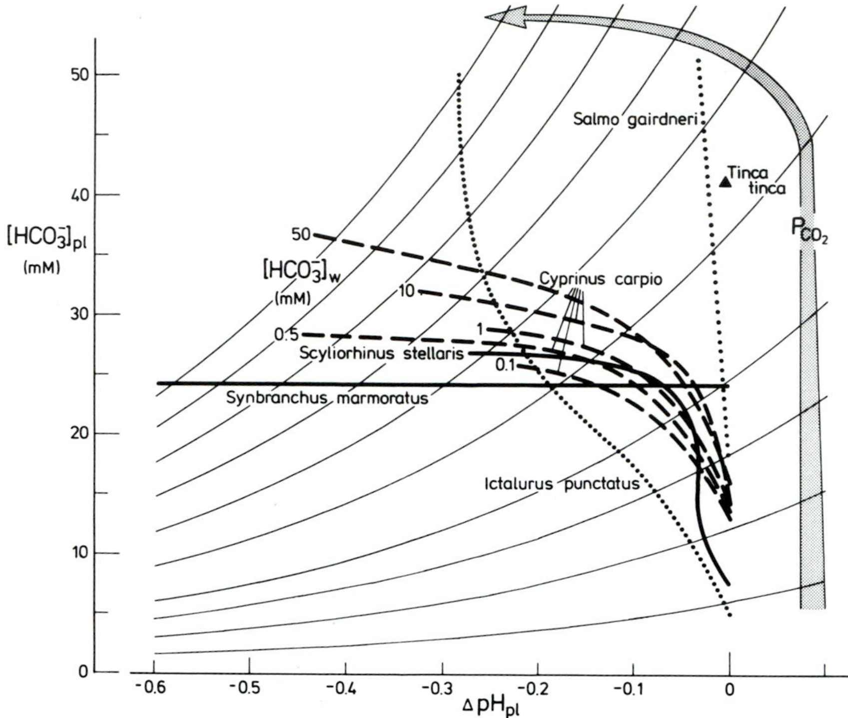

FIGURE 8. Bicarbonate concentration during hypercapnia as a function of water P_{CO_2} and $[HCO_3^-]$. In most fish species studied, the degree of pH compensation falls (indicated by no further rise in $[HCO_3^-]$ with larger shifts in pH) when certain plasma bicarbonate levels are attained (e.g., *Scyliorhinus stellaris, Cyprinus carpio,* and *Synbranchus marmoratus*), whereas in some experiments an improved capacity for pH compensation and bicarbonate accumulation is observed (*Salmo gairdneri, Ictalurus punctatus* and *Tinca tinca*). P_{CO_2} isobars are only relative (with increasing values indicated by the shaded arrow), due to the normalization of individual steady-state pH values. For references and details see text.

A factor more likely to contribute is the environmental $[Ca^{2+}]$. All experiments in which high plasma bicarbonate levels were achieved have been conducted at environmental $[Ca^{2+}]$ close to, or higher than, 2 m*M*, combined with relatively high levels of bicarbonate, in the range of 3 to 4 m*M*. The extraordinary capacity to compensate hypercapnic acidoses may, accordingly, be related to a combination of the supportive effect of high environmental bicarbonate with the sealing effect of Ca^{2+} on paracellular pathways, and the associated general reduction of unidirectional flux rates (including a reduction in the rate of transepithelial bicarbonate leakage). Even a direct effect of Ca^{2+} on the ion-transporting structures can presently not be excluded. The precise interrelationship of these factors is unclear and requires further investigation.

MECHANISMS AND SITES OF IONIC TRANSFER

EPITHELIAL ION TRANSFER

Although it may not accurately delineate the intimate action of epithelial ion-transfer mechanisms, ionic transfer in fish is generally considered to be best described as active electroneutral ion exchange processes of H^+ or NH_4^+ against Na^+, and of HCO_3^- against Cl^- (Figure 9; Maetz, 1974; Evans, 1986; Heisler 1986b). As an additional mechanism, an electrogenic H^+ pump in the apical membrane of branchial chloride cells has been postulated recently (Lin and Randall, 1991), being linked only indirectly to movement of Na^+ or other cations (Avella and Bornancin, 1989; Lin and Randall, 1991). Evidence has been provided for all of these mechanisms to operate under certain conditions (Evans, 1986; Lin and Randall, 1991). Under physiological conditions, however, obviously not all of them are involved. Recent determination of unidirectional Na^+ and Cl^- tracer ion fluxes in *Cyprinus carpio* indicate that Na^+-related mechanisms play no role in transfer of acid-base relevant ions for compensa-

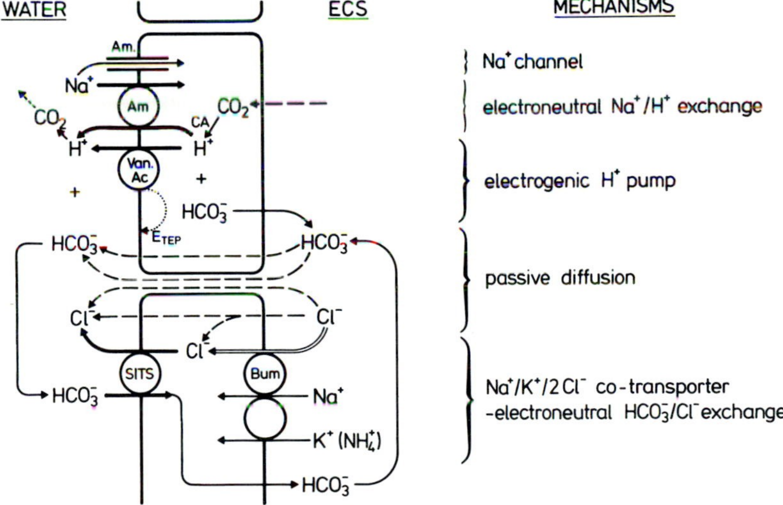

FIGURE 9. Mechanisms of transepithelial ion transfer. H^+ ions may be extruded at the apical epithelial membrane by amiloride-sensitive (Am) electroneutral ion exchange against Na^+, or by an electrogenic vanadate- (Van) and acetozolamide-sensitive (Ac) H^+ pump. As a consequence of electrogenic transfer, Na^+ may be translocated through amiloride-sensitive (Am) Na^+ channels, or Cl^- diffuses through paracellular pathways or the cytoplasm. Bicarbonate may be translocated by 1:1 electroneutral ion exchange. CA, Carbonic anhydrase; AM, Van, Ac, Bum, SITS, transfer sensitive to amiloride, vanadate, acetazolamide, bumetanide, or SITS, respectively. Dotted lines indicate the effect of the electrogenic pump on the transepithelial potential (E_{TEP}). See also text.

tion of hypercapnic acidoses (Heisler, 1990a, 1990b), whereas the unidirectional Cl^- flux was directly related. This strategy may be related to the fact that the HCO_3^-/Cl^- ion exchange is osmotically neutral, whereas the H^+/Na^+ exchange leads to accumulation of additional osmotically active molecules in the body fluids of the animal, a factor possibly also prohibiting utilization of cotransport processes (Heisler, 1989). The often-observed lack of relationship between transfer of acid-base relevant ions with Na^+ movements is in line with the proposed electrogenic proton pump (Lin and Randall, 1991). Recent studies, however, indicate that the overall transepithelial potential in trout changes during hypercapnia to more positive values, which is opposite to that expected with electrogenic transfer of H^+ in a direction out of the fish (Iwama and Heisler, 1991). Also, close correlation of acid-base relevant ion fluxes with Cl^- flux, together with the lack of appreciable movements of other ions suggests that the main fraction of transfer is performed by electroneutral HCO_3^-/Cl^- ion exchange.

Interrelation with Ammonia Elimination

The form in which ammonia is transferred across the gill epithelium of fishes is still not completely settled. A long-lasting discussion (for review: Evans, 1980, 1984, 1985, 1986; Heisler, 1984b, 1986b; Evans and Cameron, 1986) has confirmed generally that all three possible modes, non-ionic diffusion, ionic diffusion and ionic carrier-mediated active transfer associated with movement of Na^+, are actually utilized under certain conditions, with the relative contributions largely depending on composition of the environmental water and other factors.

Elimination of ammonia by either ionic diffusion (as NH_4^+) or by active ionic exchange of NH_4^+ against Na^+ implies alkalinization of the body fluids by removal of H^+ ions utilized for ionization of NH_3 upon production. Quantitatively, this would represent a tremendous alkalinizing load for the acid-base regulatory system of up to 15% of the oxygen consumption (Heisler, 1984b, 1986b, 1988a). Together with other data (see below) this implication renders ionic transfer a less likely alternative during normal conditions.

Although in certain instances a close relationship between elimination of NH_4^+ and Na^+ influx has been described, the NH_4^+/Na^+ exchange mechanism is hardly exploited during acid-base stress conditions, as indicated by usually constant rate of ammonia release even during severe acid-base perturbations (Heisler, 1984b, 1986b). Only during special conditions is the ammonia elimination enhanced, such as during induction of acid-base disturbances by ammonium chloride infusion, or by low environmental pH. Enhanced ammonia flux is then, however, likely related to non-ionic diffusion of NH_3 across the epithelium, similar to the elimination of CO_2 (Claiborne and Heisler, 1986; Heisler, 1986b, 1990b, 1993).

The gill NH_3 diffusion coefficient allows elimination of ammonia completely by non-ionic diffusion as long as the environmental conditions are normal (i.e., low NH_3 partial pressure; Cameron and Heisler, 1983, 1985). In carp and trout the elimination of ammonia is linearly proportional to the partial pressure gradient across the epithelium, as long as plasma NH_4^+ concentrations of 0.2 m*M* are not surpassed (Heisler, 1990c), strongly suggesting that the elimination takes place exclusively by non-ionic diffusion (Heisler, 1990b, 1990c, 1993). This notion is supported by the fact that no correlation existed between any one of plasma $[NH_4^+]$, effective $\Delta[NH_4^+]$ or ΔP_{NH_3} between plasma and water, and the transepithelial ammonia transfer rate at higher values of plasma $[NH_4^+]$, indicating the switch to the often-described electroneutral NH_4^+/Na^+ ion exchange mechanism. The close correlation described between ammonia transfer and unidirectional Na^+ influx (e.g., Wood, 1989) may have been obtained under conditions of reduced transepithelial ΔP_{NH_3} by elevated environmental $[NH_4^+]$. This notion has been corroborated by Evans and co-workers (Evans, 1985; Evans and More, 1988; Evans et al., 1989), demonstrating that ammonia is eliminated from the elasmobranch *Squalus acanthias* mainly by non-ionic diffusion (75%, with 25 % of passive ionic diffusion of NH_4^+).

Sites of Transepithelial Ionic Transfer

Generally, three possible sites are involved in transepithelial ionic transfer in lower vertebrates: the skin, the gills, and the kidneys.

Fish Skin

The skin is the main transfer epithelium in amphibia, in fish its role is not very well established. Although hardly based on any conclusive data, the skin is considered inactive and not taken into consideration during studies of ion transfer. In only one species of elasmobranch fish was the contribution of cutaneous exchange sites assessed to be minor (<1%; Heisler et al., 1976a); analogous experiments have never been performed in teleosts.

Fish skin is generally quoted to be poorly permeable for ions and water (Fromm, 1968), but a few studies in teleosts have indicated that 50% of the Ca^{2+} uptake of rainbow trout is performed at the skin surface (Perry and Wood, 1985), and 65% of Cl^- efflux occurs through the skin of the seawater shanny (*Blennius pholis*, Nonnotte et al., 1979). Chloride cells, which are likely responsible for transepithelial ionic transfer, occur in the skin of several fish species (Henrikson and Matoltsy, 1968; Nonnotte et al., 1979; Stiffler et al., 1986; Hwang, 1989), with particularly dense chloride cell populations found, for instance, in the opercular epithelium of *Fundulus* (Karnaky and Kinter, 1977) and *Oreochromis* (Foskett et al., 1981), and in the external skin of *Gillichthys* (Marshall, 1977). The high ionic transfer activity shown for these sites *in vitro* suggest significant involvement of the general body surface in ionic regulation.

Significant ionic transfer, however, depends on a relatively high rate of convective mixing with the bulk extracellular fluid. Build-up of ions in the interstitial fluid will result in reduced ion transfer velocity by increased electrochemical gradients, or even complete shut-down of any transfer, until the accumulated ions are removed by convective mechanisms (similar to those observed during lactacidosis in skeletal muscle tissue, perfusion/equilibrium limitation, see Heisler, 1986c). This task of convective transfer may be performed by the dense capillary network of the secondary circulation above the scales directly underlying the skin epithelium of trout and tilapia (Vogel, 1985; Satchell, 1991), and the similar capillary network beneath the epidermis of the exposed part of each scale in pike (Tysekiewicz, 1969), which is most likely related to the secondary circulation as well (Satchell, 1991).

The composition of fluid from the draining vessel of the secondary circulatory system of the teleost body surface, the "lateral cutaneous vessel" (LCV), has been studied recently in order to evaluate a possible role of this system in acid-base regulation (Ishimatsu et al., 1992). As expected for a venous vessel system the virtually erythrocyte-free LCV fluid was lower in pH and higher in P_{CO_2}, but the $[HCO_3^-]$ was the same as compared to arterial plasma. Upon environmental hypercapnia and the resulting stimulation of acid-base ionic transfer the gradients for pH and P_{CO_2} reversed, and a significant positive difference in $[HCO_3^-]$ developed between LCV fluid and arterial plasma (+2.1 m*M*), with an equimolar reduction in $[Cl^-]$ (Ishimatsu et al., 1992). These data strongly suggest involvement of the secondary circulatory system in acid-base regulation, although its quantitative role in the overall process may not be large. Based on estimates of control LCV perfusion rate (Steffensen and Lomholt, 1993) the contribution to the compensation of hypercapnia would be about 4%. The absolute transfer rate is in the same range as the relative contribution of teleost kidneys, and comparable to amphibian skin and the overall ionic transfer capacity of mammals (see below). The role of the skin may even be larger than this, taking into account that the LCV perfusion rate is likely to increase considerably during hypercapnia. In animals with rich chloride cell populations on the body surface, such as *Gillichthys* (Marshall, 1977), the fish skin may be a valuable site for ionic transfer. Further investigation is required to accurately quantify the role of this system.

Kidneys

A number of studies have addressed the role of the kidneys in fish acid-base regulation and generally have indicated minor relative contribution (typical teleosts around 5%; for review see Heisler, 1986b, 1988b). This is not related to particularly inefficient kidney mechanisms in fish. The rate of kidney transfer of about 0.007 to 0.025 (μmol ion transfer per μmol O_2 consumption) compares favorably with those of typical mammals (e.g., dog and man: 0.02 to 0.03), particularly when those are adjusted for different body temperature (0.004 to 0.006 μmol ion transfer per μmol O_2 consumption; 20°C, $Q_{10} = 2.2$). The small relative contribution of the kidney is accordingly mainly associated with the fact that gill ionic transfer in fish is facilitated at much larger rates by mechanisms exploiting the advantages of large water flow across the branchial ionic exchange surface, and the lack of appreciable ionic gradients between in- and outflowing water (see above).

The typical approach of determination of urine titratable acidity and ammonia concentration may actually underestimate the relative role of the fish kidney for acid-base regulation. According to Kampmeier (1969), the LCV of teleosts is connected to the deep "lymphatics" running alongside the kidneys at the level of the caudal extremity of the abdominal cavity (called "circumanal lymph plexus"). When methylene blue is injected into the LCV the dye is almost exclusively received by the kidneys (Ishimatsu et al., 1992), indicating that the path of lowest resistance is from the lateral cutaneous vessel through the circumanal plexus to the kidneys. The functional significance of the complicated spatial interrelationship between kidneys, circumanal plexus, and LCV of the secondary circulatory system of the body surface is still unclear (Ishimatsu et al., 1992). The circumanal vascular connection may represent a functional linkage between kidneys and the secondary surface capillary bed, with the secondary circulatory system providing convective transport of ions from the interstitial fluid of body surface transport cells to the primary system (Ishimatsu et al., 1992). This may provide a special extracellular environment for ion-transferring kidney cells and thus allow for exchange of acid-base relevant ions between primary and secondary system, which is reflected by modification of renal effluent plasma but does not affect urine composition. The evaluation of the physiological role of fish kidneys by urine analysis alone may, accordingly, prove to be inadequate.

Branchial Epithelia

The predominant role of the gills in general is well established, but the precise site of ionic transfer within the gills is still unclear. Mitochondria-rich cells (chloride cells) are generally claimed to be responsible on the basis of circumstantial evidence. With some variability, the chloride cells are located mainly at the base of the secondary lamellae on the primary filament, whereas the apical fractions of the secondary lamellae are mainly covered by so-called respiratory cells. These areas are drained by two different circulatory systems. The respiratory cells on the secondary lamellae are in close connection with the main blood stream of the primary circulation through the branchial gas exchange area. The chloride cells are mainly located in close proximity to the central venous sinus (CVS) in the primary filament of the fish gills, which drains the basal portion of the secondary lamellae as part of the secondary circulatory system in fishes (Laurent, 1984). Differences between the composition of fluids from the two circulatory circuits in the gills, together with the actual flow rates, will therefore reflect the fractional contribution of epithelia in juxtaposition to perfused areas.

The blood perfusion rate in the primary gill circulation (the respiratory, apical parts of the secondary lamellae) is close to cardiac output (Ishimatsu et al., 1988), owing to the necessity of a high rate of respiratory gas exchange. Venous-arterial (ventral, VA, to dorsal aortic, DA) differences in ionic composition will, accordingly, be small and difficult to detect. The net acid-base relevant transepithelial ionic transfer capacity of 15 μmol·kg^{-1}·min^{-1}) (Heisler, 1989) will result in VA-DA differences of less than 375 μmol/l [HCO_3^-] if it would occur exclusively in the respiratory part of the gill epithelium (cardiac output ~40 ml·kg^{-1}·min^{-1}, Neumann et al., 1983). This is less than 5% of typical control [HCO_3^-] in fish and less than 1.5% of levels attained during hypercapnia (Heisler, 1984b, 1986b). Uncertainties related to the correction for blood nonbicarbonate buffering, and to changes in bicarbonate by oxygenation of hemoglobin (Haldane effect; up to 1.5 mmol/l) render the Fick principle (ionic transfer rate = v-a concentration difference × flow rate) an inadequate approach for determination of ionic transfer in the range of the primary gill circulation.

The flow rate in the secondary circulation of the gills (CVS) is much lower (for references, see Ishimatsu et al., 1988), providing an ideal background to determine the contribution to ionic transfer of this circulatory system. Measurement of flow rate in the CVS and access to its effluent fluid, however, is extremely difficult. The CVS circuit is fed from the efferent filamental and branchial arteries and drains into the branchial vein (Laurent, 1984), with little contamination by blood from gill nutritional vessels (Ishimatsu et al., 1988). The CVS flow rate was estimated in rainbow trout to be <7 % of cardiac output, using a special microcannulation technique of the branchial vein (BV) and an endogenous indicator distribution method employing hemoglobin concentrations of CVS fluid and dorsal aortic blood (Ishimatsu et al., 1988). A subsequent study focused on quantitative assessment of acid-base relevant transfer processes in branchial epithelia in juxtaposition to the secondary circulation, utilizing CVS flow rate and dorsal aortic-branchial vein ionic concentration differences for an application of the Fick principle (Iwama et al., 1993).

During normocapnia there was no difference between CVS and dorsal aortic [HCO_3^-], although pH and P_{CO_2} were slightly higher in CVS fluid owing to the venous nature of the CVS system (Figure 10). Upon environmental hypercapnia [HCO_3^-] became significantly higher in CVS fluid than in dorsal aortic plasma, due to transfer from the environmental water (Figure 10). Since the CVS circuit in rainbow trout is fed from the efferent filamental and branchial arteries (Laurent, 1984), differences in [HCO_3^-] between DA and BV are the result of ionic transfer across the CVS-related epithelium, and, to a minor extent, to nonbicarbonate buffering. As a rough estimate, 3.1 or 2.3 mmol/kg of HCO_3^- were accumulated during the entire time of exposure (8 h), calculated, on the basis of the Fick principle, from the concentration differences after 2 or 8 h, respectively, and the relative CVS flow rate (Iwama et al., 1992). This estimate compares favorably with the amount of bicarbonate accumulated in the extracellular space, 2.4 mmol/kg ($\Delta[HCO_3^-]_{pl}$ × extracellular volume = 12 mmol/l × 0.2 l/kg), with the estimate according to the 2-h concentration difference suggesting some transfer also to the intracellular space. Since the amount of bicarbonate accumulated in the intracellular space during hypercapnia in fish is generally much smaller than that in the extracellular compartments (see above and Heisler, 1984b, 1986b), the largest fraction of additional bicarbonate

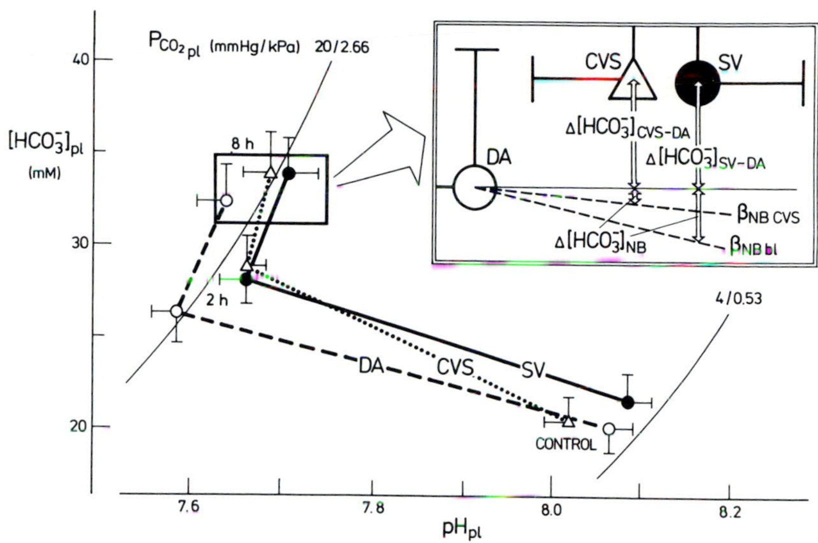

FIGURE 10. Plasma pH, P_{CO_2}, and $[HCO_3^-]$ of fluid from the central venous sinus (CVS) and blood of dorsal (DA) and ventral aorta (VA) of *Oncorhynchus mykiss* during normocapnia (control) and exposed to environmental hypercapnia (2 h and 8 h). The difference in $[HCO_3^-]$ between CVS and DA occurring under hypercapnic conditions is mainly due to transfer from the environment ($\Delta HCO_3^-{}_{CVS\text{-}DA}$), and to a small fraction due to nonbicarbonate buffering ($\Delta HCO_3^-{}_{NB}$). Differences between sinus venosus (SV) and dorsal aorta ($\Delta HCO_3^-{}_{SV\text{-}DA}$) are related to oxygenation of hemoglobin (Haldane effect). (Based on data of Iwama et al., in press). See also text.

was evidently gained by action of epithelia related to the CVS circuit (Iwama et al., in press). These data support the notion that mitochondria-rich cells at the base of the secondary lamellae are mainly responsible for ion transfer relevant for acid-base regulation in fish.

The pattern is not quite as clear as to the mechanisms involved in translocation of acid-base relevant ions. While the dorsal aortic (DA) chloride concentration was reduced equivalent to the rise in bicarbonate (as expected on the basis of a 1:1 electroneutral ion exchange of HCO_3^-/Cl^-), there were no differences in $[Cl^-]$ between DA and CVS, comparable to the differences in $[HCO_3^-]$ (Iwama et al., in press). Significance of differences in the range of 2 mmol/l may have been occluded by the variability of experimental data on the high background of about 125 mmol/l $[Cl^-]$, but the data did not indicate any sign of trend even on the basis of paired analysis. The experimental approach with separation of primary and secondary circulatory system does not allow the measurement of unidirectional efflux; the influx of Cl^- and Na^+ into the secondary circulatory system did not indicate any significant movements of neither Cl^- nor Na^+ (Iwama et al., unpublished data). These data suggest that while the acid-base relevant flux occurred at the CVS-related epithelium the correlated co- or countertransfer of ions required for electroneutrality took place at other epithelial sites. This combination is compatible with operation of an electrogenic H^+ pump, as proposed by

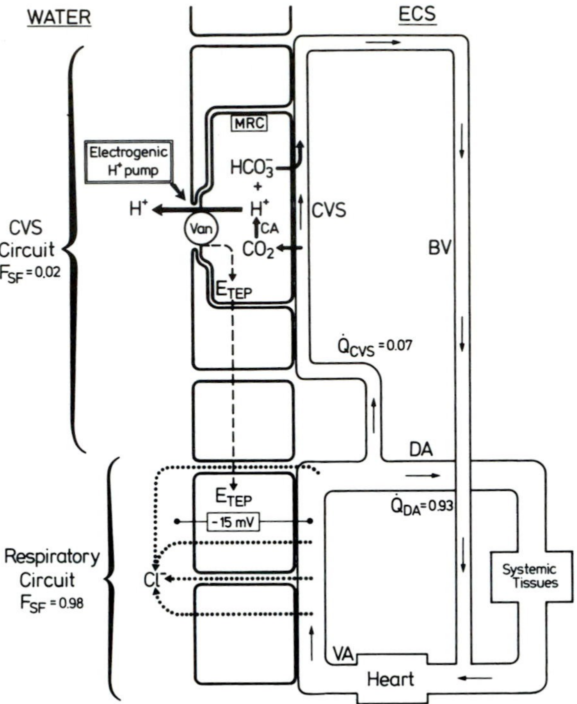

FIGURE 11. Model of ionic transfer during hypercapnia. Changes in [HCO_3^-] (see Figure 10) are effected by an electrogenic H^+ pump in the epithelium of the primary filament at the base of the secondary lamellae (which is drained at the low relative flow rate $\dot{Q}_{CVS}$ of 0.07 by the central venous sinus of the secondary branchial circulation, CVS, into the branchial vein, BV). Note: transfer of H^+ ions is bound to occur at the site of mitochondria-rich cells, whereas diffusive transfer of Cl^- following the changes in transepithelial potential will take place distributed across the epithelial surface at the site of least resistance, mainly the respiratory epithelium. E_{TEP}, transepithelial potential; F_{SF}, relative surface area of the two perfusion cicuits. See text for details and references.

Avella and Bornancin (1989) and Lin and Randall (1991). Readjustment of the transepithelial potential may, accordingly, drive passive ionic movement, reestablishing electroneutrality (Figure 11). Since this is not linked to any specific location such as carrier-mediated processes, diffusive translocation will occur at the site of least diffusive resistance, which is clearly represented by the much larger surface area of the respiratory part of the secondary lamellae in the primary circulatory system (Iwama et al., in press).

CONCLUSION

Respiratory regulation in water-breathing fish is handicapped by low P_{CO_2} in the body fluids, but this deficiency is made up for by a much larger epithelial ionic transfer capacity than in air-breathing animals. The high capacity to net gain bicarbonate from the environmental water generally facilitates normalization of the acid-base status long before the original stress factor is removed, with clear priority for intracellular vs. extracellular regulation. Although the transfer capacity of kidneys and skin in fish is comparable to that in higher vertebrates such as amphibia, the largest fraction of epithelial acid-base relevant transfer takes place at the gills, apparently by action of chloride cells located in juxtaposition to the secondary circulatory system of the central venous gill sinus. While most efficient during normal conditions the ionic transfer mechanisms of fish are largely affected by environmental ion concentrations, limiting the compensatory scope as a function of [HCO_3^-], and likely also [Ca^{2+}], in the environmental water.

REFERENCES

Avella, M. and Bornancin, M., A new analysis of ammonia and sodium transport through the gills of the freshwater rainbow trout (*Salmo gairdneri*), *J. Exp. Biol.*, 142, 155, 1989.

Børjeson, H. Some effects of high carbon dioxide tension on juvenile salmon, (*Salmo salar* L.), *Acta Univ. Upsaliensis*, 383, 3.35, 1976.

Børjeson, H. Effects of hypercapnia of the buffer capacity and haematological values in *Salmo salar* (L.) *J. Fish Biol.*, 11, 133, 1977.

Brønsted, J. N., Einige Bemerkungen über den Begriff der Säuren und Basen, *Rec. Trav. Chim. Pays-Bas*, 42, 718, 1923.

Cameron J. N., Regulation of blood pH in teleost fish, *Respir. Physiol.*, 33, 129, 1978.

Cameron, J. N., Body fluid pools, kidney function, and acid-base regulation in the freshwater catfish *Ictalurus punctatus*, *J. Exp. Biol.*, 86, 171, 1980.

Cameron, J. N., The bone compartment in a teleost fish, *Ictalurus punctatus:* size, composition and acid-base response to hypercapnia, *J. Exp. Biol.*, 117, 307, 1985.

Cameron, J. N. and Heisler, N., Studies of ammonia in the rainbow trout: physico-chemical parameters, acid-base behavior and respiratory clearance, *J. Exp. Biol.*, 105, 107, 1983.

Cameron, J. N. and Heisler, N., Ammonia transfer across fish gills: a review, in *Circulation, Respiration and Metabolism,* Gilles, R., Ed., Springer, Berlin, 1985, 91.

Cameron J. N. and Iwama, G. K., Compensation of progressive hypercapnia in channel catfish and blue crabs, *J. Exp. Biol.*, 133, 183, 1987.

Cameron, J. N. and Kormanik, G. A., Intracellular and extracellular acid-base status as a function of temperature in the freshwater channel catfish, *Ictalurus punctatus*, *J. Exp. Biol.*, 99, 127, 1982.

Claiborne, J. B. and Heisler, N., Acid-base regulation in the carp (*Cyprinus carpio*) during and after exposure to environmental hypercapnia, *J. Exp. Biol.*, 108, 25, 1984.

Claiborne, J. B. and Heisler, N., Acid-base regulation and ion transfers in the carp (*Cyprinus carpio*): pH compensation during graded long- and short-term environmental hypercapnia and the effect of bicarbonate infusion, *J. Exp. Biol.*, 126, 41, 1986.

Daxboeck, C., Barnard, D. K., and Randall, D. J., Functional morphology of the gills of bowfin *Amia calva* L. with special reference to their significance during air exposure, *Respir. Physiol.*, 43, 349, 1981.

Dejours, P., Problems of control of breathing in fishes, in *Comparative Physiology,* Bolis, L., Schmidt-Nielsen, K., and Maddressl, S. H. P., Eds., North Holland, Amsterdam, 1973, 117.

Dejours, P., *Principles of Comparative Respiratory Physiology,* North-Holland, Amsterdam, 1975, 1981.

DeLaney, R. G., Lahiri, S., and Fishman, A. P., Aestivation of the African lungfish *Protopterus aethiopicus*: cardiovascular and respiratory functions, *J. Exp. Biol.*, 61, 111, 1974.

DeLaney, R. G., Lahiri, S., Hamilton, R., and Fishman, A. P., Acid-base balance and plasma composition in the aestivating lungfish *(Protopterus), Am. J. Physiol.,* 232, R10, 1977.

Dimberg, K., High blood CO_2 levels in rainbow trout exposed to hypercapnia in bicarbonate-rich hard fresh water — a methodological verification, *J. Exp. Biol.,* 134, 463, 1988.

Eddy, F. B., Lomholt, J. P., Weber, R. E., and Johansen, K., Blood respiratory properties of rainbow trout (*Salmo gairdneri*) kept in water of high CO_2 tension, *J. Exp. Biol.,* 67, 37, 1977.

Edsall, J. T. and Wyman, J., *Biophysical Chemistry,* Academic Press, New York, 1958.

Evans, D. H., Kinetic studies of ion transport by fish gill epithelium, *Am. J. Physiol.,* 238, R224, 1980.

Evans, D. H., The role of gill permeability and transport mechanisms in euryhalinity, in *Fish Physiology,* Vol. 10B, Hoar, W. S. and Randall, D. J., Eds., Academic Press, Orlando, 1984, 315.

Evans, D. H., Modes of ammonia transport across fish gills, in *Transport Processes, Iono- and Osmoregulation,* Gilles, R. and Gilles-Baillien, M., Eds. Springer, Berlin.

Evans, D. H., The role of branchial and dermal epithelia in acid-base regulation in aquatic animals, in *Acid-Base Regulation in Animals,* Heisler, N. Ed., Elsevier, Amsterdam, 1986, 139.

Evans, D. H. and Cameron, J. N., Gill ammonia transport, *J. Exp. Zool.,* 239, 17, 1986.

Evans, D. H. and More, K. J., Modes of ammonia transport across the gill epithelium of the dogfish pup (*Squalus acanthias*)*, J. Exp. Biol.,* 138, 375, 1988.

Evans, D. H., More, K. J., and Robbins, S. L., Modes of ammonia transport across the gill epithelium of the marine teleost fish *Opsanus beta, J. Exp. Biol.,* 144, 339, 1989.

Foskett, J. K., Logsdon, C. D., Turner, T., Machen, T. E., and Bern, H. A., Differentiation of the chloride extrusion mechanism during seawater adaption of a teleost fish, the cichlid *Sarotherodon mossambius, J. Exp. Biol.,* 93, 209, 1981.

Fromm, P. O., Some quantitative aspects of ion regulation in teleosts, *Comp. Biochem. Physiol.,* 27, 865, 1968.

Fromm, P. O., A review of some physiological and toxicological responses to freshwater fish to acid stress, *Environ. Biol. Fish.,* 5, 79, 1980.

Glass, M. L., Andersen, N. A., Kruhøffer, M., Williams, E. M., and Heisler, N., Combined effects of environmental Po_2 and temperature on ventilation and blood gases in the carp *Cyprinus carpio* L., *J. Exp. Biol.,* 148, 1, 1990.

Harvey, H. W., *The Chemistry and Fertility of Sea Waters,* Cambridge University Press, London, 1974.

Heisler, N., Bicarbonate exchange between body compartments after changes of temperature in the larger spotted dogfish (*Scyliorhinus stellaris*), Respir. Physiol., 33, 145, 1978.

Heisler, N., Intracellular and extracellular acid-base regulation in the tropical freshwater teleost fish *Synbranchus marmoratus* in response to the transition from water breathing to air breathing, *J. Exp. Biol.,* 99, 9, 1982.

Heisler, N., Transepithelial ion transfer processes as mechanisms for fish acid-base regulation in hypercapnia and lactacidosis, *Can. J. Zool.,* 60, 1108, 1982b.

Heisler, N., Role of ion transfer processes in acid-base regulation with temperature changes in fish, *Am. J. Physiol.,* 246, R441, 1984a.

Heisler, N., Acid-base regulation in fishes, in *Fish Physiology,* Vol. 10A, Hoar, W. S. and Randall, D. J., Eds., Academic Press, New York, 1984b, 315.

Heisler, N., Buffering and transmembrane ion transfer processes, in *Acid-Base Regulation in Animals,* Heisler, N., Ed, Elsevier, Amsterdam, 1986a, 3.

Heisler, N., Acid-base regulation in fishes, in *Acid-Base Regulation in Animals,* Heisler, N., Ed, Elsevier, Amsterdam, 1986b, 309.

Heisler, N., Comparative aspects of acid-base regulation, in *Acid-Base Regulation in Animals,* Heisler, N., Ed., Elsevier, Amsterdam, 1986c, 397.

Heisler, N., Acid-base regulation in elasmobranch fishes, in *Physiology of Elasmobranch Fishes,* Shuttleworth, T. J., Ed., Springer, Heidelberg, 1988a, 215.

Heisler, N., Ion transfer processes as mechanisms for acid-base regulation, in *Lung Biology in Health and Disease — Comparative Pulmonary Physiology: Current Concepts,* Wood, S. C., Ed., Marcel Decker, New York, 1988b, 539.

Heisler, N., Acid-base regulation in fishes. I. Mechanisms, in *Acid Toxicity and Aquatic Animals,* Morris, R., Taylor, E. W., Brown, D. J. A., and Brown, J. A., Eds., Society of Experimental Biology Seminar Series, Cambridge University Press, London, 1989, 85.

Heisler, N., Acid-base regulation: interrelationships between gaseous and ionic exchange, in *Advances in Environmental and Comparative Physiology,* Vol. 6, *Vertebrate Gas Exchange from Environment to Cell,* Boutilier, R. G., Ed., Series: Springer, Heidelberg, 1990a, 211.

Heisler, N., Interaction between gas exchange, metabolism and ion transport in animals: an overview, *Can. J. Zool.,* 67, 2923, 1990b.

Heisler, N., Mechanisms of ammonia elimination in fishes, in *Animal Nutrition and Transport Processes. II. Transport, Respiration and Excretion,* Mellinger, J., Truchot, J. P., and Lahlou, B., Eds., S. Karger, Basel, 1990c, 137.

Heisler, N., Mode of transmembrane and transepithelial ammonia transfer, in *The Vertebrate Gas Transport Cascade: Adaptations to Environment and Mode of Life,* Bicudo, J. E., Ed., CRC Press, Boca Raton, FL, 1993.

Heisler, N. and Neumann, P., Influence of sea water pH upon bicarbonate uptake induced by hypercapnia in an elasmobranch *(Scyliorhinus stellaris) Pfluegers Arch.,* 368 (Suppl.), R19, 1977.

Heisler, N. and Neumann, P., The role of physico-chemical buffering and of bicarbonate transfer processes in intracellular pH regulation in response to changes of temperature in the larger spottted dogfish *(Scyliorhinus stellaris), J. Exp. Biol.,* 85, 99, 1980.

Heisler, N., Weitz, H., and Weitz, A. M., Hypercapnia and resultant bicarbonate transfer processes in an elasmobranch fish, *Bull. Eur. Physiopathol. Respir.,* 12, 77, 1976a.

Heisler, N., Weitz, H., and Weitz, A. M., Extracellular and intracellular pH with changes of temperature in the dogfish *Scyliorhinus stallaris, Respir. Physiol.,* 26, 249, 1976b.

Heisler, N., Neumann, P., and Holeton, G. F., Mechanisms of acid-base adjustment in dogfish *(Scyliorhinus stellaris)* subjected to long-term temperature acclimation, *J. Exp. Biol.,* 85, 89, 1980.

Heisler, N., Holeton, G. F., and Toews, D. P., Regulation of gill ventilation and acid-base status in hyperoxia-induced hypercapnia in the larger spotted dogfigh *(Scyliorhinus stellaris), Physiologist,* 24, 58 (305), 1981.

Heisler, N., Forcht, G., Ultsch, G. R., and Anderson, J. F., Acid-base regulation in response to environmental hypercapnia in two aquatic salamanders, *Siren lacertina* and *Amphiuma means, Respir. Physiol.,* 49, 141, 1982.

Heisler, N., Toews, D. P., and Holeton, G. F., Regulation of ventilation and acid-base status in the elasmobranch *Scyliorhinus stellaris* during hyperoxia-induced hypercapnia, *Resp. Physiol.,* 71, 227, 1988.

Henrikson, R. C. and Matoltsy, A. G., The fine structure of teleost epidermis. III. Club cells and other cell types, *J. Ultrastruct. Res.,* 21, 22, 1968.

Henry, R. P., Smatresk, N. J., and Cameron, J. N., The distribution of branchial carbonic anhydrase and the effects of gill and erythrocyte carbonic anhydrase inhibition in the channel catfish, *Ictalurus punctatus, J. Exp. Biol.,* 134, 201, 1988.

Hōbe, H., Wood, C. M., and Wheatly, M. G., The mechanism of acid-base and ionoregulation in the freshwater rainbow trout during environmental hyperoxia and subsequent normoxia. I. Extra- and intracellular acid-base status, *Respir. Physiol.,* 55, 139, 1984.

Holeton, G. F. and Heisler, N., Contribution of net ion transfer mechanisms to the acid-base regulation after exhausting activity in the larger spotted dogfish *(Scyliorhinus stellaris), J. Exp. Biol.,* 103, 31, 1983.

Holeton, G. F., Neumann, P., and Heisler, N., Branchial ion exchange and acid-base regulation after strenuous exercise in rainbow trout *(Salmo gairdneri), Respir. Physiol.,* 51, 303, 1983.

Hunn, J. B., Urine flow rate in freshwater salmonids: a review, *Progr. Fish-Cult,* 44, 119, 1982.

Hwang, P. P., Distribution of chloride cells in teleost larvae, *J. Morphol.,* 200, 1, 1989.

Ishimatsu, A. and Itazawa, Y., Blood oxygen levels and acid-base status following air exposure in an air-breathing fish, *Channa argus:* the role of air ventilation, *Comp. Biochem. Physiol. A* 74, 787, 1983.

Ishimatsu, A., Iwama, G. K., and Heisler, N., *In vivo* analysis of partitioning of cardiac output between systemic and central venous sinus circuits in rainbow trout: a new approach using chronic cannulation of the branchial vein, *J. Exp. Biol.,* 137, 75, 1988.

Ishimatsu, A., Iwama, G. K., Bentley, T. B., and Heisler, N., Role of the secondary circulatory system for acid-base regulation during hypercapnia in rainbow trout (*Oncorhynchus mykiss), J. Exp. Biol.,* 170, 43, 1992.

Iwama, G. K. and Heisler, N., Effect of environmental water salinity on the acid-base regulation during environmental hypercapnia in the rainbow trout (*Salmo gairdneri*), J. Exp. Biol., 158, 1, 1991.

Iwama, G. K., Ishimatsu, A., and Heisler, N., Blood flow and ion flux partitioning between central venous sinus and systemic gill compartments during exposure to environmental hypercapnia in the rainbow trout (*Oncorhynchus mykiss) Fish Physiol. Biochem.* in press.

Janssen, R. G. and Randall, D. J., The effect of changes in pH and Pco_2 in blood and water on breathing in rainbow trout, *Salmo gairdneri, Respir. Physiol.,* 25, 235, 1975.

Jensen, F. B. and Weber, R. E., Respiratory properties of tench blood and hemoglobin. Adaptation to hypoxic-hypercapnic water, *Mol. Physiol.,* 2, 235, 1982.

Kampmeier, O. F., *Evolution and Comparative Morphology of the Lymphatic System,* Charles C. Thomas, Springfield, IL, 1969.

Karnaky, K. J., Jr., and Kinter, W. B., Killifish opercular skin: a flat epithelium with a high density of chloride cells, *J. Exp. Zool.,* 199, 355, 1977.

Kirsch, R. and Nonotte, G., Cutaneous respiration in three freshwater teleosts, *Resp. Physiol.,* 29, 339, 1977.

Laurent, P., Gill internal morphology, in *Fish Physiology,* Hoar, W. S. and Randall, D. J., Eds., Vol. 10A, Academic Press, New York, 1984, 73.

Lin, H. and Randall, D. J., Evidence for the presence of an electrogenic proton pump on the trout gill epithelium, *J. Exp. Biol.,* 161, 119, 1991.

Maetz, J., Adaptation to hyper-osmotic environments, *Biochem. Biophys. Perspect. Mar. Biol.,* 1, 91, 1974.

Marshall, W. S., Transepithelial potential and short-circuit current across the isolated skin of *Gillichthys mirabilis* (Teleostei: Gobiidae), acclimated to 5% and 100% seawater, *J. Comp. Physiol. B* 114, 157, 1977.

Neumann, P., Holeton, G. F., and Heisler, N., Cardiac output and regional blood flow in gills and muscles after exhaustive exercise in rainbow trout (*Salmo gairdneri)*, J. Exp. Biol., 105, 1, 1983.

Nonotte, G. and Kirsch, R., Cutaneous respiration in seven sea-water teleosts, *Resp. Physiol.,* 35, 111, 1978.

Nonnotte, G., Nonnotte, L., and Kirsch, R., Chloride cells and chloride exchange in the skin of a sea-water teleost, the shanny (*Blennius pholis* L.), *Cell Tissue Res.,* 199, 387, 1979.

Perry, S. F. and Wood, C. M., Kinetics of branchial calcium uptake in the rainbow trout: effects of acclimation to various external calcium levels, *J. Exp. Biol.,* 116, 411, 1985.

Randall, D. J. and Cameron, J. N., Respiratory control of arterial pH as temperature changes in rainbow trout, *Am. J. Physiol.,* 225, 997, 1973.

Randall, D. J., Heisler, N., and Drees, F., Ventilatory response to hypercapnia in the larger spotted dogfish *Scyliorhinus stallaris, Am. J. Physiol.,* 230, 590, 1976.

Reeves, R. B., An imidazole alphastat hypothesis for vertebrate acid-base regulation: tissue carbon dioxide content and body temperature in bullfrogs, *Respir. Physiol.,* 14, 219, 1972.

Reeves, R. B. and Malan, A., Model studies of intracellular acid-base temperature responses in ectotherms, *Respir. Physiol.,* 28, 49, 1976.

Robin, E. D., Relationship between temperature and plasma pH and carbon dioxide tension in the turtle, *Nature,* 135, 249, 1962.

Satchell, G. H., *Physiology and Form of Fish Circulation,* Cambridge University Press, Cambridge, 1991.

Shoubridge, E. and Hochachka, P., Ethanol: novel end-product of vertebrate anaerobic metabolism, *Science,* 209, 308, 1980.

Spry, D. J., Wood, C. M., and Hodson, P. V., The effects of environmental acid on freshwater fish with particular reference to the softwater lakes in Ontario and the modifying effects of heavy metals. A literature review, *Can. Tech. Rep. Fish. Aquat. Sci.,* 999, 145, 1981.

Steffensen, J. F. and Lomholt, J. P., The secondary vascular system, in *Fish Physiology,* Vol. 12, *Cardiovascular System,* Hoar, W. S., Randall, D. J., and Farrell, A. P., Eds., Academic Press, New York, 1993.

Stiffler, D. F., Graham, J. B, Dickson, K. A., and Stockmann, W., Cutaneous ion transport in the freshwater teleost *Synbranchus marmoratus, Physiol. Zool.,* 59, 406, 1986.

Tang, Y. and Boutilier, R. G., Clearance of lactate and protons following acute lactacidosis: a comparison between seawater- and freshwater-adapted rainbow trout *(Salmo gairdneri) Exp. Biol.,* 48, 41, 1988.

Tang, Y., McDonald, D. G., and Boutilier, R. G., Acid-base regulation following exhaustive exercise: a comparison between freshwater- and seawater-adapted rainbow trout *(Salmo gairdneri), J. Exp. Biol.,* 141, 407, 1989.

Toews, D. P., Holeton, G. F., and Heisler, N., Regulation of the acid-base status during environmental hypercapnia in the marine teleost fish *Conger conger, J. Exp. Biol.,* 107, 9, 1983.

Tysekiewicz, K., Structure and vascularization of the skin of the pike *(Esox lucius* L.), *Acta Biol. Cracov. Ser.,* Zool., 12, 67, 1969.

Van den Thillart, G. and Van Waarde, A., pH changes in fish during environmental anoxia and recovery: the advantages of the ethanol pathway, in *Physiological Strategies for Gas Exchange and Metabolism,* Woakes, A. J., Grieshaber, M. K. and Bridges, C. R., Eds., Society for Experimental Biology Seminar Series 41, Society for Experimental Biology, Cambridge, U.K., 1991, 173.

Van den Thillart, G. and Van Berge-Henegouwen, M., and Kesbeke, F., Anaerobic metabolism of goldfish, *Carassius auratus* (L).: ethanol and CO_2 excretion rates and anoxia tolerance at 20, 10 and 5°C, *Comp. Biochem. Physiol.,* 76A(2), 295, 1983.

Van Slyke, D. D., On the measurement of buffer values and on the relationship of buffer value to the dissociation constant of the buffer and the concentration and the reaction of the buffer system, *J. Biol. Chem.,* 52, 525, 1922.

Vogel, W. O. P., Systemic vascular anastomoses, primary and secondary vessels in fish, and the phylogeny of lymphatics, in *Cardiovascular Shunts: Phylogenetic, Ontogenetic and Clinical Aspects,* Alfred Benzon Symposium 21, Johansen, K. and Burggren, W., Eds., Munksgaard, Copenhagen, 1985, 143.

Walsh, J. P. and Moon, T. W., The influence of temperature on extracellular and intracellular pH in the American eel, *Anguilla rostrata* (Le Suer), *Respir. Physiol.,* 50, 129, 1982.

Wilkes, P. R. H., Walker, R. L., McDonald, D. G., and Wood, C. M., Respiratory, ventilatory, acid-base and ionoregulatory physiology of the white sucker *Catostomus commersoni:* the influence of hyperoxia, *J. Exp. Biol.,* 91, 239, 1981.

Winterstein, H., Der Einfluβ der Körpertemperatur auf das Säure-Basen-Gleichewicht im Blut, *Arch. Exp. Pathol. Pharmakol.,* 223, 1, 1954.

Wood, C. M., The physiological problems of fish in acid waters, in *Acid Toxicity and Aquatic Animals,* Morris, R., Taylor, E. W., Brown, D. J. A., and Grown, J. A., Eds., Society of Experimental Biology Seminar Series, Cambridge University Press, Cambridge, 1989, 85.

Wood, C. M. and McDonald, D. G., Physiological mechanisms of acid toxicity to fish, in *Acid Rain/Fisheries,* Johnson, R. E., Ed., American Fisheries Society, Bethesda, MD, 1982, 197.

Wood, C. M. and McDonald, D. G., The physiology of acid aluminum stress in trout, *Ann. Soc. R. Zool. Belg.,* 117(S1), 399, 1987.

Wood, C. M. and Soivio, A., Environmental-effects on gill function — an introduction, *Physiol. Zool.,* 64(N1), 1, 1991.

Woodbury, J. W., Regulation of pH, in *Physiology and Biophysics,* Ruch, T. C. and Patton, H. D., Eds., W. B. Saunders, Philadelphia, 1965, 899.

13 Ammonia and Urea Metabolism and Excretion

Chris M. Wood

INTRODUCTION

General reviews of the massive literature on N-metabolism and excretion in fish have been provided by Forster and Goldstein (1969), Goldstein and Forster (1970), Campbell (1973), Watts and Watts (1973), Goldstein (1982), Perlman and Goldstein (1988), and Mommsen and Walsh (1992), the latter two specifically for elasmobranchs and teleosts, respectively. Ammonia metabolism and excretion have been reviewed by Kormanik and Cameron (1981), van Waarde (1983), Cameron and Heisler (1985), Evans and Cameron (1986), Randall and Wright (1987), and Heisler (1990). The specific interactions between ionoregulation, acid-base balance, and N-excretion have been considered by Evans (1979), Evans et al. (1982), Heisler (1984), Truchot (1987), McDonald et al. (1989), Wood (1991), and between ammonia and carbon dioxide by Wright and Randall (1987), Randall and Wright (1989), and Walsh and Henry (1992). Urea metabolism and excretion have been reviewed by Goldstein (1972), Mommsen and Walsh (1989, 1991), Griffith (1991), and Atkinson (1992).

In light of this extensive background, the present chapter will focus exclusively on ammonia and urea metabolism and excretion, summarizing those aspects which appear to be generally established and noncontroversial, while highlighting points of remaining uncertainty. Substances such as uric acid, purines, various methylamines, taurine, imidazole dipeptides, creatine, and creatinine will not be considered, because their quantitative contributions are small and their physiological roles remain poorly understood (van Waarde, 1988). Particular emphasis will be placed on environmental influences on ammonia and urea metabolism. In areas of general agreement, references to the primary literature will be selective so as to emphasize papers of key historical importance or recent impact. In areas of uncertainty, greater detail will be provided on unresolved issues, in the hope of stimulating future research.

AMMONIA CHEMISTRY

Ammonia is the most reduced and energy-efficient N-product of the biological oxidation of amino acids and proteins (19.5 kJ/g protein; Smith and

0-8493-8042-1/93/$0.00+$.50

Rumsey, 1976). Chemically, ammonia plays three different but interrelated roles in biological systems — as a buffer base, as a respiratory gas, and as a cation. In aqueous solution it behaves as a base of moderate strength with a pK $\simeq$9.0 to 10.0 in the physiological range of temperature, ionic strength, and protein content (see nomogram in Cameron and Heisler, 1983):

$$NH_3 + H^+ \rightleftarrows NH_4^+ \tag{1}$$

Thus, total ammonia (T_{Amm}) in solution will have two components — the gas NH_3 and the cation NH_4^+; the separate fractions of each can be calculated from the Henderson-Hasselbalch equation if the pH and appropriate pK are known:

$$NH_4^+ = \frac{T_{AMM}}{1 + \text{antilog (pH} - \text{pk)}} = T_{Amm} - NH_3 \tag{2}$$

The metabolic production of NH_3 will tend to raise intra- and extracellular pH by trapping H^+ ions to form NH_4^+. At the normal pH values present in the body fluids, NH_4^+ will constitute $\geq$95% of T_{Amm}. On a *net* basis, ammonia is produced as NH_3 by metabolism (in acid-base terms, this is equivalent to equimolar NH_4^+ and HCO_3^- production by the oxidation of proteins); therefore, by convention, the excretion of NH_3 is neutral with respect to acid-base balance, whereas the excretion of NH_4^+ represents the excretion of acidic equivalents. This assumption is central to techniques such as the "titration alkalinity" (Maetz, 1973; McDonald and Wood, 1981) and "ΔHCO_3^-" methods (Heisler, 1984) commonly employed for measuring *net* acid-base fluxes to the water in fish.

NH_3 is a highly soluble gas which exerts a partial pressure according to Henry's Law:

$$P_{NH3} = \frac{NH_3}{\alpha NH_3} \tag{3}$$

Cameron and Heisler (1983) have provided a tabulation of αNH_3 values for water and trout plasma at various temperatures and ionic strengths. As αNH_3 values are very high and T_{Amm} concentrations quite low in fish blood (usually 0.05 to 1 mmol·l^{-1}), P_{NH3} values are typically in the 10- to 200-µtorr range.

NH_3 solubility is approximately 1000 times that of CO_2, while its aqueous diffusion coefficient is similar to that of CO_2. Therefore the diffusivity of NH_3 through water is about 1000 times higher than that of CO_2. This statement is often misinterpreted to mean that the diffusivity of NH_3 through fish tissues is equally high. Indeed, it is commonly assumed that NH_3 is highly lipophilic and therefore diffuses even more easily through cell membranes. However, the available data, while limited, indicate that the lipid vs. water partition coefficient for NH_3 is less than 0.1 (Evans and Cameron, 1986). Diffusion through water-filled channels is likely to be much faster than diffusion through lipoprotein membranes. A recent report has even suggested that certain cell membranes in the mammalian kidney are impermeable to NH_3 (Kikeri et al., 1989).

NH_4^+, being charged and larger than NH_3, especially in the hydrated form, is normally thought to be much less diffusive and to move almost entirely through water-filled channels, particularly paracellular channels (McDonald et al., 1989). The NH_4^+ ion is considerably larger than Na^+ and slightly larger than K^+; nevertheless it appears to be more permeant through epithelia than either of these strong cations. NH_4^+ also competes with K^+ for transcellular transporters, such as $Na^+,K^+,2Cl^-$ cotransport and Na^+,K^+-ATPase; as with NH_3, its true diffusivity through biological membranes remains unknown. More important than absolute diffusivity is the relative permeability (pNH_3/pNH_4^+) of biological membranes. While there are many estimates of this ratio in the literature (e.g., Cameron and Heisler, 1985; Evans and More, 1988; Evans et al., 1989), the outstanding characteristic is their variability, with pNH_3/pNH_4^+ ratios ranging from <10:1 to >1000:1. To a certain extent, these results may be tissue specific — for example, seawater gills appear to have greater NH_4^+ permeability than freshwater gills (Goldstein et al., 1982; Evans et al., 1989).

AMMONIA INTERNAL DISTRIBUTION

If NH_3 permeability is *much* greater than NH_4^+ permeability, then the movement of ammonia will be dominated by the diffusion of NH_3 along P_{NH3} gradients. If P_{NH3} equilibrates between compartments, the distribution of T_{Amm} at steady state will be dictated by the pH gradient, as described by the theory of nonionic diffusion (Jacobs and Stewart, 1936), and the influence of electrical gradients on NH_4^+ equilibration will be negligible. The distribution of T_{Amm} across cell membranes will be a function of the intracellular (pHi) to extracellular (pHe) gradient:

$$\frac{[T_{AMM}]_i}{[T_{AMM}]_e} = \frac{1+10^{(pk-pHi)}}{1+10^{(pk-pHe)}} \tag{4}$$

This appears to be the situation in most mammalian tissues (e.g., Roos and Boron, 1981).

On the other hand, if NH_4^+ permeability is significant (though not necessarily larger than NH_3 permeability), then the influence of the trans-membrane voltage gradient on NH_4^+ dictates the steady-state distribution of T_{Amm}, as described by the Nernst equation:

$$EM = -\frac{RT}{zF}\ln\frac{[NH_4^+]_i}{[NH_4^+]_e} = -\frac{RT}{zF}\ln\frac{[T_{AMM}]_i-[NH_3]_i}{[T_{AMM}]_e-[NH_3]_e} \tag{5}$$

where R, T, z, and F have their usual meaning and E_M is the membrane potential. This distribution according to E_M will be the same as a distribution according to pHe – pHi when H^+ itself is passively distributed according to E_M; the pNH_3/pNH_4^+ ratio will exert no influence, because NH_4^+ distribution will be at equilibrium with E_M when intracellar and extracellular P_{NH3} are identical. This is the situation in mammalian red cells, where the pHe – pHi gradient is a passive Donnan ratio determined by the relatively small E_M (Roos and Boron, 1981). It also appears to be the situation in teleost red cells under truly resting conditions (Wright et al., 1988a and b). However in most other mammalian tissues, E_M is much larger, and intracellular $[H^+]$ is kept far below (i.e., pHi far above) the theoretical passive value. In this situation, the $[T_{AMM}]_i/[\dot{T}_{AMM}]_e$ ratio predicted by a dominant influence of E_M on NH_4^+ would be far greater than the ratio predicted by a dominant influence of pHe – pHi on NH_3.

Distribution according to voltage vs. distribution according to pH represent the two extreme ends of a spectrum of possible equilibrium states determined by the pNH_3/pNH_4^+ ratio. A general equation (Roos and Boron, 1981) describes the steady-state distribution of a base such as ammonia subject to simultaneous electrical and pH gradients:

$$\frac{[T_{Amm}]_i}{[T_{Amm}]_e} = \frac{[H^+]_i + K}{[H^+]_e + K} \times \frac{(pNH_3/pNH_4^+) - [F \times E_M/R \times T(1-\gamma)] \times ([H^+]_e/K)}{(pNH_3/pNH_4^+) - [F \times E_M \times \gamma/R \times T(1-\gamma)] \times ([H^+]_i/K)} \quad (6)$$

where K is the NH_3/NH_4^+ dissociation constant, and γ is $\exp(E_M F/RT)$. Figure 1 plots the steady-state $[T_{Amm}]_i/[T_{Amm}]_e$ as a function of pNH_3/pNH_4^+, assuming typical values of pHe – pHi = 0.5 units and $E_M = -90$ mV for vertebrate muscle tissue. At $pNH_3/pNH_4^+ > 100$, the ratio asymptopically approaches the theoretical minimum of about 3, where it is entirely a function of the pH gradient (as in most mammalian tissues), while at $pNH_3/pNH_4^+ < 10$, the ratio asymptopically approaches the theoretical maximum of about 35, where it is entirely a function of E_M. In between these limits, there is an approximately log/linear relationship between pNH_3/pNH_4^+ and $[T_{Amm}]_i/[T_{Amm}]_e$. Most models of ammonia distribution in fish have assumed a mammalian-type situation, with very high values of pNH_3/pNH_4^+, yielding a $[T_{Amm}]_i/[T_{Amm}]_e$ ratio of about 3 (e.g., Cameron and Heisler, 1983; Randall and Wright, 1987). On the other hand, most measurements in the muscle tissue of fish have indicated a far higher ratio (e.g., Robertson, 1975; Saha and Ratha, 1989; Tang et al., 1992; Danulat and Kempe, 1992). The data of Mommsen and Hochachka (1988) on resting rainbow trout *(Oncorhynchus mykiss)* are a notable exception.

Wright et al. (1988a, 1988b) and Wright and Wood (1988) attacked this problem in lemon sole (*Parophrys vetulus*) and rainbow trout using chronic cannulation to preserve extracellular T_{Amm} (see section below on Blood Levels of Ammonia), freeze clamping to preserve intracellular T_{Amm}, and simultaneous measurements of pHe – pHi. They found that $[T_{Amm}]_i/[T_{Amm}]_e$ was about 30 to 35 in white muscle at rest, almost exactly the ratio predicted by the membrane potential and far greater than predicted by the pHe – pHi gradient.

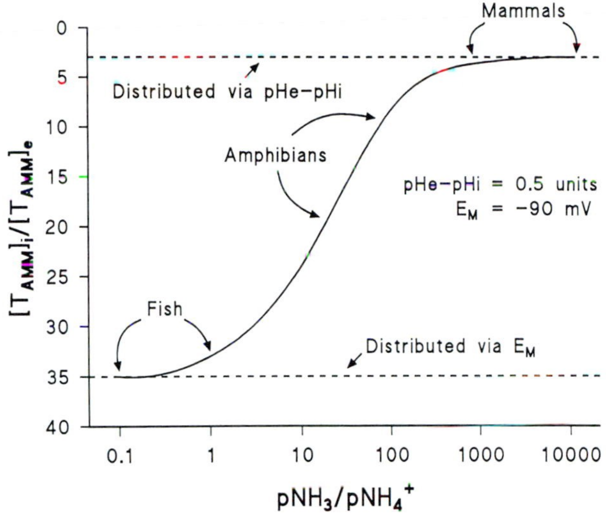

FIGURE 1. The theoretical relationship between the distribution of ammonia at equilibrium between intracellular and extracellular compartments of muscle $[T_{Amm}]i/[T_{Amm}]e$ and the relative permeability (pNH_3/pNH_4^+) of the cell membrane to NH_3 and NH_4^+ as calculated from Equation 6 of the text. A membrane potential (E_M) of −90 mV and a pHe − pHi gradient of 0.5 units have been assumed. The typical ranges of values for fish, amphibians, and mammals are indicated; fish data are taken from Wright et al. (1988b), Wright and Wood (1988), and Tang et al. (1992) and amphibian data from Wood et al. (1989a). See text for additional details.

A similar situation prevailed after exhaustive exercise in both species, as well as in the ventricular muscle and brain of the sole. Ammonia was distributed across the red blood cell membrane according to pHe − pHi under truly resting conditions, but not under stressful conditions, where catecholamine mobilization would disturb the simple Donnan distribution of H^+. Wright and co-workers concluded that pNH_3/pNH_4^+ was rather low (<20) in fish cell membranes, such that ammonia was distributed largely according to the membrane potential. They proposed a model whereby diffusive efflux of NH_3 from ICF to ECF under standing P_{NH3} gradients would be balanced by diffusive influx of NH_4^+ under standing electrochemical gradients. This dual permeability would be advantageous in moving large amounts of ammonia across the cell membranes of an ammoniotelic animal under non-steady-state conditions. This idea was supported by subsequent experiments of Wood et al. (1989a) showing that amphibians transitional between ammoniotelism and ureotelism had intracellular $[T_{Amm}]_i/[T_{Amm}]_e$ ratios (9 to 17) intermediate between those of fish and mammals.

Heisler (1990) has criticized these conclusions on theoretical grounds, most notably that the NH_4^+ "shuttle" would create a large H^+ load for the pHi regulatory mechanisms of the cells. Heisler (1990) also cites unpublished data (no values given) indicating much lower values of $[T_{Amm}]_i/[T_{Amm}]_e$ in fish and amphibians, and therefore much higher values of pNH_3/pNH_4^+. The resolution of this issue is important, for it will help establish how ammonia crosses cell membranes, how large the true intracellular reservoirs of ammonia really are in fish, and the nature of phylogenetic trends in pNH_3/pNH_4^+.

AMMONIA PRODUCTION

Ammonia is the major end product of N-metabolism in almost all osteichthyes, agnathans, and freshwater chondrichthyes, but less important than urea in marine chondrichthyes, dipnoans during estivation, and a very few teleosts living in extreme environments (Tables 1 and 3). The coelocanth (*Latimeria chalumnae*) is probably ureotelic rather than ammoniotelic, based on blood data (Table 2), but measurements of ammonia efflux (J_{Amm}) are lacking in this very rare species.

Fuels

The largest source of ammonia is catabolism of dietary or structural protein. In most studies, gut absorption efficiency of dietary N is higher than total caloric absorption efficiency, and fecal losses are low (<20%; e.g., Beamish and Thomas, 1984; Kaushik and Teles, 1985). Essential amino acids in fish are the same as those in mammals: arginine, histidine, isoleucine, leucine, lysine, methionine, phenylalanine, threonine, tryptophan, and valine (Halver and Shanks, 1960); all others can be synthesized. Fish have a remarkable capacity to utilize amino acids both as a metabolic fuel and as precursors for protein, lipid, and carbohydrate synthesis. Brown and Cameron (1991) infused channel catfish (*Ictalurus punctatus*) with a mixture of essential amino acids approximating a standard daily ration. The fish responded with twofold increases in M_{O2} and threefold increases in J_{Amm}, but only about 20% of the infused N-load was excreted. In actively growing fish fed a balanced diet, greater than 50% of dietary N-content is incorporated into structural growth, and most of the remainder is used for energy production (Atherton and Aitken, 1970; Beamish and Thomas, 1984). Dietary carbohydrate does not appear to be an important aerobic fuel, but there is nevertheless a high capacity for gluconeogesis from exogenous amino acids (Bever et al., 1981). However, dietary lipid certainly is important, especially in carnivores. If the diet is deficient in lipid, then a greater proportion of dietary protein is metabolized for energy or deaminated for conversion to fat and carbohydrate, more ammonia is excreted, and a lower percentage of dietary N is retained for growth (Atherton and Aitken, 1970).

The most important influence on J_{Amm} is the rate of dietary protein intake (Beamish and Thomas, 1984). Ammonia production increases markedly after a meal (Brett and Zala, 1975; Brown and Cameron, 1991). In fish fed to satiation, J_{Amm} may be up to 10-fold greater than rates in starved fish; once feeding is terminated, production declines rapidly, and stable levels are achieved in 5 to 10 d (Fromm, 1963; Guerin-Ancey, 1976; Kaushik and Teles, 1985). The latter represents the so-called *endogenous* fraction, required for normal body maintenance. The difference between fed and starved rates approximates the *exogenous* fraction, the portion not retained from the ingested food. All fish exhibit a marked ability to consume their own structural protein during prolonged starvation, but the size of the endogenous fraction as a proportion of metabolic rate varies depending on the extent of fat reserves (van Waarde, 1983). Glycogen reserves tend to be spared during starvation, probably because of their survival value as an anaerobic fuel during burst exercise and hypoxia.

The other important source of ammonia is the deamination of adenylates, though under steady-state conditions, *net* production from this fuel is probably negligible. However, a significant portion of the elevated J_{Amm} which occurs during burst ("anaerobic") exercise (Dobson and Hochachka, 1987; Wood, 1988) or environmental hypoxia (van Waarde et al., 1983) arises from this source through one arm of the purine nucleotide cycle (see below).

PATHWAYS

When fish are feeding and growing, absorbed amino acids in excess of those needed for protein synthesis are deaminated and then oxidized in the Kreb's cycle or converted to fat and carbohydrate. When fish are starving, or energetic expenditure exceeds intake, amino acids released by muscle proteolysis are similarly deaminated. The most important pathway (Walton and Cowey, 1977) is thought to be the transamination system first proposed by Braunstein and Bychkov (1939). Various aminotransferases transfer the amino group of L-amino acids to alpha-ketoglutarate to form glutamate, which is subsequently deaminated by glutamate dehydrogenase (Figure 2A). This occurs in many tissues, including muscle, gill, and kidney, but enzyme activities are generally highest in liver. The size of the glutamate pool in liver is large relative to other amino acids.

The other major pathway is probably the direct hydrolysis of the amide groups on glutamine and asparagine, catalyzed by glutaminase and asparaginase (Figure 2B). These enzymes occur in liver, kidney, and gills, and at least glutaminase is present in both red and white muscle (Chamberlin et al., 1991). Glutamine and asparagine (together with leucine and sometimes histidine) are preferred substrates for ammonia production by *in situ* liver preparations (Pequin and Serfaty, 1963; Vellas and Serfaty, 1974), isolated hepatocytes (van Waarde and Kesbeke, 1981), and hepatic and muscle mitochondria (Campbell et al., 1983; Chamberlin et al., 1991). Glutamine synthetase is widespread (Lund and Goldstein, 1969; Vellas and Serfaty, 1974), and serves to detoxify ammonia when internal ammonia levels rise (Levi et al., 1974; Iwata, 1989). Protection against ammonia toxicity is particularly important for brain, where glutamine synthetase activity is especially high (Lund and Goldstein, 1969; Iwata, 1989; Danulat and Kempe, 1992). Even in the absence of ammonia

TABLE 1
Rates (μmol-N·kg^{-1}·h^{-1}) and Relative Percentages of N-Excretion as Ammonia-N and Urea-N Through *"Gills"* and *"Kidney"*

Species	Medium	Gills		Kidney		Reference
		Ammonia-N	Urea-N	Ammonia-N	Urea-N	
Agnatha						
Entosphenus tridentatus[a]	Fresh water	139 (95%)	0 (0%)	6 (4%)	1 (1%)	Read (1968)
Osteichthyes						
Cyprinus carpio[a]	Fresh water	315 (88%)	27 (7%)	15 (4%)	3 (1%)	Smith (1929a)
Cyprinus carpio[b]	Fresh water	115 (82%)	11 (8%)	15 (10%)	0 (0%)	Smith (1929a)
Carassius auratus[b]	Fresh water	177 (79%)	28 (13%)	17 (7%)	2 (1%)	Smith (1929a)
Oncorhynchus mykiss[a]	Fresh water	270 (86%)	34 (11%)	4 (1%)	6 (2%)	C. M. Wood (unpublished data)
Oreochromis mossambicus[b]	Fresh water	59 (61%)	24 (25%)	0 (0%)	13 (14%)	Sayer & Davenport (1987a)
Heteropneustes fossilis[a]	Fresh water	236 (85%)	30 (11%)	1 (0%)	11 (4%)	Saha et al. (1988)
Oncorhynchus clarki henshawi[a]	10% Seawater	105 (56%)	60 (32%)	19 (10%)	4 (2%)	Wright et al. (1992)
Periophthalmus cantonensis[b]	25% Seawater	351 (47%)	174 (23%)	94 (13%)	131 (17%)	Morii et al. (1978)
Boleophthalmus pectinorostris[b]	25% Seawater	204 (61%)	46 (14%)	36 (11%)	49 (14%)	Morii et al. (1978)
Agonus cataphractus[b]	Seawater	316 (41%)	68 (9%)	329 (43%)	59 (7%)	Sayer & Davenport (1987a)
Taurulus bubalis[b]	Seawater	337 (63%)	28 (4%)	118 (20%)	76 (13%)	Sayer & Davenport (1987a)
Crenilabrus melops[b]	Seawater	263 (67%)	6 (2%)	109 (28%)	12 (3%)	Sayer & Davenport (1987a)
Blennius pholis[b]	Seawater	51 (35%)	26 (18%)	58 (39%)	12 (8%)	Sayer & Davenport (1987a)

Chondrichthyes						
Pristis microdon[a]	Fresh water	250 (18%)	759 (55%)	27 (2%)	336 (25%)	Smith & Smith (1931)
Potamotrygon sp[a]	Fresh water	981	~70	?	~30	Goldstein & Forster (1971a)
Squalus acanthias[a]	Seawater	13 (2%)	519 (91%)	0 (0%)	37 (7%)	C. M. Wood & P. A. Wright (unpublished data)
Raja erinacea[a]	Seawater	111	454	?	24	Goldstein & Forster (1971b)

Note: The percentages assume that the total of ammonia-N plus urea-N is 100%.

[a] Kidney excretion measured by urinary catheter. Excretion through skin and gastrointestinal tract is therefore included in the gill component.

[b] Kidney excretion measured by divided chamber. Excretion through skin and gastrointestinal tract is therefore largely included in the kidney component.

TABLE 2
Concentrations (mmol-N·l^{-1}) of Ammonia-N and Urea-N in the Blood Plasma

Species	Medium	Ammonia-N	Urea-N	Reference
Agnatha				
Lampetra fluviatilis[a]	Fresh water	0.38	"0"	Robertson (1954)
Myxine glutinosa[a]	Seawater	"0"	5.60	Robertson (1966)
Myxine glutinosa[a]	Seawater	1.10	—	McDonald et al. (1991)
Bdellostoma cirrhatum[a]	Seawater	0.63	"Trace"	Read (1975)
Osteichthyes				
Cyprinus carpio[a]	Fresh water	0.13	1.90	Smith (1929a)
Cyprinus carpio[a]	Fresh water	0.29	1.71	Vellas and Serfaty (1974)
Coregonus clupoides[a]	Fresh water	0.33	1.40	Robertson (1954)
Pterygoplichthys sp.[a]	Fresh water	0.04	0.80	Mangum et al. (1978)
Synbranchus marmoratus[a]	Fresh water	0.50	0.66	Mangum et al. (1978)
Heteropneustes fossilis[a]	Fresh water	0.47	1.52	Saha and Ratha (1989)
Anabas testudineus[a]	Fresh water	0.73	0.74	Saha and Ratha (1989)
Channa punctatus[a]	Fresh water	0.89	0.46	Saha and Ratha (1989)
Clarias batrachus[a]	Fresh water	0.47	1.22	Saha and Ratha (1989)
Amphipnous cuchia[a]	Fresh water	0.43	1.56	Saha and Ratha (1989)
Oreochromis nilotica[a]	Fresh water	1.04	2.32	Wood et al. (1989b)
Oncorhynchus mykiss[a]	Fresh water	0.68	4.24	Wood et al. (1989b)
Oncorhynchus mykiss[b]	Fresh water	0.20	4.40	Wood et al. (1989b)
Oncorhynchus mykiss[b]	Fresh water	0.10	5.00	Wilkie & Wood (1991)
Amia calva[b]	Fresh water	0.31	0.22	McKenzie & Randall (1990)
Oncorhynchus clarki henshawi[b]	10% Seawater[c]	0.17	8.15	Wright et al. (1993)
Periophthalmus cantonensis[a]	25% Seawater	6.70	10.00	Morii et al. (1979)
Boleophthalmus pectinirostris[a]	25% Seawater	2.30	1.30	Morii et al. (1979)
Oreochromis alcalicus grahami[a]	50% Seawater[d]	0.77	10.52	Wood et al. (1989b)
Chalcalburnus tarichi[a]	63% Seawater[e]	1.30	36.18	Danulat and Kempe (1992)

Osteichthyes				
Sicyases sanguineus[a]	Seawater	0.20	6.40	Gordon et al. (1970)
Opsanus beta[b]	Seawater	0.26	16.10	Walsh et al. (1990)
Protopterus aethiopicus[a]	Fresh water	"<0.71"	1.21	Smith (1930)
Protopterus aethiopicus[b]	Fresh water	—	8.40	Delaney et al. (1977)
Protopterus aethiopicus[b]	Estivation (13 months)	—	406	Delaney et al. (1977)
Latimeria chalumnae[a]	Seawater	—	754	Griffith et al. (1974)
Chondrichthyes				
Pristis microdon[a]	Fresh water	"Trace"	260	Smith & Smith (1931)
Potamotrygon sp.[a]	Fresh water	"0"	2.20	Gerst & Thorson (1977)
Potamotrygon motoro[a]	Fresh water	3.28	0.90	Mangum et al. (1978)
Raja diaphenes[a]	Seawater	"0"	751	Smith (1929b)
Carcharias littoralis[a]	Seawater	"Trace"	771	Smith (1929b)
Mustelis canis[a]	Seawater	"0"	761	Smith (1929b)
Squalus acanthias[a]	Seawater	0.33	672	Kormanik & Evans (1986)
Squalus acanthias[b]	Seawater	0.17	675	C.M. Wood & P.A. Wright (unpublished data)
Scyliorhinus canicula[a]	Seawater	0.66	814	Robertson (1989)
Chimaera monstrosa[a]	Seawater	0.31	664	Robertson (1976)

[a] Sampled by caudal or cardiac puncture.
[b] Sampled by in-dwelling catheters.
[c] Pyramid Lake, pH 9.4.
[d] Lake Magadi, pH 10.
[e] Lake Van, pH 9.8.

TABLE 3
Whole-Body Rates (μmol-N·kg^{-1}·h^{-1}) of Ammonia-N and Urea-N Excretion in Aquatic Osteichthyes, Osteichthyes of Alkaline Lakes, and Air-Breathing Osteichthyes

Species	Medium	Ammonia	Urea-N	OUC	References
Aquatic Osteichthyes					
Cyprinus carpio	Fresh water	330 (92%)	30	—	Smith (1929a)
Carassius auratus	Fresh water	194 (87%)	30	—	Smith (1929a)
Oncorhynchus nerka	Fresh water	521 (79%)	136	—	Brett & Zala (1975)
Oreochromis mossambicus	Fresh water	59 (61%)	37	—	Sayer & Davenport (1987a)
Oreochromis nilotica	Fresh water	763 (85%)	138	—	Wood et al (1989b)
Oncorhynchus mykiss	Fresh water	220 (87%)	32	—	Wilkie & Wood (1991)
Leptocottus armatus	Seawater	121 (74%)	43	—	Wood (1958)
Platichthys stellatus	Seawater	179 (86%)	29	—	Wood (1958)
Engraulis mordax	Seawater	293 (80%)	71	—	McCarthy & Whitledge (1972)
Engraulis ringens	Seawater	686 (75%)	229	—	McCarthy & Whitledge (1972)
Trachurus symmetricus	Seawater	271 (84%)	50	—	McCarthy & Whitledge (1972)
Agonus cataphractus	Seawater	645 (84%)	127	—	Sayer & Davenport (1987a)
Taurulus bubalis	Seawater	495 (83%)	104	—	Sayer & Davenport (1987a)
Crenilabrus melops	Seawater	372 (95%)	18	—	Sayer & Davenport (1987a)
Limanda limanda	Seawater	490 (93%)	36	—	Sayer & Davenport (1987a)
Osteichthyes of Alkaline Lakes					
Oreochromis alcalicus grahami	Lake Magadi, pH = 10, ~50% seawater	0 (0%)	7,600	Y	Randall et al (1989); Wood et al (1989b)
Oncorhynchus clarki henshawii	Pyramid Lake, pH = 9.4, ~10% seawater	123 (66%)	63	N	Wright et al. (1993)
Chalcalburnus tarichi	Lake Van, pH = 9.8, ~63% seawater	1,041 (63%)	607	N	Danulat & Kempe (1992)

Airbreathing Osteichthyes					
Amphipnous cuchia	Fresh water	171 (91%)	17	Y	Saha & Ratha (1989)
Clarias batrachus	Fresh water	421 (92%)	37	Y	Saha & Ratha (1989)
Heteropneustes fossilis	Fresh water	254 (91%)	24	Y	Saha & Ratha (1989)
Anabas testudineus	Fresh water	417 (96%)	16	Y	Saha & Ratha (1989)
Channa punctatus	Fresh water	292 (97%)	9	Y	Saha & Ratha (1989)
Amia calva	Fresh water	607 (91%)	60	—	McKenzie & Randall (1990)
Blennius pholis	Fresh water	115 (46%)	136	—	Sayer & Davenport (1987b)
Blennius pholis	Seawater	156 (79%)	42	—	Sayer & Davenport (1987b)
Sicyases sanguineus	Seawater	270 (26%)	760	—	Gordon et al. (1970)
Opsanus beta	Seawater	75 (77%)	22	Y	Walsh et al. (1990)
Periophthalmus sobrinus	40% seawater	770 (93%)	60	—	Gordon et al. (1965)
Periophthalmus sobrinus	Seawater	490 (58%)	360	—	Gordon et al. (1969)
Periophthalmus cantonensis	25% seawater	445 (59%)	305	—	Morii et al. (1978)
Periophthalmus cantonensis	Seawater	630 (33%)	1260	—	Gordon et al. (1978)
Boleophthalmus pectinorostris	25% seawater	240 (72%)	95	—	Morii et al. (1978)
Periophthalmus expeditionium	Seawater	771 (67%)	383	N	Gregory (1977)
Periophthalmus gracilis	Seawater	548 (86%)	92	N	Gregory (1977)
Scartelaos histophorus	Seawater	464 (81%)	110	N	Gregory (1977)

Note: The percentage (%) occurring as ammonia-N assumes that the total of ammonia-N plus urea-N is 100%. The presence (Y) or absence (N) of significant ornithine urea cycle activity (OUC) in the liver is noted for those species for which the information is available.

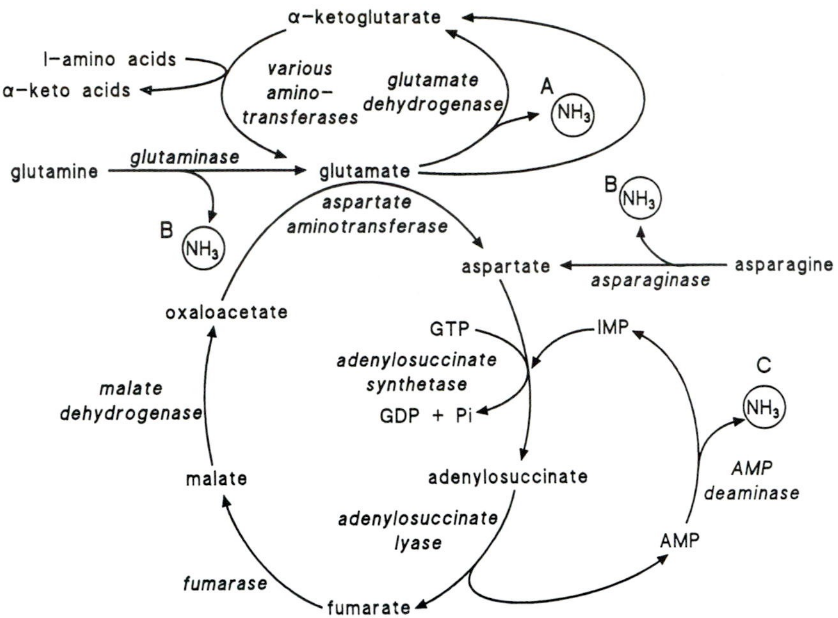

FIGURE 2. Biochemical pathways in fish for the production of ammonia from (A) *l*-amino acids (transamination), (B) amides (hydrolysis of amide groups), and (C) adenylates (purine nucleotide cycle), and their interrelationships.

loading, the infusion of glutamate into intact carp caused a significant decrease in J_{Amm}, presumably due to glutamine formation (Pequin and Serfaty, 1966).

It has been suggested that glutamine may also serve as a vehicle which carries N in the bloodstream between proteolytic and deaminating tissues, but there is little supportive evidence (van Waarde, 1988). Alanine and ammonia itself appear to be the most important transport vehicles. Spawning, starving salmon exhibited an increase in plasma ammonia and alanine but no elevation of plasma glutamine (Mommsen et al., 1980). Relative to dorsal aortic blood, there was no enrichment of glutamine in the hepatic portal blood of fasted trout, whereas alanine was again elevated (Ash et al., 1989). In elasmobranchs, glutamine is a primary substrate for urea synthesis (see section below on Urea Production), but alanine, rather than glutamine is released from muscle during starvation (Leech et al., 1979).

The purine nucleotide cycle of Lowenstein (1972) has been suggested as another deamination route, albeit energetically more expensive (Figure 2C). Here, the amino group on glutamate is transferred to oxaloacetate by aspartate aminotransferase to form aspartate and regenerate alpha-ketoglutarate. Aspartate then reacts with inosine monophosphate (IMP) and guanosine triphosphate (GTP) to form adenylosuccinate, mediated by adenylosuccinate synthetase, followed by conversion to fumarate and adenosine monophosphate (AMP), mediated by adenylosuccinate lyase. The deamination of AMP by AMP deaminase liberates ammonia and regenerates IMP, thereby completing the cycle. In addition, fumarate may be converted to malate by fumarase, and then back to oxaloacetate by malate dehydrogenase, thereby regenerating the recipient of the amino group. While the enzymes of this cycle are present in the liver, it does not appear to be the main hepatic pathway of amino acid deamination (van Waarde and Kesbeke, 1981). Enzyme activities are highest in muscle. However, in the white muscle of trout subjected to exhaustive exercise, less than 2% of the ammonia production can be traced to the depletion of aspartate because adenylosuccinate synthetase and lyase are effectively shut down (Mommsen and Hochachka, 1988). Almost all ammonia production is due to the deamination of adenylates by AMP deaminase, resulting in stoichiometrically equal accumulation of ammonia and IMP. Some of the ammonia is released into the blood and subsequently excreted at the gills (Wood, 1988), but most is retained in the muscle, which helps maintain glycolytic flux via stimulation of phosphofructokinase. At the same time, the buffer base properties of NH_3 attenuate the pHi depression of lactacidosis (Dobson and Hochachka, 1987). During recovery, AMP deaminase is shut down, while the other two enzymes replenish the adenylate pool; the required aspartate is furnished by fumarate/malate plus the resynthesis of glutamate from ammonia and alpha-ketoglutarate.

Finally, there may also be direct deamination of certain individual amino acids, such as histidine, serine, and threonine, by specific enzymes. These reactions have been detailed by van Waarde (1983). With the exception of adenylate deamination, all the above pathways are aerobic. However cyprinids such as goldfish (*Carassius auratus*) maintain unchanged excretion and body levels of ammonia under completely anoxic conditions (van den Thillart and Kesbeke, 1978; van Waversveld et al., 1989). Adenylate deamination can account for less than 50% of this anaerobic production; the major portion results from the deamination of amino acids and appears to be extrahepatic. These fish have a unique capacity to ferment carbohydrate to CO_2 and ethanol during anoxia, and a comparable fermentation of alanine and other amino acids to ethanol, CO_2, and NH_3 may serve as an anaerobic deamination mechanism.

Sites

In his classic study of N-excretion in fish, Homer Smith (1929a) concluded that ammonia was cleared directly from the blood by the gills, rather than synthesized from a precursor *in situ*, but provided no direct evidence. The first experimental attack on the problem, using the sculpin *(Myoxcephalus scorpius)*, suggested exactly the opposite — that most of the excreted ammonia was produced in the branchial tissue (Goldstein and Forster, 1961). However, a later investigation on the same species (Goldstein et al., 1964), using improved methods, confirmed Smith's original view. At least 75% of the excreted ammonia is cleared from the blood passing through the gills, even though a dynamic exchange of several amino acids occurs between blood and gill tissue (e.g., Goldstein et al., 1964; Walton and Cowey, 1977; Ogata and Murai, 1988). Only a small amount is actually produced in the gills, probably via a combination of transamination and hydrolysis of amide groups. The small amounts of ammonia excreted by the gills of ureotelic marine elasmobranchs also probably originate from blood clearance (Perlman and Goldstein, 1988).

Under normal conditions, the major production site seems to be the liver, which clearly possesses the greatest transamination capacity. The rates of ammonia production by isolated liver mitochondria incubated with realistic levels of amino acids were adequate to explain whole-body values of J_{Amm} in intact catfish (*I. punctatus*; Campbell et al., 1983). Pequin and Serfaty (1963, 1966) and Vellas and Serfaty (1974) combined measurements of ammonia concentrations in blood at different vascular sites with *in situ* hepatic perfusion in intact carp (*Cyprinus carpio*) to demonstrate that the liver was quantitatively the most important production site. The amides glutamine and asparagine were particularly potent in stimulating hepatic ammoniagenesis.

However, the liver is not the sole site of ammonia production, or under some circumstances, even the predominant site. The kidney and the lateral body musculature also make significant contributions to overall ammonia production in the carp (Vellas and Serfaty, 1974). In addition to the ammonia passing into venous blood draining the kidney, there is also passage of ammonia into the urine (see section below on Ammonia Excretion). The large production of ammonia by adenylate deamination in skeletal muscle during glycolytic exercise or hypoxia in a number of species was discussed earlier, as was the apparent shift of ammonia production to extrahepatic sites in anoxic goldfish. Total activities (i.e., weight-adjusted) of purine nucleotide cycle enzymes are greater in skeletal muscle than in liver, and total activities of asparaginase and glutaminase are greater in gill than in liver (van Waarde, 1983). Chamberlin et al. (1991) have described a high capacity for glutamine oxidation in the skeletal muscle of several species. All these observations of extrahepatic ammonia production may be significant in explaining the amazing finding of Kenyon (1967) on the eel (*Anguilla anguilla*). Kenyon reported that whole body J_{Amm} was unchanged 10 d after total hepatectomy. Clearly, this is an observation which should be confirmed.

BLOOD LEVELS OF AMMONIA

The literature contains at least several hundred reports of plasma ammonia levels, a significant fraction of which are in error. Errors arise from confusion as to units (e.g., ammonia vs. ammonia-N, NH_3 vs. T_{Amm}, μmoles vs. mmoles), analytical methodology (e.g., the use of chemical assays without deproteination, which results in great overestimation), and sampling methods (failure to freeze or assay samples immediately, which, again, results in great overestimation). Table 2 provides a selection of plasma values, chosen on the basis that these errors were likely avoided *and* that urea-N as well as ammonia-N levels were reported. However, two additional factors may have biased much of the data in Table 2 — the site and the method of blood sampling. Caudal or cardiac puncture has been the approach most commonly employed. Caudal puncture blood is usually an arterial-venous mixture, while cardiac puncture blood is venous. Venous blood contains 50 to 600% more T_{Amm} than arterial blood (Goldstein et al., 1964; Vellas and Serfaty, 1974; Walton and Cowey, 1977; Cameron and Heisler, 1983; Wright and Wood, 1985; Ogata and Murai, 1988). Furthermore, venipuncture has usually been performed on anesthetized, stunned, or restrained fish, often with air exposure. This will invariably elevate plasma T_{Amm}, due to the release of ammonia produced by adenylate deamination in hypoxemic muscle, and/or interference with branchial excretion (Wood et al., 1989b). True resting values can only be obtained via chronic indwelling catheters; such measurements are available for only a few species. Based on the limited information available (Table 2; also Cameron and Heisler, 1983; Wright and Wood, 1985; Cameron, 1986; Ogata and Murai, 1988; Heisler, 1990; Wilson and Taylor, 1992), most of which is for salmonids, the author's opinion is that *normal* arterial T_{Amm} levels are certainly less than 0.5 $mmol \cdot l^{-1}$ and usually closer to 0.1–0.2 $mmol \cdot l^{-1}$ in resting, nonfeeding fish living in flowing water. Clearly, there is a need for work on a wider range of species to substantiate or disprove this view.

A variety of endogenous and exogenous factors which alter the dynamic balance between production and excretion will affect plasma T_{Amm}. The many influences on excretion are considered below; changes in production will be discussed here. Increases in deamination rate may be caused by elevations in temperature (Maetz, 1972; Guerin-Ancey, 1976) or endocrine factors, such as the proteolytic effects of cortisol (Freeman and Idler, 1973). Unfortunately, there appear to be no data on how plasma T_{Amm} levels respond to these stimuli, though it is known that both plasma T_{Amm} and plasma cortisol levels increase during the terminal spawning migration of Pacific salmon (Mommsen et al., 1980). Acute environmental hypercapnia depressed plasma T_{Amm} in the lemon sole (*P. vetulus*; Wright et al., 1988a) and rainbow trout (*O. mykiss*; Iwama and Heisler, 1991), and lowered J_{Amm} in the toadfish (*Opsanus beta*; Evans et al., 1982) and conger eel (*Conger conger*; Toews et al., 1983). In total, these data suggest that high P_{CO2} inhibits ammonia production. Feeding, which is the single largest influence on J_{Amm} (see section above on Ammonia Production) caused plasma T_{Amm} to more than double after a meal in one study on trout where sampling occurred by caudal puncture (Kaushik and Teles, 1985). This observation should be corroborated in fish fitted with in-dwelling catheters (e.g., Ash et al., 1989). Exercise is another condition in which an excess of production over excretion can elevate blood ammonia. After exhaustive exercise in trout, plasma T_{Amm} increases 3- to 4-fold, and stays elevated for several hours (Wood, 1988; Wright and Wood, 1988; Mommsen and Hochachka, 1988). During this period, excess ammonia excretion to the water exceeds whole-body ATP depletion (Scarabello et al., 1992). Thus, the combination of increased ammonia production from amino acid deamination and adenylate deamination overwhelms the excretory mechanisms.

AMMONIA EXCRETION

Division among Sites and Possible Importance of the Skin

The predominant role of the gills in ammonia (and urea) excretion was first established by the classic experiments of Smith (1929a) on freshwater carp (*C. carpio*) and goldfish (*C. auratus*). Smith employed two different techniques to separate the contributions of gills and kidney. In the first, a rubber dam partitioned the head and gills from the rest of the fish in a "divided box", such that branchial excretion would occur into the anterior chamber, and urinary, intestinal, and skin excretion into the posterior chamber. In the second, the bladder was catheterized to collect the urinary excretion separately, while branchial, intestinal, and skin fluxes were measured together in the water. The results of the two approaches were very similar, *both alone and in combination*, indicating that only branchial and renal routes were important, and that the gills were responsible for at least 85% of whole-body J_{Amm}. These same approaches have since been duplicated on many different species.

The data in Table 1 have been chosen on the basis that both ammonia-N and urea-N excretions were measured through both routes. With very few exceptions, they are in broad agreement with Smith's conclusion as to the dominance of the gills. There is, however, an *apparent* tendency for greater percentage excretion of ammonia through the kidney in seawater than in freshwater fish. The conclusion is almost certainly an artifact of methodology. If this tendency were real, it would be most surprising, inasmuch as urine flow rate is much lower in seawater fish. Note in Table 1 that virtually all seawater data have been obtained with the divided box rather than by urinary catheterization. Thus, the skin and intestinal output would be assigned to the kidney route. However, when urine flow was collected directly by bladder catheterization in the lemon sole (*P. vetulus)*, it accounted for <1% of whole-body J_{Amm} (McDonald et al., 1982). In another marine flatfish, the dab (*Limanda limanda*), Sayer and Davenport (1987a) isolated the majority of the skin surface in a divided chamber separate from the gills, anus, and urinary papilla. They found that 48% of both whole-body J_{Amm} and total N-excretion occurred through this skin. In two amphibious mudskippers living in dilute seawater, about 15% of whole body J_{Amm} originated from the skin, based on combined divided chamber and urinary blockage experiments (Morii et al., 1978). In freshwater lamprey (*Entosphenus tridentatus*), the skin accounted for about 8% of J_{Amm} (Read, 1968). The obvious conclusion is that the skin makes a larger contribution in seawater than freshwater fish; the kidney is likely no more important (probably less important) in seawater fish. There is a need for studies on other marine and euryhaline species using urinary catheterization *in combination with* divided chambers to validate these ideas, and to characterize the mechanisms of skin excretion.

Mechanisms of Branchial Ammonia Excretion

The mechanism(s) of branchial ammonia excretion remain highly controversial (e.g., Maetz and García-Romeu, 1964; Maetz, 1972, 1973; Cameron and Heisler, 1983, 1985; Wright and Wood, 1985; Cameron, 1986; McDonald et al., 1989; Heisler, 1990). However, it seems to this reviewer that inadequate attention has been devoted to two issues. The first is the importance of gill boundary layer acidification by CO_2 excretion (and also H^+ transport processes) as a mechanism for diffusion trapping of NH_3 as NH_4^+, thereby sustaining the P_{NH3} gradient for diffusive efflux of NH_3. This was first proposed by Randall and Wright (1987) and later demonstrated by Wright et al. (1989); water passing over the gills may be acidified by up to 1.5 pH units (Wright et al., 1986; Playle and Wood, 1989; Lin and Randall, 1990). No analysis in intact fish has yet taken this into account; instead gradients are invariably defined with respect to the bulk water (e.g., Figures 4 and 5). Failure to use true gill water pH may result in large errors in P_{NH3} gradient calculations, depending on inspired pH, water buffer capacity, ventilatory flow, and fish metabolic rate (i.e., $\dot{M}_{CO2}$ and $\dot{M}_{H+}$). Furthermore, the abnormal pattern and rate of ventilatory flow, $\dot{M}_{CO2}$, and $\dot{M}_{H+}$ typical of most perfused and irrigated gill preparations make extrapolation to the intact animal problematic at best. The second issue is the simple recognition that the quantitative contribution of various mechanisms may vary greatly depending on the fish's acid-base status, the blood ammonia load, and the prevailing environmental conditions (Wright and Wood, 1985; Heisler, 1990). These are considered in greater detail in the following section.

Here it is sufficient to note that evidence exists for at least four mechanisms of ammonia extrusion (e.g., Evans and Cameron, 1986), as summarized in Figure 3: (1) NH_3 diffusion along a P_{NH3} gradient from blood to water; (2) electroneutral Na^+/NH_4^+ exchange located on the apical membranes of branchial epithelial cells; (3) a similarly located electroneutral H^+/NH_4^+ exchange; and (4) NH_4^+ diffusion along an electrochemical gradient from blood to water. If there is any general consensus, it would be that the first two are the principal mechanisms under most normal conditions, and that mechanism (4) is only important for teleosts in seawater, where cationic permeability is much higher than in fresh water. The term "Na^+/NH_4^+ exchange" is used here in a general sense, recognizing that its exact nature remains uncertain. Thus, NH_3 diffusion stoichiometrically tied to a classic Na^+/H^+ exchange (Kinsella and Aronson, 1981) or a primary H^+-ATPase/Na^+ channel (Avella and Bornancin, 1989; Lin and Randall, 1991) by a local pH gradient would also manifest as electroneutral Na^+/NH_4^+ exchange. Because metabolism produces NH_3 on a *net* basis, note that J_{Amm} resulting from mechanisms (1) and (3) will have no *net* acid-base effect, whereas J_{Amm} by

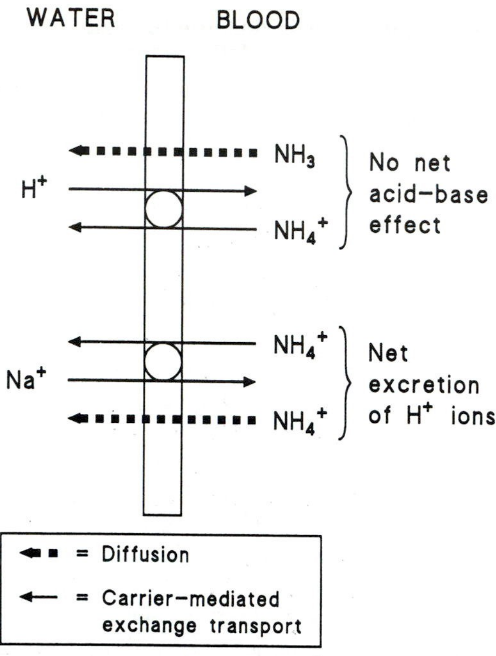

FIGURE 3. Summary of mechanisms by which ammonia may be excreted from blood to water across the mucosal surface of the gill epithelium, and their consequences for the acid-base status of the fish.

mechanisms (2) and (4) will achieve *net* excretion of H^+ ions. At least in marine fish, there is evidence of two additional mechanisms for entry through the basolateral membrane: (5) substitution of NH_4^+ for K^+ on basolateral Na^+/K^+-ATPase in teleosts (e.g., Claiborne et al., 1982; Evans et al., 1989); and (6) substitution of NH_4^+ for K^+ on basolateral Na^+,K^+,2Cl cotransporters in elasmobranchs (e.g., Evans and More, 1988). Their acid-base consequences will depend on whether NH_4^+ or NH_3 leaves across the apical membrane.

Environmental Influences on Branchial Ammonia Excretion

Low Environmental pH

An acute reduction in J_{Amm} occurs over the first few hours of exposure of freshwater fish to pH 4.0 to 5.5, resulting in an increase in plasma T_{Amm} (McDonald et al., 1983; Wright and Wood, 1985; Lin and Randall, 1990). This short-term inhibition is surprising, inasmuch as it is opposite to the predicted increase in NH_3 "diffusion trapping" by the acidic external water. Inflammatory swelling and mucification of the gill epithelium may play a role by increasing the mean blood-to-water diffusion distance and resistance. Most probably, however, a blockade of Na^+/NH_4^+ exchange is also involved, for high water $[H^+]$ is well known to competitively inhibit Na^+ uptake (Wood, 1989).

Over longer-term exposure to severe acidity, J_{Amm} recovers and becomes elevated above the control level (McDonald and Wood, 1981; McDonald, 1983). As this increase in J_{Amm} is chronic (e.g., over 81 d; Audet et al., 1988), chronic elevation of ammonia *production* must be occurring simultaneously, probably due to the catabolic effects of prolonged cortisol mobilization (Freeman and Idler, 1973; Audet et al., 1988, in press). Recovery of Na^+ uptake during this period is at best partial (McDonald et al., 1983; Audet et al., 1988), so Na^+/NH_4^+ exchange alone cannot account for the restoration of J_{Amm}. The diffusion trapping mechanism may become important at this time, in combination with elevated internal ammonia levels which would enhance any of the mechanisms shown in Figure 3. Many freshwater fish are endemic to highly acidic waters — e.g., peatbogs, tropical blackwaters — but to date there have been no investigations on N-waste excretion in such species.

High Environmental pH

The effects of water pH 8.5 to 10.5 on branchial J_{Amm} are surprisingly similar to those of low pH, and some of the same mechanisms may be involved. Exposure of salmonids to an alkaline medium causes an immediate reduction of J_{Amm} and resulting increase in plasma T_{Amm} (Figure 4A, B; Wright and

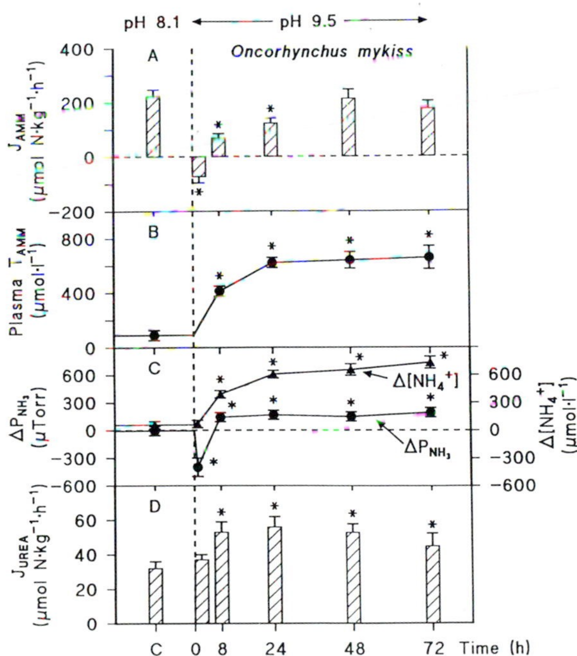

FIGURE 4. The responses of the freshwater rainbow trout (*Oncorhynchus mykiss*) to acute exposure to high environmental pH (9.5) for a 72-h period. (A) Ammonia excretion; (B) plasma total ammonia concentration; (C) the P_{NH3} gradient and the chemical gradient for NH_4^+ from blood to bulk water; and (D) urea excretion. Means ±1 SEM (N = 7 to 20); asterisks indicate significant differences from the control values. (Redrawn from Wilkie and Wood, 1991.)

Wood, 1985; Lin and Randall, 1990; Wilkie and Wood, 1991; Wright et al., 1993; Wilkie et al., 1993; Yesaki and Iwama, 1992). This is the response expected from inhibition of NH_3 diffusion. At high water pH close to or above the pK of the NH_3/NH_4^+ equilibrium, a large fraction of the ammonia immediately outside the branchial surface will be in the NH_3 form, thereby reducing the blood-to-water P_{NH3} gradient. Indeed reversal of the P_{NH3} gradient (Figure 4C) and transitory uptake of ammonia (Figure 4A) were seen by Wilkie and Wood (1991). Yesaki and Iwama (1992) reported that increases in water hardness (i.e., Ca^{2+}) help protect against the inhibition of J_{Amm} by high water pH in *O. mykiss*, but the mechanism of this curious effect is unknown. As with low pH, inflammation and mucification of the branchial surface at high pH may also contribute to the inhibition of J_{Amm}. However, more importantly, like low pH, high pH also causes a severe reduction in branchial Na^+ influx and therefore a likely reduction in Na^+/NH_4^+ exchange (Wright and Wood, 1985; Wood, 1989). Although this has been interpreted as a direct action of external alkalinity on the gill transport site, an indirect effect acting through elevated blood pH cannot be ruled out. In both the coho salmon (*O. kisutch)* and the starry flounder (*Platichthys stellatus*), J_{Amm} fell by up to 50% during an internal metabolic alkalosis induced by sodium lactate infusion, *in the absence of any change in water pH* (Milligan et al., 1991).

During prolonged exposure to high water pH, J_{Amm} fully recovers over 8 to 48 h and plasma T_{Amm} stabilizes at an elevated level (Figure 4A, B; Wilkie and Wood, 1991; Wilkie et al., 1993). The large rise in plasma T_{Amm} in itself is probably an important factor in the restoration of J_{Amm}. Note, however, that the time course of this recovery (Figure 4A) appears slower than the restoration of the P_{NH3} gradient (Figure 4C), suggesting that mechanisms in addition to NH_3 diffusion are involved. Plasma Na^+ stabilizes, but it is not yet known whether Na^+ influx recovers during prolonged exposure. Yesaki and Iwama (1992) report that amiloride inhibits the restoration of J_{Amm}, suggesting that Na^+/NH_4^+ or H^+/NH_4^+ exchanges may be involved. More work is clearly needed.

Recently, N-waste excretion has been studied in several teleosts endemic to highly alkaline lakes. The Lahontan cutthroat trout (*O. clarki henshawi)* of Pyramid Lake, Nevada (pH 9.4), has a lower overall rate of N-excretion than other salmonids, and relies on the kidney for about 15% of total J_{Amm}, an unusually high figure (Wright et al., 1993). Nevertheless, the low branchial J_{Amm} appears to occur almost entirely by diffusion of NH_3, facilitated by the high pH and P_{NH3} of plasma. There is also a greater reliance on urea-N excretion, a characteristic shared with many air-breathing fish which also have difficulty in excreting ammonia-N (Table 3). In both *O. clarki henshawi* and the cyprinid *Chalcalburnus tarichi* (Danulat and Kempe, 1992) of Lake Van, Turkey (pH 9.8), ammonia-N accounts for only about 65% of N-excretion, and urea-N for 35%, whereas the division is more normally >80% ammonia-N in most aquatic teleosts (Table 3). The extreme example of this strategy is the tilapia, *Oreochromis alcalicus grahami,* of Lake Magadi, Kenya (pH 10.0), which excretes no ammonia-N at all, and relies entirely on urea-N excretion (Randall et al., 1989; Wood et al., 1989b).

High Environmental Ammonia

The effects of high water T_{Amm} on branchial J_{Amm} are similar to those of high pH. Initially, an inhibition of ammonia excretion and/or net ammonia entry occurs; in consequence, plasma T_{Amm} increases (Figure 5A, B; Fromm and Gillette, 1968; Cameron and Heisler, 1983; Cameron, 1986; Claiborne and Evans, 1988; Wilson and Taylor, 1992; Wright et al., 1993). Entry appears to be largely due to a reversal of the normal P_{NH3} gradient from blood to water (Figure 5C), which explains the well-known protective effect of acidic environmental pH against toxicity from high environmental ammonia. Other mechanisms contributing to plasma T_{Amm} build-up could include inhibition or reversal

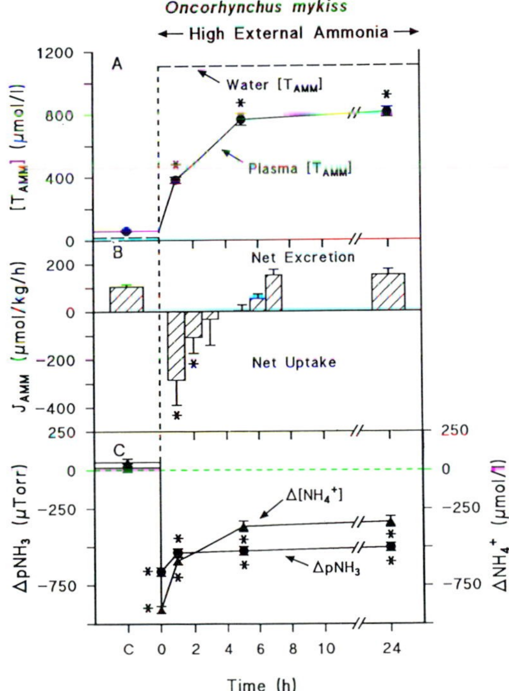

FIGURE 5. The responses of the freshwater rainbow trout (*Oncorhynchus mykiss*) to acute exposure to high environmental ammonia (~1100 $\mu mol \cdot l^{-1}$) for 24 h. (A) Plasma and water total ammonia levels; (B) ammonia excretion; (C) the P_{NH3} gradient and the chemical gradient for NH_4^+ from blood to bulk water. Means ±1 SEM (N = 16);asterisks indicate significant differences from the control values. (Unpublished data of R. W. Wilson, P. M. Wright, R. S. Munger, and C. M. Wood.)

of Na^+/NH_4^+ exchange (Maetz and García-Romeu, 1964) or NH_4^+ diffusion (Wilson and Taylor, 1992). If the increase in plasma T_{Amm} is not so large as to result in toxicity, the level stabilizes within a few hours (Figure 5A), branchial J_{Amm} resumes (Figure 5B), and a new steady state is established. Interestingly, this new stable plasma T_{Amm} may be well below water T_{Amm} (Figure 5A) such that ammonia is *apparently* excreted against both P_{NH3} and $[NH_4^+]$ gradients (Figure 5C). The explanation remains uncertain. Originally, Cameron and Heisler (1983) proposed an activation of Na^+/NH_4^+ exchange in the trout in fresh water, but provided no direct evidence. However, in a more detailed examination of the catfish (*I. punctatus*), Cameron (1986) concluded that the response was consistent with activation of H^+/NH_4^+ exchange (which is acid-base neutral), rather than Na^+/NH_4^+ exchange. Based on an analysis of blood acid-base changes, Wilson and Taylor (1992) proposed that activation of H^+/NH_4^+ exchange occurred in freshwater trout and Na^+/NH_4+ exchange in seawater trout. More work is needed. Whatever the final explanation, the phenomenon has great practical importance. The very act of confining a fish to a closed water volume so as to measure its J_{Amm} will result in a build-up of T_{Amm} in the water and blood, accompanied by a decline and perhaps a later recovery in J_{Amm}.

Salinity

The reported effects of salinity changes on J_{Amm} are surprisingly small, considering that salinity will alter the pK and αNH_3 of ammonia, the availability of Na^+ for Na^+/NH_4^+ exchange, the permeability of paracellular pathways in the gills to cations such as NH_4^+, and very likely the pH of the water. The intertidal, amphibious blenny (*Blennius pholis*) subjected to an experimental salinity cycle exhibited insignificant change in J_{Amm} but greatly increased J_{Urea} in fresh water (Sayer and Davenport, 1987b; Table 3). Gordon et al. (1965) acclimated another intertidal amphibious species (*Periophthalmus sobrinus*) to 100% and 40% seawater for several days; again J_{Amm} did not change significantly, but J_{Urea} *declined* markedly in the lower salinity! Acclimation of the ammoniotelic freshwater stingray *Potamotrygon* to 40% seawater (Gerst and Thorson, 1977) and the ureotelic marine skate *Raja erinacea* to 50% seawater (Goldstein and Forster, 1971b) again had no effect on J_{Amm}. A possible explanation for this lack of effect of salinity on J_{Amm} would be compensating changes in excretion mechanisms. For example, Evans et al. (1989) reported that NH_4^+ permeability decreased but NH_3 permeability increased in perfused head preparations when marine toadfish (*O. beta*) were acclimated to 5% seawater.

Renal Excretion

Substantial amounts of glutaminase, glutamate dehydrogenase, and aminotransferases are found in the kidneys of trout (Walton and Cowey, 1977) and other fish (King and Goldstein, 1983a, b). However, renal excretion accounts for no more than 15% of whole-body J_{Amm} in fresh water (Smith,

1929a; Wood, 1988), and an even lower percentage in seawater fish, where urine flow rate is much reduced (McDonald et al., 1982). Diuresis due to increases in glomerular filtration rate tends to elevate urinary J_{Amm} slightly (King and Goldstein, 1983b; Curtis and Wood, 1992). However, by far the largest influence on urinary J_{Amm} is systemic acid-base status. An increase in renal NH_4^+ excretion, largely by secretion, has been seen in virtually every study where a prolonged blood acidosis has been induced (e.g., McDonald and Wood, 1981; Cameron and Kormanik, 1982; Wheatly et al., 1984). This was true even in two seawater forms, the teleost *P. vetulus* (McDonald et al., 1982) and the elasmobranch *S. acanthias* (King and Goldstein, 1983a) in which the absolute magnitude of the response was too small to be of any real significance for whole-body acid-base balance. The contribution of the NH_4^+ component to net renal H^+ excretion varied from 1 to 90% among the various studies, but in all, the peak NH_4^+ excretion was virtually coincident with the time of lowest urine pH, consistent with the classic diffusion trapping model of tubule function in higher vertebrates (Pitts, 1974).

UREA CHEMISTRY

On a simple energetic basis, urea is a less efficient N-product than ammonia (15.5 kJ/g protein; Smith and Rumsey, 1976); the synthesis of urea costs at least 2 ATPs per unit N. However, under conditions where ammonia excretion is difficult, or there is a need for a relatively nontoxic osmotic effector, the benefits of urea production outweigh the energetic disadvantage. Unlike ammonia, urea exists in only one form, an uncharged molecule which can exert no direct electrical, acid-base, or gaseous effects; dissociation to ammonium cyanate is negligible in aqueous solution. Another important aspect of its much lower toxicity is the fact that urea is rather inert biochemically. The enzyme urease, which degrades urea to ammonia and CO_2, is not synthesized by vertebrates. High concentrations of urea do tend to destabilize macromolecules such as proteins. However methylamine compounds derived from choline, such as betaine, sarcosine, and trimethylamine oxide (TMAO), counteract this effect, which probably explains why these compounds are also retained by many ureosmotic regulators (Yancey and Somero, 1979).

The urea molecule is a dipole with a low lipid vs. water partition coefficient (0.00015). Passive movement through cell membranes is thought to occur almost entirely via aqueous pores. Both the aqueous diffusion coefficient and the absolute solubility of urea in water are similar to, but slightly lower (~75%) than, those of ammonia. Thus, the common observation that biological membranes are *much* less permeable to urea than ammonia must reflect a lower solubility and diffusivity of urea in lipoprotein membranes. Current uncertainties about the true values of these properties for ammonia were noted earlier.

The acid-base consequences and control of urea synthesis remain highly controversial (e.g., Campbell, 1973; Halperin et al., 1986; Meijer et al., 1990;

Atkinson, 1992). Without minimizing the important role that urea synthesis may play in acid-base adjustment under non-steady-state conditions, it must be pointed out that the common statement that "urea production results in a net formation of H^+" is misleading *under steady-state conditions*. Recall that the normal aerobic metabolism of proteins and amino acids produces equimolar amounts of NH_4^+ and HCO_3^- — i.e., NH_3 and CO_2 on a *net* basis — which is neutral with respect to acid-base balance. The *net* reaction of urea production is:

$$2NH_4^+ + 2HCO_3^- ----> NH_2 - CO - NH_2 + 3H_2O + CO_2 \quad (7)$$

The process consumes equal amounts of NH_4^+ and HCO_3^-, which represent the proportions in which they were produced, and is neutral with respect to whole animal acid-base balance (assuming CO_2 can be removed by ventilation). Ureagenesis can be considered as an ATP-driven H^+ pump, moving protons from a metabolic acid to a metabolic base, a reaction which cannot occur directly because NH_4^+ is too weak an acid (pK 9.5) to titrate HCO_3^- (pK 6.1).

UREA INTERNAL DISTRIBUTION

Most measurements suggest that intracellular concentrations of urea are roughly equivalent to extracellular concentrations in both chondrichthyes (Smith, 1936; Robertson, 1975, 1976, 1989; Forster and Goldstein, 1976) and osteichthyes (Smith, 1930; Wood et al., 1989b; Danulat and Kempe, 1992). In the dogfish (*S. acanthias*), ^{14}C-urea infused into the bloodstream slowly equilibrated into the water spaces of various body compartments (Fenstermacher et al., 1972). It is therefore surprising that Saha and Ratha (1989) report $[\text{Urea}]_i/[\text{Urea}]_e$ ratios as high as 15, with great tissue-to-tissue variability, in five Indian teleosts. All of these species are capable of both air breathing and active ureagenesis by the ornithine-urea cycle, so selective urea retention may be associated with cell volume regulation and/or ammonia detoxification during air exposure. This finding deserves further experimental analysis, especially in light of subsequent discussion on possible transport mechanisms contributing to whole-body urea retention in fish.

UREA PRODUCTION

Urea is produced by all fish but is the major end product of N-metabolism only in marine chondrichthyes, dipnoans during estivation, a very few teleosts living in extreme environments, and probably the coelacanth (Tables 1 and 3).

Pathways

The ornithine-urea cycle (OUC; Figure 6A) and uricolysis (Figure 6D) are the two basic routes of urea production; hydrolysis of dietary arginine to urea

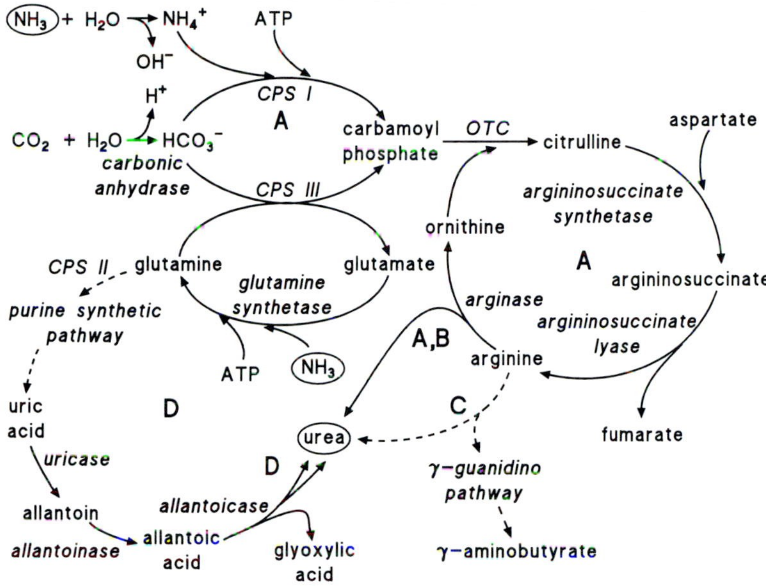

FIGURE 6. Biochemical pathways in fish for the production of urea via (A) the ornithine urea cycle; (B) arginase; (C) the gamma-guanidino urea hydrolase pathway, and (D) uricolysis, and their interrelationships.

and ornithine by arginase is one step of the OUC (Figure 6B) which may occur even if the rest of the cycle is incomplete, as in agnathans and most osteichthyes. The gamma-guanidino urea hydrolase pathway (Figure 6C) may also yield urea from arginine, and is known to be present in carp (*C. carpio*; Vellas and Serfarty, 1974) but its physiological significance remains uncertain. Arginase is ubiquitous (Read, 1968; Cvancara, 1969b; Saha and Ratha, 1989) and the uricolytic enzymes seem to be present in most fish (Brunel, 1937; Goldstein and Forster, 1965; Cvancara, 1969a) but absent in agnathans (Florkin and Duchateau, 1943; Read, 1975).

The irregular occurrence of a functional OUC is now well documented — present in all chondrichthyes, but restricted to the coelocanth, the lungfish, and a very few teleosts within the osteichthyes. Earlier it was thought that the genes for the OUC had been suppressed or deleted in teleosts (Cohen and Brown, 1960), but we now know that this is not the case. Trace levels of some or all OUC enzymes have been found in "standard" teleosts (Huggins et al., 1969; Wilson, 1973; Chiu et al., 1986; Dépêche et al., 1979). More importantly, a very active and complete hepatic OUC has now been found in several air-breathing teleosts of the Indian subcontinent (Saha and Ratha, 1987, 1989), the toadfishes (Read, 1971a; Mommsen and Walsh, 1989), and the tilapia (*O. alcalicus grahami*) of highly alkaline Lake Magadi, Kenya (Randall et al., 1989; Wood et al., 1989b; Table 3). A common circumstance linking these fish is that environmental conditions are occasionally or continually unfavorable for ammonia-N excretion. Indeed the Magadi tilapia excretes no ammonia-N at all because of the high external pH (10).

Two key features of the OUC in most fish differ from those in higher vertebrates (lungfish, amphibians, reptiles, and mammals): the glutamine dependency of the cycle (vs. ammonia dependency) and the intramitochondrial location (vs. cytosolic) of arginase. The OUC appears to have evolved as a monophyletic trait in vertebrates (Mommsen and Walsh, 1989). The fact that the coelacanth has the basic fish pattern, while the lungfish has the pattern of higher vertebrates, brings into question the pivotal role of the coelocanth in tetrapod evolution, and suggests that the lungfish led the way to higher vertebrates. This conclusion has also been supported by direct sequencing of mitochondrial DNA using the polymerase chain reaction (Meyer and Wilson, 1990).

There are five key steps in the OUC (Figure 6A), each catalysed by a specific enzyme: (1) the direct fixation of NH_4^+ and HCO_3^- together with phosphate bond energy by carbamoyl phosphate synthetase I (CPS I) which occurs in the dipnoans and higher vertebrates *or* the indirect fixation of NH_4^+ from glutamine by CPS III, which occurs in all fish possessing the OUC except the lungfish; (2) the reaction of the product, carbamoyl phosphate, with ornithine, catalyzed by ornithine carbamoyl transferase (OTC); (3) the fixation of a second amino group, donated by aspartate, under the influence of argininosuccinate synthetase; (4) cleavage to arginine and fumarate by argininosuccinate lyase; and (5) hydrolysis of arginine by arginase which produces urea and regenerates the ornithine backbone. In fish, a sixth step is essential and may be considered a functional part of the OUC: the fixation of ammonia by glutamate to form glutamine, under the influence of glutamine synthetase. The activity of this enzyme is much higher in the livers of OUC fish than non-OUC fish (Webb and Brown, 1980), and its K_m for ammonia is much lower than in higher vertebrates (Shankar and Anderson, 1985). In most fish, glutamine synthetase occurs in the mitochondria, rather than in the cytosol as in higher vertebrates. Glutamine synthetase thereby serves as a highly efficient trapping mechanism to scavenge blood ammonia and channel it directly into the OUC via mitochondrial glutamine. This likely avoids the large-scale utilization of glutamine for other purposes by cytosolic enzymes, such as CPS II which is involved in purine and pyrimidine synthesis. Other "feeder enzymes" of direct relevance to the OUC are carbonic anhydrase, to supply HCO_3^- from CO_2 within the mitochondrial matrix (Walsh et al., 1989), and aminotransferases for regulating the supply of aspartate.

Uricolysis (Figure 6D) is a three-step pathway which occurs largely in peroxisomes (Hayashi et al., 1989): (1) conversion of uric acid to allantoin by uricase; (2) subsequent hydration to allantoic acid by allantoinase; and (3) hydrolysis to glyoxylic acid plus 2 molecules of urea by allantoicase. The uric acid substrate is derived from purines via cytosolic enzymes. While purines are thought to result mainly from the metabolism of nucleic acids and high-energy phosphates, they may also be formed *de novo* by the condensation of glutamine, aspartate, and glycine. In carp (*C. carpio*), Vellas and Serfaty (1974) found that perfusion of glutamine into an *in situ* liver preparation stimulated both urea and uric acid release. The cytosolic, glutamine-dependent CPS II required for this synthesis has been documented in the liver of largemouth bass (*Micropterus salmoides*; Cao et al., 1991).

Comparisons of the rates of urea synthesis by the OUC vs. by uricolysis have been made via radiotracer techniques applied to *in vitro* liver preparations of two ureagenic species, the dogfish (*S. acanthias*; Schooler et al., 1966) and the lungfish (*Protopterus dolloi*; Forster and Goldstein, 1966) and one "standard" teleost (*C. carpio;* Vellas and Serfaty, 1974). In both ureagenic species, urea production *in vitro* by the OUC was about 100-fold faster than by uricolysis, and a preliminary *in vivo* experiment on the dogfish suggested at least a 10-fold difference. In the carp, urea was produced only by uricolysis.

SITES

For all fish that have been examined, and for all urea-producing pathways (OUC, arginolysis, uricolysis), highest enzyme activities have been found in the liver. Furthermore, incubation of liver slices, homogenates, or hepatocytes with realistic levels of appropriate substrates yielded urea production rates which appeared reasonable relative to measured whole animal J_{Urea} (Goldstein and Forster, 1965; Forster and Goldstein, 1966; Schooler et al., 1966; Funkhouser et al., 1972; Gregory, 1977; Walsh et al., 1989). In consequence, little attention has been paid to other possible production sites. For uricolysis, it is probably safe to assume that extrahepatic production is minimal; surveys of a variety of tissues (including kidney) in several teleosts have found uricolytic activity *only* in the liver (Cvancara, 1969a; Vellas and Serfaty, 1974; Saha and Ratha, 1987). However arginase activity occurred in every tissue except brain, and kidney levels were only slightly lower than those in liver (Cvancara, 1969b; Saha and Ratha, 1989). Saha and Ratha (1986, 1987, 1989) also found that the activities of most OUC enzymes were comparable in kidney and liver in several air-breathing teleosts. The only physiological study on sites of urea production appears to be that of Vellas and Serfaty (1974) on carp; they concluded that the liver was dominant, but the kidney made a small contribution via arginolysis only.

FUELS

The fuel source largely depends on the metabolic pathway. In the OUC (Figure 6A), the proximate fuels which supply amino groups are glutamine and aspartate, except in dipnoans, where ammonia-N is fixed directly. Glutamine and aspartate may originate in the diet or be synthesized from other l-amino acids. This scenario supposes that N from dietary protein is processed mainly by the OUC. However, in marine elasmobranchs, Mommsen and Walsh (1991) have suggested that the OUC supplies only the urea needed to satisfy osmoregulatory purposes, that after a meal most waste-N is excreted directly as ammonia, and that any additional urea-N excreted comes from arginolysis or uricolysis. There are no known experimental data to support these hypotheses, but virtually all work to date has been done on starved animals. Clearly, there is a need for feeding studies on ureosmotic regulators such as sharks (see Haywood, 1973).

Dietary arginine, an essential amino acid, could be a major source of urea; arginase (Figure 6B) is ubiquitous and the gamma-guanidino urea hydrolase pathway (Figure 6C) may also occur. If so, J_{Urea} might be expected to be markedly affected by feeding; urea-N excretion would represent part of the *exogenous* fraction of N-metabolism. This appears to be the situation in the seabass *(Dicentrarchus labrax*; Guerin-Ancey, 1976) and the rainbow trout (*O. mykiss*). In the latter, Kaushik (1980) reported a 75% decline in J_{Urea} during 28 d of starvation, and Beamish and Thomas (1984) found a marked variation in urea-N output with ration and diet. The other common pathway is uricolysis, in which the proximate fuel is the purine ring (Figure 6D). If purines are formed *de novo* from dietary amino acids, again a dependence of J_{Urea} on feeding might be expected. However, a more conventional purine source is the metabolism of native nucleic acids and nucleotides. Here we might expect J_{Urea} to be relatively independent of feeding, and urea-N excretion would represent part of the *endogenous* fraction. This appears to be the case in sockeye salmon *(O. nerka)*; Brett and Zala (1975) found that J_{Urea} (in contrast to J_{Amm}) remained constant throughout a diurnal feeding cycle, as well as during 22 d of starvation. The reasons for these very different patterns in congeneric salmonids are unclear.

BLOOD LEVELS OF UREA

The literature contains even more measurements of plasma urea than T_{Amm}, most of which appear to be reliable, apart from occasional confusion between urea and urea-N units. Urea is not rapidly generated or degraded by mishandling of blood samples. Concentrations are generally high relative to production/excretion rates, even in osteichthyes where urea is not a principal N-product, so biases due to sampling site or stress-induced production are minimal. The values in Table 2 have been selected on the basis of accompanying T_{Amm} data which appear reasonable and have been expressed as urea-N (i.e., urea $\times$ 2) to facilitate comparison. Additional plasma urea data can be found in heroic surveys by Holmes and Donaldson (1969), Griffith (1981), and Mangum et al. (1978).

For marine chondrichthyes, including the holocephalians (ratfish and chimaerids), living in 100% seawater, most data are in the range 600 to 800 mmol urea-N$\cdot l^{-1}$. Thus, 300 to 400 mosm$\cdot l^{-1}$, or about one third of plasma osmotic pressure is contributed by urea. Interestingly, the coelacanth (*L. chalumnae*), exhibits the same pattern. As first appreciated by Smith (1929b,

1936), this "retention" of urea, together with significant amounts of TMAO (contributing 50 to 200 mosm·l^{-1}) and somewhat higher levels of Na^+ and Cl^- (about 250 mosm·l^{-1} each) than in marine teleosts, is a strategy to raise plasma osmotic pressure slightly above that of ambient seawater. A potential drawback is destabilization of protein structure by such high levels of urea, but the simultaneous elevation of TMAO counteracts this influence; the normal 1:2 ratio of TMAO:urea in elasmobranchs (but not holocephalians; Read, 1971b) appears most effective in this regard (Yancey and Somero, 1979). Unlike the marine teleost, the urea-rich fish does not lose water across the gills, but rather gains small amounts by osmosis to allow urine formation. There is therefore no need to drink seawater with its attendant cost for excreting an extra NaCl load. Starvation, however, can impair this ability. Plasma urea and osmotic pressure levels fall gradually during fasting in marine elasmobranchs, and increase rapidly upon refeeding (Haywood, 1973; Leech et al., 1979).

Plasma urea and TMAO also fall when marine and euryhaline elasmobranchs move to more dilute media, and rise when they move to more concentrated media, but this serves as a mechanism of osmoconformity (Price, 1967). In some species, these adjustments are achieved solely by changes in clearance rates — e.g., the lemon shark, *Negaprion brevirostris* (Goldstein et al., 1968) — while in others, biosynthetic rates also change — e.g., the pyjama shark, *Poroderma africanum* (Haywood, 1973, 1974), the little skate, *R. erinacea* (Goldstein and Forster, 1971b; Payan et al., 1973), and the lipshark, *Hemiscyllium plagiosum* (Wong and Chan, 1977). Changes in clearance are largely due to alterations in renal function. Gill urea permeability is more or less unchanged such that branchial J_{Urea} remains proportional to plasma urea concentration.

Many species of elasmobranchs can inhabit fresh water, at least temporarily; Smith (1936) presented an impressive list. Urea concentrations only 20 to 50% of those in seawater residents are common. The sawfish, *Pristis microdon* (Smith and Smith, 1931), is a typical example (Table 2). The freshwater Amazonian rays (*Potamotrygon sp.*), which are completely isolated from the sea, represent the extreme. These fish are predominantly ammoniotelic (Table 1), have very low levels of plasma urea-N similar to those of osteichthyes (Table 2; Mangum et al., 1978), very low biosynthetic rates of urea, very limited capacity to reabsorb urea in the kidney, and negligible ability to raise plasma urea when challenged with salinity (Goldstein and Forster, 1971a; Gerst and Thorson, 1977). Like the marine chondrichthyes, the myxinoid Agnatha (marine hagfish) regulate a plasma osmotic pressure close to seawater. However, these animals do not retain urea but rather have very high levels of plasma Na^+ and Cl^- and the much lower plasma ureas characteristic of the petromyzonoid Agnatha (lampreys) and the osteichthyes (Table 2).

In general, plasma urea-N levels in osteichthyes are in the range of 1 to 10 mmol·l^{-1} (Table 2). Although these ammoniotelic forms produce only 5 to 40% as much urea-N as ammonia-N, they regulate internal levels which are up to 50-fold higher than plasma T_{Amm}! Even higher internal urea levels may occur during embryonic development (Dépêche et al., 1979; Griffith, 1991). The significance is unclear. Is this situation merely the result of a much lower gill permeability to urea than ammonia, or do these fish actively retain urea for some purpose? The origin of the great inter- and intraspecific variability in plasma urea levels (Table 2) is also unclear, because little is known about the factors regulating urea production and excretion in teleosts (see below). Water chemistry and salinity are influences which deserve investigation. The osmoregulatory contribution of such low levels of urea is debatable, but there appears to be a tendency for higher concentrations in seawater animals. The very lowest concentrations (<1 mmol urea-N$\cdot l^{-1}$) were reported in 10 different teleosts living in the very dilute, soft fresh water of the Amazon (Mangum et al., 1978). Very low levels were also seen in bowfin (*Amia calva*) living in similarly dilute fresh water (McKenzie and Randall, 1990). The low concentrations of calcium and the acidic pH of soft fresh water may play a role, for both will tend to increase diffusive permeability of the gill and cause diuresis (Wood, 1989). Teleosts living in alkaline lakes of intermediate salinity but very high pH (Randall et al., 1989; Wood et al., 1989b; Wright et al., 1993) exhibit high levels of plasma urea, with an extreme of 36.2 mmol urea-N$\cdot l^{-1}$ in the cyprinid *C. tarichi* of alkaline Lake Van, Turkey (pH 9.8; Danulat and Kempe, 1992). The possible interaction of acid-base status with urea biosynthesis may be an influence here.

Air exposure, which blocks ammonia excretion, and dehydration, which produces a compounding osmotic stress, may affect internal urea levels in air-breathing and intertidal species. However, the data are not entirely consistent. For several mudskippers, the balance of ammonia-N to urea-N excretion shifts strongly towards the latter immediately after reimmersion (Gordon et al., 1969, 1970, 1978). Morii et al. (1979) have disputed the results of Gordon et al. (1978) for one species, *P. cantonensis,* and have reported decreased plasma urea and J_{Urea} during and after air exposure, apparently due to hypometabolism. Davenport and Sayer (1986) also found no shift towards urea during air exposure in the blenny (*B. pholis*). Only in those air-breathers possessing a functional OUC is there an unequivocal shift. For example, the Indian air-breathing fish *H. fossilis* is predominantly ammoniotelic with typical levels of ammonia and urea in the bloodstream when living in water (Saha and Ratha, 1989; Tables 2 and 3). However, during water deprivation and/or hyperosmotic stress, urea accumulates internally, apparently through activation of the OUC. Experimental elevation of external, and therefore internal, ammonia levels produces a similar response (Saha and Ratha, 1986, 1990). The gulf toadfish (*O. beta*), which possesses a functional OUC, exhibits a similar activation of J_{Urea} in response to air exposure and hyperammonia stress, though increases in plasma urea have not been observed (Walsh et al., 1990).

The most dramatic examples of urea build-up are provided by estivating African lungfish, *Protopterus sp.*, which have an active OUC (Smith, 1930; Forster and Goldstein, 1966; Janssens and Cohen, 1966, 1968; Delaney et al.,

1977). In water, the lungfish is predominantly ammoniotelic with plasma ammonia and urea levels characteristic of other freshwater fish (Table 2). During estivation, ammonia production drops to negligible levels while urea production continues at rates actually lower than those in water due to hypometabolism. After a year's estivation, plasma urea-N levels may exceed 400 $mmol \cdot l^{-1}$, and ammonia toxicity is completely avoided. Upon return to water, urea is excreted in "spectacular quantities" (Smith, 1930), and the normal low values of J_{Urea} and plasma urea are reestablished in about a week. A similar strategy probably occurs in the South American lungfish (*Lepidosiren paradoxa*; Funkhouser et al., 1972), but not in the Australian lungfish (*Neoceratodus forsteri;* Forster and Goldstein, 1969) which does not estivate.

UREA EXCRETION

Division among Sites and Possible Importance of the Skin

The ratio of urea-N/ammonia-N in urine is routinely greater than in excretion from extrarenal sources (e.g., Goldstein and Forster, 1962; Saha et al., 1988); nevertheless, the gills predominate over the kidney as the major route of J_{Urea} in most fish (Table 1). The differences are not as clearcut as for ammonia, and the confounding influence of the skin, discussed above, may be one cause. Smith's (1929a) original work indicated that the skin was unimportant for J_{Urea} in freshwater teleosts, a conclusion which needs confirmation. However, in seawater teleosts, only the divided box approach has been used, and there appear to be no studies directly comparing branchial vs. renal J_{Urea} via urinary catheterization. Nevertheless, indirect comparisons indicate that the divided box may greatly overestimate renal J_{Urea}, and suggest that urea excretion through the skin may be quite significant in marine teleosts. Indeed Morii et al. (1978) for mudskippers, and Sayer and Davenport (1987a) for dab (*L. limanda*) concluded that the skin could account for 50% or more of whole-body J_{Urea}. More work is needed in this whole area of partitioning of J_{Urea} among branchial, renal, and skin routes. Those few teleosts in which high rates of J_{Urea} by the OUC can be induced may prove particularly interesting; toadfish (*O. beta*), for example, exhibit intermittent pulses of J_{Urea} (Walsh et al., 1990; Griffith, 1991), suggesting renal excretion via periodic micturation.

In both seawater and freshwater elasmobranchs, extrarenal excretion of urea predominates (Table 1), even though the latter forms have little capacity to reabsorb urea in the renal tubules (Goldstein and Forster, 1971a). Extrarenal J_{Urea} is commonly assumed to occur via the gills, but there are no data on possible skin contributions. Marine ureosmotic elasmobranchs reabsorb urea strongly at the kidney when in 100% seawater, but elevated renal clearance becomes the dominant pathway for dumping excess urea during adjustment to lower salinity. In the two other ureosmotic groups, excretion data are not available, but comparisons of urine and plasma indicate that urea is reabsorbed by the kidney of the holocephalian *Hydrolagus colliei* (Read, 1971b) but not the coelacanth *L. chalumnae* (Griffith et al., 1976).

Mechanisms of Branchial Urea Excretion

In teleosts, Smith (1929a) concluded that "excretion of urea by the gills can unquestionably be explained by diffusion", and this view has prevailed ever since. Indeed ^{14}C-urea has often been used as a diffusive marker for measuring gill functional surface area (e.g., Bergman et al., 1974). In ureosmotic elasmobranchs, the explanation for very low branchial permeability to urea has long been debated (*eg*. Smith, 1936; Griffith, 1991), especially in light of knowledge that active reabsorption in the kidney facilitates blood urea retention (see below). However, Boylan and colleagues performed the only detailed experiments on gill urea permeation, in the dogfish *S. acanthias* (summarized in Boylan, 1967). Possible transport mediation was evaluated by a variety of approaches: changes in temperature and plasma urea, metabolic blockers, and comparisons of urea vs. thiourea (a urea analogue poorly transported in dogfish kidney). They concluded that there was no evidence for transport mediation (inwards or outwards), and that the low permeability was simply of structural origin. Griffith (1991) has pointed to the low branchial surface area to body mass ratio in elasmobranchs (and the coelocanth) as an additional factor. Nevertheless, it remains difficult to understand how an epithelium can sustain a concentration gradient of 300 to 400 m*M* urea, while allowing the free passage of respiratory gases. Indeed even the observation (Table 2) that non-OUC teleosts maintain blood to water gradients of urea 5- to 50-fold greater than those for ammonia (despite much lower production rates) gives food for thought, as does the fact that injected ^{14}C-urea accumulated selectively in the gill chloride cells of the seawater eel, *Anguilla anguilla* (Masoni and Garcia-Romeu, 1972). Over the past decade, it has become apparent that carrier-mediated urea transport occurs in many epithelia of higher vertebrates; in most cases, earlier experiments had indicated passive diffusion alone (Marsh and Knepper, 1992).

Renal Urea Excretion

In ureosmotic marine elasmobranchs, active reabsorption of up to 95% of the filtered urea load by the kidneys has long been recognized (e.g., Smith, 1936). In teleosts, most emphasis has been placed on the kidney as an N-excretory route, and the possibility of renal conservation of urea has been overlooked. However urea, unlike other N-wastes, is not secreted by the kidneys of marine fish (Hickman and Trump, 1969), and comparison of urine and plasma data from diverse sources suggests that urea may be actively reabsorbed in a number of freshwater and marine species (e.g., Tables 1, 2). Recently this has been confirmed for rainbow trout in fresh water (Curtis and Wood, 1991, 1992) and cutthroat trout in alkaline Pyramid Lake (10% seawater; Wright et al., 1993). In both, urea concentrations in ureteral urine were less than 40% of plasma concentrations; in rainbow trout, net tubular reabsorption exceeded 70% and there was additional reabsorption across the urinary bladder. These observations help explain the apparent retention of urea in the plasma of teleosts noted earlier, reinforce the idea that it is a directed strategy, and again raise questions as to its purpose.

Nothing is known about tubular transport of urea in teleosts, but the phenomenon has been studied extensively in marine elasmobranchs. Hickman and Trump (1969) have reviewed early studies. Reabsorption is sensitive to the plasma-urine urea gradient, shows no transport maximum within the physiological range, and differs mechanistically from urea secretion in frog kidney. Transport is stoichiometrically coupled (1.6 urea:1 Na^+) to Na^+ transport (Schmidt-Nielsen et al., 1972), accepts the urea analogues methylurea and acetamide, but not thiourea (Schmidt-Nielsen and Rabinowitz, 1964), and is associated with a countercurrent exchange system (Lacy et al., 1985). During adaptation to more dilute environments, renal urea excretion increases dramatically as a mechanism to lower plasma osmotic pressure. The phenomenon is due to both an increase in glomerular filtration rate and a reduction in tubular urea transport (Goldstein and Forster, 1971a and b; Schmidt-Nielsen et al., 1972; Payan et al., 1973; Wong and Chan, 1977). The proximate controls on urea reabsorption remain unclear.

Environmental Influences on Urea Excretion

The as yet poorly understood influences of feeding and salinity on J_{Urea} have been discussed earlier.

High Environmental Ammonia

Severalfold increases in J_{Urea} in response to high external ammonia have now been documented in three of those unusual teleosts which possess a functional OUC — the Magadi tilapia, *O. alcalicus grahami* (Figure 7A; Wood et al., 1989b), the gulf toadfish, *O. beta* (Walsh et al., 1990), and the Indian air-breathing fish, *H. fossilis* (Saha and Ratha, 1986, 1990). Acid-base consequences, if any, are unknown. If ammonia enters as NH_4^+, the process will be acidifying due to the removal of "extra" HCO_3^- by urea synthesis (Equation 7), but in the more likely event of NH_3 entry, the process will be acid-base neutral. The time course of the ureagenic response is rapid (a few hours) in the tilapia (Figure 7A) and toadfish. Presumably the OUC enzymes are already present and fully active in the totally ureotelic tilapia and partially ureotelic toadfish. Entering and/or retained ammonia would be scavenged by mitochondrial glutamine synthetase, and the increased supply of glutamine would drive the OUC (Figure 6), thereby serving as a detoxification mechanism. Endocrine mediation is probably unimportant, for Mommsen et al. (1991, 1992) found a complete absence of OUC effects when a variety of hormones were applied acutely to intact toadfish or their hepatocytes. However, the ureagenic response

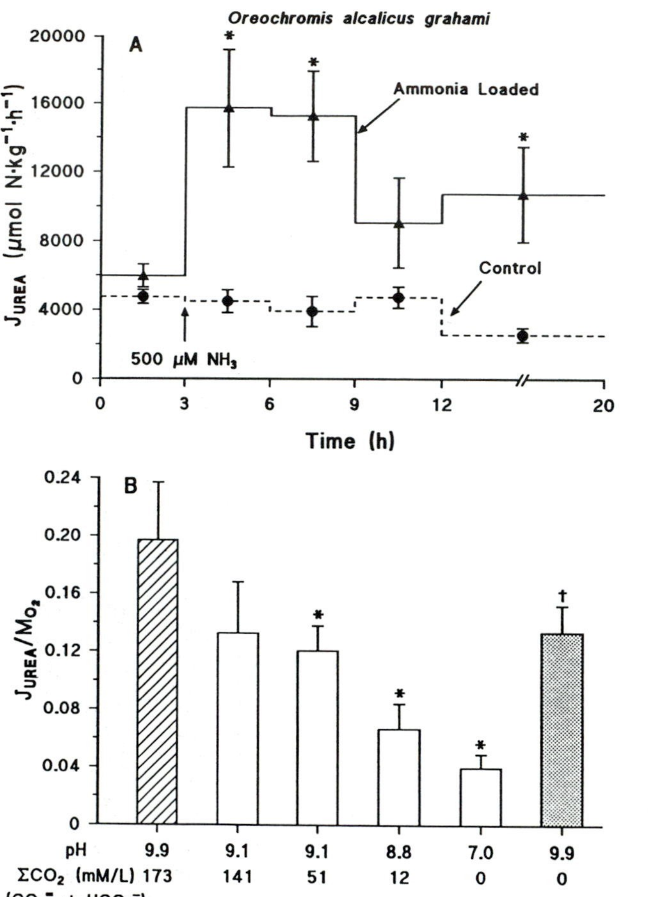

FIGURE 7. The influence of manipulations of the chemistry of Lake Magadi water on the rates of urea-N excretion in the Magadi tilapia (*Oreochromis alcalicus grahami*). (A) Responses to high external ammonia (~500 μM NH_3) in standard Lake Magadi water (pH 10). Means ±1 SEM (N = 11 to 14); asterisks indicate significant differences from the simultaneously measured flux in the control group. (Redrawn from Wood et al., 1989b.) (B) Responses to changes in the acid-base chemistry of the water, with urea excretion rates (J_{Urea}) normalized to O_2 consumption rates (M_{O2}) to correct for differences in temperature between treatment groups. Means ±1 SEM (N = 6 to 12); asterisks indicate significant differences from the control group in normal Lake Magadi water (cross-hatched bar) and dagger indicates significant difference between the two far right-hand groups. (Unpublished data of C. M. Wood, P. J. Walsh, P. L. Laurent, H. Bergman, A. Narahara, and J. N. Maina.)

to high external ammonia was rather slow (7 d) in *H. fossilis* (a normally ammoniotelic teleost), paralleling the induction of several OUC enzymes over this time course (Saha and Ratha, 1986). Similar phenomena probably occur in response to the ammonia retention accompanying air exposure (e.g., Walsh et al., 1990; Saha and Ratha, 1987, 1989), though it is not yet clear whether all species which exhibit a ureagenic response also possess the OUC (Table 3). Surprisingly there appear to have been no studies on J_{Urea} responses to ammonia loading in ureosmotic fish, an obvious area for future research. In "standard" (i.e., non-OUC) osteichthyes, the picture is currently unclear. J_{Urea} increased in both rainbow trout and goldfish during acute exposure to high ambient ammonia, though the response was much larger in the latter (Olson and Fromm, 1971). An ammoniotelic tilapia, *Oreochromis nilotica,* exhibited an elevation in J_{Urea} of marginal significance (Wood et al., 1989b). However, in recent similar experiments on trout (shown in Figure 5) and bowfin, *A. calva* (McKenzie and Randall, 1990), J_{Urea} did not change. In those instances where a ureogenic reponse does occur during high ammonia exposure, it is likely via the same mechanism as during high pH exposure.

High Environmental pH

As noted earlier, the acid-base significance of urea synthesis remains highly controversial. Nevertheless, high blood urea, high blood pH, and an increased reliance on urea as the excreted N-product (Table 3) are characteristic of teleosts endemic to highly alkaline lakes (Randall et al., 1989; Danulat and Kempe, 1992; Wright et al., 1993). It is tempting to speculate that such fish offer an example where urea synthesis really does contribute to acid-base homeostasis — by the net removal of $HCO_3^-/CO^{2-}{}_3$ which is continally entering from the water (Atkinson, 1992). However, on close inspection, the hypothesis must be qualified (Wood et al., 1989b). Ureagenesis (Equation 7) will effect *net* removal of exogenous HCO_3^- *only* if the accompanying NH_4^+ is also of exogenous origin, and does not come from the animal's own protein metabolism (which itself produces equimolar NH_4^+ and HCO_3^-). The completely ureotelic tilapia (*O. alcalicus grahami)* of Lake Magadi (pH 10, total HCO_3^- + $CO^{2-}{}_3$ = 170 to 190 m*M*), may still fit the hypothesis for it eats a cyanobacterial diet rich in nonprotein N. The idea is further supported by the dramatic inhibition of J_{Urea} which accompanied exposure to neutralized lake water (Wood et al., 1989b). Recently, the influence of the acid-base chemistry of lake water on J_{Urea} has been reexamined in greater detail (Figure 7B). The results suggest that J_{Urea} responds to both the pH and $HCO_3^-/CO^{2-}{}_3$ alkalinity of the environment. A tentative conclusion is that inhibition of urea production is proportional to the decrement of internal HCO_3^- levels caused by the various water chemistries. In toadfish hepatocytes, the rate of urea production by the OUC conformed to simple Michaelis-Menten kinetics as a function of HCO_3^- concentration in the incubation medium (Walsh et al., 1989).

As for high ammonia, there are no studies to date on ureagenic responses of ureosmotic fish to alkaline challenge. However, two "standard" teleosts, the rainbow trout (Figure 4D; Wilkie and Wood, 1991) and the Lahontan cutthroat trout (Wilkie et al., 1993) exhibited clear increases in J_{Urea} when exposed to elevated environmental pH. These responses occurred in association with marked elevations in both plasma pH and T_{Amm}. In the cutthroat trout, there was no change in liver arginase or OUC activity (which was negligible), but a significant increase in allantoicase (Wilkie et al., 1993). These results, together with data on high environmental ammonia (see above), suggest that ammonia build-up causes an increased uricolytic flux which is favored by alkalosis. The response probably serves as an ammonia-N detoxification mechanism (Figure 6), but it is unclear whether it makes any contribution to pH regulation. Interactions between acid-base and urea metabolism remain a rich area for future research.

ACKNOWLEDGMENTS

I thank Drs. P.J. Walsh, P.A. Wright, R.W. Wilson, E. Danulat, and D.H. Evans for valuable input, and Natural Sciences and Engineering Research Council of Canada (NSERC) (Canada) for financial support.

REFERENCES

Ash, R., McLean, C., and Wescott, P. A. B., Arterio-portal differences and net appearance of amino acids in hepatic portal vein blood of the trout (*Salmo gairdneri*), in *Aquaculture — A Biotechnology in Progress*, DePauw, N., Japers, E., Ackefors, H., and Wilkens, N., Eds., European Aquaculture Society, Bredene, Belgium, 1989, 801.

Atherton, W. D. and Aitken, A., Growth, nitrogen metabolism and fat metabolism in *Salmo gairdneri*, Rich., *Comp. Biochem. Physiol.*, 36, 719, 1970.

Atkinson, D. E., Functional roles of urea synthesis in vertebrates, *Physiol. Zool.*, 65, 243, 1992.

Audet, C. E., Munger, R. S., and Wood, C. M., Long term sublethal acid exposure in rainbow trout (*Salmo gairdneri*) in softwater: effects on ion exchanges and blood chemistry, *Can. J. Fish. Aquat. Sci.*, 45, 1387, 1988.

Audet, C. E. and Wood, C. M., Branchial morphological and endocrine responses of rainbow trout (*Oncorhynchus mykiss*) to a long-term sublethal acid exposure in which acclimation did not occur, *Can. J. Fish. Aquat. Sci.*, in press.

Avella, M. and Bornancin, M., A new analysis of ammonia and sodium transport through the gills of the freshwater rainbow trout (*Salmo gairdneri*), *J. Exp. Biol.*, 142, 155, 1989.

Beamish, F. W. H. and Thomas, E., Effects of dietary protein and lipid on nitrogen losses in rainbow trout, *Salmo gairdneri*, *Aquaculture* 41, 359, 1984.

Bergman, H. L., Olson, K. R., and Fromm, P. O., The effects of vasoactive agents on the functional surface area of isolated-perfused gills of rainbow trout, *J. Comp. Physiol.*, 94, 267, 1974.

Bever, K., Chenoweth, M., and Dunn, A., Amino acid gluconeogenesis and glucose turnover in kelp bass (*Paralabrax* sp.), *Am. J. Physiol.*, 240, R246, 1981.

Boylan, J. W., Gill permeability in *Squalus acanthias*, in *Sharks, Skates and Rays*, Gilbert, P. W., Mathewson, R. F., and Rall, D. P., Eds., Johns Hopkins Press, Baltimore, 1967, 197.

Braunstein, A. E. and Bychkov, S. M., A cell-free enzymatic model of l-amino acid dehydrogenase ("l-deaminase"), *Nature*, 144, 751, 1939.

Brett, J. R. and Zala, C. A., Daily pattern of nitrogen excretion and oxygen consumption of sockeye salmon *(Oncorhynchus nerka)* under controlled conditions, *J. Fish. Res. Board. Can.*, 32, 2479, 1975.

Brown, C. R. and Cameron, J. N., The induction of specific dynamic action in channel catfish by infusion of essential amino acids, *Physiol. Zool.*, 64, 276, 1991.

Brunel, A. M., Metabolisme de l'azote d'origine purique chez les poissons et chez les Batraciens. II. Catabolisme de l'azote d'origine purique chez les Téléostéens, *Bull. Soc. Chim. Biol.*, 19, 1027, 1937b.

Cameron, J. N., Responses to reversed NH_3 and NH_4^+ gradients in a teleost (*Ictalurus punctatus*), an elasmobranch (*Raja erinacea*), and a crustacean (*Callinectes sapidus*): evidence for NH_4^+/H^+ exchange in the teleost and the elasmobranch, *J. Exp. Zool.*, 239, 183, 1986.

Cameron, J. N. and Heisler, N., Studies of ammonia in the rainbow trout: physico-chemical parameters, acid-base behaviour, and respiratory clearance, *J. Exp. Biol.*, 105, 107, 1983.

Cameron, J. N. and Heisler, N., Ammonia transfer across fish gills: a review, in *Circulation, Respiration, and Metabolism,* Gilles, R., Ed., Springer-Verlag, Berlin, 1985, 91.

Cameron, J. N. and Kormanik, G. A., The acid-base responses of gills and kidneys to infused acid and base loads in the channel catfish, *Ictalurus punctatus, J. Exp. Biol.,* 99, 143, 1982.

Campbell, J. W., Nitrogen Excretion, in *Comparative Animal Physiology*, 3rd ed., Prosser, C.L., Ed., W.B.Saunders, Philadelphia, 1973, 279.

Campbell, J. W., Aster, P. L., and Vorhaben, J. E., Mitochondrial ammoniagenesis in liver of the channel catfish *Ictalurus punctatus, Am. J. Physiol.*, 244, R709, 1983.

Cao, X., Kemp, J. R., and Anderson, P. M., Subcellular localization of two glutamine dependent carbamoylphosphate synthetases and related enzymes in liver of *Micropterus salmoides* (large-mouth bass) and properties of isolated liver mitochondria: comparative relationships with elasmobranchs, *J. Exp. Zool.*, 258, 24, 1991.

Chamberlin, M. E., Glemet, H. C., and Ballantyne, J. S., Glutamine metabolism in a holostean (*Amia calva*) and teleost fish *(Salvelinus namaycush), Am. J. Physiol.*, 260, R159, 1991.

Chiu, Y. N., Austic, R. E, and Rumsey, G. L., Urea cycle activity and arginine formation in rainbow trout *(Salmo gairdneri), J. Nutr.*, 116, 1640, 1986.

Claiborne, J. B. and Evans, D. H., Ammonia and acid-base balance during high ammonia exposure in a marine teleost *(Myoxocephalus octodecimspinosus), J. Exp. Biol.*, 140, 89, 1988.

Claiborne, J. B., Evans, D. H., and Goldstein, L., Fish branchial Na^+/NH_4^+ exchange is via basolateral Na^+-K^+ activated ATPase, *J. Exp. Biol.*, 96, 431, 1982.

Cohen, P. P. and Brown, G. W., Ammonia metabolism and urea biosynthesis, in *Comparative Biochemistry*, Vol. 2, Florkin, M. and Mason, H. S., Eds., Academic Press, New York, 1960, 161.

Curtis, B. J. and Wood, C. M., The function of the urinary bladder in vivo in the freshwater rainbow, trout, *J. Exp. Biol.*, 155, 567, 1991.

Curtis, B. J. and Wood, C. M., Kidney and urinary bladder responses of freshwater rainbow trout to isosmotic NaCl and $NaHC0_3$ infusion, *J. Exp. Biol.*, 173, 181, 1992.

Cvancara, V. A., Comparative study of liver uricase activity in freshwater teleosts, *Comp. Biochem. Physiol.*, 28, 725, 1969a.

Cvancara, V. A., Studies on tissue arginase and ureogenesis in freshwater teleosts, *Comp. Biochem. Physiol.*, 30, 489, 1969b.

Danulat, E., and Kempe, S., Nitrogenous waste excretion and accumulation of urea and ammonia in *Chalcalburnus tarichi (Cyprinidae)* endemic to the extremely alkaline Lake Van (eastern Turkey), *Fish Physiol. Biochem.*, 9, 377, 1992.

Davenport, J. and Sayer, M. D. J., Ammonia and urea excretion in the amphibious teleost *Blennius pholis* (L.) in seawater and in air, *Comp. Biochem. Physiol.*, 84A, 189, 1986.

Delaney, R. G., Lahiri, S., Hamilton, R., and Fishman, A. P., Acid-base balance and plasma composition in the aestivating lungfish *(Protopterus), Am. J. Physiol.*, 232, R10, 1977.

Dépêche, J., Gilles, R., Daufresne, S., and Chiapello, H., Urea content and urea production via the ornithine-urea cycle pathway during the ontogenic development of two teleost fishes, *Comp. Biochem. Physiol.*, 63A, 51, 1979.

Dobson, G. P. and Hochachka, P., Role of glycolysis in adenylate depletion and repletion during work and recovery in teleost white muscle, *J. Exp. Biol.*, 129, 125, 1987.

Evans, D. H., Ionic exchange mechanisms in fish gills. *Comp. Biochem. Physiol.* 51A, 491, 1975.

Evans, D. H., Fish, in *Comparative Physiology of Osmoregulation in Animals,* Vol. 1., Maloiy, G.M.O., Ed., Academic Press, London, 1979, chap. 6.

Evans, D. H. and Cameron, J. N., Gill ammonia transport, *J. Exp.Zool.*, 239, 1723, 1986.

Evans, D. H. and More, K. J., Modes of ammonia transport across the gill epithelium of the dogfish pup *(Squalus acanthias), J. Exp. Biol.*, 138, 375, 1988.

Evans, D. H., More, K. J., and Robbins, S. L., Modes of ammonia transport across the gill epithelium of the marine teleost fish *Opsanus beta, J. Exp. Biol.*, 144, 339, 1989.

Fenstermacher, J., Shelden, F., Ratner, J., and Roomet, A., The blood to tissue distribution of various polar materials in the dogfish, *Squalus acanthias, Comp. Biochem. Physiol.*, 42A, 195, 1972.

Florkin, M., and Duchateau, G., Les formes du système enzymatique de l'uricolyse et l'evolution du catabolisme purique chez les animaux, *Arch. Int. Physiol.*, 53, 267, 1943.

Forster, R. P. and Goldstein, L., Urea synthesis in the lungfish: relative importance of purine and ornithine cycle pathways, *Science*, 153, 1650, 1966.

Forster, R. P. and Goldstein, L., Formation of excretory products, in *Fish Physiology*, Vol. 1, Hoar, W. S. and Randall, D. J., Eds., Academic Press, New York, 1969, 313.

Forster, R. P. and Goldstein, L., Intracellular osmoregulatory role of amino acids and urea in marine elasmobranchs, *Am. J. Physiol.*, 230, 925, 1976.

Freeman, H. C. and Idler, D. R., Effects of corticosteroids on liver transaminases in two salmonids, the rainbow trout, *Salmo gairdneri* and the brook trout, *Salvelinus fontinalis. Gen. Comp. Endocrinol.*, 20, 695, 1973.

Fromm, P. O., Studies on renal and extra-renal excretion in a freshwater teleost, *Salmo gairdneri. Comp. Biochem. Physiol.*, 10, 121, 1963.

Fromm, P. O. and Gillette, J. R., Effect of ambient ammonia on blood ammonia and nitrogen excretion of rainbow trout *(Salmo gairdneri). Comp. Biochem. Physiol.*, 26, 887, 1968.

Funkhouser, D., Goldstein, L., Forster, R. P., Urea biosynthesis in the South American lungfish, *Lepidosiren paradoxa:* relation to its ecology. *Comp. Biochem. Physiol.*, 41A, 439, 1972.

Gerst, J. W. and Thorson, T. B., Effects of saline acclimation on plasma electrolytes, urea excretion, and hepatic urea biosynthesis in a freshwater stingray *Potamotrygon* sp. Garman, 1877, *Comp. Biochem. Physiol.*, 56A, 87, 1977.

Goldstein, L., Adaptation of urea metabolism in aquatic vertebrates, in *Nitrogen Metabolism and the Environment*, Campbell, J. W. and Goldstein, L., Eds. Academic Press, New York, 1972, 55.

Goldstein, L., Gill nitrogen excretion, in *Gills*, Houlihan, D. F., Rankin, J. C., and Shuttleworth, T. J., Eds., Cambridge University Press, London, 1982, 193.

Goldstein, L. and Forster, R. P., Source of ammonia excreted by the gills of the marine teleost, *Myoxocephalus scorpius, Am. J. Physiol.*, 200, 1116, 1961.

Goldstein, L. and Forster, R. P., The relative importance of gills and kidneys in the excretion of ammonia and urea by the spiny dogfish (*Squalus acanthias*), *Bull. Mount Desert Island Biol. Lab.*, 4, 33, 1962.

Goldstein, L. and Forster, R. P., The role of uricolysis in the production of urea by fishes and other aquatic invertebrates, *Comp. Biochem. Physiol.*, 14, 567, 1965.

Goldstein, L. and Forster, R. P., Nitrogen metabolism in fishes, in *Comparative Biochemistry of Nitrogen Metabolism*, Vol. 2, Campbell, J. W., Ed., Academic Press, New York, 1970, 495.

Goldstein, L. and Forster, R. P., Urea biosynthesis and excretion in freshwater and marine elasmobranchs, *Comp. Biochem. Physiol.*, 39B, 415, 1971a.

Goldstein, L. and Forster, R. P., Osmoregulation and urea metabolism in the little skate *Raja erinacea, Am. J. Physiol.*, 220, 742, 1971b.

Goldstein, L., Forster, R. P., and Fanelli, G. M. Jr., Gill blood flow and ammonia excretion in the marine teleost, *Myoxocephalus scorpius. Comp. Biochem. Physiol.*, 12, 489, 1964.

Goldstein, L., Oppelt, W. W., and Maren, T. H., Osmotic regulation and urea metabolism in the lemon shark *Negaprion brevirostris, Am. J. Physiol.*, 215, 1493, 1968.

Goldstein, L., Claiborne, J. B., and Evans, D. E., Ammonia excretion by the gills of two marine teleost fish: the importance of NH_4^+ permeance, *J. Exp. Zool.*, 219, 395, 1982.

Gordon, M. S., Boetius, I., Boetius, I., Evans, D. H., McCarthy, R., and Oglesby, L. C., Salinity adaptation in the mudskipper fish, *Periophthalmus sobrinus, Hvalradets Skrifter Norske Videnskap Akad. Oslo*, 48, 85, 1965.

Gordon, M. S., Boetius, I., Evans, D. H., McCarthy, R., and Oglesby, L. C., Aspects of the physiology of terrestrial life in amphibious fishes. I. The mudskipper, *Periophthalmus sobrinus, J. Exp. Biol.*, 50, 141, 1969.

Gordon, M. S., Fischer, J., and Tarifeno, E., Aspects of the physiology of terrestrial life in amphibious fishes. II. The Chilean clingfish, *Sicyases sanguineus, J. Exp. Biol.*, 53, 559, 1970.

Gordon, M. S., Ng, W.W-S., and Yip, A.Y-W., Aspects of the physiology of terrestrial life in amphibious fishes. III. The Chinese mudskipper *Periopthalmus cantonensis. J. Exp. Biol.*, 72, 575, 1978.

Gregory, R. B., Synthesis and total excretion of waste nitrogen by fish of the *Periophthalmus* (mudskipper) and *Scartelaos* families, *Comp. Biochem. Physiol.*, 57A, 33, 1977.

Griffith, R. W., Composition of the blood-serum of deep-sea fishes, *Biol. Bull.*, 160, 250, 1981.

Griffith, R. W., Guppies, toadfish, lungfish, coelocanths, and frogs: a scenario for the evolution of urea retention in fishes, *Envir. Biol. Fish*, 32, 199, 1991.

Griffith, R. W., Umminger, B. L., Grant, B. F., Pang, P. K. T., and Pickford, G.E., Serum composition of the coelocanth *Latimeria chalumnae* Smith, *J. Exp. Zool.*, 187, 87, 1974.

Griffith, R. W., Umminger, B. C., Grant B. F., Pang, P. K. T., and Pickford, G.E., Composition of bladder urine of the coelocanth *Latimeria chalumnae, J. Exp. Zool.*, 196, 371, 1976.

Guerin-Ancey, O., Etude expérimentale de l'excretion azotée du bar (*Dicentrarchus labrax*) en cours de croissance. II. Effets du jeune sur l'excretion d'ammoniac et d'urée, *Aquaculture*, 9, 187, 1976.

Halperin, M. L., Chen, C. B., Cheema-Dhadli, S., West, M. L., and Jungas, R. L., Is urea formation regulated primarily by acid-base status *in vivo? Am. J. Physiol.*, 250, F605, 1986.

Halver, J. E. and Shanks, W. E., Nutrition of salmonoid fishes. VIII. Indispensable amino acids for sockeye salmon, *J. Nutr.*, 72, 340, 1960.

Hayashi, S., Fujiwara, S., and Noguchi, T., Degradation of uric acid in fish liver peroxisomes, *J. Biol. Chem.*, 264, 3211, 1989.

Haywood, G. P., Hypo-osmotic regulation coupled with reduced metabolic urea in the dogfish *Poroderma africanum*: an analysis of serum osmolarity, chloride, and urea, *Mar. Biol.*, 23, 121, 1973.

Haywood, G. P., The exchangeable ionic space and salinity effects upon ion, water, and urea turnover rates in the dogfish *Poroderma africanum, Mar. Biol.*, 26, 69, 1974

Heisler, N., Acid-base regulation in fishes, in *Fish Physiology*, Vol. 10, Gills, Part A., Hoar, W. S. and Randall, D. J., Eds., Academic Press, New York, 1984, 315.

Heisler, N., Mechanisms of ammonia elimination in fishes, in *Animal Nutrition and Transport Processes 2. Transport, Respiration, and Excretion: Comparative and Environmental Aspects, Comp. Physiol.*, Truchot, J. P. and Lahlou, B., Eds., Karger, Basel, 1990, 137.

Hickman, C. P. and Trump, B. F., The kidney, in *Fish Physiology,* Vol. 1, Hoar, W. S. and Randall, D., Eds., Academic Press, New York, 1969, 91.

Holmes, W. N. and Donaldson, E. M., The body compartments and the distribution of electrolytes, in *Fish Physiology*, Vol. 1, Hoar, W. S. and Randall, D. J., Eds., Academic Press, New York, 1969, 1.

Huggins, A. K., Skutch, G., and Baldwin, E., Ornithine-urea cycle enzymes in teleostean fishes, *Comp. Biochem. Physiol.*, 28, 587, 1969.

Iwama, G. K. and Heisler, N., Effect of environmental water salinity on acid-base regulation during environmental hypercapnia in the rainbow trout *(Oncorhynchus mykiss), J. Exp. Biol.*, 158, 1, 1991.

Iwata, K., Nitrogen metabolism in the mudskipper, *Periophthalmus cantonens*: changes in free amino acids and related compounds of various tissues under conditions of ammonia loading with special reference to its high ammonia tolerance, *Comp. Biochem. Physiol.*, 92A, 499, 1989.

Jacobs, M. H. and Stewart, D. R., The distribution of penetrating ammonium salts between cells and their surroundings, *J. Cell. Comp. Physiol.*, 7, 351, 1936.

Janssens, P. A. and Cohen, P. P., Ornithine-urea cycle enzymes in the African lungfish, *Protopterus aethiopicas, Science*, 152, 358, 1966.

Janssens, P. A. and Cohen, P. P., Nitrogen metabolism in the African lungfish, *Comp. Biochem. Physiol.*, 29, 879, 1968.

Kaushik, S. J., Influence of nutritional status on the daily patterns of nitrogen excretion in the carp *(Cyprinus carpio* L.) and the rainbow trout *(Salmo gairdneri* L.), *Reprod. Nutr. Dev.*, 20, 1751, 1980.

Kaushik, S. J. and Teles, A., Effect of digestible energy on nitrogen and energy balance in rainbow trout, *Aquaculture*, 50, 89, 1985.

Kenyon, A. J., The role of the liver in the maintenance of plasma proteins and amino acids in the eel, *Anguilla anguilla* L., with reference to amino acid deamination, *Comp. Biochem. Physiol.*, 22, 169, 1967.

Kikeri, D., Sun, A., Zeidel, M. L., and Hebert, S. C., Cell membranes impermeable to NH_3, *Nature*, 339, 478, 1989.

King, P. A. and Goldstein, L., Renal ammoniagenesis and acid excretion in the dogfish, *Squalus acanthias, Am. J. Physiol.*, 245, R581, 1983a.

King, P. A. and Goldstein, L., Renal ammonia excretion and production in goldfish, *Carassius auratus*, at low environmental pH, *Am. J. Physiol.*, 245, R590, 1983b.

Kinsella, J. L. and Aronson, P. J., Interaction of NH_4^+ and Li^+ with the renal microvillus membrane Na^+-H^+ exchanger, *Am. J. Physiol.*, 241, C220, 1981.

Kormanik, G. A. and Cameron, J. N., Ammonia excretion in animals that breathe water: a review, *Mar. Biol. Lett.*, 2, 11, 1981.

Kormanik, G. A. and Evans, D. H., The acid-base status of prenatal pups of the dogfish, *Squalus acanthias*, in the uterine environment, *J. Exp. Biol.*, 125, 173, 1986.

Lacy, E. R., Reale, E., Schlusselberg, D., Smith, W. K., and Woodward, D. J., A renal countercurrent system in marine elasmobranch fish: a computer assisted reconstruction, *Science*, 227, 1351, 1985.

Leech, A. R., Goldstein, L., Cha, C. J., and Goldstein, J. M., Alanine biosynthesis during starvation in skeletal muscle of the spiny dogfish, *Squalus acanthias, J. Exp. Zool.*, 207, 73, 1979.

Levi, G., Morisi, G., Coletti, A., and Catanzaro, R., Free amino acids in fish brain: normal levels and changes upon exposure to high ammonia concentrations in vivo, and upon incubation of brain slices, *Comp. Biochem. Physiol.*, 49A, 623, 1974.

Lin, H. and Randall, D. J., The effect of varying water pH on the acidification of expired water in rainbow trout, *J. Exp. Biol.*, 149, 149, 1990.

Lin, H. and Randall, D. J., Evidence for the presence of an electrogenic proton pump on the trout gill epithelium, *J. Exp. Biol.*, 161, 119, 1991.

Lowenstein, J. M., Ammonia production in muscle and other tissues. The purine nucleotide cycle, *Physiol. Rev.*, 52, 382, 1972.

Lund, P. and Goldstein, L., Glutamine synthetase activity in tissues of lower vertebrates, *Comp. Biochem. Physiol.*, 31, 205, 1969.

Maetz, J., Branchial sodium exchange and ammonia excretion in the goldfish *Carassius auratus*. Effects of ammonia loading and temperature changes, *J. Exp. Biol.*, 56, 601, 1972.

Maetz, J., Na^+/NH_4^+, Na^+/H^+ exchanges and NH_3 movement across the gills of *Carassius auratus, J. Exp. Biol.*, 58, 255, 1973.

Maetz, J., and García-Romeu, F., The mechanisms of sodium and chloride uptake by the gills of a freshwater fish, *Carassius auratus*. II. Evidence for NH_4^+/Na^+ and HCO_3^-/Cl^-- exchanges, *J. Gen. Physiol.*, 47, 1209, 1964.

Mangum, C. P., Haswell, M. S., Johansen, K., and Towle, D. W., Inorganic ions and pH in the body fluids of Amazon animals, *Can. J. Zool.*, 56, 907, 1978.

Marsh, D. J. and Knepper, M. A., Renal handling of urea, in *Handbook of Physiology, Section 8, Renal Physiology*, Windhager, E. E., Ed., Oxford University Press, New York, 1992, 1317.

Masoni, A. and Garciá-Romeu, F., Accumulation et excrétion de substances organiques par les cellules à chlorure de la branchie d'*Anguilla anguilla* L adapteé à l'eau de mer, *Z. Zellforsch.*, 133, 389, 1972.

McCarthy, J. J. and Whitledge, T. E., Nitrogen excretion by anchovy (*Engraulis mordax*) and jack mackerel *(Trachurus symmetricus), Fish. Bull.*, 70, 395, 1972.

McDonald, D. G., The interaction of calcium and low pH on the physiology of the rainbow trout. I. Branchial and renal net ion and H^+ fluxes, *J. Exp. Biol.*, 102, 123, 1983.

McDonald, D. G. and Wood, C. M., Branchial and renal acid and ion fluxes in the rainbow trout, *Salmo gairdneri*, at low environmental pH, *J. Exp. Biol.*, 93, 101, 1981.

McDonald, D. G, Tang, Y., and Boutilier, R. G., Acid and ion transfer across the gills of fish: mechanisms and regulation, *Can. J. Zool.*, 67, 3046, 1989.

McDonald, D. G., Walker, R. L., and Wilkes, P. R. H., The interaction of environmental calcium and low pH on the physiology of the rainbow trout, *Salmo gairdneri.* II. Branchial ionoregulatory mechanisms, *J. Exp. Biol.*, 102, 141, 1983.

McDonald, D. G., Walker, R. L., Wilkes, P. R. H., and Wood, C. M., H^+ excretion in the marine teleost *Parophrys vetulus, J. Exp. Biol.*, 98, 403, 1982.

McDonald, D. G., Cavdek, V., Calvert, C., and Milligan, C. L., Acid-base regulation in the Atlantic hagfish *Myxine glutinosa, J. Exp. Biol.*, 161, 201, 1991.

McKenzie, D. J. and Randall, D. J., Does *Amia calva* aestivate? *Fish Physiol. Biochem.*, 8, 147, 1990.

Meijer, A. J., Lamers, W. H., and Chamuleau, F. M., Nitrogen metabolism and ornithine cycle function, *Physiol. Rev.*, 70, 701, 1990.

Meyer, A. and Wilson, A. C., Origin of tetrapods inferred from their mitochondrial DNA affiliation to lungfish, *J. Molec. Evol.*, 31, 359, 1990.

Milligan, C. L., McDonald, D. G., Prior, E. T., Branchial acid and ammonia fluxes in response to alkalosis and acidosis in two marine teleosts: coho salmon *(Oncorhynchus kisutch)* and starry flounder (*Platichthys stellatus), Physiol. Zool.*, 64, 169, 1991.

Mommsen, T. P., and Hochachka, P. W., The purine nucleotide cycle as two temporally separated metabolic units: a study on trout muscle, *Metabolism*, 37, 552, 1988.

Mommsen, T. P. and Walsh, P. J., Evolution of urea synthesis in vertebrates: the piscine connection, *Science*, 243, 7275, 1989.

Mommsen, T. P. and Walsh, P. J., Urea synthesis in fishes: evolutionary and biochemical perspectives, in *Biochemistry and Molecular Biology of Fishes*, Vol. 1, Hochachka, P. W. and Mommsen, T. P., Eds., Elsevier, New York, 1991, 137.

Mommsen, T. P. and Walsh, P. J., Biochemical and environmental perspectives on nitrogen metabolism in fishes, *Experientia*, 48, 583, 1992.

Mommsen, T. P., French, C. J., and Hochachka, P. W., Sites and patterns of protein and amino acid utilization during the spawning migration of salmon, *Can. J. Zool.*, 58, 1785, 1980.

Mommsen, T. P., Danulat, E., and Walsh, P. J., Hormonal regulation of metabolism in hepatocytes of the ureogenic teleost *Opsanus beta, Fish. Physiol. Biochem.*, 9, 247, 1991.

Mommsen, T. P., Danulat, E., and Walsh, P. J., Metabolic actions of glucagon and dexamethasone in liver of the ureogenic teleost *Opsanus beta, Gen. Comp. Endocrinol.*, 85, 316, 1992.

Morii, H., Nishikata, K., and Tamura, O., Nitrogen excretion of mudskipper fish *Periophthalmus cantonensis* and *Boleophthalmus pectinorostris* in water and on land, *Comp. Biochem. Physiol.*, 60A, 189, 1978.

Morii, H., Nishikata, K., and Tamura, O., Ammonia and urea excretion from mudskipper fishes *Periophthalmus cantonensis* and *Boleophthalmus pectinorostris* transferred from land to water, *Comp. Biochem. Physiol.*, 63A, 23, 1979.

Olson, K. R. and Fromm, P. O., Excretion of urea by two teleosts exposed to different concentrations of ambient ammonia, *Comp. Biochem. Physiol.*, 40A, 999, 1971.

Ogata, H. and Murai, T., Changes in ammonia and amino acid levels in the erythrocytes and plasma of carp, *Cyprinus carpio*, during passage through the gills, *J. Fish. Biol.*, 33, 471, 1988.

Payan, P., Goldstein, L., and Forster, R. P., Gills and kidneys in ureosmotic regulation in euryhaline skates, *Am. J. Physiol.*, 224, 367, 1973.

Pequin, L. and Serfaty, A., L'excretion ammoniacale chez un téléostéen dulcicole: *Cyprinus carpio* L., *Comp. Biochem. Physiol.*, 10, 315, 1963.

Pequin, L. and Serfaty, A., Acide glutamique et excrétion azotée chez la carpe commune *(Cyprinus carpio* L.), *Comp. Biochem. Physiol.*, 18, 141, 1966.

Perlman, D. F. and Goldstein, L., Nitrogen metabolism, in *Physiology of Elasmobranch Fishes*, Shuttleworth, T. J., Ed., Springer-Verlag, Berlin, 1988, 253.

Pitts, R. F., *Physiology of the Kidney and Body Fluids,* 3rd edition, Yearbook Medical, Chicago, 1974.

Playle, R. C. and Wood, C. M., Water chemistry changes in the gill micro-environment of rainbow trout: experimental observations and theory, *J. Comp. Physiol. B.*, 159, 527, 1989.

Price, K. J., Fluctuations in two osmoregulatory components, urea and sodium chloride, of the clearnose skate, *Raja eglanteria* Busc. 1802. II. Upon natural variation of the salinity of the external medium, *Comp. Biochem. Physiol*, 23, 77, 1967.

Randall, D. J. and Wright, P.A., Ammonia distribution and excretion in fish, *Fish Physiol. Biochem.*, 3, 107, 1987.

Randall, D. J. and Wright, P. A., The interaction between carbon dioxide and ammonia excretion and water pH in fish, *Can. J. Zool.*, 67, 2936, 1989.

Randall, D. J., Wood, C. M., Perry, S. F., Bergman, H. L., Maloiy, G. M. O., Mommsen, T. P., and Wright, P. A., Urea excretion as a strategy for survival in a fish living in a very alkaline environment, *Nature*, 337, 165, 1989.

Read, L. J., A study of ammonia and urea production and excretion in the freshwater-adapted form of the Pacific lamprey, *Entosphenus tridentatus, Comp. Biochem. Physiol.*, 26, 455, 1968.

Read, L. J., The presence of high ornithine-urea cycle enzyme activity in the teleost *Opsanus tau, Comp. Biochem. Physiol.*, 39B, 409, 1971a.

Read, L. J., Chemical constituents of the body fluids and urine of the holocephalian *Hydrolagus collei, Comp. Biochem. Physiol.*, 39A, 185, 1971b.

Read, L. J., Absence of ureogenic pathways in the liver of the hagfish, *Bdellostoma cirrhatum, Comp. Biochem. Physiol.*, 51B, 139, 1975.

Robertson, J. D., The chemical composition of the blood of some aquatic chordates of the *Tunicata, Cyclostomata,* and *Osteichthyes, J. Exp. Biol.*, 31, 424, 1954.

Robertson, J. D., Osmotic constituents of the blood plasma and parietal muscle of *Myxine glutinosa* L., in *Some Contemporary Studies in Marine Science*, Barnes, H., Ed., Allen and Unwin, London, 1966, 631.

Robertson, J. D., Osmotic constituents of the blood plasma and parietal muscle of *Squalus acanthias* L., *Biol. Bull.*, 148, 303, 1975.

Robertson, J. D., Chemical composition of the body fluids and muscle of the hagfish *Myxine glutinosa* and the rabbit-fish *Chimaera monstrosa, J. Zool.*, 178, 261, 1976.

Robertson, J. D., Osmotic constituents of the blood plasma and parietal muscle of *Scyliorhinus canicula, Comp. Biochem. Physiol.*, 93A, 799, 1989.

Roos, A., and Boron, W. F., Intracellular pH., *Physiol. Rev.*, 61, 296, 1981.

Saha, N., and Ratha, B. K., Effect of ammonia stress on ureogenesis in a freshwater air-breathing teleost, *Heteropneustes fossilis,* in *Contemporary Themes in Biochemistry,* Vol. 6, Kon, O. L., Chung, M. C. M., Hwang, P. L. H., Leong, S. F., Loke, K. H., Thiyagarah, P., and Wong, P. T. H., Eds., Cambridge University Press, Cambridge, 1986, 342.

Saha, N. and Ratha, B. K., Active ureogenesis in a freshwater air-breathing teleost, *Heteropneustes fossilis, J. Exp. Zool.*, 241, 137, 1987.

Saha, N. and Ratha, B. K., Comparative study of ureogenesis in freshwater air-breathing teleosts, J. Exp. Zool., 252, 1, 1989.

Saha, N. and Ratha, B. K., Alterations in excretion pattern of ammonia and urea in a freshwater air-breathing teleost, *Heteropneustes fossilis* (Bloch) during hyper-ammonia stress, *Indian J. Exp. Biol.*, 28, 597, 1990.

Saha, N., Chakravorty, J., and Ratha, B. K., Diurnal variation in renal and extra-renal excretion of ammonia-N and urea-N in a freshwater airbreathing teleost, *Heteropneustes fossilis* (Bloch), *Proc. Indian. Acad. Sci. (Anim. Sci),* 97, 529, 1988.

Sayer, M. D. J. and Davenport, J., The relative importance of the gills to ammonia and urea excretion in five seawater and one freshwater teleost species, *J. Fish Biol.,* 31, 561, 1987a.

Sayer, M. D. J. and Davenport, J., Ammonia and urea excretion in the amphibious teleost *Blennius pholis* exposed to fluctuating salinity and pH, *Comp. Biochem. Physiol.*, 87A, 851, 1987b.

Scarabello, M., Heigenhauser, G. J. F., and Wood, C. M., Gas exchange, metabolite status, and excess post-exercise oxygen consumption after repetitive bouts of exhaustive exercise in juvenile rainbow trout, *J. Exp. Biol.*, 167, 155, 1992.

Schmidt-Nielsen, B. and Rabinowitz, L., Methylurea and acetamide: active reabsorption by elasmobranch renal tubules, *Science*, 146, 1587, 1964.

Schmidt-Nielsen, B., Truniger, B., and Rabinowitz, L., Sodium-linked urea transport by the renal tubule of the spiny dogfish, *Squalus acanthias, Comp. Biochem. Physiol.*, 42A, 13, 1972.

Schooler, J. M., Goldstein, S., Hartman, C., and Forster, R. P., Pathways of urea synthesis in the elasmobranch, *Squalus acanthias, Comp. Biochem. Physiol.*, 18, 271, 1966.

Shankar, R. A. and Anderson, P. M., Purification and properties of glutamine synthetase from liver of *Squalus acanthias, Arch. Biochem. Biophys.*, 239, 248, 1985.

Smith, H. W., The excretion of ammonia and urea by the gills of fish, *J. Biol. Chem.*, 81, 727, 1929a.

Smith, H. W., The composition of the body fluids of elasmobranchs, *J. Biol. Chem.*, 81, 407, 1929b.

Smith, H. W., Metabolism of the lung-fish, *Protopterus aethiopicus, J. Biol. Chem.*, 88, 97, 1930.

Smith, H. W., The retention and physiological role of urea in the elasmobranchii, *Biol. Rev.*, 11, 49, 1936.

Smith, H. W. and Smith, C. G., The absorption and excretion of water and salts by the elasmobranch fishes. I. Freshwater elasmobranchs, *Am. J. Physiol.*, 98, 279, 1931.

Smith, R. R. and Rumsey, G. L., Nutrient utilization by fish, in *First Int. Symp., Feed Composition, Animal Nutrient Requirements, and Computerization of Diets,* Fonnesbeck, P. V., Harris, L. E., and Kearl, L. C., Eds., Utah Agricultural Experiment Station, Utah State University, Logan, Utah, 1976, 320.

Tang, Y., Lin, H., and Randall, D. J., Compartmental distributions of carbon dioxide and ammonia in rainbow trout at rest and following exercise, and the effect of bicarbonate infusion, *J. Exp. Biol.*, 169, 235, 1992.

Toews, D. P., Holeton, G. F., and Heisler, N., Regulation of the acid-base status during environmental hypercapnia in the marine teleost fish *Conger conger, J. Exp. Biol.*, 107, 9, 1973.

Truchot, J. P., *Comparative Aspects of Extracellular Acid-Base Balance,* Springer-Verlag, Berlin, 1987.

Van den Thillart, G., and Kesbeke, F., Anaerobic production of carbon dioxide and ammonia by goldfish *Carassius auratus* (L.), *Comp. Biochem. Physiol.*, 59A, 393, 1978.

Van Waarde, A., Aerobic and anaerobic ammonia production by fish, *Comp. Biochem. Physiol.*, 74B, 675, 1983.

Van Waarde, A., Biochemistry of non-protein nitrogeneous compounds in fish including the use of amino acids for anaerobic energy production, *Comp. Biochem. Physiol.*, 91B, 207, 1988.

Van Waarde, A. and Kesbeke, F., Nitrogen metabolism in goldfish, Carassius auratus L. Influence of substrates and enzyme inhibitors on ammonia production by isolated hepatocytes, *Comp. Biochem. Physiol.*, 70B, 499, 1981.

Van Waarde, A., van den Thillart, G., and Kesbeke, F., Anaerobic energy metabolism of the European eel, *Anguilla anguilla* (L.), *J. Comp. Physiol.*, 149, 469, 1983.

Van Waversveld, J., Addink, A. D. F., and Van den Thillart, G., The anaerobic energy metabolism of goldfish determined by simultaneous direct and indirect calorimetry during anoxia and hypoxia, *J. Comp. Physiol. B.*, 159, 263, 1989.

Vellas, F. and Serfaty, A., L'ammoniaque et l'urée chez un Téléostéen d'eau douce: la Carpe *(Cyprinus carpio* L.), *J. Physiol. (Paris)*, 68, 591, 1974.

Walsh, P. J. and Henry, R. P., Carbon dioxide and ammonia metabolism and exchange, in *Biochemistry and Molecular Biology of Fishes,* Vol. 2, Hochachka, P. W. and Mommsen, T. P., Eds., Elsevier, New York, 1992, 181.

Walsh, P. J., Parent, J. J., and Henry, R. P., Carbonic anhydrase supplies biocarbonate for urea synthesis in toadfish *(Opsanus beta)* hepatocytes, *Physiol. Zool.*, 62, 1257, 1989.

Walsh, P. J., Danulat, E., and Mommsen, T. P., Variation in urea excretion in the gulf toadfish Opsanus beta, *Mar. Biology*, 106, 323, 1990.

Walton, M. J. and Cowey, C. B., Aspects of ammoniogenesis in rainbow trout, *Salmo gairdneri, Comp. Biochem. Physiol.*, 57B, 143, 1977.

Watts, R. L. and Watts, D. C., Nitrogen metabolism in fishes, *Chem. Zool.*, 8, 369, 1973.

Webb, J. T. and Brown, G. W., Glutamine synthetase: assimilatory role in liver as related to urea retention in marine chondrichthyes, *Science,* 208, 293, 1980.

Wheatly, M. G., Höbe, H., and Wood, C. M., The mechanisms of acid-base and ionoregulation in the freshwater rainbow trout during environmental hyperoxia and subsequent normoxia. II. The role of the kidney, *Respir. Physiol.*, 55, 155, 1984.

Wilkie, M. P. and Wood, C. M., Nitrogenous waste excretion, acid-base balance, and ionoregulation in rainbow trout *(Oncorhynchus mykiss)* exposed to extremely alkaline water, *Physiol. Zool.*, 64, 1069, 1991.

Wilkie, M. P., Wright, P. A., Iwama, G. K., and Wood, C. M., The physiological responses of the Lahontan cutthroat trout (*Oncorhynchus clarki henshawi*), a resident of highly alkaline Pyramid Lake (pH 9.4), to challenge at pH 10, *J. Exp. Biol.*, 175, 173, 1993.

Wilson, R. P., Nitrogen metabolism in channel catfish *Ictalurus punctatus* II. Evidence for an apparent incomplete ornithine-urea cycle, *Comp. Biochem. Physiol.*, 46B, 625, 1973.

Wilson, R. W. and Taylor, E. W., Transbranchial ammonia gradients and acid-base responses to high external ammonia in rainbow trout *(Oncorhynchus mykiss)* acclimated to different salinities, *J. Exp. Biol.*, 166, 95, 1992.

Wong, T. M. and Chan, D. K. O., Physiological adjustment to dilution of the external medium in the lip shark, *Hemiscyllium plagiosum* (Bennet). II. Branchial, renal, and rectal gland function, *J. Exp. Zool.*, 200, 85, 1977.

Wood, C. M., Acid-base and ionic exchanges at gills and kidney after exhaustive exercise in the rainbow trout, *J. Exp. Biol.,* 136, 461, 1988.

Wood, C. M., The physiological problems of fish in acid waters, in *Acid Toxicity and Aquatic Animals,* Morris, R., Taylor, E. W., Brown, D. J. A. and Brown, J. A., Eds., Cambridge University Press, Cambridge, 1989, 125.

Wood, C. M., Branchial ion and acid-base transfer in freshwater teleost fish: environmental hyperoxia as a probe, *Physiol. Zool.*, 64, 68, 1991.

Wood, C. M., Munger, R. S., and Toews, D. P., Ammonia, urea and H^+ distribution and the evolution of ureotelism in amphibians, *J. Exp. Biol.,* 144, 215, 1989a.

Wood, C. M., Perry, S. F., Wright, P. A., Bergman, H.L., and Randall, D. J., Ammonia and urea dynamics in the Lake Magadi tilapia, a ureotelic teleost fish adapted to an extremely alkaline environment, *Respir. Physiol.*, 77, 1, 1989b.

Wood, J. D., Nitrogen excretion in some marine teleosts, *Can. J. Biochem. Physiol.*, 36, 1237, 1958.

Wright, P. A. and Randall, D. J., The interaction between ammonia and carbon dioxide stores and excretion rates in fish, *Ann. Soc. R. Zool. Belg.*, Suppl. 1, 321, 1987.

Wright, P. A. and Wood, C. M., An analysis of branchial ammonia excretion in the freshwater rainbow trout: effects of environmental pH change and sodium uptake blockade, *J. Exp. Biol.*, 114, 329, 1985.

Wright, P. A. and Wood, C. M., Muscle ammonia stores are not determined by pH gradients, *Fish. Physiol. Biochem.*, 5, 159, 1988.

Wright, P. A., Heming, T., and Randall, D., Downstream pH changes in water flowing over the gills of rainbow trout, *J. Exp. Biol.*, 126, 499, 1986.

Wright, P. A., Randall, D. J., and Wood, C. M., The distribution of ammonia and H^+ between tissue compartments in lemon sole *(Parophrys vetulus)* at rest, during hypercapnia, and following exercise, *J. Exp. Biol.*, 136, 149, 1988a.

Wright, P. A., Wood, C. M., and Randall, D. J., An *in vitro* and *in vivo* study of the distribution of ammonia between plasma and red cells of rainbow trout *(Salmo gairdneri)*, *J. Exp. Biol.*, 134, 423, 1988b.

Wright, P. A., Randall, D. J., and Perry, S. F., Fish gill boundary layer: a site of linkage between carbon dioxide and ammonia excretion, *J. Comp. Physiol. B.*, 158, 627, 1989.

Wright, P. A., Iwama, G. K., and Wood, C. M., Ammonia and urea excretion in Lahontan cutthroat *(Oncorhynchus clarki henshawii)* adapted to highly alkaline Pyramid Lake (pH 9.4), *J. Exp. Biol.*, 175, 153, 1993.

Yancey, P. H. and Somero, G. N., Counteraction of urea destabilization of protein structure by methylamine osmoregulatory compounds of elasmobranch fishes, *Biochem. J.*, 183, 317, 1979.

Yesaki, T. Y. and Iwama, G. K., Survival, acid-base regulation, ion regulation, and ammonia excretion in rainbow trout in highly alkaline, hard water, *Physiol. Zool.*, 65, 763, 1992.

14 Thermal Biology

Jeffrey R. Hazel

INTRODUCTION

The thermal biology of fishes is dominated by constraints governing respiration in an aquatic environment. Because of the limited solubility of oxygen in water, gill-breathers must irrigate their respiratory surfaces with a 40 times larger volume than air-breathers in order to extract an equivalent amount of oxygen (Block, 1991a). In addition, the high heat capacity of water (3000-fold greater than air) combined with higher rates of thermal than molecular (i.e., oxygen) diffusion ensure that by the time blood in the gills is saturated with oxygen, it has also equilibrated with the water temperature (Carey, 1973). Thus, in a typical fish, the heat generated by metabolism is carried via the blood to the gills, where it is lost to the environment (Carey and Lawson, 1973); consequently, tissue temperatures are at, or within, 1°C of the ambient water temperature (Carey et al., 1971; Reynolds et al., 1976).

Variations in body temperature impact fish physiology in two primary ways (Hochachka and Somero, 1984): (1) temperature *determines* the *rate* of chemical reactions, and even modest reductions in temperature (e.g., 10°C within the biological temperature range, corresponding to a change of only ≈3% in average kinetic energy) markedly depress reaction rates (typically 2- to 3-fold); (2) Temperature *dictates* the point of *equilibrium* between the formation and disruption of the noncovalent forces which stabilize both macromolecular (e.g., biological membranes and subunit interactions between proteins) and molecular (e.g., proteins and nucleic acids) structures of physiological importance, and the binding of ligands to proteins (e.g., hormone-receptor and enzyme-substrate interactions). Structural flexibility is an obligate requirement for both protein function and the integrity of biological membranes, yet cold temperatures constrain this flexibility, thereby stabilizing less active conformations. Conversely, high temperatures promote conformational plasticity to a point where biological structure can no longer be maintained.

It follows that significant changes in body temperature pose a serious challenge to the maintenance of integrated physiological function. Among fish, this challenge is met and the effects of acute temperature change obviated in various ways. (1) Some fish, described as *endothermic* or warm-bodied, have evolved specialized mechanisms for the production and retention of metabolic heat. (2) Alternately, most fish are essentially isothermal with their environ-

0-8493-8042-1/93/$0.00+$.50

ment and cope with the problems of *poikilothermy* (i.e., variable body temperatures) by a diversity of mechanisms. The aim of this chapter is to review mechanisms which permit (1) a limited number of fish taxa to exploit endothermy and (2) fish that are poikilothermic to either avoid extremes of environmental temperature (or sustained activity at extremes of temperature), or maintain physiological function at temperature extremes.

ENDOTHERMY IN FISHES

Endothermy in fish is restricted to one group of large marine fishes, the Scrombroidei (a suborder of Perciformes including mackerels, tunas, bonitos, and billfishes [swordfish, marlin, and spearfish]), and sharks of the family Lamnidae (including the mako, porbeagle, and white sharks) (Block, 1991b). The requirements for endothermy include large body size (i.e., thermal inertia), a source of metabolic heat, and mechanisms of heat conservation. These requirements are manifest in two patterns of endothermy: (1) *whole body endothermy*, best exemplified by tunas, employs vascular heat exchangers to retain heat generated by the red oxidative swimming muscles in addition to the brain, eyes, and viscera; and (2) *regional endothermy*, displayed by billfishes, utilizes a unique thermogenic tissue (the heater organ), to elevate brain and eye temperatures.

Vascular Countercurrent Heat Exchangers and Whole-Body Endothermy

Observations that tunas exhibit higher muscle temperatures than other fish of comparable size date from the 1830s (Stevens and Neill, 1978). For example, red muscle temperature in freshly caught skipjack (*Katsuwonus pelamis*) was 9.1°C above the ambient water temperature of 25.6°C, but decreased to between 2 to 4°C above ambient in captivity (Stevens and Fry, 1971). Both skipjack and yellowfin (*Thunnus albacares*), which occur mainly in waters warmer than 20°C, maintain relatively fixed intervals between muscle and water temperature (intervals of 4 to 7 and 2.5 to 3.5°C, respectively; Carey et al., 1971), whereas free-swimming bluefin tuna (*Thunnus thynnus*), which range from the Bahamas (30°C) to the Newfoundland Coast (6°C), maintain muscle temperatures between 28 to 33°C over a range of water temperatures from 7 to 30°C (Carey and Lawson, 1973). These abilities are even more remarkable in light of the 5 to 10 times larger gill surface areas of tuna compared to other active teleosts of similar size (Bushnell and Brill, 1992). The abilities of bluefin tuna to vary the magnitude of excess muscle temperature from ± 20°C in 7°C water to only 2 to 3°C in 30°C water, and yellowfin tuna to swim significantly faster at 30 than 20°C but maintain a lower excess muscle temperature (Dizon and Brill, 1979), suggest a limited capacity for thermoregulation. Most notably, bigeye tuna (*Thunnus obesus*) vary whole-body conductivity (by two orders of magnitude) by disengaging the heat exchangers during ascent from cold to warm surface waters, and reengaging them to conserve heat when they return to depth (Holland et al., 1992). However, bluefin tuna do not increase their metabolic rate (i.e., heat production) at low temperature (Stevens and Neill, 1978), and muscle temperatures of fish in the same school vary by as much as 5°C (Carey and Lawson, 1973).

Both elevated rates of heat production and specialized mechanisms of heat retention contribute to endothermy in tunas. Maximum rates of oxygen consumption in active, free-swimming skipjack (2.5 mg O_2/g/h) are more than twice as great as those reported for other teleosts (Stevens and Carey, 1981). In addition, two distinct anatomical adaptations account for heat retention: (1) movement of the red muscle mass internally; and (2) placement of a countercurrent heat exchanger in selected regions of the systemic circulation (Block, 1991a). Typically, in fish, blood is distributed to the swimming musculature via segmental arteries that arise from a centrally located dorsal aorta, and venous blood is returned to a centrally located postcardinal vein (Figure 1A); consequently, heat produced by muscular activity is carried to the gill, where it is lost to the environment. In contrast, four cutaneous arteries (two pairs arising on each side of the body posterior to the origin of the coeliacomesenteric artery), which run just under the skin and close to the upper and lower boundaries of the red muscle mass, are the major vessels delivering blood to the musculature in large tunas (Figure 1C). Numerous small vessels (0.1 mm in diameter) arranged at right angles to the cutaneous arteries and veins penetrate into the musculature to form the *rete mirabile*, a system of parallel, intermingled arterioles and venules which constitutes a physiological barrier to the loss of metabolic heat. The mass of the rete is thickest (≈1 cm) and the vessels finest near its origin from the cutaneous vessels. Because of the large area of contact between the fine venules carrying warm, deoxygenated blood back to the gills and the small arterioles delivering cold, oxygenated blood to the musculature, and the countercurrent arrangement of arterial and venous flow, heat is readily transferred from venous to arterial vessels and thus retained within the tissue. Because the deepest muscles receive some "cold" blood derived from the dorsal aorta in addition to that derived from the rete, the highest muscle temperatures are actually displaced toward the center of the dark muscle band on either side of the fish. Nevertheless, a large volume of both the dark and light muscle mass is considerably warmer than the water temperature.

Although peripheral or cutaneous retia are present in bluefin, bigeye, and albacore tunas (Carey et al., 1971), in smaller fish such as the yellowfin, and especially the skipjack, the primary heat exchanger is located centrally (Stevens et al., 1974). In skipjack, the large dorsal aorta and postcardinal vein run along the floor of the hemal canal and give rise to numerous small vessels that project upward and intermingle to form a massive rete (in a 2-kg tuna, the rete is ≈7 mm wide with vessels extending 10 mm in length and contains ≈5% of the total blood volume), completely filling the canal (Figure 1D). Arterial blood thus passes from the dorsal aorta through the rete before entering the segmental arteries. Because the velocity of blood flow through the rete is only 1/80th of that in the dorsal aorta, heat

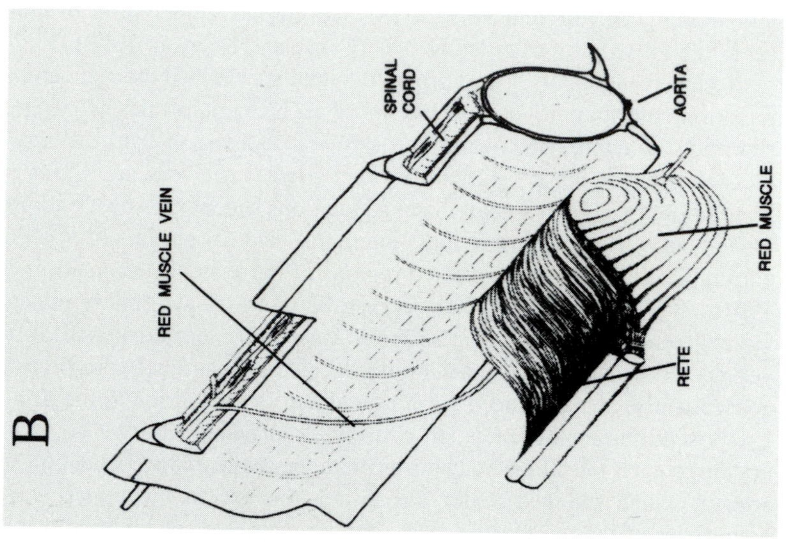
B
RED MUSCLE VEIN
SPINAL CORD
AORTA
RED MUSCLE
RETE

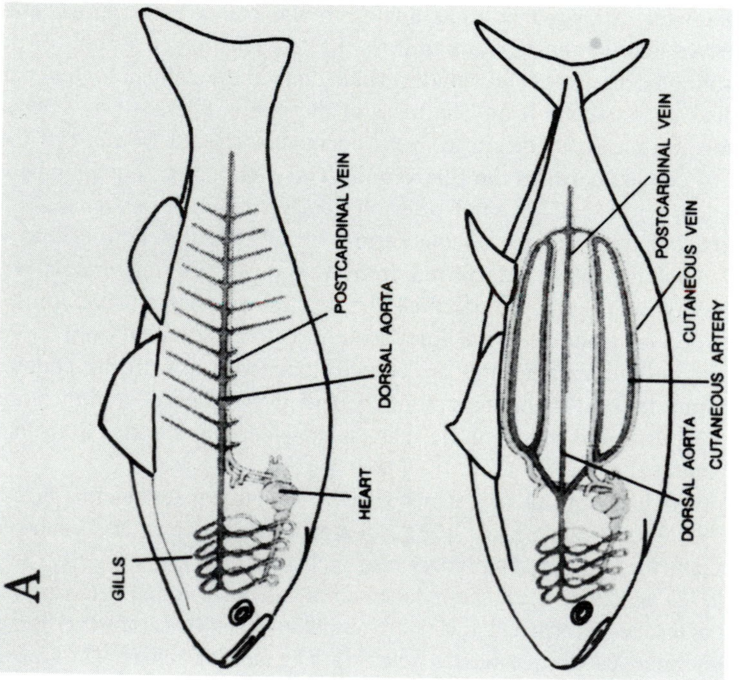
A
GILLS
HEART
DORSAL AORTA
POSTCARDINAL VEIN
DORSAL AORTA
CUTANEOUS ARTERY
CUTANEOUS ARTERY
CUTANEOUS VEIN
POSTCARDINAL VEIN

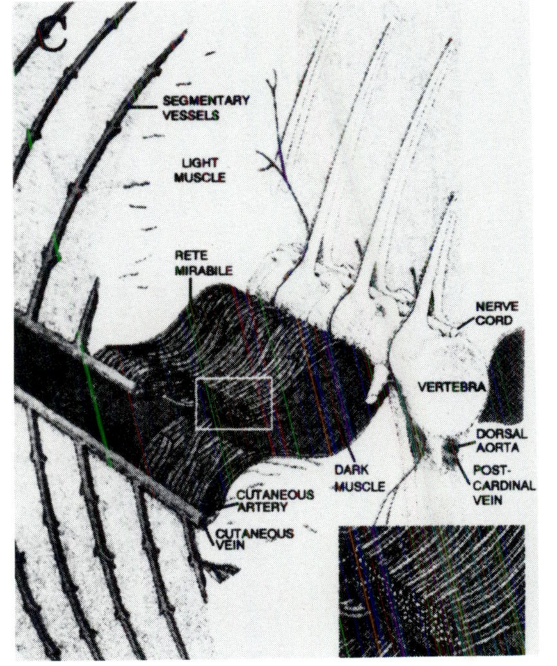

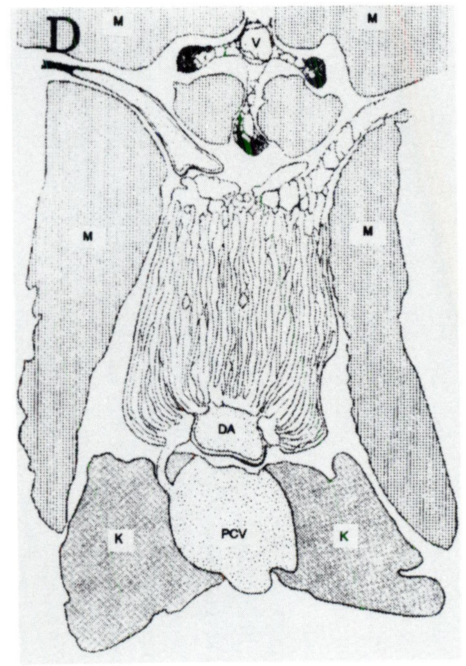

FIGURE 1. Circulatory plan and vascular anatomy of heat exchangers in endothermic fish. (A) Circulatory plans of poikilothermic (top) and endothermic (bottom) fish. (From Carey, F. G., *Sci. Am.*, 228(2), 36, 1973. With permission.) (B) The red muscle vein of the mako shark, *I. oxyrinchus*, which delivers warm blood from the red muscle to the central nervous system. (From Wolf, N. G. et al., *J. Comp. Physiol.*, 157, 709, 1988. With permission.) (C) Peripheral *rete mirabile* of the bluefin tuna, *T. thynnus*. (From Carey, F. G., *Sci. Am.*, 228(2), 36, 1973. With permission.) (D) The central rete of the skipjack tuna, *K. pelamis*. (From Stevens, E. D. et al., *J. Exp. Biol.*, 61, 145, 1974. With permission.) Symbols: M, muscle; V, vertebrae; DA, dorsal aorta; PCV, postcardinal vein; K, kidney.

exchange is highly efficient: 97% of the heat produced by the oxidative muscles is transferred to the arterial blood (Stevens and Neill, 1978).

Both tunas and mackerel sharks also possess visceral heat exchangers capable of maintaining the internal organs (but not the heart) significantly warmer (by as much as 6 to 8°C) than water temperature. In an 88-kg porbeagle shark (*Lamna nasus*), the suprahepatic rete, located on the surface of the liver, weighed 267 g, displaced a volume of 240 ml, possessed a surface area for thermal exchange of 4 m^2, and comprised 3.4% of the weight of the visceral organs (Carey et al., 1981). Telemetered data from free-swimming sharks indicate that stomach temperature is not constant. Nevertheless, visceral temperatures vary with time and circumstance, in some cases correlating with the activity of the digestive system (Carey et al., 1971). For example, stomach temperatures of bluefin tuna rise slowly following a meal and decline when feeding stops (Stevens and McLeese, 1984).

Brain and eye temperatures of tuna and lamnid sharks are also elevated above water temperature, most commonly due to the presence of a carotid or orbital rete (Stevens and Fry, 1971). For example, in bluefin tuna captured at 20°C, brain and eye temperatures averaged 27 and 26°C, respectively (compared to 30°C for muscle) and varied by less than 6°C as water temperature fluctuated between 7 and 23°C (Linthicum and Carey, 1972). Similarly, the mean excess temperatures for sharks possessing an orbital rete averaged 3 and 2.8°C for brain and eye tissues, respectively (Block and Carey, 1985). Furthermore, in lamnid sharks, a large vein (valved to direct blood flow anteriorly) drains the anterior half of the red muscle mass and transfers warm blood to the central nervous system (Figure 1B).

Regional Endothermy: The Billfish Heater Organ

Billfish are large pelagic fish with a preponderance of white muscle that stalk and sprint rather than sustaining the high levels of continuous activity typical of tuna and lamnid sharks (Carey, 1982). Accordingly, in the swordfish (*Xiphias gladius*), only brain and eye temperatures are above ambient (up to 13°C), but the cranial temperature of a free-swimming fish remained at 28°C for 36 h even though water temperature varied from 13 to 17°C (Block, 1991b). The basis for this regional endothermy is a heat-generating organ (with a color and consistency similar to liver) located at the base of the brain that is actually a bilateral modification of the superior rectus muscle (or the lateral recti in the mackerel, *Gasterochisma*; Carey, 1982). In a 120-kg swordfish, the heater organ weighs 50 g. Blood flow to the organ is derived from a branch of the carotid artery and passes through a highly developed carotid rete before continuing on to supply warm blood to the retina and brain. An insulating layer of fat prevents conductive heat loss through the surface of the head.

The extraocular muscles of the heater organ express a novel type of muscle fiber lacking contractile components and specialized for heat production (Block,

1991a). These specialized heater cells possess the highest volume of mitochondria of any cell in the animal kingdom (55 to 70%) and up to 35 nmol/g of cytochrome c, a value comparable to that of brown adipose tissue (Block, 1991b), in addition to large quantities of smooth endoplasmic reticulum (25 to 30% of cell volume). Furthermore, billfish species that experience the coldest temperatures express more of the heater phenotype than warmer-water fish (Block, 1991a).

Specialized heat-producing tissues are rare in the animal kingdom, being restricted to mammalian brown adipose tissue (BAT) and the billfish heater organ. However, unlike the situation in BAT, mitochondria of the heater organ are coupled (ADP/O ratios of 2.3; Ballantyne et al., 1992) and do not express an uncoupling protein (Block, 1987). The extensive hypertrophy of the sarcoplasmic reticulum (SR) in heater cells, in combination with high mitochondrial volumes, suggests that a cycle of calcium release and reuptake may play a major role in heat production, a hypothesis supported by the high levels of SR Ca^{2+}-ATPase (34,000 molecules per μm^2 of membrane in the blue marlin) and phospholambin (involved in the regulation of the ATPase) in heater tissue (Block, 1991a and b).

Why billfishes exhibit regional rather than whole-body endothermy is perhaps related to the organization of their musculature (Block, 1991b). Although red muscle fibers of the striped marlin (*Tetrapturus audax*) are similar in oxidative capacity to the corresponding fibers in tuna, they comprise only 5 to 6% of the total muscle mass (compared to 4–13% in tunas) and are widely dispersed throughout the myomeres, suggesting that reorganization of the epaxial musculature is a major requirement for whole-body endothermy.

The Significance of Endothermy in Fish

The presumed advantages of endothermy are several. The capacity for fast, sustained swimming has been an important development in the evolution of large predatory fish of the open ocean, for there is an abundance of food in such fast-swimming prey as squid, mackerel, and herring (Carey et al., 1971). Yet high-speed swimming is a demanding mode of locomotion given the density and viscosity of water. Nevertheless, yellowfin and wahoo (*Acanthocybium solanderi*) tuna attain speeds of up to 40 mph during sprints of 10 to 20 s (Carey, 1973). Since the shortening speed and power production of muscle fibers increase 1.6 to 3-fold for each 10°C rise in temperature (Rome, 1990), elevated power output of the locomotory muscles may be one advantage of endothermy, although this claim has been disputed (Block, 1991b). Elevated temperatures may also accelerate the myoglobin-mediated diffusion of oxygen, and thereby the rate of oxygen delivery, to mitochondria of the swimming musculature, rather than influencing muscle performance directly (Stevens and Carey, 1981). Furthermore, accelerated rates of metabolism associated with endothermy may aid in the clearance of lactate: typical teleosts require 12 to 24 h for blood lactate to return to control levels following exhaustive exercise, while only 2 to 3 h are required in skipjack and yellowfin tuna (Stevens and Neill, 1978). An additional benefit of visceral endothermy is increased digestive efficiency, which may compensate for the relatively small mass of visceral organs in warm-bodied fish (Carey et al., 1971): skipjack held at 23 to 26°C, empty their stomach contents within 12 h, which is 4 to 5 times faster than typical of temperate-water piscivorous fish of comparable size (Stevens and McLeese, 1984). Finally, insulation from large and rapid changes in body temperature may permit niche expansion. For example, even small tunas can sound for several minutes below the thermocline with only a slight depression in muscle temperature (Stevens and Neill, 1978).

ECTOTHERMY/POIKILOTHERMY IN FISHES

With the exception of the endothermic fishes, all living members of the vertebrate classes Agnatha, Chondrichthyes, and Osteichthyes are *ectothermic*, with their body temperatures being determined by the thermal characteristics of the environment. For example, mean excess tissue temperatures for both muscle and brain were ≈0.45°C in *Tilapia* weighing between 40 and 200 g, but averaged less than 0.15°C in smaller fish (Stevens and Fry, 1970). The rate at which fish change temperature when moving from one thermal environment to another depends primarily on body size, with half-lives for thermal exchange varying from 1 min in a 30-g alewife (*Alosa pseudoharengus*) to 13 min in a 3.6-kg lake trout (Spigarelli et al., 1977). Thus, body and ambient water temperatures may vary significantly over the short term because of the thermal inertia imparted by tissue heat capacity and physiological factors that differentially influence rates of warming and cooling (heating generally occurs more rapidly than cooling, presumably as a result of cardiovascular and respiratory adjustments; Crawshaw, 1976). However, in general, body temperatures of fish depart little from ambient water temperature, except when the latter is changing.

Although changes in body and ambient temperature are obligately linked, fish manage and compensate for thermal stress by a variety of mechanisms (Crawshaw et al., 1990), including: (1) behavioral adjustments; (2) physiological or autonomic responses to acute temperature change; and (3) acclimatory (within the lifetime of an individual) or adaptational (over evolutionary time) adjustments to persistently altered temperatures. The first two of these mechanisms are complementary and activated within seconds to minutes, whereas the third requires some period of exposure to altered body temperatures before becoming effective. This temporal hierarchy of compensatory responses provides a conceptual framework for discussing the thermal biology of poikilothermic fishes.

Behavioral Thermoregulation in Fishes

Although most fish are obligate poikilotherms, they do regulate their body temperature if given the opportunity to do so (Crawshaw and Hammel, 1974).

For example, green sunfish (*Lepomis cyanellus*) acclimated to 25°C, although spending 30% of their time at temperatures more than 6°C removed from the *preferred temperature* of 24.7°C, regulate their body temperature to within 3 to 5°C (Nelson and Prosser, 1979). Similarly, largemouth blackbass (*Micropterus salmoides*) precisely regulate their body temperature within the range from 29.3 to 30.9°C when placed in a gradient tank (Reynolds et al., 1976). Even endothermic fish bask in warmer surface waters as a means of regulating body temperature (Block, 1991b). Furthermore, frequency distributions of temperature data transmitted from striped bass (*Marone saxatilis*) confirm that free-swimming fish occupy progressively deeper positions in the water column of an experimental quarry as summer progresses, because surface warming and water mixing cause the preferred temperature to be located at greater depths (Coutant, 1987). Annual migrations of striped bass (from the coast of southern New England in summer to the Carolina coast in winter) and longhorn sculpin (*Myoxocephalus octodecemspinosus*) also appear to be directed movements tracking a zone of water at the preferred body temperature (Coutant, 1986; DeVries, 1982). Temperature selection behavior presumably places a fish at the temperature where it functions most efficiently and narrows the range of temperatures to which it must compensate by other means.

Fish, like all other vertebrates, initiate thermoregulatory behavior by responding to sensory input from both peripheral and central thermoreceptors (Crawshaw et al., 1990). Processing and integration of this information is accomplished in the preoptic anterior hypothalamus (Crawshaw, 1980). Intracellular recordings from preoptic neurons of the green sunfish, *Lepomis cyanellus*, have identified temperature-sensitive units which respond primarily (by an increase in firing rate) to warming (17% of the cells sampled; only 2% were cold-sensitive, and 81% did not respond to temperature) and which are randomly dispersed rather than grouped into discrete nuclei (Nelson and Prosser, 1981a and b). Similar temperature-sensitive units have also been described in trout brain (Greer and Gardner, 1974). Furthermore, heating of the rostral brainstem in both the brown bullhead, *Ictalurus nebulosus* (Crawshaw and Hammel, 1974), and sharks (Crawshaw et al., 1990) leads to the selection of lower temperatures (cooling has the opposite effect) and lesions to the preoptic area in both goldfish and green sunfish markedly impair thermoregulatory ability (Nelson and Prosser, 1979). Intracranial injections of norepinephrine (NE, 5 to 50 ng) into the nucleus preopticus periventricularis (NPP) in goldfish lead to a consistent, dose-dependent decrease in selected temperature, suggesting that the NPP is an important locus of thermoregulatory integration and that release of NE initiates movement to cooler water (Wollmuth et al., 1987). The actions of NE were mimicked by both α_1-(phenylephrine)- and α_2-(clonidine)-specific agonists and eliminated by prior injections of an α-(phentolamine) but not a β-(propranolol) antagonist, whereas β-agonists (isoproterenol) were without effect, implicating both α_1- and α_2-adrenergic receptors in the selection of cooler temperatures (Wollmuth et al., 1988). However, the failure of α-antagonists to influence temperature selection in the absence of NE suggests that NE does not exert a chronic influence on body temperature regulation. Dopamine (DA — 25 to 250 ng) also produces a dose-dependent decrease in selected temperature as does a DA antagonist (haloperidol) alone, which (unlike NE) implicates DA in the tonic regulation of body temperature in fish (Wollmuth et al., 1989).

A number of factors modify the setpoint of temperature selection in teleosts. For example, the preferred body temperature of goldfish fluctuates between 26°C during the day and 30°C at night (Reynolds et al., 1978). Growth temperature is of paramount importance: e.g., brown bullheads acclimated to temperatures of 7, 15, 24, and 32°C selected temperatures of 16, 21, 26, and 31°C, respectively (Crawshaw, 1975), and a wide variety of teleosts behave similarly (Cherry et al., 1975). In addition, the preferred temperatures of both striped bass (Coutant, 1986) and rainbow trout (Spigarelli and Thommes, 1979) decrease with age (in the case of striped bass, from 26°C for juveniles to 18–20°C for adult females). Interactions between attributes of the physical environment and temperature regulation are particularly interesting. For example, rainbow trout under conditions of normoxia, and at oxygen tensions of 13.3, 9.3, and 5.3 kPa select temperatures of 16.1, 14.9, and 12.7°C, respectively (Schurmann et al., 1991). A decreased rate of metabolism and increased affinity of hemoglobin for oxygen are among the presumed advantages of selecting a lower temperature during hypoxic exposure. Anoxic goldfish also select cooler (12.8°C) temperatures than normoxic animals (17.5°C), but unlike trout, produce ethanol as an endproduct of anaerobic metabolism. Furthermore, injections of ethanol (0.048 ng) into the NPP of goldfish lead to the immediate selection of cooler temperatures (by 8.7°C; Crawshaw et al., 1989). Finally, threespine stickleback select cooler temperatures (16°C) in fresh than salt water (18°C), perhaps to reduce metabolic and ionic stress (Crawshaw, 1980).

Physiological Adjustments to Acute Temperature Change

Autonomic adjustments associated with the regulation of body temperature in endotherms (e.g., shivering, sweating, panting, and the redistribution of blood flow), although present to a limited extent in some ectotherms, are of no major significance in fish. Nevertheless, changes in temperature of both inspired water and the general body surface alter respiratory and cardiovascular function in fish (heating and cooling produce increases and decreases, respectively, in both gill ventilation and heart rate) and may promote the maintenance of arterial oxygenation during heating, or minimize the energy expended for gill ventilation and active ion transport during cooling (Crawshaw, 1976). Warming also darkens the skin surface and increases peripheral blood flow if applied locally; however, vasomotion has not been implicated as a thermoregulatory effector in fish (Crawshaw, 1980).

Strategies for Coping with Extremes of Environmental Temperature

Although most fish are able to thermoregulate behaviorally, microhabitat selection is not effective in dealing with the long-term (i.e., diurnal, seasonal, or evolutionary) fluctuations in temperature. Consequently, fish display a variety of alternate mechanisms for coping with the extremes of environmental temperature.

Some fish simply respond to seasonal extremes of temperature (frequently correlated with other changes in the environment such as water or food availability) by entering a state of dormancy, thereby circumventing environmental extremes. For example, American eels (*Anguilla rostrata*) swim actively and feed regularly at summer temperatures (13 to 17°C), but at temperatures below 8°C, cease feeding and burrow in the mud (Walsh et al., 1983). Similarly, cooling the brainstem of brown bullheads (*Ictalurus nebulosus*) to 10°C causes fish to select the coldest temperature in a gradient tank and become torpid (Crawshaw and Hammel, 1974), while unrestrained fish at 3°C depress their activity, burrow in the mud, and enter a sleeplike state (Crawshaw et al., 1982). Most importantly, dormancy depresses the metabolic rate. In both American eels and the killifish, *Fundulus heteroclitus*, the Q_{10} for rates of oxygen consumption increases dramatically (from 2–3 to >4) at low temperature (<10°C), implying a regulated state of metabolic depression (Walsh et al., 1983; Targett, 1978). The potential energy savings derived from winter dormancy are considerable: American eels fasted for 6 months at 10°C utilize a larger portion of their lipid reserves and possess lower hepatosomatic indices than dormant eels. Thus, in some circumstances, the costs of remaining active may exceed the derived benefits or the resources available, and in such situations, thermal adaptation (see below) would be maladaptive.

More commonly, exposure to novel temperatures initiates a physiological reorganization which compensates, to varying degrees, for the direct effects of temperature change and permits similar levels of activity to be maintained at different body temperatures. Accordingly, temperature *acclimated* (temperature is the only variable altered in a laboratory experiment), *acclimatized* (referring to seasonal adjustments to the simultaneously changing attributes of a natural environment), or *adapted* fish perform best at or near the temperature of their environment. The remainder of this chapter will focus on various mechanisms of thermal compensation.

Evolutionary Adaptation to Specific Thermal Environments

Aquatic habitats vary in temperature from close to the freezing point of seawater (–1.86°C) in polar regions to temperatures in the range of 40 to 45°C in desert springs and streams, while temperate and tropical seas possess surface temperatures in the range of 10 to 20°C and ≈30°C, respectively. This temperature range far exceeds the acclimatory abilities of any single species. Thermal adaptation is best illustrated by interspecific comparisons among ecotypically similar species inhabiting diverse thermal environments. In particular, attention has focused on polar fishes and their ability to function at temperatures near 0°C.

Antifreeze Production by Polar Fishes

The colligative properties of solutes in seawater depress its freezing point to ≈–1.9°C, whereas the serum of a typical marine teleost freezes at ≈–0.7°C, indicating that, in the absence of special adaptations, fish are at risk of freezing when their temperature falls below –0.7°C (DeVries, 1982). Yet fish are abundant in the polar oceans, which remain at the freezing point of seawater and are ice-covered in their shallow waters. The near-shore waters of North temperate oceans also attain freezing temperatures during winter months.

In the absence of ice, fish can supercool by several degrees, and bottom fishes inhabiting the fjords of Labrador (e.g., *Lycodes turneri* and *Liparis koefoedi*; plasma freezing points ≈–0.97 to –1.07°C), where the water temperature is ≈–1.7°C, spend their entire lives in this metastable state (Scholander et al., 1957). Yet, many Antarctic and Arctic fishes come into direct contact with ice, but do not freeze; in fact, freezing in these fish does not occur unless they are exposed to ice at temperatures below –2.2°C (DeVries, 1988). Furthermore, the colligative properties of the serum account for only ≈40 to 50% of the freezing point depression (Cheng and DeVries, 1991): the melting point of ice in the serum of Antarctic fish is ≈–1°C (largely explained by colligative properties), however, no growth of a seed crystal is observed until the temperature is lowered to ≈–2.2°C (DeVries, 1982). This separation of melting and freezing points (by 1 to 1.6°C or more) is referred to as *thermal hysteresis* and is due to the presence of proteinaceous antifreezes; after dialysis of the serum (to remove salts and other small solutes), the freezing point rises to only –1.2°C (from ≈–2.2°C) whereas the melting point is –0.02°C (van Voorhies et al., 1978). Thus, serum antifreezes are without effect on the melting point of ice, but lower the freezing point to well below that predicted from colligative properties. Antifreeze proteins may be present at concentrations up to 10 mg/ml (Scott et al., 1986), accounting for as much as 3% of the total serum protein (Cheng and DeVries, 1991), and on a molar basis are 200 to 300 times more effective than NaCl in depressing the freezing point of water (Devries, 1988).

Antifreeze proteins produced by fish fall into two general categories, the glycoprotein antifreezes (AFGPs) and the nonglycosylated, antifreeze peptides (AFPs). At least three families of Antarctic fish belonging to the suborder Nototheniodei and the northern cod (Gadiformes) produce essentially identical AFGPs which are composed of tandem repeats of the tripeptide alanine-alanine-threonine with the disaccharide, β-D-galactopyranosyl-(1→3)-2-acetamido-2-deoxy-α-D-galacto-pyranose, linked to the threonine residue (Figure 2) (Cheng and DeVries, 1991). AFGPs are produced in at least eight different

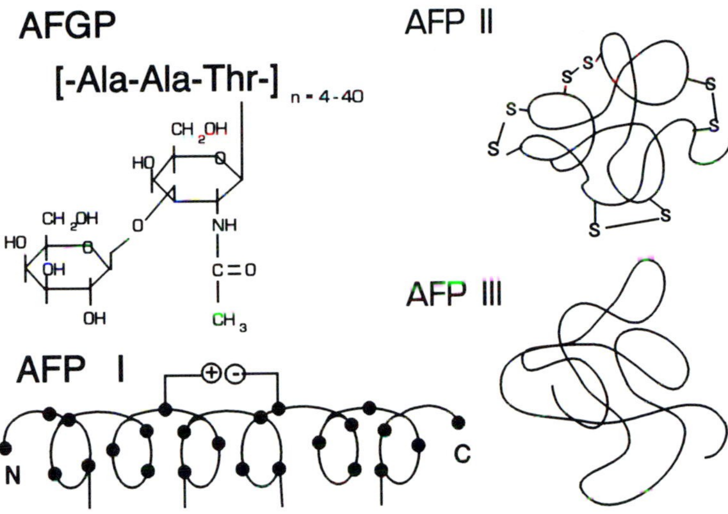

FIGURE 2. Schematic representations of the structures of antifreeze proteins (redrawn from Davies and Hew, 1990). AFGP, antifreeze glycoproteins; AFP, antifreeze proteins. Refer to text for further details.

size classes varying in molecular weight from 2.6 to 34 kDa, depending on the number of repeating glycotripeptide units; the smaller AFGPs are less effective antifreezes than the larger-size classes (Schrag et al., 1982). The *cis*-hydroxyl groups projecting from the disaccharide units are essential for function since treatment with either borate (DeVries, 1971) or lectin (Osuga and Feeney, 1978) eliminates antifreeze activity.

In contrast, AFPs occur in three, chemically distinct subtypes (Davies and Hew, 1990). Type I AFPs are present in winter flounder (*Pseudopleuronectes americanus*), Alaskan plaice (*Pleuronectes quadritaberulatus*), and short-horn sculpin (*Myoxocephalus scorpius*) and vary in molecular weight between 3 and 4 kDa. The two major AFPs in this category are each 37 amino acids in length and contain three 11-amino acid tandem repeats of the sequence Thr-X_2-Asx-X_7, in which X is usually alanine or some other amino acid that favors helical structure. The net effect of the tandem repeats is the formation of an α-helix which is amphipathic, with hydrophilic and hydrophobic amino acid residues largely sorted to different sides of the helix (Figure 2); the helical content of flounder AFPI is 85% or greater at 1°C (Davies and Hew, 1990). Furthermore, the polar amino acids appear to be essential for antifreeze function, since both chemical modification of these residues in native peptides (Duman and DeVries, 1976) and omission of these amino acids from chemically synthesized analogues (Chakrabartty et al., 1989a and b) result in reduced antifreeze activity.

Type II AFPs have been reported only in sea ravens (*Hemitripterus americanus*) of the northern hemisphere, are 14 kDa in size, and are unique in possessing a high content (8/100 amino acid residues) of cysteine, making them susceptible to inactivation by sulfhydryl reagents (Davies and Hew, 1990).

Finally, the AFPs of one Arctic (*Macrozoarces americanus*) and two Antarctic (*Rigophila dearborni* and *Austrolycicthys brachycephalus*) eel pouts (Zoarcids) comprise a third class (Type III) of AFP consisting of from 3 to 7 peptides (depending on species) all of about 6.9 kDa. Unlike other AFPs, these peptides lack both a novel amino acid composition and repeating elements in their primary structure (DeVries, 1988).

Correlations of antifreeze (AF) structure (type) with the zoogeographical distribution and phylogenetic relationships among fish indicate that closely related species inhabiting the same environment often produce different types of antifreeze, while other species thought to be in different orders and inhabiting opposite poles express essentially identical antifreezes. Paleoclimatology suggests that teleostian emergence and radiation occurred in a thermal environment 6 to 14°C warmer than present day and that two subsequent cooling events, Antarctic and Arctic glaciation (occurring 30 million and 2.5 million years ago, respectively), drove antifreeze evolution. Fish lineages which have been challenged to produce antifreeze in response to glaciation events and the types of antifreeze they produce are summarized in Figure 3 (Scott et al., 1986). Note that at least three families of Antarctic Nototheniids produce essentially identical AFGPs, consistent with their origin slightly before or during family diversification 30 to 40 million years ago. The presence of type III AFP in both Antarctic and northern hemisphere Zoarcids (a different suborder than Nototheniids) implies an independent evolution occurring sometime after suborder diversification and further suggests that northern Zoarcids arose by the transhemispheric migration of their antecedents. In addition, the nearly identical AFGPs of Northern cods (Gadiformes) and Nototheniids suggests (1) that the latter may be more closely related to the more primitive Gadiformes than the Perciformes, and (2) that the divergence between the Nototheniids and Gadiformes occurred 30 to 40 million years ago after AFGP evolution. Conversely, the expression of different AFP types in two species of northern hemisphere cottids (sculpin and sea raven) that bear a close physical resemblance to one another and share a common habitat confined to the North Atlantic, presumably reflects the more recent challenge of Arctic glaciation to AF production. Finally, the similarity of AFPs in winter flounder and the distantly related shorthorn sculpin is presumed to reflect convergent evolution, since cDNA probes to flounder AFP failed to detect any homology to sculpin liver mRNA.

For the most part, all AFPs and AFGPs depress the freezing point of water to a similar extent (on a weight basis), suggesting a common mode of interaction with ice regardless of structure. Unlike other molecules of similar size that are excluded from ice during freezing, AFPs and AFGPs are retained in the ice crystal lattice, prompting DeVries and colleagues to postulate an adsorptive mechanism of crystal growth inhibition (Cheng and DeVries, 1991). The

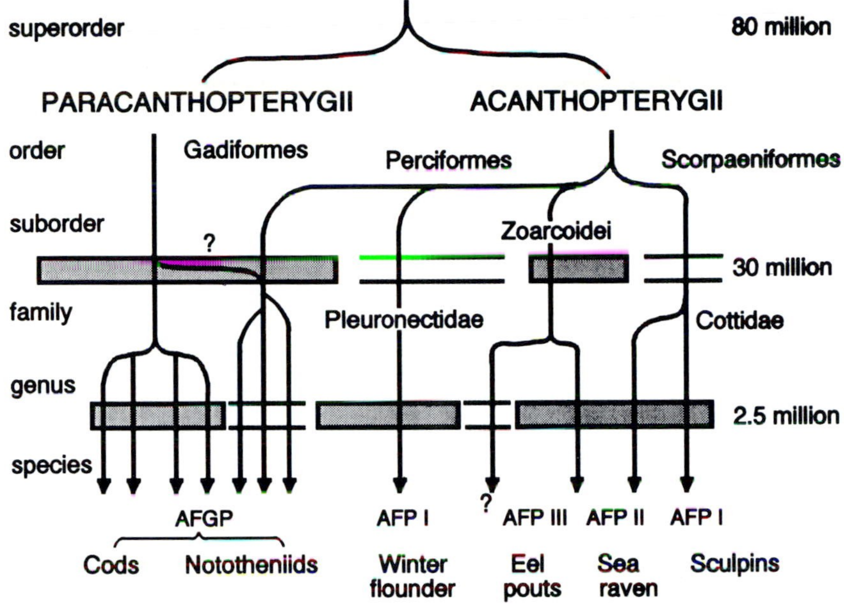

FIGURE 3. Presumed phylogenetic relationships between present-day fish that express well-characterized antifreeze proteins (redrawn from Scott et al., 1986). The band at 30 million years corresponds to the onset of Antarctic glaciation and the stippled regions correspond to lineages presumed to have developed antifreezes at that time. The similar band at 2.5 million years represents Arctic glaciation. Refer to the text and Figure 2 for a description of antifreeze types. The question marks indicate uncertainty about the antifreeze type(s) in the Antarctic eelpout, and the time of divergence of the nototheniids and cods.

adsorption of antifreeze to the ice crystal surface is presumed to involve a lattice match between hydrogen-bonding groups in the antifreeze molecule and oxygen atoms in the ice crystal. That different antifreezes preferentially adsorb to different planes of the ice crystal (Raymond et al., 1989) further supports the argument for a structural match. Furthermore, in Type I AFPs, the polar threonine residues are nearly colinear on one side of the helix and separated by a distance of 16.5Å, which closely matches the 16.7-Å repeat spacing along one plane of the ice crystal lattice (Knight et al., 1991). Adsorbed AF molecules adversely affect the thermodynamics of crystal growth by dividing the growing face into many fronts and forcing growth of the crystal to occur in regions between them. As a result, growth fronts possess larger than normal surface areas, which is thermodynamically unfavorable and retards crystal growth (Knight et al., 1991).

Although it is clear that AF molecules can adsorb to the prism faces of an ice crystal and thereby inhibit growth, it is less clear how and where they function *in situ*. In the Notothenioid, *Dissostichus mawsoni*, AFGPs are present in the peritoneal, pericardial, and extradural fluids in quantities similar to blood, but are absent from most secreted fluids, including the urine, endolymph, and aqueous and vitreous humours, indicating that AF proteins secreted by the liver may distribute passively into some fluid compartments but do not readily cross "tight" epithelia (Ahlgren et al., 1988). However, AFGPs are carried into the intestine along with the bile, where they may prevent inoculation of the intestinal contents (which are hypo-osmotic to seawater; DeVries, 1988), and AF molecules of all types appear in the urine of northern hemisphere fish with glomerular kidneys (Fletcher et al., 1989), although this finding has been disputed (Cheng and DeVries, 1991). Nevertheless, it is not possible from the systemic distribution of AF proteins to determine whether they function by modifying the growth habits and temperatures of ice crystals in the body fluids, or by preventing the propagation of ice from the environment. AF proteins do penetrate the skin and increase the effectiveness of the dermal epithelium in preventing ice propagation (Valerio et al., 1992), but this does not prevent the proteins from acting in other ways.

Evidence for Physiological Adaptation to Temperature

Although production of antifreeze proteins is the most unique adaptation to life at freezing temperatures, it is but one of many, frequently profound, adjustments in physiological systems that permit maintenance of function under extreme environmental conditions. Although numerous examples could be cited, the following discussion will focus on evidence supporting thermal adaptation in: (1) rates of metabolism; (2) neuromuscular function; and (3) the physical properties of biological membranes.

Thermal Compensation of Metabolism

The extent to which organisms are adapted to their thermal environment is evident from interspecific comparisons among polar, temperate, and tropical teleosts of metabolic rates (Scholander et al., 1953), which when compared at the respective environmental temperatures are roughly similar. Conversely, rates of metabolism are generally higher in polar than temperate fishes at temperatures near freezing. These observations describe a phenomenon referred to as "temperature compensation of metabolism". Although early studies failed to control adequately for such factors as taxonomic diversity, body size, level of activity, locomotory abilities, and general patterns of life history, all of which can significantly influence metabolic rate (Holeton, 1974), more carefully controlled comparisons between confamilial and congeneric species confirm that Antarctic fish, in particular, display higher rates of oxygen consumption than temperate species (by at least twofold) when compared at an assay temperature near 0°C (Somero, 1991). Pairwise comparisons of maximal enzyme activities between Antarctic and temperate zone teleosts of similar ecotypes confirm this trend: enzymes from both central pathways of aerobic metabolism (citrate synthase, cytochrome c oxidase) and fatty acid oxidation (e.g., carnitine palmitoyl transferase) were more active (1.5- to 27-fold) in oxidative muscles of the polar species (Crockett and Sidell, 1990). These data indicate significant metabolic cold adaptation of aerobic energy metabolism reliant on fatty fuels in Antarctic species (Sidell, 1991). Similarly, at 5°C, rates of hepatic protein synthesis in the toadfish are only one tenth of those observed for Antarctic fish (*Trematomus bernacchii* and *T. hansoni*) at –1.5°C (Haschemeyer and Mathews, 1983).

Adaptations in Neuromuscular Function

Conduction velocities determined at 0°C average nearly twofold higher (≈8 vs. 4.7 m/s) in neurons of Antarctic than temperate fish (Macdonald, 1981). Synaptic transmission at the neuromuscular junction is similarly cold adapted in the Antarctic fish, *Pagothenia borchgrevinki* (Macdonald and Montgomery, 1982). In addition, the maximal isometric tension developed at normal body temperatures by muscles of Antarctic, temperate, and tropical teleosts is similar (150 to 300 kN/m^2), indicating nearly perfect thermal compensation of force generation (Johnston and Altringham, 1985). Conversely, when compared at 0°C, maximal isometric tension is 5 to 7 times greater in muscles of Antarctic than tropical fish species (Johnston and Harrison, 1985). Furthermore, at temperatures below 10°C, the kinetics of muscle contraction are considerably faster in Antarctic than tropical species (McVean and Montgomery, 1987), presumably reflecting the 6-fold higher rates of sarcoplasmic reticular Ca^{2+} transport at 0°C in Antarctic compared to tropical species (Johnson and Johnston, 1991). In contrast, there is relatively little adaptation in unloaded contraction velocities (V_{max}), an impression reinforced by the lower maximal swimming speeds of Antarctic compared to temperate fishes (Johnson and Johnston, 1991). Thus, thermal adaptation of muscle function is reflected primarily in the ability of cold-adapted species to generate more force per ATP molecule at low temperatures than temperate and tropical forms (Altringham and Johnston, 1986).

Homeoviscous Adaptation in Biological Membranes

Temperature exerts a modifying influence on the conformation of phospholipid acyl chains, and thereby determines: (1) the phase state; (2) rates of molecular motion; and (3), the molecular order (i.e., packing arrangement) of lipids in biological membranes. Within the physiological range of temperatures, membrane lipids are present in a relatively fluid, disordered *liquid-crystalline* phase, in which the acyl chains of phospholipids are highly mobile (Figure 4) (Casal and Mantsch, 1984). As temperature drops below the physiological range, a "frozen" crystalline or *gel phase* is formed as the acyl chains pack tightly into a highly ordered, all-*trans* conformation. A region of phase separation (consisting of coexisting domains of fluid and gel-phase lipids) is frequently present in the phase transition (gel/fluid) region, which, in biological membranes, may extend over a relatively broad temperature range (10°C or more). Conversely, with warming above the normal

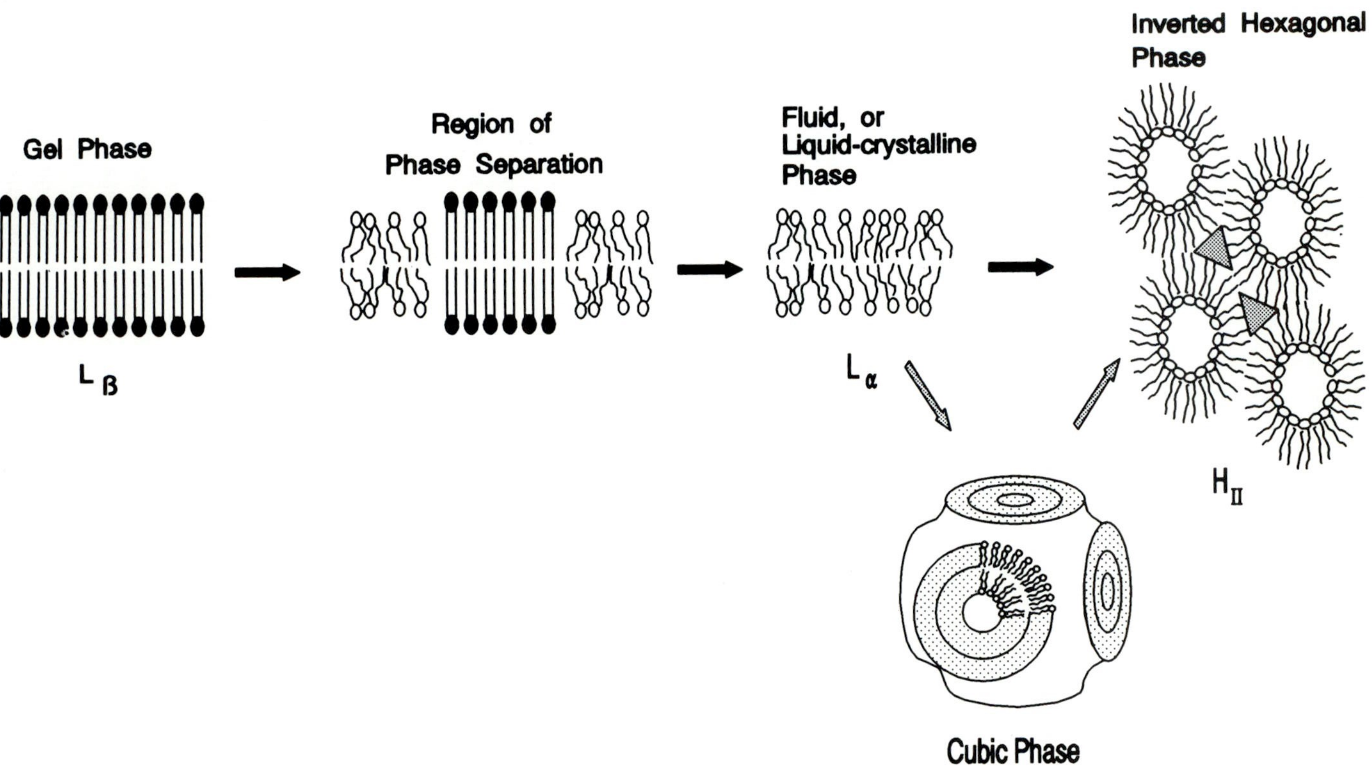

FIGURE 4. The effects of increasing temperature (reading from left to right) on the phase state and conformation of phospholipids in biological membranes.

physiological range, some lipids assume nonbilayer structures (such as hexagonal and cubic phases — Figure 4), resulting in a loss of membrane integrity (Mantsch et al., 1981). Temperature also modulates rates of molecular motion and the order of membrane constituents within a given phase, with order increasing and rates of motion decreasing as temperature declines (illustrated in Figure 8). Consequently, changes in temperature perturb membrane structure and function and pose a formidable challenge to the maintenance of integrated body function since membranes mediate a variety of essential cell activities (see Hazel and Williams, 1990 for an extensive review).

In spite of the marked temperature dependence of membrane structure and function, interspecific comparisons among teleosts from different thermal environments document a conservation of membrane fluidity (as sensed by the motion of fluorescence probes such as 1,6-diphenyl-1,3,5-hexatriene [DPH]) when measured at normal physiological temperatures. As illustrated in Figure 5A, synaptosomal membranes from Antarctic fish (of the genus *Notothenia*) and the Arctic sculpin are more fluid (as indicated by lower polarization values, reflecting a less restricted probe environment and a more disordered membrane) than membranes of the desert pupfish when measured at a common temperature, with membranes of the temperate goldfish being intermediate in fluidity (Cossins et al., 1987). However, when polarization values are compared at the respective cell or body temperatures (Figure 5B), they are roughly equivalent, illustrating the phenomenon of *homeoviscous adaptation* (HVA). Temperatures at which values of DPH fluorescence polarization are identical in synaptosomes of Antarctic fish and desert pupfish differ by 18°C, whereas body temperatures differ by ≈34°C, indicating a homeoviscous efficacy of 53% (18/34 × 100) (Cossins and Prosser, 1978). In general, adaptation to cold temperatures disorders biological membranes to an extent which only partially compensates (50 to 70%) for the direct effects of temperature on membrane fluidity (Cossins and MacDonald, 1989). Nevertheless, several lines of evidence suggest that even partial HVA may contribute to thermal compensation of membrane function. Correlations between membrane fluidity and the activity of both Na^+/K^+- and Ca^{2+}-ATPases are robust, indicating that properties of the lipid environment can modulate the activity of these enzymes (Harris, 1985; Squier et al., 1988). Furthermore, the LT_{50} (the temperature at which 50% of the native enzyme activity is denatured in a 15-min period) of Na^+/K^+-ATPase from the fluid synaptosomes of an Antarctic fish (38.4°C) is ≈9°C lower than that of 28°C-acclimated goldfish (47.7°C), implying a causal relationship between membrane fluidity and the thermal stability of membrane-associated proteins (Cossins et al., 1986). Additional evidence of thermal adaptation in biological membranes is the lower gel/fluid phase transition temperature of sarcoplasmic reticular lipids from cold-water cod (–11.6°C) than either carp or *Tilapia* (–1.8 and –5.4°C, respectively; Howell and Matthews, 1991).

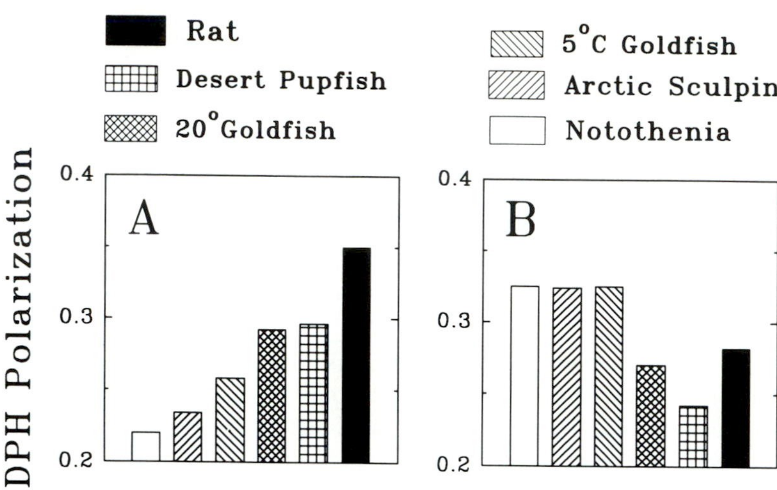

FIGURE 5. Fluorescence polarization values for DPH in synaptosomal membranes of various species of fish and a homeotherm (the rat) illustrating the phenomenon of homeoviscous adaptation. Values in A were determined at a common assay temperature of 20°C; the observed ranking of membrane disorder is inversely correlated with body or habitat temperature: Antarctic fish (–1.9°C)>Arctic fish (0°C)>goldfish (5 to 25°C)>pupfish (34°C). Values in B were determined at the respective cell or body temperatures. (From Hazel, J. R. and Williams, E. E., *Prog. Lipid Res.*, 29(3), 167, 1990. With permission from Pergamon Press, Ltd., Oxford, UK.)

Mechanisms of Thermal Adaptation

Thermal adaptation is evident at all levels of biological organization ranging from the morphological and organ/system levels down to individual cells and molecules. For example, Antarctic fish reduce red blood cell number and hemoglobin concentrations (or in the extreme case of the *Channichthyidae*, produce no red cells at all) to offset the increase in blood viscosity at low temperature and ensure adequate oxygen delivery by relative (compared to temperate fish) increases in heart size, blood volume, and cardiac output (D'Avino et al., 1991). Nevertheless, molecular mechanisms which permit rates of catalysis and the stability of biological structure to be temperature compensated have been most thoroughly documented.

Adaptive Variations in Protein Concentration and Primary Structure

Two primary adaptations of polar fishes include: (1) a relatively high intracellular concentration (i.e., an increased number of catalytic sites per unit of cell volume) of some enzymes, especially those associated with aerobic respiration; and (2) a higher (compared to temperate species) inherent catalytic activity per active site (i.e., faster-working enzymes, as reflected in higher turnover numbers [the number of moles of substrate converted to product per mole of enzyme per second — k_{cat}]; Somero, 1991). Of these, the latter has been the easiest to demonstrate. In particular, free energies of activation decline with a drop in cell temperature in comparisons among homologous enzymes from species inhabiting different thermal environments, due primarily to a decrement in the enthalpic contribution to the free energy term (Sidell and Moerland, 1989); consequently, catalytic efficiencies are inversely correlated with body temperature, as illustrated by higher values of k_{cat} for enzymes of cold- compared to warm-adapted teleosts in Figure 6. Central to the thermal compensation of enzyme-catalyzed reaction rates are adjustments in the binding of both substrates and modulators, which, although essential to the regulation of metabolic flux, is very sensitive to perturbation by temperature. Interspecific comparisons among enzyme homologues reveal that, although K_m rises steeply with temperature just above the normal physiological range, K_m values are conserved across species when compared at the normal body temperature (Figure 7). For example, K_ms for pyruvate of lactate dehydrogenase (LDH) homologues from Antarctic (*P. borchgrevinki*) and temperate fish, as well as 15 species of teleost of the family Scianidae which differ in thermal habitat, are conserved within the range of 0.1 to 0.4 m*M* when compared at their respective body temperatures, but are consistently and significantly higher in cold- than warm-adapted species when compared at a common assay temperature (Yancey and Somero,

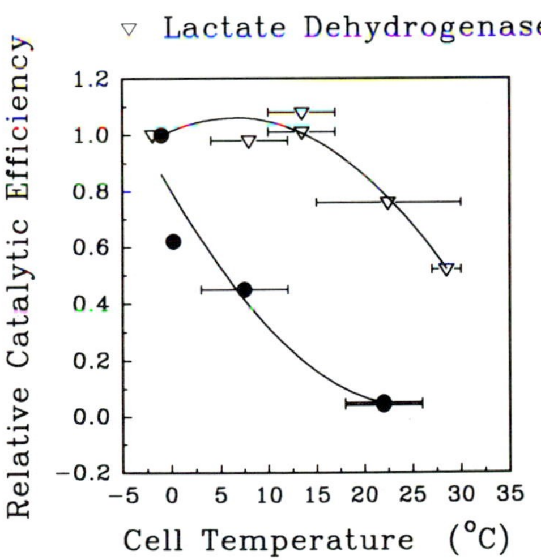

FIGURE 6. Variations in relative k_{cat} with cell or habitat temperature in homologues of lactate dehydrogenase and myosin ATPase for various species of teleost fish. Species illustrated (proceeding from left to right) are: *P. borchgrevinki*, *S. pinninger*, *P. stellatus*, *S. alascanus*, *T. thynnus*, and *P. volitans* (for LDH); *C. gunnari*, *N. neglecta*, *C. bubalis*, *D. carneus*, and *P. uniocellatus* (for myosin ATPase). The error bars indicate the temperature range of the species in question. (Data plotted are from Table 1 of Somero, 1991.)

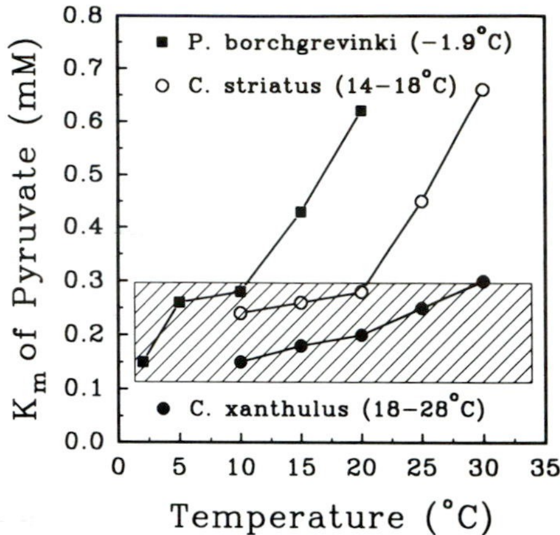

FIGURE 7. The temperature dependence of the K_m for pyruvate in homologues of lactate dehydrogenase from three species of teleost fish inhabiting different thermal environments. The cross-hatched region illustrates the conservation of K_m in different thermal environments (refer to text). (Data are combined and redrawn from Figures 3 and 4 of Somero, 1991.)

1978; Coppes and Somero, 1990). Somero speculates, based on thermodynamic arguments, that differences in binding and catalytic efficiencies among enzyme homologues reflect underlying differences in the number of weak bonds formed between the enzyme and ligand during the activation process (Somero, 1978). Namely, the disruption of relatively few bonds in the formation of products by enzymes of cold-adapted species would keep the activation energy relatively low and the catalytic rate relatively high. However, for the enzyme of a tropical species to bind ligand as strongly at high temperature as a polar homologue at 0°C, necessitates the formation of additional weak bonds that must ultimately be broken during the activation event, resulting in reduced catalytic efficiency.

Weak bonding forces are also important in stabilizing the higher levels of protein structure and proteins of warm-adapted species are generally more thermostable than those of cold-adapted species (Somero, 1991). Furthermore, globular proteins are only marginally stable at their respective cell temperatures and the energy barrier for thermal denaturation can be as low as 5 to 15 kcal/mole (Fontana, 1991). These observations document structure- function compromises in enzyme evolution, for although enzyme homologues from cold-adapted species are catalytically more efficient than their temperate and tropical counterparts, they are also more susceptible to thermal denaturation, indicating a less rigid structure. Enzymes must be rigid enough to maintain the appropriate geometries at ligand-binding sites under the prevailing thermal conditions, yet remain sufficiently flexible to undergo changes in conformation required for catalysis. Thus, the stability of protein structure reflects an evolutionary balance between catalytic efficiency and ligand binding.

Thermal adaptation of enzyme function most likely reflects variations in protein primary structure which modulate a variety of interactions (including hydrophobic, ionic, hydrogen and disulfide bonding, in addition to metal binding) that vary in significance depending on the enzyme in question (Fontana, 1991). Surprisingly, only small differences in environmental temperature are sufficient to drive these adaptations, as illustrated by comparisons among LDH homologues from congeneric teleost species isolated from one another either latitudinally (Graves and Somero, 1982) or by the Panama land bridge (Graves et al., 1983). For example, Pacific congeners, which experience average water temperatures 2 to 3°C cooler than Atlantic species, express homologues of LDH that may be considered to be cold-adapted based on their higher values of K_m and k_{cat}.

Although the development of more efficient catalysts by cold-adapted fish is an elegant example of molecular evolution, the increase in catalytic efficiency rarely exceeds twofold, whereas, as noted previously, maximal rates of enzyme activity are frequently more than twofold higher in cold- compared to warm-adapted teleosts. The greater compensation of reaction rates than catalytic efficiencies is circumstantial evidence suggesting that cold-adapted fish express higher concentrations of selected enzymes, although this has yet to be directly demonstrated.

Interactions between proteins (and protein subunits) are also temperature compensated as indicated by highly conserved values for the equilibrium constant of the globular-to-filamentous transition in actins from species adapted to widely different temperatures when measured under *in situ* conditions (Swezey and Somero, 1982, 1985). Microtubule assembly (from the $\alpha\beta$-dimers of tubulin) is of particular interest since this process is entropically driven and thus favored by high temperatures, as evidenced by the cold lability of mammalian microtubules at 0 to 4°C. Brain tubulins of Antarctic fish, however, polymerize efficiently at low temperature, indicating a stronger interaction between tubulin dimers in polar fish than homeotherms (Detrich, 1991). In addition, the critical concentration of tubulin required to drive microtubule assembly is conserved within the range from 0.04 to 0.34 mg/ml when measured at the normal cell temperatures. Surprisingly, hydrophobic interactions play a greater role in stabilizing microtubule structure in Antarctic than warm-adapted species, indicating that tubulins of Antarctic fish either form more, and/or qualitatively stronger hydrophobic interactions than those of temperate species.

Adaptations in Membrane Lipid Composition

Similarity in the physical properties of biological membranes from tropical, temperate, and polar teleosts, when compared at their respective temperatures, is attributable to *differences* in membrane lipid composition. For example, brain fatty acids of Antarctic fish are enriched in long-chain, 24-carbon polyunsaturated fatty acids (PUFA) (Morris and Schneider, 1969), and deep-water marine fish possess higher proportions of PUFA (42 vs. 33%) and lower amounts of saturated fatty acids (SFA — 17 vs. 35%) than temperate-water species (Patton, 1975). Similarly, the unsaturation of mitochondrial lipids, as reflected in the ratio of linolenic(18:3) to linoleic(18:2) acids, is higher in catfish (3.85) than rat (0.01) (Richardson and Tappel, 1962). Because a double bond introduces a bend of $\approx 30°$ into the acyl chain, unsaturated fatty acids adopt a more expanded conformation, pack less efficiently, and consequently possess lower melting points than their saturated homologues. Increased membrane unsaturation thus offsets the direct effects of low temperature on acyl chain conformation by reducing membrane order and increasing fluidity (Hazel, 1988).

Phenotypic Variation in Fluctuating Environments: Temperature Acclimation

Thermal compensation in fishes is most evident in eurythermal species that must accommodate to seasonal or diurnal fluctuations in temperature. Individual fish are able to survive and function at widely different water temperatures because many aspects of their physiology and biochemistry are altered in response to temperature change in a manner that compensates, to some degree, for the acute effects of temperature on physiological activity. Such environmentally induced phenotypic plasticity (i.e., the expression of environment-specific phenotypes) generally requires a period of adjustment (i.e., acclimation or acclimatization) that may last from days to weeks.

Evidence for Thermal Acclimation of Physiological Function

The classical evidence for thermal acclimation in fishes and other poikilotherms has been extensively reviewed (Hazel and Prosser, 1974; Hochachka and Somero, 1984) and is derived from comparisons between individuals of the same species that have been acclimated or acclimatized to different temperatures. Acclimation to temperature extremes both improves performance (as assessed by swimming ability) at the acclimation temperature (Griffiths and Alderdice, 1972) and extends the range over which normal activity can be maintained. For example, in goldfish, the upper lethal temperature is increased 1°C (up to 36.5°C) for every 3°C rise in acclimation temperature (Fry, 1967). It is beyond the scope of this chapter to exhaustively review the large body of literature in this area. Instead, selected examples will be used to illustrate the general phenomenon.

Rates of Whole Animal and Cellular Metabolism

Just as polar teleosts exhibit higher rates of metabolism than their tropical counterparts, so do individuals of a species display higher rates of oxygen consumption when acclimated to cold than warm temperatures. For example, goldfish acclimated to 10°C exhibit rates of standard and active oxygen consumption that are 1.92- and 1.7-fold higher, respectively, than those of fish acclimated to 30°C when compared at an intermediate temperature of 15°C (Kanungo and Prosser, 1959). At the cellular level, the capacity for ATP production by aerobic means is commonly elevated at cold temperatures (Hazel and Prosser, 1974). Although pathways of carbohydrate (Stone and Sidell, 1981), lipid (Sellner and Hazel, 1982), RNA/DNA (Das, 1967), and protein metabolism (Kent and Prosser, 1980) all acclimate to temperature change, the nature of these compensations is frequently highly specific. For example, catabolism of carbohydrate via the pentose phosphate pathway and the Krebs cycle is generally elevated in cold- compared to warm-acclimated fish (Helly, 1976), whereas rates of glycolysis are not (Jones and Sidell, 1982). Similarly, although fatty acid oxidation (in red muscle; Jones and Sidell, 1982) and desaturation (Sellner and Hazel, 1982) are accelerated in the cold, rates of phospholipid biosynthesis are depressed (Hazel, 1990). Thus, temperature acclimation involves a total reorganization of cellular metabolism rather than a simple up- or downregulation of general metabolic rates.

Muscle Mechanics and Swimming Performance

Because temperature exerts a large influence on the maximal shortening velocity (V_{max}) and power output of skeletal muscles (Q_{10}s ≈1.5 to 3) (Rome, 1990), locomotor capacities of fish are markedly perturbed by alterations in temperature (Sisson and Sidell, 1987). Among teleosts, cyprinids are unique in exhibiting thermal acclimation of muscle function at the cellular and biochemical level. For example, the maximum isometric tension (P_0) developed at 7°C is 1.3 (slow fibers) to 2 times (fast fibers) higher in 7°C- than 23°C-acclimated carp (*Cyprinus carpio*) and unloaded contraction velocities (V_{max}) increase 1.5-fold following cold acclimation (Johnston et al., 1990); consequently, maximum power output at 8°C was 26.5 and 18 W/kg in cold- and warm-acclimated carp, respectively (Langfeld et al., 1991). Similarly, V_{max} and P_0 values determined at 0°C were 2 to 3 times higher in skinned fibers of 11- compared to 23°C-acclimated carp (Crockford and Johnston, 1990), and slow fibers of the pectoral fin adductor muscles both developed tension and relaxed more rapidly in cold- than warm-acclimated fish (Heap et al., 1987; Bailey and Driedzic, 1990). Furthermore, the activities of both actin-activated myosin ATPase (Hwang et al., 1990, 1991) and SR Ca^{2+}-ATPase (Johnston et al., 1990) are consistently 1.5- to 4-fold higher in cold- than warm-acclimated cyprinids.

Acclimation of Membrane Fluidity

The concept of HVA extends beyond genotypic adaptation to extreme thermal environments, as is illustrated by the fluidity compensation in basolateral membranes of enterocytes from thermally acclimated carp (Figure 8). Note that movement of 30°C-acclimated carp into 10°C water orders the membrane to an

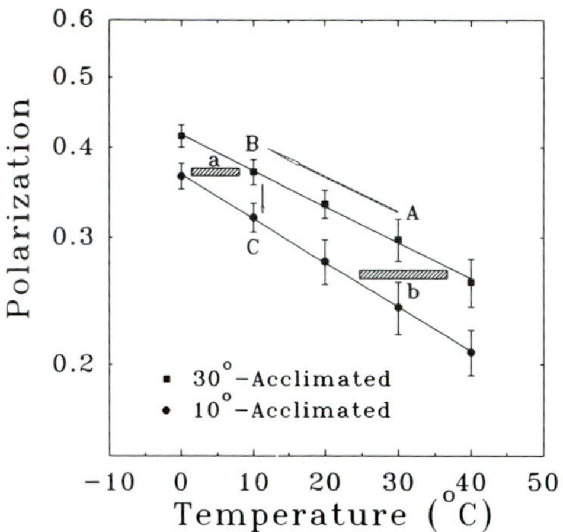

FIGURE 8. Fluorescence polarization values of DPH in basolateral membranes of carp enterocytes illustrating the phenomenon of acclimatory (or intraspecific) homeoviscous adaptation. Acute exposure of a 30°C-acclimated carp to 10°C results in a rigidification of the membrane corresponding to a rise in polarization from point A to point B. Continued exposure to 10°C results in an acclimatory fluidization of the membrane corresponding to a reduction in the polarization from point B to C. The extent of the adaptation varies from ≈50% (refer to bar a) to ≈75% (refer to bar b). (From Hazel, J. R. and Williams, E. E., *Prog. Lipid Res.*, 29(3), 167, 1990. With permission from Pergamon Press, Ltd., Oxford, UK.)

extent indicated by the rise in fluorescence polarization from point A to B. However, prolonged exposure (i.e., acclimation) to 10°C fluidizes the membrane (reflected in the leftward translation of the polarization curve on the temperature axis) to an extent represented by the interval from B to C. Since the temperatures of equivalent fluidity for membranes of the two acclimation groups differ by only 10 to 15°C, whereas the difference in acclimation temperatures is 20°C, the homeoviscous efficacy varies between 50 and 75% (Lee and Cossins, 1990). Although the efficacy of acclimatory HVA is membrane-specific and ranges from 0 to 100% (reviewed in Cossins and Sinensky, 1986), average values are slightly lower (20 to 50%) than the efficacy of genotypic HVA (50 to 70%).

The adaptive significance of acclimatory adjustments in membrane fluidity must ultimately be judged by the extent to which membrane function is temperature compensated. Acclimation of carp from 26 to 10°C shifts the discontinuity in Arrhenius plots of hepatic mitochondrial succinate oxidation (resulting in a marked decrement in enzyme activity at temperatures below the break point) from 23 to 15°C; thus, higher rates of catalytic activity are maintained at low temperature in cold- than warm-acclimated fish (Wodtke, 1976). Furthermore, the temperature at which Na^+/K^+-ATPase denatures rises from 44.9 to 47.7°C in 6°C- compared to 28°C-acclimated goldfish (Cossins et al., 1981), exhibiting a similar dependence on membrane fluidity as interspecific comparisons among polar, temperate, and desert teleosts, further strengthening the causal relationship between membrane fluidity and the thermal stability of membrane proteins.

Modulation of fluidity has also been implicated in the elevated activities of succinate dehydrogenase (Hazel, 1972), cytochrome oxidase (Wodtke, 1981), and Na^+/K^+-ATPase (Raynard and Cossins, 1991) typical of cold-acclimated fish, as well as the compensation of membrane permeability (Hazel and Schuster, 1976) and neural function (Cossins et al., 1977).

Mechanisms of Thermal Acclimation

Altered Gene Expression

Many, if not most, of the physiological adjustments responsible for thermal acclimation in fishes appear to result from altered gene expression (interpreted in the broadest sense to include modulation of transcription, translation, post-translational modification, or protein catabolism).

Temperature regulation of gene expression is clearly illustrated by the synthesis of heat shock proteins (HSPs), a highly conserved family of "stress proteins" (Misra et al., 1989) that vary in size from 25 to 90 kDa and are thought to help solubilize and refold denatured or misfolded proteins (Cheng et al., 1989). HSPs are preferentially synthesized during acute heat stress and their synthesis permits a cell (or organism) transiently to tolerate subsequent exposure to temperatures that would otherwise be lethal (Subjeck and Shyy, 1986). Transfer of rainbow trout fibroblasts from 22 to 28°C leads to both the rapid synthesis (within 6 h) of 27-, 70-, and 87-kDa HSPs and tolerance of exposure to 32°C (Mosser and Bols, 1988). In addition, both the expression of HSP 90 in brain tissue and the threshold temperature for its synthesis (32 vs. 28°C) are higher in summer- than winter-acclimatized gobys (*Gillichthys mirabilis*; Dietz and Somero, 1992), whereas Antarctic fish initiate HSP synthesis at temperatures as low as 8°C (Maresca et al., 1988).

The seasonal appearance of antifreeze proteins in north temperate fishes is an additional example of thermal modulation of gene expression. In winter flounder, AFP mRNA appears in and disappears from the liver 1 to 2 months prior to the appearance and disappearance of AFP in the serum, suggesting multiple sites of regulation (Lin, 1979). Temperature, rather than photoperiod, is the primary factor regulating the expression of AFP genes (Duman and DeVries, 1974; Fletcher et al., 1987), but cold exposure (4°C) leads to the accumulation of AFP mRNA in winter flounder only during fall and winter due to an endogenous rhythm (Price et al., 1986) of endocrine secretion, since hypophysectomized flounder continue to express AFP even in summer (Hew and Fletcher, 1979). In addition, the half-life of circulating AFGPs in Atlantic cod is temperature dependent, varying from 99.4 d at 0°C to 15.6 d at 5°C (Fletcher et al., 1987).

Modification of enzyme quantity has generally been assumed, in the absence of direct experimental evidence, to account for the elevated activities of enzymes of aerobic energy metabolism in cold-acclimated teleosts, thus providing a mechanistic explanation for the thermal compensation of cellular metabolism (Hazel and Prosser, 1974). However, increased enzyme activity does not necessarily reflect altered rates of gene transcription alone. For example, in skeletal muscle of green sunfish, the rate of cytochrome c synthesis declined by 40% following acclimation from 25 to 5°C, but the concentration of cytochrome actually increased 1.5-fold due to an even larger reduction (by 60%) in its rate of degradation (Sidell, 1977).

Temperature acclimation also varies the relative proportions of allozymes or isozymes of a particular protein expressed at a given temperature, so that the catalysts produced are those best suited to function in the prevailing thermal environment. The expression of the heart-type (B_4) isozyme of lactate dehydrogenase in liver tissue of *Fundulus heteroclitus* is the best-studied example of temperature-dependent polymorphism (Place and Powers, 1979). Two allozymes (LDH-B_4^a and LDH-B_4^b) vary in frequency as a function of latitude: the *b* allele is fixed at northern latitudes (coast of Maine with a mean water temperature of ≈10°C), whereas the *a* allele is fixed in southern populations (Florida coast, ≈25°C). Furthermore, at 10°C, the V_{max}/K_m ratio (an index of catalytic efficiency) is 1.5- to 1.7-fold higher for the *b* than *a* allozyme, indicating that the allozyme fixed at northern latitudes is a better catalyst at low temperature (Place and Powers, 1984). The greater catalytic efficiency of the *b* allozyme, coupled with a higher concentration of enzyme (reflected in 2.31- and 1.38-fold higher levels of mRNA and enzyme, respectively) in northern populations, combine to produce nearly perfect compensation of catalytic rate (Crawford and Powers, 1989). Variations in allozyme complement are also correlated with differences in carbon flux (DiMichelle et al., 1991) and rates of oxygen consumption (Paynter et al., 1991) in embryonic fish, indicating that modulation of enzyme polymorphism may alter metabolic capabilities and permit subpopulations of a species to acclimate to specific thermal environments.

The environmental induction of isozymes best suited to function at either warm or cold temperatures has long been advanced as a mechanism of thermal acclimation (Hochachka and Somero, 1984). However, many early studies failed to distinguish between genetic polymorphism and isozyme induction, and subsequent research combining thorough genetic and electrophoretic analyses has frequently failed to detect changes in isozyme repertory attributable to temperature acclimation (Wilson et al., 1973; Shaklee et al., 1977). Nevertheless, in both carp (Kurokawa and Nakano, 1991) and the temperate fish, *Leiostomus xanthurus* (Schwantes and Schwantes, 1982), the proportions of the A_2 homodimer of soluble malate dehydrogenase, which is relatively thermostable compared to the B_2 homodimer, increase (relative to the B_2 homodimer) in both heart and skeletal muscle of 30°C- compared to 15°C-acclimated fish, a pattern consistent with thermal acclimation of enzyme function; similar results have also been reported for lactate dehydrogenase isozymes in goldfish liver (Yamawaki and Tsukuda, 1979). The role of isozyme induction in the low-temperature acclimation of muscle function in cyprinid fishes is also contentious. For example, peptide maps (generated by α-chymotrypsin cleavage) of purified myosin heavy chains (SF1) (Hwang et al., 1991) and the isoforms of myosin heavy chain mRNA (Gerlach et al., 1990) differed in skeletal muscle of carp acclimated to 10 and 30°C. Other authors, however, failed to detect temperature-induced differences in myosin (or troponin subunit) isoform content in the same species of fish (Crockford and Johnston, 1990; Langfeld et al., 1991). Finally, the proportions of myosin light chain isoforms favoring higher contraction speeds have also been reported to increase at low temperature in carp muscle (Johnston et al., 1990; Langfeld et al., 1991).

Membrane Remodeling

Fish, like other poikilotherms, exploit the diversity of lipid structure to fashion membranes with physical properties appropriate to the prevailing ambient temperature. The retailoring of membrane lipid composition is probably the most ubiquitous and one of the few continuously graded cellular responses to temperature change.

In fish, many aspects of membrane lipid composition are influenced by growth temperature. A reduction in saturated fatty acids (SFA) and a corresponding increase in unsaturates (UFA) is a common consequence of cold acclimation: in fish, UFA/SFA ratios increase, on average, 1.3- to 1.4-fold for a 20°C drop in acclimation temperature, due primarily to elevated proportions of polyunsaturated fatty acids (PUFA) in winter-active species (Hazel, 1988). These acclimatory adjustments in fatty acid composition are similar to those previously noted in studies of cold adaptation when comparing temperate and polar fishes, and presumably have a similar explanation. However, results of numerous studies reveal a range of compositional adjustments in addition to the modulation of acyl chain unsaturation, and a level of detail and complexity to the acclimatory response that have not yet been examined in an interspecific context. For example, a membrane containing equal proportions of 16:0/16:0-PC (i.e., dipalmitoyl phosphatidylcholine), 18:0/18:0-PC, and 18:1/18:1-PC has the same fatty acid composition as a membrane composed of 16:0/18:0-, 16:0/18:1-, and 18:0/18:1-PC, even though the molecular species compositions are different. Since different mixtures of phospholipid molecular species sharing the same overall fatty acid composition can display markedly different properties, "molecular species remodeling" may contribute to the early stages of the acclimatory response, since no input is required from the lipid biosynthetic apparatus to simply rearrange the existing complement of fatty acids. Consistent with this hypothesis, changes in phospholipid molecular species composition occur rapidly (within 8 to 16 h) both in response to diurnal warming in the Sonoran desert teleost, *Agosia chrysogaster* (the percentage of diunsaturated species of PC declined from 7.23 to 2.46% in muscle microsomes between the cool of the morning and the heat of the afternoon, a temperature difference of 13°C; Carey and Hazel, 1989) and during acclimation from 20 to 5°C in renal plasma membranes of rainbow trout (Hazel and Landrey, 1988a).

Cold-acclimated teleosts also possess higher proportions of phosphatidylethanolamine (PE) and lower proportions of PC in their membranes than warm-acclimated conspecifics; consequently, the PC/PE ratio is positively correlated with growth temperature (Hazel and Williams, 1990). Since PE destabilizes phospholipid bilayers (by favoring formation of the hexagonal phase) whereas PC has the opposite effect, the replacement of PC by PE at cold temperatures, although not sufficient to induce formation of a nonbilayer phase, displaces the phase behavior of the membrane to a point closer to that of the bilayer/ nonbilayer transition (T_h). Consequently, although a drop in temperature increases the temperature interval between T_h and the growth temperature, compensatory adjustments in headgroup composition reduce T_h and thereby restore the proximity of the membrane to the bilayer/ nonbilayer transition point. The balance between nonbilayer- and bilayer-forming lipids (or alternately stated, the proximity to the nonbilayer phase transition) is an important determinant of membrane function (Hui and Sen, 1989), presumably explaining the temperature dependence of the PC/PE ratio in fish membranes. Furthermore, alterations in phospholipid composition are among the most rapidly occurring adjustments in membrane composition: the PC/PE ratio in renal plasma membranes of trout acclimating from 20 to 5°C declines from 1.71 to 0.78 within 8 h (Hazel and Landrey, 1988b).

Other adjustments in membrane lipid composition believed to contribute to the thermal acclimation of membrane function include modulation of membrane cholesterol and the content of ether-linked phospholipids (i.e., plasmalogens), both of which are positively correlated with growth temperature (Wodtke, 1983; Hazel, 1988). Cholesterol presumably stabilizes membranes at high temperature by virtue of its well-established role in ordering fluid-phase lipids (Yeagle, 1985), whereas plasmalogens, in addition to possessing a greater chemical stability than diacyl phospholipids, also contribute to membrane stability by reducing membrane permeability (Schwartz and Paltauf, 1977).

The temperature-induced restructuring of biological membranes undoubtedly is due, in part, to altered activities of key regulatory enzymes in lipid metabolism, as illustrated by the elevated levels of fatty-acyl CoA desaturase activity in cold-acclimated fish (Schunke and Wodtke, 1983). However, the differential effects of temperature on rates of SFA (Q_{10} = 4.3 in flounder; Hansen and Knudsen, 1981) vs. UFA biosynthesis (Q_{10} = 1.4 to 1.5 in carp liver; Schunke and Wodtke, 1983) and a direct effect of temperature on the substrate specificity of the enzymes of phospholipid turnover (Livermore and Hazel, 1988) have also been implicated in shaping the lipid composition of membranes as temperature changes.

Integrated Acclimatory Responses

While many aspects of temperature acclimation have a largely cellular explanation, compensation of some whole animal functions involves the integrated responses of more than one cell or tissue type. Thermal compensation of locomotory function in fish is a case in point. Because muscle performance is impaired at low temperature (Q_{10} of V_{max} 1.6), while the force required to swim at a given speed remains nearly constant, colder-bodied fish must activate more muscle fibers to attain a given swimming speed (Sidell and Moerland, 1989). Consequently, fish compensate for the diminished power output of their muscles at low temperature by compressing the recruitment order of muscle fiber types into a narrower range of locomotor speeds (Rome, 1990). Since only red muscle fibers are recruited at low and sustainable swimming speeds and anaerobic fibers only at higher speeds, the compression of recruitment order at low temperature is manifest in the activation of anaerobic fibers at lower absolute swimming speeds. For example, in carp, white muscle fibers are first recruited at a swimming speed of 26 cm/s at 10°C compared to 46 cm/s at 20°C. Although the compression of recruitment order enables fish to maintain locomotory performance over a moderate range of speeds, locomotion is impaired at low temperature in the sense that the maximum sustainable swimming speeds are reduced (Rome, 1990). However, even in noncyprinid fish (for which the contractile properties of individual muscle cells are not temperature compensated), swimming performance improves following a period of cold acclimation. For example, striped bass (*Marone saxatilis*) acclimated to 9°C can sustain speeds up to 2.5 body lengths per second at 15°C prior to the recruitment of anaerobic fibers (as indicated by electromyography), whereas 25°C-acclimated fish can sustain speeds of only 1.8 lengths per second at the same temperature (Sisson and Sidell, 1987). This improvement in locomotory performance is attributable solely to an increased proportion of red muscle fibers in the trunk musculature, for the cross-sectional area of red fibers in striped bass increases by 60% in cold- compared to warm-acclimated fish (Jones and Sidell, 1982). Unlike the biochemical and cellular adaptations evident only in cyprinid fishes, alterations in the proportion of red muscle mass and the compression of recruitment order are acclimatory responses common to all fish. In fact, the comparison between cyprinid and noncyprinid fishes provides a "natural experiment" to evaluate whether changes in myofibrillar proteins are necessary for enhanced locomotory performance at low temperature (Sidell and Moerland, 1989). The answer appears to be "no", since noncyprinid fishes can improve swimming performance at low temperature in the absence of changes in myofibrillar proteins.

Finally, the failure of *in vitro* acclimation (to 7 and 25°C) of cultured hepatocytes isolated from 15°C-acclimated channel catfish (*Ictalurus punctatus*) to reproduce the *in vivo* acclimatory response, provides further evidence that temperature acclimation, in some cases, may require an integrated, systemic response, possibly endocrine mediated (Koban, 1986).

Acclimation of Tissue and Cellular Morphology

Rearrangements of tissue and cell architecture also contribute to the thermal acclimation of physiological function. Hypertrophy of liver and cardiac muscle is a common adjustment to life at cold temperature. For example, the mass and protein content of the liver nearly double as a result of acclimating channel catfish from 25 to 15°C (Kent et al., 1988). The practical implication of this finding is that the total metabolic capacity of a tissue may well increase as a result of cold acclimation, even though the specific activity of diagnostic enzymes does not. In addition, cold acclimation of carp (from 30 to 10°C) increases both the mean height and breadth of villi along the entire length of the intestine, resulting in a 58% increase in the mucosal surface area (Lee and Cossins, 1988). Furthermore, the length of the intestinal villus over which Na^+-dependent alanine transport occurs is greater in 10°C- than 20°C-acclimated carp (Lee et al., 1991). Thus, variations in tissue morphology may influence intestinal function independently of the transport capacity of individual enterocytes.

Similarly, changes in cellular morphology play a central role in temperature acclimation. For example, the fraction of cell volume occupied by mitochondria rises from 2.9 to 4.5% in red muscle fibers of striped bass following acclimation from 25 to 5°C, which in combination with the hypertrophy of red muscle mass, increases the total mitochondrial volume from 2.58 to 6.73 cm^3/100 g body weight (Eggington and Sidell, 1989). This expansion in mitochondrial volume not only increases the oxidative capacity of the tissue, but also effectively reduces the diffusion distances for oxygen and other metabolites. Cold-induced proliferation of sarcoplasmic reticular surface area and volume density has also been noted in muscle fibers of goldfish (Penney and Goldspink, 1980).

Regulation of the Intracellular Environment

Regulation of the microenvironment in which macromolecules function is an additional mechanism of thermal acclimation. In fish, the pH or arterial blood (pH_e) and the intracellular pH (pH_i) of most tissues vary inversely ($\Delta pH/°C \approx -0.015$) with body temperature (Cameron and Kormanik, 1982). This pattern of pH_i regulation is believed to conserve the net charge state of proteins (α_{im}) at different temperatures, a phenomenon termed imidazole alphastat regulation (Reeves, 1977). Alphastat conservation contributes to the thermal compensation of enzyme function, because (1) the pH optima of enzymes vary with temperature similarly to pH_i (i.e., pH optima shift to more alkaline pH values as temperature drops), consequently, pH_i parallels the pH of optimal activity regardless of temperature (Hazel et al., 1978); and (2) the K_m of lactate dehydrogenase (from Antarctic, temperate, and Amazonian fishes, in addition to tuna) for pyruvate varies less ($\approx$50% less) with temperature (between 10 and 30°C) when assayed under conditions of variable ($\Delta pH/°C \approx -0.015$) than constant pH (i.e., K_m values are conserved as a consequence of imidazole alphastat regulation) (Yancey and Somero, 1978). Furthermore, recent evidence suggests that temperature-dependent patterns of acid-base regulation may account for up to 40% of the fluidization experienced in plasma membranes of trout hepatocytes during acclimation from 20 to 5°C (Hazel et al., 1992).

CONCLUSIONS

In summary, although most fish are obligately poikilothermic due to the constraints imposed by extracting oxygen from an aquatic environment, they nevertheless display a variety of adaptations that reduce the impact of temperature on their physiological function. These adaptations include (1) the capacity for general body endothermy in tuna and lamnid sharks, and regional endothermy in billfishes, which permit key body tissues to operate at a relatively stable, elevated temperature; (2) the capability to regulate body temperature behaviorally in the absence of well-developed mechanisms for heat production and retention; (3) the ability of some fish to depress, in a regulated manner, their metabolic rate and physiological function at extremes of environmental temperature (i.e., to become dormant); and (4) the proficiency of most fish to compensate for differences in body temperature through the processes of temperature adaptation (genotypic variation over evolutionary time) and temperature acclimation (phenotypic variation within the lifetime of an individual). Mechanisms of temperature adaptation and acclimation include (1) refinements to the primary structure of proteins which ensure that catalytic, binding, and assembly characteristics are conserved across the range of biological temperatures; (2) thermal modulation of gene expression, reflected either in altered amounts of the same protein or the preferential expression of thermally adapted isozymes or allozymes at different temperatures; (3) regulation of the chemical composition and physical properties of biological membranes, so that membrane fluidity and function are appropriately attuned to the prevailing thermal environment; (4) a reorganization of both tissue and cellular morphology; (5) adjustments to the composition of the intracellular environment (particularly with respect to pH, but most likely including other attributes as well); and (6) integrative adjustments that ensure the patency of physiological function at extremes of temperature.

REFERENCES

Ahlgren, J. A., Cheng, C. C., Shrag, J. D. and DeVries, A. L., Freezing avoidance and the distribution of antifreeze glycopeptides in body fluids and tissues of antarctic fish, *J. Exp. Biol.*, 137, 549, 1988.

Altringham, J. D. and Johnston, I. A., Evolutionary adaptation to temperature in fish muscle cross bridge mechanisms: tension and ATP turnover, *J. Comp. Physiol. B*, 156, 819, 1986.

Bailey, J. R. and Driedzic, W. R., Enhanced maximum frequency and force development of fish hearts following temperature acclimation, *J. Exp. Biol.*, 149, 239, 1990.

Ballantyne, J. S., Chamberlin, M. E., and Singer, T. D., Oxidative metabolism in thermogenic tissues of the swordfish and mako shark, *J. Exp. Zool.*, 261, 110, 1992.

Block, B. A., Billfish brain and eye heater: a new look at nonshivering heat production, *NIPS*, 2, 208, 1987.

Block, B. A., Evolutionary novelties: How fish have built a heater out of muscle, *Amer. Zool.*, 31, 726, 1991a.

Block, B. A., Endothermy in fish: thermogenesis, ecology and evolution, in *Biochemistry and Molecular Biology of Fishes*, vol. I, Hochachka, P. W. and Mommsen, T., Eds., Elsevier Science Publishers B. V., Amsterdam, 1991b, 269.

Block, B. A. and Carey, F. G., Warm brain and eye temperatures in sharks, *J. Comp. Physiol. B.*, 156, 229, 1985.

Bushnell, P. G. and Brill, R. W., Oxygen transport and cardiovascular responses in skipjack tuna (*Katsuwonus pelamis*) and yellowfin tuna (*Thunnus albacares*) exposed to acute hypoxia, *J. Comp. Physiol. B.*, 162, 131, 1992.

Cameron, J. N. and Kormanik, G. A., Intracellular and extracellular acid-base status as a function of temperature in the freshwater channel catfish, Ictalurus punctatus, *J. Exp. Biol.*, 99, 127, 1982.

Carey, F. G., Fishes with warm bodies, *Sci. Am.*, 228(2), 36, 1973.

Carey, F. G., A brain heater in the swordfish, *Science*, 216, 1327, 1982.

Carey, C. and Hazel, J. R., Diurnal variation in membrane lipid composition of Sonoran desert teleosts, *J. Exp. Biol.*, 147, 375, 1989.

Carey, F. G. and Lawson, K. D., Temperature regulation in free-swimming bluefin tuna, *Comp. Biochem. Physiol.*, 44A, 375, 1973.

Carey, F. G., Teal, J. M., Kanwisher, J. W., Lawson, K. D. and Beckett, J. S., Warm-bodied fish, *Am. Zool.*, 11, 137, 1971.

Carey, F. G., Teal, J. M. and Kanwisher, J. W., The visceral temperatures of mackeral sharks (lamnidea), *Physiol. Zool.*, 54(3), 334, 1981.

Casal, H. L. and Mantsch, H. H., Polymorphic phase behavior of phospholipid membranes studied by infrared spectroscopy, *Biochim. Biophys. Acta*, 779, 381, 1984.

Chakrabartty, A., Ananthanarayanan, V. S., and Hew, C. L., Structure-function relationships in a winter flounder antifreeze polypeptide. I. Stabilization of α-helical antifreeze polypeptide by charged-group and hydrophobic interactions, *J. Biol. Chem.*, 264(19), 11307, 1989a.

Chakrabartty, A., Yang, D. S. C., and Hew, C. L., Structure-function relationship in a winter flounder antifreeze polypeptide II. Alteration of the component growth rates of ice by synthetic antifreeze polypeptides, *J. Biol. Chem.*, 264(19), 11313, 1989b.

Cheng, C. C. and DeVries, A. L., The role of antifreeze glycopeptides and peptides in the freezing avoidance of cold-water fish, in *Life Under Extreme Conditions*, di Prisco, G., Ed., Springer-Verlag, Berlin, 1991, 1.

Cheng, M. Y., Hartl, F.-U., Martin, J., Pollock, R. A., Kalousek, F., Neupert, W., Hallberg, E. M., Hallberg, R. L., and Horwich, A. L., Mitochondrial heat-shock protein hsp60 is essential for assembly of proteins imported into yeast mitochondria, *Nature*, 337, 620, 1989.

Cherry, D. S., Dickson, K. L., and Cairns, J., Temperatures selected and avoided by fish at various acclimation temperatures, *J. Fish. Res. Bd. Can.*, 32(4), 485, 1975.

Coppes, Z. L. and Somero, G. N., Temperature-adaptive differences between the M_4 lactate dehydrogenases of stenothermal and eurythermal sciaenid fishes, *J. Exp. Zool.*, 254, 127, 1990.

Cossins, A. R. and MacDonald, A. G., The adaptation of biological membranes to temperature and pressure: fish from the deep and cold, *J. Bioenerg. Biomembr.*, 21(1), 115, 1989.

Cossins, A. R. and Prosser, C. L., Evolutionary adaptation of membranes to temperature, *Proc. Nat. Acad. Sci.*, 75(4), 2040, 1978.

Cossins, A. R. and Sinensky, M., Adaptation of membranes to temperature, pressure, and exogenous lipids, in *Physiology of Membrane Fluidity*, Shinitzky, M., Ed., CRC Press, Boca Raton, FL, 1986, 1.

Cossins, A. R., Friedlander, M. J., and Prosser, C. L., Correlations between behavioral temperature adaptations of goldfish and the viscosity and fatty acid composition of their synaptic membranes, *J. Comp. Physiol.*, 120, 109, 1977.

Cossins, A. R., Bowler, K. and Prosser, C. L., Homeoviscous adaptation and its effects upon membrane-bound proteins, *J. Thermal. Biol.*, 6, 183, 1981.

Cossins, A. R., Lee, J. A. C., Lewis, R. N. A. H. and Bowler, K., The adaptation to cold of membrane order and (Na^++K^+)-ATPase properties, in *Living in the Cold,* Heller, H. C., Ed., Elsevier, Amsterdam, 1986, 13.

Cossins, A. R., Behan, M. K., Jones, G. and Bowler, K., Lipid-protein interactions in the adaptive regulation of membrane function, *Biochem. Soc. Trans.*, 15(1), 77, 1987.

Coutant, C. C., Thermal niches of striped bass, *Sci. Am.*, 255, 98, 1986.

Coutant, C. C., Thermal preference: when does an asset become a liability? *Environ. Biol. Fish.*, 18(3), 161, 1987.

Crawford, D. L. and Powers, D. A., Molecular basis of evolutionary adaptation at the lactate dehydrogenase-B locus in the fish *Fundulus heteroclitus, Proc. Natl. Acad. Sci. U.S.A.*, 86, 9365, 1989.

Crawshaw, L. I., Attainment of the final thermal preferendum in brown bullheads acclimated to different temperatures, *Comp. Biochem. Physiol.*, 52A, 171, 1975.

Crawshaw, L. I., Effect of temperature change on mean body temperature and gill ventillation in carp, *Am. J. Physiol.*, 231(3), 837, 1976.

Crawshaw, L. I., Temperature regulation in vertebrates, *Annu. Rev. Physiol.*, 42, 473, 1980.

Crawshaw, L. I. and Hammel, H. T., Behavioral regulation of internal temperature in the brown bullhead, *Ictalurus nebulosis, Comp. Biochem. Physiol.*, 47A, 51, 1974.

Crawshaw, L. I., Lemons, D. E., Parmer, M., and Messing, J. M., Behavioral and metabolic aspects of low temperature dormancy in the brown bullhead, *Ictalurus nebulosus, J. Comp. Physiol.*, 148, 41, 1982.

Crawshaw, L. I., Wollmuth, L. P. and O'Connor, C. S., Intracranial ethanol and ambient anoxia elicit selection of cooler water by goldfish, *Am. J. Physiol.*, 256(25), R133, 1989.

Crawshaw, L. I., Wollmuth, L. P., O'Connor, C. S., Rausch, R. N. and Simpson, L., Body temperature regulation in vertebrates: comparative aspects and neuronal elements, in *Thermoregulation: Physiology and Biochemistry,* Schönbaum, E. and Lomax, P., Eds., Pergamon Press, New York, 1990, 209.

Crockett, E. L. and Sidell, B. D., Some pathways of energy metabolism are cold adapted in Antarctic fishes, *Physiol. Zool.*, 63(3), 472, 1990.

Crockford, T. and Johnston, I. A., Temperature acclimation and the expression of contractile protein isoforms in the skeletal muscles of the common carp *(Cyprinus carpio* L.), *J. Comp. Physiol.*, 160B, 23, 1990.

D'Avino, R., Caruso, C., Camardella, L., Schinna, M. E., Rutigliano, B., Romano, M., Carratore, V., Barra, D. and Diprisco, G., An overview of the molecular structure and functional properties of the hemoglobins of a cold-adapted Antarctic teleost, in *Life Under Extreme Conditions*, DiPrisco, G., Ed., Springer-Verlag, Berlin, 1991, 15.

Das, A.B., Biochemical changes in tissues of goldfish acclimated to high and low temperatures. II. Synthesis of proteins and RNA of subcellular fractions and tissue composition, *Comp. Biochem. Physiol.*, 21, 469, 1967.

Davies, P. L. and Hew, C. L., Biochemistry of fish antifreeze proteins, *FASEB J.*, 4, 2460, 1990.

Detrich, H. W., Cold-stable microtubules from Antarctic fish, in *Life Under Extreme Conditions,* di Prisco, G., Ed., Springer-Verlag, Berlin Heidelberg, 1991, 35.

DeVries, A. L., Glycoproteins as biological antifreeze agents in antarctic fishes, *Science*, 172, 1152, 1971.

DeVries, A. L., Biological antifreeze agents in coldwater fishes, *Comp. Biochem. Physiol.*, 73A(4), 627, 1982.

DeVries, A. L., The role of antifreeze glycopeptides and peptides in the freezing avoidance of antarctic fishes, *Comp. Biochem. Physiol.*, 90B(3), 611, 1988.

Dietz, T. J. and Somero, G. N., The threshold induction temperature of the 90-kDa heat shock protein is subject to acclimatization in eurythermal goby fishes (Genus *Gillichthys*), *Proc. Natl. Acad. Sci. U.S.A.*, 89, 3389, 1992.

DiMichelle, L., Paynter, K. T., and Powers, D. A., Evidence of lactate dehydrogenase-B allozyme effects in the teleost, *Fundulus heteroclitus, Science*, 253, 898, 1991.

Dizon, A. E. and Brill, R. W., Thermoregulation in yellowfin tuna, *Thunnus albacares, Physiol. Zool.*, 52(4), 581, 1979.

Duman, J. G. and DeVries, A. L., The effects of temperature and photoperiod on antifreeze production in cold water fishes, *J. Exp. Zool.*, 190(1), 89, 1974.

Duman, J. G. and DeVries, A. L., Isolation, characterization, and physical properties of protein antifreezes from the winter flounder, *Pseuopleuronectes americanus, Comp. Biochem. Physiol.*, 54B, 375, 1976.

Eggington, S. and Sidell, B. D., Thermal acclimation induces adaptive changes in subcellular structure of fish skeletal muscle, *Am. J. Physiol.*, 256(25), R1, 1989.

Fletcher, G. L., King, M. J., and Kao, M. H., Low temperature regulation of antifreeze glycopeptide levels in Atlantic cod *(Gadus morhua), Can. J. Zool.*, 65(2), 227, 1987.

Fletcher, G. L., King, M. J., Kao, M. H. and Shears, M. A., Antifreeze proteins in the urine of marine fish, *Fish Physiol. Biochem.*, 6(2), 121, 1989.

Fontana, A., How nature engineers protein (thermo) stability, in *Life Under Extreme Conditions*, di Prisco, G., Ed., Springer-Verlag, Berlin, 1991, 89.

Fry, F. E. J., Responces of vertebrate poikilotherms to temperature, in *Thermobiology*, Rose, A. H., Ed., Academic Press, New York, 1967, 375.

Gerlach, G-F., Turay, L., Malik, K. T. A., Lida, J., Scutt, A., and Goldspink, G., Mechanisms of temperature acclimation in the carp: a molecular biology approach, *Am. J. Physiol.*, 259(28), R237, 1990.

Graves, J.E. and Somero, G.N., Electrophoretic and functional enzymic evolution in four species of eastern pacific barracudas from different thermal environments, *Evolution*, 36(1), 97, 1982.

Graves, J. E., Rosenblatt, R. H., and Somero, G. N., Kinetic and electrophoretic differentiation of lactate dehydrogenases of teleost species-pairs from the Atlantic and Pacific coasts of Panama, *Evolution*, 37(1), 30, 1983.

Greer, G. L. and Gardner, D. R., Characterization of responses from temperature-sensitive units in trout brain, *Comp. Biochem. Physiol.*, 48A, 189, 1974.

Griffiths, J. S. and Alderdice, D. F., Effects of acclimation and acute temperature experience on the swimming speed of juvenile coho salmon, *J. Fish. Res. Bd. Can.*, 29, 251, 1972.

Hagar, A. F. and Hazel, J. R., Changes in desaturase activity and the fatty acid composition of microsomal membranes from liver tissues of thermally-acclimating rainbow trout, *J. Comp. Physiol.*, 156B, 35, 1985.

Hansen, H. O. and Knudsen, J., The influence of environmental and incubation temperature on fatty acid synthetase from flounder *(Platichthys flesus* L.), *Comp. Biochem. Physiol.*, 70B, 515, 1981.

Harris, W. E., Modulation of (Na^+, K^+)-ATPase activity by the lipid bilayer examined with dansylated phosphatidylserine, *Biochemistry*, 24(12), 2873, 1985.

Haschemeyer, A. E. V. and Mathews, R. W., Temperature dependency of protein synthesis in isolated hepatocytes of Antarctic fish, *Physiol. Zool.*, 56(1), 78, 1983.

Hazel, J. R., The effect of thermal acclimation upon succinic dehydrogenase activity from the epaxial muscle of the common goldfish *(Carassius auratus)* II. Lipid reactivation of the soluble enzyme, *Comp. Biochem. Physiol.*, 43B, 863, 1972.

Hazel, J. R., Homeoviscous adaptation in Animal Cell Membranes, in *Advances in Membrane Fluidity — Physiological Regulation of Membrane Fluidity*, Aloia, R. C., Curtain, C. C. and Gordon, L. M., Eds., Alan R. Liss, Inc., New York, 1988, 149.

Hazel, J. R., Adaptation to temperature: phospholipid synthesis in hepatocytes of rainbow trout, *Am. J. Physiol.*, 258(27), R1495, 1990.

Hazel, J. R. and Landrey, S. R., Timecourse of thermal acclimation in plasma membranes of trout kidney. II. Molecular species composition, *Am. J. Physiol.*, 255(24), 628, 1988a.

Hazel, J. R. and Landrey, S. R., Timecourse of thermal acclimation in plasma membranes of trout kidney. I. Headgroup composition, *Am. J. Physiol.*, 255(24), R622, 1988b.

Hazel, J. R. and Prosser, C. L., Molecular mechanisms of temperature compensation in poikilotherms, *Physiol. Rev.*, 54(3), 620, 1974.

Hazel, J.R. and Schuster, V.L., The effects of temperature and thermal acclimation upon the osmotic properties and nonelectrolyte permeability of liver and gill mitochondria from rainbow trout *(Salmo gairdneri), J. Exp. Zool.*, 195(3), 425, 1976.

Hazel, J. R. and Williams, E. E., The role of alterations in membrane lipid composition in enabling physiological adaptation of organisms to their physical environment, *Prog. Lipid Res.*, 29(3), 167, 1990.

Hazel, J. R., Garlick, W. S., and Sellner, P. A., The effect of assay temperature upon the pH optima of enzymes from poikilotherms: a test of the imidazole alphastat hypothesis, *J. Comp. Physiol.*, 123, 97, 1978.

Hazel, J. R., McKinley, S. J., and Williams, E. E., Thermal adaptation in biological membranes: interacting effects of temperature and pH, *J. Comp. Physiol. B*, 162, 593, 1992.

Heap, S. P., Watt, P. W. and Goldspink, G., Contractile properties of goldfish fin muscles following temperature acclimation, *J. Comp. Physiol. B*, 157, 219, 1987.

Helly, J. J., The effect of temperature and thermal distribution on glycolysis in two rockfish species *(Sebastes), Mar.Biol.*, 37, 89, 1976.

Hew, C. L. and Fletcher, G. L., The role of pituitary in regulating antifreeze protein synthesis in the winter flounder, *FEBS Lett.*, 99(2), 337, 1979.

Hochachka, P.W. and Somero, G.N., *Biochemical Adaptation*, Princeton University Press, Princeton, N.J., 1984.

Holeton, G. F., Metabolic cold adaptation of polar fish: fact or artifact, *Physiol. Zool.*, 47(3), 137, 1974.

Holland, K. N., Brill, R. W., Chang, R. K. C., Sibert, J.R. and Fournier, D.A., Physiological and behavioural thermoregulation in bigeye tuna *(Thunnus obesus), Nature*, 358, 410, 1992.

Howell, B. K. and Matthews, A. D., Environmental temperature adaptation of fish sarcoplasmic reticulum: a differential scanning calorimetric study, *Comp. Biochem. Physiol. B.*, 99(1), 175, 1991.

Hui, S-W. and Sen, A., Effects of lipid packing on polymorphic phase behavior and membrane properties, *Proc. Natl. Acad. Sci. U.S.A.*, 86, 5825, 1989.

Hwang, G. C., Watabe, S. and Hashimoto, K., Changes in carp myosin ATPase induced by temperature acclimation, *J. Comp. Physiol. B*, 160, 233, 1990.

Hwang, G. C., Ochiai, Y., Watabe, S. and Hashimoto, K., Changes of carp myosin subfragment-1 induced by temperature acclimation, *J. Comp. Physiol. B*, 161, 141, 1991.

Johnson, T. P. and Johnston, I. A., Temperature adaptation and the contractile properties of live muscle fibres from teleost fish, *J. Comp. Physiol. B*, 161, 27, 1991.

Johnston, I. A. and Altringham, J. D., Evolutionary adaption of muscle power output to environmental temperature: force-velocity characteristics of skinned fibres isolated from Antarctic, temperate and tropical marine fish, *Pfluegers. Arch.*, 405, 136, 1985.

Johnston, I. A. and Harrison, P., Contractile and metabolic characteristics of muscle fibers from antarctic fish, *J. Exp. Biol.*, 116, 223, 1985.

Johnston, I. A., Fleming, J. D., and Crockford, T., Thermal acclimation and muscle contractile properties in cyprinid fish, *Am. J. Physiol.*, 259(28), R231, 1990.

Jones, P. L. and Sidell, B. D., Metabolic responses of striped bass *(Morone saxatilis)* to temperature acclimation. II. Alterations in metabolic carbon sources and distributions of fiber types in locomotory muscle, *J. Exp. Zool.*, 219, 163, 1982.

Kanungo, M. S. and Prosser, C. L., Physiological and biochemical adpatation of goldfish to cold and warm temperatures I. Standard and active oxygen consumptions of cold- and warm-acclimated goldfish at various temperatures, *J. Cell. Comp. Physiol.*, 54(3), 259, 1959.

Kent, J. and Prosser, C. L., Effects of incubation and acclimation temperatures on incorporation of U-(^{14}C) glycine into mitochondrial protein of liver cells and slices from green sunfish, *Lepomis cyanellus, Physiol. Zool.*, 53(3), 293, 1980.

Kent, J., Koban, M., and Prosser, C. L., Cold-acclimation-induced protein hypertrophy in channel catfish and green sunfish, *J. Comp. Physiol. B*, 158, 185, 1988.

Knight, C. A., Cheng, C. C., and DeVries, A. L., Adsorption of -helical antifreeze peptides on specific ice crystal surface planes, *Biophys. J.*, 59, 409, 1991.

Koban, M., Can cultured teleost hepatocytes show temperature acclimation? *Am. J. Physiol.*, 250(19), R211, 1986.

Kurokawa, T. and Nakano, H., Changes in isozymes of soluble and mitochondrial malate dehydrogenase of carp *(Cyprinus carpio* L.) during thermal acclimation, *Comp. Biochem. Physiol. B*, 99(4), 911, 1991.

Langfeld, K. S., Crockford, T., and Johnston, I. A., Temperature acclimation in the common carp: force-velocity characteristics and myosin subunit composition of slow muscle fibres, *J. Exp. Biol.*, 155, 291, 1991.

Lee, J. A. C. and Cossins, A. R., Adaptation of intestinal morphology in the temperature-acclimated carp, *Cyprinus carpio* L, *Cell Tissue Res.*, 251, 451, 1988.

Lee, J. A. C. and Cossins, A. R., Temperature adaptation of biological membranes: differential homeoviscous responses in brush border and basolateral membranes of carp intestinal mucosa, *Biochim. Biophys. Acta,* 1026, 195, 1990.

Lee, J. A. C., James, P. S., Smith, M. W., and Cossins, A. R., Amino acid transport in the intestinal mucosa of temperature-acclimated carp, *J. Therm. Biol.*, 16(1), 7, 1991.

Lin, Y., Environmental regulation of gene expression - in vitro translation of winter flounder antifreeze messenger RNA, *J. Biol. Chem.*, 254(4), 1422, 1979.

Linthicum, D. S. and Carey, F. G., Regulation of brain and eye temperatures by the bluefin tuna, *Comp. Biochem. Physiol.*, 43A, 425, 1972.

Livermore, R. C. and Hazel, J. R., Acylation of lysophosphatidylcholine in liver microsomes of thermally acclimated trout, *Am. J. Physiol.*, 255(24), R923, 1988.

Macdonald, J. A., Temperature compensation in the peripheral nervous system: Antarctic vs temperate poikilotherms, *J. Comp. Physiol. B*, 142, 411, 1981.

Macdonald, J. A. and Montgomery, J. C., Thermal limits of neuromuscular function in an Antarctic fish, *J. Comp. Physiol. B*, 147, 237, 1982.

Mantsch, H. H., Martin, A., and Cameron, D. G., Characterization by infrared spectroscopy of the bilayer to non-bilayer phase transition of phosphatidylethanolamines, *Biochemistry*, 20, 3138, 1981.

Maresca, B., Patriarca, C., Goldenberg, C., and Sacco, M., Heat shock and cold adaptation in antarctic fishes: a molecular approach, *Comp. Biochem. Physiol.*, 90B(3), 623, 1988.

McVean, A. R. and Montgomery, J. C., Temperature compensation in myotomal muscle: Antarctic versus temperate fish, *Environ. Biol. Fish.*, 19(1), 27, 1987.

Misra, S., Zafarullah, M., Price-Haughey, J., and Gedamu, L., Analysis of stress-induced gene expression in fish cell lines exposed to heavy metals and heat shock, *Biochim. Biophys. Acta,* 1007, 325, 1989.

Morris, R. W. and Schneider, M. J., Brain fatty acids of an antarctic fish, *Trematomus bernachii, Comp. Biochem. Physiol.*, 28, 1461, 1969.

Mosser, D. D. and Bols, N. C., Relationship between heat-shock protein synthesis and thermotolerance in rainbow trout fibroblasts, *J. Comp. Physiol. B*, 158, 457, 1988.

Nelson, D. O. and Prosser, C. L., Effect of preoptic lesions on behavioral thermoregulation of green sunfish, Lepomis cyanellus, and of goldfish, *Carassius auratus, J. Comp. Physiol.*, 129, 193, 1979.

Nelson, D. O. and Prosser, C. L., Temperature-sensitive neurons in the preoptic region of the sunfish, *Am. J. Physiol.*, 241(10), R259, 1981a.

Nelson, D. O. and Prosser, C. L., Intracellular recordings from thermosensitive preoptic neurons, *Science*, 213, 787, 1981b.

Osuga, D.T. and Feeney, R.E., Antifreeze glycoproteins from Arctic fish, *J. Biol. Chem.*, 253(15), 5338, 1978.

Patton, J. S., The effect of pressure and temperature on phospholipid and triglyceride fatty acids of fish white muscle: a comparison of deepwater and surface marine species, *Comp. Biochem. Physiol.*, 52B, 105, 1975.

Paynter, K. T., DiMichelle, L., Hand, S. C., and Powers, D. A., Metabolic implications of Ldh-B genotype during early development in *Fundulus heteroclitus, J. Exp. Zool.*, 257, 24, 1991.

Penney, R. K. and Goldspink, G., Temperature adaptation of sarcoplasmic reticulum of fish muscle, *J. Therm. Biol.*, 5, 63, 1980.

Place, A.R. and Powers, D.A., Genetic variation and relative catalytic efficiencies: lactate dehydrogenase B allozyme of *Fundulus heteroclitus, Proc. Nat. Acad. Sci. U.S.A.*, 76, 2354, 1979.

Place, A. R. and Powers, D. A., Kinetic characterization of the lactate dehydrogenase (LDH-B4) allozymes of *Fundulus heteroclitus, J. Biol. Chem.*, 259(2), 1309, 1984.

Price, J. L., Gourlie, P. B., Lin, Y., and Huang, R. C. C., Induction of winter flounder antifreeze protein messenger RNA at 4 C in vivo and in vitro, *Physiol. Zool.*, 59(6), 679, 1986.

Prosser, C. L. and Nelson, D. O., The role of nervous systems in temperature adaptation of poikilotherms, *Annu. Rev. Physiol.*, 43, 281, 1981.

Raymond, J. A., Wilson, P., and DeVries, A. L., Inhibition of growth of nonbasal planes in ice by fish antifreezes, *Proc. Natl. Acad. Sci. U.S.A.*, 86, 881, 1989.

Raynard, R. S. and Cossins, A. R., Homeoviscous adaptation and thermal compensation of sodium pump of trout erythrocytes, *Am. J. Physiol.*, 260(29), R916, 1991.

Reeves, R. B., The interaction of body temperature and acid-base balance in ectothermic vertebrates, in *Annual Review of Physiology*, 39th ed., Knobil, E.K., Sonnenschein, R.R., and Edelman, I.S., Eds., Annual Reviews, Inc., Palo Alto, 1977, 559.

Reynolds, W. W., McCauley, R. W., Casterlin, M. E., and Crawsha, L. I., Body temperatures of behaviorally thermoregulating largemouth black bass *(Micropterus salmoides), Comp. Biochem. Physiol.*, 54A, 461, 1976.

Reynolds, W. W., Casterlin, M. E., Matthey, J. K., Millington, S. T., and Ostrowski, A. C., Diel patterns of preferred temperature in the goldfish, *Carassius auratus, Comp. Biochem. Physiol.*, 59A, 225, 1978.

Richardson, T. and Tappel, A.L., Swelling of fish mitochondria, *J. Cell Biol.*, 13, 45, 1962.

Rome, L. C., Influence of temperature on muscle recruitment and muscle function in vivo, *Am. J. Physiol.*, 259(28), R210, 1990.

Scholander, P. F., Flagg, W., Walters, V., and Irving, L., Climatic adaptation in arctic and tropical poikilotherms, *Physiol. Zool.*, 26, 67, 1953.

Scholander, P.F., van Dam, L., Kanwisher, J.W., Hammel, H.T., and Gordon, M.S., Supercooling and osmoregulation in arctic fish, *J. Cell. Comp. Physiol.*, 49, 5, 1957.

Schrag, J. D., O'Grady, S. M., and DeVries, A. L., Relationship of amino acid composition and molecular weight of antifreeze glycopeptides to non-colligative freezing point depression, *Biochim. Biophys. Acta*, 717, 322, 1982.

Schunke, M. and Wodtke, E., Cold-induced increase of delta-nine and delta-six desaturase activities in endoplasmic membranes of carp liver, *Biochim. Biophys. Acta*, 734, 70, 1983.

Schurmann, H., Steffensen, J. F., and Lomholt, J. P., The influence of hypoxia on the preferred temperature of rainbow trout *Oncorhynchus mykiss, J. Exp. Biol.*, 157, 75, 1991.

Schwantes, M. L. B. and Schwantes, A. R., Adaptive features of ectothermic enzymes. II. The effects of acclimation temperature on the malate dehydrogenase of the spot, *Leiostomus xanthurus, Comp. Biochem. Physiol.*, 72B, 59, 1982.

Schwartz, F. T. and Paltauf, F., Influence of the ester carbonyl oxygens of lecithin on the permeability properties of mixed lecithin-cholesterol bilayers, *Biochemistry*, 16(20), 4335, 1977.

Scott, G.K., Fletcher, G.L., and Davies, P.L., Fish antifreeze proteins: recent gene evolution, *Can. J. Fish. Aquat. Sci.*, 43(5), 1028, 1986.

Sellner, P.A. and Hazel, J.R., Desaturation and elongation of unsaturated fatty acids in hepatocytes from thermally-acclimated rainbow trout, *Arch. Biochem. Biophys.*, 213(1), 58, 1982.

Shaklee, J. B., Christiansen, J. A., Sidell, B. D., Prosser, C. L., and Whitt, G. S., Molecular aspects of temperature acclimation in fish: contributions of changes in enzyme activities and isozyme patterns to metabolic reorganization in the green sunfish, *J. Exp. Zool.*, 201(1), 1, 1977.

Sidell, B. D., Turnover of cytochrome c in skeletal muscle of green sunfish *(Lepomis cyanellus,* R.) during thermal acclimation, *J. Exp. Zool.*, 199(2), 233, 1977.

Sidell, B. D., Physiological roles of high lipid content in tissues of Antarctic fish species, in *Biology of Antarctic Fish,* di Prisco, G., Maresca, B., and Tota, B., Eds., Springer-Verlag, Berlin, 1991, 220.

Sidell, B. D. and Moerland, T. S., Effects of temperature on muscular function and locomotory performance in teleost fish, in *Advances in Comparative and Environmental Physiology*, Vol. 5, Springer-Verlag, Berlin, 1989, 116.

Sisson, J. E. and Sidell, B. D., Effect of thermal acclimation on muscle fiber recruitment of swimming striped bass *(Morone saxatilis), Physiol.Zool.*, 60(3), 310, 1987.

Somero, G. N., Temperature adaptation of enzymes: biological optimization through structure-function compromises, *Annu. Rev. Ecol. Syst.*, 9, 1, 1978.

Somero, G. N., Biochemical mechanisms of cold adaptation and stenothermality in Antarctic fish, in *Biology of Antarctic Fish*, di Prisco, G., Maresca, B., and Tota, B., Eds., Springer-Verlag, Berlin, 1991, 231.

Spigarelli, S. A. and Thommes, M. M., Temperature selection and estimated thermal acclimation by rainbow trout *(Salmo gairdneri*) in a thermal plume, *J. Fish Res. Bd. Can.*, 36(4), 366, 1979.

Spigarelli, S. A., Thommes, M. M., and Beitinger, T. L., The influence of body weight on heating and cooling of selected Lake Michigan fishes, *Comp. Biochem. Physiol.*, 56A, 51, 1977.

Squier, T. C., Bigelow, D. J., and Thomas, D. D., Lipid fluidity directly modulates the overall protein rotational mobility of the Ca-ATPase in sarcoplasmic reticulum, *J. Biol. Chem.*, 263(19), 9178, 1988.

Stevens, E. D. and Carey, F. G., One why of warm-bodied fish, *Am. J. Physiol.*, 240(9), R151, 1981.

Stevens, E. D. and Fry, F. E. J., The rate of thermal exchange in a teleost, *Tilapia mossambica, Can. J. Zool.*, 48, 221, 1970.

Stevens, E. D. and Fry, F. E. J., Brain and muscle temperature in ocean caught and captive skipjack tuna, *Comp. Biochem. Physiol.*, 38A, 203, 1971.

Stevens, E. D. and McLeese, J. M., Why bluefin tuna have warm tummies: temperature effect on trypsin and chymotrypsin, *Am. J. Physiol.*, 246(15), R487, 1984.

Stevens, E. D. and Neill, W. H., Body temperature relations of tunas, especially skipjack, in *Fish Physiology*. Vol. VII, Hoar, W.S. and Randall, D.J., Eds., Academic Press, New York, 1978, 315.

Stevens, E. D., Lam, H. M. and Kendall, J., Vascular anatomy of the counter-current heat exchanger of skipjack tuna, *J. Exp. Biol.*, 61, 145, 1974.

Stone, B.B. and Sidell, B.D., Metabolic responses of striped bass (*Morone saxatilis*) to temperature acclimation I. Alterations in carbon sources for hepatic energy metabolism, *J. Exp. Zool.*, 218, 371, 1981.

Subjeck, J. R. and Shyy, T. T., Stress protein systems of mammalian cells, *Am. J. Physiol.*, 250(19), C1, 1986.

Swezey, R. R. and Somero, G. N., Polymerization thermodynamics and structural stabilities of skeletal muscle actins from vertebrates adapted to different temperatures and hydrostatic pressures, *Biochemistry*, 21(18), 4496, 1982.

Swezey, R. R. and Somero, G. N., Pressure effects on actin self-assembly: interspecific differences in the equilibrium and kinetics of the G to F transformation, *Biochemistry*, 24(4), 852, 1985.

Targett, T. E., Respiratory metabolism of temperature acclimated Fundulus heteroclitus (L.): zones of compensation and dependence, *J. Exp. Mar. Biol. Ecol.*, 32, 197, 1978.

Valerio, P. F., Kao, M. H., and Fletcher, G. L., Fish skin: an effective barrier to ice crystal propagation, *J. Exp. Biol.*, 164, 135, 1992.

van Voorhies, W. V., Raymond, J. A., and De Vries, A. L., Glycoproteins as biological antifreeze agents in the cod *Gadus ogac* (Richardson), *Physiol. Zool.*, 51(4), 347, 1978.

Walsh, P. J., Foster, G. D., and Moon, T. W., The effects of temperature on metabolism of the american eel *Anguilla rostrata* (le Sueur): compensation in the summer and torpor in the winter, *Physiol. Zool.*, 56(4), 532, 1983.

Wilson, F. R., Whitt, G. S., and Prosser, C. L., Lactate dehydrogenase and malate dehydrogenase isozyme patterns in tissues of temperature acclimated goldfish *(Carassius auratus* L.), *Comp. Biochem. Physiol.*, 46B, 105, 1973.

Wodtke, E., Discontinuities in the Arrhenius plots of mitochondrial membrane-bound enzyme systems from a poikilotherm: acclimation temperature of carp affects transition temperatures, *J. Comp. Physiol.*, 110, 145, 1976.

Wodtke, E., Temperature adaptation of biological membranes — Compensation of the molar activity of cytochrome c oxidase in the mitochondrial energy-transducing membrane during thermal acclimation of the carp, *Biochim. Biophys. Acta*, 640, 710, 1981.

Wodtke, E., On adaptation of biomembranes to temperature: membrane dynamics and membrane functions, *J. Therm. Biol.*, 8(4), 416, 1983.

Wolf, N. G., Swift, P. R., and Carey, F. G., Swimming muscle helps warm the brain of lamnid sharks, *J. Comp. Physiol.*, 157, 709, 1988.

Wollmuth, L. P., Crawshaw, L. I., and Panayiotides-Djaferis, H., Thermoregulatory effects of intracranial norepinephrine injections in goldfish, *Am. J. Physiol.*, 253, R821, 1987.

Wollmuth, L. P., Crawshaw, L. I., and Rausch, R. N., Adrenoceptors and temperature regulation in goldfish, *Am. J. Physiol.*, 255(24), R600, 1988.

Wollmuth, L. P., Crawshaw, L. I., and Panayiotides-Djaferis, H., The effects of dopamine on temperature regulation in goldfish, *J. Comp. Physiol. B*, 159, 83, 1989.

Yamawaki, H. and Tsukuda, H., Significance of the variation in isozymes of liver lactate dehydrogenase with thermal acclimation in goldfish I. Thermostability and temperature dependency, *Comp. Biochem. Physiol.*, 62B, 89, 1979.

Yancey, P. H. and Somero, G. N., Temperature dependence of intracellular pH: its role in the conservation of pyruvate apparent Km values of vertebrate lactate dehydrogenase, *J. Comp. Physiol.*, 125, 129, 1978.

Yeagle, P. L., Cholesterol and cell membranes, *Biochim. Biophys. Acta*, 822, 267, 1985.

15 Endocrinology

Sjoerd E. Wendelaar Bonga

INTRODUCTION

Communication between cells and organs is essential for the coordination of physiological processes, including those that adapt animals in time and space to their environment. Classically, the nervous system and the endocrine system were considered complementary for communication in a complex organism. The developments of the last two decades have not only shown that both systems operate in a highly integrated way, but also that they show a substantial overlap with respect to the mechanisms and messengers they employ. Endocrine messengers have been detected in cells, tissues, and organs other than the classical endocrine glands and tissues. For instance, in the vertebrates, including fishes, somatostatins are produced in the islet tissue, the gut, as well as in the brain; in mammals, and probably other vertebrates, arginine vasotocin is not only produced in the brain but also in peripheral organs, including the gonads, and calcitonin gene expression has been established in the nervous system and in pulmonary neuroepithelial cells, in addition to the well-known locations in the C-cells of the thyroid gland or the ultimobranchials. The messengers concerned may function, depending on their site of production, as neurohormones, neurotransmitters, or classical hormones. These and similar discoveries have led to the growing awareness that the boundaries between the nervous and endocrine systems, from an endocrine point of view, are arbitrary. This has resulted in the neuroendocrine concept of regulation. It has also become clear that there is an intricate system of intercellular communication at the tissue level, with mechanisms defined as paracrine, autocrine, or intracrine interactions, that shares an important element with the neuroendocrine system: communication accomplished by chemical messengers that evoke a physiological response by binding to specific receptors (Bern, 1990a).

The new developments have been initiated and facilitated by technical innovations, such as improved separation techniques and more sensitive assay methods, and by the introduction in the neuroendocrinological research of new techniques, such as protein and gene sequencing, immunocytochemistry, and *in situ* hybridization. The latter techniques have led to an explosive growth of the number of known messenger molecules, in particular peptides and proteins. This started in the 1970s and still continues.

0-8493-8042-1/93/$0.00+$.50

Groups of these molecules appeared to be structurally related and have been classified as peptide or protein families, such as the vasopressin, proopiomelanocortin, somatostatin, neuropeptide-Y, and glucagon families, with members functioning, even in one species, as neurotransmitters, neuromodulators, hormones, and/or paracrine messengers, and occurring in a large variety of animal groups (Sherwood and Parker, 1990). Many peptide hormones long considered typical for the vertebrates have now been found in the invertebrates, and messengers first identified in invertebrates, FMRFamide-like peptides, have now been found in the vertebrate nervous system. Even steroids, catecholamines, and peptides that function as hormones in the metazoa have been identified in bacteria and protozoa, frequently in association with specific binding proteins. This marks the long evolutionary history of the genes coding for these messengers (Scharrer, 1990). Pearse and Pollack (1971) have suggested that a large group of peptide-producing endocrine cells that had APUD-characteristics (the capacity for the uptake and decarboxylation of amine precursors) have a common embryological origin, from the neural crest. This generalization turned out to be too rigorous, however, and an underestimation of the variability of these cells with respect to their embryological origin and biochemical characteristics.

In the vertebrates the peptide and protein messengers are produced in the brain and in the classical endocrine organs, as well as in many other organs including the heart, gut, skin, and the lymphoid system. The lymphoid system most clearly exemplifies the new developments in the field of endocrinology. Its cells produce a variety of peptide and protein messengers that regulate the immune response and in addition modulate the hypothalamo-pituitary adrenal/interrenal axis. In this way they represent essential elements of the integrated stress response. These messengers include cytokines typical for the immune system such as the interleukins, as well as products identical, similar, or related to classical hormones, e.g., prolactin, adrenocorticotropic hormone (ACTH), α-melanophore-stimulating hormone (α-MSH), growth hormone, and thyroid- and follicle-stimulating hormones (Weigent and Blalock, 1987). The first results on fishes have now become available.

Thus, the field of endocrinology can no longer be clearly delineated, and this also applies to the endocrinology of fishes. In this chapter we are dealing mainly with the endocrine system of fishes in the classical way. This is rather a matter of convenience than an approach that is self-evident. The chapter intends, first, to show how the recent developments in the field have changed our view on the endocrine system of fishes and, second, to identify the specific aspects of this system that are related to the aquatic habitat of fishes.

THE MESSENGERS

Hypophyseal Hormones

Like all vertebrates, fishes have a neurohypophysis which is continuous with the floor of the diencephalon, and an adenohypophysis which originates from non-neural tissues. In most fishes the adenohypophysis is differentiated into a rostral and a proximal pars distalis and a pars intermedia (Figure 1). For a detailed description of the embryology and morphology of the hypophysis in fishes one is referred to Schreibman (1986).

Neurohypophyseal Factors

In the neurohypophysis, where many axons terminate that originate from hypothalamic nuclei, storage and release of neurohormones takes place. There are also direct nervous connections between the neurohypophysis and the adenohypophysis, which are typical for fishes (with the exception of the agnathans). In the nonteleostean fishes the interdigitation of the neurohypophysis is confined to the region of the pars intermedia, a condition related to that of the higher vertebrates. In these fishes a primitive median eminence and associated portal system may be present. In the teleosts an anterior and a caudal part of the neurohypophysis can be distinguished. The caudal part is in contact with the pars intermedia. The anterior part invades the proximal pars distalis of the adenohypophysis, and a median eminence or portal system is absent or, at most, poorly developed.

Part of the hormones of the neurohypophysis are released into the general circulation. Like in other vertebrates, several of these belong to the family of vasopressin-related nonapeptides. Of the ten members of this family identified in the vertebrates, seven have been found in fishes. These are mainly produced in the preoptic nuclei. In agnathans only arginine vasotocin has been demonstrated. In chondrichthyans glumitocin, aspartocin, and valitocin have been

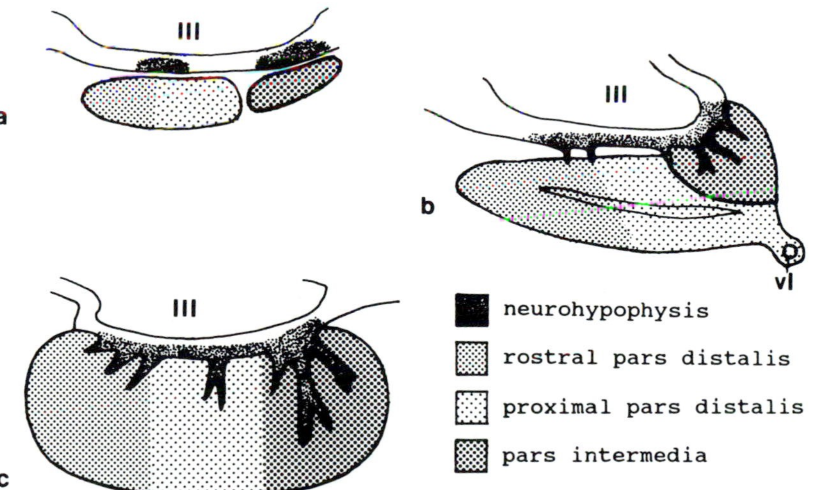

FIGURE 1. Representation of a midsagittal section of the pituitary glands of fishes: (a) agnathan (lamprey; in the hagfishes a pars intermedia is absent); (b) chondrichthyan (shark), vl, ventral lobe; (c) actinopterygian (teleost).

found. Most bony fishes have arginine vasotocin and isotocin, a peptide related to oxytocin, whereas in lungfishes arginine vasotocin, oxytocin, and mesotocin are present (Perks, 1987). The precise function of these peptides has not been clarified. Most of them have effects on blood pressure and, directly or indirectly, on fluid volume control. Diuretic and antidiuretic effects have been reported for arginine vasotocin and isotocin in teleosts. The diuretic actions may be due to the use of nonphysiologically high doses of the hormones (Balment et al., 1993). Transfer of teleosts from fresh water to seawater usually leads to depletion of the arginine vasotocin stores in the neurohypophysis, whereas the reverse occurs after transfer from seawater to lower salinities (Haruta et al., 1991). Plasma levels of this peptide in eels were higher in seawater than in freshwater eels, but this could not be confirmed for flounder and trout in which the higher levels were found in fish from fresh water (Perrott et al., 1991). Administration of arginine vasotocin elicits a spawning reflex. It further stimulates cortisol secretion in the goldfish. This effect may be mediated by ACTH, because arginine vasotocin is colocated with corticotropin-releasing hormone in hypothalamic fibers innervating the ACTH cells in some teleosts (Moons et al., 1991). Arginine vasotocin as well as isotocin promote gonadal testosterone production *in vitro* in rainbow trout testis (Rodriguez and Specker, 1991).

Another group of neuropeptides are released in the neurohypophysis and adjacent areas. These are the factors that regulate the secretion of endocrine cells of the adenohypophysis. Immunoreactivity has been reported for a large variety of peptide hormones. These include growth hormone-, corticotropin-, thyrotropin-, and gonadotropin-releasing hormones, melanophore-concentrating hormone, enkephalin, glucagon-like peptide, FMRFamide-related peptides, somatostatins (in particular somatostatin-14), galanin, neuropeptide Y, bombesin, vasointestinal peptide, and gastrin-like peptides (e.g., Batten et al., 1990; Peter et al., 1990). This list is certainly not complete. Of some of these peptides, two (gonadotropin-releasing hormone) or more (somatostatin) molecular forms have been demonstrated within one species. Many are produced in the preoptic nuclei, but several other hypothalamic nuclei are involved as well. Depending on the species, the peptides may be released into blood vessels and intercellular spaces, and function as hormones, or (with exception of the agnathans), in axons ending on the adenohypophyseal cells. Some of the peptides may also have effects outside the brain and pituitary gland. Melanophore-concentrating hormone, which inhibits α-MSH secretion via fibers that are ending on or nearby the α-MSH-cells in the pars intermedia, is also released in the general circulation and opposes the stimulating action of MSH on the skin melanophores of teleostean and holostean fish. Interestingly, melanophore-concentrating hormone was first isolated from the chum salmon hypophysis 10 years ago, and was subsequently identified in the hypothalamus of other vertebrates, including lampreys, amphibians, and mammals, where it may function mainly as a neurotransmitter and neuromodulator (Baker, 1991).

In addition to peptidergic fibers, serotonergic, noradrenergic, dopaminergic, and GABA-ergic fibers are present, which may end synaptically on, or nearby, the endocrine cells. In the higher vertebrates two or more of these factors may be colocated in the same nerve fibers. This phenomenon does also occur in fishes, e.g., colocation of corticotropin-releasing hormone with arginine vasotocin or isotocin (Fryer, 1989), or with galanin (Batten et al., 1990).

Each adenohypophyseal cell type probably receives a multiple input from the neurohypophysis, as is indicated by immunohistochemical and biochemical observations. For instance, teleost prolactin secretion *in vitro* is stimulated by corticotropin- and thyroid hormone-releasing hormone and by serotonin, and inhibited by somatostatin, urotensin I, vasointestinal peptide, and dopamine (Grau and Helms, 1990). Similar lists can be presented for growth hormone, gonadotropic hormone, and α-MSH. Although the significance of these effects *in vivo* remains to be established for most of these factors, it is clear that the endocrine cells of the pituitary gland are under a complicated multifactorial control. This control may further include paracrine mechanisms that are still poorly understood. This indicates that these cells are centers of neuroendocrine integration.

Adenohypophyseal Factors

The location of the endocrine cell types of the adenohypophysis has been studied in a large number of fish species, originally mainly with antisera raised against mammalian hormones. Later on more and more antisera against fish hormones have become available, with essentially similar results. With the exception of agnathans, fishes show a clear zonation with respect to the location of the endocrine cell types of the adenohypophysis, as is shown for teleosts and elasmobranchs (Figure 1). Glia-like stellate cells have frequently been noticed, e.g., in the rostral pars distalis and the pars intermedia of teleost. Evidence for a regulatory function of these cells, as has been found recently in higher vertebrates, remains to be demonstrated in fishes. Immunocytochemical evidence for the presence of prolactin, somatotropic, corticotropic, thyrotropic, gonadotropic, and melanotropic activity has been reported for chondrichthyan and osteichthyan fishes. For the agnathans corticotropic, thyrotropic, gonadotropic, and melanotropic cells have been reported. Some other typical adenohypophyseal hormones are released in agnathans from the neurohypophysis (see below). Osteichthyans have a unique cell type that is located in the pars intermedia: the PAS-positive or somatolactin cells. The different hormones produced by the adenohypophyseal cells will now be discussed briefly.

The Prolactin Family: Prolactin, Growth Hormone and Somatolactin

The prolactin family includes prolactin, growth hormone, somatolactin, and, in mammals, placental lactogens and proliferins.

Prolactin, the hormone that received its name from its stimulating action on milk production in mammals, has a variety of effects in the higher vertebrates, but no clearly delineated function. In fishes it is essential for hydromineral control in fresh water (Hirano, 1986). Prolactin cells are prominent in the pituitary gland of all fishes except for the agnathans. Prolactin immunoreactivity has further been demonstrated in nerve fibers in the brain. In the agnathans such fibers terminate in the neurohypophyseal region adjacent to the adenohypophysis, and it is possible that from here the hormone may enter the systemic circulation (Schreibman, 1986).

Removal of the pituitary gland of teleost fishes in fresh water leads to losses of body electrolytes that may be lethal. These losses can be reduced or completely redressed by administration of mammalian or teleost prolactin preparations. The major function of the hormone most likely is the maintenance of the permeability to water and ions of epithelia of the skin, in particular the gills, the intestine, and the renal tubules including the bladder, at the low level that is required for surviving of fishes in water with a low salinity. The substantial ionic and osmotic gradients between the body fluids and the water promote passive losses of ions and the osmotic uptake of water, mainly across the integument. For maintaining hydromineral equilibrium, the ion losses have to be kept at a level that can be compensated for by active ion uptake. The low permeability of the renal tubular epithelia is essential for the production of large volumes of hypoosmotic urine, to compensate for the osmotic water inflow. These actions of prolactin are of particular importance during migration of euryhaline fishes from seawater to fresh water (Hirano, 1986; Wendelaar Bonga and Pang, 1991).

In some teleosts, prolactin has hypercalcemic effects, but these may be related to the general function in hydromineral control. The hormone does not seem to have a specific function in calcium homeostasis (Wendelaar Bonga and Pang, 1991). Many pollutants, as well as stressors such as handling, induce hydromineral imbalance in fishes, and this may evoke a prominent response of the prolactin cells. This has been shown for several freshwater teleosts, with the exception of the salmonids. In this group the role of prolactin in hydromineral control is unclear. The function of prolactin in seawater fishes is unknown. The circulating hormone levels are much lower than in freshwater fishes. Apart from the control of the hydromineral balance in teleosts, prolactin has been implicated in the control of gonadal steroidogenesis (Singh et al., 1988) and of parental fanning and in lipid metabolism (Wendelaar Bonga and Pang, 1989). With cDNA cloning, two different forms of prolactin have been identified in several teleost species. Notable functional differences have so far not been demonstrated (Specker et al., 1985; Swennen et al., 1991).

Growth hormone, like prolactin, is present in the adenohypophysis of all fishes except for the agnathans, where growth hormone-immunoreactivity has been demonstrated in fibers in the neurohypophysis. Two structural variants of growth hormone occur in the teleosts examined. The actions of the hormone seem to be mediated, like in the higher vertebrates, by insulin-like growth factors

(IGF), which are members of the insulin family. In mammals two forms predominate, IGF-I and IGF-II. Data on fishes are scarce. Some IGF-I-like immunoreactivity has been demonstrated in the plasma of teleost and elasmobranch fishes (Daughady et al., 1985), including salmon, in which a rise was reported during adaptation to seawater (smoltification; Lindahl et al., 1985). Indications for the production of IGF have been demonstrated recently in some teleosts, for instance in salmonid liver, gills, and kidney (Sakamoto et al., 1993). Growth hormone stimulates body growth and the peripheral conversion of T4 to T3, at least in eels and salmonids (see below). In addition, the hormone promotes the development of the capacity for hypo-osmoregulation during smoltification, probably in synergism with cortisol (Madsen, 1990). It stimulates chloride cell development and Na^+/K^+-ATPase activity in the gills. Plasma growth hormone concentration and turnover increase during seawater adaptation in salmonids (Sakamoto et al., 1991). Whether growth hormone stimulates seawater adaptation in other euryhaline fish is still unclear. Growth hormone also has been implicated in the inhibitory control of antifreeze protein production in winter flounder (Idler et al., 1989) and in gonadal steroid synthesis (Van der Kraak et al., 1990).

Somatolactin — the predicted amino acid sequence of the hormone of the somatolactin cells of the pars intermedia cells has been reported recently for flounder (Ono et al., 1990). This hormone has structural homology with prolactin and growth hormone, and this resulted in the name somatolactin. Like growth hormone, it is glycosylated. Its function is unclear. The somatolactin cells of teleosts (Figure 2) respond to a variety of stimuli, such as the calcium concentration, pH, or salinity of the water, and change in background color (Wendelaar Bonga and Pang, 1989). Plasma somatolactin levels increased markedly during gonadal growth in coho salmon, and somatolactin stimulated steroidogenesis of gonadal tissue of coho salmon *in vitro* (Planas et al., 1992).

The POMC-Gene Derived Peptides: ACTH and MSH Cells

The ACTH and MSH cells, which have been identified in the pituitary gland of all groups of fishes (Schreibman, 1986), synthesize and release peptides which originate from a common precursor molecule, proopiomelanocortin (POMC). This precursor molecule, which is also found, for example, in mammalian lymphocytes (Weigent and Blalock, 1987) and in the brains of many vertebrates, including fishes (Olivereau and Olivereau, 1990), is processed in a way that is specific for each cell type. In the ACTH cells of the vertebrate pituitary gland, this results in an N-terminal peptide, in addition to ACTH and lipotropin (β-LPH) as end products. In the MSH cells these peptides are processed further: ACTH yields α-MSH and corticotropin-like intermediate lobe peptide, while β-LPH is cleaved to give β-endorphin and γ-LPH. The latter may be processed further to β-MSH, and the N-terminal fragment to γ-MSH (Figure 3). In general, fishes (there are data from lungfish, elasmobranchs, and actinopterygians) seem to follow this

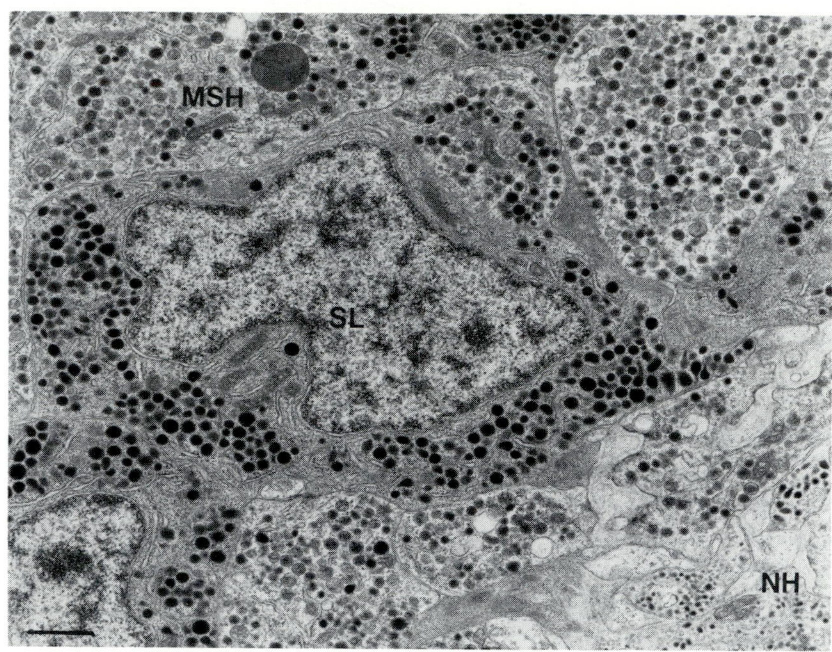

FIGURE 2. Electron micrograph of the pars intermedia of the pituitary gland of goldfish; somatolactin cell (SL) surrounded by melanotropic cells (MSH); NH, axon terminals of the neurohypophysis; bar represents 1 μm. (From Wendelaar Bonga et al., *Cell Tissue Res.*, 243, 609, 1986. With permission.)

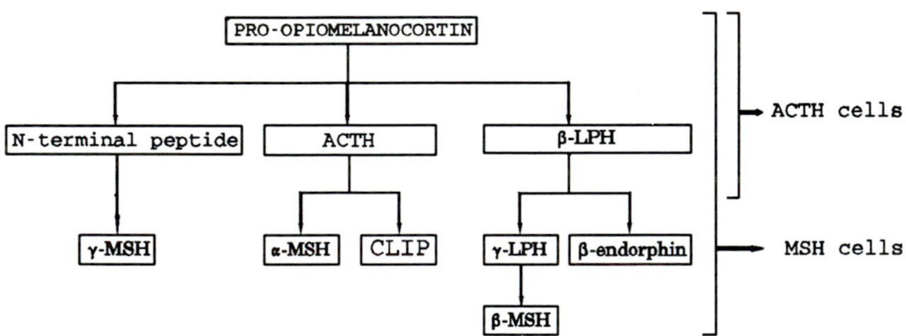

FIGURE 3. A generalized diagram of the processing of proopiomelanocortin (POMC) in ACTH and MSH cells of the vertebrates; ACTH, adrenocorticotropic hormone; LPH, lipotropin; MSH, melanophore-stimulating hormone; CLIP, corticotropin-like intermediate lobe peptide.

or a similar pattern, although γ-MSH has so far only been demonstrated in a dogfish. Two separate POMC molecules encoded by two different genes are present in chum salmon (Kawauchi, 1988). Indications for different forms of ACTH with corticotropic activity were obtained in goldfish (Fryer, 1989).

ACTH probably is the major secretagogue regulating the synthesis and release of corticosteroids in fishes, as it is in the higher vertebrates. The function of the N-terminal peptide is not known. The name MSH was derived from its first known function, i.e., the dispersion of melanin in skin melanophores in amphibians, but its function in many other vertebrate groups, including fishes, is unclear. It has been implicated in melanogenesis and background adaptation of the skin color in some teleosts (Van Eys and Peters, 1981), but in many other fishes no effects on skin pigmentation could be demonstrated. Three forms of α-MSH have been identified in carp, goldfish, and tilapia: des-acetyl, monoacetyl-, and diacetyl-α-MSH. All forms have melanotropic activity *in vitro*. Recently, in mammals, corticotropic activity has been ascribed to α-MSH, and this has also been found in teleost fishes, with di-acetyl—MSH as the most active form (Lamers et al., 1992). Balm et al. (1993) recently reported that treatment with a bacterial endotoxin inhibited α-MSH secretion *in vitro*, and that this effect may be mediated by an interleukin-1-like factor. Bayne and Levy (1991) demonstrated that ACTH stimulates the respiratory burst activity of phagocytes in trout. These data indicate for fishes a type of interaction between the hypothalamo-pituitary-interrenal axis and the immune system comparable to that established in mammals (Weigent and Blalock, 1987).

The Gonadotropin Family: Gonadotropins and Thyrotropin

This family consists of glycoproteins, each consisting of two different subunits (α and β). Within a species the hormones have identical α-subunits, while the β-subunit is hormone specific. The gonadotropins will be discussed in the chapter on reproduction. Thyroid-stimulating hormone is an important stimulator of thyroid function in fishes as it is in higher vertebrates. It has received little attention.

UROTENSINS

Jawed fishes are unique in having a neuroendocrine system almost at the end of the spinal cord: the caudal neurosecretory system. Typically, it is a ventral swelling of the cord, consisting of axon terminals forming a neurohemal organ, structurally similar to the neurohypophysis. The terminals originate from large neurons, the Dahlgren cells, that are distributed over the caudal part of the spinal cord. In the agnathans, neither a caudal neurosecretory system nor Dahlgren cells have been found. In chondrichthyans, Dahlgren cells are present, but their terminals do not form a swelling and are distributed along the periphery of the caudal spinal cord. This situation is also typical for many osteichthyans. Most teleosts have a well-developed caudal neurosecretory system: the urophysis. Two hormones predominate in the urophysis of teleosts: urotensin I and II. Urotensin I has structural homology to corticotropin-releasing hormone, and urotensin II to the somatostatins. There are subpopulations of Dahlgren cells producing urotensin I or urotensin II, or both. Immunoreactivity to these peptides has further been demonstrated in the brain, where the urotensin-like peptides may act as neurotransmitters and as hypophysiotropic factors (Bern, 1990b). Both urotensins have myotropic actions and are weakly vasoconstricting, and both have osmoregulatory effects, in particular in seawater. The effects may be partly indirect. Urotensin I stimulates interrenal steroid synthesis. Urotensin II stimulates the contraction of the teleost heart, urinary bladder, intestine, and gonadal ducts, and promotes the intestinal uptake of ions in seawater (Bern, 1990b). It further appeared to be hypoglycemic and enhanced lipogenesis in coho salmon (Sheridan et al., 1987). The release of both hormones seems to be inhibited by transfer of seawater fish to lower salinities, and stimulated by the reverse transfer (Larson and Madani, 1991). Removal of the urophysis and urotensin-producing cell bodies of the spinal cord in goldfish resulted in an increase in hypothalamic urotensin-I-like activity and in increased release of pituitary ACTH and cortisol (Woo et al., 1985). Administration of urotensin-I increased plasma cortisol in flounder (Arnold-Reed and Balment, 1989). These results indicate a complex interaction of the urophysis with the hypothalamo-pituitary-interrenal axis.

Thyroid Hormones

The functional unit of the thyroid tissue of the vertebrates is the epithelial follicle. In amphibians and higher vertebrates these follicles form compact or multilobular, single or paired glands. Contrastingly, in fishes the thyroid tissue displays a large variety of shapes. In agnathans and teleosts the follicles are distributed in the connective tissue ventral to the pharynx. In some species the follicles may be found outside this area, mostly in the head kidney. In only a few teleosts, the follicles are concentrated in one or more compact lobes or glands. A single gland, located ventromedially to the pharynx, is found in chondrichthyans, lungfishes, and crossopterygians. In lampreys, the thyroid develops from the endostyle, a ciliated pharyngeal groove present in the larva. In all jawed fishes the thyroid follicles arise from the ventral wall of the pharynx, posterior to the first pair of pharyngeal pouches. The structure of the follicles is very similar from agnathans to mammals (Dent, 1986).

The gland cells actively concentrate iodine from the blood and, after oxidation, this is bound to the tyrosyl groups of a protein produced by these cells, thyroglobulin. This protein is synthesized in the gland cells and is secreted into, and temporarily stored in, the follicular lumen, where iodination takes place. When the cells are stimulated by TSH, the iodinated thyroglobulins are taken up from the tissues by endocytosis, and partly destroyed in lysosomes. This leads to the release of the iodothyrosines, mainly tetra-iodothyrosine (thyroxin or T4) which is secreted into the blood. In the target tissues, such as liver and gills, it is deiodinated to tri-iodothyrosine (T3), the biologically more active form, under the control of hormones, including growth hormone and testosterone. Cortisol may have a reverse effect (Dent, 1986; Eales and McLatchy, 1989).

Thyroid hormones are important for the control of metabolism, growth, and development, and osmoregulation in fish as well as in higher vertebrates. They frequently act in synergism with growth hormone or cortisol. Thyroid hormones stimulate metamorphosis of flounders, comparable to their effect on amphibian metamorphosis. Increased activity of the pituitary cells producing thyroid-stimulating hormone and a peak of plasma T4 coincident with a cortisol peak, has been observed during the climax of metamorphosis (Grace de Jesus et al., 1991). Thyroid hormones of maternal origin have been identified in the yolk of unfertilized fish eggs, and these hormones may be of importance for the regulation of early development (Brown and Bern, 1989). Another important function of thyroid hormones is the triggering of migratory behavior and the control of osmoregulatory adaptation to seawater in salmonids and possibly other migratory fishes. High surges of T4 and T3 have been reported during the parr-smolt transformation in fresh water, days or weeks before a peak in branchial Na^+/K^+-ATPase activity, an indicator of seawater preadaptation, is displayed. A link has been demonstrated between the T4 peaks and the lunar cycle during this preadaptive period (Grau, 1987). Freshwater to seawater transfer of trout involves an increase of T4 and T3 metabolism, and of deiodination of T4 to T3 (de Luze et al., 1987). This has also been shown by the same authors for eels, although the data on the relationship between plasma T3 and T4 levels and seawater adaptation in nonsalmonid euryhaline species are less consistent than in salmonids.

Interrenal Hormones

The homologue in fishes of the adrenal gland of the higher vertebrates is the interrenal tissue. In adrenal glands, the medullary tissue is of neural crest origin and produces catecholamines. It is surrounded by a cortex of steroid-producing tissue, which originates from the mesoderm. The interrenal tissue of fishes shows a large heterogeneity with respect to the medullary and cortical homology, and no true medulla or cortex can be distinguished. The term interrenal is usually, although not consistently, restricted to the steroid-producing cells, whereas the catecholamine-producing elements are called chromaffin cells. In the agnathans, the interrenal cells are distributed in groups in the pronephric region. In the chondrichthyans one or more large groups of interrenal cells are spatially separated from small groups of chromaffin cells, which are located on the surface of the kidneys. In most teleost fishes (Figure 4) both cell types are intermingled and located around the posterior cardinal veins in the pronephric part of the kidneys, the head kidneys (Chester Jones and Phillips, 1986).

The Interrenal Steroid-Producing Cells

Low plasma levels of cortisol, corticosterone, and deoxycortisol have been demonstrated in agnathans. The major corticosteroid produced in most gnathostomes is cortisol. In elasmobranchs it is 1α-hydroxycorticosterone. Only traces of other steroids such as corticosterone (most groups) or aldosterone

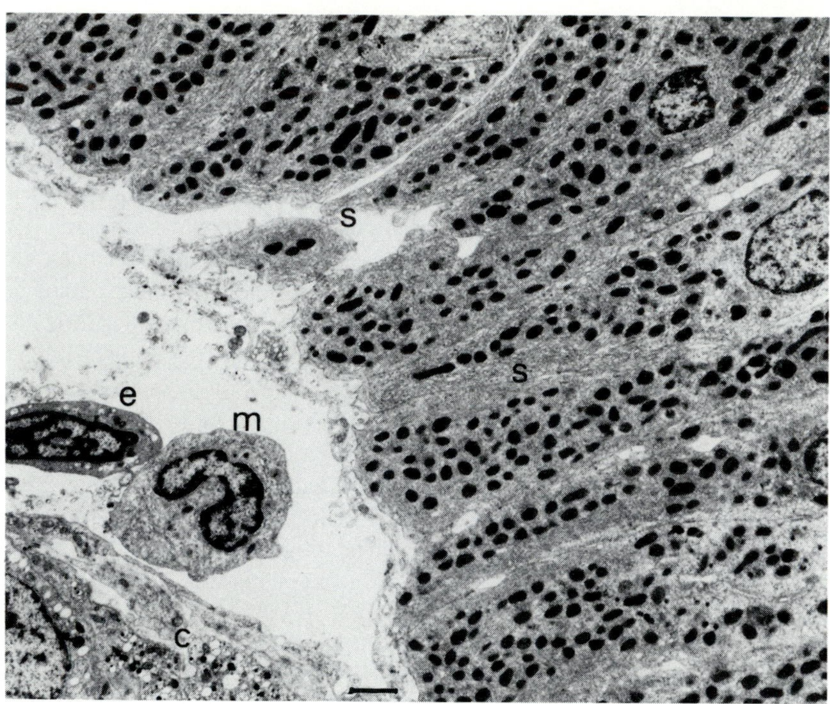

FIGURE 4. Electron micrograph of the wall of a branch of the posterior cardinal veins, showing steroid-producing interrenal cells (s) and chromaffin cells (c) of the teleost *Oreochromis mossambicus*; e, erythrocyte; m, monocyte-like leukocyte; bar represents 1 μm.

(lungfishes) have been detected (Henderson and Kime, 1987). Cortisol is involved in energy metabolism, ion regulation, and in the responses to stressors.

Cortisol is hyperglycemic by stimulating glycolysis and gluconeogenesis from protein and lipid sources. It activates key enzymes for hepatic intermediary metabolism, as was shown in chondrichthyans and teleosts (Vijayan et al., 1991). The ionic regulatory function of corticosteroids is very important because the osmoregulatory problems faced by most fishes are substantial. Freshwater fishes are hyperregulators, and thus face hyperhydration and diffusive ion losses. Osmotic and ionic homeostasis is maintained by the active uptake of ions through the gut and, in particular, across the gills. The marine jawless fishes are osmoconformers. The composition of the main ions of their blood plasma reflects that of seawater. In the marine chondrichthyans the blood plasma is slightly hyperosmotic because of a high content of urea. The relatively low salt levels of the plasma are maintained by the secretion of sodium and chloride by the rectal salt gland. Seawater teleost fishes are hyporegulators; they maintain their plasma osmotic value at a level about one third that of seawater, i.e., only slightly higher than that of freshwater fishes. The resulting osmotic water losses are compensated by drinking seawater. The water uptake is driven by a $Na^+/K^+/Cl^-$ cotransporter in the intestinal epithelium. The high uptake of monovalent ions is compensated by the excretion of these ions via the chloride cells in the gills (see Chapter 11, this volume). The control of the ion-regulatory processes is dependent to a high degree on corticosteroids and this most likely holds for all fishes, even though data on fishes other than teleosts are limited (Henderson and Kime, 1987). Corticosteroid injections decrease salt losses in lampreys when in fresh water, and stimulate salt excretion by the rectal glands in chondrichthyans. In both freshwater and seawater teleost fishes ionic regulation is controlled by cortisol, which is known to stimulate the proliferation of chloride cells and to promote the Na^+/K^+-ATPase activity of these cells, the driving force for monovalent ion transport in both fresh water and seawater. Cortisol secretion is stimulated transiently during migration or transfer of euryhaline fishes from fresh water to seawater, and this has led to the designation of cortisol as a seawater adaptation hormone. This is no longer tenable because cortisol secretion is also enhanced during seawater-to-freshwater transfer: it is indispensable as a gluco- and mineralocorticoid in freshwater as well as seawater fishes (Mayer-Gostan et al., 1987; Laurent and Perry, 1990).

Cortisol secretion is stimulated in response to many stressful stimuli, varying from handling, confinement, and other forms of disturbance, to toxicants and water acidification. This has led to the recognition of cortisol as an important stress hormone in freshwater as well as seawater fishes, like it is in the terrestrial vertebrates (Schreck et al., 1989). This function of cortisol can be related directly to the circumstance that stressful stimuli induce hydromineral disturbances and require the mobilization of energy reserves (Bonga and Lock, 1992). Cortisol probably mediates the immune suppression and the inhibition of growth and reproduction that may occur during prolonged and severe stress (Carragher et al., 1989). Enhanced cortisol secretion is often associated with gonadal maturation (Schreck et al., 1989). A remarkable and still not fully understood phenomenon is the hyperplasia of the interrenal cells during sexual maturation, for instance in some salmonid species during and after migration to their spawning grounds, which may end in large-scale post-spawning mortality (Henderson and Kime, 1987).

The many actions of cortisol are reflected by the complexity of the mechanisms that control its secretion. In addition to ACTH, the major secretagogue, and α-MSH (see above), hormones such as growth hormone, thyroxin, angiotensin II, arginine vasotocin, and catecholamines may have corticotropic activities (Schreck et al., 1989). Factors such as the N-terminal peptide of POMC (see above) and factors produced by the immune system may modulate cortisol secretion directly or by affecting other regulatory components of the hypothalamo-pituitary-interrenal axis (Balm et al., 1993).

The Catecholamines

The chromaffin cells synthesize epinephrine and norepinephrine as the main end products. In addition, important amounts of dopamine may be produced. The significance of the dopamine release is unknown. In teleosts, epinephrine and norepinephrine are produced in specific cell types. Distinct dopamine cells seem to be absent. In eels the colocalization of catecholamines with peptides such as methionine enkephalin has been demonstrated. In fishes, the bulk of the circulating catecholamines is released from chromaffin cells associated with the cardiovascular system, in particular those located in the head kidney region. This contrasts with the situation in mammals, where there is also a substantial "overflow" from sympathetic nerves to the blood (Epple et al., 1989). Catecholamines in fishes have effects on the blood circulation, respiration, and energy balance, like in other vertebrates. Hyperglycemic effects of epinephrine and norepinephrine have been reported for all major groups of fishes, although the significance of the data has been doubted because of the high doses used in many experiments (Epple et al., 1989). In teleosts, epinephrine and norepinephrine stimulate both glycogenolysis and gluconeogenesis, effects mediated by β-adrenoceptors. The relative potency of both catecholamines differs between species (Danulat and Mommsen, 1990). Typical for fishes is the important role of catecholamines, in particular epinephrine, for the control of the vasculature of the gills (Mayer-Gostan et al., 1987). This vasculature has a higher sensitivity to catecholamines than the systemic vessels. Typically, administration of epinephrine produces a biphasic response: an initial vasoconstriction that can be blocked by α-adrenoreceptor antagonists is followed by a more prolonged vasodilation that can be inhibited by β-antagonists (Wahlqvist, 1981). The vasodilating effect that has been produced, for example, in agnathans, chondrichthyans, and teleosts promotes perfusion of the gill lamellae and thus stimulates oxygen uptake. The oxygen supply of the body tissues is further promoted by the stimulatory effects on ventilatory movements and by direct effects on the erythrocyte volume and intracellular pH. Upon β-adrenergic activation of the Na^+/H^+ antiporter in the erythrocyte membrane, which leads to a rise in intracellular pH, the affinity and capacity of oxygen binding to hemoglobin is enhanced (Aota et al., 1990). The above effects are of primary importance for the rapid mobilization of energy during exposure to a variety of stressors. The catecholamines are important stress hormones in fish, like in the higher vertebrates (Mazeaud and Mazeaud, 1981; Epple et al., 1989).

Several effects of catecholamines on osmoregulation have been reported. They promote ion exchanges across the gills, probably as an indirect effect of their stimulatory action on gill perfusion, which increases the area of the gill lamellae that is in direct contact with the water. In seawater teleost fishes chloride secretion by the chloride cells can be inhibited by α-, and stimulated by β-adrenergic mechanisms (Mayer-Gostan et al., 1987).

In teleost fishes the release of catecholamines is under the control of cholinergic nerve fibers and of humoral factors. In this respect, the teleosts are intermediate between the cyclostomes, which are lacking nervous innervation of the interrenal steroid-producing and chromaffin cells, and mammals, in which the adrenomedullary secretion of catecholamines is fully dependent on cholinergic innervation. Hypoxia and blood acidosis strongly increase ventilatory movements, and this effect is mediated by catecholamines. High P_{CO2} and low blood pH are important stimulants of catecholamine secretion (Aota et al., 1990). Other stimulants seem to be the hormones themselves: at least in eels an increase of the blood level of one of the three catecholamines stimulates the release of the other two. This effect was not influenced by removal of the brain and spinal cord destruction (Epple et al., 1989). The blood-borne factors may control the catecholamine secretion under normal conditions, whereas the cholinergic innervation may mediate the rapid responses of these cells to stressors: sectioning of the nervous supply to the chromaffin cells resulted in a decreased ability to respond to stressors (Wahlqvist and Nilsson, 1980).

Natriuretic Peptides

A new and exciting development in vertebrate endocrinology is the discovery of the natriuretic peptide family, a group of closely related peptides of about 3 kDa. They were detected in the early 1980s, and are produced in several tissues, in particular the brain, where they function as neurotransmitters, and in cardiac tissue. The cardiac peptides, synthesized and released in atrial as well as (in the lower vertebrates) ventricular cells, function as hormones. The amino acid sequences of the peptides from the three sources may differ slightly, and they are called A-type natriuretic peptide (first found in the atrium) B-type and C-type natriuretic peptides (first identified in the brain) and V-type natriuretic peptide. The latter is a novel type, isolated from the eel ventricle (Takei et al., 1991). In mammals the natriuretic peptides have potent natriuretic, diuretic, and vasodilatory actions, which result in a decrease of blood pressure. They act directly, by decreasing tubular sodium resorption and increasing the glomerular filtration rate in the kidneys and by dilating contracted arteries, as well as indirectly, by inhibiting hormones such as aldosterone, vasopressin, or angiotensin. Synthesis and release of the cardiac peptides are controlled by factors connected with an increase in blood volume such as atrial and ventricular stretch and increased plasma Na^+ or osmolarity. In addition, various hormones including glucocorticoids, thyroid hormone, angiotensin II, and endothelins (the vasoconstrictive peptides produced by the vascular endothelium) may also control secretion (Evans, 1990, in press).

The presence of natriuretic peptides in fish has been demonstrated in brains, cardiac tissues and blood plasma of agnathans, chondrichthyans, and teleosts. The available literature has recently been reviewed extensively by Evans (1990; in press) and references for the data mentioned in the following summary can be found in these reviews. Unlike mammals, in fish the ventricular production seems to be higher than the contribution of the atria. Most data have been obtained by immunocytochemistry and radioimmunoassays with mammalian antisera. Only recently, A-type and C-type natriuretic peptides of two teleosts and two sharks have been sequenced. The first results indicate that, because of amino acid sequence differences, mammalian antisera underestimate the actual natriuretic peptide content of teleost tissues and body fluids.

The intestine, gills, kidneys, and cardiovascular system have been identified as the main targets in fish. In the intestine of seawater flounder, mammalian atrial natriuretic peptide inhibits the active uptake of Na^+ and Cl^-, the driving force for intestinal water uptake. As this water uptake is essential to offset the dehydration of seawater teleosts, this effect seems to counteract the regulation of water balance in seawater fish (O'Grady et al., 1985). Contrastingly, the promotion of salt secretion in seawater fish points to a function in ion regulation that is essential for survival in seawater; the Cl^- extrusion by gills of seawater killifish was stimulated (Scheide and Zadunaisky, 1988), like the salt secretion by the shark rectal gland. The last observation has been confirmed in a shark, with extracts from atrial and ventricular shark tissues. That atrial natriuretic peptides are more important in seawater than in freshwater fish is indicated by the rise of plasma atrial natriuretic peptide levels when euryhaline freshwater fish are exposed to higher salinities, and the reduction of these levels when marine teleosts are acclimated to dilute seawater (Evans et al., 1989, Evans, in press).

In the kidneys, high doses of mammalian atrial natriuretic peptides produce natriuresis and a slight diuresis. Urine flow was reduced in a shark, an effect mediated by the glomerular filtration rate. Antidiuretic activity has further been shown with homologous atrial natriuretic peptide and ventricular natriuretic peptide in freshwater eel. Drinking was reduced at low doses of atrial natriuretic peptide in both freshwater and seawater eels (Takei and Balment, 1993).

Mammalian atrial natriuretic peptides have no effect or produce tachycardia when tested on fish hearts. Mammalian as well as homologous atrial natriuretic peptides induce hypotension in eel and shark. In general, preconstricted teleost vessels *in vitro* respond with relaxation to mammalian or fish atrial natriuretic peptides (Evans, in press). An interesting observation is that the potency of eel and killifish was less than that of rat atrial natriuretic peptide in causing relaxation of aortic rings of toadfish (Evans et al., 1989). Because fish natriuretic peptides have a much higher potency in homologous assays, this indicates substantial interspecies differences between fish. Recently it was reported that human atrial natriuretic peptide increases plasma cortisol levels in seawater flounder, and stimulates cortisol secretion by interrenal tissue *in vitro*. The latter effect could only be produced in tissue from seawater-acclimated fish, and not freshwater fish, and has recently been confirmed with homologous atrial natriuretic peptide in eels for cortisol secretion *in vivo* (Takei and Balment, 1993).

In conclusion, the function of natriuretic peptides in fishes seems more connected with salt regulation than volume regulation and this function seems more important in seawater fishes than in freshwater fishes (Evans, in press). The data are limited, however.

Calcitonin

Calcitonin is produced in the ultimobranchial glands. These occur in all vertebrates except the agnathans, which are lacking calcitonin, and the mammals, where the calcitonin-producing cells are distributed in the thyroid gland. In fishes the gland is unpaired and located in the connective tissue sheets around the heart. The gland consists of endocrine cells, originating from the neural crest, and of nonglandular cells which show structural resemblance to the stellate cells of the pituitary gland. The function of calcitonin in fishes has not been fully clarified. In mammals and birds calcitonin has some hypocalcemic potency, particularly in young animals. The hormone redresses the hypercalcemia resulting from the episodic intestinal calcium uptake after a meal, by stimulating calcium deposition in the skeleton. It further protects the skeleton against excessive demineralization during pregnancy and lactation, by inhibiting osteoclastic activity and osteocytic osteolysis (Wendelaar Bonga and Pang, 1991).

Calcitonins from sharks and some teleosts have been isolated and sequenced and are very potent in mammalian assay systems. Salmon calcitonin is of therapeutic value for man and is used as an analgesic and for treatment of osteoporosis and other bone diseases. However, calcitonin has no clear hypocalcemic potency in fishes. It promotes bone formation by osteoblasts, and this action may be the main function of calcitonin in fishes (Wendelaar Bonga and Lammers, 1982). The bone of fishes is either cellular, with limited numbers of encapsulated osteocytes, or acellular, with osteoblasts and osteocytes limited to the periphery of the bone elements. Osteoclasts have been described, but they are probably scarce in most fishes. Effects of calcitonin on these cells have not been reported.

Calcitonin secretion is stimulated during gonadal maturation in female fishes. This process involves the production in the liver of calcium-containing phospholipoproteins (vitellogenins) and the subsequent transport of these proteins to the ovaries. Whether the high calcitonin secretion during ovarian development, which is under the control of estrogens, is connected with the protection of the skeleton against excessive demineralization, like in mammals, is not known. Demineralization of both cellular and acellular bone is possible and has been demonstrated, for instance, by experimental induction of gonadal maturation by high doses of estrogens. A specific role for calcitonin in gonadal maturation has also been suggested (Björnsson et al., 1989), but direct evidence is lacking. The interesting possibility has been presented that calcitonin, like other hormones that show peaks in plasma levels during ovarian maturation (see above), penetrates the eggs and becomes involved in early development of the embryos (Brown and Bern, 1989; Bern, 1990a). The factors controlling calcitonin secretion, with exception of estrogens, have not been identified in fishes.

Immunoreactivity to calcitonin — and to calcitonin gene-related peptide, originating from the calcitonin gene by alternative RNA splicing — has been found in the nervous system, including the neurohypophysis, of different classes of fishes. The peptides may function as neurotransmitters and neuromodulators rather than hormones in these areas (Kaneko et al., 1989).

Stanniocalcin

The corpuscles of Stannius are typical for holosteans and teleosts. In the last 10 years it has become clear that these glands produce a hormone that is unique for these osteichthyan groups and that prevents hypercalcemia. The corpuscles are usually paired and located in or on the kidneys. In holosteans and some teleosts tens of small corpuscles may be present. They arise from the nephric ducts during embryonic development. Surgical removal of the Stannius corpuscles (stanniectomy) causes a pronounced, up to threefold increase of the plasma calcium concentration. This is accounted for by an increase in the protein-bound and the ionic calcium fraction of the blood plasma. This rise takes place in both freshwater and seawater fishes, and is correlated with the water calcium concentration. This indicates that stanniectomy leads to an increase of the inflow of calcium from the water, and this has indeed been demonstrated. The rise of plasma calcium in stanniectomized fish can be redressed by injection of homogenates of the glands (Wendelaar Bonga and Pang, 1991). The active principle, originally called hypocalcin or teleocalcin, has been identified and sequenced in recent years and is now named stanniocalcin (STC). It is a glycoprotein without clear structural homology with other known hormones (Butkus et al., 1987; Wagner and Friesen, 1989). The Stannius corpuscles of many species contain two cell types, one of which produces stanniocalcin; the function of the second cell type is unclear (Figure 5).

Fishes obtain most of their calcium from the water, unlike terrestrial vertebrates, for which food is the only source. In freshwater teleosts the calcium is transported by the chloride cells in the gills. Both in fresh water and seawater, calcium ions enter these cells passively, down the electrochemical gradient, via calcium channels in the apical cell membrane. Subsequently these ions are actively extruded across the basolateral cell membrane of these cells into the blood. This extrusion, against a high electrochemical gradient, is facilitated by ATP-dependent calcium pumps. Stanniocalcin controls this process, probably by acting as a calcium-channel blocker and in this way limiting the passive entry of calcium into the chloride cells (Flik, 1990). Freshwater as well as seawater fishes have access to a continuous supply of calcium, and their main problem seems to be the limitation of calcium entry unless the water calcium levels are very low. This contrasts with terrestrial animals, which are fully dependent on the food as their — often limited — calcium supply. This

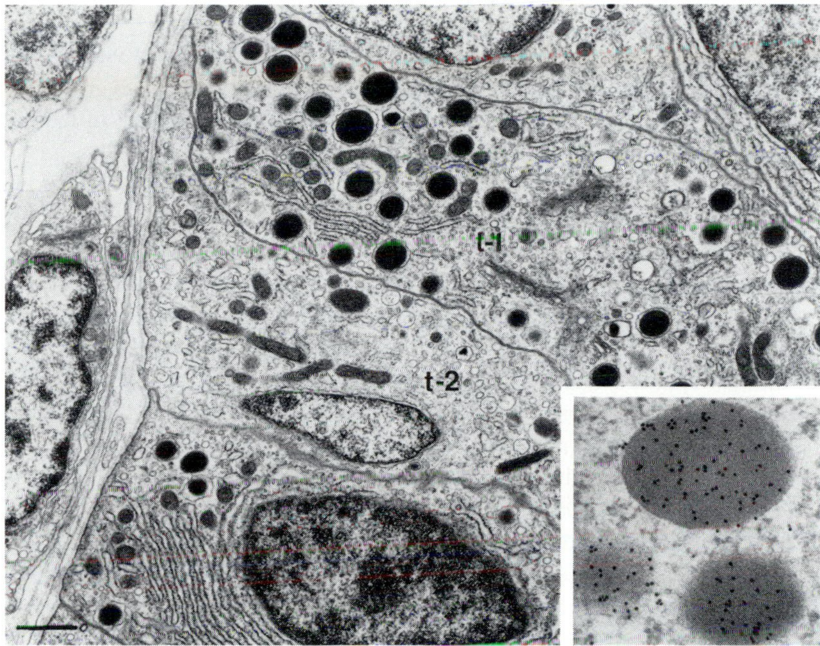

FIGURE 5. Electron micrograph of Stannius corpuscle of goldfish, showing type 1 cells (t-1), producing stanniocalcin, and type 2 cells (t-2) of unknown function; inset, secretory granules showing immunoreactivity to trout stanniocalcin (immunogold labeling); bar represents 1 μm.

difference between the aquatic and terrestrial conditions explains why the holostean and teleostean fishes depend for the control of their extracellular calcium on the hypocalcemic action of stanniocalcin. Contrastingly, in the terrestrial animals the endocrine factor dominating the control of extracellular calcium level is parathyroid hormone (PTH), which is a fast-acting and very potent hypercalcemic hormone. Parathyroid glands do not occur in fishes, and probably have evolved during the water-to-land transition of the early vertebrates (Wendelaar Bonga and Pang, 1991). Apparently, a PTH-like hypercalcemic hormone is not necessary in fishes, whereas terrestrial vertebrates do not require a potent hypocalcemic hormone like stanniocalcin (STC). One important question remains to be answered, and this is how the chondrichthyans and all other fishes that have no corpuscles of Stannius are able to maintain their plasma calcium levels, in particular in seawater.

Although true parathyroid glands are absent in fishes, immunoreactive PTH-, PTH-related peptide-, and STC-like activities were detected in coho salmon pituitary gland. Cytochemically, PTH- and STC-like immunoreactive material was localized in the neurohypophysis and preoptic area of platyfish brain. The authors suggested that this might reflect the presence of novel hypophysiotropic neuroendocrine systems (Fraser et al., 1991).

Vitamin D_3

The livers of marine teleosts are well known as rich stores of vitamin D_3 and have been applied for human use to prevent and treat rickets. Nevertheless, the function of vitamin D_3 and its metabolites in fishes is unclear. In the higher vertebrates, $1,25(OH)_2D_3$ is the most active metabolite that stimulates intestinal calcium uptake, bone remodeling, and calcium reabsorption in the kidney. Vitamin D_3 hydroxylation has been demonstrated in teleosts and lungfishes. Teleosts are able to produce $1,25(OH)_2D_3$ as well as $25(OH)D_3$ and $24,25(OH)_2D_3$, and high-affinity binding proteins for these metabolites have been demonstrated in the blood plasma in a few species. In blood plasma of lungfish and chondrichthyans, the presence of $25(OH)D_3$ has been demonstrated. Vitamin D_3 as well as $1,25(OH)_2D_3$ increased active intestinal calcium uptake in eels *in vivo* and *in vitro*. These substances were inactive when tested in perfused Atlantic cod, whereas $25(OH)D_3$ stimulated, and $24,25(OH)_2D_3$ decreased active calcium uptake in this preparation. In eel and cod $1,25(OH)_2D_3$ were hypercalcemic, whereas the other metabolites had no effect (Fenwick et al., 1984; Sundell and Björnsson, 1990; Sundell et al., 1993). Thus, although many aspects of the function of vitamin D_3 metabolites are still unclear, now there is biochemical and physiological evidence that the vitamin D_3 system is operating in fishes.

Pancreatic Hormones

In most vertebrates the pancreas contains islets of endocrine tissue among the trypsin-producing exocrine gland cells. The major hormones produced in most vertebrates are insulin, glucagon, pancreatic polypeptide, somatostatin, and pancreastatin. The physiological significance of the association of the endocrine and exocrine pancreatic tissue is unclear and does not seem to be essential. The analysis of the development of the islet tissue in fishes has contributed substantially to this notion. The distribution of the islet tissue in fishes is extremely variable. In the cephalochordate *Amphioxus*, which is missing a true pancreas, the immunoreactivity to insulin, glucagon, somatostatin, and pancreastatin is restricted to the open-type endocrine gut cells (see next section). The same situation is found in the ammocete larvae of lampreys, but in addition some insulin cells show an islet-like organization outside the gut epithelium. In adult agnathans, where the exocrine pancreas is represented by secretory cells in the intestinal epithelium, small islets are found in a ring-like association around the bile duct, as in hagfishes, or as a few relatively large islets, partly associated with the liver, in lampreys. In the gnathostomes, the islet cells evolved as a cell mass that mostly becomes associated with exocrine pancreatic tissue and gives rise to many small islets (e.g., in chondrichthyans), a few large islets (in most teleosts), or even a single large islet, such as in the swordtail fish. In many actinopterygian fishes the islet tissue is associated with some exocrine pancreatic tissue, forming large Brockmann bodies. In Polypteridae and Chondrostei, the pancreatic tissue with the islets is partially included in the liver, while in some teleosts, islet tissue is associated with the spleen. Thus, the structural organization of the islet tissue is very variable in fishes and an association with the exocrine pancreatic tissue is not always present (Epple and Brinn, 1987).

In the islets of agnathans, immunoreactivity to insulin and two somatostatins has been demonstrated. No immunoreactivity to glucagon, glucagon-like peptide, or pancreatic polypeptide could be found. However, pancreatic polypeptide is a member of the family of neuropeptide-like peptides, and in lampreys immunoreactivity to several members of this family (neuropeptide Y, peptide XY, and anglerfish peptide Y) has been demonstrated. The reactive peptides were located in the same cells (Cheung et al., 1991). In the islets of gnathostomes immunoreactivity to insulin, glucagon, glucagon-like peptide, somatostatins, and pancreatic polypeptide was found (Epple and Brinn, 1987; Conlon et al., 1988). The specificity of the pancreatic polypeptide immunoreactivity has been doubted and, like agnathans, it probably is due mainly to peptide Y and related peptides of the neuropeptide Y family (Blomqvist et al., 1992). In rainbow trout and coho salmon glucagon and glucagon-like peptide — the products of a single gene coding for the preproglucagon sequence — are colocalized in one cell type (Nozaki et al., 1988). Another member of the glucagon family, oxyntomodulin, has been identified in the islet tissue of a gar pike (Pollock et al., 1988). Immunoreactivity to pancreastatin, a recently identified pancreatic hormone of mammals that inhibits glucose-induced insulin release, has been demonstrated in the islets of a hagfish, a ray, a holocephalian, and a teleost (Reinecke et al., 1991).

The functions of the islet hormones in fishes are incompletely known. The available evidence indicates that these are comparable to some extent with those in mammals. In the agnathans, removal of the islet tissue was shown to increase, and exogenous insulin to reduce, blood sugar levels, but the reports are not always consistent. However, isletectomy in the lamprey, *Geotria australis*, the only known vertebrate in which it is possible to remove all the pancreatic islet tissue without damaging other organs, was followed by a sharp rise in plasma glucose. The main source of the elevated blood glucose is unclear. Whereas in most fishes and the higher vertebrates the liver forms the main target for insulin, the small liver of lampreys forms an exception. The main energy store of these animals, the muscles of the body wall, are also rather insensitive to insulin (Epple et al., 1992). Insulin has been shown to be hypoglycemic in a number of gnathostome fishes. However, removal of the islet tissue by pancreatectomy is, in contrast to mammals where this operation is fatal, not followed by a dramatic hyperglycemia. The absence of this effect has been ascribed to the presence of insulin cells in the gut (in the chondrichthyans) or to the simultaneous removal of insulin and glucagon cells. At least in eels the hypoglycemic action of insulin is more balanced by the hyperglycemic action of glucagon than in mammals. Insulin stimulates glucose uptake by muscle and liver and promotes gluconeogenesis and lipogenesis in these organs. An important function of insulin, at least in teleosts, is the stimulation of amino acid incorporation in tissue proteins, and amino acids are potent activators of insulin secretion (Epple and Brinn, 1987).

The understanding of the function of other islet hormones is even less than that of insulin. Glucagon causes hepatic glycogenolysis and lipolysis. It has been shown further to promote ion extrusion by the chloride cells in the gills of seawater teleosts. The first data on glucagon-like peptide show that it also is a potent stimulator of glycogenolysis. It has further gluconeogenetic and lipolytic activity. It seems to be more potent than glucagon. In fishes, in contrast to land vertebrates, it does not seem to have clear stimulatory effects on the secretion of other islet hormones such as insulin (Plisetskaya et al., 1989; Mommsen and Moon, 1989).

Somatostatins inhibit the release of insulin, glucagon, and pancreastatin in mammals, and there are indications that this holds for at least insulin and glucagon in teleosts. They further induce lipolysis and hyperglycemia (Sheridan et al., 1987). Hepatic glycogenolysis could be induced by somatostatin-14 and -25 in trout liver *in vitro* (Eilertson et al., 1991), indicating that the hyperglycemic effect of the hormone is direct, and not mediated by the inhibition of insulin secretion. Somatostatin has been shown to inhibit the ion secretion by chloride cells of seawater teleosts. The functions of the neuropeptide Y-like and pancreastatin-like factors in fishes are not known.

Gastro-Intestinal Hormones

The gastrointestinal tract of the vertebrates is an important source of hormones that control many important processes related to food digestion and energy metabolism. The hormones are produced by cells widely distributed among the normal epithelial cells of the gut. Although these endocrine cells have originally been proposed to originate from the neural crest, they are now considered to be true endodermal cells. They seem to have the capacity to perceive signals from the gut lumen and/or from the blood, and to respond to these signals by the release of hormones. Two structural types of secretory cells are distinguished: the open and closed types (Figure 6). Both types are in contact with the basal lamina. The open-type cells are also in contact with the gut lumen and are considered chemosensory cells. The closed cells have lost contact with the lumen and only seem to be able to transduce signals from the serosal side of the epithelium. The endocrine products of both types are released at the lateral and basal cell membranes (Vigna, 1986).

The hormones may diffuse to the blood circulation or they may influence neighboring cells by local diffusion (paracrine regulation). These neighboring cells may include other endocrine cells. For instance, in the mammalian stomach somatostatin secretion inhibits the release of gastrin. Both hormones are produced by separate cell types located in the epithelium of the stomach. To our knowledge paracrine effects have yet to be demonstrated in fishes. The neuropeptides produced by the gastrointestinal cells include the major hor-

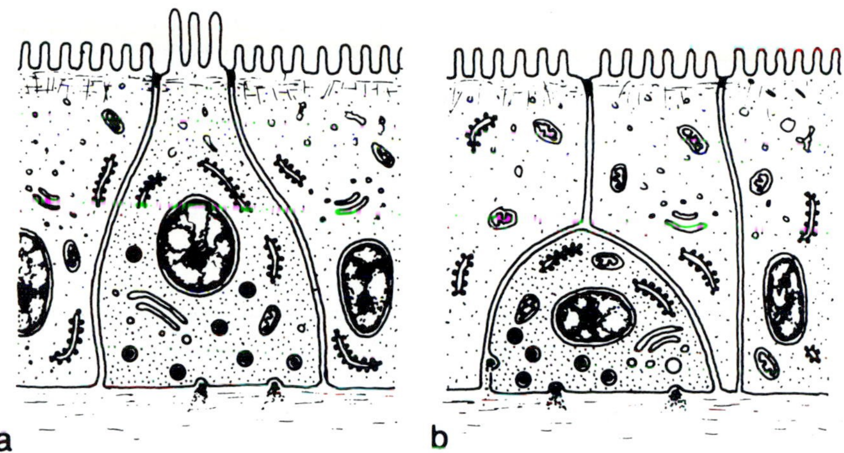

FIGURE 6. Gut endocrine cells of the open type (a) and of the closed type (b). Cells of both types are in contact with the basal lamina; hormonal messengers are released by exocytosis of secretory granules; the open-type endocrine cells extend microvilli-like processes, probably with a receptor function, into the gut lumen; cells of the closed type are lacking this specialization.

mones of the pancreatic islets: glucagon, glucagon-like peptide, somatostatin, pancreatic polypeptide, and, in agnathans and chondrichthyans, insulin. Recently, pancreastatin has also been demonstrated (Nozaki et al., 1988; Reinecke et al., 1991). This circumstance, together with the common embryological origin of the gastrointestinal and the pancreatic endocrine cells, has initiated the concept of the gastro-entero-pancreatic (GEP) endocrine system, which includes the endocrine cells of the gut and the islets. Although this name is less appropriate for fishes because of the absence in several fish species of a close association of the islets with the pancreatic tissue, the underlying idea has derived much support from comparative and developmental studies in fishes (Epple and Brinn, 1987; see also above). Immunoreactivity to many other types of gastrointestinal neuropeptides of mammals have also been demonstrated in fishes. These include gastrin/cholecystokinin, bombesin, enkephalin, vasointestinal peptide, tachykinins, neuropeptide Y-like peptides (in particular peptide YY, Blomqvist et al., 1992), neurotensin, secretin, gastric inhibitory peptide, and serotonin (Bjenning and Holmgren, 1988; Burckhardt-Holm and Holmgren, 1989). This list is certainly incomplete. Several of the mammalian gut hormones have not been studied extensively, and undoubtedly other hormones still await identification.

The study of the function of the gut hormones is hampered by the circumstance that the classical endocrine approach for function analysis (extirpation and replacement therapy) is not feasible for the endocrine system of the gut because of its diffuse nature. It has been suggested that the endocrine function of these cells is important for the adjustment of the release of digestive secretions to the amount and composition of the nutrients in the digestive tract. Some peptides may stimulate (gastrin/cholecystokinin, bombesin, substance P, serotonin) or inhibit (vasointestinal peptide) gut motility, or have excitatory effects on gastric epithelia (enkephalin, neurotensin) and increase gastric acid secretion (bombesin), but experimental evidence is limited (Bjenning and Holmgren, 1988).

Neuropeptides of the Gills

In addition to the gastrointestinal system, other organs may contain dispersed neuroendocrine cells. Such cells, producing bioamines and peptides frequently located in the same cells, are particularly common in the respiratory tract of terrestrial vertebrates (Scheuermann, 1987). A local, paracrine function has been ascribed to these cells, although a more hormonal action cannot be excluded. Such cells have also been described in pulmonary structures in fishes. For instance, serotonin-producing cells occur in the pulmonary epithelium of lungfish and in the swim bladder of the air-breathing *Polypterus* (Zaccone et al., 1992). Serotonin-producing cells have been described in the gills of chondrichthyans, actinopterygians (Figure 7), and the African lungfish (Laurent, 1984; Zaccone et al., 1993). Some of the open-type cells exhibited immunoreactivity for metenkephalin and leu-enkephalin. These cells responded by increased release of their secretory material to environmental hypoxia and reduction of the water calcium concentration. Leu-enkephalin-positive cells were further found in the gills of a lamprey. We have shown neuropeptide Y immunoreactivity in closed-type cells in teleost gill (Figure 7). It has been suggested that these cells modulate gill functions and that they may act as chemoreceptor cells (Bailly et al., 1992).

Angiotensin

Angiotensin II is the main product of the renin-angiotensin system, in fishes as well as in the higher vertebrates. It originates from a precursor molecule (angiotensinogen) that is processed by renin, produced mainly in the kidneys by the juxtaglomerular cells, to angiotensin I, and by converting enzyme, produced in several organs, to angiotensin II.

A renin-angiotensin system has been found in all major groups of fishes, with the exception of the agnathans (Sokabe and Ogawa, 1974). Until recently it was also thought to be absent in chondrichthyans, but now there is good evidence for its presence and functioning in elasmobranchs. Similar to other fishes, juxtaglomerular cells showing renin-like immunoreactivity could be demonstrated in the kidneys of several elasmobranchs (Lacy and Reale, 1989) and renin-like and angiotensin I converting enzyme-like activities were found in a dogfish. Both activities were present in the kidneys, whereas converting enzyme-like activity was further found in gills, spleen, and brain (Uva et al., 1992).

In mammals, angiotensin II has effects on blood pressure, aldosterone secretion, and drinking behavior. In fishes, angiotensin II increases systemic

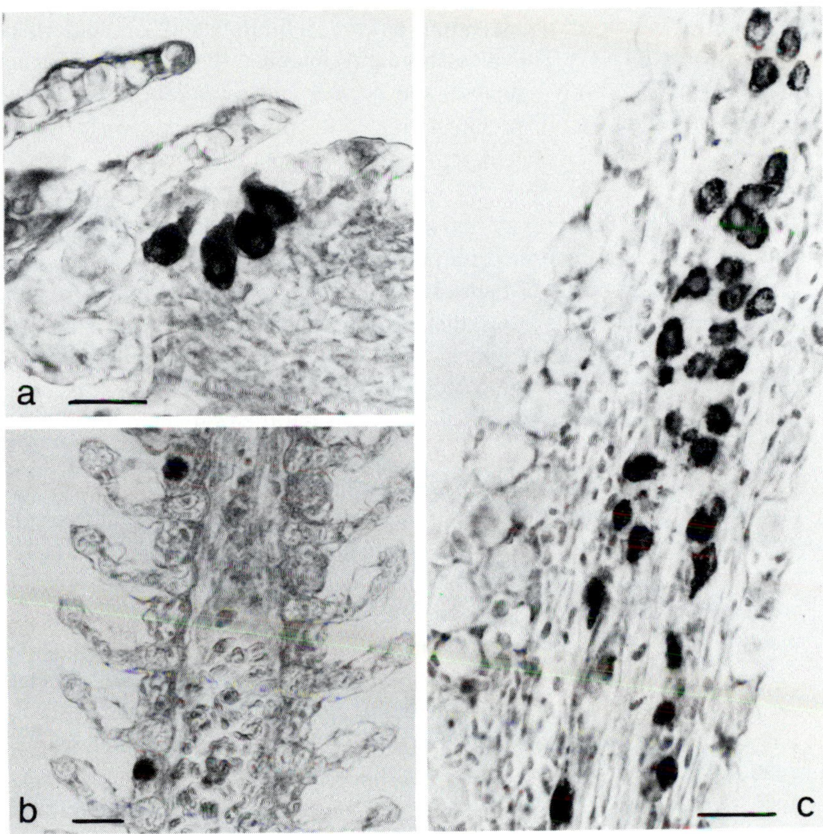

FIGURE 7. Light micrographs of immunoreactive neuroendocrine cells in the gills of fishes; PAP method; bars represent 20 μm. (a) Cluster of met-5-enkephalin-positive neuroendocrine cells in the filamental epithelium of brown trout; (b) neuropeptide Y-positive neuroendocrine cells in the filamental epithelium of the teleost *Oreochromis mossambicus* (G. Geeraedts, A. Coenen, and S.E. Wendelaar Bonga, unpublished); (c) serotonin-positive neuroendocrine cells in the basal area of the filamental epithelium of the catfish *Ictalurus nebulosus* (a and c from Zaccone et al., *Acta Zool.*, 73, 177, 1992. With permission from Pergamon Press, Oxford, UK).

blood pressure, an effect that is indirect and mediated by catecholamines (Nishimura, 1987). In an elasmobranch, administration of angiotensin II stimulated the secretion of 1α-hydroxycorticosterone (Hazon and Henderson, 1985), and in teleosts that of cortisol (Perrott and Balment, 1990). Angiotensin II further initiates drinking, particularly in seawater teleost fishes (Hazon et al., 1989). Adaptation of euryhaline teleosts to increased water salinity requires renal antidiuresis effected by reduction of the glomerular filtration rate. This can be effected by angiotensin II, and this may therefore be an essential hormone for adaptation to seawater (Brown et al., 1990). Recovery from pharmacologically induced hypotension in flounders could be inhibited by injection of captopril, a converting enzyme inhibitor, and this also reduced plasma cortisol levels. This indicates a physiological function for endogenous angiotensin II in body fluid volume control and interrenal steroidogenesis (Perrott and Balment, 1990).

Melatonin

In fishes, the pineal organ (epiphysis) receives and integrates light stimuli, and this input is transformed into electrical and hormonal responses. In this respect the pineal organ of fishes is comparable with that of most amphibians and reptiles, and contrasts with that of birds and mammals. In the latter groups the receptor function of the pineal organ is present only during early development. Subsequently the pineal organ reduces to a gland mainly consisting of neuroendocrine transducer cells that receive information about the light regimen via the eyes.

The pineal organ projects from the roof of the diencephalon. In most fishes the organ forms a pineal complex with another projection of the diencephalic root, the parapineal organ. In teleosts it is rudimentary, and most adult teleosts only have a pineal organ. It has a lumen that in most fishes is in open connection, via the pineal stalk, with the third ventricle. The tissue of the pineal wall consists of pinealocytes, glial cells, and neurons. The pinealocytes of fishes are true photoreceptor cells with a lamellated outer segment that contains the photopigments and that is comparable with the retinal cones of the eyes. The outer segments project into the pineal lumen and are connected, via an inner segment that is rich in mitochondria, with the cell body that contains the nucleus. The cell body forms the secretory part of the cell. It may produce neurotransmitters as well as hormonal messengers. One or more cytoplasmic processes from these cells may be synaptically connected with other cells. The skin may be less pigmentated and the skull less calcified in the area overlying the pineal. This pineal window facilitates the illumination of the organ.

The major hormone of the pineal organ is melatonin, an indole amine. In the higher vertebrates several other potential messengers have been identified, including peptides. Melatonin-like material has been demonstrated in the cell bodies of the pineal photoreceptor cells of fishes and is most likely secreted from the basal parts of these cells. In this respect these cells are comparable to the photoreceptor cells of the retina, that also have the capacity to secrete melatonin. At least in the higher vertebrates most of the melatonin present in the circulation originates from the pineal organ, and this seems also applicable to fishes. Retinal melatonin may have a paracrine function (Falcón and Collin, 1991). Different types of photoreceptor cells may be present in the pineal organ, and not all of these seem to secrete melatonin (Tamotsu et al., 1990).

The pineal plays a crucial role in adaptation to the light regime. The secretion of melatonin — predominantly in the dark phase — mediates the adjustment of behavioral and physiological processes to the daily light-dark cycle and the annual cycle in the daily light period. Virtually all body processes are influenced, from mitotic activity to locomotory activity (including migration), skin pigmentation, or the timing of reproduction. In fishes, the pineal organ functions as an endogenous oscillator, showing a circadian rhythmicity — also *in vitro* (Iigo et al., 1991) — which is adjusted via the photic input of the pineal. In the retina of pike the melatonin content showed a peak during the dark phase, similar to the pineal gland (Falcón and Collin, 1991). The functional relationship between melatonin production in the retina and in the pineal in fishes is unknown. Van Eys and Wendelaar Bonga (1981) have shown that, for the control of skin pigmentation in a teleost, the photic input of both the eyes and the pineal organ is essential.

CONCLUSIONS

The above survey shows that our knowledge of the neuroendocrine system of fishes is expanding rapidly. The system shares its general pattern with that of all other vertebrates. The developments in the endocrine research follow those in mammals. More and more messenger molecules are identified, and it appears that these messengers are not only produced in the brain and the classical endocrine system but also in many other organs. The same messenger molecule, or closely related variants, may be produced in different organs in the same organism and, depending on the location, function as neurotransmitter, neuromodulator, hormone, or paracrine messenger.

These conclusions are based on a limited set of data, particularly when it is taken into account that the fishes are a very diverse group of animals, and, with over 20,000 representatives, comprise more than half of the vertebrate species. As far as the identification of the messengers is concerned, originally this has been almost exclusively based on immunoreactivity to antisera raised against mammalian hormones and on mammalian assay systems. This implies that hormones unique for fish may have escaped attention. This may, furthermore, have caused an overestimation of the similarities between the fish and mammalian hormones. The structural analysis of more and more fish hormones has shown that this indeed has taken place and that the fish equivalents of mammalian hormones often show significant structural differences not only between fishes and higher vertebrates, but also between various groups of fishes, between species of the same systematic category, or between different production sites in the same species. This structural diversity is further expressed in the production of different variants of one hormone by the same cell type, for instance in the case of prolactin, somatostatin, growth hormone, or MSH. The diversity is already immense, even though the neuroendocrine system of only a small number of fishes has been explored.

For most species this exploration has been superficial, and restricted to the accumulation of data on the location of immunoreactivity to one or more hormone antisera. More detailed analysis of the structure of the messengers and analysis of their function is badly needed for many of the known fish hormones. This also holds for the receptors of these hormones, an important aspect of neuroendocrine communication that has been neglected in this chapter. It should be kept in mind that hormone receptors have been evolved in parallel with the hormones concerned, and therefore the receptors may be expected to show a similar structural diversity as the hormones.

Although the neuroendocrine system of fishes conforms to the general pattern of the vertebrates, there are some interesting differences that are more fundamental than the many structural variants of the messenger molecules or the structural organization of the neuroendocrine tissues. First, these concern both the function of hormones that are present in fishes as well as the terrestrial vertebrates, and, second, the presence of hormones that are typical for fishes. Not surprisingly, most aspects that are unique for fishes are related, directly or indirectly, to their aquatic habitat. Fishes have a very close interaction with the ambient water. As mentioned above, there is an intensive exchange of monovalent and divalent ions, related to physiological functions such as water and ion regulation, respiration, acid-base balance, and excretion of waste products. The physiological importance and complexity of these exchanges are reflected by the large number of hormones that are implicated in the control of these substances. Prolactin, growth hormone, cortisol, glucagon, and somatostatin are only a few examples of hormones with important ionic regulatory functions and effects in fishes and not, or to a limited extent, in the terrestrial vertebrates. Messengers with ionic regulatory capacities that are absent, at least as true hormones, in the terrestrial vertebrates are the urotensins and stanniocalcin. Somatolactin and melanophore concentrating hormone probably also belong to this category. Some other endocrine messengers found in mammals have not been demonstrated in fishes. This may be caused by lack of research efforts or specific detection techniques rather than the absence of fish equivalents of these hormones. However, there is good evidence for the absence of one hormone, PTH, and the near-absence of another, aldosterone. The appearance of PTH in the land vertebrates has been connected with the change during the water-to-land transition of the vertebrates, which implied the loss of water as a readily accessible calcium source. The absence of aldosterone in most fishes (traces of this corticosteroid have been found in some, mainly nonteleostean species; aldosterone may have physiological significance only in lungfishes) is more difficult to understand, because it is the mineralocorticoid of the higher vertebrates. This function is exerted by cortisol or, possibly, corticosterone in fishes, which are also the main glucocorticoids in these animals. Perhaps the regulation of energy metabolism and the control of ion regulation are so closely linked in purely aquatic animals that these functions can be combined easily in one hormone.

This chapter shows that fish endocrinologists have contributed substantially to our knowledge of the evolution of the vertebrate neuroendocrine system.

Their work is also valuable for improving fish farming, by increasing survival, reproduction, and growth rate of fish. These applied aspects of fish endocrinology will certainly become more and more important in the near future.

ACKNOWLEDGMENT

The author is grateful to Mrs. E.M. Jansen-Hoorweg for assistance during the preparation of the manuscript.

REFERENCES

Aota, S., Holmgren, K. D., Gallagher, P., and Randall, D. J., A possible role for catecholamines in the ventilatory responses associated with internal acidosis or external hypoxia in rainbow trout, *Oncorhynchus mykiss, J. Exp. Biol.*, 151, 57, 1990.

Arnold-Reed, D. E. and Balment, R. J., Steroidogenic role of the caudal neurosecretory system in the flounder, *Platichthys flesus, Gen. Comp. Endocrinol.*, 76, 267, 1989.

Bailly, Y., Dunel-Erb, S., and Laurent, P., The neuroepithelial cells of the fish gill filament: indolamine-immunocytochemistry and innervation, *Anat. Rec.*, 233, 143, 1992.

Baker, B. I., Melanin-concentrating hormone: a general vertebrate neuropeptide, *Int. Rev. Cytol.*, 126, 1, 1991.

Balm, P. H. M., Pepels, P., Van Lieshout, E., and Wendelaar Bonga, S. E., Neuroimmunological regulation of αMSH release in tilapia *(Oreochromis mossambicus), Fish Physiol. Biochem.*, 11, 125, 1993.

Balment, R. J., Warne, J. M., Tierney, M., and Hazon, N., Arginine vasotocin and fish osmoregulation, *Fish Physiol. Biochem.*, 11, 1993.

Batten, T. F. C., Cambre, M. L., Moons, L., and Vandesande, F., Comparative distribution of neuropeptide-immunoreactive systems in the brain of the green molly, *Poecillia latipinna, J. Comp. Neurol.*, 302, 893, 1990.

Bayne, C. J. and Levy, S., The respiratory burst of rainbow trout, Oncorhynchus mykiss (Walbaum), phagocytes is modulated by sympathetic neurotransmitters and the neuropeptide ACTH, *J. Fish Biol.*, 38, 609, 1991.

Bern, H. A., The "new" endocrinology: its scope and its impact, *Am. Zool.*, 30, 877, 1990a.

Bern, H. A., The caudal neurosecretory system: quest and bequest, in *Progress in Comparative Endocrinology*, Epple, A., Scanes, C. G. and Stetson, M. H., Eds., Wiley-Liss, New York, 1990b, 242.

Bjenning, C. and Holmgren, S., Neuropeptides in the fish gut. An immunohistochemical study of evolutionary patterns, *Histochemistry*, 88, 155, 1988.

Björnsson, B. T., Haux, C., Bern, H. A., and Deftos, L. J., 17β-estradiol increases plasma calcitonin levels in salmonid fish, *Endocrinology*, 125, 1754, 1989.

Blomqvist, A. G., Söderberg, C., Lundell, I., Milner, R. J., and Larhammer, D., Strong evolutionary conservation of neuropeptide Y: sequences of chicken, goldfish, and Torpedo marmorata DNA clones, *Proc. Natl. Acad. Sci. U.S.A.*, 89, 2350, 1992.

Brown, C. L. and Bern, H. A., Hormones in early development, with special reference to teleost fishes, in *Hormones in Maturation, Aging and Senescence of the Neuroendocrine System*, Schreibman, M. P., and Scanes, C., Academic Press, New York, 1989, 289.

Brown, J. A., Gray, C. J., and Taylor, S. M., Direct effects of angiotensin II on glomerular ultrastructure in the rainbow trout, *Salmo gairdneri, Cell Tiss. Res.*, 260, 315, 1990.

Burckhardt-Holm, P. and Holmgren, S., A comparative study of neuropeptides in the intestine of two stomachless teleosts *(Poecilia reticulata, Leucisces idus melanotus)* under conditions of feeding and starvation, *Cell Tiss. Res.*, 255, 245, 1989.

Butkus, A., Roche, P. J., Fernley, R. T., Haralambidis, J., Penschow, J. D., Ryan, J. B., Trahair, J. F., Tregéar, G.W., and Coghlan, J.P., Purification and cloning of a corpuscles of Stannius protein from *Anguilla australis, Mol. Cell Endocrinol.*, 54, 123, 1987.

Carragher, J. F., Sumpter, J. P., Pottinger, T. G., and Pickering, A. D., The deleterious effects of cortisol implantation on reproductive function in two species of trout, *Salmo trutta* L. and *Salmo gairdneri* Richardson, *Gen. Comp. Endocrinol.*, 76, 310, 1989.

Chester Jones, I. and Phillips, J. G., The adrenal and interrenal glands, in *Vertebrate Endocrinology: Fundamentals and Biomedical Implications*, Vol. 1, Pang, P. K. T., and Schreibman, M. P., Eds., Academic Press, Orlando, 1986, 319.

Cheung, R., Andrews, P. C., Plisetskaya, E. M., and Youson, J. H., Immunoreactivity to peptides belonging to the pancreatic polypeptide family (NPY, aPY, PP, PYY) and to glucagon-like peptide in the endocrine pancreas and anterior intestine of adult lampreys, *Petromyzon marinus*: an immunohistochemical study, *Gen. Comp. Endocrinol.*, 81, 51, 1991.

Conlon, J. M., Deacon, C. F., Hazon, N., Henderson, I. W., and Thim, L., Somatostatin-related and glucagon-related peptides with unusual structural features from the European eel *(Anguilla anguilla), Gen. Comp. Endocrinol.*, 72, 181, 1988.

Danulat, E. and Mommsen, T. P., Norepinephrine: a potent activator of glycogenolysis and gluconeogenesis in rockfish hepatocytes, *Gen. Comp. Endocrinol.*, 78, 12, 1990.

Daughady, W. H., Kapadia, M., Yanow, C. E., Fabrick, K., and Mariz, I. K., Insulin-like growth factor I and II of non-mammalian sera, *Gen. Comp. Endocrinol.*, 59, 316, 1985.

de Luze, A., Leloup, J., Redding, M., and Leloup-Hatey, J., Changes in thyroid function and plasma cortisol during seawater acclimation in eel *(Anguilla anguilla), Gen. Comp. Endocrinol.*, 66, 29, 1987.

Dent, J., The thyroid gland, in *Vertebrate Endocrinology: Fundamentals and Biomedical Implications*, Vol. 1, Pang, P. K. T., and Schreibman, M. P., Eds., Academic Press, Orlando, 1986, 175.

Eales, J. G. and McLatchy, D. L., The relationship between T_3 production and energy balance in salmonids and other teleosts, *Fish Physiol. Biochem.*, 7, 289, 1989.

Eilertson, C. D., O'Connor, P. K., and Sheridan, M. A., Somatostatin-14 and somatostatin-25 stimulate glycogenolysis in rainbow trout, *Oncorhynchus mykiss*, liver incubated *in vitro*: a systemic role for somatostatins, *Gen. Comp. Endocrinol.*, 82, 192, 1991.

Epple, A. and Brinn, J.E., *The Comparative Physiology of the Pancreatic Islets*, Springer-Verlag, Berlin, 1987.

Epple, A., Hathaway, C. B., and Nibbio, B., Circulatory catecholamines in the eel: origins and functions, *Fish Physiol. Biochem.*, 7, 123, 1989.

Epple, A., Cake, M. H., Potter, J. C., and Tajbakhsh, M., Impact of complete isletectomy on plasma glucose in the southern hemisphere lamprey *Geotria australis, Gen. Comp. Endocrinol.*, 86, 284, 1992.

Evans, D. H., An emerging role for a cardiac peptide hormone in fish osmoregulation, *Ann. Rev. Physiol.*, 52, 43, 1990.

Evans, D. H., The roles of natriuretic peptide hormones (NP) in fish osmoregulation, in *Advances in Environmental and Comparative Physiology — Mechanisms of Systemic Regulation in Lower Vertebrates II: Acid-Base Regulation, Ion Transfer, and Metabolism*, Heisler, N., Ed., Springer Verlag, Heidelberg, in press.

Evans, D. H., Chipouras, E., and Payne, J. A., Immunoreactive atriopeptin in the plasma of fishes: a potential role in gill hemodynamics, *Am. J. Physiol.*, 257, R939, 1989.

Falcón, J. and Collin, J.-P., Pineal-retinal relationships: rhythmic biosynthesis and immunocytochemical localization of melatonin in the retina of the pike, *(Esox lucius), Cell Tiss. Res.*, 265, 601, 1991.

Fenwick, J. C., Smith, K., Smith, J., and Flik, G., Effect of various vitamin D analogs on plasma calcium and phosphorus and intestinal calcium absorption in fed and unfed American eels, *Anguilla rostrata, Gen. Comp. Endocrinol.*, 55, 398, 1984.

Flik, G., Hypocalcin physiology, in *Progress in Comparative Endocrinology*, Epple, A., Scanes, C. G., and Stetson, M. H., Eds., Wiley-Liss, New York, 1990, 534.

Fraser, R. A., Kaneko, T., Pang, P. K. T., and Harvey, S., Hypo- and hypercalcemic peptides in fish pituitary glands, *Am. J. Physiol.*, 260, R622, 1991.

Fryer, J. N., Neuropeptides regulating the activity of goldfish corticotropes and melanotropes, *Fish Physiol. Biochem.*, 7, 21, 1989.

Grace de Jesus, E., Hirano, T., Inui, Y., Changes in cortisol and thyroid hormone concentrations during early development and metamorphosis in the Japanase flounder *Paralichthys olivaceus, Gen. Comp. Endocrinol.*, 82, 369, 1991.

Grau, E. G., Thyroid hormones, in *Vertebrate Endocrinology: Fundamentals and Biomedical Implications*, Vol. 2, Academic Press, Orlando, 1987, 85.

Grau, E. G. and Helms, L. M. H., The tilapia prolactin cell: twenty-five years of investigation, in *Progress in Comparative Endocrinology*, Epple, A., Scanes, C. G. and Stetson, M. H., Eds., Wiley-Liss, New York, 1990, 534.

Haruta, K., Yamashita, T., and Kawashima, S., Changes in arginine vasotocin content in the pituitary of the medaka *(Oryzias latipes)* during osmotic stress, *Gen. Comp. Endocrinol.*, 83, 327, 1991.

Hazon, N. and Henderson, I.W., Factors affecting the secretory dynamics of 1α-hydroxycorticosterone in the dogfish, *Scylliorhinus canicula, Gen. Comp. Endocrinol.*, 59, 50, 1985.

Hazon, N., Balment, R. J., Perrott, M., and O'Toole, L. B., The renin-angiotensin system and vascular and dipsogenic regulation in elasmobranchs, *Gen. Comp. Endocrinol.*, 74, 230, 1989.

Henderson, I. W. and Kime, D. E., The adrenal cortical steroids, in *Vertebrate Endocrinology: Fundamentals and Biomedical Implications*, Vol. 2., Pang, P.K.T. and Schreibman, M.P., Eds., Academic Press, Orlando, 1987, 121.

Hirano, T., The spectrum of prolactin actions in teleosts, in *Comparative Endocrinology, Developments and Directives*, Ralph, C.L., Ed., Alan Liss, Inc., New York, 1986, 53.

Idler, D. R., Fletcher, G. L., Belkhode, S., King, M. J., and Hwang, S.J., Regulation of antifreeze protein production in winter flounder: a unique function for growth hormone, *Gen. Comp. Endocrinol.*, 74, 327, 1989.

Iigo, M., Kezuka, H., Aida, K., and Hanyu, I., Circadian rhythms of melatonin secretion from superfused goldfish *(Carassius auratus)* pineal glands in vitro, *Gen. Comp. Endocrinol.*, 83, 152, 1991.

Kaneko, T., Harvey, S., Kline, L. W., and Pang, P. K. T., Localization of calcium regulatory proteins in fish, *Fish Physiol. Biochem.*, 7, 337, 1989.

Kawauchi, H., The melanotropic peptides: structure and chemistry, in *The Melanotropic Peptides*, Vol. 1, Hadley, M. E., Ed., CRC Press, Boca Raton, FL, 1988, 39.

Lacy, E. R. and Reale, E., Granulated peripolar epithelial cells in the renal corpuscle of marine elasmobranch fish, *Cell Tiss. Res.*, 257, 61, 1989.

Lamers, A. E., Flik, G., Atsma, W., and Wendelaar Bonga, S. E., A role for di-acetyl α-melanocyte-stimulating hormone in the control of cortisol release in the teleost *Oreochromis mossambicus, J. Endocrinol.*, 135, 285, 1992.

Larson, B. A. and Madani, Z., Increased urotensin I and II immunoreactivity in the urophysis of *Gillichthys mirabilis* transferred to low salinity water, *Gen. Comp. Endocrinol.*, 83, 379, 1991.

Laurent, P., Gill internal morphology, in *Fish Physiology*, Vol. XA, Hoar, W.S., and Randall, D.J., Eds., Academic Press, New York, 1984, 73.

Laurent, P. and Perry, S. F., Effects of cortisol on gill chloride cell morphology and ionic uptake in the freshwater trout, *Salmo gairdneri, Cell Tissue Res.*, 259, 429, 1990.

Lindahl, K. I., Sara, V., Fridberg, G., and Nishimiya, N., The presence of somatomedin in the baltic salmon, *Salmo salar*, with special reference to smoltification, *Aquaculture,* 45, 177, 1985.

Madsen, S.S., The role of cortisol and growth hormone in seawater adaptation and development of hypoosmoregulatory mechanisms in sea trout parr *(Salmo trutta trutta), Gen. Comp. Endocrinol.*, 79, 1, 1990.

Mayer-Gostan, N., Wendelaar Bonga, S. E., Balm, P. H. M., Mechanisms of hormone actions on gill transport, in *Vertebrate Endocrinology: Fundamentals and Medical Implications*, Vol. 2, Pang P. K. T. and Schreibman, M. P., Eds., Academic Press, Orlando, 1987, 211

Mazeaud, M. M. and Mazeaud, F., Adrenergic responses to stress in fish, in *Stress and Fish*, Pickering, A.D., Ed., Academic Press, London, 1981, 49.

Mommsen, T. P. and Moon, T., Metabolic actions of glucagon-family hormones in liver, *Fish Physiol. Biochem.*, 7, 279, 1989.

Moons, L., Cambre, M., Marovoet, S., Batten, T. F. C., Vanderhaeghen, J.J., Ollevier, F., and Vandesande, F., Peptidergic innervation of the adrenocorticotropic hormone (ACTH)- and growth hormone (GH)-producing cells in the pars distalis of the sea bass *(Dicentrarchus labrax). Gen. Comp. Endocrinol.,* 83, 327, 1991.

Nishimura, H., Role of the renin-angiotensin system in osmoregulation, in *Vertebrate Endocrinology: Fundamentals and Biomedical Implications*, Vol. 2, Pang, P. K. T., and Schreibman, M.P., Eds., Academic Press, Orlando, 1987, 157.

Nozaki, M., Miyata K., Oota, Y., Gorbman, A., and Plisetskaya, E.M., Colocalization of glucagon-like peptide and glucagon immunoreactivities in pancreatic islets and intestine of salmonids, *Cell Tissue Res.*, 253, 371, 1988.

O'Grady, S. M., Field, N., Nash, N. T., and Rao, M. C., Atrial natriuretic factor inhibits Na-K-Cl cotransport in teleost intestine, *Am. J. Physiol.*, 249, C531, 1985.

Olivereau, M. and Olivereau, J. M., Corticotropin-like immunoreactivity in the brain and pituitary of three teleost species (goldfish, trout and eel), *Cell Tissue Res.*, 262, 115, 1990.

Ono, M., Takayama, Y., Rand-Weaver, M., Sakata, S., Yasunaga, T., Noso, T., and Kawauchi, H., CDNA cloning of somatolactin, a new pituitary protein related to growth hormone and prolactin, *Proc. Natl. Acad. Sci.* U.S.A., 87, 4330, 1990.

Pearse, A. G. E. and Pollack, J. M., Neural crest origin of the endocrine polypeptide (APUD) cells of the gastrointestinal tract and pancreas, *Gut*, 12, 783, 1971.

Perks, A. M., The neurohypophysis: the place of comparative studies in biology and medicine, in *Vertebrate Endocrinology: Fundamentals and Biomedical Implications,* Vol. 2, Pang, P. K. T., and Schreibman, M. P., Eds., Academic Press, Orlando, 1987, 9.

Perrott, M. N. and Balment, R. J., The renin-angiotensin system and the regulation of plasma cortisol in the flounder, *Platichthys flesus, Gen. Comp. Endocrinol.*, 78, 414, 1990.

Perrott, M. N., Carrick, S., and Balment, R. J., Pituitary and plasma arginine vasotocin levels in teleost fish, *Gen. Comp. Endocrinol.*, 83, 68, 1991.

Peter, R. E., Yu, K.-L., Marchant, T. A., and Rosenblum, P. M., Direct neural regulation of the teleost neurohypophysis, *J. Exp. Zool. Suppl.*, 4, 84, 1990.

Planas, J. V., Swanson, P., Rand-Weaver, M., and Dickhoff, W. W., Somatolactin stimulates *in vitro* gonadal steroidogenesis in coho salmon, *Oncorhynchus kisutch, Gen. Comp. Endocrinol.*, 87, 1, 1992.

Plisetskaya, E.M., Ottolenghi, C., Sheridan, M.A., Mommsen, T.P., and Gorbman, A., Metabolic effects of salmon glucagon and glucagon-like peptide in coho and chinook salmon, *Gen. Comp. Endocrinol.*, 73, 205, 1989.

Pollock, H. G., Kimmel, J. R., Ebner, K. E., Hamilton, J. W., Rouse, J. B., Lance, V., and Rawitch, A.B., Isolation of alligator gar *(Lepisosteus spatula*) glucagon, oxyntomodulin, and glucagon-like peptide: amino acid sequence of oxyntomodulin and glucagon-like peptide, *Gen. Comp. Endocrinol.*, 69, 133, 1988.

Reinecke, M., Hoög, A., Östenson, C.-G., Egendic, S., Grimelius, L., and Falkmer, S., Phylogenetic aspects of pancreastatin- and chromogranin-like immunoreactive cells in the gastro-entero-pancreatic neuroendocrine system of vertebrates, *Gen. Comp. Endocrinol.*, 83, 167, 1991.

Rodriguez, M. and Specker, J. L., In vitro effects of arginine vasotocin on testosterone production by testes of rainbow trout *(Oncorhynchus kisutch), Gen. Comp. Endocrinol.*, 83, 249, 1991.

Sakamoto, T., Iwata, M., and Hirano, T., Kinetic studies of growth hormone and prolactin during adaptation of coho salmon, *Oncorhynchus kisutch*, to different salinities, *J. Endocrinol.*, 130, 425, 1991

Sakamoto, T., McCormick, S. D., and Hirano, T., Osmoregulatory actions of growth hormone and its mode of action in salmonids: a review, *Fish Physiol. Biochem.*, 11, 155, 1993.

Scharrer, B., The neuropeptide saga, *Am. Zool.*, 30, 887, 1990.

Scheide, J. I. and Zadunaisky, J. A., Effect of atriopeptin II on isolated opercular epithelium of *Fundulus heteroclitus, Am. J. Physiol.*, 254, R27, 1988.

Scheuermann, D. W., Morphology and cytochemistry of the endocrine epithelial system in the lung, *Int. Rev. Cytol.*, 106, 35, 1987.

Schreck, C. B., Bradford, C., Fitzpatrick, M. S., and Patiño, R., Regulation of the interrenal of fishes: non-classical control mechanisms, *Fish Physiol. Biochem.*, 7, 259, 1989.

Schreibman, M. P., The pituitary gland, in *Vertebrate Endocrinology: Fundamentals and Biomedical Implications*, Vol. 1, Pang, P. K. T. and Schreibman, M. P., Eds., Academic Press, Orlando, 1986, 11.

Sheridan, M. A., Plisetskaya, E. M., Bern, H. A., and Gorbman, A., Effects of somatostatin-25 and urotensin II on lipid and carbohydrate metabolism of coho salmon, *Oncorhynchus kisutch, Gen. Comp. Endocrinol.*, 66, 405, 1987.

Sherwood, N. M. and Parker, D. B., Neuropeptide families: an evolutionary perspective, *J. Exp. Zool. Suppl.*, 4, 63, 1990.

Singh, H., Griffith, R. W., Takahashi, A., Kawauchi, H., Thomas, P., and Stegeman, J. J., Regulation of gonadal steroidogenesis in *Fundulus heteroclitus* by recombinant salmon growth hormone and purified salmon prolactin, *Gen. Comp. Endocrinol.*, 72, 144, 1988.

Sokabe, H. and Ogawa, M., Comparative studies of the juxtaglomerular apparatus, *Int. Rev. Cytol.*, 37, 271, 1974.

Specker, J. L., King, D. S., Nishioka, R. S., Shiharata, K., Yamaguchi, K., and Bern, H.A., Isolation and characterization of two prolactins released *in vitro* by the pituitary of a cichlid fish, *Proc. Natl. Acad. Sci. U.S.A.*, 82, 7490, 1985.

Sundell, K. and Björnsson, B.T., Effects of vitamin D_3, 25(OH) vitamin D_3, 24,25$(OH)_2$ vitamin D_3 and 1,25$(OH)_2$ vitamin D_3 on the *in vitro* intestinal calcium absorption in the marine teleost, Atlantic cod *(Gadus morhua), Gen. Comp. Endocrinol.*, 78, 74, 1990.

Sundell, K., Norman, A. W., and Björnsson, A. T., Effects of 1,25$(OH)_2$ vitamin D_3 and 25(OH) vitamin D_3 on calcium homeostasis in the Atlantic cod, *Gadus morhua*, a marine teleost, in press.

Swennen, D., Rentier-Delrue, F., Auperin, B., Prunet, P., Flik, G., Wendelaar Bonga, S. E., Lion, M., and Martial, J. A., Production and purification of biologically active recombinant tilapia *(Oreochromis niloticus)* prolactins, *J. Endocrinol.*, 131, 219, 1991.

Takei, Y., Takahashi, A., Watanabe, T. X., Nakajima, K., and Sakakibara, S., A novel natriuretic peptide isolated from eel cardiac ventricles, *Febs Lett.*, 282, 317, 1991.

Takei, Y. and Balment, R. J., Biochemistry and physiology of a family of eel natriuretic peptides, *Fish Physiol. Biochem.*, 11, 183, 1993.

Tamotsu, S., Korf, H. W., Morita, Y., and Oksche, A., Immunocytochemical localization of serotonin and photoreceptor-specific proteins (rod-opsin, s-antigen) in the pineal complex of the river lamprey, *Lampetra japonica*, with special reference to photoneuroendocrine cells, *Cell Tiss. Res.*, 262, 205, 1990.

Uva, B., Masini, M. A., Hazon, N., O'Toole, L. B., Henderson, I. W., and Ghiani, P., Renin and angiotensin converting enzyme in elasmobranchs, *Gen. Comp. Endocrinol.*, 86, 407, 1992.

Van der Kraak, G., Rosenblum, P. M., and Peter, R. E., Growth hormone-dependent potentiation of gonadotropin-stimulated steroid production by ovarian follicles of the goldfish, *Gen. Comp. Endocrinol.*, 79, 233, 1990.

Van Eys, G. J. J. M. and Peters, P. T. W., Evidence for a direct regulatory role of α-MSH in the morphological background adaptation of the cichlid teleost *Sarotherodon mossambicus, Cell. Tissue Res.*, 217, 361, 1981.

Van Eys, G. J. J. M. and Wendelaar Bonga, S. E., Structural changes in the pars intermedia of the cichlid teleost *Sarotherodon mossambicus* as a result of background adaptation and illumination. III. The pineal organ, *Cell Tiss. Res.*, 220, 561, 1981.

Vigna, S. R., Gastrointestinal tract, in *Vertebrate Endocrinology: Fundamentals and Biomedical Implications*, Vol. 1, Pang, P.K.T., and Schreibman, M. P., Eds., Academic Press, Orlando, 1986, 261.

Vijayan, M. M., Ballantyne, J. S., and Leatherland, J. F., Cortisol-induced changes in some aspects of the intermediary metabolism of *Salvelinus fontinalis, Gen. Comp. Endocrinol.*, 82, 476, 1991.

Wagner, G. F. and Friesen, H. G., Studies on the structure and physiology of salmon teleocalcin, *Fish Physiol. Biochem.*, 7, 367, 1989.

Wahlqvist, I., Branchial vascular effects of catecholamines released from the head kidney of the Atlantic cod, *Gadus morhua, Mol. Physiol.*, 1, 235, 1981.

Wahlqvist, I. and Nilsson, S., Adrenergic control of the cardiovascular system of the Atlantic cod, *Gadus morhua*, during 'stress', *J. Comp. Physiol.*, 137, 145, 1980.

Weigent, D. A. and Blalock, J. E., Interactions between the neuroendocrine and immune systems: common hormones and receptors, *Immunol. Rev.*, 100, 79, 1987.

Wendelaar Bonga, S. E. and Lammers, P. I., Effects of calcitonin on ultrastructure and mineral content of bone and scales of the cichlid teleost *Sarotherodon mossambicus, Gen. Comp. Endocrinol.*, 48, 60, 1982.

Wendelaar Bonga, S. E. and Lock, R. A. C., Toxicants and osmoregulation in fish, *Neth. J. Zool*, 42, 183, 1992.

Wendelaar Bonga, S. E. and Pang, P. K. T., Pituitary hormones, in *Vertebrate Endocrinology: Fundamentals and Biochemical Implications,* Vol. 3, Pang, P. K. T. and Schreibman, M. P., Eds., Academic Press, Orlando, 1989, 105.

Wendelaar Bonga, S. E. and Pang, P. K. T., Control of calcium regulating hormones in the vertebrates: parathyroid hormone, calcitonin, prolactin, and stanniocalcin, *Int. Rev. Cytol.*, 128, 139, 1991.

Wendelaar Bonga, S. E., Van der Mëy, J. C. A., and Flik, G., Response of PAS-positive cells of the pituitary pars intermedia in the teleost *Carassius auratus* to acid water, *Cell Tissue Res.*, 243, 609, 1986.

Woo, N. Y. S., Hontela, A., Fryer, J. N., Kobayashi, Y., and Lederis, K., Activation of hypothalamo-hypophysial-interrenal system by urophysectomy in goldfish, *Am. J. Physiol.*, 238, R197, 1985.

Zaccone, G., Lauwerijns, J. M., Fasulo, S., Tagliafierro, G., Ainis, L., and Licata, A., Immunocytochemical localization of serotonin and neuropeptides in the neuroepithelial endocrine cells of teleost and lung fish gills, *Acta Zool.*, 73, 177, 1992.

Zaccone, G., Fasulo, S., and Ainis, L., Gross anatomy, histology and immunohistochemistry of respiratory organs of air-breathing and teleost fishes with particular reference to the neuroendocrine cells and their relationship to the lung and the gill as endocrine organs, in *The Respiratory System of Non-Mammalian Vertebrates,* Pastor, L. M., Ed., Secretariado de Publicaciones e Intercambio Cientifico, Universidad de Murcia, 1993.

16 Reproductive Physiology

J. Michael Redding and Reynaldo Patiño

INTRODUCTION

Fishes are an astonishingly diverse class of animals with an equally diverse array of reproductive strategies and functions. A complete treatment of our topic should then consider reproduction in agnathan, chondrichthyan, and osteichthyan fishes. Most species are gonochoristic (dioecious), each individual possessing a single sexual phenotype that is stable after its initial appearance. Less common, hermaphroditic (intersexual, ambisexual) species develop male and female gonads in the same individual; in some cases, a reversal of gonadal sex occurs during development, while in others both male and female germ cells exist concurrently in the same gonad. There is at least one report of a self-fertilizing hermaphrodite. Unisexual gynogenetic species in which the genome of the progeny is entirely of maternal origin also occur, albeit rarely. Strategies for bearing progeny range from oviparity to placental viviparity in females with complementary modes of external or internal fertilization by males. This remarkable diversity makes difficult the formulation of general statements about fish reproduction, but at the same time it also makes for exciting research opportunities in comparative biology. Most of the knowledge available for fish reproduction, and consequently our present review, derives from studies of teleostean species. However, where possible we attempt to include information from other phyletic groups.

ENVIRONMENTAL PHYSIOLOGY

Environmental and Reproductive Cycles

In fishes, as in other organisms, natural selection has favored the evolution of phenotypic traits causally linked to greater relative rates of survival and reproduction. Reproductive processes often exhibit endogenous rhythms controlled by internal "biological clocks" and entrained by environmental cues (*zeitgebers*) (Baggerman, 1990). The timing of reproduction during optimal environmental conditions would have obvious consequences for survival of the progeny. Environmental conditions typically are predicated on diurnal, lunar, and annual astronomical cycles, and on regional geographic characteristics. Variability of environmental conditions during the year depends, in part, on geographic latitude. For teleost fishes of the temperate zone, highly predictable

0-8493-8042-1/93/$0.00+$.50

correlations are often observed for seasonal and reproductive cycles. In tropical zones, seasonal changes of environment are less extreme, and many fishes exhibit extended or continuous reproductive patterns; however, even for these species seasonal changes in reproductive activity have been observed (Lam, 1983). The environmental control of fish reproduction was summarized recently in Munro et al. (1990); hence, it will not be discussed further here.

Physiological Transduction of Environmental Cues

Transduction of information inherent in environmental cues, such as changes in photoperiod or specific odors, begins with the activation of specific sensory modalities of the fish, such as photoreceptors in the retina or chemoreceptors in the olfactory epithelium. These sensory structures convert environmental stimuli into electrochemical signals, which are then propagated via sensory neurons to specific sites within the fish's central nervous system. In the central nervous system these signals are processed and interpreted according to their physiological relevance. If appropriate, specific responses are initiated within and outside the central nervous system, leading to pertinent physiological and behavioral effects.

Transduction processes in fish reproduction are likely to involve diverse and complex neurophysiological phenomena. For example, activation of certain autonomic pathways may shift metabolic processes toward gonadal growth at the expense of somatic growth. Likewise, transduction of certain environmental cues may initiate migratory and searching behavior to locate mates or suitable spawning habitat. Space limitation does not allow elaboration of this complex topic; however, some neurological structures and functions are noteworthy in that they are likely to directly affect endocrine pathways that are known to regulate reproduction in fishes.

Many transduction processes related to reproduction undoubtedly involve changes in hypothalamic secretion of gonadotropin-releasing hormone (GnRH) or the modification of its effects on the pituitary gland (Peter, 1983). Other important neural pathways appear to alter the activity of the pituitary directly (Peter et al., 1990). Some of the neurotransmitters and neurohormones that affect either hypothalamic or pituitary function *in vitro* are dopamine and norepinephrine (Yu and Peter, 1992), serotonin (Somoza and Peter, 1991), arginine vasotocin (Groves and Batten, 1986), neuropeptide Y (Breton et al., 1989), and gamma-amino-butyric acid (Kah et al., 1991). The concentrations and turnover of dopamine and serotonin in brain tissue change during the annual reproductive cycles of some fishes (Saligaut et al., 1992), perhaps reflecting the direct or indirect dependence of neurophysiological processes on environmental conditions.

The pineal gland and the terminal nerve of the brain are other putative transduction systems. The pineal of some fishes is a photoreceptive organ and, as in mammals, it may affect hypothalamic reproductive functions via its secretion of serotonin and melatonin (Reiter, 1991). The terminal nerve in the brain of fishes is linked anatomically to the olfactory system, the hypothalamus, the pituitary, and, perhaps, the retina and pineal gland (Krishna et al., 1992). Moreover, it also produces GnRH and could therefore play a direct, prominent role in transduction of olfactory cues into reproductive functions (Demski, 1989).

ENDOCRINE PHYSIOLOGY

This section reviews the physiology of the brain-pituitary-gonadal axis and its constitutive elements (Figure 1). Most reproductive functions, from gametogenesis to behavior, are dependent upon the regulatory activity of this endocrine axis. Experimental disruption of any part of this axis usually results in profound disturbances of the reproductive process. This endocrine axis, like all other systems in an organism, undergoes developmental changes from birth to senescence. Thus, one should be aware that results obtained for one developmental stage may not hold true for other stages. Although most research on the endocrine regulation of reproduction has focused on sexually mature fishes, some important developmental studies have been conducted, notably on the platyfish *Xiphophorus maculatus* (Schreibman et al., 1991). In the platyfish, significant genetic variability is associated with the developmental timing of reproductive functions.

Hypothalamic and Pituitary Function

The hypothalamus lies at the base of the brain in close proximity to the pituitary gland below it. It is a structure rich with neurons whose secretory products are transported to the pituitary via a specialized portal blood supply, found in osteichthyan fishes and all higher vertebrate classes (Van Oordt and Peute, 1983). Alternately, in many but not all teleosts, neurons originating in the hypothalamus may penetrate into all parts of the pituitary, allowing for direct neural control of pituitary function. However, neither specialized portal system nor direct innervation are found in the agnathan pituitary or in the reproductively active ventral lobe of the chondrichthyan pituitary; thus, the assumption of hypothalamic regulation of pituitary function in these fishes is dubious (Dodd, 1983; Gorbman, 1983).

Immunocytochemical and biochemical analyses show that the hypothalamus and pituitary of fishes contain one or more of five known forms of piscine GnRH, a peptide hormone composed of ten amino acids (King and Millar, 1991). The phyletic distribution of GnRH forms in fishes and other vertebrates provides an intriguing perspective on the evolution of hormones. Agnathans possess one identified form of GnRH, and perhaps other similar molecules as yet unidentified, while chondrichthyan and osteichthyan fishes have one or more of the remaining four types of GnRH (Sherwood and Lovejoy, 1989).

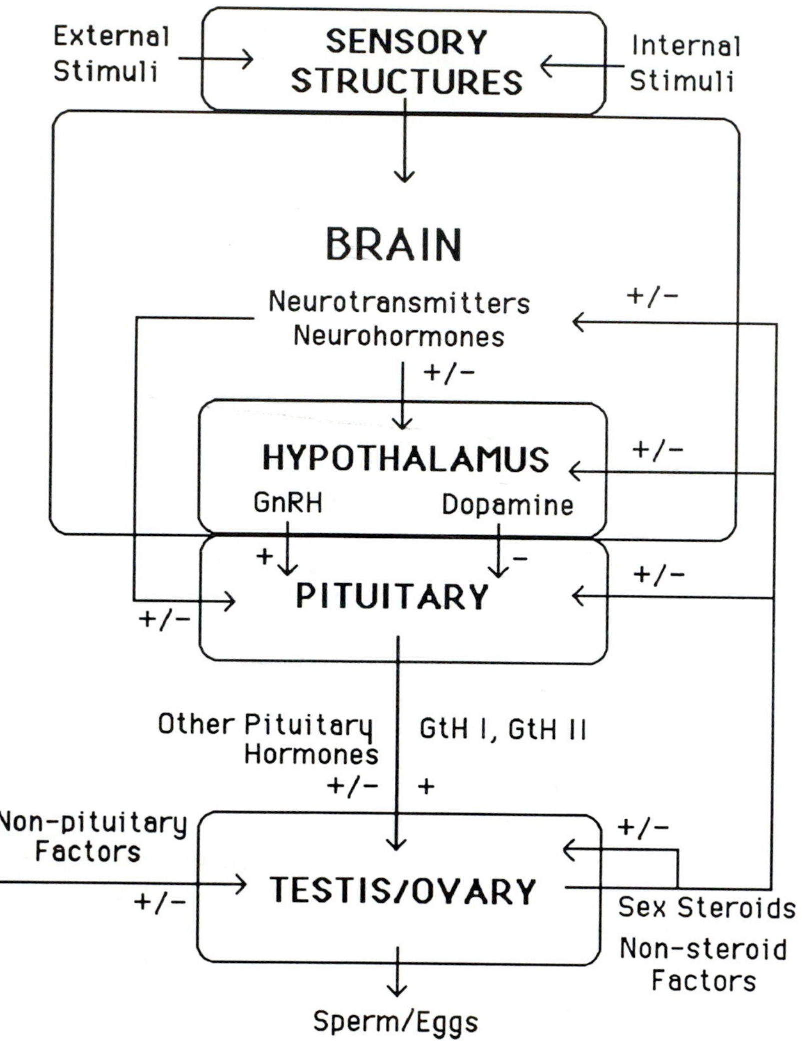

FIGURE 1. Schematic representation of the hypothalamic-pituitary-gonadal axis in fishes. Sensory stimuli originating internally or externally are transduced by the brain so that the activities of specific neurohormones and neurotransmitters are appropriately modified. As a consequence, the secretion rates of hypothalamic hormones (gonadotropin-releasing hormone [GnRH], dopamine) are changed. Pituitary gonadotropin (GtH I, GtH II) release is stimulated (+) by GnRH and inhibited (–) by dopamine; various other neurohormones may directly affect the release of gonadotropin. Gonadotropin stimulates gametogenesis and steroidogenesis by the testis or ovary. Other pituitary hormones and factors from non-pituitary sources may also regulate gonadal function. Sex steroids and perhaps other gonadal factors exert both positive and negative feedback control on the brain, hypothalamus, pituitary, and the gonad itself.

Although the presence of GnRH-like molecules is well documented in fishes, the actual function of these molecules has been studied in very few species. Changes in the concentrations of hypothalamic GnRH of the sea lamprey, *Petromyzon marinus,* and goldfish, *Carassius auratus,* are correlated to the release of pituitary gonadotropic hormones (gonadotropins, GtH) and gonadal recrudescence (increased gametogenic activity) (Fahien and Sower, 1990; Yu et al., 1991a). Experimental treatment of fishes, mostly salmonids and cyprinids, with exogenous GnRH-like molecules increases blood concentrations of GtH and sex steroids and promotes ovulation or spermiation (Donaldson and Hunter, 1983). This knowledge has been used successfully to control the timing of reproduction in commercial pisciculture.

Release of GtH by pituitary glands cultured *in vitro* increases in response to GnRH, mimicking *in vivo* responses (J. P. Chang et al., 1990). Curiously, the goldfish brain contains two forms of GnRH, each of which stimulates the release of pituitary GtH *in vitro* by different intracellular pathways, suggesting the existence of multiple redundant regulatory paths toward the same functional endpoint (Chang and Jobin, 1991). As is true for many hormonal systems, GnRH exerts its effects on the pituitary via one or more types of specific receptors located on membranes of gonadotropic cells (i.e., cells that produce GtH) (Weil et al., 1992).

Although GnRH seems to be the primary regulator of GtH release, other molecules found in the hypothalamus and pituitary also affect the release of GtH directly or indirectly (see section above on Physiological Transduction of Environmental Cues). Thus, norepinephrine stimulates GnRH release from the hypothalamus (but not pituitary), whereas serotonin stimulates and dopamine inhibits its release from both sites *in vitro* (Yu et al., 1991b; Yu and Peter, 1992). (Note that pituitary GnRH presumably originates in hypothalamic neurons, some of which may terminate within the pituitary.) Release of GtH from the pituitary *in vitro* is inhibited by dopamine but stimulated by norepinephrine, serotonin, vasotocin, and neuropeptide Y (Groves and Batten, 1986; Breton et al., 1989; Peter et al., 1990; Somoza and Peter, 1991). Such multihormonal control of GtH release undoubtedly allows for very precise regulatory control. In this regard, it is interesting to note that simultaneous administration of GnRH-like molecules with antagonists of dopamine have proved more effective at inducing ovulation in some fishes than GnRH alone (Lin et al., 1991). However, species-specific differences may exist for GtH regulatory mechanisms, since dopamine does not inhibit GtH release in sciaenid teleosts (Copeland and Thomas, 1989).

Biologists and fish culturists have long known that pituitary tissue contains gonadotropic substances, *i.e.*, factors that stimulate the growth and development of ovarian or testicular tissues. Pituitary morphology changes during gonadal maturation. Hypophysectomy (ablation of the pituitary gland) typically causes gonadal regression (decreased gametogenesis) and inhibition of gonadal steroidogenesis, and replacement therapy with crude or purified pituitary extracts restores gonadal function. Numerous studies have correlated increased pituitary and blood concentrations of GtH with gonadal development, maturation, or spawning activity in male and female fishes. Moreover, treatment of fishes with exogenous GtH-like molecules (or molecules like GnRH which stimulate the release of GtH) typically promotes reproductive processes, particularly during periods when the fishes might be environmentally "primed" and ready to mature and spawn naturally. The cellular source, biochemical purification, and biological characterization of GtH and its receptors in target tissues have been topics of intense interest and recent rapid progress, evidenced in seminal reviews by Idler and Ng (1983), Kawauchi et al. (1989), and Ishii (1991).

Fish endocrinologists have attempted to identify two types of pituitary GtH in fishes corresponding to the two forms present in most other vertebrates, namely luteinizing hormone and follicle-stimulating hormone. Although some have long argued for the presence of two types of GtH (Idler and Ng, 1983), it has been only recently that researchers have confirmed their existence in chum salmon, *Oncorhynchus keta,* and carp, *Cyrinus carpio* (Kawauchi et al., 1989; Van der Kraak et al., 1992). Identification of the two types of GtH in fishes, currently known as GtH I and GtH II, has been followed by several reports concerning the duality and ontogeny of gonadotropic cells and their secretions relative to reproductive functions. Knowledge of the amino acid sequence of GtH I and GtH II in salmon has led to molecular cloning of genes for subunits of the GtH molecules (Sekine et al., 1989; Y. S. Chang et al., 1990).

Gametogenesis and steroidogenesis in salmonids are promoted by GtH, though the relative contribution of each form of GtH seems to vary depending on the developmental stage of the fishes (Suzuki et al., 1988; Swanson, 1991). GtH I is structurally and functionally comparable to mammalian follicle-stimulating hormone and is the predominant GtH found in the pituitary and blood of fishes whose gonads are undergoing active growth and gametogenesis. In contrast, GtH II appears similar to luteinizing hormone, predominating during times of final gonadal maturation and spawning.

The actions of GtH on gonadal tissue are mediated by specific membrane-bound receptors. Analysis of GtH receptors in fishes has been limited (Van der Kraak, 1983; Kanamori and Nagahama, 1988). Recently, Yan et al. (1991) demonstrated tissue- and ligand-specific binding for both GtH I and GtH II in salmon ovarian and testicular tissue. These authors postulate the existence of both shared and separate binding sites for the two types of GtH.

Gonadal Structure and Function

Anatomy

Knowledge of the anatomy of the gonads is essential to understand gonadal physiology. Hoar (1969) wrote a seminal treatise on gonadal embryology, anatomy, and phylogeny in fishes. More recent and exhaustive reviews have been written for agnathan (Gorbman, 1983), chondrichthyan (Dodd, 1983), and teleostean (Nagahama, 1983) fishes.

Gonads consist of germ cells, which produce gametes, and of somatic cells, which support, nourish, and regulate the development of germ cells. Gonoducts are present in most species to carry gametes to their appropriate internal or external destinations. Basic similarities in the organization of the gonads are seen throughout the vertebrates. However, there are also some structural variations among species which reflect either general phyletic patterns or species-specific adaptations to the environment.

Sex Differentiation

Gonadal sex differentiation in fishes is a complex and labile process, controlled ultimately by specific sex-determining genes. Interactions between the genome and variable environmental factors, such as temperature or social status, may determine sex in some species (Conover and Fleisher, 1986; Shapiro, 1988). Gonadal differentiation is highly sensitive to sex-inducing substances produced by the somatic elements of developing gonads and, perhaps, by other tissues (Yamamoto, 1969; Adkins-Regan, 1987). Thus, indifferent, bipotential germ cells would differentiate into spermatogonia in response to andro-inducers, whereas those same cells would yield oogonia in the presence of gyno-inducers. The source and nature of these putative inducers is not known definitively. Prevailing evidence seems to support a role for sex steroid hormones as the proximate sex inducers in fishes. Steroid synthesizing enzymes and their products, including androgens, estrogens, and progestins, are present before gonadal differentiation in those few species that have been studied. The concentration of these steroids in tissues and yolk changes during the period of gonadal differentiation, reflecting utilization of residual steroids of maternal origin as well as synthesis of new steroids by embryonic tissue (Feist et al., 1990). Most compelling, however, are the numerous studies showing reversal of expected sexual phenotype in many fishes by the judicious application of exogenous androgens (to masculinize) or estrogens (to feminize) during the sensitive period before or during gonadal differentiation (Hunter and Donaldson, 1983). In some species, androgens may induce a paradoxical feminization (Goudie et al., 1983); however, this effect may be indicative of *in situ* conversion of androgens into estrogens (Piferrer and Donaldson, 1991).

Although the central event in sexual differentiation is the dedication of male or female germ cell lines, sex-specific development must also occur in the somatic part of the gonads and other body structures. Indeed, the structure and function of the nervous system in mammals reflects sexual dimorphism due, in part, to steroid-induced differentiation during early development (Selcer and Leavitt, 1991). Aromatase, the enzyme that converts circulating testosterone to estradiol, has high activity in brain tissue of goldfish where its action is likely to be critical for the differentiation and function of the nervous system (Callard, et al., 1990). Thus, many physiological processes and behavioral patterns associated with reproduction undoubtedly reflect underlying sex-specific mechanisms.

Sex differentiation and sex reversal in hermaphroditic fishes present a provocative and problematic contrast to our generalizations about sex-stable gonochoristic fishes (Chan and Yeung, 1983). Clearly, a single hermaphroditic individual possesses the genetic potential to express either sexual phenotype at some time during its life. Thus, typical genetic mechanisms of sex differentiation are inadequate to explain sexual phenotype in these fishes. Virtually nothing is known about the physiological regulation of natural sex reversal. Treatment with sex steroids does not consistently induce sex reversal in these fishes. In some coral reef species, changes in social status may initiate sex reversal via visual or chemical cues (Shapiro, 1988).

Testis

Great diversity of testicular morphology among fishes precludes brief summarization. Grier (1981), Lofts (1987), and Callard (1991) provide comprehensive reviews of the anatomy and function of the fish testis. All testis types contain germ cells, in synchronous or variable stages of development, and a complement of somatic cells specialized for the physical support and regulation of spermatogenesis, including Sertoli (sustentacular, intralobular, cyst) and Leydig (interstitial, interlobular) cells. Sertoli cells are found in direct association with germ cells, which they support physically and nurture by modifying the chemical microenvironment. Leydig cells are typically interspersed in the connective tissue surrounding the Sertoli-germ cell units; their primary function is to produce steroids needed for gametogenesis and expression of secondary sex characteristics.

Testicular tissue in agnathan fishes exists as a single cord, and in hagfish it is common for this cord to contain ovarian tissue too (Gorbman, 1983). Germ cells develop in follicles which eventually rupture to release sperm into the body cavity, and from there to the exterior via an unmodified opening in the cloacal wall. Spermatogenesis in lampreys is a single synchronous event, since the fishes die after spawning. Hagfish testes contain germ cells at all stages of maturity, and thus spawning is likely to be repetitive.

In elasmobranchs, like *Squalus acanthias*, the testes are paired structures within which the germ cells and their supportive Sertoli cells divide and develop together in numerous, discrete, spherical units called spermatocysts (ampullae) (Dodd, 1983). Interestingly, Leydig cells are not active in *Squalus*, their function probably subsumed by the Sertoli cells (Pudney and Callard, 1984). Synchronous waves of spermatocysts arise in a germinal ridge and mature in an orderly progression as evidenced by the zonate appearance of the testis. Mature spermatocysts open to release cohorts of sperm into a contiguous system of efferent ducts in which they are transported to external intromittent organs for internal insemination of females. Seasonal, cyclic spawning is typical in elasmobranchs.

The teleostean testis, typically gonochoristic but sometimes hermaphroditic, is usually paired bilaterally and has a lobular or tubular organization within which germ cells develop partially imbedded in Sertoli cells (Nagahama, 1983). The relationship between Sertoli and germ cells may be less intimate in some species such as *Fundulus* (Selman and Wallace, 1986). According to Grier (1981), the teleost testis may be classified into two major morphotypes based on the distribution of spermatogonia. In most teleosts, the actively mitotic spermatogonia are organized in lobules located along the entire length of a given tubular segment, but in atheriniform fishes (e.g., *Fundulus*) they are located only at the terminal ends of tubules. Mature sperm are released into a central lumen which eventually leads to efferent ducts. Efferent ducts terminate at the urogenital opening, which may be modified as an intromittent organ in some species. Sperm production in some teleosts is a single synchronous event, whereas in others it is cyclic or even continuous.

Ovary

The ovaries of adult fishes generally are paired structures attached to the body cavity on either side of the dorsal mesentery. Exceptions to this rule are found in lampreys (Hoar, 1969) and some teleosts (Nagahama, 1983), where the two ovaries fuse into a single structure during development, and in hagfish (Hoar, 1969) and some elasmobranchs (Dodd, 1983), where only one ovary develops in adults. Oviducts occur in most teleosts; however, they are absent in lampreys and hagfish, and lost secondarily during development in certain elasmobranchs and teleosts (e.g., salmonids) (Hoar, 1969).

Three patterns of ovarian development are generally observed in fishes: synchronous, group synchronous, and asynchronous (Wallace and Selman, 1981; Gorbman, 1983). For the synchronous pattern, all oocytes develop in unison, leading to a single, terminal ovulatory event as seen, for example, in Pacific salmon, lampreys, and catadromous eels. For the group synchronous pattern, two or more distinct clutches of oocytes exist concurrently, each clutch at a different developmental stage. This pattern allows for multiple, distinct ovulatory events that are typically entrained to seasonal, lunar, or diurnal cycles. Finally, ovaries showing the asynchronous pattern contain oocytes at all stages of maturity, allowing for protracted or continuous ovulation.

The structure of the growing ovarian follicle is remarkably similar in most fishes (Figure 2). The developing oocyte is located in the center of the follicle and is surrounded by follicle cells. The follicle cell layer generally consists of an inner sublayer, the granulosa cell layer, and one or two outer sublayers of theca cells. The theca and granulosa layers are separated by a basement membrane. Between the surface of the oocyte and the granulosa cell layer there is an acellular layer, the zona radiata. Microvilli, originating from both the oocyte and granulosa cells, penetrate the zona radiata through so-called pore canals. Moreover, cytoplasmic connections (gap junctions) between granulosa cells and the oocyte have been identified in fish follicles (Kessel et al., 1985;

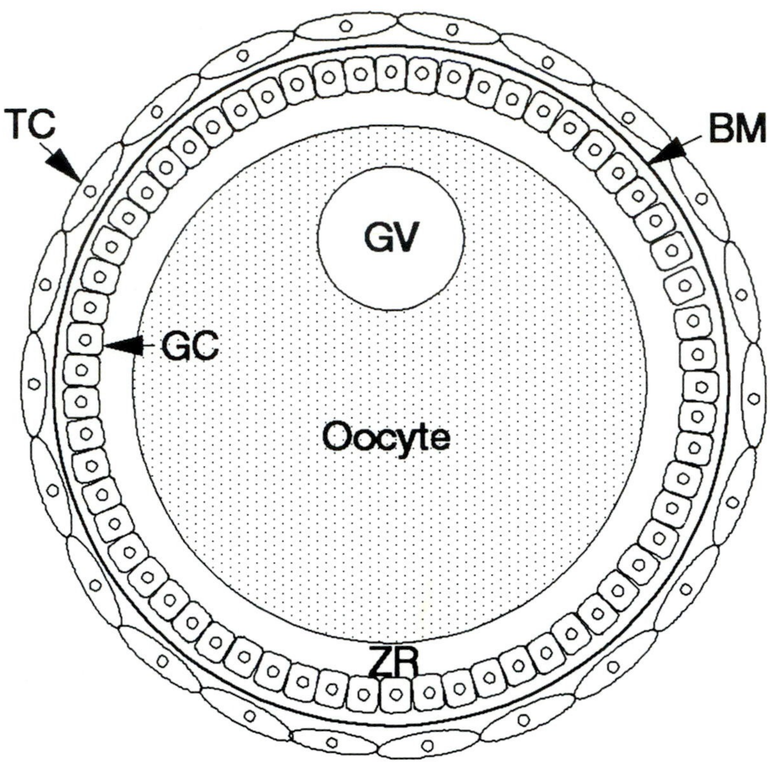

FIGURE 2. Generalized diagram of a full-grown, immature ovarian follicle in fishes. The acellular layer immediately surrounding the oocyte is called the zona radiata (ZR). This zona is in turn surrounded by a layer of granulosa cells (GC). Cellular processes of oocyte and granulosa cell origin penetrate the zona radiata through pore canals. The outer follicle layer is composed of theca cells (TC), which is separated from the granulosa cell layer by a basement membrane (BM). The germinal vesicle (GV) is shown in an eccentric position (after onset of germinal vesicle migration) in this diagram, but in many species the GV remains centrally located even after full growth of the oocyte.

York, 1993). These gap junctions form a metabolic syncytium between granulosa cells and the oocyte, and are likely to be involved in the transfer of nutrients and chemical messengers between the interconnected cells.

Gonadal Hormones

Sex Steroids

Fish testes and ovaries are capable of synthesizing a variety of steroid molecules which regulate diverse functions, including gametogenesis in the gonads, secretory activity of the hypothalamus and pituitary, the expression of various secondary sex characteristics and behaviors, and general metabolism (Fostier et al., 1983). Steroidogenic activity in the gonads is regulated primarily by GtH, although hormones and other factors from both gonadal and nongonadal tissues undoubtedly contribute to this regulation (Aida, 1988; Van Der Kraak, 1991; Callard, 1992). Numerous studies have correlated fluctuations of sex steroid concentrations in blood and gonadal tissue with cycles of gonadal regression and recrudescence (e.g., Kobayashi et al., 1988b; Barry et al., 1990b). Analyses of gonadal ultrastructure, histochemistry, and steroid metabolism corroborate observations of cyclic reproductive activity. Most compellingly, experimental manipulation of the steroidal environment, achieved by treatment with exogenous steroids or by manipulation of the hypothalamic-pituitary-gonadal axis, confirms the central role of steroids in the reproductive process.

Leydig cells appear to be major sites of androgen synthesis in the testis, but Sertoli and other cell types may augment or even replace Leydig cell steroid synthesis, as in *Squalus* (Pudney and Callard, 1984). Major androgens produced by testicular tissue vary between species and developmental stages but may include: testosterone, 11-ketotestosterone, and androstenedione (Fostier et al., 1983; Loir, 1990; Bourne, 1991). The testis of some fishes may also produce progesterone, 17α-hydroxy-4-pregnen-3-one, 17α,20β-dihydroxy-4-pregnen-3-one, 11-deoxycorticosterone, and perhaps small amounts of estrogens (Fostier et al., 1987; Barry et al., 1993).

In the ovary, the primary sites for steroid synthesis are the theca and granulosa cells of the ovarian follicle. These cells may produce 17β-estradiol, estrone, 17α,20β,21-trihydroxy-4-pregnen-3-one, and all the steroids listed above for the testis, again depending on the species and developmental stage (Fostier et al., 1983; Trant and Thomas, 1989). Steroids in both male and female fishes are often conjugated to other molecules, usually glucuronides. Steroid conjugates may be inactive metabolites, ready for disposal or reserved for possible reactivation, but some may also function in pheromonal communication before and during spawning (see section below on Spawning).

The known effects of gonadal steroids are quite diverse. These effects are mediated by the interaction of the steroids with specific, high-affinity receptors typically located in the cytoplasm or nuclear chromatin of the target cells (Callard and Callard, 1987). When complexed with their steroid ligands, such receptors cause changes in the transcriptional rates of specific genes and induce tissue-specific responses at the cellular level. Recently, receptors for maturational steroids have been identified in plasma membranes of fish oocytes, where they appear to regulate nongenomic intracellular processes (see section below on *Cell and Molecular Mechanisms*).

In immature fishes, steroids probably play a major role in the differentiation of the gonads and other sexually dimorphic tissues (see section above on Anatomy, *Sex Differentiation*). In mature fishes, gametogenesis and spawning in both sexes are regulated by steroids acting directly on germ cells or indirectly on somatic elements of the gonads (see below). For example, in male carp a shift in steroid production from androgens to progestins may regulate spermiation during spawning activity (Barry et al., 1990a). In female fishes, estrogens and progestins (or progestin-like steroids) regulate the vitellogenic and maturational phases of oogenesis, respectively. Androgens may also regulate oogenesis. For example, in coho salmon, *Oncorhynchus kisutch*, testosterone concentration in the blood correlates closely with maturational state of the female (Fitzpatrick et al., 1987). In Atlantic croaker, *Micropogonias undulatus*, testosterone seems to accelerate the early stages of oocyte maturation and to retard the final stages; thus, high periovulatory levels of testosterone may effectively synchronize the timing of final maturation and ovulation in this species (Patiño and Thomas, 1990a, 1990b).

Indirect control of gametogenesis by steroids occurs via positive and negative feedback on the endocrine cells of the hypothalamus and pituitary (Goos, 1987; Trudeau et al., 1991). The developmental stage of the fish determines whether the feedback will be positive or negative. Thus, in immature or sexually regressed fishes, sex steroids are likely to promote GnRH and GtH release leading to gonadal recrudescence and maturation (Weil and Marcuzzi, 1990a, 1990b). However, at other times the feedback control is inhibitory, since gonadectomy stimulates and steroid replacement inhibits GtH release (Habibi et al., 1989; Kobayashi and Stacey, 1990). Known mechanisms of steroid feedback include modulation of GnRH and GtH production, pituitary sensitivity to GnRH, and central transduction systems (see section above on Physiological Transduction of Environmental Cues).

There is much evidence supporting a major role for sex steroids in many behavioral events linked to reproduction, including development and recognition of secondary sex characteristics, pheromonal attraction, spawning, and parental care (Liley and Stacey, 1983). Sex steroids also affect processes that may indirectly affect reproductive performance. Thus, androgens or estrogens may alter intestinal uptake of amino acids (Habibi et al., 1983), growth rate (Gannam and Lovell, 1991), plasma lipid dynamics (Wiegand and Peter, 1980), hepatic P-450 enzyme activity (Pajor et al., 1990), sound production (Fine et al., 1982), electroreception (Meyer and Zakon, 1982), and temperature selection (Johansen and Cross, 1980).

Other Gonadal Hormones

Other hormones with nonsteroidal chemistry are secreted by the gonads and regulate various local events associated with reproduction. Most well studied of these hormones are the prostaglandins, a family of cyclopentane fatty acids found in most tissues and affecting a diverse array of functions. In certain fishes, maturational steroids may induce the formation of specific prostaglandins which, in turn, stimulate ovulation via calcium-mediated events (Goetz et al., 1991). Prostaglandins and their metabolites may, like steroid hormones, function as pheromones in fishes (see section below on Spawning and Fertilization). Recently, Ge et al. (1991) demonstrated the existence of inhibin-like proteins in ovarian fluid of goldfish. These proteins stimulated release of pituitary GtH *in vitro*, an effect opposite to that which occurs in mammals. Based on results from mammalian models, the topic of nonsteroidal, gonadal hormones is likely to attract much future attention.

Other Reproductive Hormones

Multiple targets and effects of hormones seem to be the rule, not the exception. Thus, it is not surprising to observe that many different hormones can affect reproductive processes in fishes. However, it is necessary to assess the physiological importance of such effects. What seems possible in an experimental context may not be relevant to the animal, especially when excessively high dosages of hormones are employed. Given this caveat, the following few examples may provoke the reader's interest.

Since gametogenesis and spawning are energetically demanding processes, it is plausible that all hormones regulating intermediary metabolism (e.g., pancreatic, thyroid, growth, and corticosteroid hormones) could potentially influence reproductive functions. Changes in circulating concentrations of metabolic hormones have been correlated to seasonal cycles in fishes, suggesting their possible involvement in reproductive processes (Leatherland, 1987; Dickhoff et al., 1989). Thyroid hormones may play a synergistic role in GtH-induced steroidogenesis (Cyr and Eales, 1988) and final oocyte maturation (Dickhoff et al., 1989; Weber et al., 1992), and promote survival and growth of the progeny (Brown et al., 1989). Growth hormone and a related, newly discovered hormone, somatolactin, from the pituitary gland are steroidogenic in testicular and ovarian tissue, although their potency is much lower than that of GtH (Singh et al., 1988; Planas et al., 1992). Insulin-like growth factors can stimulate spermatogonial cell proliferation in *Squalus* (DuBois and Callard, in press). Chronic, stress-induced corticosteroid production typically suppresses reproductive processes; however, exposure to cortisol during the final stages of follicular development may potentiate or directly trigger oocyte maturation in some fishes (Jalabert, 1976; Patiño and Thomas, 1990a).

Prolactin is a pituitary hormone whose function in fishes is usually regarded as osmoregulatory, not reproductive. However, in this capacity it may directly and indirectly regulate important functions upon which reproduction depends, like calcium mobilization and diadromous migration to spawning areas (Mazzi and Vellano, 1987). Various effects of prolactin on gonadal function (Singh et al., 1988) and parental behavior have also been noted. Arginine vasotocin, found in the brain and pituitary, stimulates testosterone production by rainbow trout, *Oncorhynchus mykiss,* testis *in vitro,* supporting results from mammalian research (Rodriguez and Specker, 1991). Hormone-induced maturation of goldfish oocytes *in vitro* may be suppressed by GnRH, an effect seemingly contrary to its stimulatory effect on GtH production (Habibi et al., 1988).

GAMETOGENESIS

Gametogenesis refers to the process of mitotic proliferation, meiotic division, and differentiation of germ cells from diploid, primary germ cells (gonia) to haploid, mature gametes, either spermatozoa in males or ova in females. Spermatogenesis and oogenesis are remarkably different processes. Thus, meiosis is uninterrupted during spermatogenesis, whereas it is interrupted twice during oocyte development. Second, while the spermatocyte normally undergoes a decrease in size during meiosis, the oocyte in most nonmammalian species grows enormously during this period. Third, final maturation of sperm occurs only after the conclusion of meiosis, but oocyte maturation occurs concomitantly with meiosis. Finally, although each spermatogonium entering meiosis ultimately yields four spermatozoa, each oogonium yields only one functional ovum.

SPERMATOGENESIS

Spermatogenesis in fishes is fundamentally similar to that found in other vertebrates (Nagahama, 1983; Callard, 1991). Differences between taxa are superficial, typically relating to the morphology of the testis and spermatozoa, the number of spermatogonial divisions before entry into meiosis, and the timing between successive developmental stages. Characteristic cytological changes during spermatogenesis include mitotic proliferation of spermatogonial stem cells and the subsequent differentiation of some, but not all, into committed primary spermatocytes. The first meiotic division marks the conversion of primary spermatocytes into secondary spermatocytes which, in turn, undergo a second meiotic division to form spermatids. During spermiogenesis the spermatids mature, develop a flagellum, and separate from their supportive Sertoli cells to become spermatozoa. Spermatozoa are collected in the efferent tubules, from which they are expelled during spermiation.

Spermatogenesis is synchronized so that clones of spermatogonia and their associated Sertoli cells develop in a strict spatial and temporal relationship within the tubules or spermatocysts of the testis (see above). Such synchrony likely reflects the coupling of germ cells via intercellular bridges and the influence of coordinating hormonal signals originating outside and within the testis (Billard et al., 1982; Callard, 1992). Hormonal regulatory mechanisms for spermatogenesis include GtH-induced spermatogonial proliferation and testicular steroidogenesis in Leydig or, in some species, Sertoli cells. Spermatogenesis in mammals requires high intratesticular concentrations of androgens, achieved in part by the presence of androgen-induced steroid binding proteins produced by Sertoli cells. Androgen binding proteins have been found in *Squalus* testis (Mak and Callard, 1987). Recently, Miura et al. (1991) showed that for immature testis of Japanese eel, *Anguilla japonica,* cultured *in vitro* and initially containing only premitotic spermatogonia, the presence of

11-ketotestosterone allowed complete spermatogenesis during a 21-day period. Testosterone, but not other androgens, promoted differentiation of spermatogonia in *Fundulus* testis *in vitro* (Cochran, 1992). Germ cells do not appear to be direct targets of steroid action, but androgen and estrogen receptors are found in Sertoli cells of *Squalus* spermatocysts during early, but not late, stages of spermatogenesis, further supporting the notion of stage-specific steroid regulation of spermatogenesis (Callard, 1992).

Oogenesis

By definition, an oogonium becomes an oocyte upon entering meiosis. In all vertebrate species, the first meiotic block of oogenesis occurs at prophase I. Oocyte meiosis is released from this block during maturation, but it is once again arrested when it reaches metaphase II. The mature oocyte, or ovum, is then ready for fertilization. Release from the second block and completion of the second meiotic division is triggered by fertilization of the egg (Masui, 1985).

Vitellogenic Growth

Most growth of the vertebrate oocyte occurs during the period of first meiotic arrest. Oocyte growth in nonmammalian species is due to the uptake of circulating vitellogenin, which is then modified by, and deposited as yolk in, the oocyte (Wallace et al., 1987). Vitellogenin is produced by the liver in response to estradiol, which in turn is produced by ovarian follicle cells in response to increased GtH levels in blood (Nagahama, 1983; Ho, 1991). In salmonids, production of follicular estradiol is most likely regulated by GtH I, since there is little if any GtH II produced during ovarian recrudescence (Swanson, 1991).

Circulating vitellogenin seems to penetrate the follicular cell layer through intercellular channels between granulosa cells and then reach the oocyte via the pore canals of the zona radiata (Abraham et al., 1984). The appearance of this extracellular pathway of channels and pore canals is developmentally regulated. Uptake of vitellogenin by the rainbow trout oocyte can be induced only by GtH I, not GtH II (Tyler, 1991); this uptake occurs via micropinocytosis and appears to involve binding of vitellogenin to specific receptors on the oocyte surface.

Oocyte Maturation

Strictly speaking, oocyte maturation comprises the period between prophase I-release and metaphase II-arrest. However, morphological criteria, such as germinal vesicle (i.e., nucleus) migration (GVM), are widely used as indices of the onset of maturation. Various aspects of oocyte maturation in fishes were recently reviewed by Nagahama (1987), Detlaff (1988) and Jalabert et al. (1991).

Hormonal Control

The gonadotropic control of oocyte maturation in fishes is perhaps best understood for salmonids. In these species, pituitary GtH (most likely GtH II) induces the production of oocyte maturation-inducing substance (MIS, not to be confused with Mullerian-inhibiting substance in mammals) by the ovarian follicle cells in a two-step process (Nagahama and Yamashita, 1989). In the first step, GtH induces the synthesis of 17α-hydroxy-4-pregnen-3-one by the theca cells. In the second step, which is also under the control of GtH, this steroid precursor is converted by the granulosa cells to 17α,20β-dihydroxy-4-pregnen-3-one, the MIS of amago salmon, *Oncorhynchus rhodurus,* and perhaps many other teleosts. Trant et al. (1986) positively identified a related MIS, 17α,20β,21-trihydroxy-4-pregnen-3-one, in isolated follicle cultures of Atlantic croaker. There is strong evidence that MIS is a steroid in all fish and amphibian species examined to date (Jalabert et al., 1991). After its production by granulosa cells, MIS traverses the zona radiata to contact the oocyte, triggering its maturation. This two-cell model for the synthesis of MIS may not be applicable to all fishes, since in *Fundulus* the granulosa cells are capable of producing maturational steroid without any precursor contribution from theca cells (Petrino et al., 1989).

Prevailing models suggest that the maturation of mammalian oocytes is largely regulated by a putative oocyte-maturation inhibitor (OMI) originating in the follicle cell. According to this model, oocyte maturation is preceded by uncoupling of gap junctions between follicle cells. Such uncoupling blocks the transfer of OMI to the oocyte, thereby permitting the resumption of meiosis. One major observation supporting this notion is that mammalian oocytes resume meiosis spontaneously upon defolliculation. In fish and amphibian follicles, removal of the follicle cell layer can also result in spontaneous oocyte maturation (Vilain et al., 1980; Greeley et al., 1987). Thus, the existence of OMI has also been postulated for nonmammalian ovarian follicles (Jalabert et al., 1991), although its prominence in the regulation of oocyte maturation is questionable.

Morphological Changes

In general, the most obvious morphological changes of the oocyte nucleus during maturation are the onset or resumption (depending on the species) of GVM toward the periphery of the animal pole and the subsequent germinal vesicle breakdown (GVBD). Research with goldfish suggests that the mechanisms of GVM and GVBD involve disruptions or alterations of the cytoskeletal network (Lessman et al., 1988). Cytoplasmic changes observed during GVM and GVBD include the coalescence of lipid droplets and yolk globules and increased oocyte translucency (Goetz, 1983). In Atlantic croaker, the earliest pre-GVM change following GtH stimulation is a slight darkening of the cytoplasm around the nucleus (R. Patiño, unpublished data). In some teleosts, especially in marine species, GVBD is accompanied by marked hydration resulting in swelling of the oocyte (Wallace and Selman, 1981; Greeley et al., 1991). Mechanisms of hydration are not well understood, but evidence suggests the involvement of Na^+/K^+-ATPase-dependent ion regulation (LaFleur and Thomas, 1991).

Cell and Molecular Mechanisms

Cellular Interactions

Following its synthesis by follicular cells, MIS travels across the zona radiata to reach the oocyte and induce its maturation. The prevailing notion of MIS transport involves simple extracellular and transmembrane diffusion along a concentration gradient between the follicle cells and the oocyte (Schuetz, 1985). However, follicular granulosa cells and the oocyte are physically coupled through their microvillar-gap junctions (see above). In *Xenopus laevis* this coupling is known to be developmentally and hormonally regulated (Browne et al., 1979; Browne and Werner, 1984). Moreover, results of recent studies at our laboratory have strongly implicated the gap junctions in the process of oocyte maturation. Thus, chemical inhibitors of gap junctions, such as heptanol and octanol, completely but reversibly blocked GtH- and MIS-induced GVBD in *Xenopus* intrafollicular oocytes (Patiño and Purkiss, unpublished data). This blockage did not alter GtH-induced steroidogenesis in follicles or responsiveness of denuded (defolliculated) oocytes to MIS, indicating that the blockage of GtH-induced GVBD was a specific effect of the inhibitors on gap junctions. Similar effects of these inhibitors on MIS-induced maturation were observed with follicle-enclosed oocytes of spotted seatrout, *Cynoscion nebulosus* (York, 1993). It is relevant to note that the existence of a MIS has also been postulated for mammalian ovaries, and that this MIS seems to require intact gap junctions for its transfer to the oocyte (Fagbohun and Downs, 1991). These observations suggest that gap junctions are essential conduits for intrafollicular trasport of MIS to the oocyte in vertebrate ovaries. Moreover, our results with fishes and amphibians provide evidence for the transfer of a specific message (i.e., steroidal MIS) between follicle cells and the oocyte during maturation, presenting a viable new conceptual model of intercellular steroid transport within the vertebrate ovary.

MIS Receptor and Its Membrane Transduction Pathways

In oocytes of fishes and amphibians, external application of MIS induces maturation while internal application (by microinjection) is not effective (Nagahama, 1987). This early observation was widely interpreted to impute the existence of a novel steroid receptor for MIS on the oocyte membrane. However, previous studies with the amphibian MIS receptor reported an atypically low affinity (K_D, 10^{-7} to 10^{-6} *M*) for MIS, similar to the affinity of nonspecific steroid binding sites (Sadler and Maller, 1982; Blondeau and Baulieu, 1984); thus, the existence of specific MIS receptors in oocyte membranes has been questioned (Smith, 1989). Recently, specific, high-affinity (K_D, 10^{-9} *M*) receptors for MIS were demonstrated in oocyte membranes from spotted seatrout and *Xenopus* (Patiño and Thomas, 1990c; Z. Liu and R. Patiño, unpublished data). Based on competition studies, the specificity of these receptors is not as precise as that of classical intracellular steroid receptors. However, MIS is produced within the ovarian follicle itself and may be delivered to its receptors in the oocyte membrane via direct pathways, i.e., gap junctions. Thus, the MIS receptor may not require high specificity, since normally it will be exposed only to a relatively narrow spectrum of steroids produced by the ovarian follicle at a defined stage of development.

Given the possible role of gap junctions in the transfer of MIS to the oocyte, it would be of interest to determine whether MIS receptors and gap junctions are colocalized in the oocyte membrane. Moreover, since gap junctions allow passage of MIS directly into the oocyte cytoplasm, the orientation of the ligand-binding region of the MIS receptor may be intracellular, i.e., on the inner layer of the plasma membrane. A similar conclusion was reached by Cartaud et al. (1984), who found that synthetic maturational steroids that were unable to penetrate the outer layer of the plasma membrane could not induce maturation of *Xenopus* oocytes.

The mechanism by which the binding of MIS to its receptor induces oocyte maturation is controversial. Smith (1989) concluded that there are at least two pathways involved: one involving a decrease of intracellular cyclic adenosine monophosphate (cAMP), and the other a decrease of 1,2-diacylglycerol (DAG) in the membrane. In this model, binding of MIS to its receptor would inhibit the production of both cAMP and DAG, thereby inhibiting the activities of cAMP-dependent protein kinase A and DAG-dependent protein kinase C. Depending on the specific stage of oocyte differentiation, inhibition of one or both kinases could trigger subsequent cytoplasmic events leading to oocyte maturation. Specific intermediate steps beyond inhibition of kinases remain to be clarified.

Intracellular Mechanisms

A key event of oocyte maturation is the activation of maturation-promoting factor (MPF) in oocyte cytosol (Smith, 1989). In all animals examined to date, including fishes (Yamashita et al., 1992), MPF is composed of two protein subunits: (1) a protein exhibiting MPF kinase activity; and (2) a regulatory protein known as cyclin B. Activation of MPF kinase is induced by MIS, and this induction requires the synthesis of new protein. However, microinjection of activated MPF can induce GVBD even in the absence of protein synthesis (Smith, 1989). Thus, the requirement for protein synthesis during oocyte maturation is related to early cytoplasmic events occurring after MIS stimulation but before activation of MPF kinase. In carp oocytes, *de novo* production of cyclin B may be necessary for the activation of MPF kinase (Nagahama, 1991).

Oocyte Maturational Competence

Maturational competence can be studied at different levels in the ovarian follicle. One can look at the competence of the follicle cells to produce MIS in response to maturational GtH, or at the competence of the oocyte to mature in response to MIS stimulation.

Appearance and Hormonal Regulation

Oocyte maturational or meiotic competence has somewhat different definitions in different vertebrates. In fishes and amphibians, it has been defined as the ability of the oocyte to mature in the presence of MIS. In these animals, competence appears either gradually during the final stages of oocyte growth or after completion of growth (Sadler and Maller, 1983; Jalabert and Fostier, 1984; Patiño and Thomas, 1990a, 1990b). Recent evidence obtained from *in vivo* and *in vitro* studies of teleosts clearly shows that the acquisition of maturational competence depends on GtH stimulation (Kobayashi et al., 1988a; Zhu et al., 1989; Patiño and Thomas, 1990a, 1990b; Thomas and Patiño, 1991; Degani and Boker, 1992). However, GtH-induced maturational competence in Atlantic croaker is not mediated by MIS, unlike GtH-induced GVBD. Thus, Patiño and Thomas (1990b) proposed a two-stage model for the GtH control of final maturation in sciaenid teleosts: a steroid-independent "priming stage" during which maturational competence develops, followed by a MIS-dependent "GVBD stage". Further studies are required to determine the general applicability of this model.

In mammals, acquisition of oocyte competence to resume meiosis (i.e., undergo GVBD) does not necessarily imply competence to complete the first meiotic division (i.e., progress to metaphase II) (Eppig, 1991). Rather, these competencies develop sequentially during oocyte differentiation. Little is known about this topic in nonmammalian vertebrates.

Cell and Molecular Mechanisms

The acquisition of oocyte maturational competence requires *de novo* RNA and protein synthesis (Patiño and Thomas, 1990b). Although the specific nature and localization of these newly synthesized follicular products are not known with certainty, a possible candidate for one of the new proteins is the MIS receptor. Supporting this prospect, the ovarian concentration of MIS receptor in spotted seatrout increases during GnRH-induced GVM *in vivo* and during GtH induction of maturational competence *in vitro* (Patiño and Thomas, 1990c; Thomas and Patiño, 1991). Similarly, MIS receptor concentration in the *Xenopus* oocyte may increase during growth (Sadler and Maller, 1983; Z. Liu and R. Patiño, unpublished data) and after GtH stimulation *in vitro* (Z. Liu and R. Patiño, unpublished data). In the *Xenopus* follicle, GtH induces gap junctional contact and permeability between the follicle cells and oocyte (Browne et al., 1979; Browne and Werner, 1984). Thus, the establishment of gap junction pathways for MIS transport to the oocyte may also be involved in the acquisition of maturational competence (see section above on *Cell and Molecular Mechanisms*).

Ovulation

Ovulation refers to the expulsion of the mature oocyte from its follicle. During ovulation there is a separation of microvillar processes between the follicle cells and the oocyte. This separation is followed by the rupture of the follicle cell layer and the expulsion of the egg. Proteolytic enzymes originating in the follicle cells may be involved in the rupture of the follicle, while contraction of specialized theca cells may aid in the actual expulsion of the egg (Goetz, 1983; Detlaff, 1988).

In most vertebrates, ovulation is induced by GtH and normally follows maturation of the oocyte. However, although maturation and ovulation share the same primary hormonal stimulus (i.e., GtH), each event seems to be mediated by different mechanisms. Production of MIS or other follicular steroids in response to GtH may be necessary for ovulation in some species, but the site of action of these steroids may be extrafollicular (Detlaff, 1988; Goetz et al., 1991). Increased synthesis of prostaglandins also appears to be a general requirement for GtH-induced ovulation in fishes and most other vertebrates (Goetz et al., 1991).

SPAWNING AND FERTILIZATION

Spawning

Spawning is the culmination of complex behavioral and physiological events leading to the fusion of gametes. It may involve either the coordinated release of gametes into the external environment or an internal copulatory event. Spawning depends not only on gametogenesis, but also on behaviors such as prespawning migration, habitat selection, nest building, mate selection, and courtship. Reproductive physiology and behavior are clearly interactive and appear to share common regulatory pathways in the nervous and endocrine systems. Thus, environmental stimuli, including behavioral cues from other fishes, may elicit primary neural and hormonal responses within an individual that, in turn, regulate secondary behavioral and physiological responses. Behavioral physiology of fish reproduction has been reviewed exhaustively (Liley and Stacy, 1983); however, recent exciting work on sex pheromones merits mention.

Sex pheromones are externally secreted chemicals that elicit specific reproductive responses in a conspecific individual. In fishes, at least some of these pheromones also serve as hormones (or derivatives thereof) within the individual that produces them, hence they are termed "hormonal pheromones" (Stacey, 1991). The obvious advantage of such a system is its potential to coordinate internal functions with appropriate external events (e.g., gametogenesis and mate attraction). Pheromones released by male fishes may be steroids or their metabolites, and may function to attract females or cause them to ovulate (van Den Hurk and Resink, 1992), or to coordinate reproductive events in other males (Sherwood et al., 1991). Steroids released by female goldfish, in particular 17α,20β-dihydroxy-4-pregnen-3-one, decrease turnover of the neurotransmitter dopamine in the pituitary, increase blood GtH concentration, and increase milt volume in males (Dulka et al., 1987, 1992). Prostaglandins, potent ovulatory hormones in female goldfish, may also function as pheromonal stimulants of male spawning behavior (Sorenson et al., 1988). The fish's olfactory system mediates responses to pheromones, as indicated by electrical activity in the olfactory nerve and by the lack of responses in fish with lesions in the olfactory tract.

FERTILIZATION

Sperm Activation

Fish spermatozoa are typically immotile and inactive while resident within the testis. Motility of sperm is initiated after spermiation into the aqueous environment or into the reproductive tract of the female. It is generally thought that motility is suppressed by specific chemical factors in the testis or seminal fluid (Stoss, 1983). High concentrations of cations (e.g., K^+), acidic pH, and isotonic conditions appear to suppress motility in some species. Thus, activation of the sperm may occur when suppressive factors are diluted, the pH becomes alkaline, and the osmolality shifts to hypotonic or hypertonic in freshwater or marine species, respectively. Recent research suggests that 17α,20β-dihydroxy-4-pregnen-3-one, produced by the testis, promotes the acquisition of sperm motility in salmon, and that this effect is mediated by increased pH in the sperm duct and elevated cAMP content of the sperm (Miura et al., 1992).

Egg-Sperm Interaction

The egg envelope (chorion, vitelline envelope, zona radiata) surrounds the ovulated egg in teleosts and contains a small opening over the animal pole, known as the micropyle. The micropyle allows the sperm direct access to the egg surface, and may also be the source of chemotaxic proteins that lead sperm toward the entry point (Kudo, 1988).

Fertilization of animal eggs by more than one sperm (polyspermy) is generally lethal. Therefore, most animal eggs have developed a variety of mechanisms to prevent its occurrence. Unlike eggs of many other animals, fish eggs do not exhibit electrical "fast-blocks" to polyspermy (Kudo, 1988). Instead, fish eggs rely largely on the impermeability of the egg envelope to sperm. However, in some species the diameter of the micropyle is wide enough to permit the simultaneous entry of more than one sperm. Therefore, in these fishes other preventative mechanisms have evolved. For example, a single "sperm entry site" has been described on the surface of eggs at the base of the micropyle in species such as carp and medaka, *Oryzias latipes* (Kudo, 1988). This structure seems important for the attachment of the sperm to the egg surface, and it disappears following the entry of the first sperm. In some fishes a "fertilization cone" forms at the base of the micropyle after attachment of the first sperm, thus blocking further access to the egg surface and extruding supernumerary spermatozoa out from the micropylar canal (Kudo, 1988). In medaka, formation of the fertilization cone occurs within 15 s of fertilization (Iwamatsu et al., 1991). Formation of the fertilization cone is probably independent of the cortical reaction (see section below on Egg Activation) (Kudo, 1988). Finally, in fishes such as rainbow trout and carp, the postfertilization perivitelline fluid contains a sperm-agglutinating substance from cortical alveoli that prevents polyspermy after retraction of the fertilization cone (Kudo, 1988).

Egg Activation

Contact of the sperm with the egg surface leads to marked changes in the structure and function of the egg. These changes, including those for the prevention of polyspermy (see above), are referred to as egg activation (Gilkey, 1981). Among fishes, egg activation has been most thoroughly examined in the medaka. The earliest reactions of the egg surface to the attachment of the sperm include increased ion permeability of the membrane followed by a change in membrane potential (Ito and Shimamoto, 1989). Seemingly as a result of these electrochemical changes, production of inositol triphosphate increases and triggers the release of calcium stored within the cortical cytoplasm (Iwamatsu et al., 1988). This temporary increase in calcium ("calcium transient") spreads through the cortical cytoplasm in a wave-like fashion beginning at the sperm's point of contact. The calcium transient appears to be both necessary and sufficient for activation of the egg (Gilkey, 1981).

During egg activation cortical alveoli close to the inside surface of the oocyte membrane rupture to release their contents into the perivitelline space, a process known as cortical reaction (Gilkey, 1981). In medaka, the leading edge of the calcium transient just precedes the rupture of cortical alveoli. Cortical alveolar material increases the osmolality of the perivitelline fluid and draws in extracellular water, thus increasing the perivitelline volume. Polymerization of proteins in the egg envelope hardens it into a "fertilization envelope" (Iuchi et al., 1991). Also, the outermost layer of the egg envelope disappears and is replaced by a new layer consisting of material released by the cortical alveoli (Kudo and Inoue, 1988).

The hardened fertilization envelope provides protection for the embryo against physical injury. Moreover, it has a bactericidal action, not found in the egg envelope and probably derived from the cortical alveoli (Kudo and Inoue, 1989). Thus, an important function of egg activation in fishes is protection of the embryo from injury and pathogens.

Other major events of egg activation include completion of the second meiotic division and the extrusion of the second polar body. Molecular mechanisms underlying egg release from the second meiotic block have been reviewed elsewhere (Masui, 1991). Ultimately, male and female pronuclei form and fuse to create a diploid zygote nucleus. Embryonic development proceeds in the external environment or within the maternal reproductive tract, depending on the particular strategy adopted by the species.

MATERNAL-EMBRYONIC RELATIONSHIPS

In most fishes, oviparity persists as a primitive yet effective mode of reproduction. In oviparous fishes, ovulation is followed by expulsion of the eggs into the external environment, there to be fertilized by the sperm of one or more attending male fish. Embryonic development proceeds with no direct connection to the mother (or father) aside from the various, and sometimes exotic, forms of direct parental care that may occur in some species. Viviparity is an alternative mode of reproduction, found in some taxonomically advanced fishes, in which fertilization and full embryonic development occur within the maternal reproductive tract, ultimately leading to the birth of free-living progeny. Wourms and Lombardi (1992) have estimated that viviparous strategies have evolved in about 55% of extant species of sharks and rays and 2 to 3% of teleostean fishes. Between the two extremes, some fishes employ internal fertilization followed by variable periods of development within the mother until the embryo, still contained within an egg envelope, is expelled to the external environment some time before hatching.

Evolutionary, morphological, ecological, and physiological aspects of viviparity in fishes have been exhaustively reviewed by others (Wourms et al., 1988; Wourms and Lombardi, 1992). Central to the study of viviparity are considerations of the degree of metabolic autonomy by the developing embryo. The anatomical connections between maternal and embryonic tissues may be rudimentary, providing little more than a safe environment for development, or they may be highly specialized to both house and nourish the embryo, as seen in placental sharks (Hamlett, 1989). Thus, some embryonic fishes require only those nutrients stored within the yolk during vitellogenesis. In other fishes, nutritional requirements not met by the yolk are supplied by nutrient-rich, uterine secretions that are ingested or absorbed by the embryo, or by nutrients transported across placental membranes. Metabolic waste produced by the embryo must be stored, flushed from the uterus, or processed through maternal tissues. Osmotic conditions within the uterine environment may vary and appear to correlate with the ability of the embryo to osmoregulate independently (Kormanik, 1992). Specialized mechanisms of respiratory gas exchange allow for adequate oxygenation of the embryo. Immunological factors must be specialized in the mother or embryo to prevent rejection of an embryo carrying foreign antigens. The hormonal regulation of reproductive functions also appears to be specialized in viviparous fishes (Callard et al., 1992).

CONCLUSIONS

It is our hope that the above discussion, although shaped by our personal biases, will provide a satisfying overview of current research on the subject of fish reproductive physiology. Our discussion is not comprehensive, and practitioners in specific fields will undoubtedly find deficits in our coverage. Obviously, most research on reproductive physiology has focused on the function of the gonads and their regulation by the endocrine system. However, the reader should appreciate that reproduction is the ultimate objective of all living things. As such, one might muse that all life processes in fishes and in other organisms are in some way pertinent to their reproductive physiology.

ACKNOWLEDGMENTS

We thank our colleagues Gloria V. Callard, Martin S. Fitzpatrick, and Carl B. Schreck for their reviews of earlier versions of this manuscript and for their useful suggestions. Original research cited in this chapter was supported by National Institutes of Health Grant HD277729 (R.P.) and U.S. Department of Agriculture Grant 91-37203-6680 (R.P.).

REFERENCES

Abraham, M., Hilge, V., Lison, S., and Tibika, H., The cellular envelope of oocytes in teleosts, *Cell Tissue Res.*, 235, 403, 1984.

Adkins-Regan, E., Hormones and sexual differentiation, in *Hormones and Reproduction in Fishes, Amphibians, and Reptiles*, Norris, D. O. and Jones, R. E., Eds., Plenum Press, New York, 1987, chap. 1.

Aida, K., A review of plasma hormone changes during ovulation in cyprinid fishes, *Aquaculture*, 74, 11, 1988.

Baggerman, B., Sticklebacks, in *Reproductive Seasonality in Teleosts: Environmental Influences*, Munro, A. D., Scott, A. P., and Lam, T. J., Eds., CRC Press, Boca Raton, FL, 1990, chap. 5.

Barry, T., Thomas, P., and Callard, G. V., Stage-related production of 21-hydroxylated progestins by the dogfish *(Squalus acanthias)* testis, *J. Exp. Zool.*, 265, 522, 1993.

Barry, T. P., Aida, K., Okumura, T., and Hanyu, I., The shift from C19 to C21 steroid synthesis in spawning male common carp, *Cyprinus carpio*, is regulated by the inhibition of androgen production by progestogens produced by spermatozoa, *Biol. Reprod.*, 43, 105, 1990a.

Barry, T. P., Santos, A. J. G., Furukawa, K., Aida, K., and Hanyu, I., Steroid profiles during spawning in male common carp, *Gen. Comp. Endocrinol.*, 80, 223, 1990b.

Billard, R., Fostier, A., Weil, C., and Breton, B., Endocrine control of spermatogenesis in teleost fish, *Can. J. Fish. Aquat. Sci.*, 39, 65, 1982.

Blondeau, J. P. and Baulieu, E. E., Progesterone receptor characterized by photoaffinity labelling in the plasma membrane of Xenopus laevis oocytes, *Biochem. J.*, 219, 785, 1984.

Bourne, A., Androgens, in *Vertebrate Endocrinology: Fundamentals and Biomedical Implications*, Vol 4 B, Pang, P. K. T. and Schreibman, M. P., Eds., Academic Press, San Diego, 1991, chap. 11.

Breton, B., Mikolajczyk, T., Danger, J. M., Gonnet, F., SaintPierre, S., and Vaudry, H., Neuropeptide Y (NPY) modulates in vitro gonadotropin release from rainbow trout pituitary glands, *Fish Physiol. Biochem.*, 7, 77, 1989.

Brown, C. L., Doroshov, S. I., Cochran, M. D., and Bern, H. A., Enhanced survival in striped bass fingerlings after maternal triiodothyronine treatment, *Fish Physiol. Biochem.*, 7, 295, 1989.

Browne, C. L. and Werner, W., Intercellular junctions between the follicle cells and oocytes of *Xenopus laevis, J. Exp. Zool.*, 230, 105, 1984.

Browne, C. L., Wiley, H. S., and Dumont, J. N., Oocyte follicle cell gap junctions in *Xenopus laevis* and the effects of gonadotropins on their permeability, *Science*, 203, 182, 1979.

Callard, G. V., Spermatogenesis, in *Vertebrate Endocrinology: Fundamentals and Biomedical Implications*, Vol 4 A, Pang, P. K. T. and Schreibman, M. P., Eds., Academic Press, San Diego, 1991, chap. 6.

Callard, G. V., Autocrine and paracrine role of steroids during spermatogenesis: Studies in *Squalus acanthias* and *Necturus maculosus, J. Exp. Zool.*, 261, 132, 1992.

Callard, I. P. and Callard, G. V., Sex steroid receptors and nonreceptor binding proteins, in *Hormones and Reproduction in Fishes, Amphibians, and Reptiles*, Norris, D. O. and Jones, R. E., Eds., Plenum Press, New York, 1987, chap. 12.

Callard, G. V., Schlinger, B., and Pasmanik, M., Nonmammalian vertebrate models in studies of brain-steroid interactions, *J. Exp. Zool. Suppl.*, 4, 6, 1990.

Callard, I. P., Fileti, L.A., Perez, L. E., Sorbera, L. A., Giannoukos, G., Klosterman, L. L., Tsang, P., and McCracken, J. A., Role of the corpus luteum and progesterone in the evolution of vertebrate viviparity, *Am. Zool.*, 32, 264, 1992.

Cartaud, A., Marcher, K., and Ozon, R., Digitoxigenin, a digitalis steroid, induces meiotic maturation of *Xenopus laevis* oocytes, *J. Steroid Biochem.*, 21, 101, 1984.

Chan, S. T. H. and Yeung, W. S. B., Sex control and sex reversal in fish under natural conditions, in *Fish Physiology*, Vol. IX B, Hoar, W. S., Randall, D. J., and Donaldson, E. M., Eds., Academic Press, New York, 1983, chap. 4.

Chang, J. P. and Jobin, R. M., Two native GnRH peptides stimulate gonadotropin release in goldfish by different second messenger components, in *Proc. Fourth Int. Symp. Reprod. Physiol. Fish*, Scott, A. P., Sumpter, J. P., Kime, D. E., and Rolfe, M. S., Eds., Fish Symp 91, Sheffield, 1991, 221.

Chang, J. P., Cook, H., Freedman, G. L., Wiggs, A. J., Somoza, G.M., De Leeuw, R., and Peter, R. E., Use of a pituitary cell dispersion method and primary culture system for the studies of gonadotropin-releasing hormone action in the goldfish, *Carassius auratus*. I. Initial morphological, static, and cell column perifusion studies, *Gen. Comp. Endocrinol.*, 77, 256, 1990.

Chang, Y. S., Huang, C. J., Huang, F. L., Liu, C. S., and Lo, T. B., Purification, characterization, and molecular cloning of gonadotropin subunits of silver carp *(Hypophthalmichthys molitrix), Gen. Comp. Endocrinol.*, 78, 23, 1990.

Cochran, R. C., *In vivo* and *in vitro* evidence for the role of hormones in fish spermatogenesis, *J. Exp. Zool.*, 261, 143, 1992.

Conover, D. O. and Fleisher, M. H., Temperature-sensitive period of sex determination in the Atlantic silverside, *Menidia menidia, Can. J. Fish. Aquat. Sci.*, 43, 514, 1986.

Copeland, P. A. and Thomas, P., Control of gonadotropin release in the Atlantic croaker *(Micropogonias undulatus)*: evidence for a lack of dopaminergic inhibition, *Gen. Comp. Endocrinol.*, 74, 474, 1989.

Cyr, D. G. and Eales, J. G., In vitro effects of thyroid hormones on gonadotropin-induced estradiol-17β secretion by ovarian follicles of rainbow trout, *Salmo gairdneri, Gen. Comp. Endocrinol.*, 69, 80, 1988.

Degani, G. and Boker, R., Sensitivity to maturation-inducing steroids and gonadotropin in the oocytes of blue gourami Trichogaster trichopterus (Anabantidae, Pallas, 1770), *Gen. Comp. Endocrinol.*, 85, 430, 1992.

Demski, L. S., Pathways for GnRH control of elasmobranch reproductive physiology and behavior, *J. Exp. Zool. Suppl.*, 2, 4, 1989.

Dettlaff, T. A., Development of the mature egg organization in amphibians, fish, and starfish during the concluding stages of oogenesis, in the period of maturation, in *Oocyte Growth and Maturation*, Dettlaff, T. A., and Vassetzky, S. G., Eds., Consultants Bureau, New York, 1988, chap. 3.

Dickhoff, W. W., Yan, L., Plisetskaya, E. M., Sullivan, C. V., Swanson, P., Hara, A., and Bernard, M. G., Relationship between metabolic and reproductive hormones in salmonid fish, *Fish Physiol. Biochem.*, 7, 147, 1989.

Dodd, J. M., Reproduction in cartilaginous fishes (Chondrichthyes), in *Fish Physiology*, Vol. IX A, Hoar, W. S., Randall, D. J., and Donaldson, E. M., Eds., Academic Press, New York, 1983, chap. 2.

Donaldson, E. M. and Hunter, G. A., Induced final maturation, ovulation, and spermiation in cultured fish, in *Fish Physiology*, Vol. IX B, Hoar, W. S., Randall, D. J., and Donaldson, E. M., Eds., Academic Press, New York, 1983, chap. 7.

DuBois, W. and Callard, G. V., Culture of intact Sertoli/germ cell units and isolated Sertoli cells from *Squalus* testis. II. Stimulatory effects of insulin and IGFI on DNA synthesis in premeiotic stages, *J. Exp. Zool.*, in press.

Dulka, J. G., Stacey, N. E., Sorensen, P. W., and Van Der Kraak, G. J., A steroid sex pheromone synchronizes male-female spawning readiness in goldfish, *Nature*, 325, 251, 1987.

Dulka, J. G., Sloley, B. D., Stacey, N. E., and Peter, R. E., A reduction in pituitary dopamine turnover is associated with sex pheromone-induced gonadotropin secretion in male goldfish, *Gen. Comp. Endocrinol.*, 86, 496, 1992.

Eppig, J. J., Mammalian oocyte development *in vivo* and *in vitro*, in *Elements of Mammalian Fertilization*, Vol. 1, Wassarman, P. M., Ed., CRC Press, Boca Raton, FL, 1991, chap. 3.

Fagbohun, C. F. and Downs, S. M., Metabolic coupling and ligand-stimulated meiotic maturation in the mouse oocyte-cumulus-cell complex, *Biol. Reprod.*, 45, 851, 1991.

Fahien, C. M. and Sower, S. A., Relationship between brain gonadotropinreleasing hormone and final reproductive period of the adult male sea lamprey, *Petromyzon marinus, Gen. Comp. Endocrinol.*, 80, 427, 1990.

Feist, G., Schreck, C. B., Fitzpatrick, M. S., and Redding, J. M., Sex steroid profiles of coho salmon *(Oncorhynchus kisutch)* during early development and sexual differentiation, *Gen. Comp. Endocrinol.*, 80, 299, 1990.

Fine, M. L., Keefer, D. A., and Leichnetz, G. R., Testosterone uptake in the brainstem of a sound-producing fish, *Science*, 215, 1265, 1982.

Fitzpatrick, M. S., Redding, J. M., Ratti, F. D., and Schreck, C. B., Plasma testosterone concentration predicts the ovulatory response of coho salmon *(Oncorhynchus kisutch)* to gonadotropin-releasing hormone analog, *Can. J. Fish. Aquat. Sci.*, 44, 1351, 1987.

Fostier, A., Jalabert, B., Billard, R., Breton, B., and Zohar, Y., The gonadal steroids, in *Fish Physiology*, Vol. IX A, Hoar, W. S., Randall, D. J., and Donaldson, E. M., Eds., Academic Press, NewYork, 1983, chap. 7.

Fostier, A., Le Gac, F., and Loir, M., Steroids in male reproduction, in *Proc. Third Int. Symp. Reprod. Physiol. Fish*, Idler, D. R., Crim, L. W., and Walsh, J. M., Eds., Memorial University of Newfoundland, St. John's, 1987, 239.

Gannam, A. L. and Lovell, R. T., Effects of feeding 17α-methyltestosterone, 11-ketotestosterone, 17β-estradiol, and 3,5,3′-triiodothyronine to channel catfish, *Ictalurus punctatus, Aquaculture*, 92, 377, 1991.

Ge, W., Peter, R. E., Vaughan, J., Rivier, J., and Vale, W., Inhibin/activin-like protein-mediated feedback between the pituitary and ovary in goldfish, in *Proc. Fourth Int. Symp. Reprod. Physiol. Fish,* Scott, A. P., Sumpter, J. P., Kime, D. E., and Rolfe, M. S., Eds., Fish Symp 91, Sheffield, 1991, 25.

Gilkey, J. C., Mechanisms of fertilization in fishes, *Am. Zool.,* 21, 359, 1981.

Goetz, F. W., Hormonal control of oocyte final maturation and ovulation, in *Fish Physiology,* Vol. IX B, Hoar, W. S., Randall, D. J., and Donaldson, E. M., Eds., Academic Press, New York, 1983, chap. 3.

Goetz, F. W., Berndtson, A. K., and Ranjan, M., Ovulation: Mediators at the ovarian level, in *Vertebrate Endocrinology: Fundamentals and Biomedical Implications,* Vol, 4 A, Pang, P. K. T. and Schreibman, M. P., Eds., Academic Press, San Diego, 1991, chap. 4.

Goos, H. J. T., Steroid feedback on pituitary gonadotropin secretion, in *Proc. Third Int. Symp. Reprod. Physiol. Fish,* Idler, D. R., Crim, L. W., and Walsh, J. M., Eds., Memorial University of Newfoundland, St. John's, 1987, 16.

Gorbman, A., Reproduction in cyclostome fishes and its regulation, in *Fish Physiology,* Vol. IX A, Hoar, W. S., Randall, D. J., and Donaldson, E. M., Eds., Academic Press, New York, 1983, chap. 1.

Goudie, C. A., Redner, B. D., Simco, B. A., and Davis, K. B., Feminization of channel catfish by oral administration of steroid sex hormones, *Trans. Am. Fish. Soc.,* 112, 670, 1983.

Greeley, M. S., Begovac, P. C., and Wallace, R. A., Removal of enveloping follicle cells can trigger resumption of meiotic maturation in *Fundulus heteroclitus* oocytes, *J. Exp. Zool.,* 244, 177, 1987.

Greeley, M. S., Hols, H., and Wallace, R. A., Changes in size, hydration and low molecular weight osmotic effectors during meiotic maturation of *Fundulus* oocytes *in vivo, Comp. Biochem. Physiol.,* 100A, 639, 1991.

Grier, H., Cellular organization of the testis and spermatogenesis in fishes, *Am. Zool.,* 21, 345, 1981.

Groves, D. J. and Batten, T. F. C., Direct control of the gonadotroph in a teleost, *Poecilia latipinna.* II. Neurohormones and neurotransmitters, *Gen. Comp. Endocrinol.,* 62, 315, 1986.

Habibi, H., Ince, B. W., and Matty, A. J., Effects of 17α-methyltestosterone and 17β-oestradiol on intestinal transport and absorption of L(^{14}C)-leucine *in vitro* in rainbow trout*(Salmo gairdneri), J. Comp. Physiol.,* 151, 247, 1983.

Habibi, H.R., Van der Kraak, G., Bulanski, E., and Peter, R.E., Effect of teleost GnRH on reinitiation of oocyte meiosis in golfish, *in vitro, Am. J. Physiol.,* 255, R268, 1988.

Habibi, H. R., De Leeuw, R., Nahorniak, C. S., Goos, H. J. T., and Peter, R. E., Pituitary gonadotropin-releasing hormone (GnRH) receptor activity in goldfish and catfish: seasonal and gonadal effects, *Fish Physiol. Biochem.,* 7, 109, 1989.

Hamlett, W. C., Evolution and morphogenesis of the placenta in sharks, *J. Exp. Zool. Suppl.,* 2, 35, 1989.

Ho, S.-M., Vitellogenesis, in *Vertebrate Endocrinology: Fundamentals and Biomedical Implications,* Vol. 4 A, Pang, P. K. T. and Schreibman, M. P., Eds., Academic Press, San Diego, 1991, chap. 3.

Hoar, W. S., Reproduction, in *Fish Physiology,* Vol. III, Hoar, W. S. and Randall, D. J., Eds., Academic Press, New York, 1969, chap. 1.

Hunter, G. A. and Donaldson, E. M., Hormonal sex control and its application to fish culture, in *Fish Physiology,* Vol. IX B, Hoar, W. S., Randall, D. J., and Donaldson, E. M., Eds., Academic Press, New York, 1983, chap. 5.

Idler, D. R. and Ng, T. B., Teleost gonadotropins: Isolation, biochemistry and function, in *Fish Physiology,* Vol. IX A, Hoar, W. S., Randall, D. J., and Donaldson, E. M., Eds., Academic Press, New York, 1983, chap. 5.

Ishii, S., Gonadotropins, in *Vertebrate Endocrinology: Fundamentals and Biomedical Implications,* Vol. 4 B, Pang, P. K. T. and Schreibman, M. P., Eds., Academic Press, San Diego, 1991, chap. 9.

Ito, I. and Shimamoto, K., Fertilization potential of the medaka egg, *Dev. Growth Differ.*, 31, 55, 1989.

Iuchi, I., Masuda, K., and Yamagani, K., Change in component proteins of the egg envelope (chorion) of rainbow trout during hardening, *Dev. Growth Differ.*, 33, 85, 1991.

Iwamatsu, T., Yoshimoto, Y., and Hiramoto, Y., Mechanism of Ca^{2+} release in medaka eggs microinjected with inositol 1,4,5-triphosphate and Ca^{2+}, *Dev. Biol.*, 129, 191, 1988.

Iwamatsu, T., Onitake, K., Yoshimoto, Y., and Hiramoto, Y., Time sequence of early events in fertilization of medaka eggs, *Dev. Growth Differ.*, 33, 479, 1991.

Jalabert, B., *In vitro* oocyte maturation and ovulation in rainbow trout *(Salmo gairdneri)*, northern pike *(Esox lucius)*, and goldfish *(Carassius auratus)*, *J. Fish. Res. Board Can.*, 33, 974, 1976.

Jalabert, B. and Fostier, A., The follicular sensitivity *in vitro* to maturation-inducing hormones in rainbow trout, *Salmo gairdneri*: role of oestradiol-17β, *Aquaculture*, 43, 1, 1984.

Jalabert, B., Fostier, A., Breton, B., and Weil, C., Oocyte maturation in vertebrates, in *Vertebrate Endocrinology: Fundamentals and Biomedical Implications*, Vol. 4 A, Pang, P. K.T. and Schreibman, M. P., Eds., Academic Press, San Diego, 1991, chap. 2.

Johansen, P. H. and Cross, J. A., Effects of sexual maturation and sex steroid hormone treatment on the temperature preference of the guppy, *Poecilia reticulata* (Peters), *Can. J. Zool.*, 58, 586, 1980.

Kah, O., Trudeau, V. L., Sloley, B. D., Marinoli, M. G., Chang, J. P., Yu, K. L. and Peter, R. E., Implication of GABA in the neuroendocrine regulation of gonadotrophin release in the goldfish *(Carassius auratus)*, in *Proc. Fourth Int. Symp. Reprod. Physiol. Fish*, Scott, A. P., Sumpter, J. P., Kime, D. E., and Rolfe, M. S., Eds., FishSymp 91, Sheffield, 1991, 57.

Kanamori, A. and Nagahama, Y., Developmental changes in the properties of gonadotropin receptors in the ovarian follicles of amago salmon *(Oncorhynchus rhodurus)* during oogenesis, *Gen. Comp. Endocrinol.*, 72, 210, 1988.

Kawauchi, H., Suzuki, K., Itoh, H., Swanson, P., Naito, N., Nagahama, Y., Nozaki, M., Nakai, Y., and Itoh, S., The duality of teleost gonadotropins, *Fish Physiol. Biochem.*, 7, 29, 1989.

Kessel, R. G., Tung, H. N., Roberts, R., and Beams, H. W., The presence and distribution of gap junctions in the oocyte-follicle cell complex of the zebrafish, *Brachydanio rerio*, *J. Submicrosc. Cytol.*, 17, 239, 1985.

King, J. A. and Millar, R. P., Gonadotropin-releasing hormones, in *Vertebrate Endocrinology: Fundamentals and Biomedical Implications*, Vol. 4 B, Pang, P. K. T. and Schreibman, M. P., Eds., Academic Press, San Diego, 1991, chap. 8.

Kobayashi, M. and Stacey, N. E., Effects of ovariectomy and steroid hormone implantation on serum gonadotropin levels in female goldfish, *Zool. Sci.*, 7, 715, 1990.

Kobayashi, M., Aida, K., Furukawa, K., Law, Y. K., Moriwaki, T., and Hanyu, I., Development of sensitivity to maturation-inducing steroids in the oocytes of the daily spawning teleost, the kisu *Sillago japonica*, *Gen. Comp. Endocrinol.*, 72, 264, 1988a.

Kobayashi, M., Aida, K., and Hanyu, I., Hormone changes during the ovulatory cycle in goldfish, *Gen. Comp. Endocrinol.*, 69, 301, 1988b.

Kormanik, G. A., Ion and osmoregulation in prenatal elasmobranchs: Evolutionary implications, *Am. Zool.*, 32, 294, 1992.

Krishna, N. S. R., Subhedar, N., and Schreibman, M. P., FMRFamide-like immunoreactive nervus terminalis innervation to the pituitary in the catfish, *Clarias batrachus* (Linn.): demonstration by lesion and immunocytochemical techniques, *Gen. Comp. Endocrinol.*, 85, 111, 1992.

Kudo, S., Polyspermy-preventing mechanisms in fish eggs, in *Regeneration and Development*, Inoue, S. et al., Eds., Okada Printing and Publishing, Maebashi, Japan, 1988, 395.

Kudo, S. and Inoue, M., Structure and functions of the fertilization envelope in fish eggs, in *Regeneration and Development*, Inoue, S. et al., Eds., Okada Printing and Publishing, Maebashi, Japan, 1988, 411.

Kudo, S. and Inoue, M., Bacterial action of fertilization envelope extract from eggs of the fish *Cyprinus carpio* and *Plecoglossus altivelis*, *J. Exp. Zool.*, 250, 219, 1989.

LaFleur, G. J. and Thomas, P., Evidence for a role of Na^+,K^+-ATPase in the hydration of Atlantic croaker and spotted seatrout oocytes during final maturation, *J. Exp. Zool.*, 258, 126, 1991.

Lam, T. J., Environmental influences on gonadal activity in fish, in *Fish Physiology*, Vol. IX B, Hoar, W. S., Randall, D. J., and Donaldson, E. M., Eds., Academic Press, New York, 1983, chap. 2.

Leatherland, J. F., Thyroid hormones and reproduction, in *Hormones and Reproduction in Fishes, Amphibians and Reptiles*, Norris, D. O. and Jones, R. E., Eds., Plenum Press, New York, 1987, chap. 14.

Lessman, C. A., Habibi, H. R., and Macrae, T. H., Effect of microtubule reactive drugs on steroid and centrifugation-induced germinal vesicle migration during goldfish oocyte meiosis, *Biol. Cell*, 64, 293, 1988.

Liley, N. R. and Stacey, N. E., Hormones, pheromones, and reproductive behavior in fish, in *Fish Physiology*, Vol. IX B, Hoar, W. S., Randall, D. J., and Donaldson, E. M., Eds., Academic Press, New York, 1983, chap. 1.

Lin, H. R., Zhou, X. J., Van Der Kraak, G., and Peter, R. E., Effects of gonadotropin-releasing hormone agonists and dopamine antagonists on gonadotropin secretion and ovulation in Chinese loach, *Paramisgurnus dabryanus, Aquaculture*, 95, 139, 1991.

Lofts, B., Testicular function, in *Hormones and Reproduction in Fishes, Amphibians, and Reptiles*, Norris, D. O. and Jones, R. E., Eds., Plenum Press, New York, 1987, chap. 10.

Loir, M., Trout steroidogenic testicular cells in primary culture I. Changes in free and conjugated androgen and progestagen secretions: effects of gonadotropin, serum, and lipoproteins, *Gen. Comp. Endocrinol.*, 78, 374, 1990.

Mak, P. and Callard, G.V., A novel steroid binding protein in the testis of the dogfish *Squalus acanthias, Gen. Comp. Endocrinol.*, 68, 104, 1987.

Masui, Y., Meiotic arrest in animal oocytes, in *Biology of Fertilization*, Vol. 1, Metz, C. B. and Monroy, A., Eds., Academic Press, Orlando, 1985, chap. 8.

Masui, Y., The role of "cytostatic factor (CSF)" in the control of oocyte cell cycles: a summary of 20 years of study, *Dev. Growth Differ.*, 33, 543, 1991.

Mazzi, V. and Vellano, C., Prolactin and reproduction, in *Hormones and Reproduction in Fishes, Amphibians, and Reptiles*, Norris, D. O. and Jones, R. E., Eds., Plenum Press, New York, 1987, chap. 4.

Meyer, J. H. and Zakon, H. H., Androgens alter the tuning of electroreceptors, *Science*, 217, 635, 1982.

Mirua, T., Yamauchi, K., Takahashi, H., and Nagahama, Y., Hormonal induction of all stages of spermatogenesis in vitro in the male Japanese eel *(Anguilla japonica), Proc. Natl. Acad. Sci. U.S.A.*, 88, 5774, 1991.

Miura, T., Yamauchi, K., Takahashi, H., and Nagahama, Y., The role of hormones in the acquisition of sperm motility in salmonid fish, *J. Exp. Zool.*, 261, 359, 1992.

Munro, A. D., Scott, A. P., and Lam, T. J., Eds., *Reproductive Seasonality in Teleosts: Environmental Influences*, CRC Press, Boca Raton, FL, 1990.

Nagahama, Y., The functional morphology of teleost gonads, in *Fish Physiology*, Vol. IX A, Hoar, W. S., Randall, D. J., and Donaldson, E. M., Eds., Academic Press, New York, 1983, chap. 6.

Nagahama, Y., Endocrine control of oocyte maturation, in *Hormones and Reproduction in Fishes, Amphibians, and Reptiles*, Norris, D. O., and Jones, R. E., Eds., Plenum Press, New York, 1987, chap. 7.

Nagahama, Y., Molecular aspect of final oocyte maturation in fish, in *Proc. Third MAFF Homeostasis Workshop on the Endocrine Control of Reproduction and Early Development in Fish,* Mie, Japan, 1991, 31.

Nagahama, Y. and Yamashita, M., Mechanisms of synthesis and action of 17α,20β-dihydroxy-4-pregnen-3-one, a teleost maturation-inducing substance, *Fish Physiol. Biochem.*, 7, 193, 1989.

Pajor, A. M., Stegeman, J. J., Thomas, P., and Woodin, B. R., Feminization of the hepatic microsomal cytochrome P-450 system in brook trout by estradiol, testosterone, and pituitary factors, *J. Exp. Zool.*, 253, 51, 1990.

Patiño, R. and Thomas, P., Induction of maturation of Atlantic croaker oocytes by 17α,20β,21-trihydroxy-4-pregnen-3-one *in vitro*: consideration of some biological and experimental variables, *J. Exp. Zool.*, 255, 97, 1990a.

Patiño, R. and Thomas, P., Effects of gonadotropin on ovarian intrafollicular processes during the development of oocyte maturational competence in a teleost, the Atlantic croaker: evidence for two distinct stages of gonadotropic control of final oocyte maturation, *Biol. Reprod.*, 43, 818, 1990b.

Patiño, R. and Thomas, P., Characterization of membrane receptor activity for 17α,20β,21-trihydroxy-4-pregnen-3-one in ovaries of spotted seatrout *(Cynoscion nebulosus), Gen. Comp. Endocrinol.*, 78, 204, 1990c.

Peter, R. E., The brain and neurohormones in teleost reproduction, in *Fish Physiology,* Vol. IX A, Hoar, W. S., Randall, D. J., and Donaldson, E. M., Eds., Academic Press, New York, 1983, chap. 3.

Peter, R. E., Yu, K., Marchant, T. A., and Rosenblum, P. M., Direct neural regulation of the teleost adenohypophysis, *J. Exp. Zool. Suppl.*, 4, 84, 1990.

Petrino, T. R., Greeley, M. S., Selman, K., Lin, Y.W., and Wallace, R. A., Steroidogenesis in *Fundulus heteroclitus*. II. Production of 17α-hydroxy-20β-dihydroprogesterone, testosterone, and 17β-estradiol by various components of the ovarian follicle, *Gen. Comp. Endocrinol.*, 76, 230, 1989.

Piferrer, F. and Donaldson, E. M., Dosage-dependent differences in the effect of aromatizable and nonaromatizable androgens on the resulting phenotype of coho salmon *(Oncorhynchus kisutch), Fish Physiol. Biochem.*, 9, 145, 1991.

Planas, J. V., Swanson, P., Rand-Weaver, M., and Dickhoff, W. W., Somatolactin stimulates *in vitro* gonadal steroidogenesis in coho salmon, *Oncorhynchus kisutch, Gen. Comp. Endocrinol.*, 87, 1,1992.

Pudney, J. and Callard, G. V., Identification of Leydiglike cells in the interstitium of the shark testis *(Squalus acanthias), Anat. Rec.*, 209, 311, 1984.

Reiter, R. J., The pineal gland: Reproductive interactions, in *Vertebrate Endocrinology: Fundamentals and Biomedical Implications*, Vol. 4 B, Pang, P. K. T. and Schreibman, M. P., Eds., Academic Press, San Diego, 1991, chap 15.

Rodriguez, M. and Specker, J. L., *In vitro* effects of arginine vasotocin on testosterone production by testes of rainbow trout *(Oncorhynchus mykiss), Gen. Comp. Endocrinol.,* 83, 249, 1991.

Sadler, S. E. and Maller, J. L., Identification of a steroid receptor on the surface of *Xenopus* oocytes by photoaffinity labelling, *J. Biol. Chem.*, 257, 355, 1982.

Sadler, S. E. and Maller, J. L., The development of competence for meiotic maturation during oogenesis in *Xenopus laevis, Dev. Biol.*, 98, 165, 1983.

Saligaut, C., Salbert, G., Bailhache, T., Bennani, S., and Jego, P., Serotonin and dopamine turnover in the female rainbow trout *(Oncorhynchus mykiss)* brain and pituitary: changes during the annual reproductive cycle, *Gen. Comp. Endocrinol.*, 85, 261, 1992.

Schreibman, M. P., Holtzman, S., and Margolis-Nunno, H., Maturation and aging of the reproductive system, in *Vertebrate Endocrinology: Fundamentals and Biomedical Implications*, Vol. 4 A, Pang, P. K. T. and Schreibman, M. P., Eds., Academic Press, San Diego, 1991, chap. 7.

Schuetz, A. W., Local control mechanisms during oogenesis and folliculogenesis, in *Developmental Biology. A Comprehensive Synthesis,* Vol. 1, Browder, L. W., Ed., Plenum Press, New York, 1985, chap. 1.

Sekine, S., Saito, A., Itoh, H., Kawauchi, H., and Itoh, S., Molecular cloning and sequence analysis of chum salmon gonadotropin cDNAs, *Proc. Natl. Acad. Sci. U.S.A.*, 86, 8645, 1989.

Selcer, K. W. and Leavitt, W. W., Estrogens and progestins, in *Vertebrate Endocrinology: Fundamentals and Biomedical Implications,* Vol. 4 B, Pang, P. K. T. and Schreibman, M. P., Eds., Academic Press, San Diego, 1991, chap. 10.

Selman, K. and Wallace, R. A., Gametogenesis in Fundulus heteroclitus, *Am. Zool.*, 26, 173, 1986.

Shapiro, D. Y., Behavioral influences on gene structure and other new ideas concerning sex change in fishes, *Environ. Biol. Fishes,* 23, 283, 1988.

Sherwood, N. M. and Lovejoy, D. A., The origin of the mammalian form of GnRH in primitive fishes, *Fish Physiol. Biochem.*, 7, 85, 1989.

Sherwood, N. M., Kyle, A., Kreiberg, H., Warby, C. M., Magnus, T. H., Carolsfeld, J., and Price, W. S., Partial characterisation of a spawning pheromone in the herring *Clupea harengus pallasi, Can. J. Zool.*, 69, 93, 1991.

Singh, H., Griffith, R. W., Takahashi, A., Kawauchi, H., Thomas, P., and Stegeman, J., Regulation of gonadal steroidogenesis in *Fundulus heteroclitus* by recombinant salmon growth hormone and purified salmon prolactin, *Gen. Comp. Endocrinol.*, 72, 144, 1988.

Smith, L. D., The induction of oocyte maturation: transmembrane signaling events and regulation of the cell cycle, *Development*, 107, 685, 1989.

Somoza, G. M. and Peter, R. E., Effects of serotonin on gonadotropin and growth hormone release from *in vitro* perifused goldfish pituitary fragments, *Gen. Comp. Endocrinol.*, 82, 103, 1991.

Sorenson, P. W., Hara, T. J., Stacey, N. E., and Goetz, F. W., F prostaglandins function as potent olfactory stimulants comprising the postovulatory female sex pheromone in goldfish, *Biol. Reprod.,* 39, 1039, 1988.

Stacey, N., Hormonal pheromones in fish: status and prospects, in *Proc. Fourth Int. Symp. Reprod. Physiol. Fish,* Scott, A. P., Sumpter, J. P., Kime, D. E., and Rolfe, M. S., Eds., FishSymp 91, Sheffield, 1991, 177.

Stoss, J., Fish gamete preservation and spermatozoan physiology, in *Fish Physiology,* Vol. IX B, Hoar, W. S., Randall, D. J., and Donaldson, E. M., Eds., Academic Press, New York, 1983, chap. 6.

Suzuki, K., Nagahama, Y., and Kawauchi, H., Steroidogenic activities of two distinct salmon gonadotropins, *Gen. Comp. Endocrinol.*, 71, 452, 1988.

Swanson, P., Salmon gonadotropins: reconciling old and new ideas, in Proc. Fourth Int. Symp. Reprod. Physiol. Fish, Scott, A. P., Sumpter, J. P., Kime, D. E., and Rolfe, M. S., Eds., FishSymp 91, Sheffield, 1991, 2.

Thomas, P. and Patiño, R., Changes in 17α,20β,21-trihydroxy-4-pregnen-3-one membrane receptor concentrations in ovaries of spotted seatrout during final oocyte maturation, in *Proc. Fourth Int. Symp. Reprod. Physiol. Fish,* Scott, A. P., Sumpter, J. P., Kime, D. E., and Rolfe, M. S., Eds., FishSymp 91, Sheffield, 1991, 122.

Trant, J. M. and Thomas, P., Isolation of a novel maturation-inducing steroid produced *in vitro* by ovaries of Atlantic croaker, *Gen. Comp. Endocrinol.*, 75, 397, 1989.

Trant, J. M., Thomas, P., and Shackleton, C. H. L., Identification of 17α,20β,21-trihydroxy-4-pregnen-3-one as the major ovarian steroid produced by the teleost *Micropogonias undulatus* during final oocyte maturation, *Steroids*, 47, 89, 1986.

Trudeau, V. L., Sloley, B. D., Wong, A. O. L., and Peter, R. E., Mechanisms of sex steroid negative and positive feedback control of gonadotropin (GtH) secretion in teleosts, in *Proc. Fourth Int. Symp. Reprod. Physiol. Fish,* Scott, A. P., Sumpter, J. P., Kime, D. E., and Rolfe, M. S., Eds., FishSymp 91, Sheffield, 1991, 224.

Tyler, C., Vitellogenesis in salmonids, in *Proc. Fourth Int. Symp. Reprod. Physiol. Fish,* Scott, A. P., Sumpter, J. P., Kime, D. E., and Rolfe, M. S., Eds., FishSymp 91, Sheffield, 1991, 295.

Van Den Hurk, R. and Resink, J. W., Male reproductive system as sex pheromone producer in teleost fish, *J. Exp. Zool.*, 261, 204, 1992.

Van Der Kraak, G., An introduction to gonadotropin receptor studies in fish, in *Fish Physiology,* Vol. IX A, Hoar, W. S., Randall, D. J., and Donaldson, E. M., Eds., Academic Press, New York, 1983,chap. 9.

Van Der Kraak, G., Role of calcium in the control of steroidogenesis in preovulatory ovarian follicles of the goldfish, *Gen. Comp. Endocrinol.*, 81, 268, 1991.

Van Der Kraak, G., Suzuki, K., Peter, R. E., Itoh, H., and Kawauchi, H., Properties of common carp gonadotropin I and gonadotropin II, *Gen. Comp. Endocrinol.*, 85, 217, 1992.

Van Oordt, P. G. W. J. and Peute, J., The cellular origin of pituitary gonadotropins in teleosts, in *Fish Physiology*, Vol. IX A, Hoar, W. S., Randall, D. J., and Donaldson, E. M., Eds., Academic Press, New York, 1983, chap. 4.

Vilain, J. P., Moreau, M., and Guerrier, P., Uncoupling of oocyte-follicle cells triggers reinitiation of meiosis in amphibian oocytes, *Dev. Growth Differ.*, 22, 687, 1980.

Wallace, R. A. and Selman, K., Cellular and dynamic aspects of oocyte growth in teleosts, *Am. Zool.*, 21, 325, 1981.

Wallace, R. A., Selman, K., Greeley, M. S., Begovac, P. C., Lin, Y.W. P., and McPherson, R., Current status of oocyte growth, in *Proc. Third Int. Symp. Reprod. Physiol. Fish*, Idler, D. R., Crim, L. W., and Walsh, J. M., Eds., Memorial University of Newfoundland, St. John's, 1987, 167.

Weber, G. W., Okimoto, D. K., Richman, N. H., and Grau, E. G., Patterns of thyroxine and triiodothyronine in serum and follicle-bound oocytes of the tilapia, *Oreochromis mossambicus*, during oogenesis, *Gen. Comp. Endocrinol.*, 85, 392, 1992.

Wiegand, M. D. and Peter, R. E., Effects of sex steroids on plasma lipids in the goldfish, *Carassius auratus, Can. J. Zool.*, 58, 967, 1980.

Weil, C. and Marcuzzi, O., Cultured pituitary cell GtH response to GnRH at different stages of rainbow trout oogenesis and influence of steroid hormones, *Gen. Comp. Endocrinol.*, 79, 483, 1990a.

Weil, C. and Marcuzzi, O., Cultured pituitary cell GtH response to GnRH at different stages of rainbow trout spermatogenesis and influence of steroid hormones, *Gen. Comp. Endocrinol.*, 79, 492, 1990b.

Weil, C., Crim, L. W., Wilson, C. E., and Cauty, C., Evidence of GnRH receptors in cultured pituitary cells of the winter flounder *(Pseudopleuronectes americanus* W.), *Gen. Comp. Endocrinol.*, 85, 156, 1992.

Wourms, J. P. and Lombardi, J., Reflections on the evolution of piscine viviparity, *Am. Zool.*, 32, 276, 1992.

Wourms, J. P., Grove, B. D., and Lombardi, J., The maternal-embryonic relationship in viviparous fishes, in *Fish Physiology*, Vol. XI B, Hoar, W. S. and Randall, D. J., Eds., Academic Press, San Diego, 1988, chap. 1.

Yamamoto, T., Sex differentiation, in *Fish Physiology*, Vol. III, Hoar, W. S. and Randall, D. J., Eds., Academic Press, New York, 1969, chap. 3.

Yamashita, M., Fukada, S., Yoshikuni, M., Bulet, P., Hirai, T., Yamaguchi, A., Lou, Y.H., Zhao, Z., and Nagahama, Y., Purification and characterization of maturation-promoting factor in fish, *Dev.Biol.*, 149, 8, 1992.

Yan, L., Swanson, P., and Dickhoff, W. W., Binding of gonadotropins (GtH I and GtH II) to coho salmon gonadal membrane preparations, *J. Exp. Zool.*, 258, 221, 1991.

York, W. S., Follicle Cell-Oocyte Gap Junctional Coupling and Oocyte Maturation in Sciaenid Fishes, M.S. thesis, Texas Technological University, Lubbock, TX, 1993.

Yu, K. L. and Peter, R. E., Adrenergic and dopaminergic regulation of gonadotropin-releasing hormone release from goldfish preoptic-anterior hypothalamus and pituitary *in vitro, Gen. Comp. Endocrinol.*, 85, 138, 1992.

Yu, K. L., Peng, C., and Peter, R. E., Changes in brain levels of gonadotropin-releasing hormone and serum levels of gonadotropin and growth hormone in goldfish during spawning, *Can. J. Zool.*, 69, 182, 1991a.

Yu, K. L., Rosenblum, P. M., and Peter, R. E., *In vitro* release of gonadotropin-releasing hormone from the brain preoptic-anterior hypothalamic region and pituitary of female goldfish, *Gen. Comp. Endocrinol.*, 81, 256, 1991b.

Zhu, Y., Aida, K., Furukawa, K., and Hanyu, I., Development of sensitivity to maturation-inducing steroids and gonadotropins in the oocytes of the tobinumeri-dragonet, *Rapomucenus beniteguri*, Callionymidae (Teleostei), *Gen. Comp. Endocrinol.*, 76, 250, 1989.

17 Coloration and Chromatophores

Ryozo Fujii

COLORATION

Color is the sensation that results from the stimulation of the retina by light of certain wavelengths, and many animals, including many species of fish, are known to perceive colors. When we discuss color, we inevitably refer to what we, as humans, see. However, the range of perceivable wavelengths of light has been shown to differ to some extent among different species.

Many fishes are reported to have color vision with spectral ranges similar to that of humans, while some are totally color-blind. We know too that the photic conditions around fish differ considerably among their various habitats (Lythgoe, 1988). To adapt to these conditions, fish must have acquired color vision that is appropriate to their specific survival strategies (Fernald, Chapter 6 in this volume). Recent work on some freshwater fish indicates that their retinas can detect even near-ultraviolet rays (Harosi and Hashimoto, 1983). Bearing in mind the significance of such an ability, we must be careful in our assessment of the colors seen by fish.

The mechanisms responsible for manifestation of colors in fish are fundamentally similar to those in terrestrial organisms. The coloration of the skin is generated by the absorption, scattering, and reflection of light of certain wavelengths. The colored pigmentary substances absorb light, while minute structures with refractive indices that are very different from those of their surrounding medium (the cytosol in most cases) scatter or reflect light.

Let us consider an example of the coloration of a fish in water. In clear water, light is attenuated by scattering and absorption by water molecules, in a manner dependent on the wavelength. Maximal transmission occurs at about 460 nm. The deeper the water, therefore, the more the bluish light is dominant. Under such photic conditions the skin, or some parts of it, naturally reflects small amounts of light at longer wavelengths, and, therefore, it looks darker. For example, the brilliant blue spots on the red dorsolateral surface of the red sea bream, *Pagrus major*, which normally lives at depths more than 20 m, should be much more conspicuous against a very dark red background than when we see them in an aquarium or on a table. Thus, the ethological impli-

0-8493-8042-1/93/$0.00+$.50

cations of coloration or color patterns should be discussed in the context of the photic environment.

As in the case in many other groups of animals, the body colors of fish are predominantly dependent on the presence of special cells in the skin, called chromatophores, that contain colored pigments, light-scattering or light-reflecting organelles. In poikilothermic vertebrates, these cells can be classified into five groups, as discussed below in the section on Chromatophores.

Since an apparently identified hue may be the result of different distributions of spectral energy, measurements of spectral reflectance are sometimes required. If they are equipped with a reflectance measurement device, conventional spectrophotometers can be used for this purpose. With recent developments in electronic engineering, as well as in optoelectronics, direct rapid measurements of spectral reflectance from skin have become possible, for example by use of a multichannel photodiode array apparatus (Kasukawa et al., 1987). Color photographic prints of high quality can also be adequately and easily assessed by this method.

PIGMENTS

For detailed descriptions of pigmentary substances responsible for the coloration of animals, including fish, the reader is referred to the excellent monographs by Needham (1974) and Fox (1979). In this section, only a brief description of the major pigments that are detectable in the skin of fish is presented.

MELANINS

Of the pigments that color the skin of animals, melanins are also the most widely distributed in fish. Brown to black tones generally result from the presence of melanins. They are highly polymerized compounds derived from the amino acid tyrosine. They are synthesized and deposited in the dark organelles of the melanophores, which are called the melanosomes. The first steps in the synthesis are the oxidation of tyrosine to dopa and the oxidation of dopa to dopaquinone by tyrosinase. These compounds are then oxidatively polymerized and decarboxylated to form melanins. Incidentally, pheomelanins, which are related polymers with S-containing moieties derived from cysteine, are not known in fish. Thus, fish have only "eumelanins", being different in this respect from poikilothermic vertebrates.

Mature melanosomes are retained in the dermal melanophores for long periods of time, normally playing an active role in rapid changes in color via the motile activity of these cells. By contrast, the melanosomes produced in epidermal melanophores are often observed to be transferred to epidermal cells (keratinocytes) to increase the darkness of the skin (see the discussion of morphological color changes, later in this chapter), although there are cases in which the movement of organelles results in rapid changes in color, as it does in dermal melanophores.

Carotenoids

The beautiful and flamboyant coloration exhibited by many fish is mostly due to the presence of carotenoid pigments. These are polyene pigments that are insoluble in water and they are deposited in carotenoid vesicles in the xanthophores or erythrophores. There are many molecular species, but all of them are derived from the prey or come directly from plant sources. Sometimes plant carotenoids are modified to yield pigments that are found only in animals.

Results, mainly from Japan, indicate that tunaxanthin is a rather common pigment in marine fish. This carotenoid is especially abundant in yellow-colored fishes. Astaxanthin, by contrast, seems to be dominant in red marine fish. Lutein is also widely found in many marine species. Carotenoids commonly found in freshwater fish include beta-carotene, lutein, taraxanthin, astaxanthin, tunaxanthin, alpha- and beta-doradexanthins, and zeaxanthin.

Pteridines

Pteridines are compounds with a pteridine nucleus (pyrimido[4,5-*b*]pyradine) as their basic skeleton. They are present in many species of invertebrates. Only recently were several molecular species in this category found in the skins of poikilothermic vertebrates, playing active roles in the generation of bright colors, namely, yellow, red, and brown, as do the carotenoids. In contrast to carotenoids, pteridines are soluble in water. The cell organelles that contain pteridines are called "pterinosomes". To describe such organelles, however, it is safer to use the terms "xanthosomes" or "erythrosomes", at least until chemical identification of the contents of the organelles has been made. Xanthophores and erythrophores contain these organelles.

Purines

Some purine bases play an indispensable role in the generation of color in fish. Among the purines, guanine apparently predominates, and extraordinarily large amounts of guanine can be found in the silvery belly skin of most species of fish. Being barely soluble in water, purines are usually found in crystalline form in iridophores, and they act efficiently to reflect light (see below).

CHROMATOPHORES

Integumentary colors are primarily dependent on the presence of the pigment cells in the skin, namely the chromatophores. Derived from the neural crest, all chromatophores in fish and other vertebrates have a common embryonic origin. Chromatophores are now categorized as "paraneurons." In poikilothermic vertebrates, they can be classified into five groups: melanophores, erythrophores, xanthophores, leucophores, and iridophores (Fujii, 1969, 1993; Fujii and Oshima, 1986).

In the skin tissues of fish, chromatophores are present mostly in the dermis, and pigment cells of all the above-mentioned types can be found there. Epidermal chromatophores are found rather sporadically. Most of them are melanophores, but xanthophores that are in the epidermis were recently described for the first time in an Antarctic blenny, the emerald rockcod, *Trematomus bernacchii* (Obika and Meyer-Rochow, 1990). A further survey may provide information about the occasional existence of bright-colored chromatophores in the epidermis of other fishes.

For survival, fish need the ability to change their hues and colors rapidly and effectively. Thus, the chromatophores must also react rapidly and effectively. If the chromatophores are in the epidermis, where there are no capillaries, hormonal effects should be delayed and weak. Controlling nerve fibers can more easily reach and control the chromatophores when the latter are in the dermis. Conversely, in fish, the pigmentary effector system has evolved predominantly in the dermis to meet the demands for speed and efficiency of changes in pigmentation.

Sharing their common ontogenetic origin with nerve cells, chromatophores are fundamentally dendritic cells, and several dendritic processes emanate from the perikaryon. In contrast to the neurons, however, the processes develop parallel to the plane of the skin, for effective revelation of the pigmentary colors when the skin is viewed from outside. Dendritic chromatophores include the melanophores, the xanthophores, the erythrophores, and the leucophores. Pigmentary organelles (chromatosomes) migrate centripetally (aggregation) or centrifugally (dispersion) in response to various signals received by the cells (Plates 1, 2, and 5). Having no dendrites, by contrast, most iridophores assume a round, oval, or multilateral shape, although in some cases they, too, have dendritic processes (see below). Such differences in configuration are related to the physiological characteristics of each cell.

In the dendritic chromatophores, a number of microtubules have been found emanating from the perikaryon into the dendritic processes. Apparently, they function as cytoskeletal elements that maintain the stellate shape of the cells. In addition, recent studies strongly suggest an active role for these microtubules in the centripetal and centrifugal translocation of melanosomes.

In the skin of many fish, there are wide extracellular spaces of rather uniform width, composed mainly of collagen fibrils. The dermal chromatophores are usually present below this compact collagenous layer, being not in direct contact with the bottom of the epidermis. Parallel collagen fibrils form a thin sheet. Several sheets are arranged as lamellae, but the fibrils within alternating sheets run approximately perpendicular to those in adjacent ones. Resembling plywood, the lamellar structure apparently reinforces the thin integument. In very small fish, such as guppies, neon tetras, and many aquarium fish, no such lamellar structure is found, but there is always a thin collagenous layer. As crucial but fragile cells, chromatophores have to be protected by a strong covering.

---

*Plates 1 to 5 follow page 546.

The laminated collagenous structure can also be assigned another indispensable role. Its architecture closely resembles that of the stroma (substantia propria) of the vertebrate cornea. The latter, of course, has to be transparent to light in addition to being mechanically rigid. If the transparency of the structure that overlies the chromatophores is low, the state of the chromatophores cannot be appreciated easily by an observer, because both entering and reflected light is scattered *en route*. The remarkable degree of architectural order in the lamellar gridwork must, therefore, be of great importance to the animal in the accomplishment of effective and visible chromatic responses (Fujii, 1993).

Melanophores

Melanophores are the most commonly found dendritic chromatophores in fish. With deposits of melanoprotein inside them, the chromatosomes within melanophores are called melanosomes. Melanophores are found almost everywhere in the skin where any sign of a dark shade is recognizable. Melanosomes are round or slightly ellipsoidal, with a diameter of about 0.5 μm. Melanosomes in which melanization has not been completed are usually referred to as premelanosomes. Being rather common inclusions in epidermal melanophores, these premature organelles usually assume more slender shapes than the mature ones (Fujii, 1993). In dermal melanophores, by contrast, premelanosomes are only rarely detected, perhaps because of the low rate of turnover of the organelles in the dermal cells. A single limiting membrane encloses the melanoprotein deposits.

As compared to other types of chromatophore, the melanophores usually exhibit higher motile activities: in response to nervous and hormonal stimuli, the rapid aggregation or dispersion of melanosomes takes place (Plate 1). Thus, the melanophores play a leading role in the rapid changes in color exhibited by many species of fish. A number of microtubules that radiate from the center of the cell into the dendritic processes are thought to be responsible for the translocation of melanosomes, as well as for maintaining the dendritic configuration of the cell as a cytoskeleton.

Under the electron microscope, Turner et al. (1975) observed the fine structure of the melanophores of black goldfish and they described how melanosomes develop as multivesicular bodies, with melanization proceeding around the small vesicles within them. On the basis of their observations of the melanophores of the medaka (*Oryzias latipes*), Nakajima and Obika (1986) reported that there are two distinct types of premelanosome in the melanophore: multivesicular premelanosomes, which develop into melanosomes with electron-lucent hollows in the central core; and another type with a highly organized, fibrous structure. Nakajima and Obika proposed that a mature melanosome is formed by the fusion of small, unmelanized vesicles or fibrillar premelanosomes with a preformed melanosome.

In paler mutants or varieties of normally dark species, such as the "orange-red" and "white" forms, of the medaka, melanophores without melanization can be detected. Such colorless melanophores are called "amelanotic melanophores". Although the organelles corresponding to the melanosomes are devoid of melanin deposits, they can aggregate or disperse in response to hormonal and nervous stimuli (Sugimoto et al., 1985).

Melanophores also contain nuclei and mitochondria, in addition to other organelles. Very frequently, two nuclei can be found in a single melanophore, but the biological significance of this phenomenon is still unclear.

Xanthophores and Erythrophores

Xanthophores and erythrophores are the dendritic chromatophores that contribute the yellowish and reddish components to the coloration of fish skin. They are usually a little smaller than the melanophores, but their overall configuration and morphological features are fundamentally the same as those of melanophores. Usually, they are present in the dermis. After examining the presence of epidermal xanthophores of an Antarctic blenny, *Trematomus*, Obika and Meyer-Rochow (1990) reported that these cells are motile, as are many dermal xanthophores, and that they respond, at least to theophylline, by dispersion of pigment.

The compounds that absorb light of colors that are complementary to cellular colorations are the carotenoids and pteridines. Pheomelanins, which are frequently responsible for the yellowish or reddish coloration of homeothermic vertebrates, have not been found in lower vertebrates, including fish. Organelles containing such pigments are normally called carotenoid vesicles (carotenoid droplets) or pterinosomes (Matsumoto, 1965), respectively. Many xanthophores and erythrophores contain both carotenoid vesicles and pterinosomes. Cases of the same organelle containing both pigments have not been reported. Prior to complete chemical identification of the included pigment, it may be safest to call yellowish and reddish organelles "xanthosomes" and "erythrosomes", respectively.

Usually, the dimensions of pterinosomes are similar to those of melanosomes. Being limited by a single membrane, they either contain a concentric lamellar structure or fine filaments (Matsumoto, 1965). The matrix of the carotenoid vesicles usually appears homogeneously electron-dense. With only a barely visible limiting membrane, they sometimes look like cytoplasmic oil droplets, as they do in the erythrophores of the squirrelfish *Holocentrus ascensionis* (Byers and Porter, 1977) and of *Trematomus* (Obika and Meyer-Rochow, 1990). The vesicles are, however, very homogeneous in terms of size, at least within a given species. This homogeneity suggests that there is no fusion or division of vesicles, even during vigorous movement within the cell. On the surface of the vesicle, in addition, there should be a site for some motor protein molecules that interact with their counterpart proteins. More detailed studies of the possible existence of a limiting membrane are needed.

Containing fewer cytoskeletal components, the cytoplasm of the bright-colored chromatophores usually appears clearer than that of the melanophores. Recent studies on *Trematomus* by Obika and Meyer-Rochow (1990) also indicate that microfilaments, intermediate filaments, and microtubules are rather rare inclusions. However, we can cite the erythrophores of *Holocentrus* (Byers and Porter, 1977) and of the swordtail, *Xiphophorus helleri* (Matsumoto, 1979), as exceptional cases, in which a number of microtubules can be found radiating from the center of cells. The erythrophores of *Holocentrus* are known for the extraordinarily high speed of their motile responses. In general, however, the rates of translocation of chromatosomes in erythrophores and xanthophores are rather low. There may be a positive correlation between the speed of the response and the density of microtubules. Plate 3 illustrates the pigment-aggregating response to melatonin of erythrophores on the belly skin of the neon tetra, *Paracheirodon innesi* .

LEUCOPHORES

In the dermis of certain species of fish, there are dendritic chromatophores known as leucophores. If we look through the vertebrate classes, however, we find chromatophores of this type only in fairly limited groups within the Osteichthyes. Sound descriptions are presently available for only a few species of Cyprinodontiformes, including the killifish *Fundulus heteroclitus*, the medaka, and the guppy, *Poecilia reticulata*. Further surveys will doubtlessly lead to their identification in related groups.

In contrast to the light-absorbing chromatophores described above, leucophores are light-reflecting chromatophores. Often discussed with another category of light-reflecting chromatophores, the iridophores, the leucophores have sometimes been described as iridophores. Leucophores should, however, be distinguished from iridophores because the mechanism of reflection of light by the former is entirely different from that by the latter (Fujii, 1993). The cellular organelles responsible for the cellular reflectivity are called leucosomes, and they scatter visible light in all directions. Enclosed by a single limiting membrane, each leucosome is a globular organelle, and is similar in size to other pigmentary inclusions in light-absorbing chromatophores. These organelles appear hollow in *Oryzias* leucophores (Obika, 1988), while those in *Fundulus* cells contain fibrous or electron-dense material (Menter et al., 1979).

Although the numbers of dendritic processes are usually smaller than those in melanophores and other light-absorbing chromatophores, the size of leucophores is similar to, and sometimes even larger than, that of other dendritic chromatophores. Their architectural features are fundamentally similar to those of other chromatophores. Like other dendritic chromatophores, the leucophores contain microtubules and several 10-nm filaments that run parallel to the axis of the dendrites (Obika, 1988).

It should be pointed out here that a leucophore is very frequently found to be located concentrically just beneath a melanophore in some species, including *Fundulus* and *Oryzias* (Plate 1). No ontogenetic explanation for this organization is available, nor is its physiological significance understood. To date, neither the localization nor the chemical nature of the substance(s) that are responsible for the light-reflecting properties of leucosomes has been determined. Further studies are needed to clarify these issues.

Like the chromatosomes in light-absorbing chromatophores, leucosomes move centripetally or centrifugally within the leucophores. In general, the direction of their movement is opposite to that seen in light-absorbing chromatophores (Plate 1, D–F). That is, the leucophore and the other dendritic chromatophores behave in quite the opposite manner in response to various stimuli. However, when we consider the contribution of the light-reflecting and absorbing chromatophores to the darkness of the skin, these reciprocal movements can be easily understood.

Iridophores

Iridophores are commonly found, wherever there are whitish areas of skin. On whitish and silvery belly surfaces, in particular, they are so densely distributed that there is practically no space for other types of chromatophore. They contain light-reflecting crystals that are responsible for the reflection of light from the skin. The large crystalline platelets tend to form parallel stack(s) in the cytoplasm (Denton and Nicol, 1966; Land, 1972). In addition, the spacing between adjacent platelets is strikingly uniform within a stack. Each platelet has been shown to be a monoclinic crystal composed mainly of guanine, but platelets often also contain purines other than guanine, namely, hypoxanthine and uric acid (Fujii, 1993).

The platelets are highly refractive, having refractive indices (*NB*) of no less than 1.83. The presence of such highly refractive structures in the cytoplasm renders the cells highly light-reflective. If we adopt the value of 1.37 for the refractive index of the cytoplasm (*NA*), the reflectivity at the surface of crystal can be calculated to be about 0.021. Light is reflected for the most part at the wide planar, parallel surfaces of the platelets. The reflection of light by iridophores can, thus, be explained primarily by the geometric optics.

When the platelets pile up, a much higher reflectivity can be attained as a result of the multiple thin-film interference phenomenon (Huxley, 1968; Land, 1972). A general diagram explaining the system is given in Figure 1A. In the so-called ideal system, the optical thickness (i.e., the refractive index multiplied by the actual thickness) of the platelets and that of the cytoplasmic sheets between the adjacent platelets are almost the same, being about one-quarter of the wavelength of the spectral peak of light reflected from the cell. The peak wavelength (*L-max*) is given as follows:

$$L\text{-}max = 4 \cdot NA \cdot DA = 4 \cdot NB \cdot DB \qquad (1)$$

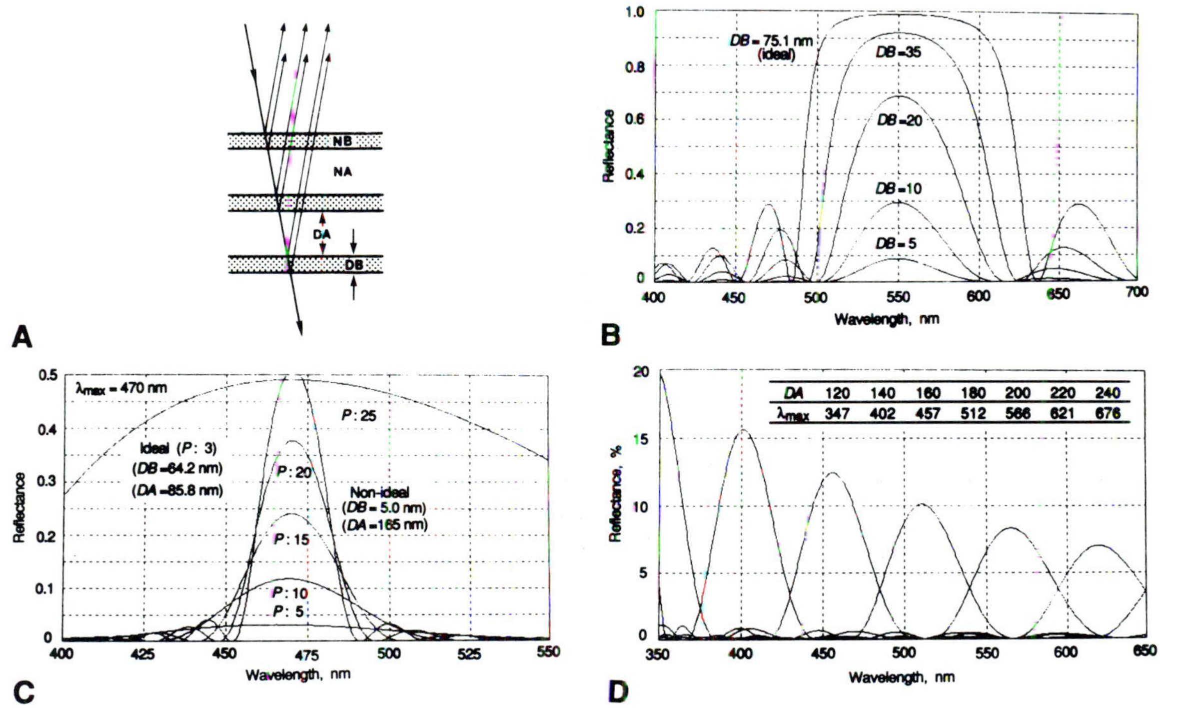

FIGURE 1. (A) Diagram explaining the interference of light rays in a thin-film reflection system. *NA* and *NB*, refractive indices of the optically clear and opaque strata in the pile, respectively; *DA* and *DB*, actual thicknesses of the optically clear and dense strata, respectively. (B) Spectra of the light reflected from a pile of reflecting platelets, computed and graphically displayed so that the principal peak is at 550 nm (yellowish green). The number of platelets was set at 10. When the system is ideal, *DB* is calculated to be 75.1 nm. (C) Spectra of the light reflected from a pile of reflecting platelets in a typical nonideal system, computed to have principal peaks at 470 nm (cobalt blue). Here, *DB* was set at 5 nm. The number of platelets in a pile was changed from 5 to 25 in increments of 5. With increases in the number, the peak heights increase, while the widths decrease. The broad peak for an ideal system (*DB* = 64.2 nm) with 3 platelets is shown for reference. (D) Shift of spectral peak reflected from a pile of platelets, when *DA* is changed from 120 to 240 nm, in increments of 20 nm, while *DB* is fixed at 5 nm. Simulated for iridophores of the blue damselfish type. Values of *DA* and of corresponding *L-max* are shown at the top. (From Fujii, R., *Int. Rev. Cytol.*, 143, 191, 1993. With permission.)

where *DB* and *DA* are the actual thicknesses of the platelet and the cytoplasmic sheet mentioned above. In this system, the spectral peak becomes very wide, and the reflectivity becomes very high, often resulting in the very high reflectance that resembles that of a mirror. Denton and Nicol (1966) explained the silvery side and belly skin of many teleosts, including oceanic migratory fishes, as being the result of the ideal multilayer thin-film interference phenomenon.

Light penetrating the cell is reflected at the wide surfaces of the flat reflecting platelets by specular reflection, and the interference between light waves, reflected at the parallel surfaces of several crystals in a stack, should be primarily responsible for the reflectivity of the skin. The random scattering of light mentioned previously may be only partly involved in the generation of silvery glitter. The reflectivity or the hue generated by interference of rays of light in such alternating layers is called the "interference color," and represents the "physical" or "structural" coloration.

The iridophores of many fish do not exhibit motile responses to stimulation of any kind. However, they play an integral part in the generation of high reflectivity, apparent as the whitish or metallic sides and belly skin that is indispensable to many fish (Denton and Nicol, 1966; Land 1972).

Iridophores in some teleostean species have been found to be motile, and studies of this phenomenon have made remarkable progress during the past decade. As can be seen from the diagrams of three representative kinds of motile iridophore, the essence of the motility is the change in the distance between adjacent platelets in a pile (Figure 2, A–C). Thus, the optical treatment of the multilayer interference system in the motile cells should be different from that of an ideal system (Equation 1). The spectral peak wavelength (*L-max*) can be given as follows:

$$L\text{-}max = 2(NA \cdot DA + NB \cdot DB) \tag{2}$$

Here again, *NA* and *NB* are assumed to be 1.37 and 1.83. Figure 1B shows the simulated spectral reflectance curves, obtained when we tentatively place the spectral peak at 550 nm and set the number of platelets in the pile of platelets at 10. The highest and widest peak in the figure is the spectral curve computed for the ideal case. With the decrease in *DB*, both the height and the width of the primary spectral peak become smaller. That is, the thinner the platelets are, the lower is the reflectance. The progressive narrowing of peaks indicates that the color becomes purer, i.e., of higher chroma, with decrease in *DB*. In such a way, the brilliant colors displayed by many tropical and coral-reef species are realized, sometimes rivaling fluorescent colors. Note that all the curves in Figure 1 (B, C, and D) are the results of computations based on the mathematical treatments by Huxley (1968) for multilayer interference phenomena of the nonideal type.

The higher purity of the color generated in the nonideal system can be seen more clearly in Figure 1C, in which sharp spectral reflectance curves for piles

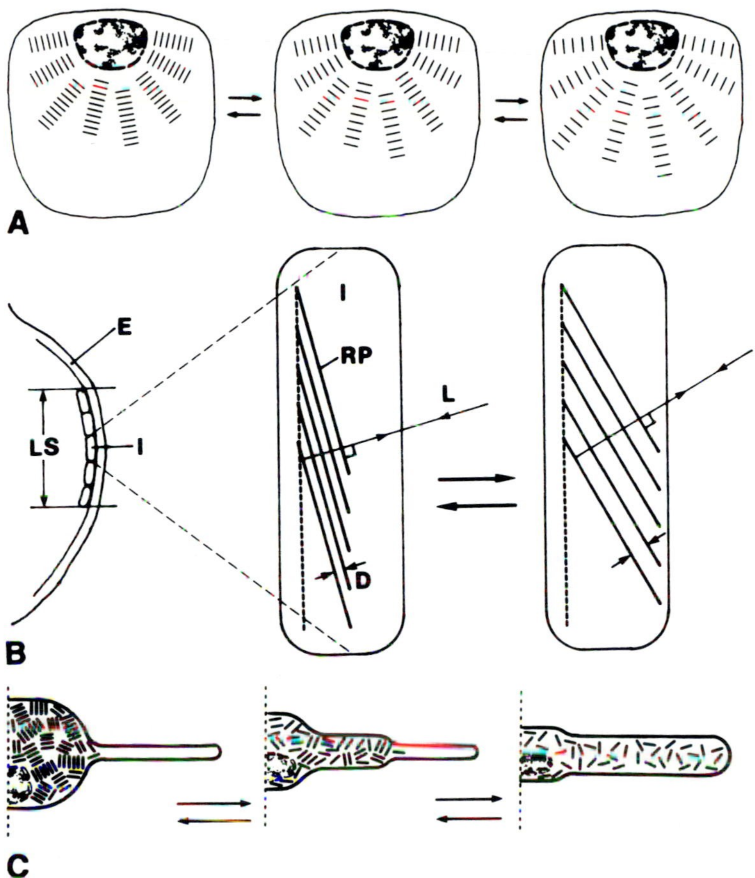

FIGURE 2. Diagrams showing the nature of motile responses of iridophores of three types, as viewed in a vertical section through the skin. (A) Blue damselfish type, (B) neon tetra type. The drawing on the left indicates the location of the iridophores that constitute the lateral blue-green stripe. D, Distance between platelets; E, epidermis; I, iridophore; L, incident and reflected light rays; RP, reflecting platelet; LS, zone of longitudinal stripe. (C) Goby type. In each series, a rightward progression makes the cell reflect light of longer wavelengths, while leftward progression makes cells reflect light of shorter wavelengths. (A and B from Fujii, R., *Int. Rev. Cytol.*, 143, 191, 1993. With permission.)

of varying numbers of 5-nm platelets are shown. Here, the position of the peaks is set at 470 nm, which corresponds to the color of schooling blue damselfish. The spectral peak becomes higher with increases in the number of platelets. It is also clear that no less than 30 platelets are required to attain a reflectance of 0.5, whereas, in the ideal system, the same effect is attainable with only 3 platelets.

Many species of fish possess iridophores in which a number of very thin platelets pile up in regular fashion to manifest the nonideal interference phenomenon (Fujii, 1993). These iridophores are often motile, playing a predominant role in color changes through their motile activities that vary the distance between platelets.

In the iridophores of the blue damselfish, *Chrysiptera cyanea*, or of related fish, a number of piles of very thin platelets are arranged in the cytoplasm so that their axes radiate from the nucleus, which is situated in the apical part of the cytoplasm (Figure 2A). In a given pile, the distance between platelets is very uniform. Stimulation of chromatic nerves induces a shift in the spectral reflectance peak towards longer wavelength. Adrenergic agonists have the same effects. Conversely, the cotransmitter substance (see below) induces a reverse shift in the peak. Apparently, the simultaneous changes in the distances between platelets in a pile result in the changes in light-reflecting characteristics of the cell.

Figure 1D illustrates a simulation, for the cells of the blue damselfish, of the spectral changes when *DA* is changed. Actual spectral changes measured with the skin of live fish and with excised pieces of skin fit the simulation very closely (Fujii, 1993). In addition, it should be emphasized here that the change in hue is continuous during the cycle of the response. The tubulin-dynein system has been shown to be responsible for the motility (Oshima and Fujii, 1987).

In the blue-green damselfish, *Chromis viridis*, iridophores with similar morphology to those of *Chrysiptera* can be categorized as either motile or nonmotile. The former type normally assumes a bluish hue, while the latter displays mainly yellowish tints (Plate 4). The motile and nonmotile iridophores are found assembled above the cell body and the surrounding dendrite zone, respectively, of the melanophore. The greenish hue of the skin is due to the combined effect of the colors from different minute areas. In this way, some fish generate their beautiful intermediate coloration by making good use of the divisionistic technique (see below).

Motile iridophores very similar to those of damselfishes may exist in fish that belong to other groups. Very recently, the common surgeonfish, *Paracanthurus hepatus,* was found to possess such iridophores (Fujii, 1993). Further surveys may lead to identification of other examples, but we do not expect very many.

Another interesting case in which changes in *DA* result in a change in hue was found in popular tropical fish that belong to Characidae, Characiformes, which are represented by neon tetras (*Paracheirodon innesi*) and related fish (Lythgoe and Shand, 1982). During the daytime, the longitudinal stripe of these species is blue-green, but it becomes dark violet at night. Each iridophore contains rows of parallel platelets. As shown schematically in Figure 2B, simultaneous changes in the inclination of platelets in a cell result in an alteration in the distance between them, with a resultant change in spectral reflectance (Nagaishi et al., 1990). Light or an adrenergic stimulus induces the rightward response in the figure, whereas darkness or the cessation of sympathetic pulses brings about the reverse response.

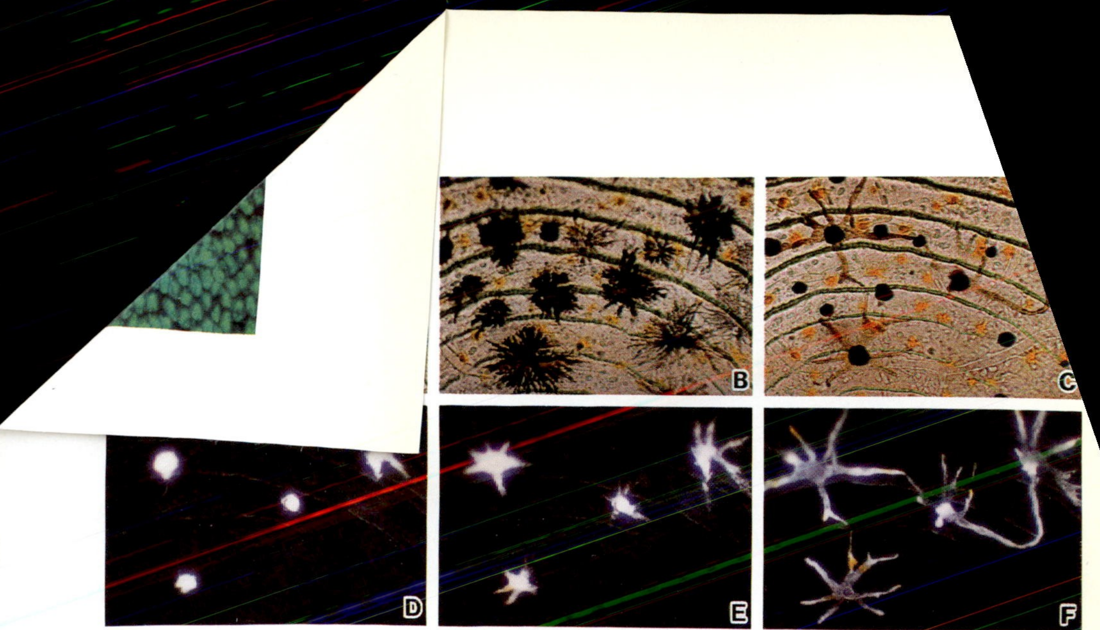

Plate 1. Photomicrographs showing motile responses of melanophores and leucophores on a scale of the medaka, *Oryzias latipes,* viewed from dermal side. Plates A–C photographed with normal transmission optics; D–F photographed with epi-illumination. A and D equilibrated in physiological saline. Melanosomes are totally dispersed; whereas leucosomes are completely aggregated. Note that the perikarya of leucophores lie concentrically under those of melanophores. B and E, 1 and 2 min after the application of K^+-rich saline (K^+: 50 m*M*), respectively. Melanosomes and leucosomes are in the process of aggregating and dispersing, respectively. C and F, 4 min after the application of K^+-rich saline; melanosomes are totally aggregated, while leucosomes are dispersed. Magnification × 150.

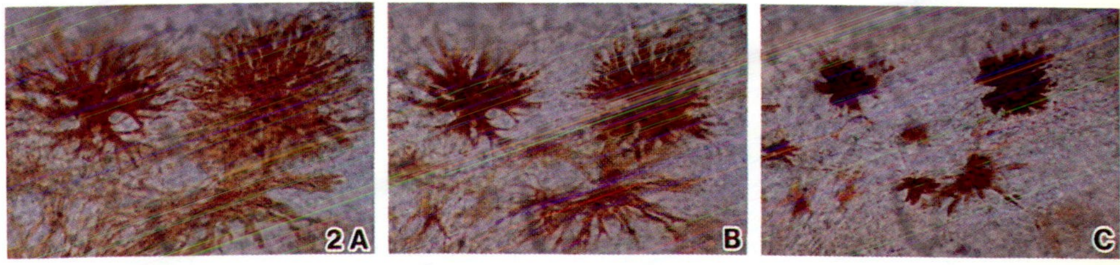

Plate 2. Responses of erythrophores on the lateroventral skin of the posterior trunk of the neon tetra, *Paracheirodon innesi* (transmission optics). Plate A equilibrated in physiological saline; erythrosomes are totally dispersed. B and C, 6 and 15 min after the application of 10 n*M* melatonin; erythrosomes are gradually aggregating. Magnification × 150.

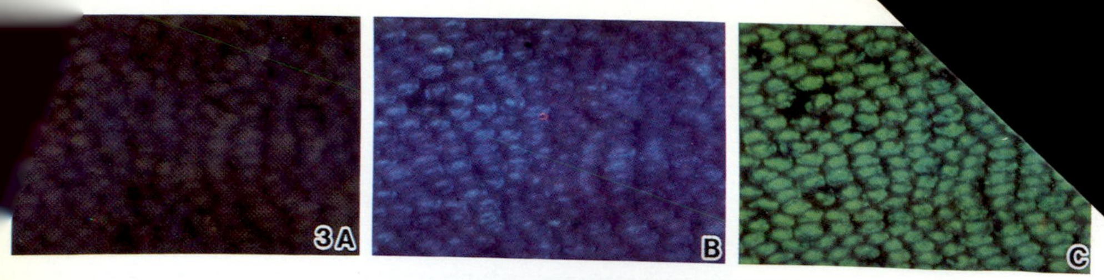

Plate 3. Responses of motile iridophores on a piece of split dorsal fin of the blue damselfish, *Chrysiptera cyanea* (epi-illumination optics). A, in physiological saline; B, after 3-min stimulation of controlling nerves by 0.5-Hz pulses of 10 V and 1.0-msec duration; C, 3 min after the application of 5 μ*M* norepinephrine. Magnification × 250. (From Kasukawa, H. et al., *Zool. Sci.*, 4, 243, 1987. With permission of the Zoological Society of Japan.)

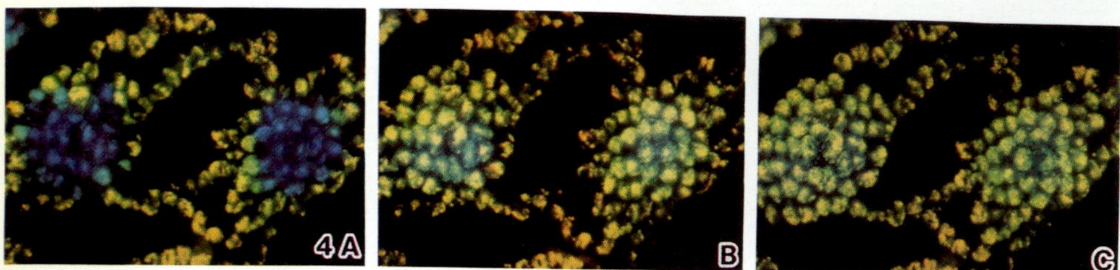

Plate 4. Responses of motile iridophores on a piece of split dorsal fin from a blue-green damselfish, *Chromis viridis* (epi-illumination). A, 5 min after the application of 5 μ*M* isoproterenol; B, equilibrated in physiological saline; C, 5 min after the application of 5 μ*M* norepinephrine. Blue cells, motile iridophores; yellow cells, nonmotile iridophores. Magnification × 180. (From Fujii, R. et al., *Zool. Sci.*, 6, 477, 1989. With permission of the Zoological Society of Japan.)

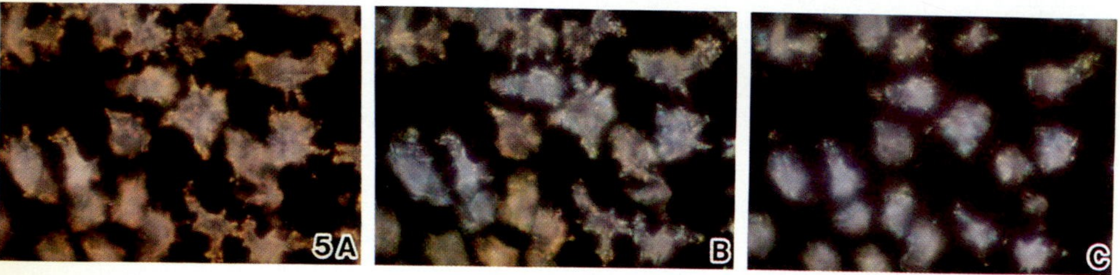

Plate 5. Responses of iridophores on an isolated scale of the dark sleeper goby, *Odontobutis obscura obscura* (epi-illumination). Plate A equilibrated in 1 μ*M* norepinephrine for 120 min; reflected platelets are dispersed throughout the cytoplasm. B and C, 30 and 120 min after treatment with 1 μ*M* alpha-MSH; platelets are gradually aggregating in the perikarya. Magnification × 240. (From Fujii, R. et al., *Zool. Sci.*, 8, 461, 1991. With permission of the Zoological Society of Japan.)

While studying some freshwater gobiid fishes, including the dark sleeper, *Odontobutis obscura obscura*, Iga and Matsuno (1986) found motile iridophores of the dendritic type, in which small light-reflecting platelets of about 60 nm in thickness aggregate and disperse. Fujii et al. (1991) reported that the cells become bluish in color with the progression of aggregation of the platelets. The same cells looked yellowish when the platelets were widely dispersed in the cytoplasm (Plate 5). The bluish tone is due to the formation of piles of platelets during the aggregation of the platelets (Figure 2C). Such delicate changes in hue probably have some ethological significance, for example, in intraspecific communication.

COMBINED EFFECTS OF PIGMENTARY COLORS

Many different colored varieties of the medaka, *Oryzias latipes*, the results of spontaneous mutations, are known and are being maintained for experimental use (Yamamoto, 1975). The wild, "brown-black" type, for instance, has dermal melanophores, xanthophores and leucophores that are distributed rather homogeneously over the dorsolateral skin. All these cells are dendritic, and the dispersion or aggregation of chromatosomes naturally results in the augmentation or reduction of the color of the pigments that they contain. Each kind of chromatophore is controlled somewhat differently by the actions of the complicated and elaborate endocrine and nervous systems (see section below on Regulation of Motility). Thus, the fish are able to adapt to background hues over a wide range and to cope with various ethological encounters by delivering chromatic signals.

When chromatophores of different types overlap one another, the subtractive mixture of component pigmentary colors may explain the macroscopically perceived colors. This phenomenon resembles the mixing of paints. When differently colored cells are placed side by side, the colors are produced by the additive mixture of colors, as in offset printing. The body colors of many fishes are the results of the integration of these effects.

In the erythrophores of xiphophorine poecilds, such as the swordtail (*Xiphophorus helleri*) and platyfish (*X. maculatus*), two kinds of pigmentary organelle, namely, brownish-red pterinosomes and yellow carotenoid vesicles, coexist (Matsumoto, 1965). Representing a rare example of bichromatic chromatophores in vertebrates, these erythrophores function to generate an orange-red hue.

Brilliant yellow-to-red colors are often attained by lining the backs of xanthophores or erythrophores with a layer of iridophores. Many such examples can be enumerated, especially among red varieties of fancy goldfish, colored carp, and tropical coral-reef fishes. As another example, the vivid orange-red portions, separated by transverse white bands, of the clown anemone fish (e.g., many species of *Amphiprion*) are produced by such a mechanism. The white bands are apparently due to dense populations of iridophores.

Even over the course of the long history of their evolution, vertebrates, including fish, have not generated any molecular species of pigment for production of skin colors with shorter wavelengths than yellow. Thus, fishes must produce violet through greenish hues by combinatorial effects when such hues are required.

In many amphibian and reptilian species, xanthophores, iridophores, and melanophores often form orderly strata in the dermis, from the surface to the bottom, for the generation of greenish hues. This architecture is called the "dermal chromatophore unit" (Bagnara and Hadley, 1973). In fish, such units are not common, but a simpler layered structure is sometimes found that generates colors that cannot be produced by the available pigments alone.

A wonderful system, in which motile iridophores play a leading role, is the bilayered construction found in the blue damselfish, *Chrysiptera cyanea*, and related pomacentrids. Round motile iridophores are arranged in a monolayer just below the epidermis. The characteristic blue coloration due to the light-reflecting characteristics of the iridophores is augmented by a layer of melanophores that lines the former cells (Kasukawa et al., 1987; Fujii et al., 1989). Absorbing light that has passed through the layer of iridophores, the melanophores function to increase the purity of the color, i.e., to give a higher chroma in the Munsell color system. The beautiful fluorescent-like colors from violet to bluish-green displayed by damselfish are due to the activities of the iridophores.

When the color of the skin is primarily due to the light-absorbing, bright-colored chromatophores, namely, the xanthophores or the erythrophores, a lining composed of a whitish layer increases the brightness (chroma) of the hue of its pigment. When the iridophores are predominantly responsible for the hue, through the nonideal interference phenomenon, a lining of light-absorbing melanin makes the reflectance spectrum sharper, increasing the chroma. Many examples of both aposematic and recognition coloration may be generated by such mechanisms.

Intermediate tints are frequently realized by the horizontal, mosaic distribution of chromatophores of different colors. In the dermis of the scalycheek damselfish, *Pomacentrus lepidogenys*, for example, there are two types of iridophore: dendritic and round (Kasukawa and Oshima, 1987). Several iridophores of each type assemble to form islets with a diameter of no more than 100 μm. The dendritic iridophores are mostly overlaid by xanthophores. The combination generates a bright yellowish tone. The round iridophores reflect light with a relatively narrow spectral peak from blue to green. A mosaic of two types of differently colored patch brings about the macroscopically perceived yellowish-green hue, characteristic of this species of fish. The mechanism of coloration is fundamentally analogous to the divisionistic techniques employed by French neo-impressionists to produce brighter colors.

In the blue-green damselfish, *Chromis viridis*, a more complex system of chromatic effectors is operative (Fujii et al., 1989). A chromatophore unit, consisting of a melanophore, several small iridophores, and, on occasion, a few small xanthophores, in that order from the bottom to the top, is present in the dermis. Being on the perikaryon of the melanophore, the motile iridophores reflect light from blue to yellow as a result of the interference phenomenon, while the nonmotile iridophores over the dendrite region of the melanophore display yellowish tints. When the motile iridophores reflect blue light, the combined effect of the blue and the yellow of the nonmotile cells leads to the production of the characteristic green color.

Visual acuity of teleosts is relatively low, especially when the eyes are small (Chapter 6). Thus, the conspecifics or individuals of other species may perceive the mixed color in the same way that we do. In any case, fish clearly succeeded in inventing the divisionistic technique much earlier than the French pointillists.

PATTERNS OF COLORATION

Many fish exhibit beautiful patterns of coloration in addition to brilliant tints. Among classes of fish, the Osteichthyes are the most gorgeous group. Their vertical, horizontal, or sometimes oblique stripe(s) are remarkable, but they represent only simple examples of their coloration. Thus, innumerable linguistic expressions are needed to describe such patterns, such as spotted, particolored, gradation of various colors, to name only a few. The many illustrated books about tropical fishes allow us to admire and wonder at the variety of these colors and patterns. The ethological implications of the patterns and their changes is discussed briefly later in this chapter.

In vertebrates, chromatoblasts, arising from the neural crest, migrate, proliferate and differentiate into chromatophores. As discussed below (section on Ethological Implications), ventral skins are generally pale. Working with amphibian materials, Fukuzawa and his associates (e.g., Fukuzawa and Bagnara, 1989) found that melanoblasts are present in ventral skin, but their differentiation into melanophores is blocked by a factor that they named "melanization inhibiting factor (MIF)". MIF could only be detected in the ventral skin. Presumably, such a factor is involved in the dorsolateral gradient of pigmentation in fish too.

Where a certain color dominates within an area of the skin, a colony of chromatophores of that color is usually seen, although in some cases the combined effect of different kinds of chromatophore is observed. Red speckles on fancy colored carp and black stripes on *Pterophyllum* angelfish are good examples of aggregates of erythrophores and melanophores, respectively. Unfortunately, however, practically no information is presently available about the mechanism responsible for the segregation of specified kinds of chromatophore in the skin. Further intensive studies on the dynamic behavior of chromatoblasts, as well as on the mechanisms responsible for revealing and suppressing phenotypic characteristics of chromatophores, are needed if we are to elucidate the mechanisms of such pattern formation.

It should be noted that the pigmentary patterns are not always stationary: in many cases, the patterns undergo changes. Ontogenetic changes are naturally brought about by morphological color changes. Very slow modifications for adaptation to environmental conditions are also commonly executed by morphological means.

Faster changes in patterns, which are also frequently observed, are considered to be physiological color changes. For instance, circadian changes in pattern of coloration of the pencil fish (e.g., *Nannostomus beckfordi*) are dependent on the pineal gland, which secretes melatonin during the night (Reed, 1968; see below). More rapid changes needed for adapting to patterns of backgrounds or for intraspecific communication are, by contrast, realized mainly through the activities of the nervous systems. For example, flatfish are well known for their rapid adaptability to background patterns (Parker, 1948). Recent studies have indicated that the changes are due to differental neural commands to chromatophores or to groups of chromatophores, and that there is a limited number of preset patterns. On the basis of the coarseness of the texture of the background, the central nervous system "selects" an appropriate pattern (Fujii and Oshima, 1986). Studies on the chromatic adaptation of the common freshwater goby, *Rhinogobius brunneus*, to black and white checkerboard backgrounds also indicated that numerous nerve fibers control integumentary chromatophores differentially and very systematically (Naitoh et al., 1985).

Tilapias have been widely studied with respect to the communicatory functions of various pigmentary patterns, mainly among conspecifics (Baerends and Baerends van Roon, 1950). Working with a mouth-brooder tilapia, *Haplochromis burtoni*, Muske and Fernald (1987) showed that nervous cues control the very rapid appearance or disappearance of the facial stripe, the "eyebar", which signals territorial ownership and aggressive intent in territorial males.

COLOR CHANGES

Physiological and Morphological Color Changes

Changes in hue and pattern of coloration result either from the motile activities of chromatophores or from changes in the amount of pigmentary materials deposited in the skin.

The color changes due to motile activities of chromatophores are called "physiological color changes." Being primarily controlled by the nervous and/or endocrine system (see below), such changes are rather rapid, and are involved in the faster chromatic adaptation of fish. Accompanied by changes in the net amount of pigmentary materials in the skin, the "morphological color changes" usually proceed more slowly. When present, epidermal melanophores play a very large role in depositing melanin in epidermal cells. No other types of chromatophore are known to produce pigment that is transferred into epidermal cells.

Increases or decreases in the numbers of dermal chromatophores and sometimes in numbers of dermal melanophores should also influence integumentary coloration. Such phenomena should be due to the proliferation of chromatoblasts or to the degradation of terminally differentiated pigment cells, and may be accompanied by a net increase or decrease in the amount of pigment (Parker, 1948). Working with *Oryzias*, Sugimoto recently counted the number of chromatophores after a prolonged period of adaptation to a dark or a white background. He found that dark adaptation gave rise to apparent increases and decreases in the numbers of melanophores and the leucophores, respectively, and that white adaptation caused the reverse changes (Sugimoto, 1993). Concurrent changes in the amount of pigments were also confirmed. No analogous quantitative studies have yet been performed on epidermal melanophores.

In many species, such as the blue angelfish (*Pomacanthus semicirculatus*), the pattern of pigmentation changes remarkably during growth. The so-called "nuptial" coloration can be observed during the breeding season in many fish. Such ontogenetic changes are the result of morphological changes.

Pigmentation due to the deposition of carotenoids in erythrophores or xanthophores is certainly influenced by dietary conditions because animals cannot synthesize carotenoids. For example, when red sea bream (*Pagrus major*) are raised commercially, the fish are given carotenoid-rich feed to enhance their color and raise their commercial value.

Motility of Chromatophores

Aggregation of pigment in the perikaryon and dispersion throughout the cytoplasm are the basis of the cellular motility of an ordinary dendritic chromatophore (Plates 1, 2, and 5). To date, a number of investigations have been performed on the mechanism of the motility; see reviews by Obika (1986) and Fujii (1993).

Among the hypotheses about the mechanism of chromatophore motility, one hypothesis has been widely accepted, namely, that the microtubules that radiate from the cell center towards the periphery of the dendritic processes, play an active role. A sliding force between the surface of chromatosomes and the microtubules would cause the movement of chromatosomes in a centripetal or centrifugal direction. Presumably, a set of motor proteins, for example, cytoplasmic dynein and kinesin, are involved in the centripetal (retrograde) and centrifugal (orthograde) translocation of pigment, respectively. Studies on the mechanisms of iridophore motility have only recently been initiated, but the results obtained to date indicate that the motility is dependent primarily on microtubules.

For a deeper understanding of the mechanisms of chromatophore motility, precise descriptions of the states of the chromatophores, as well as the ability to record changes in these states are indispensable. Classical but still useful methods, as well as more modern ones, have been reviewed previously by the author (Fujii, 1969, 1993). For detailed descriptions of photometric methods, the reader can refer to the many original papers, such as that by Oshima and Fujii (1984).

Regulation of Motility

Motile activities of chromatophores are regulated by the endocrine and/or the nervous systems. In vertebrates, the endocrine system is almost universally operative. Among the classes of vertebrates, however, the Osteichthyes constitute a group in which the nervous system has evolved mechanisms for the regulation of rapid and quickly reversible changes in coloration. A review of this field has been published (Fujii and Oshima, 1986), so only a brief discussion is presented here.

Parker (1948) suggested that the chromatophores of elasmobranchs are regulated by both the endocrine and the nervous systems. Until now, however, no reliable data are available about the possible role of the nervous system in lower classes of fish, including Chondrichthyes. Presumably, hormonal regulation is exclusively responsible for the motility of their chromatophores. Among vertebrates higher than fishes, incidentally, only a few lizards, including chameleons, are known to have neurally controlled chromatophores.

Various factors are known to influence the chromatic state of animals. As might naturally be supposed, the colors of fish are strongly influenced by the photic environment. The information perceived by the lateral eyes and/or the pineal is processed in the central nervous system which transmits efferent signals to chromatophores via hormonal and/or nervous pathways (Figure 3).

The major hormones that regulate chromatophore movements in fish are melanophore-stimulating hormone (MSH) from the intermediate lobe of the pituitary, melanin-concentrating hormone (MCH) from the posterior lobe, melatonin from the pineal gland, and epinephrine, which may possibly originate in the chromaffin cells correspond to the adrenal medullary cells of higher vertebrates. MSH is known to darken skin, whereas MCH and melatonin have the reverse effect. Being secreted during the night-time hours, melatonin is mainly concerned with circadian changes in coloration.

Upon excitement of an animal, a bolus of epinephrine is secreted from the chromaffin cells. If the adrenoceptors of the chromatophores are mainly of the alpha type, instantaneous blanching takes place. If the adrenoceptors are of the beta type, a rapid darkening is seen. These phenomena are called "excitement pallor" and "excitement darkening", respectively, although their ethological implications have not yet been adequately explained.

Apparently, the nervous system of fish has evolved to allow faster chromatic adaptation than can be achieved via the hormonal system. The sympathetic division of the autonomic nervous system has been shown to be involved in aggregation of pigment in dendritic chromatophores (Figure 3). Norepineph-

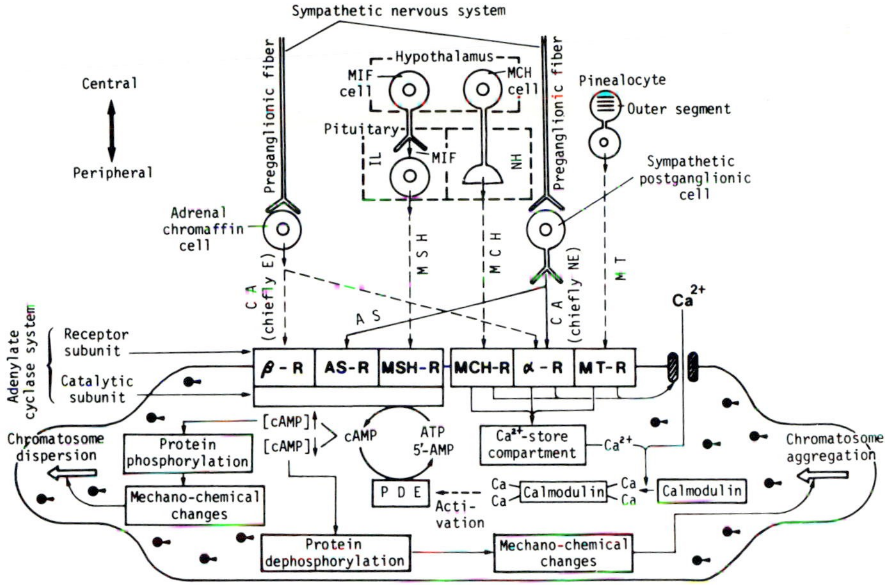

FIGURE 3. Diagram showing the system for controlling motile activities of light-absorbing chromatophores in teleosts. AS, Adenosine; AS-R, adenosine receptor; CA, catecholamines; E, epinephrine; IL, intermediate lobe of pituitary; MCH, melanin-concentrating hormone; MCH-R, MCH receptor; MIF, MSH-release inhibiting factor; MSH, melanophore-stimulating hormone; MSH-R, MSH receptor; MT, melatonin; MT-R, melatonin receptor; NE, norepinephrine; NH, neurohypophysis; PDE, phosphodiesterase; α-R, alpha-adrenoceptor; β-R, beta-adrenoceptor. (From Fujii, R. and Oshima, N., *Zool. Sci.*, 3, 13, 1986. With permission of the Zoological Society of Japan.)

rine is responsible for the peripheral neurotransmission, being mediated by alpha-adrenoceptors. Contrary to the earlier suggestion by Parker (1948), the parasympathetic division of the autonomic nervous system has not been shown to take part in the regulation of chromatophores.

Recently, it was discovered that, together with norepinephrine, ATP is liberated from chromatophore nerve terminals (Fujii and Oshima, 1986; Fujii, 1993). ATP is successively dephosphorylated by exoenzymes to adenosine (Figure 4). By interacting with adenosine receptors on the pigment cell membrane, adenosine acts to disperse pigment. Although liberated concurrently, the true transmitter, norepinephrine, acts first. The amine is quickly removed by reuptake into the nerve terminals, or it is inactivated by catechol *O*-methyltransferase (COMT) or by monoamine oxidase (MAO). After the cessation of sympathetic pulses, adenosine survives for a longer time and disperses pigment, reversing the effect of the true transmitter, norepinephrine. In this way, adenosine, as a functioning cotransmitter, acts to disperse chromatosomes rapidly, thereby enabling fish to change their coloration very quickly. Thus, the singly innervated chromatophores of teleosts have acquired

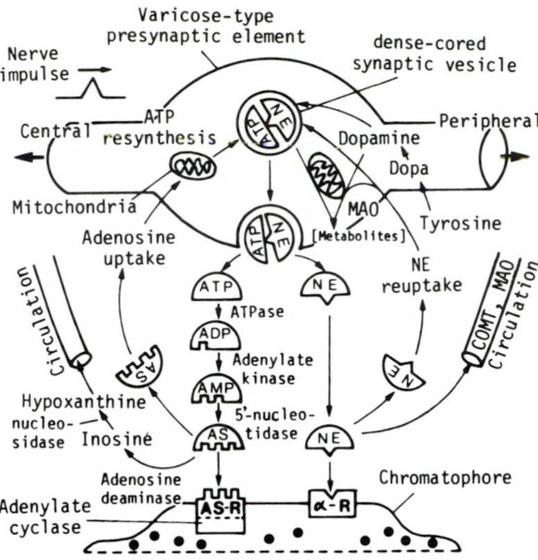

FIGURE 4. Dynamics of a catecholaminergic true transmitter and a purinergic cotransmitter liberated from a sympathetic postganglionic fiber that controls chromatophores in a teleost. AS, Adenosine; AS-R, adenosine receptor; COMT, catechol *O*-methyltransferase; MAO, monoamine oxidase; NE, norepinephrine; α-R: alpha-adrenoceptor. (From Fujii, R. and Oshima, N., *Zool. Sci.,* 3, 13, 1986. With permission of the Zoological Society of Japan.)

the ability to react very quickly, resembling in this regard reciprocally innervated effectors.

Receptors for signaling ligands, namely, hormones and neurotransmitters, are found on the cell membrane of chromatophores. Among the principal receptors, the alpha-adrenoceptors and receptors for MCH and melatonin mediate the aggregation of chromatosomes, while beta-adrenoceptors and receptors for MSH and adenosine mediate their dispersion (Figure 3). The extracellular signals are translated by the receptor system into an increase or a decrease in the levels of intracellular second messengers. Until recently, three kinds of second messenger were known to take part in the motile responses of chromatophores, namely, cyclic adenosine-3′,5′-monophosphate (cAMP), Ca^{2+} ions, and inositol 1,4,5-trisphosphate (IP_3) (Fujii, 1993). Adrenoceptors of the $alpha_1$ type cause the breakdown of phosphatidyl inositol in the membrane to IP_3, which in turn acts to release Ca^{2+} ions from the sites of intracellular storage. Stimulation of beta-adrenoceptors, or of receptors for MSH or adenosine activates the adenylate cyclase system to increase intracellular levels of cAMP. Through the stimulation of their specific receptors, melatonin and MCH are thought to inhibit the cyclase system, with resultant aggregation of chromatosomes via decreases in levels of cAMP. Other hormonal substances, autacoids, or local hormones may have some modifying influences on the actions of major ligands (Fujii and Oshima, 1986). The changes in the levels of second messengers are thought to induce the phosphorylation of certain proteins that are more directly coupled with the motile activities of chromatophores.

It is astonishing how large a number of ligands are involved in the control of chromatophores in fish. An explanation of why fish possess such a complicated system for regulating chromatophores may be found in the importance of chromatic behaviors in their ethological encounters.

ETHOLOGICAL IMPLICATIONS

CRYPTIC COLORATION

During the course of their evolution over more than 400 million years, fish have developed a large variety of systems for chromatic adaptation, as illustrated in picture books of fish. The ethological significance of coloration and colored patterns has been discussed in earlier publications, to which the reader can refer (Cott, 1940; Needham, 1974). In this section, an attempt is made to provide a brief summary of current concepts related to these phenomena.

The "cryptic (concealing) colorations" are the body colors and patterns that enable the animal to blend into the background. In this way, a predator can approach its prey without being noticed. The prey, too, can easily escape from predators, with the latter phenomenon being commonly known as "protective coloration". These phenomena usually involve resemblance, not only of body hue, but also of color pattern to the background. In this way, for example, gobies, flatfish, and many others on the substratum adapt so well to their environment that, even with utmost care, we can barely locate them.

In tropical waters, for example, around coral reefs, many fish have showy, distinct bands and spots which exemplify the "disruptive coloration" that is considered to be suitable for adaptation to a heterogenously contrasted environment. The normally conspicuous, lateral eye is very frequently covered by either a longitudinal or a transverse dark stripe, the "eye-bar". The crucial organ is thus protected from attack.

The countershading achieved by the contrast of darker dorsal and whitish ventral skin is also an effective chromatic adaptation and examples both in freshwater and in marine fishes are innumerable. A modified phenomenon seen among oceanic migratory fish is worth mentioning. In the dorsal to lateral skin, the presence of melanophores backed by a layer of iridophores generates the dark bluish tone of the integument. When seen from the top, the body color merges into the marine blue. In the iridophores that are present all over the silvery skin, very large light-reflecting platelets lie practically parallel to the sagittal plane of the swimming individual, thus rendering the fish equivalent to a mirror (Denton and Nicol, 1966). When observed from above, the platelets reflect only faint blue light from the deep, and the same platelets reflect the light from the sky when observed from below. In both cases, the body merges well into the background (Denton and Nicol, 1966; Land, 1972).

When an animal exhibits a color, color pattern, form, or posture, such that it resembles another organism, either an animal or a plant, or sometimes a lifeless object, the phenomenon is commonly called "mimicry". When the animal becomes inconspicuous as a result, the phenomenon is called "concealing mimicry", or more appropriately "mimesis". Among the parameters involved in mimesis, color is considered to be of special importance, and this aspect of the phenomenon has been designated "chromatic mimesis". A few examples of "chromatic mimesis" are as follows: leaf fish (e.g., *Monocirrhus polyacanthus*) mimic withered leaves; the leafy sea dragon, (e.g., *Phycodurus eques*) mimics algae, and anglers (e.g., *Lophius litulon*) mimic rocks.

Aposematic Coloration

The purpose of "aposematic coloration" is also to increase the fitness of the fish but, in this case, the increase in fitness results from the possessor advertising its presence. Thus, the term "advertisement coloration" is also frequently used. Among the phenomena included into this category, "warning coloration" serves to inform predators that the possessor of such coloration is dangerous. Warning coloration becomes useful after the aggressor has learned from experience that the possessor tastes bad or dangerous in other ways. By contrast, "threatening coloration", another type of aposematic coloration, may have a defensive function that is independent of a predator's prior experience. That is, it is based on an innate negative response to certain signals. The large "eyespot markings" frequently seen around the posterior parts of many butterfly fish are representative of "threatening coloration". However, they may more appropriately be considered as deflecting devices to which the attention of the predators is directed. By being attacked in less critical regions of its body, the possessor could achieve a higher rate of survival.

When mimicry, which we also discussed above, is not concealing, but is used to advertise a fish as one that resembles a fish that predators have learned to avoid, the phenomenon is called "mimicry" in a broad sense. Such warning coloration may have been developed as a special type of strategy. By disguising itself as dangerous, a fish that is not actually dangerous deceives its predators. This is the so-called "Batesian mimicry" (Wickler, 1968). One interesting example of such mimicry has been described, involving the blue streak cleaner wrasse, *Labroides dimidiatus*, which cleans larger fish (Wickler, 1968). In addition to its characteristic dancing behavior during its approach to the fish to be cleaned, its conspicuous longitudinal stripe must be an important signal. The aggressive striated saber-toothed blenny, *Aspidontus taeniatus*, mimics both the color patterns and the swimming behavior of the cleaner wrasse and snaps at larger fish. A more complex example involves the territorial male of the mouth-brooder cichlid, *Haplochromis burtoni*. It has an anal fin, on which a few round, yellow patches resembling the eggs are seen. At the moment when a female, who already holds her released eggs, tries to suck the yellow patches into her mouth, the male ejaculates, and the eggs are fertilized.

Recognition Coloration

Although defined as a kind of aposematic coloration, "recognition coloration" plays rather different roles from those described above. Except in a few special cases, such as the yellow patches on the fin of a male *Haplochromis* mentioned above, the colorations described so far have been essentially part of the fishes' strategies to avoid predators. Recognition coloration is exclusively for mutual identification among individuals of the same species. That is to say, the coloration functions as a releaser or key stimulus for intraspecific social behavior. The colors and the patterns in this category function to signal aggressive intent, to attract mates, or to communicate with young. Such functions are more common among highly visual, social teleosts.

The well-known "nuptial coloration" is an example of a display for conspecifics, and it normally appears in the male forms during the breeding season. It functions, of course, to attract females. Tinbergen (1965) showed that the conspicuous red color of the abdomen of the stickleback, *Gasterosteus aculeatus*, is an effective releaser for fighting, in addition to its role in attracting females.

Being induced ontogenetically or by morphological color changes, many instances of coloration and color patterns may safely be said to be static. As mentioned before, however, the patterns sometimes vary remarkably according to environment, ethological conditions, or to mood. Studies have been carried out with several African cichlids, examining their territorial behavior and conspecific communication (e.g., Baerends and Baerends, 1950). Here, in Figure 5, the color patterns expressed by a freshwater serranid, *Siniperca kawamebari*, during agonistic behaviors are included as an example (Kohda and Watanabe, 1982).

Under certain ethological conditions or in the dark, the peak of the spectral reflectance from the skin of the blue damselfish, *Chrysipetra cyanea*, shifts to the near ultraviolet (UV) region (Kasukawa et al., 1987). The spectral reflectance peak from the blue-green stripe of the neon tetra, *Paracheirodon innesi*, also changes to reflect UV light (Nagaishi et al., 1990). Thus, some fish, including those mentioned above, may employ the reflection of UV light to create certain visual signals or releasers, presumably for the purpose of conspecific communication.

PERSPECTIVES

Our understanding of the chemistry of colored substances in the skin and the physiological characteristics of the chromatophores that contain these substances is increasing rapidly. However, mechanisms underlying the manifestation of color as well as the formation of patterns in the skin have not been fully clarified. Mixed effects of multiple pigments or of multiple types of chromatophore have to be investigated in more detail. Many issues remain to be investigated even by conventional morphological and histological methods. Application of currently developing histochemical or immunocytochemical

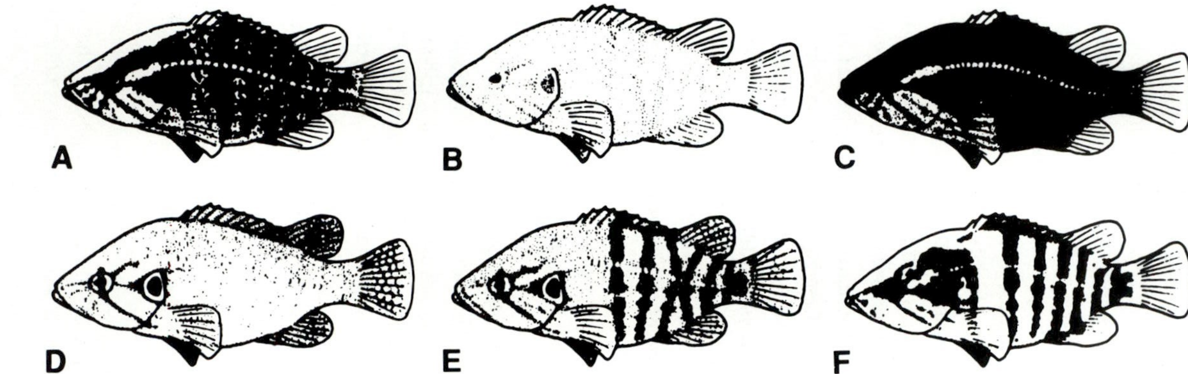

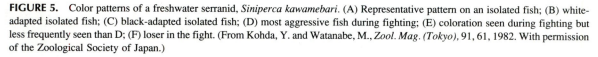

FIGURE 5. Color patterns of a freshwater serranid, *Siniperca kawamebari*. (A) Representative pattern on an isolated fish; (B) white-adapted isolated fish; (C) black-adapted isolated fish; (D) most aggressive fish during fighting; (E) coloration seen during fighting but less frequently seen than D; (F) loser in the fight. (From Kohda, Y. and Watanabe, M., *Zool. Mag. (Tokyo)*, 91, 61, 1982. With permission of the Zoological Society of Japan.)

techniques will surely also provide many important results, useful for enhancing our understanding of the significance of coloration, color patterns, and their changes during ontogeny and social encounters.

Both professional and amateur aquarists have, for many years, been making immense efforts to generate more beautiful, fancy strains of some domestic species by mating alone. Naturally, the species have to be limited to proliferative ones, such as the goldfish, *Carassius auratus*, the guppy, *Poecilia reticulata*, and the discus, *Symphisodon* spp.. Unfortunately, past attempts to obtain favorable mutants by applying mutagenic agents, such as ionizing radiation and ultraviolet light, to germ cells, have been unsuccessful. However, techniques of genetic engineering have now developed to such an extent that they may be applicable to alteration of phenotypic characteristics, for example, via introduction of foreign genes. By means of microinjection or electroporation techniques, the genes for expression of desirable colors or patterns from other species, not only within the same genus, but even from species belonging to other genera, families, orders, classes, or even to other phyla of animals, can now be transferred. Of course, such techniques are not so easily applied. Recently, however, the medaka, (Ozato et al., 1989), and the zebra danio, *Brachydanio rerio* (Powers, 1989) have been shown to be very convenient material for such experiments: these small fish are easy to breed, and their life cycle is short enough for speedy confirmation of phenotypic expression, as well as of transformed genotype. Another advantage is that the oocyte or egg is sufficiently large to allow micromanipulation.

The medaka should be an especially convenient material for generating fundamental and useful information about obtaining modifications of coloration, and color patterns. There already exist a number of established genetic varieties, with various chromatic characteristics (Yamamoto, 1975). Very recently, for example, DNA encoding tyrosinase, the key enzyme for melanization, cloned from the mouse, was injected into unfertilized eggs of the orange-red variety of this species, and transgenic dark individuals with an active mammalian enzyme were obtained (Matsumoto et al., 1991). The melanization was observed exclusively in amelanotic melanophores. These and improved techniques will surely provide many transgenic varieties with desirable hues and patterns.

Given the demand in the world today and in the future for food products, the importance of cultivating fish is growing all the time. Trials with several species of freshwater and marine fish are being conducted in several countries, and some have already turned out to be quite successful. In order to extend these undertakings to meet the demands and to increase the species available, surveys of the migration of cultured fish in the hydrosphere are needed, which naturally include the pursuit of fishes with artificial markings. Markings that have little or no influence on swimming ability as well as on ethological status are ideal. Slightly modified color patterns may be convenient for such purposes. Such chromatically modified transgenic fishes may provide us with more reliable information about migratory behaviors, rates of survival, and so on.

The commercial value of their stock is a very important problem for commercial fisheries. Preservation of the color of fish has improved with quick freezing immediately after capture, but many problems still remain to be solved. For instance, some commercially cultured fishes are subject to changes in color before they are forwarded to market. Consider, in this regard, the aquaculture of the red sea bream or red porgy, *Pagrus major,* which is a favorite food in Japan. With its beautiful crimson color, it decorates Japanese tables on many festive occasions. During growth in the fish preserves near the surface of the sea, sunshine melanizes the skin. By shading the preserves, such darkening can be minimized. However, in order to generate a more natural hue, in addition, the bream are now commonly given carotenoid-rich feed (red crustacean wastes and/or some algae) for a certain period of time prior to shipment.

In many species that are appreciated for their beauty or as food, higher commercial value may be due to more attractive colors and patterns. Currently developing biotechnological methods should be applicable to the selective breeding of favorable stock without too much difficulty in the near future.

ACKNOWLEDGMENTS

I thank Drs. N. Oshima and M. Sugimoto for discussion and help during the preparation of the manuscript, and Mr. H. Hayashi for technical assistance. This chapter was supported by grants from the Ministry of Education, Science and Culture of Japan.

REFERENCES

Bagnara, J.T. and Hadley, M.E., *Chromatophores and Color Change*, Prentice-Hall, Englewood Cliffs, NJ, 1973.

Baerends, G. and Baerends-van Roon, J.M., An introduction to the ethology of cichlid fishes, *Behaviour, Suppl. I*, 1, 1950.

Byers, H.R. and Porter, K.R., Transformation in the structure of the cytoplasmic ground substance in erythrophores during pigment aggregation and dispersion. I. A study using whole-cell preparations in stereo high voltage electron microscopy, *J. Cell Biol.*, 75, 541, 1977.

Cott, H.B., *Adaptive Colouration in Animals*, Methuen & Co., London, 1940.

Denton, E.J. and Nicol, J.A.C., A survey of reflectivity in silvery teleosts, *J. Mar. Biol. Assoc. U.K.*, 46, 685, 1966.

Fox, D.L., *Biochromy*, University of California Press, Berkeley, 1979.

Fujii, R., Chromatophores and pigments, in *Fish Physiology*, Vol. III, Hoar, W.S. and Randall, D.J., Eds., Academic Press, New York, 1969, 307.

Fujii, R., Cytophysiology of fish chromatophores, *Int. Rev. Cytol.*, 143, 191, 1993.

Fujii, R., and Oshima, N., Control of chromatophore movements in teleost fishes, *Zool.Sci.*, 3, 13, 1986.

Fujii, R., Kasukawa, H., Miyaji, K., and Oshima, N., Mechanism of skin coloration and its changes in blue-green damselfish, *Chromis viridis*, *Zool. Sci.*, 6, 477, 1989.

Fujii, R., Hayashi, H., Toyohara, J., and Nishi, H., Analysis of the light reflection from motile iridophores of the dark sleeper, *Odontobutis obscura obscura*, *Zool. Sci.*, 8, 461, 1991.

Fukuzawa, T. and Bagnara, J.T., Control of melanoblast differentiation in amphibia by a melanocyte stimulating hormone, a serum melanization factor, and a melanization inhibiting factor, *Pigment Cell Res.*, 2, 171, 1989.

Harosi, F.I. and Hashimoto, Y., Ultraviolet visual pigment in a vertebrate: a tetrachromatic cone system in the dace, *Science*, 222, 1021, 1983.

Huxley, A.F., A theoretical treatment of the reflexion of light by multilayer structures, *J. Exp. Biol.*, 48, 227, 1968.

Iga, T. and Matsuno, A., Motile iridophores of a freshwater goby, *Odontobutis obscura*, *Cell Tissue Res.*, 244, 165, 1986.

Kasukawa, H. and Oshima, N., Divisionistic generation of skin hue and the change of shade in the scalycheek damselfish, *Pomacentrus lepidogenys*, *Pigment Cell Res.*, 1, 152, 1987.

Kasukawa, H., Oshima, N., and Fujii, R., Mechanism of light reflection in blue damselfish iridophore, *Zool. Sci.*, 4, 243, 1987.

Kohda, Y. and Watanabe, M., Agonistic behavior and color pattern in a Japanese freshwater serranid fish, *Coreoperca kawamebari*, *Zool. Mag. (Tokyo)*, 91, 61, 1982.

Land, M.F., The physics and biology of animal reflectors, *Prog. Biophys. Mol. Biol.*, 24, 75, 1972.

Lythgoe, J.N., Light and vision in the aquatic environment, in *Sensory Biology of Aquatic Animals*, Atema, J., Fay, R.R., Popper, A.N. and Tavolga, W.N., Eds., Springer-Verlag, New York, 1988, 57.

Lythgoe, J.N. and Shand, J., Changes in spectral reflexions from the iridophores of the neon tetra, *J. Physiol.*, 325, 23, 1982.

Matsumoto, J., Studies on fine structure and cytochemical properties of erythrophores in swordtail, *Xiphophorus helleri*, with special reference to their pigment granules (pterinosomes), *J. Cell Biol.*, 27, 493, 1965.

Matsumoto, J., Morphological and physiological events associated with pigment translocation in swordtail fish erythrophores, in *Pigment Cell*, Vol. 4, Klaus, S.N., Ed., S. Karger, Basel, 1979, 56.

Matsumoto, J., Akiyama, T., Hirose, Eu., Nakamura, M.,Yamamoto, H. and Takeuchi, T., Expression and transmission of the melanogenic phenotypes in the orange-colored mutant of medaka fish introduced with mouse tyrosinase gene, *Zool. Sci.*, 9, 1081, 1991.

Menter, D.G., Obika, M., Tchen, T.T. and Taylor, J.D., Leucophores and iridophores of *Fundulus heteroclitus*: biophysical and ultrastructural properties, *J. Morphol.*, 160, 103, 1979.

Muske, L. and Fernald, R.D., Control of a teleost social signal: neural basis for differential expression of a color pattern, *J. Comp. Physiol*, 160, 89, 1987.

Nagaishi, H., Oshima, N. and Fujii, R., Light-reflecting properties of the iridophores of the neon tetra, *Paracheirodon innesi*, *Comp. Biochem. Physiol.*, 95A, 337, 1990.

Naitoh, T., Morioka, A., and Omura, Y., Adaptation of a common freshwater goby, yoshinobori, *Rhinogobius brunneus* Temminck et Schlegel to various backgrounds including those containing different sizes of black and white checkerboard squares, *Zool. Sci.*, 2, 59, 1985.

Nakajima, Y. and Obika, M., Growth and maturation of melanosomes in the melanophores of a teleost, *Oryzias latipes*, *Cell Tissue Res.*, 244, 279, 1986.

Needham, A.E., *Significance of Zoochromes*, Springer-Verlag, Berlin, 1974.

Obika, M., Intracellular transport of pigment granules in fish chromatophores, *Zool. Sci.*, 3, 1, 1986.

Obika, M., Ultrastructure and physiological response of leucophores of the medaka *Oryzias latipes*, *Zool. Sci.* 5, 311, 1988.

Obika, M. and Meyer-Rochow, V.B., Dermal and epidermal chromatophores of the Antarctic teleost *Trematomus bernacchi*, *Pigment Cell Res.*, 3, 33, 1990.

Oshima, N. and Fujii, R., A precision photoelectric method for recording chromatophore responses *in vitro*, *Zool. Sci.*, 1, 545, 1984.

Oshima, N. and Fujii, R., Motile mechanism of blue damselfish (*Chrysiptera cyanea*) iridophores, *Cell Motility & Cytoskeleton*, 8, 85, 1987.

Ozato, K., Inoue, K., and Wakamatsu, Y., Transgenic fish: biological and technical problems, *Zool. Sci.*, 6, 445, 1989.

Parker, G.H., *Animal Colour Changes and Their Neurohumours*, Cambridge University Press, Cambridge, 1948.

Powers, D.A., Fish as model systems, *Science*, 246, 352, 1989.

Reed, B.L., The control of circadian pigment changes in the pencil fish: a proposed role for melatonin, *Life Sci.*, 7(II), 961, 1968.

Sugimoto, M., Morphological color changes in the medaka, *Oryzias latipes*, after prolonged background adaptation — I. Changes in the population and morphology of melanophores. *Comp. Biochem. Physiol.*, 104A, 513, 1993.

Sugimoto, M., Oshima, N., and Fujii, R., Mechanisms controlling motile responses of amelanotic melanophores in the medaka, *Oryzias latipes*, *Zool. Sci.*, 2, 317, 1985.

Tinbergen, N., *Social Behaviour in Animals*, Methuen, London, 1965

Turner, W.A., Taylor, J.D., and Tchen, T.T., Melanosome formation in goldfish: the role of multivesicular bodies, *J. Ultrastruct. Res.*, 51, 16, 1975.

Wickler, W., *Mimicry in Plants and Animals*, G. Weidenfeld and Nicolson, London, 1968.

Yamamoto, T., *Medaka (killifish) — Biology and Strains*, Keigaku, Tokyo, 1975.

Systematic Index

D

I

J

K

L

M

N

O

Y

Z

General Index

A

B

C

D

E

F

G

I

J

K

L

M

N

O

P

R

S

T

U

V

W

X

Z